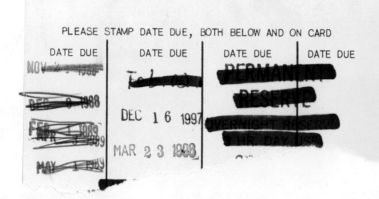

Applied Mathematical Sciences
Volume 75

Applied Mathematical Sciences

1. *John:* Partial Differential Equations, 4th ed.
2. *Sirovich:* Techniques of Asymptotic Analysis.
3. *Hale:* Theory of Functional Differential Equations, 2nd ed.
4. *Percus:* Combinatorial Methods.
5. *von Mises/Friedrichs:* Fluid Dynamics.
6. *Freiberger/Grenander:* A Short Course in Computational Probability and Statistics.
7. *Pipkin:* Lectures on Viscoelasticity Theory.
9. *Friedrichs:* Spectral Theory of Operators in Hilbert Space.
11. *Wolovich:* Linear Multivariable Systems.
12. *Berkovitz:* Optimal Control Theory.
13. *Bluman/Cole:* Similarity Methods for Differential Equations.
14. *Yoshizawa:* Stability Theory and the Existence of Periodic Solution and Almost Periodic Solutions.
15. *Braun:* Differential Equations and Their Applications, 3rd ed.
16. *Lefschetz:* Applications of Algebraic Topology.
17. *Collatz/Wetterling:* Optimization Problems.
18. *Grenander:* Pattern Synthesis: Lectures in Pattern Theory, Vol I.
20. *Driver:* Ordinary and Delay Differential Equations.
21. *Courant/Friedrichs:* Supersonic Flow and Shock Waves.
22. *Rouche/Habets/Laloy:* Stability Theory by Liapunov's Direct Method.
23. *Lamperti:* Stochastic Processes: A Survey of the Mathematical Theory.
24. *Grenander:* Pattern Analysis: Lectures in Pattern Theory, Vol. II.
25. *Davies:* Integral Transforms and Their Applications, 2nd ed.
26. *Kushner/Clark:* Stochastic Approximation Methods for Constrained and Unconstrained Systems
27. *de Boor:* A Practical Guide to Splines.
28. *Keilson:* Markov Chain Models—Rarity and Exponentiality.
29. *de Veubeke:* A Course in Elasticity.
30. *Sniatycki:* Geometric Quantization and Quantum Mechanics.
31. *Reid:* Sturmian Theory for Ordinary Differential Equations.
32. *Meis/Markowitz?* Numerical Solution of Partial Differential Equations.
33. *Grenander:* Regular Structures: Lectures in Pattern Theory, Vol. III.
34. *Kevorkian/Cole:* Perturbation methods in Applied Mathematics.
35. *Carr:* Applications of Centre Manifold Theory.
36. *Bengtsson/Ghil/Källén:* Dynamic Meteorology: Data Assimilation Methods.
37. *Saperstone:* Semidynamical Systems in Infinite Dimensional Spaces.
38. *Lichtenberg/Lieberman:* Regular and Stochastic Motion.
39. *Piccini/Stampacchia/Vidossich:* Ordinary Differential Equations in R^n.
40. *Naylor/Sell:* Linear Operator Theory in Engineering and Science.
41. *Sparrow:* The Lorenz Equations: Bifurcations, Chaos, and Strange Attractors.
42. *Guckenheimer/Holmes:* Nonlinear Oscillations, Dynamical Systems and Bifurcations of Vector Fields.
43. *Ockendon/Tayler:* Inviscid Fluid Flows.
44. *Pazy:* Semigroups of Linear Operators and Applications to Partial Differential Equations.
45. *Glashoff/Gustafson:* Linear Optimization and Approximation: An Introduction to the Theoretical Analysis and Numerical Treatment of Semi-Infinite Programs.
46. *Wilcox:* Scattering Theory for Diffraction Gratings.
47. *Hale et al.:* An Introduction to Infinite Dimensional Dynamical Systems—Geometric Theory.
48. *Murray:* Asymptotic Analysis.
49. *Ladyzhenskaya:* The Boundary-Value Problems of Mathematical Physics.
50. *Wilcox:* Sound Propagation in Stratified Fluids.
51. *Golubitsky/Schaeffer:* Bifurcation and Groups in Bifurcation Theory, Vol. I.
52. *Chipot:* Variational Inequalities and Flow in Porous Media.
53. *Majda:* Compressible Fluid Flow and Systems of Conservation Laws in Several Space Variables.
54. *Wasow:* Linear Turning Point Theory.

(continued in back)

R. Abraham
J.E. Marsden
T. Ratiu

Manifolds, Tensor Analysis, and Applications

Second Edition

Springer-Verlag
New York Berlin Heidelberg
London Paris Tokyo

Ralph Abraham
Department of Mathematics
University of California—Santa Cruz
Santa Cruz, CA 95064
U.S.A.

J.E. Marsden
Department of Mathematics
University of California—Berkeley
Berkeley, CA 94720
U.S.A.
and
Department of Mathematics
Cornell University
Ithaca, NY 14853-7901
U.S.A.

Tudor Ratiu
Department of Mathematics
University of California—Santa Cruz
Santa Cruz, CA 95064
U.S.A.

AMS Subject Classifications (1980): 34, 58, 70, 76, 93

Library of Congress Cataloging-in-Publication Data
Abraham, Ralph
Manifolds, tensor analysis, and applications, Second Edition

(Applied Mathematical Sciences; v. 75)
Bibliography: p. 631
Includes index.
1. Global analysis (Mathematics) 2. Manifolds (Mathematics) 3. Calculus of tensors.
I. Marsden, Jerrold E. II. Ratiu, Tudor S. III. Title. IV. Series.
QA614.A28 1983514.382-1737 ISBN 0-387-96790-7

First edition published by Addison-Wesley Publishing Company © 1983

Printed and bound by R.R. Donnelley and Sons, Harrisonburg, Virginia
Printed in the United States of America.

9 8 7 6 5 4 3 2 1

ISBN 0-387-96790-7 Springer-Verlag New York Berlin Heidelberg
ISBN 3-540-96790-7 Springer-Verlag Berlin Heidelberg New York

Preface

The purpose of this book is to provide core material in nonlinear analysis for mathematicians, physicists, engineers, and mathematical biologists. The main goal is to provide a working knowledge of manifolds, dynamical systems, tensors, and differential forms. Some applications to Hamiltonian mechanics, fluid mechanics, electromagnetism, plasma dynamics and control theory are given in Chapter 8, using both invariant and index notation. The current edition of the book does not deal with Riemannian geometry in much detail, and it does not treat Lie groups, principal bundles, or Morse theory. Some of this is planned for a subsequent edition. Meanwhile, the authors will make available to interested readers supplementary chapters on Lie Groups and Differential Topology and invite comments on the book's contents and development.

Throughout the text supplementary topics are given, marked with the symbols ☞ and ✍. This device enables the reader to skip various topics without disturbing the main flow of the text. Some of these provide additional background material intended for completeness, to minimize the necessity of consulting too many outside references.

We treat finite and infinite-dimensional manifolds simultaneously. This is partly for efficiency of exposition. Without advanced applications, using manifolds of mappings, the study of infinite-dimensional manifolds can be hard to motivate. Chapter 8 gives a hint of these applications. In fact, some readers may wish to skip the infinite-dimensional case altogether. To aid in this we have separated into supplements some of the technical points peculiar to the infinite-dimensional case. Our own research interests lean toward physical applications, and the choice of topics is partly molded by what is useful for this kind of research. We have tried to be as sympathetic to our readers as possible by providing ample examples, exercises, and applications. When a computation in coordinates is easiest, we give it and do not hide things behind complicated invariant notation. On the other hand, index-free notation sometimes provides valuable geometric and computational insight so we have tried to simultaneously convey this flavor.

The prerequisites required are solid undergraduate courses in linear algebra and advanced calculus. At various points in the text contacts are made with other subjects, providing a good way for students to link this material with other courses. For example, Chapter 1 links with point-set topology, parts of Chapter 2 and 7 are connected with functional analysis, Section 4.3 relates to ordinary differential equations, Chapter 3 and Section 7.5 are linked to differential topology and algebraic topology, and Chapter 8 on applications is connected with applied mathematics, physics, and engineering.

This book is intended to be used in courses as well as for reference. The sections are, as far as possible, lesson sized, if the supplementary material is omitted. For some sections, like 2.5, 4.2, or 7.5, two lecture hours are required. A standard course for mathematics graduate students could omit Chapter 1 and the supplements entirely and do Chapters 2 through 7 in one semester with the possible exception of Section 7.4. The instructor could then assign certain supplements for reading and choose among the applications of Chapter 8 according to taste. A shorter course, or a course advanced undergraduates, probably should omit all supplements, spend about two lectures on Chapter 1 for reviewing background point set topology, and cover Chapters 2 through 7 with the exception of Sections 4.4, 7.4, 7.5 and all the material relevant to volume elements induced by metrics, the Hodge star, and codifferential operators in Sections 6.2, 6.4, 6.5, and 7.2. A more applications oriented course could skim Chapter 1, review without proofs the material of Chapter 2, and cover Chapters 3 to 8 omitting the supplementary material and Sections 7.4 and 7.5. For such a course the instructor should keep in mind that while Sections 8.1 and 8.2 use only elementary material, Section 8.3 relies heavily on the Hodge star and codifferential operators, and Section 8.4 consists primarily of applications of Frobenius' theorem dealt with in Section 4.4.

The notation in the book is as standard as conflicting usages in the literature allow. We have had to compromise among utility, clarity, clumsiness, and absolute precision. Some possible notations would have required too much interpretation on the part of the novice while others, while precise, would have been so dressed up in symbolic decorations that even an expert in the field would not recognize them.

In a subject as developed and extensive as this one, an accurate history and crediting of theorems is a monumental task, especially when so many results are folklore and reside in private notes. We have indicated some of the important credits where we know of them, but we did not undertake this task systematically. We hope our readers will inform us of these and other shortcomings of the book so that, if necessary, corrected printings will be possible. The reference list at the back of the book is confined to works actually cited in the text. These works are cited by author and year like this: deRham [1955].

During the preparation of the book, valuable advice was provided by Malcolm Adams, Morris Hirsch, Charles Pugh, Alan Weinstein, and graduate students in mathematics, physics and engineering at Berkeley, Santa Cruz and elsewhere. Our other teachers and collaborators from whom we learned the material and who inspired, directly and indirectly, various portions of the text are too numerous to mention individually, so we hereby thank them all collectively. We have taken the opportunity in this edition to correct some errors kindly pointed out by our readers and to rewrite numerous sections. This book was typeset on a Macintosh using Mathwriter (Cooke Publications Inc, Ithaca, N.Y.); we thank Connie Calica, Dotty Hollinger, Marnie MacElhiny and Esther Zack for their invaluable help with the typing.

We intend this book to be an evolving project. That is, we invite corrections and comments from our readers to be incorporated into future printings. We are currently preparing some supplementary chapters and plan to include a differential topology and Lie groups chapter in the next printing—space permitting. Meanwhile, if you wish to see these chapters, we will be happy to send them to you in exchange for your comments.

February, 1988

RALPH ABRAHAM
JERROLD E. MARSDEN
TUDOR RATIU

Contents

Contents

Background Notation

The reader is assumed to be familiar with the usual notations of set theory such as $\in$, $\subset$, $\cup$, $\cap$ and with the concept of a mapping. If A and B are sets and if f: $A \to B$ is a mapping, we write $a \mapsto f(a)$ for the effect of the mapping on the element of $a \in A$; "iff" stands for "if and only if" (= "if" in definitions). Other notations we shall use without explanation include the following:

◆	end of an example or remark
■	end of a proof
▼	proof of a lemma is done, but the proof of the theorem goes on
$\mathbb{R}$, $\mathbb{C}$	real, complex numbers
$\mathbb{Z}$, $\mathbb{Q}$	integers, rational numbers
$A \times B$	Cartesian product
$\mathbb{R}^n$, $\mathbb{C}^n$	Euclidean n-space, complex n-space
$(x^1, \ldots, x^n) \in \mathbb{R}^n$	point in $\mathbb{R}^n$
$A \subset B$	set theoretic containment (means same as $A \subseteq B$)
$A \setminus B$	set theoretic difference
I or Id	identity map
$f^{-1}(B)$	inverse image of B under f
$\Gamma_f = \{(x, f(x)) \mid x \in \text{domain of } f\}$	graph of f
inf A	infinimum (greatest lower bound) of the set $A \subset \mathbb{R}$
sup A	supremum (least upper bound) of $A \subset \mathbb{R}$
$e_1, \ldots, e_n$	basis of an n-dimensional vector space
ker T, range T	kernel and range of a linear transformation T
$D_r(m)$	open ball about m of radius r
$B_r(m)$	closed ball of radius r (also denoted $\overline{D}_r(m)$).

Chapter 1
Topology

The purpose of this chapter is to introduce just enough topology for later requirements. It is assumed that the reader has had a course in advanced calculus and so is acquainted with open, closed, compact, and connected sets in Euclidean space (see for example Marsden [1974a] and Rudin [1976]). If this background is weak, the reader may find the pace of this chapter too fast. If the background is under control, the chapter should serve to collect, review, and solidify concepts in a more general context. Readers already familiar with point set topology can safely skip this chapter.

A key concept in manifold theory is that of a differentiable map between manifolds. However, manifolds are also topological spaces and differentiable maps are continuous. Topology is the study of continuity in a general context; it is therefore appropriate to begin with it. Topology often involves interesting excursions into pathological spaces and exotic theorems. Such excursions are deliberately minimized here. The examples will be ones most relevant to later developments, and the main thrust will be to obtain a working knowledge of continuity, connectedness, and compactness.

We shall take for granted the usual logical structure of analysis without much comment, except to recall one of the basic axioms that is in common use and an equivalent result. These will be used occasionally in the text.

Axiom of choice *If* $\mathcal{S}$ *is a collection of nonempty sets, then there is a function*

$$\chi : \mathcal{S} \to \bigcup_{S \in \mathcal{S}} S \quad \textit{such that} \quad \chi(S) \in S \quad \textit{for every} \quad S \in \mathcal{S}.$$

The function χ chooses one element from each $S \in \mathcal{S}$ and is called a ***choice function.*** Even though this statement seems self-evident, it has been shown to be equivalent to a number of nontrivial statements, using other axioms of set theory. To discuss them, we need a few definitions. An ***order*** on a set A is a binary relation, usually denoted by "$\leq$" satisfying the following conditions:

$a \leq a$	(***reflexivity***)
$a \leq b$ and $b \leq a$ implies $a = b$	(***antisymmetry***), and
$a \leq b$ and $b \leq c$ implies $a \leq c$	(***transitivity***).

An ordered set A is called a ***chain*** if for every $a, b \in A$, $a \neq b$ we have $a \le b$ or $b \le a$. The set A is said to be ***well ordered*** if it is a chain and every nonempty subset B has a first element; i.e., there exists an element $b \in B$ such that $b \le x$ for all $x \in B$. An ***upper bound*** $u \in A$ of a chain $C \subset A$ is an element for which $c \le u$ for all $c \in C$. A ***maximal element*** m of an ordered set A is an element for which there is no other $a \in A$ such that $m \le a$, $a \neq m$; in other words $x \le m$ for all $x \in A$ that are comparable to m. We state the following without proof.

Theorem *Given other axioms of set theory, the following statements are equivalent:*

(i) *The axiom of choice.*

(ii) **Product Axiom** *If* $\{A_i\}_{i\in I}$ *is a collection of nonempty sets then the product space* $\prod_{i\in I} A_i = \{(x_i) \mid x_i \in A_i\}$ *is nonempty.*

(iii) **Zermelo's Theorem** *Any set can be well ordered.*

(iv) **Zorn's Theorem** *If* A *is an ordered set for which every chain has an upper bound (i.e.,* A *is* ***inductively ordered****), then* A *has at least one maximal element.*

§1.1 *Topological Spaces*

Abstracting ideas about open sets in $\mathbb{R}^n$ leads to the notion of a topological space.

1.1.1 Definition *A* ***topological space*** *is a set* S *together with a collection* $\mathcal{O}$ *of subsets called* ***open sets*** *such that*

T1 $\varnothing \in \mathcal{O}$ *and* $S \in \mathcal{O}$;

T2 *if* $U_1, U_2 \in \mathcal{O}$, *then* $U_1 \cap U_2 \in \mathcal{O}$;

T3 *the union of any collection of open sets is open.*

A basic example is the real line. We choose $S = \mathbb{R}$, with $\mathcal{O}$ consisting of all sets that are unions of open intervals. As exceptional cases, the empty set $\varnothing \in \mathcal{O}$ and $\mathbb{R}$ itself belong to $\mathcal{O}$. Thus **T1** holds. For **T2**, let U_1 and $U_2 \in \mathcal{O}$; to show that $U_1 \cap U_2 \in \mathcal{O}$, we can suppose that $U_1 \cap U_2 \neq \varnothing$. If $x \in U_1 \cap U_2$, then x lies in an open interval $]a_1, b_1[\subset U_1$ and also in the interval $]a_2, b_2[\subset U_2$. We can write $]a_1, b_1[\cap]a_2, b_2[=]a, b[$ where $a = \max(a_1, a_2)$ and $b = \min(b_1, b_2))$. Thus $x \in]a, b[\subset U_1 \cap U_2$. Hence $U_1 \cap U_2$ is the union of such intervals, so is open. Finally, **T3** is clear by definition.

Similarly, $\mathbb{R}^n$ may be topologized by declaring a set to be open if it is a union of open rectangles. An argument similar to the one just given for $\mathbb{R}$ shows that this is a topology, called the ***standard topology*** on $\mathbb{R}^n$.

The ***trivial topology*** on a set S consists of $\mathcal{O} = \{\varnothing, S\}$. The ***discrete topology*** on S is defined by $\mathcal{O} = \{A \mid A \subset S\}$; i.e., $\mathcal{O}$ consists of all subsets of S.

Topological spaces are specified by a pair $(S, \mathcal{O})$; we shall, however, simply write S if there is no danger of confusion.

1.1.2 Definition *Let* S *be a topological space. A set* $A \subset S$ *will be called* ***closed*** *if its complement* $S \setminus A$ *is open. The collection of closed sets is denoted* $\mathcal{C}$.

For example, the closed interval $[0, 1] \subset \mathbb{R}$ is closed as it is the complement of the open set $]-\infty, 0[\cup]1, \infty[$.

1.1.3 Proposition *The closed sets in a topological space satisfy:*

C1 $\varnothing \in \mathcal{C}$ *and* $S \in \mathcal{C}$;

C2 *if* $A_1, A_2 \in \mathcal{C}$ *then* $A_1 \cup A_2 \in \mathcal{C}$;

C3 *the intersection of any collection of closed sets is closed.*

Proof **C1** follows from **T1** since $\varnothing = S \setminus S$, $S = S \setminus \varnothing$. The relations

$$S \setminus (A_1 \cup A_2) = (S \setminus A_2) \cap (S \setminus A_2) \quad \text{and} \quad S \setminus \left(\bigcap_{i \in I} B_i\right) = \bigcup_{i \in I} (S \setminus B_i)$$

for $\{B_i\}_{i \in I}$ a family of closed sets show that **C2, C3** are equivalent to **T2, T3**, respectively. ■

Closed rectangles in $\mathbb{R}^n$ are closed sets, as are closed balls, one-point sets, and spheres. Not every set is either open or closed. For example, the interval $[0, 1[$ is neither an open nor a closed set. In a discrete topology on S any set $A \subset S$ is both open and closed, whereas in the trivial topology any $A \neq \varnothing$ or S is neither.

Closed sets can be used to introduce a topology just as well as open ones. Thus, if $\mathcal{C}$ is a collection satisfying **C1-C3** and $\mathcal{O}$ consists of the complements of sets in $\mathcal{C}$, then $\mathcal{O}$ satisfies **T1-T3.**

1.1.4 Definition *An* ***open neighborhood*** *of a point* u *in a topological space* S *is an open set* U *such that* $u \in U$. *Similarly, for a subset* A *of* S, U *is an open neighborhood of* A *if* U *is open and* $A \subset U$. *A* ***neighborhood*** *of a point (or a subset) is a set containing some open neighborhood of the point (or subset).*

Examples of neighborhoods of $x \in \mathbb{R}$ are $]x - 1, x + 3]$, $]x - \varepsilon, x + \varepsilon[$ for any $\varepsilon > 0$, and $\mathbb{R}$ itself; only the last two are open neighborhoods. The set $[x, x + 2[$ contains the point x but is not one of its neighborhoods. In the trivial topology on a set S, there is only one neighborhood of any point, namely S itself. In the discrete topology any subset containing p is a neighborhood of the point $p \in S$, since $\{p\}$ is an open set.

1.1.5 Definition *A topological space is called* ***first countable*** *if for each* $u \in S$ *there is a sequence* $\{U_1, U_2, \ldots\} = \{U_n\}$ *of neighborhoods of* u *such that for any neighborhood* U *of* u, *there is an integer* n *such that* $U_n \subset U$. *A subset* $\mathcal{B}$ *of* $\mathcal{O}$ *is called a* ***basis*** *for the topology, if each open set is a union of elements in* $\mathcal{B}$. *The topology is called* ***second countable*** *if it has a countable basis.*

Most topological spaces of interest to us will be second countable. For example $\mathbb{R}^n$ is second countable since it has the countable basis formed by rectangles with rational side length and centered at points all of whose coordinates are rational. Clearly every second-countable space is also first countable, but the converse is false. For example if S is an infinite noncountable set, the discrete topology is not second countable, but S is first countable, since $\{p\}$ is a neighborhood of $p \in S$. The trivial topology on S is second countable (see Exercises **1.1I, 1.1J** for more interesting counter-examples).

1.1.6 Lindelöf's Lemma *Every covering of a set* A *in a second countable space* S *by a family of open sets* U_a *(that is* $\bigcup_a U_a \supset A$*) contains a countable subcollection also covering* A.

Proof Let $\mathcal{B} = \{B_n\}$ be a countable basis for the topology of S. For each $p \in A$ there are indices n and α such that $p \in B_n \subset U_\alpha$. Let $\mathcal{B}' = \{B_n \mid$ there exists an α such that $B_n \subset U_\alpha\}$. Now let $U_{\alpha(n)}$ be one of the U_α that includes the element B_n of $\mathcal{B}'$. Since $\mathcal{B}'$ is a covering of A, the countable collection $\{U_{\alpha(n)}\}$ covers A. ■

1.1.7 Definition *Let* S *be a topological space and* $A \subset S$. *The* ***closure*** *of* A, *denoted* cl(A) *is the intersection of all closed sets containing* A. *The* ***interior*** *of* A, *denoted* int(A) *is the union of all open sets contained in* A. *The* ***boundary*** *of* A, *denoted* bd(A) *is defined by*

$$\mathrm{bd}(A) = \mathrm{cl}(A) \cap \mathrm{cl}(S \setminus A).$$

By **C3**, cl(A) is closed and by **T3**, int(A) is open. Note that as bd(A) is the intersection of closed sets, bd(A) is closed, and $\mathrm{bd}(A) = \mathrm{bd}(S \setminus A)$

On $\mathbb{R}$, for example, $\mathrm{cl}([0, 1[) = [0, 1]$, $\mathrm{int}([0, 1[) =]0, 1[$, and $\mathrm{bd}([0, 1[) = \{0, 1\}$. The reader is assumed to be familiar with examples of this type from advanced calculus.

1.1.8 Definitions *A subset* A *of* S *is called* ***dense*** *in* S *if* $\mathrm{cl}(A) = S$, *and is called* ***nowhere dense*** *if* $S \setminus \mathrm{cl}(A)$ *is dense in* S. *The space* S *is called* ***separable*** *if it has a countable dense subset. A point in* S *is called an* ***accumulation point*** *of the set* A *if each of its neighborhoods contains a point of* A *other than itself. The set of accumulation points of* A *is called the* ***derived set*** *of* A *and is denoted by* der(A). *A point of* A *is said to be* ***isolated*** *if it has a neighborhood in* S *containing no other points of* A *than itself.*

The set $A = [0,1[\cup \{2\}$ in $\mathbb{R}$ has the element 2 as its only isolated point, its interior is $\mathrm{int}(A) =]0, 1[$, $\mathrm{cl}(A) = [0, 1] \cup \{2\}$ and $\mathrm{der}(A) = [0, 1]$. In the discrete topology on a set S, $\mathrm{int}\{p\} = \mathrm{cl}\{p\} = \{p\}$, for any $p \in S$.

Since the set $\mathbb{Q}$ of rational numbers is dense in $\mathbb{R}$ and is countable, $\mathbb{R}$ is separable. Similarly $\mathbb{R}^n$ is separable. A set S with the trivial topology is separable since $\mathrm{cl}\{p\} = S$ for any $p \in S$. But $S = \mathbb{R}$ with the discrete topology is not separable since $\mathrm{cl}(A) = A$ for any $A \subset S$. Any second-countable space is separable, but the converse is false; see Exercises **1.1I, 1.1J**.

1.1.9 Proposition *Let* S *be a topological space and* $A \subset S$. *Then*

(i) $u \in cl(A)$ *iff for every neighborhood* U *of* u, $U \cap A \neq \varnothing$;

(ii) $u \in int(A)$ *iff there is a neighborhood* U *of* u *such that* $U \subset A$;

(iii) $u \in bd(A)$ *iff for every neighborhood* U *of* u, $U \cap A \neq \varnothing$ *and* $U \cap (S \setminus A) \neq \varnothing$.

Proof (i) $u \notin cl(A)$ iff there exists a closed set $C \supset A$ such that $u \notin C$. But this is equivalent to the existence of a neighborhood of u not intersecting A, namely $S \setminus C$. (ii) and (iii) are proved in a similar way. ■

1.1.10 Proposition *Let* A, B *and* A_i, $i \in I$ *be subsets of* S.

(i) $A \subset B$ *implies* $int(A) \subset int(B)$, $cl(A) \subset cl(B)$, *and* $der(A) \subset der(B)$;

(ii) $S \setminus cl(A) = int(S \setminus A)$, $S \setminus int(A) = cl(S \setminus A)$, *and* $cl(A) = A \cup der(A)$;

(iii) $cl(\varnothing) = int(\varnothing) = \varnothing$, $cl(S) = int(S) = S$, $cl(cl(A)) = cl(A)$ *and* $int(int(A)) = int(A)$;

(iv) $cl(A \cup B) = cl(A) \cup cl(B)$, $der(A \cup B) = der(A) \cup der(B)$, $int(A \cup B) \supset int(A) \cup int(B)$;

(v) $cl(A \cap B) \subset cl(A) \cap cl(B)$, $der(A \cap B) \subset der(A) \cap der(B)$, $int(A \cap B) = int(A) \cap int(B)$;

(vi) $cl(\bigcup_{i\in I} A_i) \supset \bigcup_{i\in I} cl(A_i)$, $cl(\bigcap_{i\in I} A_i) \subset \bigcap_{i\in I} cl(A_i)$,
$int(\bigcup_{i\in I} A_i) \supset \bigcup_{i\in I} int(A_i)$, $int(\bigcap_{i\in I} A_i) \subset \bigcap_{i\in I} int(A_i)$.

Proof (i), (ii), and (iii) are consequences of the definition and of Proposition **1.1.9.** Since for each $i \in I$, $A_i \subset \bigcup_{i\in I} A_i$, by (i) $cl(A_i) \subset cl(\bigcup_{i\in I} A_i)$ and hence $\bigcup_{i\in I} cl(A_i) \subset cl(\bigcup_{i\in I} A_i)$. Similarly, since $\bigcap_{i\in I} A_i \subset A_i \subset cl(A_i)$ for each $i \in I$, it follows that $\bigcap_{i\in I} cl(A_i)$ is a subset of the closet set $\bigcap_{i\in I} cl(A_i)$; thus by (i) $cl(\bigcap_{i\in I} A_i) \subset cl(\bigcap_{i\in I} cl(A_i)) = \bigcap_{i\in I} (cl(A_i))$.The other formulas of (vi) follow from these and (ii). This also proves all the other formulas in (iv) and (v) except the ones with equalities. Since $cl(A) \cup cl(B)$ is closed by **C2** and $A \cup B \subset cl(A) \cup cl(B)$, it follows by (i) that $cl(A \cup B) \subset cl(A) \cup cl(B)$ and hence equality by (vi). The formula $int(A \cap B) = int(A) \cap int(B)$ is a corollary of the previous formula via (ii). ■

The inclusions in the above proposition can be strict. For example, if we let $A =]0,1[$ and $B = [1, 2[$, then one finds $cl(A) = der(A) = [0, 1]$, $cl(B) = der(B) = [1, 2]$, $int(A) =]0, 1[$, $int(B) =]1, 2[$, $A \cup B =]0, 2[$, and $A \cap B = \varnothing$, and therefore $int(A) \cup int(B) =]0,1[\cup]1, 2[\neq]0, 2[= int(A \cup B)$, and $cl(A \cap B) = \varnothing \neq \{1\} = cl(A) \cap cl(B)$. Let $A_n =]-1/n, 1/n[$, $n = 1, 2,\ldots$; then $\bigcap_{n\geq 1} A_n = \{0\}$, $int(A_n) = A_n$ for all n, and $int(\bigcap_{n\geq 1} A_n) = \varnothing \neq \{0\} = \bigcap_{n\geq 1} int(A_n)$. Dualizing this via (ii) gives $\bigcup_{n\geq 1} cl(\mathbb{R} \setminus A_n) = \mathbb{R} \setminus \{0\} \neq \mathbb{R} = cl(\bigcup_{n\geq 1}(\mathbb{R} \setminus A_n))$. If $A \subset B$, there is, in general, no relation between the sets bd(A) and bd(B). For example, if $A = [0, 1]$ and $B = [0, 2]$, $A \subset B$, yet we have $bd(A) = \{0, 1\}$ and $bd(B) = \{0, 2\}$.

1.1.11 Definition *Let* S *be a topological space and* $\{u_n\}$ *a sequence of points in* S. *The sequence is said to* **converge** *if there is a point* $u \in S$ *such that for every neighborhood* U *of* u, *there is an* N *such that* $n \geq N$ *implies* $u_n \in U$. *We say that* u_n **converges** *to* u, *or* u *is a* **limit point** *of* $\{u_n\}$.

For example, the sequence $\{1/n\}$ in $\mathbb{R}$ converges to 0. It is obvious that limit points of sequences u_n of distinct points are accumulation points of the set $\{u_n\}$. In a first countable topological space any accumulation point of a set A is a limit of a sequence of elements of A. Indeed, if $\{U_n\}$ denotes the countable collection of neighborhoods of $a \in \mathrm{der}(A)$ given by definition **1.1.5**, then choosing for each n an element $a_n \in U_n \cap A$ such that $a_n \neq a$, we see that $\{a_n\}$ converges to a. We have proved the following.

1.1.12 Proposition *Let* S *be a first-countable space and* $A \subset S$. *Then* $u \in \mathrm{cl}(A)$ *iff there is a sequence of points of* A *that converges to* u *(in the topology of* S*).*

It should be noted that a sequence can be divergent and still have accumulation points. For example $\{2, 0, 3/2, -1/2, 4/3, -2/3, \ldots\}$ does not converge but has both 1 and -1 as accumulation points. In arbitrary topological spaces, limit points of sequences are in general *not* unique. For example, in the trivial topology of S any sequence converges to all points of S. In order to avoid such situations several ***separation axioms*** have been introduced, of which the three most important ones will be mentioned.

1.1.13 Definition *A topological space* S *is called* ***Hausdorff*** *if each two distinct points have disjoint neighborhoods (that is, with empty intersection). The space* S *is called* ***regular*** *if it is Hausdorff and if each closed set and point not in this set have disjoint neighborhoods. Similarly,* S *is called* ***normal*** *if it is Hausdorff and if each two disjoint closed sets have disjoint neighborhoods.*

Most standard spaces in analysis are normal. The discrete topology on any set is normal, but the trivial topology is not even Hausdorff. It turns out that "Hausdorff" is the necessary and sufficient condition for uniqueness of limit points of sequences in first countable spaces (see Exercise **1.1E**). Since in Hausdorff space single points are closed (Exercise **1.1F**), we have the implications: normal $\Rightarrow$ regular $\Rightarrow$ Hausdorff. Counterexamples for each of the converses of these implications are given in Exercises **1.1I** and **1.1J**.

1.1.14 Proposition *A regular second-countable space is normal.*

Proof Let A and B be two disjoint closed sets in S. By regularity, for every point $p \in A$ there are disjoint open neighborhoods U_p of p and U_B of B. Hence $\mathrm{cl}(U_p) \cap B = \varnothing$. Since $\{U_p \mid p \in A\}$ is an open covering of A, by the Lindelöf lemma (**1.1.6**), there is a countable collection $\{U_k \mid k = 1, 2, \ldots\}$ covering A. Thus $\bigcup_{k\geq 1} U_k \supset A$ and $\mathrm{cl}(U_k) \cap B = \varnothing$.

Similarly, find a family $\{V_k\}$ such that $\bigcup_{k\geq 0} V_k \supset B$ and $\mathrm{cl}(V_k) \cap A = \varnothing$. Then the sets $G_{n+1} = U_{n+1} \setminus \bigcup_{k=0,1,\ldots,n} \mathrm{cl}(V_k)$, $H_n = V_n \setminus \bigcup_{k=0,1,\ldots,n} \mathrm{cl}(U_k)$, $G_0 = U_0$, are open and $G = \bigcup_{n\geq 0} G_n \supset A$, $H = \bigcup_{n\geq 0} H_n \supset B$ are also open and disjoint. ■

In the remainder of this book Euclidean n-space $\mathbb{R}^n$ will be understood to have the standard topology (unless explicitly stated to the contrary).

Exercises

1.1A Let $A = \{(x, y, z) \in \mathbb{R}^3 \mid 0 \le x < 1 \text{ and } y^2 + z^2 \le 1\}$. Find int(A).

1.1B Show that any finite set in $\mathbb{R}^n$ is closed.

1.1C Find the closure of $\{1/n \mid n = 1,2,\ldots\}$ in $\mathbb{R}$.

1.1D Let $A \subset \mathbb{R}$. Show that $\sup(A) \in \mathrm{cl}(A)$ where $\sup(A)$ is the supremum (l.u.b.) of A.

1.1E Show that a first countable space is Hausdorff iff all sequences have at most one limit point.

1.1F (i) Prove that in a Hausdorff space, single points are closed.
(ii) Prove that a topological space is Hausdorff iff the intersection of all closed neighborhoods of a point equals the point itself.

1.1G Show that in a Hausdorff space S the following are equivalent; (i) S is regular; (ii) for every point $p \in S$ and any of its neighborhoods U, there exists a closed neighborhood V of p such that $V \subset U$; (iii) for any closed set A, the intersection of all of the closed neighborhoods of A equals A.

1.1H (i) Show that if $\mathcal{V}(p)$ denotes the set of all neighborhoods of $p \in S$, then the following are satisfied:

V1 if $A \supset U$ and $U \in \mathcal{V}(p)$, then $A \in \mathcal{V}(p)$;

V2 every finite intersection of elements in $\mathcal{V}(p)$ is an element of $\mathcal{V}(p)$;

V3 p belongs to all elements of $\mathcal{V}(p)$;

V4 if $V \in \mathcal{V}(p)$ then there is a set $U \in \mathcal{V}(p)$, $U \subset V$ such that for all $q \in U$, $U \in \mathcal{V}(q)$.

(ii) If for each $p \in S$ there is a family $\mathcal{V}(p)$ of subsets of S satisfying **V1-V4**, prove that there is a unique topology $\mathcal{O}$ on S such that for each $p \in S$, the family $\mathcal{V}(p)$ is the set of neighborhoods of p in the topology $\mathcal{O}$. (*Hint:* Prove uniqueness first and then define elements of $\mathcal{O}$ as being subsets $A \subset S$ satisfying: for each $p \in A$, we have $A \in \mathcal{V}(p)$.)

1.1I Let $S = \{p = (x,y) \in \mathbb{R}^2 \mid y \ge 0\}$ and let $D_\varepsilon(p) = \{q \mid \|q - p\| < \varepsilon\}$ denote the usual ε-disk about p in the plane $\mathbb{R}^2$. Define

$$B_\varepsilon(p) = \begin{cases} D_\varepsilon(p) \cap S, \text{ if } p = (x, y) \text{ with } y > 0 \\ \{(x, y) \in D_\varepsilon(p) \mid y > 0\} \cup \{p\}, \text{ if } p = (x, 0). \end{cases}$$

Prove the following:

(i) $\mathcal{V}(p) = \{U \subset S \mid \text{there exists } B_\varepsilon(p) \subset U\}$ satisfies **V1-V4** of Exercise **1.1H**. Thus S becomes a toplogical space.

(ii) S is first countable.

(iii) S is Hausdorff.

(iv) S is separable. (*Hint :* The set $\{(x, y) \in S \mid x, y \in \mathbb{Q}, y > 0\}$ is dense in S.)

(v) S is not second countable (*Hint :* Assume the contrary and get a contradiction by looking at the points $(x, 0)$ of S.)

(vi) S is not regular. (*Hint:* Try to separate the point $(x_0, 0)$ from the set $\{(x, 0) \mid x \in \mathbb{R}\} \setminus \{(x_0, 0)\}$.)

1.1J With the same notations as in the preceding exercise, except changing $B_\varepsilon(p)$ to

$$B_\varepsilon(p) = \begin{cases} D_\varepsilon(p) \cap S, \text{ if } p = (x, y) \text{ with } y > 0 \\ \\ \{(x, y) \in D_\varepsilon(p) \mid y > 0\} \cup \{p\}, \text{ if } p = (x, 0), \end{cases}$$

show that (i)-(v) of **1.1I** remain valid and that

(vi) S is regular ; (*Hint:* Use Exercise **1.1G**.)

(vii) S is not normal. (*Hint* : Try to separate the set $\{(x, 0) \mid x \in \mathbb{Q}\}$ from the set $\{(x, 0) \mid x \in \mathbb{R} \setminus \mathbb{Q}\}$.)

1.1K Prove the following properties of the boundary operation and show by example that each inclusion cannot be replaced by equality.

Bd1 $\mathrm{bd}(A) = \mathrm{bd}(S \setminus A)$;

Bd2 $\mathrm{bd}(\mathrm{bd}(A) \subset \mathrm{bd}(A)$;

Bd3 $\mathrm{bd}(A \cup B) \subset \mathrm{bd}(A) \cup \mathrm{bd}(B) \subset \mathrm{bd}(A \cup B) \cup A \cup B$;

Bd3 $\mathrm{bd}(\mathrm{bd}(\mathrm{bd}(A))) = \mathrm{bd}(\mathrm{bd}(A))$.

Properties **Bd1-Bd4** may be used to characterize the topology.

1.1L Let p be a polynomial in n variables $z_1, \ldots, z_n$ with complex coefficients. Show that $p^{-1}(0)$ has open dense complement. (*Hint:* If p vanishes on an open set of $\mathbb{C}^n$, then all its derivatives also vanish and hence all its coefficients are zero.)

1.1M Show that a subset $\mathcal{B}$ of $\mathcal{O}$ is a basis for the topology of S if and only if the following three conditions hold:

B1 $\varnothing \in \mathcal{B}$;

B2 $\bigcup_{B \in \mathcal{B}} B = S$;

B3 if $B_1, B_2 \in \mathcal{B}$, then $B_1 \cap B_2$ is a union of elements of $\mathcal{B}$.

§1.2 *Metric Spaces*

One of the common ways to form a topological space is through the use of a distance function, also called a (topological) metric. For example, on $\mathbb{R}^n$ the standard distance

$$d(\mathbf{x}, \mathbf{y}) = \left(\sum_{i=1}^{n} (x_i - y_i)^2 \right)^{1/2}$$

between $\mathbf{x} = (x_1, ..., x_n)$ and $\mathbf{y} = (y_1, ..., y_n)$ can be used to construct the open disks and from them the topology. The abstraction of this proceeds as follows.

1.2.1 Definition *Let* M *be a set. A* ***metric*** *(also called a* ***topological metric****) on* M *is a function* $d : M \times M \to \mathbb{R}$ *such that for all* $m_1, m_2, m_3 \in M$,

M1 $d(m_1, m_2) = 0$ *iff* $m_1 = m_2$ *(definiteness);*

M2 $d(m_1, m_2) = d(m_2, m_1)$ *(symmetry); and*

M3 $d(m_1, m_3) \le d(m_1, m_2) + d(m_2, m_3)$ *(triangle inequality).*

A ***metric space*** *is the pair* (M, d); *if there is no danger of confusion, just write* M *for* (M, d).

Taking $m_1 = m_3$ in **M3** shows that d is necessarily a non-negative function. It is proved in advanced calculus courses (and is geometrically clear) that the standard distance on $\mathbb{R}^n$ satisfies **M1-M3**. The topology determined by a metric is defined as follows.

1.2.2 Definition *For* $\varepsilon > 0$ *and* $m \in M$, *the* ***open*** ***ε-ball*** ***(or disk)*** *about* m *is defined by*

$$D_\varepsilon(m) = \{m' \in M \mid d(m', m) < \varepsilon\} ,$$

and the ***closed*** ***ε-ball*** *is defined by*

$$B_\varepsilon(m) = \{m' \in M \mid d(m', m) \le \varepsilon\} .$$

The collection of subsets of M *that are unions of such disks is the* ***metric topology*** *of the metric space* (M, d). *Two metrics on a set are called* ***equivalent*** *if they induce the same metric topology.*

1.2.3 Proposition

(i) *The description of open sets given in the preceding definition is a topology.*

(ii) *A set* $U \subset M$ *is open iff for each* $m \in U$ *there is an* $\varepsilon > 0$ *such that* $D_\varepsilon(m) \subset U$.

Proof (i) **T1** and **T3** are clearly satisfied. To prove **T2,** it suffices to show that the intersection of two disks is a union of disks, which in turn is implied by the fact that any point in the intersection of two disks sits in a smaller disk included in this intersection. To verify this, suppose that $p \in D_\varepsilon(m) \cap D_\delta(n)$ and let $0 < r < \min(\varepsilon - d(p, m), \delta - d(p, n))$. Hence $D_r(p) \subset D_\varepsilon(m) \cap D_\delta(n)$, since for any $x \in D_r(p)$, $d(x, m) \le d(x, p) + d(p, m) < r + d(p, m) < \varepsilon$, and similarly $d(x, n) < \delta$.

(ii) By definition of the metric topology, a set V is a neighborhood of $m \in M$ iff there exists a disk $D_\varepsilon(m) \subset V$. Thus the statement in the theorem is equivalent to $U = \operatorname{int}(U)$. ∎

Every set M can be made into a metric space by the ***discrete metric*** defined by setting $d(m, n) = 1$ for all $m \neq n$. The metric topology of M is the discrete topology.
A ***pseudometric*** on the set M is a function $d : M \times M \to \mathbb{R}$ that satisfies **M2, M3,** and

PM1 $d(m, m) = 0$ for all m.

Thus the distance between distinct points can be zero for a pseudometric. The pseudometric topology is defined exactly as the metric space topology. Any set M can be made into a pseudometric space by the ***trivial pseudometric*** : $d(m, n) = 0$ for all $m, n \in M$; the pseudometric topology on M is the trivial topology. Note that a pseudometric space is Hausdorff iff it is a metric space.

If M is a metric space (or pseudometric space) and $u \in M$, $A \subset M$, we define $d(u, A) = \inf\{d(u,v) \mid v \in A\}$ if $A \neq \varnothing$, and $d(u, \varnothing) = \infty$. The ***diameter*** of a set $A \subset M$ is defined by $\operatorname{diam}(A) = \sup\{d(u, v) \mid u, v \in A\}$. A set is called ***bounded*** if its diameter is finite.

Clearly metric spaces are first-countable and Hausdorff; in fact:

1.2.4 Proposition *Every metric space is normal.*

Proof Let A and B be closed, disjoint subsets of M, and let

$$U = \{u \in M \mid d(u, A) < d(u, B)\} \quad \text{and} \quad V = \{v \in M \mid d(v, A) > d(v, B)\} .$$

It is verified that U and V are open, disjoint and $A \subset U$, $B \subset V$. ∎

1.2.5 Definition *Let* M *be a metric space with metric* d *and* $\{u_n\}$ *a sequence in* M. *Then* $\{u_n\}$ *is a* ***Cauchy sequence*** *if for all real* $\varepsilon > 0$, *there is an integer* N *such that* $n, m \ge N$ *implies* $d(u_n, u_m) < \varepsilon$. *The space* M *is called* ***complete*** *if every Cauchy sequence converges.*

A sequence $\{u_n\}$ *converges to* u *iff for every* $\varepsilon > 0$ *there is an integer* N *such that* $n \ge N$ *implies* $d(u_n, u) < \varepsilon$. This follows readily from the definitions **1.1.11** and **1.2.2**. *A convergent sequence* $\{u_n\}$ *is a Cauchy sequence.* To see this, let $\varepsilon > 0$ be given. Choose N such that $n \ge N$ implies $d(u_n, u) < \varepsilon/2$. Thus, $n, m \ge N$ implies $d(u_n, u_m) \le d(u_n, u) + d(u, u_m) < \varepsilon/2 + \varepsilon/2 =$

ε by the triangle inequality. Completeness requires that, conversely, every Cauchy sequence converges. A basic fact about $\mathbb{R}^n$ is that with the standard metric, it is complete. The proof is found in any textbook on advanced calculus.

1.2.6 Contraction Mapping Theorem *Let* M *be a complete metric space and* $f : M \to M$ *a mapping. Assume there is a constant* k, *where* $0 \le k < 1$ *such that*

$$d(f(m), f(n)) \le k\, d(m, n) ,$$

for all $m, n \in M$; *such an* f *is called a* ***contraction***. *Then* f *has a unique fixed point; i.e., there exists a unique* $m_* \in M$ *such that* $f(m_*) = m_*$.

Proof Let m_0 be an arbitrary point of M and define recursively $m_{i+1} = f(m_i)$, $i = 0, 1, 2, \ldots$. Induction shows that $d(m_i, m_{i+1}) \le k^i\, d(m_0, m_1)$, so that for $i < j$,

$$d(m_i, m_j) \le (k^i + \ldots + k^{j-1}) d(m_0, m_1) .$$

For $0 \le k < 1$, $1 + k + k^2 + k^3 + \ldots$ is a convergent series, and so $k^i + k^{i+1} + \ldots + k^{j-1} \to 0$ as $i, j \to \infty$. This shows that the sequence $\{m_i\}$ is Cauchy and thus by completeness of M it converges to a point m_*. Since

$$d(m_*, f(m_*)) \le d(m_*, m_i) + d(m_i, f(m_i)) + d(f(m_i), f(m_*)) < (1 + k) d(m_*, m_i) + k^i\, d(m_0, m_1)$$

is arbitrarily small, it follows that $m_* = f(m_*)$, thus proving the existence of a fixed point of f. If m' is another fixed point of f, then $d(m', m_*) = d(f(m'), f(m_*)) \le k d(m', m_*)$, which, by virtue of $0 \le k < 1$, implies $d(m', m_*) = 0$, so $m' = m_*$. Thus we have uniqueness. ■

We note that the condition $k < 1$ is necessary, for if $M = \mathbb{R}$ and $f(x) = x + 1$, then $k = 1$, but f has no fixed point (see also Exercise **1.5E**).

At this point the true significance of the Contraction Mapping Theorem cannot be demonstrated. When applied to the right spaces, however, it will yield the Inverse Function Theorem (Chapter 2) and the Basic Existence Theorem for Differential Equations (Chapter 4). A hint of this is given in Exercise **1.2I**.

Exercises

1.2A Let $d((x_1, y_1), (x_2, y_2)) = \sup(|x_1 - x_2|, |y_1 - y_2|)$. Show that d is a metric on $\mathbb{R}^2$ and is equivalent to the standard metric.

1.2B Let $f(x) = \sin(1/x)$, $x > 0$. Find the distance between the graph of f and (0, 0).

1.2C Show that every separable metric space is second countable.

1.2D Show that every metric space has an equivalent metric in which the diameter of the space is 1. (*Hint*: Consider $d_1(m, n) = d(m, n)/(1 + d(m, n))$.)

1.2E In a metric space M, let $\mathcal{U}(m) = \{U \subset M \mid \text{there exists } \varepsilon > 0 \text{ such that } D_\varepsilon(m) \subset U\}$. Show that $\mathcal{U}(m)$ satisfies **V1-V4** of Exercise **1.1H**. This shows how the metric topology can be defined in an alternative way starting from neighborhoods.

1.2F In a metric space show that $\mathrm{cl}(A) = \{u \in M \mid d(u, A) = 0\}$.

*Exercises **1.2G-I** use the notion of continuity from calculus (see Section 1.3).*

1.2G Let M denote the set of continuous functions $f : [0,1] \to \mathbb{R}$ on the interval [0, 1]. Show that

$$d(f, g) = \int_0^1 | f(x) - g(x) | \, dx$$

is a metric.

1.2H Let M denote the set of continuous functions $f : [0,1] \to \mathbb{R}$. Set

$$d(f, g) = \sup\{ \, | f(x) - g(x) | \mid 0 \le x \le 1 \} \, .$$

(i) Show that d is a metric on M.
(ii) Show that $f_n \to f$ in M iff f_n converges *uniformly* to f.
(iii) By consulting theorems on uniform convergence from your advanced calculus text, show that M is a complete metric space.

1.2I Let M be as in the previous exercise and define $T : M \to M$ by

$$T(f)(x) = a + \int_0^x K(x, y) \, f(y) \, dy \ ,$$

where a is a constant and K is a continuous function of two variables. Let

$$k = \sup\left\{ \int_0^x | K(x, y) | \, dy \;\middle|\; 0 \le x \le 1 \right\}$$

and suppose $k < 1$. Prove the following:

(i) T is a contraction.

(ii) Deduce the existence of a unique solution of the integral equation

$$f(x) = a + \int_0^x K(x, y)\, f(y)\, dy \ .$$

(iii) Taking a special case of (ii), prove the "existence of e^x."

§1.3 *Continuity*

1.3.1 Definition *Let* S *and* T *be topological spaces and* $\varphi : S \to T$ *be a mapping. We say that* φ *is* ***continuous*** *at* $u \in S$ *if for every neighborhood* V *of* $\varphi(u)$ *there is a neighborhood* U *of* u *such that* $\varphi(U) \subset V$. *If, for every open set* V *of* T, $\varphi^{-1}(V) = \{u \in S \mid \varphi(u) \in V\}$ *is open in* S, φ *is* ***continuous.*** *(Thus,* φ *is continuous if* φ *is continuous at each* $u \in S$*). If the map* $\varphi : S \to T$ *is a* ***bijection*** *(that is, one-to-one and onto), and both* φ *and* φ^{-1} *are continuous,* φ *is called a* ***homeomorphism*** *and* S *and* T *are said to be* ***homeomorphic***.

For example, notice that any map from a discrete topological space to any topological space is continuous. Similarly, any map from an arbitrary topological space to the trivial topological space is continuous. Hence the identity map from the set S topologized with the discrete topology to S with the trivial topology is bijective and continuous, but its inverse is not continuous, hence it is not a homeomorphism.

It follows from **1.3.1** by taking complements and using the set theoretic identity $S\backslash\varphi^{-1}(A) = \varphi^{-1}(T\backslash A)$ that $\varphi : S \to T$ is continuous iff the inverse image of every closed set is closed.

1.3.2 Proposition *Let* S, T *be topological spaces and* $\varphi: S \to T$. *The following are equivalent:*

(i) φ *is continuous;*

(ii) $\varphi(\mathrm{cl}(A)) \subset \mathrm{cl}(\varphi(A))$ *for every* $A \subset S$;

(iii) $\varphi^{-1}(\mathrm{int}(B)) \subset \mathrm{int}(\varphi^{-1}(B))$ *for every* $B \subset T$.

Proof If φ is continuous, then $\varphi^{-1}(\mathrm{cl}(\varphi(A))$ is closed. But $A \subset \varphi^{-1}(\mathrm{cl}(\varphi(A)))$ and hence $\mathrm{cl}(A) \subset \varphi^{-1}(\mathrm{cl}(\varphi(A)))$, or $\varphi(\mathrm{cl}(A)) \subset \mathrm{cl}(\varphi(A))$. Conversely, let $B \subset T$ be closed and $A = \varphi^{-1}(B)$. Then $\mathrm{cl}(A) \subset \varphi^{-1}(B) = A$, so A is closed. A similar argument shows that (ii) and (iii) are equivalent. ■

This proposition combined with **1.1.12** (or a direct argument) gives the following.

1.3.3 Corollary *Let* S *and* T *be topological spaces with* S *first countable and* $\varphi : S \to T$. *The map* φ *is continuous iff for every sequence* $\{u_n\}$ *converging to* u, $\{\varphi(u_n)\}$ *converges to* $\varphi(u)$, *for all* $u \in S$.

1.3.4 Proposition *The composition of two continuous maps is a continuous map.*

Proof If $\varphi_1 : S_1 \to S_2$ and $\varphi_2 : S_2 \to S_3$ are continuous maps and if U is open in S_3, then $(\varphi_2 \circ \varphi_1)^{-1}(U) = \varphi_1^{-1}(\varphi_2^{-1}(U))$ is open in S_1 since $\varphi_2^{-1}(U)$ is open in S_2 by continuity of φ_2

and hence its inverse image by φ_1 is open in S_1, by continuity of φ_1. ∎

1.3.5 Corollary *The set of all homeomorphisms of a topological space to itself forms a group under composition.*

Proof Composition of maps is associative and has for identity element the identity mapping. Since the inverse of a homeomorphism is a homeomorphism by definition, and since for any two homeomorphisms φ_1, φ_2 of S to itself, the maps $\varphi_1 \circ \varphi_2$ and $(\varphi_1 \circ \varphi_2)^{-1} = \varphi_2^{-1} \circ \varphi_1^{-1}$ are continuous by **1.3.4**, the corollary follows. ∎

1.3.6 Proposition *The space of continuous maps* $f : S \to \mathbb{R}$ *forms an algebra under pointwise addition and multiplication.*

Proof We have to show that if f and g are continuous, then so are $f + g$ and fg. Let $s_0 \in S$ be fixed and $\varepsilon > 0$. By continuity of f and g at s_0, there exists an open set U in S such that $| f(s) - f(s_0) | < \varepsilon/2$, and $| g(s) - g(s_0) | < \varepsilon/2$ for all $s \in U$. Then

$$| (f + g)(s) - (f + g)(s_0) | \leq | f(s) - f(s_0) | + | g(s) - g(s_0) | < \varepsilon .$$

Similarly, for $\varepsilon > 0$, choose a neighborhood V of s_0 such that $| f(s) - f(s_0) | < \delta$, $| g(s) - g(s_0)| < \delta$ for all $s \in V$, where δ is any positive number satisfying $(\delta + | f(s_0) |)\delta + | g(s_0) | \delta < \varepsilon$. Then

$$\begin{aligned} | (fg)(s) - (fg)(s_0) | &\leq | (f(s) | \; | g(s) - g(s_0) | + | f(s) - f(s_0) | \; | g(s_0) | \\ &< (\delta + | f(s_0) |)\delta + \delta | g(s_0) | < \varepsilon . \end{aligned}$$

Therefore, $f + g$ and fg are continuous at s_0. ∎

Continuity is defined by requiring that *inverse images* of open (closed) sets are open (closed). In many situations it is important to ask whether the *image* of an open (closed) set is open (closed).

1.3.7 Definition *A map* $\varphi : S \to T$, *where* S *and* T *are topological spaces, is called* ***open*** *(resp.,* ***closed****) if the image of every open (resp., closed) set in* S *is open (resp., closed) in* T.

Thus a homeomorphism is a bijective continuous open (closed) map. It should be noted, however, that not every open (closed) map is closed (open). Examples and applications of this notion will be given in the next section. For the moment, note that the identity map of S topologized with the trivial and discrete topologies on the domain and range, respectively, is not continuous but is both open and closed.

Let us now turn our attention to continuous maps between metric spaces. For these spaces, continuity may be expressed in terms of ε's and δ's familiar from calculus.

1.3.8 Proposition *Let* (M_1, d_1) *and* (M_2, d_2) *be metric spaces, and* $\varphi : M_1 \to M_2$ *a given mapping. Then* φ *is continuous at* $u_1 \in M_1$ *iff for every* $\varepsilon > 0$ *there is a* $\delta > 0$ *such that* $d_1(u_1, u_1') < \delta$ *implies* $d_2(\varphi(u_1), \varphi(u_1')) < \varepsilon$.

Proof Let φ be continuous at u_1 and consider $D^2_\varepsilon(\varphi(u_1))$, the ε–disk at $\varphi(u_1)$ in M_2. Then there is a δ–disk $D^1_\delta(u_1)$ in M_1 such that $\varphi(D^1_\delta(u_1)) \subset D^2_\varepsilon(\varphi(u_1))$ (**1.3.1**); i.e., $d(u_1, u_1') < \delta$ implies $d_2(\varphi(u_1), \varphi(u_1')) < \varepsilon$. Conversely, assume this latter condition is satisfied and let V be a neighborhood of $\varphi(u_1)$ in M_2. Choosing an ε–disk $D^2_\varepsilon(\varphi(u_1)) \subset V$ there exists $\delta > 0$ such that $\varphi(D^1_\delta(u_1)) \subset D^2_\varepsilon(\varphi(u_1))$ by the foregoing argument. Thus φ is continuous at u_1. ■

In a metric space we also have the notions of uniform continuity and uniform convergence.

1.3.9 Definition

(i) *Let* (M_1, d_1) *and* (M_2, d_2) *be metric spaces and* $\varphi : M_1 \to M_2$. *We say* φ *is* ***uniformly continuous*** *if for every* $\varepsilon > 0$ *there is a* $\delta > 0$ *such that* $d_1(u, v) < \delta$ *implies* $d_2(\varphi(u), \varphi(v)) < \varepsilon$.

(ii) *Let* S *be a set,* M *a metric space,* $\varphi_n : S \to M$, $n = 1, 2, \ldots$, *and* $\varphi : S \to M$ *be given mappings. We say* $\varphi_n \to \varphi$ ***uniformly*** *if for every* $\varepsilon > 0$ *there is an* N *such that* $d(\varphi_n(u), \varphi(u)) < \varepsilon$ *for all* $n \geq N$ *and all* $u \in S$.

For example, a map satisfying $d(\varphi(u), \varphi(v)) \leq Kd(u, v)$ for a constant K is uniformly continuous. Uniform continuity and uniform convergence ideas come up in the construction of a metric on the space of continuous maps. This is considered next.

1.3.10 Proposition *Let* M *be a topological space and* (N, d) *a complete metric space. Then the collection* $C(M, N)$ *of all bounded continuous maps* $\varphi : M \to N$ *forms a complete metric space with the metric*

$$d^0(\varphi, \psi) = \sup\{d(\varphi(u), \psi(u)) \mid u \in M\} .$$

Proof It is readily verified that d^0 is a metric. Convergence of a sequence $f_n \in C(M, N)$ to $f \in C(M, N)$ in the metric d^0 is the same as ***uniform convergence***, as is readily checked. (See Exercise **1.2H**.) Now, if $\{f_n\}$ is a Cauchy sequence in $C(M, N)$, then $\{f_n(x)\}$ is Cauchy for each $x \in M$ since $d(f_n(x), f_m(x)) \leq d^0(f_n, f_m)$. Thus f_n converges pointwise, defining a function $f(x)$. We must show that $f_n \to f$ uniformly and that f is continuous. First, given $\varepsilon > 0$, choose N such that $d^0(f_n, f_m) < \varepsilon/2$ if $n, m \geq N$. Second, for any $x \in M$, pick $N_x \geq N$ so that $d(f_m(x), f(x)) < \varepsilon/2$ if $m \geq N_x$. Thus with $n \geq N$ and $m \geq N_x$,

$$d(f_n(x), f(x)) \leq d(f_n(x), f_m(x)) + d(f_m(x), f(x)) < \varepsilon/2 + \varepsilon/2 = \varepsilon ,$$

so $f_n \to f$ uniformly. The reader can similarly verify that f is continuous (see Exercise **1.3F**; look in any advanced calculus text such as Marsden [1974a] for the case of $\mathbb{R}^n$ if you get stuck). ∎

Exercises

1.3A Show that a map $\varphi : S \to T$ between the topological spaces S and T is continuous iff for every set $B \subset T$, $\text{cl}(\varphi^{-1}(B)) \subset \varphi^{-1}(\text{cl}(B))$. Show that continuity of φ does *not* imply any inclusion relations between $\varphi(\text{int } A)$ and $\text{int}(\varphi(A))$.

1.3B Show that a map $\varphi : S \to T$ is continuous and closed if for every subset $U \subset S$, $\varphi(\text{cl}(U)) = \text{cl}(\varphi(U))$.

1.3C Show that compositions of open (closed) mappings are also open (closed) mappings.

1.3D Show that $\varphi :]0, \infty[\to]0, \infty[$ defined by $\varphi(x) = 1/x$ is continuous but not uniformly continuous.

1.3E Show that if d is a pseudometric on M, then the map $d(\cdot, A) : M \to \mathbb{R}$, for $A \subset M$ a fixed subset, is continuous.

1.3F If S is a topological space, T a metric space, and $\varphi_n : S \to T$ a sequence of continuous functions uniformly convergent to a mapping $\varphi : S \to T$, then φ is continuous.

§1.4 *Subspaces, Products, and Quotients*

This section concerns the construction of new topological spaces from old ones. The first basic operation of this type we consider is the formation of subset topologies.

1.4.1 Definition *If* A *is a subset of a topological space* S *with topology* $\mathcal{O}$, *the* ***relative topology*** *on* A *is defined by* $\mathcal{O}_A = \{U \cap A \mid U \in \mathcal{O}\}$.

The following identities show that $\mathcal{O}_A$ *is indeed a topology*:

(1) $\varnothing \cap A = \varnothing$, $S \cap A = A$;

(2) $(U_1 \cap A) \cap (U_2 \cap A) = (U_1 \cap U_2) \cap A$; and

(3) $\bigcup_\alpha (U_\alpha \cap A) = (\bigcup_\alpha U_\alpha) \cap A$.

Thus, for example, the topology on $S^{n-1} = \{x \in \mathbb{R}^n \mid d(x, 0) = 1\}$ is the relative topology induced from $\mathbb{R}^n$; i.e., a neighborhood of a point $x \in S^{n-1}$ is a subset of S^{n-1} containing the set $D_\varepsilon(x) \cap S^{n-1}$ for some $\varepsilon > 0$. Note that an open (closed) set in the relative topology of A is in general *not* open (closed) in S. For example, $D_\varepsilon(x) \cap S^{n-1}$ is open in S^{n-1} but it is neither open nor closed in $\mathbb{R}^n$. However, if A is open (closed) in S, then any open (closed) set in the relative topology is also open (closed) in S.

If $\varphi : S \to T$ is a continuous mapping, then the restriction $\varphi \mid A : A \to T$ is also continuous in the relative topology. The converse is false. For example, the mapping $\varphi : \mathbb{R} \to \mathbb{R}$ defined by $\varphi(x) = 0$ if $x \in \mathbb{Q}$ and $\varphi(x) = 1$ if $x \in \mathbb{R} \backslash \mathbb{Q}$ is discontinuous, but $\varphi \mid \mathbb{Q} : \mathbb{Q} \to \mathbb{R}$ is a constant mapping and is thus continuous.

1.4.2 Definition *Let* S *and* T *be topological spaces and* $S \times T = \{(u, v) \mid u \in S$ *and* $v \in T\}$. *The* ***product topology*** *on* $S \times T$ *consists of all subsets that are unions of sets which have the form* $U \times V$, *where* U *is open in* S *and* V *is open in* T. *Thus, these* ***open rectangles*** *form a basis for the topology.*

Products of more than two factors can be considered in a similar way; it is straightforward to verify that the map $((u, v), w) \mapsto (u, (v, w))$ is a homeomorphism of $(S \times T) \times Z$ onto $S \times (T \times Z)$. Similarly, one sees that $S \times T$ is homeomorphic to $T \times S$. Thus one can take products of any number of topological spaces and the factors can be grouped in any order; we simply write $S_1 \times \cdots \times S_n$ for such a finite product. For example, $\mathbb{R}^n$ has the product topology of $\mathbb{R} \times \cdots \times \mathbb{R}$ (n times). Indeed, using the ***maximum metric*** $d(x, y) = \max_{1 \leq i \leq n} (\mid x^i - y^i \mid)$, which is equivalent to the standard one, we see that the ε-disk at $\mathbf{x}$ coincides with the set $]x^1 - \varepsilon, x^1 + \varepsilon[\times \cdots \times]x^n - \varepsilon, x^n + \varepsilon[$. For generalizations to infinite products see Exercise

1.4K, and to metric spaces see Exercise **1.4N**.

1.4.3 Proposition *Let* S *and* T *be topological spaces and denote by* $p_1 : S \times T \to S$ *and* $p_2 : S \times T \to T$ *the canonical projections:* $p_1(s, t) = s$, *and* $p_2(s, t) = t$. *Then*

(i) p_1 *and* p_2 *are open mappings; and*

(ii) *a mapping* $\varphi : X \to S \times T$, *where* X *is a topological space, is continuous iff both the maps* $p_1 \circ \varphi : X \to S$ *and* $p_2 \circ \varphi : X \to T$ *are continuous.*

Proof (i) follows directly from the definitions.

(ii) φ is continuous iff $\varphi^{-1}(U \times V)$ is open in X, for $U \subset S$ and $V \subset T$ open sets. Since $\varphi^{-1}(U \times V) = \varphi^{-1}(U \times T) \cap \varphi^{-1}(S \times V) = (p_1 \circ \varphi)^{-1}(U) \cap (p_2 \circ \varphi)^{-1}(V)$, the assertion follows. ■

In general, the maps p_i, $i = 1, 2$, are *not* closed. For example, if $S = T = \mathbb{R}$ the set $A = \{(x, y) \mid xy = 1, x > 0\}$ is closed in $S \times T = \mathbb{R}^2$, but $p_1(A) =]0, \infty[$ which is not closed in S.

1.4.4 Proposition *A topological space* S *is Hausdorff iff the* ***diagonal*** *which is defined by* $\Delta_S = \{(s, s) \mid s \in S\} \subset S \times S$ *is a closed subspace of* $S \times S$, *with the product topology.*

Proof It is enough to remark that S is Hausdorff iff for every two distinct points $p, q \in S$ there exist neighborhoods U_p, U_q of p, q, respectively, such that $(U_p \times U_q) \cap \Delta_S = \varnothing$. ■

In a number of places later in the book we are going to form new topological spaces by collapsing old ones. We define this process now and give some examples.

1.4.5 Definition *Let* S *be a set. An* ***equivalence relation*** $\sim$ *on* S *is a binary relation such that for all* $u, v, w \in S$,

(i) $u \sim u$ *(**reflexivity**);*

(ii) $u \sim v$ *iff* $v \sim u$ *(**symmetry**); and*

(iii) $u \sim v$ *and* $v \sim w$ *implies* $u \sim w$ *(**transitivity**).*

The ***equivalence class*** *containing* u, *denoted* [u], *is defined by* $[u] = \{v \in S \mid u \sim v\}$. *The set of equivalence classes is denoted* $S/\sim$, *and the mapping* $\pi : S \to S/\sim$ *defined by* $u \mapsto [u]$ *is called the* ***canonical projection.***

Note that S is the disjoint union of its equivalence classes. The collection of subsets U of $S/\sim$ such that $\pi^{-1}(U)$ is open in S is a topology because

(1) $\pi^{-1}(\varnothing) = \varnothing$, $\pi^{-1}(S/\sim) = S$;

(2) $\pi^{-1}(U_1 \cap U_2) = \pi^{-1}(U_1) \cap \pi^{-1}(U_2)$; and

(3) $\pi^{-1}(\bigcup_\alpha U_\alpha) = \bigcup_\alpha \pi^{-1}(U_\alpha)$.

1.4.6 Definition *Let* S *be a topological space and* ~ *an equivalence relation on* S. *Then the collection of sets* $\{U \subset S/\!\sim \mid \pi^{-1}(U)$ *is open in* $S\}$ *is called the* ***quotient topology*** *on* $S/\!\sim$.

1.4.7 Examples

A The Torus Consider $\mathbb{R}^2$ and the relation ~ defined by

$$(a_1, a_2) \sim (b_1, b_2) \text{ if } a_1 - b_2 \in \mathbb{Z} \text{ and } a_2 - b_2 \in \mathbb{Z}$$

($\mathbb{Z}$ denotes the integers). Then $\mathbb{T}^2 = \mathbb{R}^2/\!\sim$ is called the *2-torus*. In addition to the quotient topology, it inherits a group structure by setting $[(a_1, a_2)] + [(b_1, b_2)] = [(a_1, a_2) + (b_1, b_2)]$. The n-dimensional torus $\mathbb{T}^n$ is defined in a similar manner.

The torus $\mathbb{T}^n$ may be obtained in two other ways. First, let $\square$ be the unit square in $\mathbb{R}^2$ with the subspace topology. Define ~ by $\mathbf{x} \sim \mathbf{y}$ iff any of the following hold:

(i) $\mathbf{x} = \mathbf{y}$;

(ii) $x_1 = y_1$, $x_2 = 0$, $y_2 = 1$;

(iii) $x_1 = y_1$, $x_2 = 1$, $y_2 = 0$;

(iv) $x_2 = y_2$, $x_1 = 0$, $y_1 = 1$; or

(v) $x_2 = y_2$, $x_1 = 1$, $y_1 = 0$,

as indicated in Figure 1.4.1. Then $\mathbb{T}^2 = \square/\!\sim$. Second, define $\mathbb{T}^2 = S^1 \times S^1$, also shown in Figure 1.4.1.

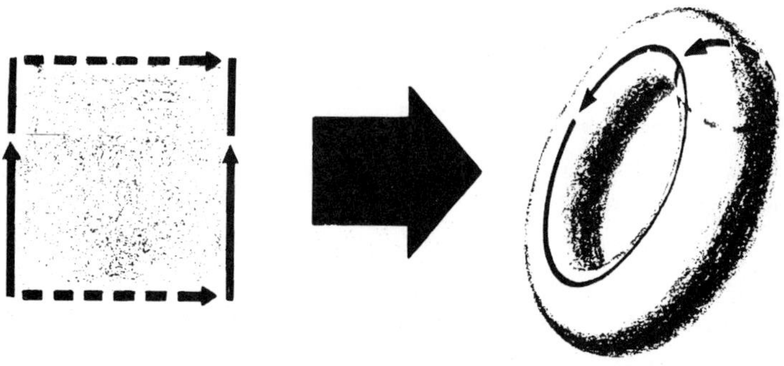

Figure 1.4.1

B The Klein bottle is obtained by reversing one of the orientations on $\square$, as indicated in Figure 1.4.2. Then $\mathbb{K} = \square/\!\sim$ (the equivalence relation indicated) is the ***Klein bottle***. Although it is realizable as a subset of $\mathbb{R}^4$, it is convenient to picture it in $\mathbb{R}^3$ as shown. In a sense we will make precise in Chapter 6, one can show that $\mathbb{K}$ is not "orientable." Also note that $\mathbb{K}$ does not inherit a group structure from $\mathbb{R}^2$, as did $\mathbb{T}^2$.

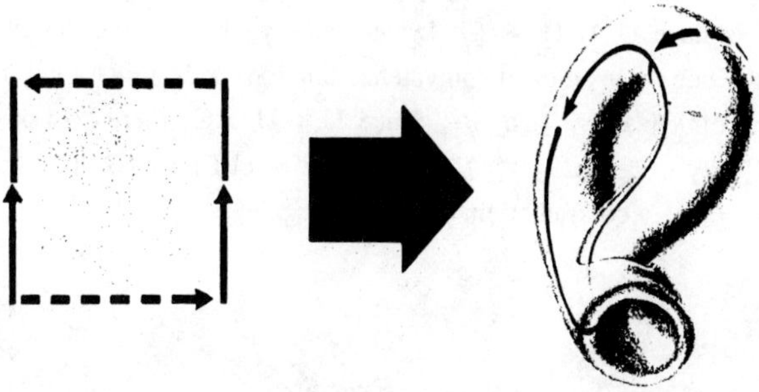

Figure 1.4.2

C **Projective Space** On $\mathbb{R}^n\backslash\{0\}$ define $\mathbf{x} \sim \mathbf{y}$ if there is a nonzero real constant λ such that $\mathbf{x} = \lambda\mathbf{y}$. Then $(\mathbb{R}^n\backslash\{0\})/\sim$ is called ***real projective*** $(n-1)$-***space*** and is denoted by $\mathbb{RP}^{n-1}$. Alternatively, $\mathbb{RP}^{n-1}$ can be defined as S^{n-1} (the unit sphere in $\mathbb{R}^n$) with antipodal points $\mathbf{x}$ and $-\mathbf{x}$ identified. (It is easy to see that this gives a homeomorphic space.) One defines ***complex projective space*** $\mathbb{CP}^{n-1}$ in an analogous way where now λ is complex. ♦

1.4.8 Proposition *Let* $\sim$ *be an equivalence relation on the topological space* S *and* $\pi : S \to S/\sim$ *the canonical projection. A map* $\varphi : S/\sim \to T$, *where* T *is another topological space, is continuous iff* $\varphi \circ \pi : S \to T$ *is continuous.*

Proof φ is continuous iff for every open set $V \subset T$, $\varphi^{-1}(V)$ is open in $S/\sim$, i.e., iff the set $(\varphi \circ \pi)^{-1}(V)$ is open in S. ■

1.4.9 Definition *The set* $\Gamma = \{(s, s') \mid s \sim s'\} \subset S \times S$ *is called the* **graph** *of the equivalence relation* $\sim$. *The equivalence relation is called* **open (closed)** *if the canonical projection* $\pi : S \to S/\sim$ *is open (closed).*

We note that $\sim$ is open (closed) iff for any open (closed) subset A of S the set $\pi^{-1}(\pi(A))$ is open (closed). As in **1.4.8**, for an open (closed) equivalence relation $\sim$ on S, a map $\varphi : S/\sim \to T$ is open (closed) iff $\varphi \circ \pi : S \to T$ is open (closed). In particular, if $\sim$ is an open (closed) equivalence relation on S and $\varphi : S/\sim \to T$ is a bijective continuous map, then φ is a homeomorphism iff $\varphi \circ \pi$ is open (closed).

1.4.10 Proposition *If* $S/\sim$ *is Hausdorff, then the graph* Γ *of* $\sim$ *is closed in* $S \times S$. *If the equivalence relation* $\sim$ *is open and* Γ *is closed (as a subset of* $S \times S$*), then* $S/\sim$ *is Hausdorff.*

Proof If $S/\sim$ is Hausdorff, then $\Delta_{S/\sim}$ is closed by **1.4.4** and hence $\Gamma = (\pi \times \pi)^{-1}(\Delta_{S/\sim})$ is closed on $S \times S$, where $\pi \times \pi : S \times S \to (S/\sim) \times (S/\sim)$ is given by $(\pi \times \pi)(x, y) = ([x], [y])$.

Assume that Γ is closed and $\sim$ is open. If $S/\sim$ is not Hausdorff then there are distinct points $[x], [y] \in S/\sim$ such that for any pair of neighborhoods U_x and U_y of $[x]$ and $[y]$, respectively, we have $U_x \cap U_y \neq \varnothing$. Let V_x and V_y be any open neighborhoods of x and y, respectively. Since $\sim$ is an open equivalence relation, $\pi(V_x) = U_x$ and $\pi(V_y) = U_y$ are open neighborhoods of $[x]$ and $[y]$ in $S/\sim$. Since $U_x \cap U_y \neq \varnothing$, there exist $x' \in V_x$ and $y' \in V_y$ such that $[x'] = [y']$; i.e., $(x', y') \in \Gamma$. Thus $(x, y) \in \text{cl}(\Gamma)$ by **1.1.9**(i). As Γ is closed, $(x, y) \in \Gamma$, i.e., $[x] = [y]$, a contradiction. ■

Exercises

1.4A Show that the sequence $x_n = 1/n$ in the topological space $]0, 1]$ (with the relative topology from $\mathbb{R}$) does not converge.

1.4B If $\varphi : S \to T$ is continuous and T is Hausdorff, show that the graph of f, $\Gamma_f = \{(s, f(s)) \mid s \in S\}$ is closed in $S \times T$.

1.4C Let X and Y be topological spaces with Y Hausdorff. Show that for any continuous maps $f, g : X \to Y$, the set $\{x \in X \mid f(x) = g(x)\}$ is closed. (*Hint*: Consider the mapping $x \mapsto (f(x), g(x))$ and use **1.4.4**.) Thus, if $f(x) = g(x)$ at all points of a dense subset of X, then $f = g$.

1.4D Define a ***topological manifold*** to be a space locally homeomorphic to $\mathbb{R}^n$. Find a topological manifold that is not Hausdorff. (*Hint:* Consider $\mathbb{R}$ with "extra origins.")

1.4E Show that a mapping $\varphi : S \to T$ is continuous iff the mapping $s \mapsto (s, f(s))$ of S to the graph $\Gamma_f = \{(s, f(s)) \mid s \in S\} \subset S \times S$ is a homeomorphism of S with Γ_f (give Γ_f the subspace topology induced from the product topology of $S \times T$).

1.4F Show that every subspace of a Hausdorff (resp., regular) space is Hausdorff (resp., regular). Conversely, if each point of a topological space has a closed neighborhood that is Hausdorff (resp., regular) in the subspace topology, then the topological space is Hausdoff (resp., regular). (*Hint:* use Exercise **1.1F** and **1.1G**.)

1.4G Show that a product of topological spaces is Hausdorff iff each factor is Hausdorff.

1.4H Let S, T be topological spaces and $\sim, \approx$ be equivalence relations on S and T, respectively. Let $\varphi : S \to T$ be continuous such that $s_1 \sim s_2$ implies $\varphi(s_1) \approx \varphi(s_2)$. Show that the induced mapping $\hat{\varphi} : S/\sim \to T/\approx$ is continuous.

1.4I Let S be a Hausdorff space and assume there is a continuous map $\sigma : S/\sim \to S$ such that $\pi \circ \sigma = i_{S/\sim}$, the identity. Show that $S/\sim$ is Hausdorff and $\sigma(S/\sim)$ is closed in S.

1.4J Let M and N be metric spaces, N complete, and $\varphi : A \to N$ be uniformly continuous (A with the induced metric topology). Show that φ has a unique extension $\varphi : \mathrm{cl}(A) \to N$ that is uniformly continuous.

1.4K Let S be a set, T_α a family of topological spaces, and $\varphi_\alpha : S \to T_\alpha$ a family of mappings. Let $\mathcal{B}$ be the collection of finite intersections of sets of the form $\varphi_\alpha^{-1}(U_\alpha)$ for U_α open in T_α. The ***initial topology*** on S given by the family $\varphi_\alpha : S \to T_\alpha$ has as basis the collection $\mathcal{B}$. Show that this topology is characterized by the fact that any mapping $\varphi : R \to S$ from a topological space R is continuous iff all $\varphi_\alpha \circ \varphi : R \to T_\alpha$ are continuous. Show that the subspace and product topologies are initial topologies. Define the product of an arbitrary infinite family of topological spaces and describe the topology.

1.4L Let T be a set and $\varphi_\alpha : S_\alpha \to T$ a family of mappings, S_α topological spaces with topologies $\mathcal{O}_\alpha$. Let $\mathcal{O} = \{U \subset T \mid \varphi_\alpha^{-1}(U) \in \mathcal{O}_\alpha$ for each $\alpha\}$. Show that $\mathcal{O}$ is a topology on T, called the ***final topology*** on T given by the family $\varphi_\alpha : S_\alpha \to T$. Show that this topology is characterized by the fact that any mapping $\varphi : T \to R$ is continuous iff $\varphi \circ \varphi_\alpha : S_\alpha \to R$ are all continuous. Show that the quotient topology is a final topology.

1.4M Show that in a complete metric space a subspace is closed iff it is complete.

1.4N Show that a product of two metric spaces is also a metric space by finding at least three equivalent metrics. Show that the product is complete if each factor is complete.

§1.5 Compactness

Some basic theorems of calculus, such as "every real valued continuous function on [a,b] attains its maximum and minimum" implicitly use the fact that [a,b] is compact. The general theory of compactness is described in this section.

1.5.1 Definition *Let* S *be a topological space. Then* S *is called* ***compact*** *if for every covering of* S *by open sets* U_α *(that is* $\bigcup_\alpha U_\alpha = S$*) there is a finite subcollection of the* U_α *also covering* S. *A subset* $A \subset S$ *is called* ***compact*** *if* A *is compact in the relative topology. A subset* A *is called* ***relatively compact*** *if* cl(A) *is compact. A space is called* ***locally compact*** *if it is Hausdorff and each point has a relatively compact neighborhood.*

1.5.2 Proposition (i) *If* S *is compact and* $A \subset S$ *is closed, then* A *is compact.*
(ii) *If* $\varphi : S \to T$ *is continuous and* S *is compact, then* $\varphi(S)$ *is compact.*

Proof (i) Let $\{U_\alpha\}$ be an open covering of A. Then $\{U_\alpha, S \setminus A\}$ is an open covering of S and hence contains a finite subcollection of this covering also covering S. The elements of this collection, except $S \setminus A$, cover A.

(ii) Let $\{U_\alpha\}$ be an open covering of $\varphi(S)$. Then $\{\varphi^{-1}(U_\alpha)\}$ is an open covering of S and thus by compactness of S a finite subcollection $\{\varphi^{-1}(U_{\alpha(i)}) \mid i = 1,\ldots, n\}$, covers S. But then $\{U_{\alpha(i)}\}$, $i = 1,\ldots, n$ covers $\varphi(S)$ and thus $\varphi(S)$ is compact. ■

In a Hausdorff space, compact subsets are closed (exercise). Thus if S is compact, T is Hausdorff and φ is continuous, then φ is closed; if φ is also bijective, then it is a homeomorphism.

1.5.3 Proposition *A product space* $S \times T$ *is compact iff both* S *and* T *are compact.*

Proof. In view of **1.5.2** all we have to show is that if S and T are compact, so is $S \times T$. Let $\{A_\alpha\}$ be a covering of $S \times T$ by open sets. Each A_α is the union of sets of the form $U \times V$ with U and V open in S and T, respectively. Let $\{U_\beta \times V_\beta\}$ be a covering of $S \times T$ by open rectangles. If we show that there exists a finite subcollection of $U_\beta \times V_\beta$ covering $S \times T$, then clearly also a finite subcollection of $\{A_\alpha\}$ will cover $S \times T$.

A finite subcollection of $\{U_\beta \times V_\beta\}$ is found in the following way. Fix $s \in S$. Since the set $\{s\} \times T$ is compact, there is a finite collection

$$U_s \times V_{\beta_1}, \ldots, U_s \times V_{\beta_{i(s)}}$$

covering it. If $U_s = \bigcap_{j=1,\ldots,i(s)} U_{\beta_j}$, then U_s is open, contains s, and

$$U_s \times V_{\beta_1}, \ldots, U_s \times V_{\beta_{i(s)}}$$

covers $\{s\} \times T$. Let $W_s = U_s \times T$; then the collection $\{W_s\}$ is an open covering of $S \times T$ and if we show that only a finite number of these W_s cover $S \times T$, then since

$$W_s = \bigcup_{j=1,\ldots,i(s)} (U_s \times V_{\beta_j}),$$

it follows that a finite number of $U_\beta \times V_\beta$ will cover $S \times T$. Now look at $S \times \{t\}$, for $t \in T$ fixed. Since this is compact, a finite subcollection $W_{s_1}, \ldots, W_{s_k}$ covers it. But then

$$\bigcup_{j=1,\ldots,k} W_{s_i} = S \times T. \qquad \blacksquare$$

As we shall see shortly in **1.5.8**, $[-1, 1]$ is compact. Thus $\mathbb{T}^1$ is compact. It follows from **1.5.3** that the torus $\mathbb{T}^2$, and inductively $\mathbb{T}^n$, are compact. Thus if $\pi : \mathbb{R}^2 \to \mathbb{T}^2$ is the canonical projection we see that $\mathbb{T}^2$ is compact without $\mathbb{R}^2$ being compact; i.e., the converse of **1.5.2**(ii) is false. Nevertheless it sometimes occurs that one does have a converse; this leads to the notion of a proper map discussed in Exercise **1.5J**.

1.5.4 Bolzano-Weierstrass Theorem *If S is a compact Hausdorff space, then every sequence has a convergent subsequence. The converse is also true in metric and second-countable Hausdorff spaces.*

Proof Suppose S is compact and $\{u_n\}$ contains no convergent subsequences. We may assume that the poins of the sequence are distinct. Then $\mathrm{cl}(\{u_n\}) = \{u_n\}$ is compact and each u_n has a neighborhood U_n that contains no other u_m, for otherwise u_n would be a limit of a subsequence. Thus $\{U_n\}$ is an open covering of the compact subset $\{u_n\}$ which contains no finite subcovering, a contradiction.

Let S be second countable, Hausdorff, and such that every sequence has a convergent subsequence. If $\{U_\alpha\}$ is an open covering of S, by the Lindelöf lemma there is a countable collection $\{U_n \mid n = 1, 2 \ldots\}$ also covering S. Thus we have to show that $\{U_n \mid n = 1, 2, \ldots\}$ contains a finite collection covering S. If this is not the case, the family $\{S \setminus \bigcup_{i=1,\ldots,n} U_i\}$ consists of closed nonempty sets and $S \setminus \bigcup_{i=1,\ldots,n} U_i \supset S \setminus \bigcup_{i=1,\ldots,m} U_i$ for $m \geq n$. Choose $p_n \in S \setminus \bigcup_{i=1,\ldots,n} U_i$. If $\{p_n \mid n = 1, 2, \ldots\}$ is infinite, by hypothesis is contains a convergent subsequence; let its limit point be denoted p. Then $p \in S \setminus \bigcup_{i=1,\ldots,n} U_i$ for all n, contradicting the fact that $\{U_n \mid n = 1, 2, \ldots\}$ covers S. Thus $\{p_n \mid n = 1, 2, \ldots\}$ must be a finite set; i.e., for all $n \geq N$, $p_n = p_N$. But then again $p_N \in S \setminus \bigcup_{i=1,\ldots,n} U_i$ for all n, contradicting the fact that $\{U_n \mid n = 1, 2, \ldots\}$ covers S. Hence S is compact.

Let S be a metric space such that every sequence has a convergent subsequence. If we show that S is separable, then since S is a metric space it is second countable (Exercise **1.2C**), and by the preceding paragraph, it will be compact. Separability of S is proved in two steps.

First we show that for any $\varepsilon > 0$ there is a *finite* set of points $\{p_1,...,p_n\}$ such that $S = \bigcup_{i=1,...,n} D_\varepsilon(p_i)$. If this were false, there would exist an $\varepsilon > 0$ such that no finite number of ε-disks cover S. Let $p_1 \in S$ be arbitrary. Since $D_\varepsilon(p_1) \neq S$, there is a point $p_2 \in S \setminus D_\varepsilon(p_1)$. Since $D_\varepsilon(p_1) \cup D_\varepsilon(p_2) \neq S$, there is also a point $p_3 \in S \setminus (D_\varepsilon(p_1) \cup D_\varepsilon(p_2))$, etc. The sequence $\{p_n \mid n=1,2,...\}$ is infinite and $d(p_i, p_j) \geq \varepsilon$. But this sequence has a convergent subsequence by hypothesis, so this subsequence must be Cauchy, contradicting $d(p_i, p_j) \geq \varepsilon$ for all i, j.

Second, we show that the existence for every $\varepsilon > 0$ of a finite set $\{p_1,..., p_{n(\varepsilon)}\}$ such that $S = \bigcup_{i=1,..n(\varepsilon)} D_\varepsilon(p_i)$ implies S is separable. Let A_n denote this finite set for $\varepsilon = 1/n$ and let $A = \bigcup_{n\geq 0} A_n$. Thus A is countable and it is easily verified that cl(A) = S. ■

A property that came up in the preceding proof turns out to be important.

1.5.5 Definition *Let* S *be a metric space. A subset* $A \subset S$ *is called* ***totally bounded*** *if for any* $\varepsilon > 0$ *there exists a finite set* $\{p_1,..., p_n\}$ *in* S *such that* $A \subset \bigcup_{i=1,..n} D_\varepsilon(p_i)$.

1.5.6 Corollary *A metric space is compact iff it is complete and totally bounded. A subset of a complete metric space is relatively compact iff it is totally bounded.*

Proof The previous proof shows that compactness implies total boundedness. As for compactness implying completeness, it is enough to remark that in this context, a Cauchy sequence having a convergent subsequence is itself convergent. Conversely, if S is complete and totally bounded, let $\{p_n \mid n = 1,2, ...\}$ be a sequence in S. By total boundedness, this sequence contains a Cauchy subsequence, which by completeness, converges. Thus S is compact by the Bolzano-Weierstrass theorem. The second statement now readily follows. ■

1.5.7 Proposition *In a metric space compact sets are closed and bounded.*

Proof This is a particular case of the previous corollary but can be easily proved directly. If A is compact, it can be finitely covered by ε-disks: $A = \bigcup_{i=1,..n} D_\varepsilon(p_i)$. Thus

$$\mathrm{diam}(A) \leq \sum_{i=1}^{n} \mathrm{diam}(D_\varepsilon(p_i)) = 2n\varepsilon . \qquad ■$$

From **1.5.2** and **1.5.6**, if S is compact and $\varphi : S \to \mathbb{R}$ is continuous, then $\varphi(S)$ is closed and bounded. Thus φ attains its sup and inf (see Exercise **1.1D**).

1.5.8 Heine-Borel Theorem *In* $\mathbb{R}^n$ *a closed and bounded set is compact.*

Proof By **1.5.2**(i) it is enough to show that closed bounded rectangles are compact in $\mathbb{R}^n$, which in turn is implied via **1.5.3** by the fact that closed bounded intervals are compact in $\mathbb{R}$. To show that $[-a, a]$, $a > 0$ is compact, it suffices to prove (by **1.5.6**) that for any given $\varepsilon > 0$, $[-a, a]$ can be finitely covered by intervals of the form $]p - \varepsilon, p + \varepsilon[$, since we are accepting completeness of $\mathbb{R}$. Let n be a positive integer such that $a < n\varepsilon$. Let $t \in [-a, a]$ and k be the largest (positive or negative) integer satisfying $k\varepsilon \leq t$. Then $-n \leq k \leq n$ and $k\varepsilon \leq t < (k + 1)\varepsilon$. Thus any point $t \in [-a, a]$ belongs to an interval of the form $]k\varepsilon - \varepsilon, k\varepsilon + \varepsilon[$, where $k = -n, \ldots, 0, \ldots, n$ and hence $\{]k\varepsilon - \varepsilon, k\varepsilon + \varepsilon[\mid k = 0, \pm 1, \ldots, \pm n\}$ is a finite covering of $[-a, a]$. ■

This theorem is also proved in virtually every textbook on advanced calculus. As is known from calculus, continuity of a function on an interval $[a, b]$ implies uniform continuity. The generalization to metric spaces is the following.

1.5.9 Proposition *A continuous mapping* $\varphi : M_1 \to M_2$, *where* M_1 *and* M_2 *are metric spaces and* M_1 *is compact, is uniformly continuous.*

Proof The metrics on M_1 and M_2 are denoted by d_1 and d_2. Fix $\varepsilon > 0$. Then for each $p \in M_1$, by continuity of φ there exists $\delta_p > 0$ such that if $d_1(p, q) < \delta_p$, then $d_2(\varphi(p), \varphi(q)) < \varepsilon/2$. Let

$$D_{\delta_1/2}(p_1), \ldots, D_{\delta_n/2}(p_n)$$

cover the compact space M_1 and let $\delta = \min\{\delta_1/2, \ldots, \delta_n/2\}$. Then if $p, q \in M_1$ are such that $d_1(p, q) < \delta$, there exists an index i, $1 \leq i \leq n$, such that $d_1(p, p_i) < \delta_i/2$ and thus

$$d_1(p_i, q) \leq d_1(p_i, p) + d_1(p, q) < (\delta_i/2) + \delta \leq \delta_i.$$

Thus $d_2(\varphi(p), \varphi(q)) \leq d_2(\varphi(p), \varphi(p_i)) + d_2(\varphi(p_i)), \varphi(q)) < \varepsilon$. ■

A useful application of Corollary **1.5.6** concerns relatively compact sets in $C(M, N)$, for metric spaces (M, d_M) and (N, d_N) with M compact and N complete.

1.5.10 Definition *A subset* $\mathcal{F} \subset C(M,N)$ *is called* ***equicontinuous*** *at* $m_0 \in M$, *if given* $\varepsilon > 0$, *there exists* $\delta > 0$ *such that whenever* $d_M(m, m_0) < \delta$, *we have* $d_N(\varphi(m), \varphi(m_0)) < \varepsilon$ *for every* $\varphi \in \mathcal{F}$ *(*δ *is independent of* φ*).* $\mathcal{F}$ *is called* ***equicontinuous****, if it is equicontinuous at every point in* M.

1.5.11 Arzela-Ascoli Theorem *Let* (M, d_M) *and* (N, d_N) *be metric spaces, with* M *compact and* N *complete. A set* $\mathcal{F} \subset C(M, N)$ *is relatively compact iff it is equicontinuous and all the sets* $\mathcal{F}(m) = \{\varphi(m) \mid \varphi \in \mathcal{F}\}$ *are relatively compact in* N.

Proof If $\mathcal{F}$ is relatively compact, it is totally bounded and hence so are all the sets $\mathcal{F}(m)$. Since N is complete, by Corollary **1.5.6** the sets $\mathcal{F}(m)$ are relatively compact. Let $\{\varphi_1, ..., \varphi_n\}$ be the centers of the ε-disks covering $\mathcal{F}$. Then there exists $\delta > 0$ such that if $d_M(m, m') < \delta$, we have $d_N(\varphi_i(m), \varphi_i(m')) \leq \varepsilon/3$, for $i = 1, ..., n$ and hence if $\varphi \in \mathcal{F}$ is arbitrary, φ lies in one of the ε-disks whose center, say, is φ_i, so that

$$d_N(\varphi(m), \varphi(m')) \leq d_N(\varphi(m), \varphi_i(m)) + d_N(\varphi_i(m), \varphi_i(m')) + d_N(\varphi_i(m'), \varphi(m')) < \varepsilon.$$

This shows that $\mathcal{F}$ is equicontinuous.

Conversely, since C(M, N) is complete, by Corollary **1.5.6** we need only show that $\mathcal{F}$ is totally bounded. For $\varepsilon > 0$, find a neighborhood U_m of $m \in M$ such that for all $m' \in U_m$, $d_N(\varphi(m), \varphi(m')) < \varepsilon/4$ for all $\varphi \in \mathcal{F}$ (this is possible by equicontinuity). Let $U_{m(1)}, ..., U_{m(n)}$ be a finite collection of these neighborhoods covering the compact space M. By assumption each $\mathcal{F}(m)$ is relatively compact, hence $\mathcal{F}(m(1)) \cup ... \cup \mathcal{F}(m(n))$ is also relatively compact, and thus totally bounded. Let $D_{\varepsilon/4}(x_1), ..., D_{\varepsilon/4}(x_k)$ cover this union. If $\mathcal{A}$ denotes the set of all mappings $\alpha : \{1, ..., n\} \to \{1, ..., k\}$, then $\mathcal{A}$ is finite and

$$\mathcal{F} = \bigcup_{\alpha \in \mathcal{A}} \mathcal{F}_\alpha, \text{ where } \mathcal{F}_\alpha = \{\varphi \in \mathcal{F} \mid d_N(\varphi(m(i)), x_{\alpha(i)}) < \varepsilon/4, \text{ for all } i = 1, ..., n\}.$$

But if $\varphi, \psi \in \mathcal{F}_\alpha$ and $m \in M$, then $m \in D_{\varepsilon/4}(x_i)$ for some i, and thus

$$d_N(\varphi(m), \psi(m)) \leq d_N(\varphi(m), \varphi(m(i)) + d_N(\varphi(m(i)), x_{\alpha(i)}) + d_N(x_{\alpha(i)}, \psi(m(i))) + d_N(\psi(m(i)), \psi(m)) < \varepsilon;$$

i.e., the diameter of $\mathcal{F}_\alpha$ is $\leq \varepsilon$, so $\mathcal{F}$ is totally bounded. ∎

Combining this with the Heine-Borel theorem, we get the following.

1.5.12 Corollary *If* M *is a compact metric space, a set* $\mathcal{F} \subset C(M, \mathbb{R}^n)$ *is relatively compact iff it is equicontinuous and uniformly bounded (i.e.,* $\|\varphi(m)\| \leq$ *constant for all* $\varphi \in \mathcal{F}$ and $m \in M$).

The following example shows how to use the Arzela-Ascoli theorem.

1.5.13 Example *Let* $f_n : [0, 1] \to \mathbb{R}$ *be continuous and be such that* $|f_n(x)| \leq 100$ *and the derivatives* f_n' *exist and are uniformly bounded on* $]0, 1[$. *Prove* f_n *has a uniformly convergent subsequence.*

We verify that the set $\{f_n\}$ is equicontinuous and bounded. The hypothesis is that $|f_n'(x)| \leq M$ for a constant M. Thus by the mean-value theorem,

$$|f_n(x) - f_n(y)| \leq M|x - y|,$$

so given ε we can choose $\delta = \varepsilon/M$, independent of x, y, and n. Thus $\{f_n\}$ is equicontinuous. It is bounded because $\| f_n \| = \sup_{0 \le x \le 1} | f_n(x) | \le 100$. ♦

Exercises

1.5A Show that a topological space S is compact iff every family of closed subsets of S whose intersection is empty contains a finite subfamily whose intersection is empty.

1.5B Show that every compact metric space is separable. (*Hint:* Use total boundedness.)

1.5C Show that the space of exercise **1.1I** is not locally compact. (*Hint*: Look at the sequence (1/n, 0).)

1.5D (i) Show that every closed subset of a locally compact space is locally compact.
(ii) Show that $S \times T$ is locally compact if both S and T are locally compact.

1.5E Let M be a compact metric space and $T : M \to M$ a map satisfying $d(T(m_1), T(m_2)) < d(m_1, m_2)$ for $m_1 \neq m_2$. Show that T has a unique fixed point.

1.5F Let S be a compact topological space and ~ an equivalence relation on S, so that S/~ is compact. Prove that the following conditions are equivalent (cf. **1.4.10**):

(i) The graph C of ~ is closed in $S \times S$;

(ii) ~ is a closed equivalence relation;

(iii) S/~ is Hausdorff.

1.5G Let S be a Hausdorff space that is locally homeomorphic to a locally compact Hausdorff space (that is, for each $u \in S$, there is a neighborhood of u homeomorphic, in the subspace topology, to an open subset of a locally compact Hausdorff space). Show that S is locally compact. In particular, Hausdorff spaces locally homeomorphic to $\mathbb{R}^n$ are locally compact. Is the conclusion true without the Hausdorff assumption?

1.5H Let M_3 be the set of all 3×3 matrices with the topology obtained by regarding M_3 as $\mathbb{R}^9$. Let $\mathbf{SO}(3) = \{A \in M_3 \mid A \text{ is orthogonal and } \det A = 1\}$.

(i) Show that $\mathbf{SO}(3)$ is compact.

(ii) Let $P = \{Q \in \mathbf{SO}(3) \mid Q \text{ is symmetric}\}$ and let $\varphi : \mathbb{RP}^2 \to \mathbf{SO}(3)$ be given by $\varphi(\ell)$ = the rotation by π about the line $\ell \subset \mathbb{R}^3$. Show that φ maps the space $\mathbb{RP}^2$ homeomorphically onto $P \setminus \{\text{Identity}\}$.

1.5I Let $f_n : [a,b] \to \mathbb{R}$ be uniformly bounded continuous functions. Set

$$F_n(x) = \int_a^x f_n(t)\,dt\,, \qquad a \le x \le b\,.$$

Prove that F_n has a uniformly convergent subsequence.

1.5J Let X and Y be topological spaces, Y be first countable, and $f : X \to Y$ be a continuous map. The map f is called ***proper***, if $f(x_n) \to y$ implies the existence of a convergent subsequence $\{x_{n(i)}\}$, $x_{n(i)} \to x$ such that $f(x) = y$.

(i) Show that f is a closed map.

(ii) Show that if Y is locally compact, f is proper if and only if the inverse image by f of every compact set in Y is a compact set in X.

(iii) Show that if f is proper and Y is locally compact, then X is also locally compact. (We have defined properness only when Y is first countable. The same definition and properties of proper maps hold for general Y if in the definition "sequence" is replaced by "net.")

(iv) Show that the composition of two proper maps is again proper.

(v) Show that any continuous map defined on a compact space is proper.

§1.6 *Connectedness*

Three types of connectedness treated in this section are arcwise connectedness, connectedness, and simple connectedness.

1.6.1 Definition *Let* S *be a topological space and* $I = [0, 1] \subset \mathbb{R}$. *An* ***arc*** φ *in* S *is a continuous mapping* $\varphi : I \to S$. *If* $\varphi(0) = u$, $\varphi(1) = v$, *we say* φ *joins* u *and* v; S *is called* ***arcwise connected*** *if every two points in* S *can be joined by an arc in* S. *A space* S *is called* ***locally arcwise connected*** *if for each point* $x \in S$ *and each neighborhood* U *of* x, *there is a neighborhood* V *of* x *such that any pair of points in* V *can be joined by an arc in* U.

For example, $\mathbb{R}^n$ is arcwise and locally arcwise connected: any two points of $\mathbb{R}^n$ can be joined by the straight line segment connecting them. A set $A \subset \mathbb{R}^n$ is called ***convex*** if this property holds for any two of its points. Thus convex sets in $\mathbb{R}^n$ are arcwise and locally arcwise connected. A set with the trivial topology is arcwise and locally arcwise connected, but in the discrete topology it is neither (unless it has only one point).

1.6.2 Definition *A topological space* S *is* ***connected*** *if* $\varnothing$ *and* S *are the only subsets of* S *that are both open and closed. A subset of* S *is* ***connected*** *if it is connected in the relative topology. A* ***component*** A *of* S *is a nonempty connected subset of* S *such that the only connected subset of* S *containing* A *is* A *itself;* S *is called* ***locally connected*** *if each point has a connected neighborhood. The* ***components*** *of a subset* $T \subset S$ *are the components of* T *in the relative topology of* T *in* S.

For example, $\mathbb{R}^n$ and any convex subset of $\mathbb{R}^n$ are connected and locally connected. The union of two disjoint open convex sets is disconnected but is locally connected; its components are the two convex sets. The trivial topology is connected and locally connected, whereas the discrete topology is neither: its components are all the one-point sets.

Connected spaces are characterized by the following.

1.6.3 Proposition *The following are equivalent:*

(i) S *is not connected;*

(ii) *there is a nonempty proper subset of* S *that is both open and closed;*

(iii) S *is the disjoint union of two nonempty open sets;*

(iv) S *is the disjoint union of two nonempty closed sets.*

The sets in (iii) *or* (iv) *are said to* ***disconnect*** S.

Proof To prove that (i) implies (ii), assume there is a nonempty proper set A that is both open and closed. Then $S = A \cup (S\backslash A)$ with A, $S\backslash A$ open and nonempty. Conversely, if $S = A \cup B$ with A, B open and nonempty, then A is also closed, and thus A is a proper nonempty set of S that is both open and closed. The equivalences of the remaining assertions are similarly checked. ■

1.6.4 Proposition *If* $f : S \to T$ *is a continuous map of topological spaces and* S *is connected (resp., arcwise connected) then so is* $f(S)$.

Proof Let S be arcwise connected and consider $f(s_1), f(s_2) \in f(S) \subset T$. If $c : I \to S$, $c(0) = s_1$, $c(1) = s_2$ is an arc connecting s_1 to s_2, then clearly $f \circ c : I \to T$ is an arc connecting $f(s_1)$ to $f(s_2)$; i.e., $f(S)$ is arcwise connected. Let S be connected and assume $f(S) \subset U \cup V$, where U and V are open and $U \cap V = \varnothing$. Then $f^{-1}(U)$ and $f^{-1}(V)$ are open by continuity of f, $f^{-1}(U) \cup f^{-1}(V) = f^{-1}(U \cup V) \supset f^{-1}(f(S)) = S$, and $f^{-1}(U) \cap f^{-1}(V) = f^{-1}(\varnothing) = \varnothing$, thus contradicting connectedness of S by **1.6.3**. Hence $f(S)$ is connected. ■

We shall prove that arcwise connected spaces are connected. We shall use the following.

1.6.5 Lemma *The only connected sets of* $\mathbb{R}$ *are the intervals (finite, infinite, open, closed, or half-open).*

Proof Let us prove that $[a, b[$ is connected; all other possibilities have identical proofs. If not, $[a, b[= U \cup V$ with U, V nonempty disjoint closed sets in $[a, b[$. Assume that $a \in U$. If $x = \sup(U)$, then $x \in U$ since U is closed in $[a, b[$, and $x < b$ since $V \neq \varnothing$. But then $]x, b[\subset V$ and, since V is closed, $x \in V$. Hence $x \in U \cap V$, a contradiction.

Conversely, let A be a connected set of $\mathbb{R}$. We claim that $[x, y] \subset A$ whenever $x, y \in A$, which implies that A is an interval. If not, there exists $z \in [x, y]$ with $z \notin A$. But in this case $]-\infty, z[\cup A$ and $]z, \infty[\cup A$ are open nonempty sets disconnecting A. ■

1.6.6 Proposition *If* S *is arcwise connected then it is connected.*

Proof If not, there are nonempty, disjoint open sets U_0 and U_1 whose union is S. Let $x_0 \in U_0$ and $x_1 \in U_1$ and let φ be an arc joining x_0 to x_1. Then $V_0 = \varphi^{-1}(U_0)$ and $V_1 = \varphi^{-1}(U_1)$ disconnect $[0, 1]$. ■

A standard example of a space that is connected but is not arcwise connected nor locally connected, is

$$\{(x, y) \in \mathbb{R}^2 \mid x > 0 \text{ and } y = \sin(1/x)\} \cup \{(0, y) \mid -1 < y < 1\} .$$

1.6.7 Proposition *If a space is connected and locally arcwise connected, it is arcwise connected. In particular, a space locally homeomorphic to* $\mathbb{R}^n$ *is connected iff it is arcwise connected.*

Proof Fix $x \in S$. The set $A = \{y \in S \mid y$ can be connected to x by an arc$\}$ is nonempty and open since S is locally arcwise connected. For the same reason, $S\backslash A$ is open. Since S is connected we must have $S\backslash A = \varnothing$; so $A = S$, i.e., S is arcwise connected. ■

1.6.8 Intermediate Value Theorem *Let* S *be a connected space and* $f : S \to \mathbb{R}$ *be continuous. Then* φ *assumes every value between any two values* $f(u)$ *and* $f(v)$.

Proof Suppose $f(u) < a < f(v)$ and f does not assume the value a. Then $U = \{u_0 \mid f(u_0) < a\}$ is both open and closed. ■

An alternative proof uses the fact that $f(S)$ is connected in $\mathbb{R}$ and therefore is an interval.

1.6.9 Proposition *Let* S *be a topological space and* $B \subset S$ *be connected.*

(i) *If* $B \subset A \subset \mathrm{cl}(B)$, *then* A *is connected.*

(ii) *If* B_α *is a family of connected subsets of* S *and* $B_\alpha \cap B \neq \varnothing$, *then* $B \cup (\bigcup_\alpha B_\alpha)$ *is connected.*

Proof If A is not connected, A is the disjoint union of $U_1 \cap A$ and $U_2 \cap A$ where U_1 and U_2 are open in S. Then from **1.1.9**(i), $U_1 \cap B \neq \varnothing$, $U_2 \cap B \neq \varnothing$, so B is not connected. We leave (ii) as an exercise. ■

1.6.10 Corollary *The components of a topological space are closed. Also,* S *is the disjoint union of its components. If* S *is locally connected, the components are open as well as closed.*

1.6.11 Proposition *Let* S *be a first countable compact Hausdorff space and* $\{A_n\}$ *a sequence of closed, connected subsets of* S *with* $A_n \subset A_{n-1}$. *Then* $A = \bigcap_{n \geq 1} A_n$ *is connected.*

Proof As S is normal, if A is not connected, A lies in two disjoint open subsets U_1 and U_2 of S. If $A_n \cap (S\backslash U_1) \cap (S\backslash U_2) \neq \varnothing$ for all n, then there is a sequence $u_n \in A_n \cap (S\backslash U_1) \cap (S\backslash U_2)$ with a subsequence converging to u. As A_n, $S\backslash U_1$, and $S\backslash U_2$ are closed sets, $u \in A \cap (S\backslash U_1) \cap (S\backslash U_2)$, a contradiction. Hence some A_n is not connected. ■

Finally we consider the notion of simple connectivity.

1.6.12 Definition *Let* S *be a topological space and* $c : [0, 1] \to S$ *a continuous map such that* $c(0) = c(1) = p \in S$. *We call* c *a* ***loop*** *in* S *based at* p. *The loop* c *is called* ***contractible*** *if there is a continuous map* $H : [0, 1] \times [0, 1] \to S$ *such that* $H(t, 0) = c(t)$ *and* $H(0, s) = H(1, s) = H(t, 1) = p$ *for all* $t \in [0, 1]$. (See Figure 1.6.1.)

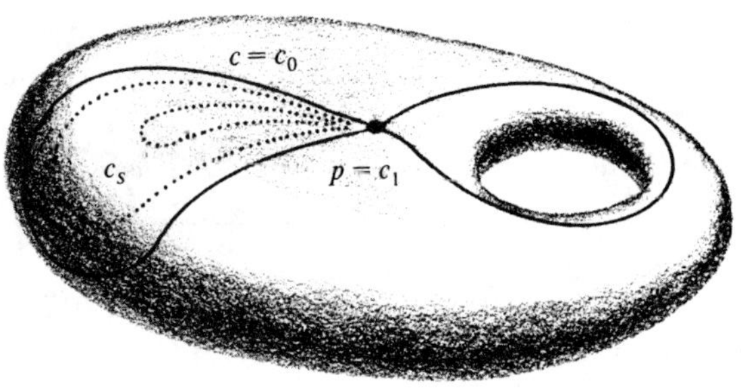

Figure 1.6.1

We think of $c_s(t) = H(t, s)$ as a family of arcs connecting $c_0 = c$ to c_1, a constant arc; see Figure 1.6.1. Roughly speaking, a loop is contractible when it can be shrunk continuously to p by loops beginning and ending at p. The study of loops leads naturally to homotopy theory. In fact, the loops at p can, by successively traversing them, be made into a group called the ***fundamental group***; see Exercise **1.6F.**

1.6.13 Definition *A space* S *is* ***simply connected*** *if* S *is connected and every loop in* S *is contractible.*

In the plane $\mathbb{R}^2$ there is an alternative approach to simple connectedness, by way of the Jordan Curve Theorem; namely, that every simple (nonintersecting) loop in $\mathbb{R}^2$ divides $\mathbb{R}^2$ (that is, its complement has two components). The bounded component of the complement is called the interior, and a subset A of $\mathbb{R}^2$ is simply connected iff the interior of every loop in A lies in A.

We close this section with an optional theorem sometimes used in Riemannian geometry (to show that a Riemannian manifold is second countable) that illustrates the interplay between various notions introduced in this chapter.

1.6.14 Alexandroff's Theorem *An arcwise connected locally compact metric space is separable and hence is second countable.*

Proof (Pfluger [1957]) Since the metric space M is locally compact, each $m \in M$ has compact neighborhoods that are disks. Denote by r(m) the least upper bound of the radii of such disks. If $r(m) = \infty$, since every metric space is first countable, M can be written as a countable union of compact disks. But since each compact metric space is separable (Exercise **1.5B**), these disks and also their union will be separable, and so the proposition is proved in this case. If $r(m_0) < \infty$, since $r(m) \le r(m_0) + d(m, m_0)$, we see that $r(m) < \infty$ for all $m \in M$. By the preceding argument, if we show that M is a countable union of compact sets, the proposition is proved. Then second countability will follow from Exercise **1.2C**.

To show that M is a countable union of compact sets, define the set G_m by $G_m = \{m' \in M \mid d(m', m) \le r(m)/2\}$. These G_m are compact neighborhoods of m. Fix $m(0) \in M$ and put $A_0 = G_{m(0)}$, and, inductively, define

$$A_{n+1} = \bigcup \{G_m \mid m \in A_n\} .$$

Since M is arcwise connected, every point $m \in M$ can be connected by an arc to $m(0)$, which in turn is covered by finitely many G_m. This shows that $M = \bigcup_{n \ge 0} A_n$. Since A_0 is compact, all that remains to be shown is that the other A_n are compact. Assume inductively that A_n is compact and let $\{m(i)\}$ be an infinite sequence of points in A_{n+1}. There exists $m(i)' \in A_n$ such that $m(i) \in G_{m(i)'}$. Since A_n is assumed to be compact there is a subsequence $m(i_k)'$ that converges to a point $m' \in A_n$. But

$$\begin{aligned} d(m(i), m') &\le d(m(i), m(i)') + d(m(i)', m') \\ &\le \frac{r(m(i)')}{2} + d(m(i)', m') \le \frac{r(m')}{2} + \frac{3d(m(i)', m')}{2} . \end{aligned}$$

Hence for i_k big enough, all $m(i_k)$ are in the compact set $\{n \in M \mid d(n, m') \le 3r(m')/2\}$, so $m(i_k)$ has a subsequence converging to a point m. The preceding inequality shows that $m \in A_{n+1}$. By the Bolzano-Weierstrass Theorem, A_{n+1} is compact. ■

Exercises

1.6A Let M be a topological space and $H : M \to \mathbb{R}$ continuous. Suppose $e \in H(M)$. Then show $H^{-1}(e)$ divides M; that is, $M \backslash H^{-1}(e)$ has at least two components.

1.6B Let $O(3)$ be the set of orthogonal 3×3 matrices. Show that $O(3)$ is not connected and that it has two components.

1.6C Show that $S \times T$ is connected (locally connected, arcwise connected, locally arcwise connected) iff both S and T are. (*Hint*: For connectedness write $S \times T = \bigcup_{t \in T} [(S \times \{t\}) \cup (\{s_0\} \times T)]$ for $s_0 \in S$ fixed and use **1.6.9**(ii).)

1.6D Show that S is locally connected iff every component of an open set is open.

1.6E Show that the quotient space of a connected (locally connected, arcwise connected) space is also connected (etc.). (*Hint:* For local connectedness use Exercise **1.6D** and show that the inverse image by π of a component of an open set is a union of components.)

1.6F (i) Let S and T be topological spaces. Two continuous maps $f, g : T \to S$ are called ***homotopic*** if there exists a continuous map $F : [0, 1] \times T \to S$ such that $F(0, t) = f(t)$ and $F(1, t) = g(t)$ for all $t \in T$. Show that homotopy is an equivalence relation.

(ii) Show that S is simply connected if and only if any two continuous paths c_1, $c_2 : [0, 1] \to S$ satisfying $c_1(0) = c_2(0)$, $c_1(1) = c_2(1)$ are homotopic, via a homotopy which preserves the end points, i.e., $F(s, 0) = c_1(0) = c_2(0)$ and $F(s, 1) = c_1(1) = c_2(1)$.

(iii) Define the ***composition*** $c_1 * c_2$ of two paths $c_1, c_2 : [0, 1] \to S$ satisfying $c_1(1) = c_2(0)$ by

$$(c_1 * c_2)(t) = \begin{cases} c_1(2t) & \text{if } t \in [1, 1/2] \\ c_2(2t-1) & \text{if } t \in [1/2, 1] \end{cases} .$$

Show that this composition, when defined, induces an associative operation on endpoints preserving homotopy classes of paths.

(iv) Fix $s_0 \in S$ and consider the set $\pi_1(S, s_0)$ of endpoint fixing homotopy classes of paths starting and ending at s_0. Show that $\pi_1(S, s_0)$ is a group: the identity element is given by the class of the constant path equal to s_0 and the inverse of c is given by the class of $c(1-t)$.

(v) Show that if S is arcwise connected, then $\pi_1(S, s_0)$ is isomorphic to $\pi_1(S, s)$ for any $s \in S$. $\pi_1(S)$ will denote any of these isomorphic groups.

(vi) Show that if S is arcwise connected, then S is simply connected iff $\pi_1(S) = 0$.

§1.7 *Baire Spaces*

The Baire condition on a topological space is fundamental to the idea of "genericity" in differential topology and dynamical systems; see Kelley [1975] and Choquet [1969] for additional information.

1.7.1 Definition *Let* X *be a topological space and* $A \subset X$ *a subset. Then* A *is called* ***residual*** *if* A *is the intersection of a countable family of open dense subsets of* X. *A space* X *is called a* ***Baire space*** *if every residual set is dense. A set* $B \subset X$ *is called a* ***first category set*** *if* $B \subset \bigcup_{n\geq 1} C_n$ *where* C_n *is closed with* $\text{int}(C_n) = \varnothing$. *A* ***second category set*** *is a set which is not of the first category.*

A set $B \subset X$ is called ***nowhere dense*** if $\text{int}(\text{cl}(B)) = \varnothing$, so that $X \backslash A$ is residual iff A is the union of a countable collection of nowhere dense closed sets, i.e., iff $X \backslash A$ is of first category. Clearly, a countable intersection of residual sets is residual.

In a Baire space X, if $X = \bigcup_{n \geq 1} C_n$ where C_n are closed sets, then $\text{int}(C_n) \neq \varnothing$ for some n. For if all $\text{int}(C_n) = \varnothing$, then $O_n = X \backslash C_n$ are open, dense, and we have $\bigcap_{n \geq 1} O_n = X \backslash \bigcup_{n \geq 1} C_n = \varnothing$ contradicting the definition of Baire space. In other words, *Baire spaces are of second category.*

1.7.2 Proposition *Let* X *be a* ***locally Baire space****; that is, each point* $x \in X$ *has a neighborhood* U *such that* cl(U) *is a Baire space. Then* X *is a Baire space.*

Proof Let $A \subset X$ be residual, $A = \bigcap_{n \geq 1} O_n$, where $\text{cl}(O_n) = X$. Then if U is an open set for which cl(U) is a Baire space, from the equality $A \cap \text{cl}(U) = \bigcap_{n \geq 1}(O_n \cap \text{cl}(U))$ and the density of $O_n \cap \text{cl}(U)$ in cl(U) (if $u \in \text{cl}(U)$ and $u \in O$, O open in X, then $O \cap U \neq \varnothing$, and therefore $O \cap U \cap O_n \neq \varnothing$), it follows that $A \cap \text{cl}(U)$ is residual in cl(U) hence dense in cl(U), that is, $\text{cl}(A) \cap \text{cl}(U) = \text{cl}(U)$ so that $\text{cl}(U) \subset \text{cl}(A)$. Therefore $X = \text{cl}(A)$. ■

1.7.3 Baire Category Theorem *Complete pseudometric and locally compact spaces are Baire spaces.*

Proof Let X be a complete pseudometric space. Let $U \subset X$ be open and $A = \bigcap_{n\geq 1} O_n$ be residual. We must show $U \cap A \neq \varnothing$. Since $\text{cl}(O_n) = X$, $U \cap O_n \neq \varnothing$ and so we can choose a disk of diameter less than one, say V_1, such that $\text{cl}(V_1) \subset U \cap O_1$. Proceed inductively to obtain $\text{cl}(V_n) \subset U \cap O_n \cap V_{n-1}$, where V_n has diameter $<1/n$. Let $x_n \in \text{cl}(V_n)$. Clearly $\{x_n\}$ is a Cauchy sequence, and by completeness has a convergent subsequence with limit point x. Then x

$\in \bigcap_{n\geq1} \text{cl}(V_n)$ and so $U \cap (\bigcap_{n\geq1} O_n) \neq \emptyset$; i.e., A is dense in X. If X is a locally compact space the same proof works with the following modifications: V_n are chosen to be relatively compact open sets, and $\{x_n\}$ has a convergent subsequence since it lies in the compact set $\text{cl}(V_1)$. ■

To get a feeling for this theorem, let us prove that the set of rationals $\mathbb{Q}$ cannot be written as a countable intersection of open sets. For suppose $\mathbb{Q} = \bigcap_{n\geq1} O_n$. Then each O_n is dense in $\mathbb{R}$, since $\mathbb{Q}$ is, and so $C_n = \mathbb{R} \setminus O_n$ is closed and nowhere dense. Since $\mathbb{R} \cup (\bigcup_{n\geq1} C_n)$ is a complete metric space (as well as a locally compact space), it is of second category, so $\mathbb{Q}$ or some C_n should have nonempty interior. But this is impossible.

The notion of category can lead to interesting restrictions on a set. For example in a nondiscrete Hausdorff space, any countable set is first category since the one-point set is closed and nowhere dense. Hence in such a space *every second category set is uncountable.* In particular, nonfinite complete pseudometric and locally compact spaces are uncountable.

Exercises

1.7A Let X be a Baire space. Show that

(i) X is a second category set;

(ii) if $U \subset X$ is open, then U is Baire.

1.7B Let X be a topological space. A set is called an $\mathcal{F}_\sigma$ if it is a countable union of closed sets, and is called a $\mathcal{G}_\delta$ if it is a countable intersection of open sets. Prove that the following are equivalent:

(i) X is a Baire space;

(ii) any first category set in X has a dense complement;

(iii) the complement of every first category $\mathcal{F}_\sigma$ -set is a dense $\mathcal{G}_\delta$ -set;

(iv) for any countable family of closed sets $\{C_n\}$ satisfying $X = \bigcup_{n\geq1} C_n$, the open set $\bigcup_{n\geq1} \text{int}(C_n)$ is dense in X. (*Hint:* First show that (ii) is equivalent to (iv). For (ii) implies (iv), let $U_n = C_n \setminus \text{int}(C_n)$ so that $\bigcup_{n\geq1} U_n$ is a first category set and therefore $X \setminus \bigcup_{n\geq1} U_n$ is dense and included in $\bigcup_{n\geq1} \text{int}(C_n)$. For the converse, assume X is not Baire so that $A = \bigcap_{n\geq1} U_n$ is not dense, even though all U_n are open and dense. Then $X = \text{cl}(A) \cup \{X \setminus U_n \mid n=1,2,\ldots\}$. Put $F_0 = \text{cl}(A)$, $F_n = X \setminus U_n$ and show that $\text{int}(F_n) = \text{int}(\text{cl}(A))$ which is not dense.)

1.7C Show that *there is a residual set* E *in the metric space* $C([0, 1], \mathbb{R})$ *such that each* $f \in E$ *is not differentiable at any point.* Do this by following the steps below.

(i) Let E_ε denote the set of all $f \in C([0, 1], \mathbb{R})$ such that for every $x \in [0, 1]$,

$$\operatorname{diam}\left\{ \frac{f(x+h)-f(x)}{h} \;\middle|\; \frac{\varepsilon}{2} < |h| < \varepsilon \right\} > 1$$

for $\varepsilon > 0$. Show that E_ε is open and dense in $C([0, 1], \mathbb{R})$. (*Hint:* For any polynomial $p \in C([0, 1], \mathbb{R})$, show that $p + \delta \cos(kx) \in E_\varepsilon$ for δ small and δk large.)

(ii) Show that $E = \bigcap_{n\geq 1} E_{1/n}$ is dense in $C([0, 1], \mathbb{R})$. (*Hint:* Use the Baire Category Theorem.)

(iii) Show that if $f \in E$, then f has no derivative at any point.

1.7D Prove the following: In a complete metric space (M, d) with no isolated points, no countable dense set is a G_δ-set. (*Hint:* Suppose $E = \{x_1, x_2, \ldots\}$ is dense in M and is also a G_δ set, i.e., $E = \bigcap_{n>0} V_n$ with V_n open, $n = 1, 2, \ldots$. Conclude that V_n is dense in M. Let $W_n = V_n \setminus \{x_1, \ldots, x_n\}$. Show that W_n is dense in M and that $\bigcap_{n>0} W_n = \varnothing$. This contradicts the Baire property.)

Chapter 2

BANACH SPACES AND DIFFERENTIAL CALCULUS

Manifolds have enough structure to allow differentiation of maps between them. To set the stage for these concepts requires a development of differential calculus in linear spaces from a geometric point of view. The goal of this chapter is to provide this perspective.

Perhaps the most important theorem for later use is the Implicit Function Theorem. A fairly detailed exposition of this topic will be given with examples appropriate for use in manifold theory. The basic language of tangents, the derivative as a linear map, and the chain rule, while elementary, are important for developing geometric and analytic skills necessary for mastering manifold theory.

The main goal is to develop the theory of finite-dimensional manifolds. However, it is instructive and efficient to do the infinite-dimensional theory simultaneously. To avoid being sidetracked by infinite-dimensional technicalities at this stage, some functional analysis background and other topics special to the infinite-dimensional case are presented in supplements. With this arrangement readers who wish to concentrate on the finite-dimensional theory can do so with a minimum of distraction.

§2.1 *Banach Spaces*

It is assumed the reader is familiar with the concept of an abstract vector space over the real or complex numbers. Banach spaces are vector spaces with additional structure. While most of this book is concerned with finite-dimensional spaces, much of the theory is really no harder in the general case, and the infinite-dimensional case is actually essential for certain applications. Thus it makes sense to work in the setting of Banach spaces. In addition, although the primary concern is with real Banach spaces, the basic concepts for complex spaces are introduced with little extra effort.

2.1.1 Definition *A* ***norm*** *on a real (complex) vector space* E *is a mapping from* E *into the real numbers,* $\|\cdot\| : \mathrm{E} \to \mathbb{R};\ \mathrm{e} \mapsto \|\mathrm{e}\|$, *such that*

N1 $\|\mathrm{e}\| \geq 0$ *for all* $\mathrm{e} \in \mathrm{E}$ and $\|\mathrm{e}\| = 0$ *iff* $\mathrm{e} = 0$ *(positive definiteness);*

N2 $\|\lambda \mathrm{e}\| = |\lambda|\ \|\mathrm{e}\|$ *for all* $\mathrm{e} \in \mathrm{E}$ *and* $\lambda \in \mathbb{R}$ *(homogeneity);*

N3 $\|\mathrm{e}_1 + \mathrm{e}_2\| \leq \|\mathrm{e}_1\| + \|\mathrm{e}_2\|$ *for all* $\mathrm{e}_1, \mathrm{e}_2 \in \mathrm{E}$ *(triangle inequality).*

The pair $(E, \|\cdot\|)$ *is sometimes called a* ***normed space.*** *If there is no danger of confusion, we sometimes just say "*E *is a normed space." To distinguish different norms, different notations are sometimes used, e.g.,* $\|\cdot\|_E$, $\|\cdot\|_1$, $|||\cdot|||$, *etc., for the norm.*

The triangle inequality **N3** has the following important consequence:

$$|\,\|e_1\| - \|e_2\|\,| \leq \|e_1 - e_2\| \quad \text{for all} \quad e_1, e_2 \in E\,,$$

which is proved in the following way:

$$\|e_2\| = \|e_1 + (e_2 - e_1)\| \leq \|e_1\| + \|e_1 - e_2\|\,,$$
$$\|e_1\| = \|e_2 + (e_1 - e_2)\| \leq \|e_2\| + \|e_1 - e_2\|\,,$$

so that both $\|e_2\| - \|e_1\|$ and $\|e_1\| - \|e_2\|$ are smaller than or equal to $\|e_1 - e_2\|$.

If **N1** in **2.1.1** is replaced by

N1' $\|e\| \geq 0$ *for all* $e \in E$ *and* $\|e\| = 0$ *implies* $e = 0$,

the mapping $\|\cdot\| : E \to \mathbb{R}$ is called a ***seminorm.*** For example, $\mathbb{R}^n$ with standard norm

$$\|x\| = (x_1^2 + \cdots + x_n^2)^{1/2},$$

where $x = (x_1, \ldots, x_n)$, is a normed space. Actually, the standard norm on $\mathbb{R}^n$ comes from a more special structure defined as follows.

2.1.2 Definition *An* ***inner product*** *on a real vector space* E *is a mapping* $\langle \cdot, \cdot \rangle : E \times E \to \mathbb{R}$ *which we denote* $(e_1, e_2) \mapsto \langle e_1, e_2 \rangle$ *such that*

I1 $\langle e, e_1 + e_2 \rangle = \langle e, e_1 \rangle + \langle e, e_2 \rangle$;

I2 $\langle e, \alpha e_1 \rangle = \alpha \langle e, e_1 \rangle$;

I3 $\langle e_1, e_2 \rangle = \langle e_2, e_1 \rangle$;

I4 $\langle e, e \rangle \geq 0$ *and* $\langle e, e \rangle = 0$ *iff* $e = 0$.

The ***standard inner product*** on $\mathbb{R}^n$ is

$$\langle \cdot, \cdot \rangle = \sum_{i=1}^{n} x_i y_i\,,$$

and **I1 - I4** are readily checked.

For vector spaces over the complex numbers, the definition is modified slightly as follows.

2.1.2' Definition *A* ***complex inner product*** *(or a* ***Hermitian inner product****) on a complex vector* *space* E *is a mapping*

$$\langle \cdot, \cdot \rangle : \mathrm{E} \times \mathrm{E} \to \mathbb{C}$$

such that the following conditions hold:

CI1 $\langle e, e_1 + e_2 \rangle = \langle e, e_1 \rangle + \langle e, e_2 \rangle$;

CI2 $\langle \alpha e, e_1 \rangle = \alpha \langle e, e_1 \rangle$;

CI3 $\langle e_1, e_2 \rangle = \overline{\langle e_2, e_1 \rangle}$ *(so* $\langle e, e \rangle$ *is real)*;

CI4 $\langle e, e \rangle \geq 0$ *and* $\langle e, e \rangle = 0$ *iff* $e = 0$.

These properties are to hold for all $e, e_1, e_2 \in E$ and $\alpha \in \mathbb{C}$; $\overline{}$ denotes complex conjugation. Note that **CI2** and **CI3** imply that $\langle e_1, \alpha e_2 \rangle = \overline{\alpha} \langle e_1, e_2 \rangle$. Properties **CI1 - CI3** are also known in the literature under the name ***sesquilinearity.*** As is customary, for a complex number α we shall denote by

$$\operatorname{Re} \alpha = (\alpha + \bar{\alpha})/2 \ , \quad \operatorname{Im} \alpha = (\alpha - \bar{\alpha})/2 \ , \quad |\alpha| = (\alpha \bar{\alpha})^{1/2}$$

its real and imaginary parts and its absolute value. The ***standard inner product*** on the product space $\mathbb{C}^n = \mathbb{C} \times \cdots \times \mathbb{C}$ is defined by

$$\langle x, w \rangle = \sum_{i=1}^{n} z_i \bar{w}_i \ ,$$

and **CI1-CI4** are readily checked. Also $\mathbb{C}^n$ is a normed space with

$$\| z \|^2 = \sum_{i=1}^{n} | z_i |^2 \ .$$

In $\mathbb{R}^n$ or $\mathbb{C}^n$, property **N3** is a little harder to check directly. However, as we shall show in Proposition **2.1.4, N3** follows from **I1-I4** or **CI1-CI4**.

In a (real or complex) inner product space E, two vectors $e_1, e_2 \in E$ are called ***orthogonal*** and we write $e_1 \perp e_2$ provided $\langle e_1, e_2 \rangle = 0$. For a subset $A \subset E$, the set $A^\perp$ defined by $A^\perp = \{ e \in E \mid \langle e, x \rangle = 0 \text{ for all } x \in A \}$ is called the ***orthogonal complement*** of A. Two sets A, B $\subset$ E are called ***orthogonal*** and we write $A \perp B$ if $\langle A, B \rangle = 0$; that is, $e_1 \perp e_2$ for all $e_1 \in A$ and $e_2 \in B$.

2.1.3 Cauchy-Schwartz Inequality *In a (real or complex) inner product space,*

$$|\langle e_1, e_2 \rangle| \leq \langle e_1, e_1 \rangle^{1/2} \langle e_2, e_2 \rangle^{1/2}.$$

Equality holds iff e_1, e_2 *are linearly dependent.*

Proof It suffices to prove the complex case. If $\alpha, \beta \in \mathbb{C}$ we have

$$0 \leq \langle \alpha e_1 + \beta e_2, \alpha e_1 + \beta e_2 \rangle = |\alpha|^2 \langle e_1, e_2 \rangle + 2\mathrm{Re}(\alpha\bar{\beta}\langle e_1, e_2 \rangle) + |\beta|^2 \langle e_2, e_2 \rangle .$$

If we set $\alpha = \langle e_2, e_2 \rangle$, and $\beta = -\langle e_1, e_2 \rangle$, then this becomes

$$0 \leq \langle e_2, e_2 \rangle^2 \langle e_1, e_1 \rangle - 2\langle e_2, e_2 \rangle |\langle e_1, e_2 \rangle|^2 + |\langle e_1, e_2 \rangle|^2 \langle e_2, e_2 \rangle ,$$

and so

$$\langle e_2, e_2 \rangle |\langle e_1, e_2 \rangle|^2 \leq \langle e_2, e_2 \rangle^2 \langle e_1, e_1 \rangle .$$

If $e_2 = 0$, equality results in the statement of the proposition and there is nothing to prove. If $e_2 \neq 0$, the term $\langle e_2, e_2 \rangle$ in the preceding inequality can be cancelled since $\langle e_2, e_2 \rangle \neq 0$ by **CI4.** Taking square roots yields the statement of the proposition. Finally, equality results if and only if $\alpha e_1 + \beta e_2 = \langle e_2, e_2 \rangle e_1 - \langle e_1, e_2 \rangle e_2 = 0$. ■

2.1.4 Proposition *Let* $(E, \langle \cdot, \cdot \rangle)$ *be a (real or complex) inner product space and set* $\| e \| = \langle e, e \rangle^{1/2}$. *Then* $(E, \| \cdot \|)$ *is a normed space.*

Proof **N1** and **N2** are straightforward verifications. As for **N3**, the Cauchy-Schwartz inequality and the obvious inequality $\mathrm{Re}(\langle e_1, e_2 \rangle) \leq |\langle e_1, e_2 \rangle|$ imply

$$\begin{aligned} \| e_1 + e_2 \|^2 &= \| e_1 \|^2 + 2\,\mathrm{Re}(\langle e_1, e_2 \rangle) + \| e_2 \|^2 \leq \| e_1 \|^2 + 2\,|\langle e_1, e_2 \rangle| + \| e_2 \|^2 \\ &\leq \| e_1 \|^2 + 2\| e_1 \| \, \| e_2 \| + \| e_2 \|^2 = (\| e_1 \| + \| e_2 \|)^2 . \end{aligned}$$ ■

Some other useful facts about inner products are given next.

2.1.5 Proposition *Let* $(E, \langle \cdot\,, \cdot \rangle)$ *be an inner product space and* $\| \cdot \|$ *the corresponding norm. Then*

(i) *(Polarization)*

$$4\langle e_1, e_2 \rangle = \begin{cases} \| e_1 + e_2 \|^2 - \| e_1 - e_2 \|^2 + i \| e_1 + e_2 \|^2 - i \| e_1 - ie_2 \|^2 & \text{if } E \text{ is complex} , \\ \\ \| e_1 + e_2 \|^2 - \| e_1 - e_2 \|^2 & \text{if } E \text{ is real} . \end{cases}$$

(ii) *(Parallelogram law)*

$$2 \| e_1 \|^2 + 2 \| e_2 \|^2 = \| e_1 + e_2 \|^2 + \| e_1 - e_2 \|^2 .$$

Proof (i) $\| e_1 + e_2 \|^2 - \| e_1 - e_2 \|^2 + i \| e_1 + ie_2 \|^2 - i \| e_1 - ie_2 \|^2$

$$= \| e_1 \|^2 + 2\mathrm{Re}(\langle e_1, e_2 \rangle) + \| e_2 \|^2 - \| e_1 \|^2 + 2\mathrm{Re}(\langle e_1, e_2 \rangle) - \| e_2 \|^2 +$$

$$+ i \| e_1 \|^2 + 2i\mathrm{Re}(\langle e_1, ie_2 \rangle) + i \| e_2 \|^2 - i \| e_1 \|^2 + 2i\mathrm{Re}(\langle e_1, ie_2 \rangle) - i \| e_2 \|^2$$
$$= 4\mathrm{Re}(\langle e_1, e_2 \rangle) + 4i\mathrm{Re}(-i \langle e_1, e_2 \rangle)$$
$$= 4\mathrm{Re}(\langle e_1, e_2 \rangle) + 4i\, \mathrm{Im}(\langle e_1, e_2 \rangle) = 4 \langle e_1, e_2 \rangle.$$

The real case is proved in a similar way.

(ii) $$\| e_1 + e_2 \|^2 + \| e_1 - e_2 \|^2 = \| e_1 \|^2 + 2\mathrm{Re}(\langle e_1, e_2 \rangle) + \| e_2 \|^2 + \| e_1 \|^2 - 2\mathrm{Re}(\langle e_1, e_2 \rangle) e_2 \|^2$$
$$= 2 \| e_1 \|^2 + 2 \| e_2 \|^2. \quad \blacksquare$$

Not all norms come from an inner product. For example in $\mathbb{R}^n$,

$$||| x ||| = \sum_{i=1}^{n} | x_i |$$

is a norm, but there is no inner product with this as norm, since this norm fails to satisfy the parallelogram law (see Exercise **2.1A** for a discussion).

2.1.6 Proposition *Let* $(E, \| \cdot \|)$ *be a normed (or a seminormed) space and define* $d(e_1, e_2) = \| e_1 - e_2 \|$. *Then* (E, d) *is a metric (pseudometric) space.*

Proof The only non-obvious verification is the triangle inequality. By **N3** we have

$$d(e_1, e_3) = \| e_1 - e_3 \| = \|(e_1 - e_2) + (e_2 - e_3) \| \le \| e_1 - e_2 \| + \| e_2 - e_3 \|$$
$$= d(e_1, e_2) + d(e_2, e_3) \ . \quad \blacksquare$$

Thus we have the following hierarchy of generality:

more general $\Longrightarrow$

inner product spaces	$\subset$	normed spaces	$\subset$	metric spaces	$\subset$	topological spaces.

$\Longleftarrow$ more special

Since inner product and normed spaces are metric spaces, we can use the concepts from Chapter 1. In a normed space, **N1** and **N2** imply that the maps $(e_1, e_2) \mapsto e_1 + e_2$, $(\alpha, e) \mapsto \alpha e$ of $E \times E \to E$, and $\mathbb{C} \times E \to E$, respectively, are continuous. Hence for $e_0 \in E$, $\alpha_0 \in \mathbb{C}$, $\alpha_0 \neq 0$ fixed, the mappings $e \mapsto e_0 + e$, $e \mapsto \alpha_0 e$ are homeomorphisms. Thus U is a neighborhood of the origin iff $e + U = \{e + x \mid x \in U\}$ is a neighborhood of $e \in E$. In other words, all the neighborhoods of $e \in E$ are sets that contain translates of disks centered at the origin. This constitutes a complete description of the topology of a normed vector space $(E, \| \cdot \|)$.

Finally, note that the inequality $|\,\|e_1\| - \|e_2\|\,| \le \|e_1 - e_2\|$ implies that the norm is uniformly continuous on E. In inner product spaces, the Cauchy-Schwartz inequality implies the continuity of the inner product as a function of two variables.

2.1.7 Definition *Let* $(E, \|\cdot\|)$ *be a normed space. If the corresponding metric* d *is complete, we say* $(E, \|\cdot\|)$ *is a* ***Banach space.*** *If* $(E, \langle\cdot,\cdot\rangle)$ *is an inner product space whose corresponding metric is complete, we say* $(E, \langle\cdot,\cdot\rangle)$ *is a* ***Hilbert space.***

For example, it is proven in books on advanced calculus that $\mathbb{R}^n$ is complete. Thus, $\mathbb{R}^n$ with the standard norm is a Banach space and with the standard inner product is a Hilbert space. Not only is the standard norm on $\mathbb{R}^n$ complete, but so is the nonstandard one

$$|||\, x\, ||| = \sum_{i=1}^{n} |x_i|.$$

However, it is equivalent to the standard one in the following sense.

2.1.8 Definition *Two norms on a vector space* E *are* ***equivalent*** *if they induce the same topology on* E.

2.1.9 Proposition *Two norms* $\|\cdot\|$ *and* $|||\cdot|||$ *on* E *are equivalent iff there is a constant* M *such that* $M^{-1}|||\, e\, ||| \le \|e\| \le M|||\, e\, |||$ *for all* $e \in E$.

Proof Let

$$B_r^1(x) = \{y \in E \mid \|y - x\| \le r\}\ , \quad B_r^2(x) = \{y \in E \mid |||\, y - x\, ||| \le r\}$$

denote the two closed disks of radius r centered at $x \in E$ in the two metrics defined by the norms $\|\cdot\|$ and $|||\cdot|||$, respectively. Since neighborhoods of an arbitrary point are translates of neighborhoods of the origin, the two topologies are the same iff for every $R > 0$, there are constants $M_1, M_2 > 0$ such that

$$B_{M_1}^2(0) \subset B_R^1(0) \subset B_{M_2}^2(0).$$

The first inclusion says that if $|||\, x\, ||| \le M_1$, then $\|x\| \le R$, i.e., if $|||\, x\, ||| \le 1$, then $\|x\| \le R/M_1$. Thus, if $e \ne 0$, then

$$\left\| \frac{e}{|||\, e\, |||} \right\| = \frac{\|e\|}{|||\, e\, |||} \le \frac{R}{M_1},$$

that is, $\|e\| \le (R/M_1)|||\, e\, |||$ for all $e \in E$. Similarly, the second inclusion is equivalent to the assertion that $(M_2/R)\,|||\, e\, ||| \le \|e\|$ for all $e \in E$. Thus the two topologies are the same if there exist constants $N_1 > 0$, $N_2 > 0$ such that

$$N_1 ||| e ||| \leq \| e \| \leq N_2 ||| e |||$$

for all $e \in E$. Taking $M = \max(N_2, 1/N_1)$ gives the statement of the proposition. ■

If E and F are normed vector spaces, the map $\| \cdot \| : E \times F \to \mathbb{R}$ defined by $\| (e, e') \| = \| e \| + \| e' \|$ is a norm on $E \times F$ inducing the product topology. Equivalent norms on $E \times F$ are $(e, e') \mapsto \max(\| e \|, \| e' \|)$ and $(e, e') \mapsto (\| e \|^2 + \| e' \|^2)^{1/2}$. The normed vector space $E \times F$ is usually denoted by $E \oplus F$ and called the ***direct sum*** of E and F. Note that $E \oplus F$ is a Banach space iff both E and F are. These statements are readily checked.

2.1.10 Proposition. *Let* E *be a finite-dimensional real or complex vector space. Then*

(i) *there is a norm on* E;

(ii) *all norms on* E *are equivalent;*

(iii) *all norms on* E *are complete.*

Proof Let $e_1, \ldots, e_n$ denote a basis of E, where n is the dimension of E.

(i) A norm on E is given, for example, by

$$\| e \| = \sum_{i=1}^{n} | a^i | ,$$

where

$$e = \sum_{i=1}^{n} a^i e_i .$$

(ii) Let $\| \cdot \|'$ be any other norm on E. If

$$e = \sum_{i=1}^{n} a^i e_i \quad \text{and} \quad f = \sum_{i=1}^{n} b^i e_i ,$$

the inequality

$$| \| e \|' - \| f \|' | \leq \| e - f \|' \leq \sum_{i=1}^{n} | a^i - b^i | \, \| e_i \|'$$

$$\leq \max_{1 \leq i \leq n} \{ \| e_i \|' \} \, ||| (a^1, \ldots, a^n) - (b^1, \ldots, b^n) |||$$

shows that the map

$$(x^1, \ldots, x^n) \in \mathbb{C}^n \mapsto \left\| \sum_{i=1}^{n} x^i e_i \right\|' \in [0, \infty[$$

is continuous with respect to the $|||\cdot|||$-norm on $\mathbb{C}^n$. Since the set $S = \{x \in \mathbb{C}^n \mid |||x||| = 1\}$ is closed and bounded, it is compact. The restriction of this map to S is a continuous, strictly positive function, so it attains its minimum M_1 and maximum M_2 on S; that is,

$$0 < M_1 \leq \left\| \sum_{i=1}^{n} x^i e_i \right\|' \leq M_2$$

for all $(x^1, \ldots, x^n) \in \mathbb{C}^n$ such that $|||(x^1, \ldots, x^n)||| = 1$. Thus

$$M_1 \, |||(x^1, \ldots, x^n)||| \leq \left\| \sum_{i=1}^{n} x^i e_i \right\|' \leq M_2 \, |||(x^1, \ldots, x^n)||| ,$$

i.e., $M_1 \| e \|' \leq \| e \|' \leq M_2 \| e \|$, where

$$e = \sum_{i=1}^{n} x^i e_i .$$

Taking $M = \max(M_2, 1/M_1)$, Proposition **2.1.9** shows that $\|\cdot\|$ and $\|\cdot\|'$ are equivalent.

(iii) It is enough to observe that

$$(x^1, \ldots, x^n) \in \mathbb{C}^n \mapsto \sum_{i=1}^{n} x^i e_i \in E$$

is a norm-preserving map (i.e., an isometry) between $(\mathbb{C}^n, |||\cdot|||)$ and $(E, \|\cdot\|)$. ■

The unit spheres for the three common norms on $\mathbb{R}^2$ are shown in Figure 2.1.1.

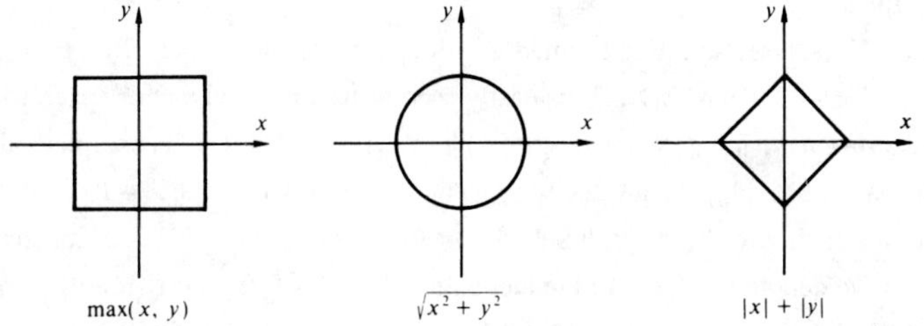

Figure 2.1.1

The foregoing proof shows that compactness of the unit sphere in a finite-dimensional space is crucial. This fact is exploited in the following supplement.

☞ SUPPLEMENT 2.1A
A Characterization of Finite-Dimensional Spaces

2.1.11 Proposition *A normed vector space is finite dimensional iff it is locally compact. (Local compactness is equivalent to the fact that the closed unit disk is compact.)*

Proof If E is finite dimensional, the previous proof of (iii) shows that E is locally compact. Conversely, assume the closed unit disk $B_1(0) \subset E$ is compact and let $\{D_{1/2}(x_i) \mid i = 1, \ldots, n\}$ be a finite cover with open disks of radii 1/2. Let $F = \text{span}\{x_1, \ldots, x_n\}$. Then F is finite dimensional, hence homeomorphic to $\mathbb{C}^k$ (or $\mathbb{R}^k$) for some $k \leq n$, and thus complete. Being a complete subspace of the metric space $(E, \|\cdot\|)$, it is closed. We shall prove F = E.

Suppose the contrary, that is, there exists $v \in E$, $v \notin F$. Since $F = \text{cl}(F)$, the number $d = \inf\{\| v - e \| \mid e \in F\}$ is strictly positive. Let $r > 0$ be such that $B_r(v) \cap F \neq \varnothing$. Note that the set $B_r(v) \cap F$ is closed and bounded in the finite-dimensional space F, and thus is compact. Since $\inf\{ \| v - e \| \mid e \in F\} = \inf\{\| v - e \| \mid e \in B_r(v) \subset F\}$ and the continuous function defined by $e \in B_r(v) \cap F \mapsto \| v - e \| \in]0, \infty[$ attains its minimum, there is a point $e_0 \in B_r(v) \cap F$ such that $d = \| v - e_0 \|$. But then there is a point x_i such that

$$\left\| \frac{v - e_0}{\|v - e_0\|} - x_i \right\| < \frac{1}{2},$$

so that

$$\| v - e_0 - (\| v - e_0 \|) x_i \| < \frac{1}{2} \| v - e_0 \| = \frac{d}{2}.$$

Since $e_0 + \| v - e_0 \| x_i \in F$, we get $\| v - e_0 - (\| v - e_0 \|) x_i \| \geq d$, which is a contradiction. ■ ✍

2.1.12 Examples

A Let X be a set and F a normed vector space. Define the set $B(X, F) = \{f : X \to F \mid \sup_{x \in X} \| f(x) \| < \infty\}$. Then $B(X, F)$ is easily seen to be a normed vector space with respect to the so-called ***sup-norm***, $\| f \|_\infty = \sup_{x \in X} \| f(x) \|$. We prove that if F is complete, then $B(X, F)$ is a Banach space. Let $\{f_n\}$ be a Cauchy sequence in $B(X, F)$, i.e., $\| f_n - f_m \|_\infty < \varepsilon$ for $n, m \geq N(\varepsilon)$. Since for each $x \in X$, $\| f(x) \| \leq \| f \|_\infty$, it follows that $\{f_n(x)\}$ is a Cauchy sequence in F, whose limit we denote by $f(x)$. In the inequality $\| f_n(x) - f_m(x) \| < \varepsilon$ for all $n, m \geq N(\varepsilon)$, let $m \to \infty$ and get $\| f_n(x) - f(x) \| \leq \varepsilon$ for $n \geq N(\varepsilon)$, i.e., $\|f_n - f\|_\infty \leq \varepsilon$ for $n \geq N(\varepsilon)$. This shows that $f_n - f \in B(X, F)$, i.e., that $f \in B(X, F)$, and that $\| f_n - f \|_\infty \to 0$ as $n \to \infty$. As a particular case, we get the Banach space c_b consisting of all bounded real sequences $\{a_n\}$ with the norm $\| \{a_n\} \|_\infty = \sup_n | a_n |$.

B If X is a topological space, the space $CB(X, F) = \{f : X \to F \mid f$ is continuous, $f \in B(X, F)\}$ is closed in $B(X, F)$. Thus if F is Banach, so is $CB(X, F)$. In particular, $C(X, F) = \{f : X \to F \mid f \text{ continuous}\}$, for X compact and F Banach, is a Banach space. For example, the vector space $C([0, 1], \mathbb{R}) = \{f : [0, 1] \to \mathbb{R} \mid f \text{ is continuous}\}$ is a Banach space with the norm $\| f \|_\infty = \sup\{| f(x) | \mid x \in [0, 1]\}$.

C (For readers with some knowledge of measure theory.) Consider the space of real valued square integrable functions defined on an interval $[a, b] \subset \mathbb{R}$, i.e., functions f that satisfy

$$\int_a^b | f(x)|^2 \, dx < \infty.$$

The function

$$\| \cdot \| : f \mapsto \left(\int_a^b | f(x) |^2 \, dx \right)^{1/2}$$

is, strictly speaking, not a norm on this space; for example, if

$$f(x) = \begin{cases} 0 & \text{for } x \neq a , \\ 1 & \text{for } x = a , \end{cases}$$

then $\| f \| = 0$, but $f \neq 0$. However, $\| \cdot \|$ does become a norm if we identify functions which differ only on a set of measure zero in $[a, b]$, i.e., which are equal almost everywhere. The resulting vector space of equivalence classes $[f]$ will be denoted $L^2\,[a, b]$. With the norm of the equivalence class $[f]$ defined as

$$\| [f] \| = \left(\int_a^b | f(x) |^2 \, dx \right)^{1/2} ,$$

$L^2\,[a, b]$ is an (infinite dimensional) Banach space. The only nontrivial part of this assertion is the completeness; this is proved in books on measure theory, such as Royden [1968]. As is customary, $[f]$ is denoted simply by f. In fact, $L^2\,[a, b]$ is a Hilbert space with

$$\langle f, g \rangle = \int_a^b f(x)\, g(x) \, dx .$$

If we use square integrable complex-valued functions we get a complex Hilbert space $L^2([a, b], \mathbb{C})$ with

$$\langle f, g \rangle = \int_a^b f(x)\, \overline{g(x)} \, dx .$$

D The space $L^p([a, b])$ may be defined for each real number $p \geq 1$ in an analogous fashion to $L^2\,[a, b]$. Functions $f : [a, b] \to \mathbb{R}$ satisfying

$$\int_a^b | f(x)|^p \, dx < \infty$$

are considered equivalent if they agree almost everywhere. $L^p([a, b])$ is then defined to be the vector space of equivalence classes of functions equal almost everywhere.

$$\| \cdot \|_p : L^p\,[a, b] \to \mathbb{R} \text{ given by } [f] \to \left(\int_a^b | f(x)|^p \, dx \right)^{1/p}$$

defines a norm, called the L^p-norm, which makes $L^p\,[a, b]$ into an (infinite dimensional) Banach space.

E Denote by $C([a, b], \mathbb{R})$ the set of continuous real valued functions on $[a, b]$. With the L^1-norm, $C([a, b], \mathbb{R})$ is not a Banach space. For example, the sequence of continuous functions f_n shown in Figure 2.1.2 is a Cauchy sequence in the L^1-norm but does not have a continuous limit function. On the other hand, with the sup norm

$$\| f \| = \sup_{x \in [0,1]} | f(x) | \,,$$

$C([0, 1])$ is complete, i.e., it is a Banach space, as in Example **B**. ♦

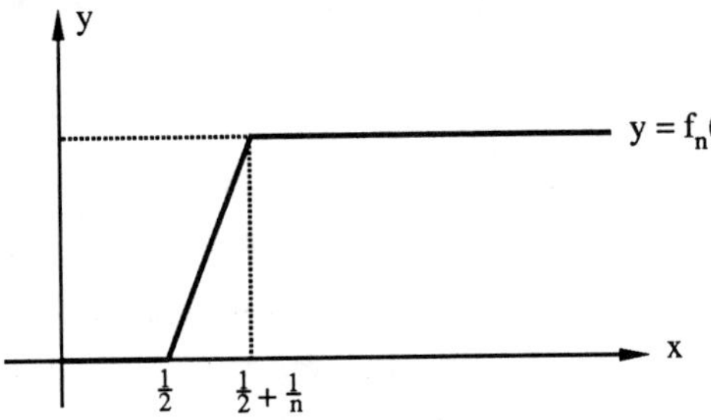

Figure 2.1.2. f_n *converges in* L^1, *but not in* C.

As in the case of vector spaces, quotient spaces of normed vector spaces play a fundamental role.

2.1.13 Proposition *Let* E *be a normed vector space,* F *a closed subspace,* E/F *the quotient vector space, and* $\pi : E \to E/F$ *the* ***canonical projection*** *defined by* $\pi(e) = [e] = e + F \in E/F$.

(i) *The mapping* $\| \cdot \| : E/F \to \mathbb{R}$,

$$\| [e] \| = \inf\{\| e + v \| \mid v \in F\}$$

defines a norm on E/F.

(ii) π *is continuous and the topology on* E/F *defined by the norm coincides with the quotient topology. In particular,* π *is open.*

(iii) *If* E *is a Banach space, so is* E/F.

Proof (i) Clearly $\| [e] \| \geq 0$ for all $[e] \in E/F$ and $\| [0] \| = \inf\{ \| v \| \mid v \in F\} = 0$. If $\| [e] \| = 0$, then there is a sequence $\{v_n\} \subset F$ such that $\lim_{n \to \infty} \| e + v_n \| = 0$. Thus $\lim_{n \to \infty} v_n = -e$ and since F is closed, $e \in F$; i.e., $[e] = 0$. Thus **N1** is verified and the necessity of having F closed becomes apparent. **N2** and **N3** are straightforward verifications.

(ii) Since $\| [e] \| \leq \| e \|$, it is obvious that $\lim_{n \to \infty} e_n = e$ implies $\lim_{n \to \infty} \pi(e_n) = \lim_{n \to \infty} [e_n] = [e]$ and hence π is continuous. Translation by a fixed vector is a homeomorphism. Thus to show that the topology of E/F is the quotient topology, it suffices to show that if $[0] \in U$ and $\pi^{-1}(U)$ is a neighborhood of zero in E, then U is a neighborhood of [0] in E/F. Since $\pi^{-1}(U)$ is a neighborhood of zero in E, there exists a disk $D_r(0) \subset \pi^{-1}(U)$. But then $\pi(D_r(0)) \subset U$ and $\pi(D_r(0)) = \{[e] \mid e \in D_r(0)\} = \{[e] \mid \| [e] \| < r\}$, so that U is a neighborhood of [0] in E/F.

(iii) Let $\{[e_n]\}$ be a Cauchy sequence in E/F. We may assume without loss of generality that $\| [e_n] - [e_{n+1}] \| \leq 1/2^n$. Inductively, we find points $e'_n \in [e_n]$ such that $\| e'_n - e'_{n+1} \| < 1/2^n$. Thus $\{e'_n\}$ is Cauchy in E so it converges to, say, $e \in E$. Continuity of π implies that $\lim_{n \to \infty} [e_n] = [e]$. ■

The ***codimension*** of F in E is defined to be the dimension of E/F. We say F is of ***finite codimension*** if E/F is finite dimensional.

2.1.14 Definition *The closed subspace* F *of the Banach space* E *is said to be* ***split,*** *or* ***complemented,*** *if there is a closed subspace* $G \subset E$ *such that* $E = F \oplus G$.

☞ SUPPLEMENT 2.1B
Split Subspaces

Definition **2.1.14** implicitly asks that the topology of E coincide with the product topology of $F \oplus G$. We shall show in Supplement **2.2C** that this topological condition can be dropped; i.e., F is split iff E is the algebraic direct sum of F and the closed subspace G.

We note that if $E = F \oplus G$ then G is isomorphic to E/F. However, F need not split for E/F to be a Banach space, as we proved in **2.1.13**. In finite dimensional spaces, any subspace is closed and splits; however, in infinite dimensions this is false. For example, let $E = L^p(S^1)$ and let $F = \{f \in E \mid f(n) = 0 \text{ for } n < 0\}$, where

$$f(n) = \frac{1}{2\pi}\int_{-\pi}^{\pi} f(\theta)e^{-in\theta}\, d\theta$$

is the nth Fourier coefficient of f. Then F is closed in E, splits in E for $1 < p < \infty$ by a theorem of M. Riesz (Rudin [1966], Theorem 17.26) but does not split in E for $p = 1$ (Rudin [1973], Example 5.19). The same result holds if $E = C^0(S^1, \mathbb{C})$ and F has the same definition.

Another example worth mentioning is $E = \ell^\infty$, the Banach space of all bounded sequences, and $F = c_0$, the subspace of ℓ^∞ consisting of all sequences convergent to zero. The subspace $F = c_0$ is closed in $E = \ell^\infty$, but does not split. However, c_0 splits in any *separable* Banach space which contains it isomorphically as a closed subspace by a theorem of Sobczyk; see Veech [1971]. If *every* subspace of a Banach space is complemented, the space must be isomorphic to a Hilbert space by a result of J. Lindenstrauss and L. Tzafriri [1971]. Supplement **2.2B** gives some general criteria useful in nonlinear analysis for a subspace to be split. But the simplest situation occurs in Hilbert spaces.

2.1.15 Proposition *If* E *is a Hilbert space and* F *a closed subspace, then* $E = F \oplus F^\perp$. *Thus every closed subspace of a Hilbert space splits.*

The proof of this theorem is done in three steps, the first two being important results in their own rights.

2.1.16 Existence and Uniqueness of Minimal Norm Elements in Closed Convex Sets *If* C *is a closed convex set (i.e.,* $x, y \in C$ *and* $0 \le t \le 1$ *implies* $tx + (1-t)y \in C$*) in* E, *then there exists a unique* $e_0 \in C$ *such that* $\| e_0 \| = \inf\{\| e \| \mid e \in C\}$.

Proof Let $\sqrt{d} = \inf\{\| e \| \mid e \in C\}$. Then there exists a sequence $\{e_n\}$ satisfying the inequality $d \le \| e_n \|^2 < d + 1/n$; hence $\| e_n \|^2 \to d$. Since $(e_n + e_m)/2 \in C$, C being convex, it follows that $\| (e_n + e_m)/2 \|^2 \ge d$. By the parallelogram law,

$$\left\| \frac{e_n - e_m}{2} \right\|^2 = 2\left\| \frac{e_n}{2} \right\|^2 + 2\left\| \frac{e_m}{2} \right\|^2 - \left\| \frac{e_n + e_m}{2} \right\|^2$$

$$< \frac{d}{2} + \frac{1}{2n} + \frac{d}{2} + \frac{1}{2m} - d = \frac{1}{2}\left(\frac{1}{n} + \frac{1}{m}\right);$$

that is, $\{e_n\}$ is a Cauchy sequence in E. Let $\lim_{n\to\infty} e_n = e_0$. Continuity of the norm implies that $\sqrt{d} = \lim_{n\to\infty} \| e_n \| = \| e_0 \|$, and so the existence of an element of minimum norm in C is proved.

Finally, if f_0 is such that $\| e_0 \| = \| f_0 \| = \sqrt{d}$, the parallelogram law implies

$$\left\| \frac{e_0 - f_0}{2} \right\|^2 = 2\left\| \frac{e_0}{2} \right\|^2 + 2\left\| \frac{f_0}{2} \right\|^2 - \left\| \frac{e_0 + f_0}{2} \right\|^2 \le \frac{d}{2} + \frac{d}{2} - d = 0 \,;$$

i.e., $e_0 = f_0$. ■

2.1.17 Lemma *Let* $F \subset E$, $F \neq E$ *be a closed subspace of* E. *Then there exists a nonzero element* $e_0 \in E$ *such that* $e_0 \perp F$.

Proof Let $e \in E$, $e \notin F$. The set $e - F = \{e - v \mid v \in F\}$ is convex and closed, so by the previous lemma it contains a unique element $e_0 = e - v_0 \in e - F$ of minimum norm. Since F is closed and $e \notin F$, it follows that $e_0 \neq 0$. We shall prove that $e_0 \perp F$.

Since e_0 is of minimal norm in $e - F$, for any $v \in F$ and $\lambda \in \mathbb{C}$ (resp., $\mathbb{R}$), we have

$$\| e_0 \| = \| e - v_0 \| \le \| e - v_0 + \lambda v \| = \| e_0 + \lambda v \| \;,$$

i.e., $2\mathrm{Re}(\lambda\langle v, e_0 \rangle) + |\lambda|^2 \| v \|^2 \ge 0$.

If $\lambda = a\langle e_0, v \rangle$, $a \in \mathbb{R}$, $a \neq 0$, this becomes $a \,|\langle v, e_0 \rangle|^2(2 + a\| v \|^2) \ge 0$ for all $v \in F$, and $a \in \mathbb{R}$, $a \neq 0$. This forces $\langle v, e_0 \rangle = 0$ for all $v \in F$, since if $-2/\| v \|^2 < a < 0$, the preceding expression is negative. ■

Proof of Proposition 2.1.15 It is easy to see that $F^\perp$ is closed (Exercise **2.1C**). We now show that $F \oplus F^\perp$ is a closed subspace. If $\{e_n + e'_n\} \subset F \oplus F^\perp$, $\{e_n\} \subset F$, $\{e'_n\} \subset F^\perp$, the relation

$$\| (e_n + e'_n) - (e_m + e'_m) \|^2 = \| e_n - e'_n \|^2 + \| e_m - e'_m \|^2$$

shows that $\{e_n + e'_n\}$ is Cauchy iff both $\{e_n\} \subset F$ and $\{e'_n\} \subset F^\perp$ are Cauchy. Thus if $\{e_n + e'_n\}$ converges, then there exist $e \in F$, $e' \in F^\perp$ such that $\lim_{n\to\infty} e_n = e$, $\lim_{n\to\infty} e'_n = e'$. Thus $\lim_{n\to\infty} (e_n + e'_n) = e + e' \in F \oplus F^\perp$.

If $F \oplus F^\perp \neq E$, then by the previous lemma there exists $e_0 \in E$, $e_0 \notin F \oplus F^\perp$, $e_0 \neq 0$, $e_0 \perp (F \oplus F^\perp)$. Hence $e_0 \in F^\perp$ and $e_0 \in F$ so that $\langle e_0, e_0 \rangle = \| e_0 \|^2 = 0$; i.e., $e_0 = 0$, a contradiction. ■ ✍

Exercises

2.1A Show that a normed space is an inner product space iff the norm satisfies the parallelogram law. (*Hint:* Use the polarization identities over $\mathbb{R}$ and $\mathbb{C}$ to guess the corresponding inner-products.) Conclude that if $n \ge 2$, $|||\, x \,||| = \Sigma\, | x^i |$ on $\mathbb{R}^n$ does not arise from an inner product.

2.1B Let c_0 be the space of real sequences $\{a_n\}$ such that $a_n \to 0$ as $n \to \infty$. Show that c_0 is a closed subspace of the space c_b of bounded sequences (see Example **2.1.12A**) and conclude that c_0 is a Banach space.

2.1C Let E_1 be the set of all C^1 functions $f: [0, 1] \to \mathbb{R}$ with the norm

$$\|f\| = \sup_{x \in [0,1]} |f(x)| + \sup_{x \in [0,1]} |f'(x)| .$$

(i) Prove that E_1 is a Banach space.

(ii) Let E_0 be the space of C^0 maps $f: [0, 1] \to \mathbb{R}$, as in **2.1.12**. Show that the inclusion map $E_1 \to E_0$ is compact; i.e., the unit ball in E_1 has compact closure E_0. (*Hint:* Use the Arzela-Ascoli Theorem.)

2.1D Let $(E, \langle \cdot, \cdot \rangle)$ be an inner product space and A, B subsets of E. Define the ***sum*** of A and B by $A + B = \{a + b \mid a \in A, b \in B\}$. Show that:

(i) $A \subset B$ implies $B^\perp \subset A^\perp$;

(ii) $A^\perp$ is a closed subspace of E;

(iii) $A^\perp = (\text{cl}(\text{span}(A)))^\perp$, $(A^\perp)^\perp = \text{cl}(\text{span}(A))$;

(iv) $(A + B)^\perp = A^\perp \cap B^\perp$;

(v) $(\text{cl}(\text{span}(A)) \cap \text{cl}(\text{span}(B)))^\perp = A^\perp + B^\perp$ (*not* necessarily a direct sum).

2.1E A sequence $\{e_n\} \subset E$, where E is an inner product space, is said to be ***weakly convergent*** to $e \in E$ iff all the numerical sequences $\langle v, e_n \rangle$ converge to $\langle v, e \rangle$ for all $v \in E$. Let

$$\ell^2(\mathbb{C}) = \left\{ \{a_n\} \mid a_n \in \mathbb{C} \text{ and } \sum_{n=1}^{\infty} |a_n|^2 < \infty \right\}$$

and put

$$\langle \{a_n\}, \{b_n\} \rangle = \sum_{n=1}^{\infty} a_n \bar{b}_n .$$

Show that:

(i) in any inner product space, convergence implies weak convergence;

(ii) $\ell^2(\mathbb{C})$ is an inner product space;

(iii) the sequence $(1, 0, 0, ...)$, $(0, 1, 0, ...)$, $(0, 0, 1, ...)$, ... is not convergent but is weakly convergent to 0 in $\ell^2(\mathbb{C})$.

Note. $\ell^2(\mathbb{C})$ is in fact complete, so it is a Hilbert space. The ambitious reader can attempt a direct proof or consult a book on real analysis such as Royden [1968].

2.1F Show that a normed vector space is a Banach space iff every absolutely convergent series is convergent. (A series $\Sigma_{n=1}^{\infty} x_n$ is called ***absolutely convergent*** if $\Sigma_{n=1}^{\infty} \|x_n\|$ converges.)

2.1G Let E be a Banach space and $F_1 \subset F_2 \subset E$ be closed subspaces such that F_2 splits in E. Show that F_1 splits in E iff F_1 splits in F_2.

2.1H Let F be closed in E of finite codimension. Show that if G is a subspace of E

containing F, then G is closed.

2.1I Let E be a Hilbert space. A set $\{e_i\}_{i \in I}$ is called ***orthonormal*** if $\langle e_i, e_j \rangle = \delta_{ij}$, the Kronecker delta. An orthonormal set $\{e_i\}_{i \in I}$ is a ***Hilbert basis*** if $\mathrm{cl}(\mathrm{span}\{e_i\}_{i \in I}) = E$.

(i) Let $\{e_i\}_{i \in I}$ be an orthonormal set and $\{e_{i(1)}, \ldots, e_{i(n)}\}$ be any finite subset. Show that

$$\sum_{j=1}^{n} |\langle e, e_{i(j)} \rangle|^2 \leq \| e \|^2$$

for any $e \in E$. (*Hint:* $e' = e - \Sigma_{j=1,\ldots,n} \langle e, e_{i(j)} \rangle e_{i(j)})$ is orthogonal to all $\{e_{i(j)} \mid j = 1, \ldots, n\}$.)

(ii) Deduce from (i) that for any positive integer n, the set $\{i \in I \mid |\langle e, e_i \rangle| > 1/n\}$ has at most $n \| e \|^2$ elements. Hence at most countably many $i \in I$ satisfy $\langle e, e_i \rangle \neq 0$, for any $e \in E$.

(iii) Show that any Hilbert space has a Hilbert basis. (*Hint:* Use Zorn's Lemma and **2.1.17.**)

(iv) If $\{e_i\}_{i \in I}$ is a Hilbert basis in E, $e \in E$, and $\{e_{i(j)}\}$ is the (at most countable) set such that $\langle e, e_{i(j)} \rangle \neq 0$, show that

$$\sum_{j=1}^{\infty} |\langle e, e_{i(j)} \rangle|^2 = \| e \|^2 .$$

(*Hint:* If $e' = \Sigma_{j=1,\ldots,\infty} \langle e, e_{i(j)} \rangle e_{i(j)} \rangle$, show that $\langle e_i, e - e' \rangle = 0$ for all $i \in I$ and then use maximality of $\{e_i\}_{i \in I}$.)

(v) Show that E is separable iff any Hilbert basis is at most countable. (*Hint:* For the "if" part, show that the set

$$\left\{ \sum_{k=1}^{n} \alpha_n e_n \;\middle|\; \alpha_k = a_k + i b_k \text{, where } a_k \text{ and } b_k \text{ are rational} \right\}$$

is dense in E. For the "only if" part, show that since $\| e_i - e_j \|^2 = 2$, the disks of radius $1/\sqrt{2}$ centered at e_i are all disjoint.)

(vi) If E is a separable Hilbert space, it is algebraically isomorphic either with $\mathbb{C}^n$ or $\ell^2(\mathbb{C})$ ($\mathbb{R}^n$ or $\ell^2(\mathbb{R})$), and the algebraic isomorphism can be chosen to be norm preserving.

§2.2 *Linear and Multilinear Mappings*

This section deals with various aspects of linear and multilinear maps between Banach spaces. We begin with a study of continuity and go on to study spaces of continuous linear and multilinear maps and some related fundamental theorems of linear analysis.

2.1.2 Proposition *Let* $A : E \to F$ *be a linear map of normed spaces. Then* A *is continuous if and only if there is a constantt* $M > 0$ *such that*

$$\| Ae \|_F \le M \| e \|_E \quad \textit{for all } e \in E \ .$$

Proof Continuity of A at $e_0 \in E$ means that for any $r < 0$, there exists $\rho > 0$ such that

$$A(e_0 + B_\rho(0_E)) \subset Ae_0 + B_r(0_F)$$

(0_F denotes the zero element in E and $B_s(0)_E$ denotes the closed disk of radius s centered at the origin in E). Since A is linear, this is equivalent to: if $\| e \| \le \rho$, then $\| Ae \| \le r$. If $M = r/\rho$, continuity of A is thus equivalent to the following: $\| e \|_E \le 1$ implies $\| Ae \|_F \le M$, which in turn is the same as: there exists $M > 0$ such that $\| Ae \|_F \le M\| e \|_E$, which is seen by choosing the vector $e/\| e \|_E$ in the preceding implication. ■

Because of this proposition one says that a continuous linear map is ***bounded.***

2.2.2 Proposition *If* E *is finite dimensional and* $A : E \to F$ *is linear, then* A *is continuous.*

Proof Let $\{e_1, ..., e_n\}$ be a basis for E. Letting $M_1 = \max(\| Ae_1 \|, ..., \| Ae_n \|)$ and setting $e = a^1 e_1 + \cdots + a^n e_n$, we see that

$$\begin{aligned} \| Ae \| &= \| a^1 Ae_1 + \cdots + a^n Ae_n \| \\ &\le | a^1 | \| Ae_1 \| + \cdots + | a^n | \| Ae_n \| \le M_1(| a^1 | + \cdots + | a^n |) \ . \end{aligned}$$

Since E is finite dimensional, all norms on it are equivalent. Since $||| e ||| = \Sigma | a^i |$ is a norm, it follows that $||| e ||| \le C\| e \|$ for a constant C. Let $M = M_1 C$ and use **2.2.1.** ■

2.2.3 Definition *If* E *and* F *are normed spaces and* $A : E \to F$ *is a continuous linear map, let the **operator norm** of* A *be defined by*

$$\| A \| = \sup\left\{ \frac{\| Ae \|}{\| e \|} \;\middle|\; e \in E, e \neq 0 \right\}$$

(which is finite by **2.2.1**). *Let* L(E, F) *denote the space of all continuous linear maps of* E *to* F. *If* $F = \mathbb{C}$ *(resp.* $\mathbb{R}$), *then* $L(E, \mathbb{C})$ *(resp.* $L(E, \mathbb{R})$) *is denoted by* E^* *and is called the* ***complex*** *(resp.* ***real****)* ***dual space*** *of* E. *(It will always be clear from the context whether* L(E, F) *or* E^* *means the real or complex linear maps or dual space; in most of the work later in this book it will mean the real case.*

A straightforward verification gives the following equivalent definitions of $\| A \|$:

$$\begin{aligned} \| A \| &= \inf\{ M > 0 \mid \| Ae \| \leq M \| e \| \text{ for all } e \in E \} \\ &= \sup\{ \| Ae \| \mid \| e \| \leq 1 \} = \sup\{ \| Ae \| \mid \| e \| = 1 \} . \end{aligned}$$

In particular, $\| Ae \| \leq \| A \| \| e \|$.

If $A \in L(E, F)$ and $B \in L(F, G)$, where E, F, and G are normed spaces, then

$$\| (B \circ A)(e) \| = \| B(A(e)) \| \leq \| B \| \| Ae \| \leq \| B \| \| A \| \| e \| ,$$

and so

$$\| (B \circ A) \| \leq \| B \| \| A \| .$$

Equality does not hold in general. A simple example is obtained by choosing $E = F = G = \mathbb{R}^2$, $A(x, y) = (x, 0)$, and $B(x, y) = (0, y)$, so that $B \circ A = 0$ and $\| A \| = \| B \| = 1$.

2.2.4 Proposition L(E, F) *with the norm just defined is a normed space. It is a Banach space if* F *is.*

Proof Clearly $\| A \| \geq 0$ and $\| 0 \| = 0$. If $\| A \| = 0$, then for any $e \in E$, $\| Ae \| \leq \| A \| \| e \| = 0$, so that $A = 0$ and thus **N1** (see Definition **2.1.1**) is verified. **N2** and **N3** are also straightforward to check.

Now let F be a Banach space and $\{A_n\} \subset L(E, F)$ be a Cauchy sequence. Because of the inequality $\| A_n e - A_m e \| \leq \| A_n - A_m \| \| e \|$ for each $e \in E$, the sequence $\{A_n e\}$ is Cauchy in F and hence is convergent. Let $Ae = \lim_{n \to \infty} A_n e$. This defines a map $A : E \to F$, which is evidently linear. It remains to be shown that A is continuous and $\| A_n - A \| \to 0$.

If $\varepsilon > 0$ is given, there exists a natural number $N(\varepsilon)$ such that for all $m, n \geq N(\varepsilon)$ we have $\| A_n - A_m \| < \varepsilon$. If $\| e \| \leq 1$, this implies $\| A_n e - A_m e \| < \varepsilon$, and now letting $m \to \infty$, it follows that $\| A_n e - Ae \| \leq \varepsilon$ for all e with $\| e \| \leq 1$. Thus $A_n - A \in L(E, F)$, hence $A \in L(E, F)$ and $\| A_n - A \| \leq \varepsilon$ for all $n \geq N(\varepsilon)$; i.e., $\| A_n - A \| \to 0$. ■

If a sequence $\{A_n\}$ converges to A in L(E, F) in the sense that $\| A_n - A \| \to 0$, i.e., if $A_n \to A$ in the norm topology, we say $A_n \to A$ ***in norm***. This phrase is necessary since other topologies on L(E, F) are possible. For example, we say that $A_n \to A$ ***strongly*** if $A_n e \to Ae$ for each $e \in E$. Since $\| A_n e - Ae \| \le \| A_n - A \| \, \| e \|$, norm convergence implies strong convergence. The converse is false as the following example shows. Let

$$E = \ell^2(\mathbb{R}) = \left\{ \{a_n\} \;\middle|\; \sum_{n=1}^{\infty} a_n^2 < \infty \right\}$$

with inner product

$$\langle \{a_n\}, \{b_n\} \rangle = \sum_{n=1}^{\infty} a_n b_n \; .$$

Let

$$e_n = (0, \ldots, 0, 1, 0 \cdots) \in E \;, \quad F = \mathbb{R} \;, \quad \text{and} \quad A_n = \langle e_n, \cdot \rangle \in L(E, F) \;,$$

where the 1 in e_n is in the nth slot. The sequence $\{A_n\}$ is not Cauchy in the operator norm since $\| A_n - A_m \| = \sqrt{2}$, but if $e = \{a_m\}$, $A_n(e) = \langle e_n, e \rangle = a_n \to 0$, i.e., $A_n \to 0$ strongly. If both E and F are finite dimensional, strong convergence implies norm convergence. (To see this, choose a basis $e_1, \ldots, e_n$ of E and note that strong convergence is equivalent to $A_k e_i \to Ae_i$ as $k \to \infty$ for $i = 1, \ldots, n$. Hence $\max_i \| Ae_i \| = |||\, A \,|||$ is a norm yielding strong convergence. But all norms are equivalent in finite dimensions.)

☞ SUPPLEMENT 2.2A
Dual Spaces

Recall from elementary linear algebra that the dual space of a finite dimensional vector space of dimension n also has dimension n and so the space and its dual are isomorphic. For general Banach spaces this is no longer true. However, it is true for Hilbert space.

2.2.5 Riesz Representation Theorem *Let* E *be a real (resp. complex) Hilbert space. The map* $e \mapsto \langle \cdot, e \rangle$ *is a linear (resp. antilinear) norm-preserving isomorphism of* E *with* E^*; *for short,* $E \cong E^*$. *(A map* $A : E \to F$ *between complex vector spaces is called* ***antilinear*** *if we have the identities* $A(e + e') = Ae + Ae'$, *and* $A(\alpha e) = \overline{\alpha} Ae$.)

Proof Let $f_e = \langle \cdot, e \rangle$. Then $\| f_e \| = \| e \|$ and thus $f_e \in E^*$. The map $A : E \to E^*$, $Ae = f_e$ is clearly linear (resp. antilinear), norm preserving, and thus injective. It remains to prove surjectivity.

Let $f \in E^*$ and $\ker f = \{ e \in E \mid f(e) = 0 \}$. $\ker f$ is a closed subspace in E. If $\ker f = E$, then $f = 0$ and $f = A(0)$ so there is nothing to prove. If $\ker f \neq E$, then by lemma **2.1.17** there exists $e \neq 0$ such that $e \perp \ker f$. Then we claim that $f = A(f(e)e/\| e \|^2)$. Indeed, any $v \in E$ can be written as

$$v = v - \frac{f(v)}{f(e)} e + \frac{f(v)}{f(e)} e \quad \text{and} \quad v - \frac{f(v)}{f(e)} e \in \ker f \,. \;\blacksquare$$

Thus, in a real Hilbert space E every continuous linear function $\ell : E \to \mathbb{R}$ can be written

$$\ell(e) = \langle e, e_0 \rangle$$

for some $e_0 \in E$ and $\| \ell \| = \| e_0 \|$.

In a general Banach space E we do not have such a concrete realization of E^*. However, one should *not* always attempt to identify E and E^*, even in finite dimensions. In fact, distinguishing these spaces is fundamental in tensor analysis.

We have a canonical map $i : E \to E^{**}$ defined by $i(e)(\ell) = \ell(e)$. (Pause and look again at this strange but natural formula: $i(e) \in E^{**} = (E^*)^*$, so $i(e)$ is applied to the element $\ell \in E^*$.) It is easy to check that i is norm preserving. One calls E ***reflexive*** if i is onto. Hilbert spaces are reflexive, by **2.2.5**. For example, let $V = L^2(\mathbb{R}^n)$ with inner product

$$\langle f, g \rangle = \int_{\mathbb{R}^n} f(x)\, g(x)\, dx \ ,$$

and let $\alpha : L^2(\mathbb{R}^n) \to \mathbb{R}$ be a continuous linear functional. Then the Riesz representation theorem guarantees that there exists a unique $g \in L^2(\mathbb{R}^n)$ such that

$$\alpha(f) = \int_{\mathbb{R}^n} g(x)\, f(x)\, dx = \langle g, f \rangle$$

for all $f \in L^2(\mathbb{R}^n)$.

In general, if E is not a Hilbert space and we wish to represent a linear functional α in the form of $\alpha(f) = \langle g, f \rangle$, we must regard $g(x)$ as an element of the dual space E^*. For example, let $E = C^0(\Omega, \mathbb{R})$, where $\Omega \subset \mathbb{R}^n$. Each $x \in \Omega$ defines a linear functional E_x : $C^0(\Omega, \mathbb{R}) \to \mathbb{R}$; $f \mapsto f(x)$. This linear functional cannot be represented in the form $E_x(f) = \langle g, f \rangle$ and, indeed, is not continuous in the L^2 norm. Nevertheless, it is customary and useful to write such linear maps as if $\langle\ ,\ \rangle$ were the L^2 inner product. Thus one writes, symbolically,

$$E_{x_0}(f) = \int_{\Omega} \delta(x - x_0) f(x)\, dx,$$

which defines the ***Dirac delta function*** at x_0; i.e., $g(x) = \delta(x - x_0)$.

Next we shall discuss integration of vector valued functions. We shall require the following.

2.2.6 Linear Extension Theorem *Let* E, F, *and* G *be normed vector spaces where*

(i) $F \subset E$;

(ii) G *is a Banach space*; *and*

(iii) $T \in L(F, G)$.

Then the closure cl(F) *of* F *is a normed vector subspace of* E *and* T *can be uniquely extended to a map* $\mathcal{T} \in L(\mathrm{cl}(F), G)$. *Moreover*, we have the equality $\| T \| = \| \mathcal{T} \|$.

Proof The fact that cl(F) is a linear subspace of E is easily checked. Note that if $\mathcal{T}$ exists it is unique by continuity. Let us prove the existence of $\mathcal{T}$. If $e \in \mathrm{cl}(F)$, we can write $e = \lim_{n \to \infty} e_n$, where $e_n \in F$, so that

$$\| Te_n - Te_m \| \leq \| T \| \| e_n - e_m \| ,$$

which shows that the sequence $\{Te_n\}$ is Cauchy in the Banach space G. Let $\mathcal{T}e = \lim_{n \to \infty} Te_n$. This limit is independent of the sequence $\{e_n\}$, for if $e = \lim e'_n$, then

$$\| Te_n - Te'_n \| \leq \| T \| (\| e_n - e \| + \| e - e'_n \|) ,$$

which proves that $\lim_{n \to \infty} (Te_n) = \lim_{n \to \infty} (Te'_n)$. It is simple to check the linearity of $\mathcal{T}$. Since $Te = \mathcal{T}e$ for $e \in F$ (because $e = \lim_{n \to \infty} e$), $\mathcal{T}$ is an extension of T. Finally,

$$\| \mathcal{T}e \| = \left\| \lim_{n \to \infty} (Te_n) \right\| = \lim_{n \to \infty} \| Te_n \| \leq \| T \| \lim_{n \to \infty} \| e_n \| = \| T \| \| e \|$$

shows that $\mathcal{T} \in L(\mathrm{cl}(F), G)$ and $\| \mathcal{T} \| \leq \| T \|$. The inequality $\| T \| \leq \| \mathcal{T} \|$ is obvious since $\mathcal{T}$ extends T. ■

As an application of this lemma we define a Banach space valued integral that will be of use later on. Fix the closed interval $[a, b] \subset \mathbb{R}$ and the Banach space E. A map $f : [a, b] \to E$ is called a ***step function*** if there exists a partition $a = t_0 < t_1 < t_2 \cdots < t_n = b$ such that f is constant on each interval $[t_i, t_{i+1}[$. Using the standard notion of a refinement of a partition, it is clear that the sum of two step functions and the scalar multiples of step functions are also step functions. Thus the set $\mathbf{S}([a, b], E)$ of step functions is a vector subspace of B([a, b], E), the Banach space of all bounded functions (see Example **2.1.12**). The integral of a step function f is defined by

$$\int_a^b f = \sum_{i=0}^{n} (t_{i+1} - t_i) f(t_i) .$$

It is easily verified that this definition is independent of the partition. Also note that

$$\left\| \int_a^b f \right\| \leq \int_a^b \| f \| \leq (b - a) \| f \|_\infty$$

where

$$\| f \|_\infty = \sup_{a \leq t b} | f(t) | ;$$

that is,

$$\int_a^b : \mathbf{S}([a, b], E) \to E$$

is continuous and linear. By the linear extension theorem, it extends to a continuous linear map

$$\int_a^b \in L(\mathrm{cl}(\mathbf{S}\,([a, b], E)), E).$$

2.2.7 Definition *The linear map*

$$\int_a^b$$

is called the ***Cauchy-Bochner*** *integral.*

Note that

$$\left\| \int_a^b f \right\| \le \int_a^b \| f \| \le (b - a)\| f \|_\infty \ .$$

The usual properties of the integral such as

$$\int_a^b f = \int_a^c f + \int_c^b f, \ \text{and} \ \int_a^b f = -\int_b^a f$$

are easily verified since they clearly hold for step functions.

The space $\mathrm{cl}(\mathbf{S}([a,b], E)$ contains enough interesting functions for our purposes, namely

$$C^0([a, b], E) \subset \mathrm{cl}(\mathbf{S}([a, b], E)) \subset B([a, b], E) \ .$$

The first inclusion is proved in the following way. Since $[a,b]$ is compact, each $f \in C^0([a, b], E)$ is uniformly continuous. For $\varepsilon > 0$, let $\delta > 0$ be given by uniform continuity of f for $\varepsilon/2$. Then take a partition $a = t_0 < \cdots < t_n = b$ such that $| t_{i+1} - t_1 | < \delta$ and define a step function g by $g \,|\, [t_i, t_{i+1}[= f(t_i)$. Then the ε-disk $D_\varepsilon(f)$ in $B([a, b], E)$ contains g.

Finally, note that if E and F are Banach spaces, $A \in L(E, F)$, and $f \in \mathrm{cl}(\mathbf{S}([a, b], E))$, we have $A \circ f \in \mathrm{cl}(\mathbf{S}([a, b], F))$ since $\| A \circ f_n - A \circ f \| \le \| A \| \, \| f_n - f \|_\infty$ where f_n are step functions in E. Moreover,

$$\int_a^b A \circ f = A\left(\int_a^b f\right)$$

since this relation is obtained as the limit of the same (easily verified) relation for step functions. The reader versed in Riemann integration should notice that this integral for $E = \mathbb{R}$ is less general than the Riemann integral; i.e., the Riemann integral exists also for functions outside of $\mathrm{cl}(\mathbf{S}([a, b], \mathbb{R}))$. For purposes of this book, however, this integral will suffice.

Next we turn to multilinear mappings. If $E_1, \ldots, E_k$ and F are linear spaces, a map

$$A : E_1 \times \cdots \times E_k \to F$$

is called ***k-multilinear*** if $A(e_1, ..., e_k)$ is linear in each argument separately. Linearity in the first argument means that

$$A(\lambda e_1 + \mu f_1, e_2, ..., e_k) = \lambda A(e_1, e_2, ..., e_k) + \mu A(f_1, e_2, ..., e_k) .$$

We shall study multilinear mappings in detail in our study of tensors. They also come up in the study of differentiation, and we shall require a few facts about them for that purpose.

2.2.8 Definition *The space of continuous* k*-multilinear maps of* $E_1, ..., E_k$ *to* F *is denoted* $L(E_1, ..., E_k; F)$. *If* $E_i = E$, $1 \le i \le k$, *this space is denoted* $L^k(E, F)$.

As in **2.2.1**, a k-multilinear map A is continuous if and only if there is an $M > 0$ such that

$$\| A(e_1, ..., e_k) \| \le M \| e_1 \| \cdots \| e_k \|$$

for all $e_i \in E_i$, $1 \le i \le k$. We set

$$\| A \| = \sup \left\{ \frac{\| A(e_1, \ldots, e_k) \|}{\| e_1 \| \cdots \| e_k \|} \;\middle|\; e_1, \ldots, e_k \neq 0 \right\}$$

which makes $L(E_1, ..., E_k; F)$ into a normed space that is complete if F is. Again $\| A \|$ can also be defined as

$$\begin{aligned} \| A \| &= \inf\{ M > 0 \mid \| A(e_1, ..., e_n) \| \le M \| e_1 \| \cdots \| e_n \| \} \\ &= \sup\{ \| A(e_1, ..., e_n) \| \mid \| e_1 \| \le 1, ..., \| e_n \| \le 1 \} \\ &= \sup\{ \| A(e_1, ..., e_n) \| \mid \| e_1 \| = \cdots = \| e_n \| = 1 \} . \end{aligned}$$

2.2.9 Proposition *There are (natural) norm-preserving isomorphisms*

$$\begin{aligned} L(E_1, L(E_2, \ldots, E_k; F)) &\cong L(E_1, \ldots, E_k; F) \cong L(E_1, \ldots, E_{k-1}; L(E_k, F)) \\ &\cong L(E_{i_1}, \ldots, E_{i_k}; F) \end{aligned}$$

where $(i_1, ..., i_k)$ *is a permutation of* $(1, ..., k)$.

Proof For $A \in L(E_1, L(E_2, ..., E_k; F))$, define $A' \in L(E_1, ..., E_k; F)$ by

$$A'(e_1, ..., e_k) = A(e_1)(e_2, ..., e_k) .$$

The association $A \mapsto A'$ is clearly linear and $\| A' \| = \| A \|$. The other isomorphisms are proved similarly. ■

In a similar way, we can identify $L(\mathbb{R}, F)$ (or $L(\mathbb{C}, F)$ if F is complex) with F: to $A \in L(\mathbb{R}, F)$ we associate $A(1) \in F$; again $\| A \| = \| A(1) \|$. As a special case of **2.2.9** note that $L(E, E^*) \cong L^2(E, \mathbb{R})$ (or $L^2(E; \mathbb{C})$, if E is complex). This isomorphism will be useful when we consider second derivatives.

We shall need a few facts about the permutation group on k elements. The information we cite is obtainable from virtually any elementary algebra book. The ***permutation group*** on k elements, denoted S_k, consists of all bijections $\sigma : \{1, ..., k\} \to \{1, ..., k\}$ together with the structure of a group under composition. Clearly, S_k has order $k!$. Letting $(\mathbb{R}, \times)$ denote $\mathbb{R}\backslash\{0\}$ with the multiplicative group structure, we have a homomorphism $\text{sign} : S_k \to (\mathbb{R}, \times)$. That is, for $\sigma, \tau \in S_k$, $\text{sign}(\sigma \circ \tau) = (\text{sign}\, \sigma)(\text{sign}\, \tau)$. The image of "sign" is the subgroup $\{-1, 1\}$, while its kernel consists of the subgroup of ***even permutations.*** Thus, a permutation σ is ***even*** when $\text{sign}\, \sigma = +1$ and is ***odd*** when $\text{sign}\, \sigma = -1$. A ***transposition*** is a permutation that swaps two elements of $\{1, ..., k\}$, leaving the remainder fixed. An even (odd) permutation can be written as the product of an even (odd) number of transpositions. The group S_k ***acts*** on the space $L^k(E; F)$; i.e., each $\sigma \in S_k$ defines a map $\sigma : L^k(E; F) \to L^k(E; F)$ by $(\sigma A)(e_1, ..., e_k) = A(e_{\sigma(1)}, ..., e_{\sigma(k)})$. Note that $(\tau\sigma)A = \tau(\sigma A)$ for all $\tau, \sigma \in S_k$. Accordingly, $A \in L^k(E, F)$ is called ***symmetric*** (***antisymmetric***) if for any permutation $\sigma \in S_k$, $\sigma A = A$ (resp. $\sigma A = (\text{sign}\, \sigma)A$.)

2.2.10 Definition *Let* E *and* F *be normed vector spaces. Let* $L^k_s(E; F)$ *and* $L^k_a(E; F)$ *denote the subspaces of symmetric and antisymmetric elements of* $L^k(E; F)$. *Write* $S^0(E, F) = F$ *and* $S^k(E, F) = \{p : E \to F \mid p(e) = A(e, ..., e)$ *for some* $A \in L^k(E; F)\}$. *We call* $S^k(E, F)$ *the* ***space of homogeneous polynomials of degree*** k *from* E *to* F.

Note that $L^k_s(E; F)$ and $L^k_a(E; F)$ are closed in $L^k(E; F)$; thus if F is a Banach space, so are $L^k_s(E; F)$ and $L^k_a(E; F)$. The antisymmetric maps $L^k_a(E; F)$ will be studied in detail in Chapter 7. For technical purposes later in this chapter we will need a few facts about $S^k(E, F)$ which are given in the following supplement.

☞ SUPPLEMENT 2.2B
Homogeneous Polynomials

2.2.11 Proposition (i) $S^k(E, F)$ *is a normed vector space with respect to the following norm:*

$$\| f \| = \inf\{M > 0 \mid \| f(e) \| \le M \| e \|^k\} = \sup\{\| f(e) \| \mid \| e \| \le 1\} = \sup\{\| f(e) \| \mid \| e \| = 1\} .$$

It is complete if F *is.*

(ii) *If* $f \in S^k(E, F)$ *and* $g \in S^n(F, G)$, *then* $g \circ f \in S^{kn}(E, G)$ *and* $\| g \circ f \| \le \| g \| \, \| f \|$.

(iii) *(Polarization). The mapping* $' : L^k(E, F) \to S^k(E, F)$ *defined by* $A'(e) = A(e, ..., e)$ *restricted to* $L^k_s(E; F)$ *has an inverse* $` : S^k(E, F) \to L^k_s(E, F)$ *given by*

$$`f(e_1, \ldots, e_k) = \frac{1}{k!} \frac{\partial^k}{\partial t_1 \cdots \partial t_k}\bigg|_{t=0} f(t_1 e_1 + \cdots + t_k e_k) .$$

(Note that $f(t_1 e_1 + \cdots + t_k e_k)$ *is a polynomial in* $t_1, \ldots, t_k$, *so there is no problem in understanding what the derivatives on the right hand side mean).*

(iv) *For* $A \in L^k(E,F)$, $\| A' \| \le \| A \| \le (k^k/k!) \| A' \|$, *which implies the maps* $'$, $`$ *are continuous.*

Proof (i) and (ii) are proved exactly as for $L(E, F) = S^1(E, F)$.

(iii) For $A \in L^k_s(E; F)$ we have

$$A'(t_1 e_1 + \cdots + t_k e_k) = \sum_{a_1 + \cdots + a_j = k} \frac{k!}{a_1! \cdots a_j!} t_1^{a_1} \cdots t_j^{a_j} A(e_1, \ldots, e_1, \ldots, e_j, \ldots, e_j) ,$$

where each e_i appears a_i times, and

$$\frac{\partial^k}{\partial t_1 \cdots \partial t_k}\bigg|_{t=0} t_1^{a_1} \cdots t_j^{a_j} = \begin{cases} 1, & \text{if } k = j \\ 0, & \text{if } k \ne j . \end{cases}$$

It follows that

$$A(e_1, \ldots, e_k) = \frac{1}{k!} \frac{\partial^k}{\partial t_1 \cdots \partial t_k} A'(t_1 e_1 + \cdots + t_k e_k) ,$$

and for $j \ne k$,

$$\frac{\partial^j}{\partial t_1 \cdots \partial t_j} A'(t_1 e_1 + \cdots + t_k e_k) = 0 .$$

This means that $`(A') = A$ for any $A \in L^k_s(E, F)$.

Conversely, if $f \in S^k(E, F)$, then

$$(`f)'(e) = `f(e, \ldots, e) = \frac{1}{k!} \frac{\partial^k}{\partial t_1 \cdots \partial t_k}\bigg|_{t=0} f(t_1 e + \cdots + t_k e)$$

$$= \frac{1}{k!} \frac{\partial^k}{\partial t_1 \cdots \partial t_k}\bigg|_{t=0} (t_1 + \cdots + t_k)^k f(e) = f(e) .$$

(iv) $\| A'(e) \| = \| A(e, \ldots, e) \| \le \| A \| \, \| e \|^k$, so $\| A' \| \le \| A \|$. To prove the other inequality, note that if $A \in L^k_s(E; F)$, then

$$A(e_1, \ldots, e_k) = \frac{1}{k! 2^k} \sum \varepsilon_1 \cdots \varepsilon_k A'(\varepsilon_1 e_1 + \cdots + \varepsilon_k e_k)$$

where the sum is taken over all the 2^k possibilities $\varepsilon_1 = \pm 1, \ldots, \varepsilon_k = \pm 1$. Put $\| e_1 \| = \cdots = \| e_k \| = 1$ and get

$$\| A'(\varepsilon_1 e_1 + \cdots + \varepsilon_k e_k) \| \le \| A' \| \, \| \varepsilon_1 e_1 + \cdots + \varepsilon_k e_k \|^k$$
$$\le \| A' \| (| \varepsilon_1 | \, \| e_1 \| + \cdots + | \varepsilon_k | \, \| e_k \|)^k = \| A' \| \, k^k ,$$

whence

$$\| A(e_1, \ldots, e_k) \| \le \frac{k^k}{k!} \| A' \| ,$$

i.e.,

$$\| A \| \le \frac{k^k}{k!} \| A' \| . \quad \blacksquare$$

Let $E = \mathbb{R}^n$, $F = \mathbb{R}$, and $e_1, \ldots, e_n$ be the standard basis in $\mathbb{R}^n$. For $f \in S^k(\mathbb{R}^n, \mathbb{R})$, set

$$c_{a_1 \cdots a_n} = f(e_1, \ldots, e_1, \ldots, e_n, \ldots e_n),$$

where each e_i appears a_i times. If $e = t_1 e_1 + \cdots + t_n e_n$, the proof of (iii) shows that

$$f(e) = \grave{f}(e, \ldots, e) = \sum_{a_1 + \cdots + a_n = k} c_{a_1 \ldots a_n} t_1^{a_1} \ldots t_n^{a_n};$$

i.e., f is a homogeneous polynomial of degree k in $t_1, \ldots, t_n$ in the usual algebraic sense.

The constant $k^k/k!$ in (iv) is the best possible, as the following example shows. Write elements of $\mathbb{R}^k$ as $\mathbf{x} = (x^1, \ldots, x^k)$ and introduce the norm $|||\, (x^1, \ldots, x^k) \,||| = | x^1 | + \cdots + | x^k |$. Define $A \in L^k_s(\mathbb{R}^k, \mathbb{R})$ by

$$A(\mathbf{x}_1, \ldots, \mathbf{x}_k) = \frac{1}{k!} \sum x_{i_1}{}^1 \cdots x_{i_k}{}^k,$$

where $\mathbf{x}_i = (x_i{}^1, \ldots, x_i{}^k) \in \mathbb{R}^k$ and the sum is taken over all permutations of $\{1, \ldots, k\}$. It is easily verified that $\| A \| = 1/k!$, and $\| A' \| = 1/k^k$; i.e., $\| A \| = (k^k/k!) \, \| A' \|$. Thus, except for $k = 1$, the isomorphism $'$ is not norm preserving. (This is a source of annoyance in the theory of formal power series and infinite-dimensional holomorphic mappings.) ✍

☞ SUPPLEMENT 2.2C
The Three Pillars of Linear Analysis

The three fundamental theorems of linear analysis are the *Hahn-Banach theorem*, the *open mapping theorem*, and the *uniform boundedness principle*. This supplement gives the classical proofs of these three fundamental theorems and derives some corollaries that will be used later. In finite dimensions these corollaries are all "obvious."

2.2.12 Hahn-Banach Theorem *Let* E *be a real or complex vector space,* $\|\cdot\| : E \to \mathbb{R}$ *a seminorm, and* $F \subset E$ *a subspace. If* $f \in F^*$ *satisfies* $|f(e)| \le \|e\|$ *for all* $e \in F$, *then there exists a linear map* $f' : E \to \mathbb{R}$ *(or* $\mathbb{C}$*) such that* $f' | F = f$ *and* $|f'(e)| \le \|e\|$ *for all* $e \in E$.

Proof *Real Case.* First we show that $f \in F^*$ can be extended with the given property to $F \oplus \operatorname{span}\{e_0\}$, for a given $e_0 \notin F$. For $e_1, e_2 \in F$ we have

$$f(e_1) + f(e_2) = f(e_1 + e_2) \le \|e_1 + e_2\| \le \|e_1 + e_0\| + \|e_2 - e_0\| ,$$

so that

$$f(e_2) - \|e_2 - e_0\| \le \|e_1 + e_0\| - f(e_1) ,$$

and hence

$$\sup\{f(e_2) - \|e_2 - e_0\| \mid e_2 \in F\} \le \inf\{\|e_1 + e_0\| - f(e_1) \mid e_1 \in F\} .$$

Let $a \in \mathbb{R}$ be any number between the sup and inf in the preceding expression and define $f' : F \oplus \operatorname{span}\{e_0\} \to \mathbb{R}$ by $f'(e + te_0) = f(e) + ta$. It is clear that f' is linear and that $f' | F = f$. To show that $|f'(e + te_0)| \le \|e + te_0\|$, note that by the definition of a,

$$f(e_2) - \|e_2 - e_0\| \le a \le \|e_1 + e_0\| - f(e_1) ,$$

so that by multiplying the second inequality by $t \ge 0$ and the first by $t < 0$, we get the desired result.

Second, one verifies that the set $\mathcal{S} = \{(G, g) \mid F \subset G \subset E,\ G$ is a subspace of E, $g \in G^*$, $g | F = f$, and $|g(e)| \le \|e\|$ for all $e \in G\}$ is inductively ordered with respect to the ordering

$$(G_1, g_1) \le (G_2, g_2) \text{ iff } G_1 \subset G_2 \text{ and } g_2 | G_1 = g_1 .$$

Thus by Zorn's Lemma there exists a maximal element (F_0, f_0) of $\mathcal{S}$.

Third, using the first step and the maximality of (F_0, f_0), one concludes that $F_0 = E$.

Complex Case. Let $f = \operatorname{Re} f + i \operatorname{Im} f$ and note that complex linearity implies that $(\operatorname{Im} f)(e) = -(\operatorname{Re} f)(ie)$ for all $e \in F$. By the real case, Re f extends to a real linear continuous map $(\operatorname{Re} f) : E \to \mathbb{R}$, such that $|(\operatorname{Re} f)'(e)| \le \|e\|$ for all $e \in E$. Define $f' : E \to \mathbb{C}$ by $f'(e) = (\operatorname{Re} f)'(e) - i(\operatorname{Re} f)'(ie)$ and note that f is complex linear and $f' | F = f$.

To show that $|f'(e)| \le \|e\|$ for all $e \in E$, write $f'(e) = |f'(e)| \exp(i\theta)$, so complex linearity of f' implies $f'(e \cdot \exp(-i\theta)) \in \mathbb{R}$, and hence

$$|f'(e)| = f'(e \cdot \exp(-i\theta)) = (\operatorname{Re} f)'(e \cdot \exp(-i\theta)) \le \|e \cdot \exp(-i\theta)\| = \|e\| . \quad \blacksquare$$

2.2.13 Corollary *Let* $(E, \|\cdot\|)$ *be a normed space,* $F \subset E$ *a subspace, and* $f \in F^*$ *(the topological dual). Then there exists* $f' \in E^*$ *such that* $f' | F = f$ *and* $\|f'\| = \|f\|$.

Proof We can assume $f \neq 0$. Then $|||e||| = \|f\|\|e\|$ is a norm on E and $|f(e)| \leq \|f\| \cdot \|e|$ $= |||e|||$ for all $e \in F$. Applying the preceding theorem we get a linear map $f' : E \to \mathbb{R}$ (or $\mathbb{C}$) with the properties $f' | F = f$ and $|f'(e)| \leq |||e|||$ for all $e \in E$. This says that $\|f'\| \leq \|f\|$, and since f′ extends f, it follows that $\|f\| \leq \|f'\|$; i.e., $\|f'\| = \|f\|$ and $f' \in E^*$. ■

Applying the corollary to the linear function $ae \mapsto a$, for $e \in E$ a fixed element, we get the following.

2.1.14 Corollary *Let* E *be a normed vector space and* $e \neq 0$. *Then there exists* $f \in E^*$ *such that* $f(e) \neq 0$. *In other words if* $f(e) = 0$ *for all* $f \in E^*$, *then* $e = 0$; i.e., E^* *separates points of* E.

2.2.15 Open Mapping Theorem of Banach and Schauder *Let* E *and* F *be Banach spaces and suppose* $A \in L(E, F)$ *is onto. Then* A *is an open mapping.*

Proof To show A is an open mapping, it suffices to prove that $A(cl(D_1(0)))$ contains a disk centered at zero in F. Let $r > 0$. Since $E = \bigcup_{n\geq 1} D_{nr}(0)$, it follows that $F = \bigcup_{n\geq 1}(A(D_{nr}(0)))$ and hence $\bigcup_{n\geq 1} cl(A(D_{nr}(0))) = F$. Completeness of F implies that at least one of the sets $cl(A(D_{nr}(0)))$ has a nonempty interior by the Baire Category Theorem **1.7.2**. Because the mapping $e \in E \mapsto ne \in E$ is a homeomorphism, we conclude that $cl(A(D_r(0)))$ contains some open set $V \subset F$. We shall prove that the origin of F is in $\text{int}\{cl[A(D_r(0))]\}$ for some $r > 0$. Continuity of $(e_1, e_2) \in E \times E \mapsto e_1 - e_2 \in E$ assures the existence of an open set $U \subset E$ such that $U - U = \{e_1 - e_2 \mid e_1, e_2 \in U\} \subset D_r(0)$. Choose $r > 0$ such that $D_r((0)) \subset U$. Then $cl(A(D_r(0))) \supset cl(A(U) - A(U)) \supset cl(A(U)) - cl(A(U)) \supset V - V$. But $V - V = \bigcup_{e \in V}(V - e)$ is open and clearly contains $0 \in F$. It follows that there exists a disk $D_t(0) \subset F$ such that $D_t(0) \subset cl(A(D_r(0)))$.

Now let $\varepsilon(n) = 1/2^{n+1}$, $n = 0, 1, 2, \ldots$, so that $1 = \Sigma_{n\geq 0}\, \varepsilon(n)$. By the foregoing result for each n there exists an $\eta(n) > 0$ such that $D_{\eta(n)}(0) \subset cl(A(D_{\varepsilon(n)}(0))$. Clearly $\eta(n) \to 0$. We shall prove that $D_{\eta(0)} \subset A(cl(D_1(0)))$. For $v \in D_{\eta(0)}(0) \subset cl(A(D_{\varepsilon(0)}(0)))$ there exists $e_0 \in D_{\varepsilon(0)}(0)$ such that $\|v - Ae_0\| < \eta(1)$ and thus $v - Ae_0 \in cl(A(D_{\varepsilon(1)}(0)))$, so there exists $e_1 \in D_{\varepsilon(1)}(0)$ such that $\|v - Ae_0 - Ae_1\| < \eta(2)$, etc. Inductively one constructs a sequence $e_n \in D_{\eta(n)}$ such that $\|v - Ae_0 - \ldots - Ae_n\| < \eta(n+1)$. The series $\Sigma_{n\geq 0}\, e_n$ is convergent because

$$\Big\| \sum_{i=n+1}^{m} e_i \Big\| \leq \sum_{i=n+1}^{m} 1/2^{i+1}, \qquad \sum_{n=0}^{\infty} 1/2^{n+1} = 1,$$

and E is complete. Let $e = \Sigma_{n\geq 0}\, e_n \in E$. Thus

$$Ae = \sum_{n=0}^{\infty} Ae_n = v\ ,$$

and

$$\|e\| \le \sum_{n=0}^{\infty} \|e_n\| \le \sum_{n=0}^{\infty} \frac{1}{2^{n+1}} = 1 \; ;$$

i.e., $v \in D_{\eta(0)}(0)$ implies $v = Ae$, $\|e\| \le 1$; that is, $D_{\eta(0)}(0) \subset A(\mathrm{cl}(D_1(0)))$. ■

An immediate consequence is the following.

2.2.16 Banach's Isomorphism Theorem *A continuous linear isomorphism of Banach spaces is a homeomorphism.*

Thus, if F and G are closed subspaces of the Banach space E and E is the *algebraic* direct sum of F and G, then the mapping $(e, e') \in F \times G \mapsto e + e' \in E$ is a continuous isomorphism, and hence a homeomorphism; i.e., $E = F \oplus G$; this proves the comment at the beginning of Supplement **2.1B.**

2.2.17 Closed Graph Theorem *Let* E, F *be Banach spaces. A linear map* $A : E \to F$ *is continuous iff its graph* $\Gamma_A = \{(e, Ae) \in E \times F \mid e \in E\}$ *is a closed subspace of* $E \oplus F$.

Proof It is readily verified that Γ_A is a linear subspace of $E \oplus F$. If $A \in L(E, F)$, then Γ_A is closed (see Exercise **1.4B**). Conversely, if Γ_A is closed, then it is a Banach subspace of $E \oplus F$, and since the mapping $(e, Ae) \in \Gamma_A \mapsto e \in E$ is a continuous isomorphism, its inverse $e \in E \mapsto (e, Ae) \in \Gamma_A$ is also continuous by **2.2.16**. Since $(e, Ae) \in \Gamma_A \mapsto Ae \in F$ is clearly continuous, so is the composition $e \mapsto (e, Ae) \mapsto Ae$. ■

The Closed Graph Theorem is often used in the following way. To show that a linear map $A : E \to F$ is continous for E and F Banach spaces, it suffices to show that if $e_n \to 0$ and $Ae_n \to e'$, then $e' = 0$.

2.2.18 Corollary *Let* E *be a Banach space and* F *a closed subspace of* E. *Then* F *is split iff there exists* $P \in L(E, E)$ *such that* $P \circ P = P$ *and* $F = \{e \in E \mid Pe = e\}$.

Proof If such a P exists, then clearly $\ker P$ is a closed subspace of E that is an algebraic complement of F; any $e \in E$ is of the form $e = e - Pe + Pe$ with $e - Pe \in \ker P$ and $Pe \in F$.

Conversely, if $E = F \oplus G$, define $P : E \to E$ by $P(e) = e_1$, where $e = e_1 + e_2$, $e_1 \in F$, $e_2 \in G$. P is clearly linear, $P^2 = P$, and $F = \{e \in E \mid Pe = e\}$, so all there is to show is that P is continuous. Let $e_n = e_{1n} + e_{2n} \to 0$ and $P(e_n) = e_{1n} \to e'$; i.e., $-e_{2n} \to e'$, and since F and G are closed this implies that $e' \in F \cap G = \{0\}$. By the closed graph theorem, $P \in L(E, E)$. ■

2.2.19 Fundamental Isomorphism Theorem *Let* $A \in L(E, F)$ *be surjective where* E *and* F *are Banach spaces. Then* $E/\ker A$ *and* F *are isomorphic Banach spaces.*

Proof The map $[e] \mapsto Ae$ is bijective and continuous (since its norm is $\leq \| A \|$), so it is a homeomorphism. ■

A sequence of maps

$$\cdots \to E_{i-1} \xrightarrow{A_i} E_i \xrightarrow{A_{i+1}} E_{i+1} \to \cdots$$

of Banach spaces is said to be ***split exact*** if for all i, $\ker A_{i+1} = \text{range } A_i$ and both $\ker A_i$ and range A_i split. With this terminology, **2.2.18** can be reformulated in the following way: *If* $0 \to G \to E \to F \to 0$ *is a split exact sequence of Banach spaces, then* E/G *is a Banach space isomorphic to* F *(thus* $F \cong G \oplus F$).

2.2.20 Uniform Boundedness Principle of Banach and Steinhaus *Let* E *and* F *be normed vector spaces, with* E *complete, and let* $\{A_i\}_{i \in I} \subset L(E, F)$. *If for each* $e \in E$ *the set* $\{\| A_i e \|\}_{i \in I}$ *is bounded in* F, *then* $\{\| A_i \|\}_{i \in I}$ *is a bounded set of real numbers.*

Proof Let $\varphi(e) = \sup\{\| A_i e \| \mid i \in I\}$ and note that

$$S_n = \{e \in E \mid \varphi(e) \leq n\} = \bigcap_{i \in I} \{e \in E \mid \| A_i e \| \leq n\}$$

is closed and $\bigcup_{n \geq 1} S_n = E$. Since E is a complete metric space, the Baire Category Theorem **1.7.2** says that some S_n has nonempty interior; i.e., there exist $r > 0$ and $e_0 \in E$ such that $\varphi(e) \leq M$, for all $e \in \text{cl}(D_r(e_0))$, where $M > 0$ is come constant.

For each $i \in I$, and $\| e \| = 1$, we have $\| A_i(re + e_0) \| \leq \varphi(re + e_0) \leq M$, so that

$$\| A_i e \| = \frac{1}{r} \| A_i(re + e_0 - e_0) \| \leq \frac{1}{r} \| A_i(re + e_0) \| + \frac{1}{r} \| A_i e_0 \| \leq \frac{1}{r} (M + \varphi(e_0)) ,$$

i.e., $\| A_i \| \leq (M + \varphi(e_0))/r$ for all $i \in I$. ■

2.2.21 Corollary *If* $\{A_n\} \subset L(E, F)$ *is a strongly convergent sequence (i.e.,* $\lim_{n \to \infty} A_n e = Ae$ *exists for every* $e \in E$), *then* $A \in L(E, F)$.

Proof A is clearly a linear map. Since $\{A_n e\}$ is convergent, it is a bounded set for each $e \in E$, so that by **2.2.20**, $\{\| A_n \|\}$ is bounded by, say, $M > 0$. But then

$$\| Ae \| = \lim_{n \to \infty} \| A_n e \| \leq \lim_{n \to \infty} \sup \| A_n \| \| e \| \leq M \| e \| ;$$

i.e., $A \in L(E, F)$. ■ ✍

Exercises

2.2A If $E = \mathbb{R}^n$ and $F = \mathbb{R}^m$ with the standard norms, and $A : E \to F$ is a linear map, show that (i) $\| A \|$ is the square root of the absolute value of the largest eigenvalue of AA^T, where A^T is the transpose of A and (ii) if $n, m \geq 2$, this norm does not come from an inner product. (*Hint*: Use Exercise **2.1A**.)

2.2B Let $E = \mathbb{R}^n$ and $F = \mathbb{R}^n$ with the standard norms and $A, B \in L(E, F)$. Let $\langle A, B \rangle = \text{trace}(AB^T)$. Show that this is an inner product on $L(E, F)$.

2.2C Show that the map

$$L(E, F) \times L(F, E) \to \mathbb{R}; \qquad (A, B) \mapsto \text{trace}(AB)$$

gives a (natural) isomorphism $L(E, F)^* \cong L(F, E)$.

2.2D Let E, F, G be Banach spaces, $A : D \subset E \to F$, and $B : D \subset E \to G$ two closed operators with the same domain D. (See Supplement **7.4A** for a full account of closed operators; all that is needed here is the definition). Show there exist constants $M_1, M_2 > 0$ such that

$$\| Ae \| \leq M_1(\| Be \| + \| e \|) \quad \text{and} \quad \| Be \| \leq M_2(\| Ae \| + \|e \|)$$

for all $e \in E$. (*Hint:* Norm $E \oplus G$ by $\| (e,g) \| = \| e \| + \| g \|$ and define $T : \Gamma_B \to G$ by $T(e, Be) = Ae$. Use the Closed Graph Theorem to show that $T \in L(\Gamma_B, G)$.)

2.2E **Linear transversality** Let E, F be Banach spaces, $F_0 \subset F$ a closed subspace, and $T \in L(E, F)$. T is said to be ***transversal*** to F_0, if $T^{-1}(F_0)$ splits in E and $T(E) + F_0 = \{Te + f \mid e \in E, f \in F\} = F$. Prove the following.

(i) T is transversal to F_0 iff $\pi \circ T \in L(E, F/F_0)$ is surjective with split kernel; here $\pi : F \to F/F_0$ is the projection.

(ii) If $\pi \circ T \in L(E, F/F_0)$ is surjective and F_0 has finite codimension, then $\ker(\pi \circ T)$ has the same codimension and T is transversal to F_0. (*Hint:* Use the algebraic isomorphism $T(E)/(F_0 \cap T(E)) \cong (T(E) + F_0)/F_0$ to show $E/\ker(\pi \circ T) \cong F/F_0$; now use **2.2.18**.)

(iii) If $\pi \circ T \in L(E, F/F_0)$ is surjective and if ker T and F_0 are finite dimensional, then $\ker(\pi \circ T)$ is finite dimensional and T is transversal to F_0. (*Hint:* Use the exact sequence $0 \to \ker T \to \ker(\pi \circ T) \to F_0 \cap T(E) \to 0$.)

2.2F Let E and F be Banach spaces. Prove the following.

(i) If $f \in \text{cl}(\mathcal{S}([a, b], L(E, F)))$ and $e \in E$, then

$$\int_a^b f(t)e\,dt = \left(\int_a^b f(t)\,dt\right)(e).$$

(*Hint:* $T \mapsto Te$ is in $L(L(E, F), F)$.)

(ii) If $f \in \mathrm{cl}(\mathcal{S}([a, b], \mathbb{R})$ and $v \in F$, then

$$\int_a^b f(t)v\,dt = \left(\int_a^b f(t)\,dt\right)(v).$$

(*Hint:* $t \mapsto$ multiplication by t in F is in $L(\mathbb{R}, L(F, F))$; apply (i).)

(iii) Let X be a topological space and $f : [a, b] \times X \to E$ be continuous. Then the mapping

$$g : X \to E,\ g(x) = \int_a^b f(t, x)dt$$

is continuous. (*Hint:* for $t \in \mathbb{R}$, $x' \in X$ and $\varepsilon > 0$ given, $\| f(s, x) - f(t, x') \| < \varepsilon$ if $(s, x) \in U_1 \times U_{x',t}$; use compactness of $[a, b]$ to find $U_{x'}$ as a finite intersection and such that $\| f(t, x) - f(t, x') \| < \varepsilon$ for all $t \in [a, b]$, $x \in U_{x'}$.)

2.2G Show that the Banach Isomorphism Theorem is false for normed incomplete vector spaces in the following way. Let E be the space of all polynomials over $\mathbb{R}$ normed as follows:

$$\| a_0 + a_1 x + \cdots + a_n x^n \| = \max\{| a_0 |, \ldots, | a_n |\}.$$

(i) Show that E is not complete.

(ii) Define $A : E \to E$ by

$$A\left(\sum_{i=0}^{n} a_i x^i\right) = a_0 + \sum_{i=1}^{n} \frac{a_i}{i} x^i$$

and show that $A \in L(E, E)$. Prove that $A^{-1} : E \to E$ exists.

(iii) Show that A^{-1} is not continuous.

2.2H Let E and F be Banach spaces and $A \in L(E, F)$. If $A(E)$ has finite codimension, show that it is closed. (*Hint:* If F_0 is an algebraic complement to $A(E)$ in F, show there is a continuous linear isomorphism $E/\ker A \cong F/F_0$; compose its inverse with $E/\ker A \to A(E)$.)

2.2I **Symmetrization operator** Define $\mathrm{Sym}^k : L^k(E, F) \to L^k(E, F)$, by

$$\operatorname{Sym}^k A = \frac{1}{k!} \sum_{a \in S_k} \sigma A ,$$

where $(\sigma A)(e_1, \ldots, e_k) = A(e_{\sigma(1)}, \ldots, e_{\sigma(k)})$. Show that:

(i) $\operatorname{Sym}^k(L^k(E, F)) = L^k_s(E, F)$.

(ii) $(\operatorname{Sym}^k)^2 = \operatorname{Sym}^k$.

(iii) $\| \operatorname{Sym}^k \| \le 1$.

(iv) If F is Banach, then $L^k_s(E, F)$ splits in $L^k(E, F)$. (*Hint:* Use **2.2.18.**)

(v) $(\operatorname{Sym}^k A)' = A'$.

2.2J Show that a k-multilinear map continuous in each argument separately is continuous. (*Hint:* for $k = 2$: If $\| e_1 \| \le 1$, then $\| A(e_1, e_2) \| \le \| A(\cdot, e_2) \|$, which by the uniform boundedness principle implies that $\| A(e_1, \cdot) \| \le M$ for $\| e_1 \| \le 1$.)

2.2K (i) Prove the following theorem of Mazur and Ulam [1932] following the steps below (see Banach [1955], p. 166): *Every isometric surjective mapping* $\varphi : E \to F$ *such that* $\varphi(0) = 0$ *is a linear map.* Here E and F are normed vector spaces; φ being isometric means $\| \varphi(x) - \varphi(y) \| = \| x - y \|$ for all $x, y \in E$.

(a) Fix $x_1, x_2 \in E$ and define

$$H_1 = \left\{ x \;\middle|\; \| x - x_1 \| = \| x - x_2 \| = \frac{1}{2} \| x_1 - x_2 \| \right\},$$

$$H_n = \left\{ x \in H_{n-1} \;\middle|\; \| x - z \| \le \frac{1}{2} \operatorname{diam}(H_{n-1}),\ z \in H_{n-1} \right\}.$$

Show that $\operatorname{diam}(H_n) \le \dfrac{1}{2^{n-1}} \operatorname{diam}(H_1) \le \dfrac{1}{2^{n-1}} \| x_1 - x_2 \|$.

Conclude that if $\bigcap_{n \ge 1} H_n \ne \varnothing$, then it consists of one point only.

(b) Show by induction that if $x \in H_n$, then $x_1 + x_2 - x \in H_n$.

(c) Show that $(x_1 + x_2)/2 = \bigcap_{n \ge 1} H_n$. (*Hint:* Show inductively that $(x_1 + x_2)/2 \in H_n$ using (b).)

(d) From (c) deduce that

$$\varphi\left(\frac{1}{2} (x_1 + x_2) \right) = \frac{1}{2} (\varphi(x_1) + \varphi(x_2)) .$$

Use $\varphi(0) = 0$ to conclude φ is linear.

(ii) (P. Chernoff [1970]). In the Mazur-Ulam Theorem, drop the assumption that φ is onto, but request that φ be homogeneous: $\varphi(tx) = t\varphi(x)$ for all $t \in \mathbb{R}$ and $x \in E$.

(a) A normed vector space is ***strictly convex*** if equality holds in the triangle inequality only for colinear points. Show that if F is strictly convex, then φ is linear. (*Hint:*

$$\| \varphi(x) - \varphi(y) \| = \left\| \varphi(x) - \varphi\left(\frac{x+y}{2}\right) \right\| + \left\| \varphi(y) - \varphi\left(\frac{x+y}{2}\right) \right\|$$

and

$$\left\| \varphi(x) - \varphi\left(\frac{x+y}{2}\right) \right\| = \left\| \varphi(y) - \varphi\left(\frac{x+y}{2}\right) \right\| .$$

Show that

$$\varphi\left(\frac{x+y}{2}\right) = \frac{1}{2}\,(\varphi(x) + \varphi(y)\ .)$$

(b) Show that, in general, the assumption on φ being onto is necessary by considering the following counterexample. Let $E = \mathbb{R}^2$ and $F = \mathbb{R}^3$, both with the max norm. Define $\varphi : E \to F$ by

$$\begin{aligned} \varphi(a, b) &= (a, b, \sqrt{ab}),\ a, b > 0\ ; \\ \varphi(-a, b) &= (-a, b, -\sqrt{ab}),\ a, b > 0\ ; \\ \varphi(a, -b) &= (a, -b, -\sqrt{ab}),\ a, b > 0\ ; \\ \varphi(-a, b) &= (-a, -b, -\sqrt{ab}),\ a, b > 0\ . \end{aligned}$$

Show that φ is not linear, φ is homogeneous, φ is an isometry, and $\varphi(0, 0) = (0, 0, 0)$. (*Hint:* Prove the inequality $|\alpha\beta - \gamma\delta| \le \max(|\alpha^2 - \gamma^2|, |\beta^2 - \delta^2|)$.)

2.2L Let E be a complex n dimensional vector space.

(i) Show that the set of all operators $A \in L(E, E)$ which have n distinct eigenvalues is open and dense in E. (*Hint:* Let p be the characteristic polynomial of A, i.e., $p(\lambda) = \det(A - \lambda I)$, and let $\mu_1, \ldots, \mu_{n-1}$ be the roots of p'. Then A has multiple eigenvalues iff $p(\mu_1) \ldots p(\mu_{n-1}) = 0$. The last expression is a symmetric polynomial in $\mu_1, \ldots, \mu_{n-1}$, and so is a polynomial in the coefficients of p' and therefore is a polynomial q in the entries of the matrix of A in a basis. Show that $q^{-1}(0)$ is the set of complex $n \times n$ matrices which have multiple eigenvalues; $q^{-1}(0)$ has open dense complement by Exercise **1.1.L**.)

(ii) Prove the ***Cayley-Hamilton Theorem***: *If* p *is the characteristic polynomial of* $A \in L(E, E)$, *then* $p(A) = 0$. (*Hint:* If the eigenvalues of A are distinct, show that the matrix of A in the basis of eigenvectors $e_1, \ldots, e_n$ is diagonal. Apply $A, A^2, \ldots, A^{n-1}$. Then show that for any polynomial q the matrix of q(A) in the same basis is diagonal with entries $q(\lambda_i)$, where λ_i are the eigenvalues of A. Finally, let q =

p. If A is general, apply (i).)

2.2M Let E be a normed real (resp. complex) vector space.

(i) Show that $\lambda : E \to \mathbb{R}$ (resp. $\mathbb{C}$) is continuous if and only if $\ker \lambda$ is closed. (*Hint:* Let $e \in E$ satisfy $\lambda(e) = 1$ and choose a disk D of radius r centered at e such that $D \cap (e + \ker \lambda) = \varnothing$. Then $\lambda(x) \neq 1$ for all $x \in D$. Show that if $x \in D$ then $\lambda(x) < 1$. If not, let $\alpha = \lambda(x)$, $|\alpha| > 1$. Then $\| x/\alpha \| < r$ and $\lambda(x/\alpha) = 1$.)

(ii) Show that if F is a closed subspace of E and G is a finite dimensional subspace, then $G + F$ is closed. (*Hint:* Assume G is one dimensional and generated by g. Write any $x \in G + F$ as $x = \lambda(x)g + f$ and use (i) to show λ is continuous on $G + F$.)

2.2N Let F be a Banach space.

(i) Show that if E is a finite dimensional subspace of F, then E is split. (*Hint:* Define $P : F \to F$ by $P(x) = \Sigma_{i = 1, \ldots, n}\, e^i(x)e_i$, where $\{e_1, \ldots, e_n\}$ is a basis of E and $\{e^1, \ldots, e^n\}$ is a dual basis, i.e., $e^i(e_j) = \delta_{ij}$. Then use Corollary **2.2.18**.)

(ii) Show that if E is closed and finite codimensional, then it is split.

(iii) Show that if E is closed and contains a finite codimensional subspace G of F, then it is split.

(iv) Let $\lambda : F \to \mathbb{R}$ be a linear discontinuous map and let $E = \ker \lambda$. Show that the codimension of E is 1 and that E is not closed. Thus finite codimensional subspaces of F are not necessarily closed. Compare this with (i) and (ii) and with **2.2H**.

2.2O Let E and F be Banach spaces and $T \in L(E, F)$. Define $T^*: F^* \to E^*$ by $\langle T^*\beta, e \rangle = \langle \beta, Te \rangle$ for $e \in E$, $\beta \in F^*$. Show that:

(i) $T^* \in L(F^*, E^*)$ and $T^{**} | E = T$.

(ii) $\ker T^* = T(E)^\circ = \{\beta \in F^* \mid \langle \beta, T e \rangle = 0$ for all $e \in E\}$ and $\ker T = (T^*(F^*))^\circ = \{e \in E \mid \langle T^*\beta, e \rangle = 0$ for all $\beta \in F^*\}$.

(iii) If $T(E)$ is closed, then $T^*(F^*) = (\ker T)^\circ$. (*Hint:* The induced map $E/\ker T \to T(E)$ is a Banach space isomorphism; let S be its inverse. If $\lambda \in (\ker T)^\circ$, define the element $\mu \in (E/\ker T)^*$ by $\mu([e]) = \lambda(e)$. Let $\nu \in F^*$ denote the extension of $S^*(\mu) \in (T(E))^*$ to $\nu \in F^*$ with the same norm and show that $T^*(\nu) = \lambda$.)

(iv) If $T(E)$ is closed, then $\ker T^*$ is isomorphic to $(F/T(E))^*$ and $(\ker T)^*$ is isomorphic to $E^*/T^*(F^*)$.

§2.3 *The Derivative*

For a differentiable function $f : U \subset \mathbb{R} \to \mathbb{R}$, the usual interpretation of the derivative at a point $u_0 \in U$ is the slope of the line tangent to the graph of f at u_0. To generalize this, we interpret $\mathbf{D}f(u_0) = f'(u_0)$ as a linear map acting on the vector $(u - u_0)$. Then we can say that $\mathbf{D}f(u_0)$ is the unique linear map from $\mathbb{R}$ into $\mathbb{R}$ such that the mapping

$$g : U \to \mathbb{R} \quad \text{given by} \quad u \mapsto g(u) = f(u_0) + \mathbf{D}f(u_0) \cdot (u - u_0)$$

is tangent to f at u_0 (see Figure 2.3.1). This version can be extended to vector valued mappings of a vector variable as follows.

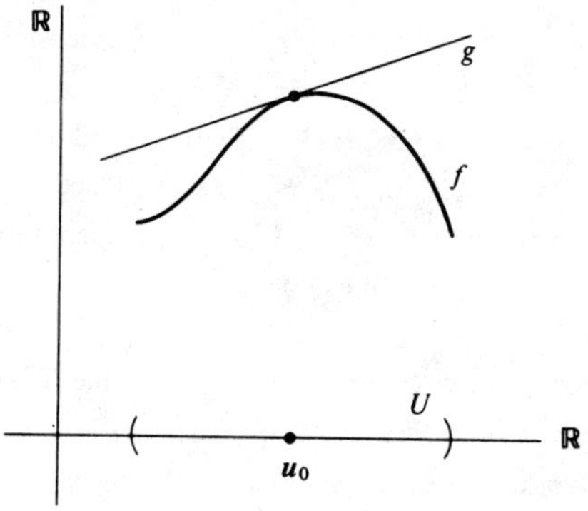

Figure 2.3.1

2.3.1 Definition *Let* E, F *be normed vector spaces, with maps* $f, g : U \subset E \to F$ *where* U *is open in* E. *We say* f *and* g *are* ***tangent*** *at the point* $u_0 \in U$ *if*

$$\lim_{u \to u_0} \frac{\|f(u) - g(u)\|}{\|u - u_0\|} = 0 ,$$

where $\|\cdot\|$ *represents the norm on the appropriate space.*

2.3.2 Proposition *For* $f : U \subset E \to F$ *and* $u_0 \in U$ *there is at most one* $L \in L(E, F)$ *such that the map* $g_L : U \subset E \to F$ *given by* $g_L(u) = f(u_0) + L(u - u_0)$ *is tangent to* f *at* u_0.

Proof Let L_1 and $L_2 \in L(E, F)$ satisfy the conditions of the proposition. If $e \in E$, $\| e \| = 1$, and $u = u_0 + \lambda e$ for $\lambda \in \mathbb{R}$ (or $\mathbb{C}$), then for λ small and $u \in U$, we have

$$\| L_1 e - L_2 e \| = \| L_1(u - u_0) - L_2(u - u_0) \| / \| u - u_0 \|$$

$$\leq \frac{\|f(u) - f(u_0) - L_1(u - u_0)\|}{\|u - u_0\|} + \frac{\|f(u) - f(u_0) - L_2(u - u_0)\|}{\|u - u_0\|} .$$

As $\lambda \to 0$, the right hand side approaches zero so that $\| (L_1 - L_2)e \| = 0$ for all $e \in E$ satisfying $\| e \| = 1$; therefore, $\| L_1 - L_2 \| = 0$ and thus $L_1 = L_2$. ■

2.3.3 Definition *If, in* **2.3.2**, *there is such an* $L \in L(E, F)$, *we say* f *is* ***differentiable at*** u_0, *and define the* ***derivative*** *of* f *at* u_0 *to be* $\mathbf{D}f(u_0) = L$. *The evaluation of* $\mathbf{D}f(u_0)$ *on* $e \in E$ *will be denoted* $\mathbf{D}f(u_0) \cdot e$. *If* f *is differentiable at each* $u_0 \in U$, *the map*

$$\mathbf{D}f : U \to L(E, F); \; u \mapsto \mathbf{D}f(u)$$

is called the ***derivative*** *of* f. *Moreover, if* $\mathbf{D}f$ *is a continuous map (where* $L(E, F)$ *has the norm topology), we say* f *is of class* C^1 *(or is* ***continuously differentiable****). Proceeding inductively we define*

$$\mathbf{D}^r f := \mathbf{D}(\mathbf{D}^{r-1} f) : U \subset E \to L^r(E, F)$$

if it exists, where we have identified $L(E, L^{r-1}(E, F))$ *with* $L^r(E, F)$ *(see* **2.2.9***). If* $\mathbf{D}^r f$ *exists and is norm continuous, we say* f *is of* ***class*** C^r.

We shall reformulate the definition of the derivative with the aid of the somewhat imprecise but very convenient ***Landau symbol*** : $o(e^k)$ *will denote a continuous function of* e *defined in a neighborhood of the origin of a normed vector space* E, *satisfying* $\lim_{e \to 0}(o(e^k)/\| e \|^k) = 0$. The collection of these functions forms a vector space. Clearly $f : U \subset E \to F$ is differentiable at $u_0 \in U$ iff there exists a linear map $\mathbf{D}f(u_0) \in L(E, F)$ such that

$$f(u_0 + e) = f(u_0) + \mathbf{D}f(u_0) \cdot e + o(e) .$$

Let us use this notation to show that *if* $\mathbf{D}f(u_0)$ *exists, then* f *is continuous at* u_0:

$$\lim_{e \to 0} f(u_0 + e) = \lim_{e \to 0} (f(u_0) + \mathbf{D}f(u_0) \cdot e + o(e)) = f(u_0) .$$

2.3.4 Linearity of the Derivative *Let* $f, g : U \subset E \to F$ *be* r *times differentiable mappings and* a *a real (or complex) constant. Then* af *and* $f + g : U \subset E \to F$ *are* r *times differentiable with*

$$\mathbf{D}^r(f + g) = \mathbf{D}^r f + \mathbf{D}^r g \quad \textit{and} \quad \mathbf{D}^r(af) = a\mathbf{D}^r f .$$

Proof If $u \in U$ and $e \in E$, then

$$f(u+e) = f(u) + \mathbf{D}f(u)\cdot e + o(e) \quad \text{and} \quad g(u+e) = g(u) + \mathbf{D}g(u)\cdot e + o(e)\ ,$$

so that adding these two relations yields

$$(f+g)(u+e) = (f+g)(u) + (\mathbf{D}f(u) + \mathbf{D}g(u))\cdot e + o(e)\ .$$

The case $r > 1$ follows by induction. Similarly $af(u+e) = af(u) + a\mathbf{D}f(u)\cdot e + ao(e) = af(u) + a\mathbf{D}f(u)\cdot e + o(e)$. ■

2.3.5 Derivative of a Cartesian Product *Let* $f_i : U \subset E \to F_i$, $1 \le i \le n$, *be* r *times differentiable mappings. Then* $f = f_1 \times \cdots \times f_n : U \subset E \to F_1 \times \cdots \times F_n$ *defined by* $f(u) = (f_1(u), ..., f_n(u))$ *is* r *times differentiable and*

$$\mathbf{D}^r f = \mathbf{D}^r f_1 \times \cdots \times \mathbf{D}^r f_n\ .$$

Proof For $u \in U$ and $e \in E$, we have

$$\begin{aligned} f(u+e) &= (f_1(u+e), ..., f_n(u+e)) \\ &= (f_1(u) + \mathbf{D}f_1(u)\cdot e + o(e), ..., f_n(u) + \mathbf{D}f_n(u)\cdot e + o(e)) \\ &= (f_1(u), ..., f_n(u)) + (\mathbf{D}f_1(u), ..., \mathbf{D}f_n(u))\cdot e + (o(e), ..., o(e)) \\ &= f(u) + \mathbf{D}f(u)\cdot e + o(e)\ , \end{aligned}$$

the last equality follows using the sum norm in $F_1 \times \cdots \times F_n$:

$$\|(o(e), ..., o(e))\| = \|o(e)\| + \cdots + \|o(e)\|\ ,$$

so $(o(e), ..., o(e)) = o(e)$. ■

Notice from the definition that for $L \in L(E, F)$, $\mathbf{D}L(u) = L$ for any $u \in E$. It is also clear that the derivative of a constant map is zero.

Usually all our spaces will be real and linearity will mean real-linearity. In the complex case, differentiable mappings are the subject of analytic function theory, a subject we shall not pursue in this book (see Exercise **2.3F** for a hint of why there is a relationship with analytic function theory).

In addition to the foregoing approach, there is a more traditional way to differentiate a function $f : U \subset \mathbb{R}^n \to \mathbb{R}^m$. We write out f in component form using the following notation: $f(x^1, ..., x^n) = (f^1(x^1, ... x^n), ..., f^m(x^1, ... x^n))$ and compute partial derivaties, $\partial f^j/\partial x^i$ for $j = 1, ..., m$ and $i = 1, ..., n$, where the symbol $\partial f^j/\partial x^i$ means that we compute the usual derivative of f^j with respect to x^i while keeping the other variables $x^1, ..., x^{i-1}, x^{i+1}, ..., x^n$ fixed.

For $f : \mathbb{R} \to \mathbb{R}$, $\mathbf{D}f(x)$ is just the linear map "multiplication by df/dx," i.e., $df/dx = \mathbf{D}f(x)\cdot 1$. This fact, which is obvious from the definitions, can be generalized to the following theorem.

2.3.6 Proposition *Suppose that* $U \subset \mathbb{R}^n$ *is an open set and that* $f : U \to \mathbb{R}^m$ *is differentiable. Then the partial derivatives* $\partial f^j/\partial x^i$ *exist, and the matrix of the linear map* $Df(x)$ *with respect to the standard bases in* $\mathbb{R}^n$ *and* $\mathbb{R}^m$ *is given by*

$$\begin{bmatrix} \dfrac{\partial f^1}{\partial x^1} & \dfrac{\partial f^1}{\partial x^2} & \cdots & \dfrac{\partial f^1}{\partial x^n} \\ \dfrac{\partial f^2}{\partial x^1} & \dfrac{\partial f^2}{\partial x^2} & \cdots & \dfrac{\partial f^2}{\partial x^n} \\ \vdots & \vdots & & \vdots \\ \dfrac{\partial f^m}{\partial x^1} & \dfrac{\partial f^m}{\partial x^2} & \cdots & \dfrac{\partial f^m}{\partial x^n} \end{bmatrix}$$

where each partial derivative is evaluated at $x = (x^1, \ldots, x^n)$. *This matrix is called the* ***Jacobian matrix*** *of* f.

Proof By the usual definition of the matrix of a linear mapping from linear algebra, the $(j, i)^{\text{th}}$ matrix element a^j_i of $Df(x)$ is given by the jth component of the vector $Df(x) \cdot e_i$, where $e_1, \ldots, e_n$ is the standard basis of $\mathbb{R}^n$. Letting $y = x + he_i$, we see that

$$\frac{\| f(y) - f(x) - Df(x)(y - x) \|}{\| y - x \|} = \frac{1}{|h|} \| f(x^1, \ldots, x^i + h, \ldots, x^n) - f(x^1, \ldots, x^n) - h\mathbf{D} f(x) e_i \|$$

approaches zero as $h \to 0$, so the jth component of the numerator does as well; i.e.,

$$\lim_{h \to 0} \frac{1}{|h|} \left| f^j(x^1, \ldots, x^i + h, \ldots, x^n) - f^j(x^1, \ldots, x^n) - ha^j_i \right| = 0,$$

which means that $a^j_i = \partial f^j/\partial x^i$. ■

In computations one can usually compute the Jacobian matrix easily, and this proposition then gives Df. In some books, Df is called the ***differential*** or the ***total derivative*** of f.

2.3.7 Example Let $f : \mathbb{R}^2 \to \mathbb{R}^3$, $f(x, y) = (x^2, x^3y, x^4y^2)$. Then $Df(x, y)$ is the linear map whose matrix in the standard basis is

$$\begin{bmatrix} \frac{\partial f^1}{\partial x} & \frac{\partial f^1}{\partial y} \\ \frac{\partial f^2}{\partial x} & \frac{\partial y^2}{\partial y} \\ \frac{\partial f^3}{\partial x} & \frac{\partial f^3}{\partial y} \end{bmatrix} = \begin{bmatrix} 2x & 0 \\ 3x^2y & x^3 \\ 4x^3y^2 & 2x^4y \end{bmatrix}$$

where $f^1(x, y) = x^2$, $f^2(x, y) = x^3y$, $f^3(x,y) = x^4y^2$. ♦

One should take special note when $m = 1$, in which case we have a real-valued function of n variables. Then **Df** has the matrix

$$\left[\frac{\partial f}{\partial x^1}, \ldots, \frac{\partial f}{\partial x^n}\right]$$

and the derivative applied to a vector $e = (a^1, \ldots, a^n)$ is

$$\mathbf{D}f(x) \cdot e = \sum_{i=1}^{n} \frac{\partial f}{\partial x^i} a^i .$$

It should be emphasized that **Df** assigns a linear mapping to each $x \in U$ and the definition of $\mathbf{D}f(x)$ is independent of the basis used. If we change the basis from the standard basis to another one, the matrix elements will of course change. If one examines the definition of the matrix of a linear transformation, it can be seen that the columns of the matrix relative to the new basis will be the derivative $\mathbf{D}f(x)$ applied to the new basis in $\mathbb{R}^n$ with this image vector expressed in the new basis in $\mathbb{R}^m$. Of course, the linear map $\mathbf{D}f(x)$ itself does not change from basis to basis. In the case $m = 1$, $\mathbf{D}f(x)$ is, in the standard basis, a $1 \times n$ matrix. The vector whose components are the same as those of $\mathbf{D}f(x)$ is called the ***gradient*** of f, and is denoted grad f or ∇f. Thus for $f : U \subset \mathbb{R}^n \to \mathbb{R}$,

$$\text{grad } f = \left[\frac{\partial f}{\partial x^1}, \ldots, \frac{\partial f}{\partial x^n}\right]$$

(Sometimes it is said that grad f is just **Df** with commas inserted!) The formation of gradients makes sense in a general inner product space as follows.

2.3.8 Definition (i) *Let* E *be a normed space and* $f : U \subset E \to \mathbb{R}$ *be differentiable so that* $\mathbf{D}f(u) \in L(E, \mathbb{R}) = E^*$. *In this case we sometimes write* $\mathbf{d}f(u)$ *for* $\mathbf{D}f(u)$ *and call* $\mathbf{d}f$ *the **differential** of* f. *Thus* $\mathbf{d}f : U \to E^*$.

(ii) *If* E *is a Hilbert space, the **gradient** of* f *is the map*

$$\text{grad } f = \nabla f : U \to E \quad \textit{defined by} \quad \langle \nabla f(u), e \rangle = \mathbf{d}f(u) \cdot e ,$$

where $\mathbf{d}f(u) \cdot e$ *means the linear map* $\mathbf{d}f(u)$ *applied to the vector* e.

Note that the existence of $\nabla f(u)$ requires the Riesz Representation Theorem (see **2.2.5**). The notation $\delta f/\delta u$ instead of $(\text{grad } f)(u) = \nabla f(u)$ is also in wide use, especially in the case in which E is a space of functions. See Supplement **2.4C** below.

2.3.9 Example Let $(E, \langle\, ,\rangle)$ be a real inner product and $f(u) = \|u\|^2$. Since $\|u\|^2 = \|u_0\|^2 + 2\langle u_0, u - u_0\rangle + \|u - u_0\|^2$, we obtain $\mathbf{d}f(u_0) \cdot e = 2\langle u_0, e\rangle$ and thus $\nabla f(u) = 2u$. Hence f is of class C^1. But since $\mathbf{D}f(u) = 2\langle u, \cdot\rangle \in E^*$ is a continuous linear map in $u \in E$, it follows that $\mathbf{D}^2 f(u) = \mathbf{D}f \in L(E, E^*)$ and thus $\mathbf{D}^k f = 0$ for $k \geq 3$. Thus f is of class C^∞. The mapping f considered here is a special case of a polynomial mapping (see **2.2.10**). ♦

We close this section with the Fundamental Theorem of Calculus in real Banach spaces. First a bit of notation. If $\varphi : U \subset \mathbb{R} \to F$ is differentiable, then $\mathbf{D}\varphi(t) \in L(\mathbb{R}, F)$. The space $L(\mathbb{R}, F)$ is isomorphic to F by $A \mapsto A(1)$, $1 \in \mathbb{R}$; note that $\|A\| = \|A(1)\|$. We denote

$$\varphi'(t) = \frac{d\varphi}{dt} = \mathbf{D}\varphi(t) \cdot 1,\ 1 \in \mathbb{R}\,.$$

$$\varphi'(t) = \lim_{h \to 0} \frac{\varphi(t+h) - \varphi(t)}{h}$$

and φ is differentiable iff φ' exists.

2.3.10 Fundamental Theorem of Calculus (i) *If* $g : [a, b] \to F$ *is continuous, for* F *a real normed space, the map*

$$f : (a, b) \to F \quad \textit{defined by} \quad f(t) = \int_a^t g(s)ds$$

is differentiable and we have $f' = g$.

(ii) *If* $f : [a, b] \to F$ *is continuous, is differentiable on* $]a, b[$ *and* f' *extends to a continuous map on* $[a, b]$, *then*

$$f(b) - f(a) = \int_a^b f'(s)\, ds.$$

Proof (i) Let $t_0 \in\ ]a, b[$. Since the integral is linear and continuous,

$$\|f(t_0 + h) - f(t_0) - hg(t_0)\| = \left\| \int_{t_0}^{t_0+h} (g(s) - g(t_0))\, ds \right\| \leq$$

$$\leq |h| \sup\{\|g(s) - g(t_0)\| \mid t_0 \leq s \leq t_0 + h\}\ .$$

The sup on the right-hand side $\to 0$ as $|h| \to 0$ by continuity of g at t_0.

(ii) Let

$$h(t) = \left(\int_a^t f'(s)\, ds \right) - f(t).$$

By (i), $h'(t) = 0$ on $]a, b[$ and h is continuous on $[a, b]$. If for some $t \in [a, b]$, $h(t) \neq h(a)$, then by the Hahn-Banach Theorem there exists $\alpha \in F^*$ such that $(\alpha \circ h)(t) \neq (\alpha \circ h)(a)$. Moreover, $\alpha \circ h$ is differentiable on $]a, b[$ and its derivative is zero (Exercise **2.3D**). Thus by elementary calculus, $\alpha \circ h$ is constant on $[a, b]$, a contradiction. Hence $h(t) = h(a)$ for all $t \in [a, b]$. In particular, $h(a) = h(b)$. ■

Exercises

2.3A Let $B : E \times F \to G$ be a continuous bilinear map of normed spaces. Show that B is C^∞ and that

$$DB(u, v)(e, f) = B(u, f) + B(e, v) .$$

2.3B Show that the derivative of a map is unaltered if the spaces are renormed with equivalent norms.

2.3C If $f \in S^k(E, F)$, show that

$$D^k f(0)(e_1, \ldots, e_k) = \frac{\partial^k}{\partial t_1 \cdots \partial t_k} f(t_1 e_1 + \cdots + t_k e_k)\Big|_{t_i = 0},\ i = 1, \ldots, k$$

and

$$D^i f(0) = 0 \text{ for } i = 1, \ldots, k-1 .$$

2.3D Let $f : U \subset E \to F$ be a differentiable (resp. C^r) map and $A \in L(F, G)$. Show that $A \circ f : U \subset E \to G$ is differentiable (resp. C^r) and $D^r(A \circ f)(u) = A \circ D^r f(u)$. (*Hint:* Use induction.)

2.3E Let $f : U \subset E \to F$ be r times differentiable and $A \in L(G, E)$. Show that

$$D^i(f \circ A)(v) \cdot (g_1, \ldots, g_i) = D^i f(Av) \cdot (Ag_1, \ldots, Ag_i)$$

exists for all $i \leq r$, where $v \in A^{-1}(U)$, and $g_1, \ldots, g_i \in G$. Generalize to the case where A is an affine map.

2.3F (i) Show that a *complex* linear map $A \in L(\mathbb{C}, \mathbb{C})$ is necessarily of the form $A(z) = \lambda z$, for some $\lambda \in \mathbb{C}$.

(ii) Show that the matrix of $A \in L(\mathbb{C}, \mathbb{C})$, when A is regarded as a *real* linear map in

$L(\mathbb{R}^2, \mathbb{R}^2)$, is of the form

$$\begin{bmatrix} a & -b \\ b & a \end{bmatrix}.$$

(*Hint:* $\lambda = a + ib$.)

(iii) Show that a map $f : U \subset \mathbb{C} \to \mathbb{C}$, $f = g + ih$, $g, h : U \subset \mathbb{R}^2 \to \mathbb{R}$ is *complex* differentiable iff the ***Cauchy-Riemann equations***

$$\frac{\partial g}{\partial x} = \frac{\partial h}{\partial y}, \quad \frac{\partial g}{\partial y} = -\frac{\partial h}{\partial x}$$

are satisfied. (*Hint:* Use (ii) and **2.3.6.**)

2.3G Let $(E, \langle\, ,\rangle)$ be a *complex* inner product space. Show that the map $f(u) = \|u\|^2$ is not differentiable. Contrast this with **2.3.9.** (*Hint:* $\mathbf{D}f(u)$, if it exists, should equal $2\mathrm{Re}(\langle u, \cdot \rangle)$.)

2.3H Show that the matrix of $\mathbf{D}^2 f(x) \in L^2(\mathbb{R}^n, \mathbb{R})$ for $f : U \subset \mathbb{R}^n \to \mathbb{R}$, is given by

$$\begin{bmatrix} \dfrac{\partial^2 f}{\partial x^1 \partial x^1} & \dfrac{\partial^2 f}{\partial x^1 \partial x^2} & \cdots & \dfrac{\partial^2 f}{\partial x^1 \partial x^n} \\ \vdots & \vdots & & \vdots \\ \dfrac{\partial^2 f}{\partial x^n \partial x^1} & \dfrac{\partial^2 f}{\partial x^n \partial x^2} & \cdots & \dfrac{\partial^2 f}{\partial x^n \partial x^n} \end{bmatrix}$$

(*Hint:* Apply **2.3.6.** Recall from linear algebra that the matrix of a bilinear mapping $B \in L(\mathbb{R}^n, \mathbb{R}^m; \mathbb{R})$ is given by the entries $B(e_i, f_j)$ (first index = row index, second index = column index), where $\{e_1, \ldots, e_n\}$ and $\{f_1, \ldots, f_m\}$ are ordered bases of $\mathbb{R}^n$ and $\mathbb{R}^m$, respectively.)

§2.4 *Properties of the Derivative*

In this section some of the fundamental properties of the derivative are developed. These properties are analogues of rules familiar from elementary calculus. Let us begin by strengthening the fact that differentiability implies continuity.

2.4.1 Proposition *Suppose* $U \subset E$ *is open and* $f : U \to F$ *is differentiable on* U. *Then* f *is continuous. In fact, for each* $u_0 \in U$ *there is a constant* $M > 0$ *and a* $\delta_0 > 0$ *with the property that* $\| u - u_0 \| < \delta_0$ *implies* $\| f(u) - f(u_0) \| \leq M \| u - u_0 \|$. *(This is called the* ***Lipschitz property.****)*

Proof

$$\begin{aligned} \big| \| f(u) - f(u_0) \| - \| \mathbf{D}f(u_0) \cdot (u - u_0) \| \big| &\leq \| f(u) - f(u_0) - \mathbf{D}f(u_0) \cdot (u - u_0) \| \\ &= \| o(u - u_0) \| \leq \| u - u_0 \| \end{aligned}$$

for $\| u - u_0 \| \leq \delta_0$, where δ_0 is some positive constant depending on u_0; this holds since

$$\lim_{u \to u_0} \frac{o(u - u_0)}{\| u - u_0 \|} = 0 .$$

Thus,

$$\| f(u) - f(u_0) \| \leq \| \mathbf{D}f(u_0) \cdot (u - u_0) \| + \| u - u_0 \| \leq (\| \mathbf{D}f(u_0) \| + 1) \| u - u_0 \|$$

for $\| u - u_0 \| \leq \delta_0$. ■

Perhaps the most important rule of differential calculus is the chain rule. To facilitate its statement, the notion of the tangent of a map is introduced. The text will begin conceptually distinguishing *points* in U from *vectors* in E. At this point it is not so clear that the distinction is important, but it will help with the transition to manifolds in Chapter 3.

2.4.2 Definition *Suppose* $f : U \subset E \to F$ *is of class* C^1. *Define the* ***tangent*** *of* f *to be the map*

$$Tf : U \times E \to F \times F \quad \textit{given by} \quad Tf(u, e) = (f(u), \mathbf{D}f(u) \cdot e)$$

where we recall that $\mathbf{D}f(u) \cdot e$ *denotes* $\mathbf{D}f(u)$ *applied to* $e \in E$ *as a linear map. If* f *is of class* C^r, *define* $T^r f = T(T^{r-1} f)$ *inductively.*

From a geometric point of view, T is more "natural" than $\mathbf{D}$. If we think of (u, e) as a vector with base point u, and vector part e then $(f(u), \mathbf{D}f(u) \cdot e)$ is the image vector *with its base point* f(u), as in Fig. 2.4.1. Another reason for this is its simple behavior under composition, as given in the next theorem.

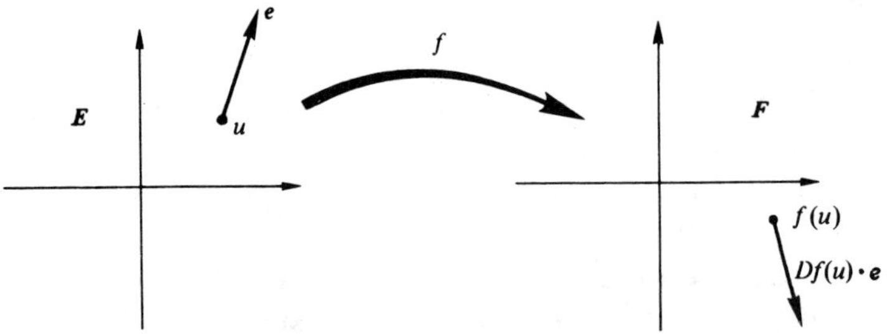

Figure 2.4.1

2.4.3 C^r Composite Mapping Theorem *Suppose* $f: U \subset E \to V \subset F$ *and* $g: V \subset F \to G$ *are differentiable (respectively* C^r*) maps. Then the composite* $g \circ f: U \subset E \to G$ *is also differentiable (respectively* C^r*) and*

$$T(g \circ f) = Tg \circ Tf,$$

(respectively, $T^r(g \circ f) = T^r g \circ T^r f$ *). The formula* $T(g \circ f) = Tf \circ Tf$ *is equivalent to the chain rule in terms of* $\mathbf{D}$:

$$\mathbf{D}(g \circ f)(u) = \mathbf{D}g(f(u)) \circ \mathbf{D}f(u).$$

Proof Since f is differentiable at $u \in U$ and g is differentiable at $f(u) \in V$, we have

$$f(u + e) = f(u) + \mathbf{D}f(u) \cdot e + o(e) \quad \text{for} \quad e \in E$$

and for $v = f(u)$ we have $g(v + w) = g(v) + \mathbf{D}g(v) \cdot w + o(w)$. Thus

$$\begin{aligned}(g \circ f)(u + e) &= g(f(u) + \mathbf{D}f(u) \cdot e + o(e)) \\ &= (g \circ f)(u) + \mathbf{D}g(f(u)) \cdot (\mathbf{D}f(u) \cdot e) + \mathbf{D}g(f(u))(o(e)) + o(\mathbf{D}f(u) \cdot e + o(e)).\end{aligned}$$

For e in a neighborhood of the origin,

$$\frac{\| \mathbf{D}f(u) \cdot e + o(e) \|}{\| e \|} \leq \left(\| \mathbf{D}f(u) \| + \frac{\|o(e)\|}{\|e\|} \right) \leq M$$

for some constant $M > 0$, and

$$\|\mathbf{D}g(f(u)) \cdot o(e)\| \leq \|\mathbf{D}g(f(u)\| \; \|o(e)\|.$$

Thus

$$\frac{\|o(\mathbf{D}f(u)\cdot e+o(e))\|}{\|e\|}=\frac{\|(o(\mathbf{D}f(u)\cdot e+o(e))\|}{\|\mathbf{D}f(u)\cdot e+o(e)\|}\cdot\frac{\|\mathbf{D}f(u)\cdot e+o(e)\|}{\|e\|}\le M\frac{\|(o(\mathbf{D}f(u)\cdot e+o(e))\|}{\|\mathbf{D}f(u)\cdot e+o(e)\|}.$$

Hence we conclude

$$\mathbf{D}g(f(u))\cdot(o(e))+o(\mathbf{D}f(u)\cdot e+o(e))=o(e)$$

and thus

$$\mathbf{D}(g\circ f)(u)\cdot e=\mathbf{D}g(f(u))\cdot(\mathbf{D}f(u)\cdot e).$$

Denote by $\varphi : L(F, G)\times L(E, F)\to L(E, G)$ the bilinear mapping $\varphi(B, A) = B\circ A$ and note that $\varphi\in L(L(F, G), L(E, F); L(E, G))$ since $\|B\circ A\|\le\|B\|\,\|A\|$; i.e., $\|\varphi\|\le 1$. Let $(\mathbf{D}g\circ f)\times\mathbf{D}f : U\to L(F, G)\times L(E, F)$ be defined by $[(\mathbf{D}g\circ f)\times\mathbf{D}f](u) = (\mathbf{D}g(f(u)), \mathbf{D}f(u))$; notice that this map is continuous if f and g are of class C^1. Therefore the composite function $\varphi\circ((\mathbf{D}g\circ f)\times\mathbf{D}f) = \mathbf{D}(g\circ f) : U\to L(E, G)$ is continuous if f and g are C^1, that is, $g\circ f$ is C^1. Inductively suppose f and g are C^r. Then $\mathbf{D}g$ is C^{r-1}, so $\mathbf{D}g\circ f$ is C^{r-1} and thus the map $(\mathbf{D}g\circ f)\times\mathbf{D}f$ is C^{r-1} (see **2.3.5**). Since φ is C^∞ (Exercise **2.3A**), again the inductive hypothesis forces $\varphi\circ((\mathbf{D}g\circ f)\times\mathbf{D}f) = \mathbf{D}(g\circ f)$ to be C^{r-1}; i.e., $g\circ f$ is C^r.

The formula $T^r(g\circ f) = T^r g\circ T^r f$ is a direct verification for $r = 1$ using the chain rule, and the rest follows by induction. ■

If $E=\mathbb{R}^m$, $F=\mathbb{R}^n$, $G=\mathbb{R}^p$ and $f=(f^1,\ldots,f^n)$, $g=(g^1,\ldots,g^p)$, where $f^i : U\to\mathbb{R}$ and $g^j : V\to\mathbb{R}$, by **2.3.6** the chain rule becomes

$$\begin{bmatrix}\frac{\partial(g\circ f)^1(x)}{\partial x^1} & \cdots & \frac{\partial(g\circ f)^1(x)}{\partial x^m}\\ \vdots & & \vdots\\ \frac{\partial(g\circ f)^p(x)}{\partial x^1} & \cdots & \frac{\partial(g\circ f)^p(x)}{\partial x^m}\end{bmatrix}=\begin{bmatrix}\frac{\partial g^1(f(x))}{\partial y^1} & \cdots & \frac{\partial g^1(f(x))}{\partial y^n}\\ \vdots & & \vdots\\ \frac{\partial g^p(f(x))}{\partial y^1} & \cdots & \frac{\partial g^p(f(x))}{\partial y^n}\end{bmatrix}\cdot\begin{bmatrix}\frac{\partial f^1(x)}{\partial x^1} & \cdots & \frac{\partial f^1(x)}{\partial x^m}\\ \vdots & & \vdots\\ \frac{\partial f^n(x)}{\partial x^1} & \cdots & \frac{\partial f^n(x)}{\partial x^m}\end{bmatrix}$$

which, when read componentwise, becomes the usual chain rule from calculus:

$$\frac{\partial(g\circ f)^j(x)}{\partial x^i}=\sum_{k=1}^{n}\frac{\partial g^j(f(x))}{\partial y^k}\frac{\partial f^k(x)}{\partial x^i}\ ,\quad i=1,\ldots,m\ .$$

The chain rule applied to $B\in L(F_1, F_2; G)$ and $f_1\times f_2 : U\subset E\to F_1\times F_2$ yields the following.

2.4.4 The Leibniz Rule *Let* $f_i : U \subset E \to F_i$, $i = 1, 2$, *be differentiable (respectively* C^r*) maps and* $B \in L(F_1, F_2; G)$. *Then the mapping* $B(f_1, f_2) = B \circ (f_1 \times f_2) : U \subset E \to G$ *is differentiable (respectively* C^r*) and*

$$\mathbf{D}(B(f_1, f_2))(u) \cdot e = B(\mathbf{D}f_1(u) \cdot e, f_2(u)) + B(f_1(u), \mathbf{D}f_2(u) \cdot e).$$

In the case $F_1 = F_2 = \mathbb{R}$ and B is multiplication, **2.4.4** reduces to the usual product rule for derivatives. Leibniz' rule can easily be extended to multilinear mappings (Exercise **2.4C**).

The first of several consequences of the chain rule involves the directional derivative.

2.4.5 Definition *Let* $f : U \subset E \to F$ *and let* $u \in U$. *We say that* f ***has a derivative in the direction*** $e \in E$ *at* u *if*

$$\frac{d}{dt} f(u + te)\Big|_{t=0}$$

exists. We call this element of F *the **directional derivative of*** f ***in the diurection*** e *at* u.

Sometimes a function all of whose directional derivatives exist is called ***Gâeaux differentiable***, whereas a function differentiable in the sense we have defined is called ***Fréchet differentiable.*** The latter is stronger, according to the following. (See also Exercise **2.4J**.)

2.4.6 Proposition *If* f *is differentiable at* u, *then the directional derivatives of* f *exist at* u *and are given by*

$$\frac{d}{dt} f(u + te)\Big|_{t=0} = \mathbf{D}f(u) \cdot e\,.$$

Proof A ***path*** in E is a map from I into E, where I is an open interval of $\mathbb{R}$. Thus, if c is differentiable, for $t \in I$ we have $\mathbf{D}c(t) \in L(\mathbb{R}, E)$, by definition. Recall that we identify $L(\mathbb{R}, E)$ with E by associating $\mathbf{D}c(t)$ with $\mathbf{D}c(t) \cdot 1$ $(1 \in \mathbb{R})$. Let

$$\frac{dc}{dt}(t) = \mathbf{D}c(t) \cdot 1\,.$$

For $f : U \subset E \to F$ of class C^1 we consider $f \circ c$, where $c : I \to U$. It follows from the chain rule that

$$\frac{d}{dt}(f(c(t))) = \mathbf{D}(f \circ c)(t) \cdot 1 = \mathbf{D}f(c(t)) \cdot \frac{dc}{dt}\,.$$

The proposition follows by choosing $c(t) = u + te$, where $u, e \in E$, $I =]-\lambda, \lambda[$, and λ is sufficiently small. ■

For $f : U \subset \mathbb{R}^n \to \mathbb{R}$, the directional derivative is given in terms of the standard basis $\{e_1, \ldots, e_n\}$ by

$$\mathbf{D}f(u) \cdot e = \frac{\partial f}{\partial x^1} x^1 + \cdots + \frac{\partial f}{\partial x^n} x^n,$$

where $e = x^1 e_1 + \cdots + x^n e_n$. This follows from **2.3.6** and **2.4.6**.

The formula in **2.4.6** is sometimes a convenient method for computing $\mathbf{D}f(u) \cdot e$. For example, let us compute the differential of a homogeneous polynomial of degree 2 from E to F. Let $f(e) = A(e, e)$, where $A \in L^2(E; F)$. By the chain and Leibniz rules,

$$\mathbf{D}f(u) \cdot e = \frac{d}{dt} A(u + te, u + te)\Big|_{t=0} = A(u, e) + A(e, u)\,.$$

If A is symmetric, then $\mathbf{D}f(u) \cdot e = 2A(u, e)$.

One of the basic tools for finding estimates is the following.

2.4.7 Proposition *Let* E *and* F *be real Banach spaces,* $f : U \subset E \to F$ *a* C^1*-map,* $x, y \in U$, *and* c *a* C^1 *arc in* U *connecting* x *to* y; *i.e.,* c *is a continuous map* $c : [0, 1] \to U$, *which is* C^1 *on* $]0, 1[$, $c(0) = x$, *and* $c(1) = y$. *Then*

$$f(y) - f(x) = \int_0^1 \mathbf{D}f(c(t)) \cdot c'(t)\, dt\,.$$

If U *is convex and* $c(t) = (1 - t)x + ty$, *then*

$$f(y) - f(x) = \int_0^1 \mathbf{D}f((1 - t)x + ty) \cdot (y - x)\, dt = \left(\int_0^1 \mathbf{D}f((1 - t)x + ty)\, dt \right) \cdot (y - x)\,.$$

Proof If $g(t) = (f \circ c)(t)$, the chain rule implies $g'(t) = \mathbf{D}f(c(t)) \cdot c'(t)$ and the fundamental theorem of calculus gives

$$g(1) - g(0) = \int_0^1 g'(t)\, dt\,,$$

which is the first equality. The second equality for U convex and $c(t) = (1 - t)x + ty$ is Exercise **2.2F**(i). ■

2.4.8 Mean Value Inequality *Suppose* $U \subset E$ *is convex and* $f : U \subset E \to F$ *is* C^1. *Then for all* $x, y \in U$

$$\| f(y) - f(x) \| \le [\sup_{0 \le t \le 1} \| \mathbf{D}f((1 - t)x + ty) \|]\ \| x - y \|.$$

Thus, if $\| \mathbf{D}f(u) \|$ *is uniformly bounded on* U *by a constant* $M > 0$, *then for all* $x, y \in U$

$$\| f(y) - f(x) \| \le M \| y - x \|.$$

If $F = \mathbb{R}$, *then* $f(y) - f(x) = \mathbf{D}f(c) \cdot (y - x)$ *for some* c *on the line joining* x *to* y.

Proof The inequality follows directly from **2.4.7**. The last assertion follows from the intermediate value theorem as in elementary calculus. ■

2.4.9 Corollary *Let* $U \subset E$ *be an open set; then the following are equivalent:*

(i) U *is connected;*

(ii) *every differentiable map* $f : U \subset E \to F$ *satisfying* $\mathbf{D}f = 0$ *on* U *is constant.*

Proof If $U = U_1 \cup U_2$, $U_1 \cap U_2 = \varnothing$, where U_1 and U_2 are open, then the mapping

$$f(u) = \begin{cases} 0, & \text{if } u \in U_1 \\ e, & \text{if } u \in U_2 \end{cases}$$

where $e \in F$, $e \neq 0$ is a fixed vector, has $\mathbf{D}f = 0$, yet is not constant.

Conversely, assume that U is connected and $\mathbf{D}f = 0$. Then f is in fact C^∞. Let $u_0 \in U$ be fixed and consider the set $S = \{u \in U \mid f(u) = f(u_0)\}$. Then $S \neq \varnothing$ (since $u_0 \in S$), $S \subset U$, and S is closed since f is continuous. We shall show that S is also open. If $u \in S$, consider $v \in D_r(u) \subset U$ and apply **2.4.8** to get

$$\| f(u) - f(v) \| \leq \sup \{ \| \mathbf{D}f((1-t)u + tv) \| \mid t \in [0, 1]\} \| u - v \| = 0;$$

i.e., $f(v) = f(u) = f(u_0)$ and hence $D_r(u) \subset S$. Connectedness of U implies $S = U$. ■

If f is Gâteaux differentiable and the Gâteaux derivative is in $L(E, F)$; i.e., for each $u \in V$ there exists $G_u \in L(E, F)$ such that

$$\frac{d}{dt} f(u + te)\Big|_{t=0} = G_u e,$$

and if $u \mapsto G_u$ is continuous, we say f is C^1-***Gâteaux***. The mean value inequality holds, replacing C^1 everywhere by "C^1-Gâteaux" and the identical proofs work. When studying differentiability the following is often useful.

2.4.10 Corollary *If* $f : U \subset E \to F$ *is* C^1*-Gâteaux then it is* C^1 *and the two derivatives coincide.*

Proof Let $u \in U$ and work in a disk centered at u. Proposition **2.4.7** gives

$$\| f(u + e) - f(u) - G_u e \| = \left\| \left(\int_0^1 (G_{u+te} - G_u)\, dt \right) e \right\| \leq \sup\{ \| G_{u+te} - G_u \| \mid t \in [0, 1]\} \| e \|$$

and the sup converges to zero as, $e \to 0$, by uniform continuity of the map $t \in [0, 1] \mapsto G_{u+te} \in L(E, F)$. This says that $\mathbf{D}f(u) \cdot e$ exists and equals $G_u e$. ■

It will be convenient to consider partial derivatives in this context. We shall discuss only functions of two variables, the generalization to n variables being obvious.

2.4.11 Definition *Let* $f: U \to F$ *be a mapping defined on the open set* $U \subset E_1 \oplus E_2$ *and let* $u_0 = (u_{01}, u_{02}) \in U$. *The derivatives of the mappings* $v_1 \mapsto f(v_1, u_{02})$, $v_2 \mapsto f(u_{01}, v_2)$, where $v_1 \in E_1$ *and* $v_2 \in E_2$, *if they exist, are called* ***partial derivatives*** *of* f *at* $u_0 \in U$ *and are denoted by* $\mathbf{D}_1 f(u_0) \in L(E_1, F)$, $\mathbf{D}_2 f(u_0) \in L(E_2, F)$.

2.4.12 Proposition *Let* $U \subset E_1 \oplus E_2$ *be open and* $f: U \to F$.

(i) *If* f *is differentiable, then the partial derivatives exist and are given by*

$$\mathbf{D}_1 f(u) \cdot e_1 = \mathbf{D}f(u) \cdot (e_1, 0) \quad \textit{and} \quad \mathbf{D}_2 f(u) \cdot e_2 = \mathbf{D}f(u) \cdot (0, e_2).$$

(ii) *If* f *is differentiable, then*

$$\mathbf{D}f(u) \cdot (e_1, e_2) = \mathbf{D}_1 f(u) \cdot e_1 + \mathbf{D}_2 f(u) \cdot e_2 .$$

(iii) f *is of class* C^r *iff* $\mathbf{D}_i f: U \to L(E_i, F)$, $i = 1, 2$ *both exist and are of class* C^{r-1}.

Proof (i) Let $j_u^1 : E_1 \to E_1 \oplus E_2$ be defined by $j_u^1(v_1) = (v_1, u_2)$, where $u = (u_1, u_2)$. Then j_u^1 is C^∞ and $\mathbf{D}j_u^1(u_1) = J_1 \in L(E_1, E_1 \oplus E_2)$ is given by $J_1(e_1) = (e_1, 0)$. By the chain rule,

$$\mathbf{D}_1 f(u) = \mathbf{D}(f \circ j_u^1)(u_1) = \mathbf{D}f(u) \circ J_1,$$

which proves the first relation in (i). One similarly defines j_u^2, J_2 and proves the second relation.

(ii) Let $P_i(e_1, e_2) = e_i$, $i = 1, 2$ be the canonical projections. Then compose the relation $J_1 \circ P_1 + J_2 \circ P_2 = \text{identity}$ on $E_1 \oplus E_2$ with $\mathbf{D}f(u)$ on the left and use (i).

(iii) Let $\Phi_i \in L(L(E_1 \oplus E_2, F), L(E_i, F))$ and $\Psi_i \in L(L(E_i, F), L(E_1 \oplus E_2, F))$ be defined by $\Phi_i(A) = A \circ J_i$ and $\Psi_i(B_i) = B_i \circ P_i$, $i = 1, 2$. Then (i) and (ii) become

$$\mathbf{D}_i f = \Phi_i \circ \mathbf{D}f \ , \quad \mathbf{D}f = \Psi_1 \circ \mathbf{D}_1 f + \Psi_2 \circ \mathbf{D}_2 f \ .$$

This shows that if f is differentiable, then f is C^r iff $\mathbf{D}_1 f$ and $\mathbf{D}_2 f$ are C^{r-1}. Thus to conclude the proof we need to show that if $\mathbf{D}_1 f$ and $\mathbf{D}_2 f$ exist and are continuous, then $\mathbf{D}f$ exists. By **2.4.7** applied consecutively to the two arguments, we get

$$\begin{aligned}
f(u_1 + e_1, u_2 + e_2) - f(u_1, u_2) - \mathbf{D}_1 f(u_1, u_2) \cdot e_1 - \mathbf{D}_2 f(u_1, u_2) \cdot e_2 \\
= f(u_1 + e_1, u_2 + e_2) - f(u_1, u_2 + e_2) - \mathbf{D}_1 f(u_1, u_2) \cdot e_1 \\
+ f(u_1, u_2 + e_2) - f(u_1, u_2) - \mathbf{D}_2 f(u_1, u_2) \cdot e_2 \\
= \left(\int_0^1 (\mathbf{D}_1 f(u_1 + te_1, u_2 + e_2) - \mathbf{D}_1 f(u_1, u_2))\, dt \right) \cdot e_1 \\
+ \left(\int_0^1 (\mathbf{D}_2 f(u_1, u_2 + te_2) - \mathbf{D}_2 f(u_1, u_2) dt \right) \cdot e_2.
\end{aligned}$$

Taking norms and using in each term the obvious inequality $\| e_1 \| \le \| e_1 \| + \| e_2 \| \equiv \|(e_1, e_2)\|$, we see that

$$\| f(u_1 + e_1, u_2 + e_2) - f(u_1, u_2) - \mathbf{D}_1 f(u_1, u_2)\cdot e_1 - \mathbf{D}_2 f(u_1, u_2) \cdot e_2 \|$$

$$\le \Big(\sup_{0\le t\le 1} \| \mathbf{D}_1 f(u_1 + te_1, u_2 + e_2) - \mathbf{D}_1 f(u_1, u_2 + e_2) \|$$

$$+ \sup_{0\le t\le 1} \| \mathbf{D}_2 f(u_1, u_2 + te_2) - \mathbf{D}_2 f(u_1, u_2) \| \Big) \| (e_1, e_2) \| .$$

Both sups in the parentheses converge to zero as $(e_1, e_2) \to (0, 0)$ by continuity of the partial derivatives. ■

If $E_1 = E_2 = \mathbb{R}$ and $\{e_1, e_2\}$ is the standard basis in $\mathbb{R}^2$ we see that

$$\frac{\partial f}{\partial x}(x, y) = \lim_{h\to 0} \frac{f(x + h, y) - f(x, y)}{h} = \mathbf{D}_1 f(x, y) \cdot e_1 \in F.$$

Similarly, $(\partial f/\partial y)(x, y) = \mathbf{D}_2 f(x, y) \cdot e_2 \in F$. Define inductively higher derivatives

$$\frac{\partial^2 f}{\partial x^2} = \frac{\partial}{\partial x}\left(\frac{\partial f}{\partial x}\right), \quad \frac{\partial^2 f}{\partial x \partial y} = \frac{\partial}{\partial y}\left(\frac{\partial f}{\partial x}\right), \text{ etc.}$$

2.4.13 Example As an application of the formalism just introduced we shall prove that for $f : U \subset \mathbb{R}^2 \to F$

$$\mathbf{D}^2 f(u) \cdot (v, w) = v^1 w^1 \frac{\partial^2 f}{\partial x^2}(u) + v^1 w^2 \frac{\partial^2 f}{\partial y \partial x}(u) + v^2 w^1 \frac{\partial^2 f}{\partial x \partial y}(u) + v^2 w^2 \frac{\partial^2 f}{\partial y^2}(u),$$

$$= (v^1, v^2) \begin{bmatrix} \dfrac{\partial^2 f}{\partial x^2}(u) & \dfrac{\partial^2 f}{\partial y \partial x}(u) \\ \dfrac{\partial^2 f}{\partial x \partial y}(u) & \dfrac{\partial^2 f}{\partial y^2}(u) \end{bmatrix} \begin{pmatrix} w^1 \\ w^2 \end{pmatrix}$$

where $u \in U$, $v, w \in \mathbb{R}^2$, $v = v^1 e_1 + v^2 e_2$, $w = w^1 e_1 + w^2 e_2$, and $\{e_1, e_2\}$ is the standard basis of $\mathbb{R}^2$. To prove this, note that by definition,

$$\mathbf{D}^2 f(u) \cdot (v, w) = \mathbf{D}(\mathbf{D}f)(\cdot) \cdot w)(u) \cdot v$$

Applying the chain rule to $\mathbf{D}f(\cdot) \cdot w = T_w : A \in L(\mathbb{R}^2, F) \mapsto A \cdot w \in F$, the above equals

$$\mathbf{D}(\mathbf{D}f(\cdot) \cdot w)(u) \cdot v$$

$$= \mathbf{D}(\mathbf{D}_1 f(\cdot) \cdot w^1 e_1 + \mathbf{D}_2 f(\cdot) \cdot w^2 e_2)(u) \cdot v \qquad \text{(by } \mathbf{2.4.12}\text{(ii))}$$

$$= \mathbf{D}\left(w^1 \frac{\partial f}{\partial x} + w^2 \frac{\partial f}{\partial y}\right)(u) \cdot v$$

$$= w^1\left[\mathbf{D}_1\left(\frac{\partial f}{\partial x}\right)(u) \cdot v^1 e_1 + \mathbf{D}_2\left(\frac{\partial f}{\partial x}\right)(u) \cdot v^2 e_2\right] + w^2\left[\mathbf{D}_1\left(\frac{\partial f}{\partial y}\right)(u) \cdot v^1 e_1 + \mathbf{D}_2\left(\frac{\partial f}{\partial y}\right)(u) \cdot v^2 e_2\right]$$

$$= v^1 w^1 \frac{\partial^2 f}{\partial x^2}(u) + v^2 w^1 \frac{\partial^2 f}{\partial x \partial y}(u) + v^1 w^2 \frac{\partial^2 f}{\partial y \partial x}(u) + v^2 w^2 \frac{\partial^2 f}{\partial y^2}(u) . \ \blacklozenge$$

For computation of higher derivatives, let us note that by repeated application of **2.4.6**,

$$\mathbf{D}^r f(u) \cdot (e_1, \ldots, e_r) = \frac{d}{dt_r} \cdots \frac{d}{dt_1}\left\{ f\left(u + \sum_{i=1}^{r} t_i e_i \right)\right\}\Bigg|_{t_1 = \cdots = t_r = 0} .$$

In particular for $f : U \subset \mathbb{R}^m \to \mathbb{R}^n$ the components of $\mathbf{D}^r f(u)$ in terms of the standard basis are

$$\frac{\partial^r f}{\partial x^{i_1} \cdots \partial x^{i_r}}, \qquad 0 \le i_k \le r .$$

Thus f is of class C^r iff all its r-th order partial derivatives exist and are continuous.

2.4.14 Proposition *(L. Euler) If* $f : U \subset E \to C^r$, *then* $\mathbf{D}^r f(u) \in L^r_s(E, F)$; *i.e.,* $\mathbf{D}^r f(u)$ *is symmetric.*

Proof First we prove the result for $r = 2$. Let $u \in U$, $v, w \in E$ be fixed; we want to show that $\mathbf{D}^2 f(u) \cdot (v, w) = \mathbf{D}^2 f(u) \cdot (w, v)$. To this, define the linear map $a : \mathbb{R}^2 \to E$ by $a(e_1) = v$, and $a(e_2) = w$, where e_1 and e_2 are the standard basis vectors of $\mathbb{R}^2$. For $(x, y) \in \mathbb{R}^2$, then $a(x,y) = xv + yw$. Now define the affine map $A : \mathbb{R}^2 \to E$ by $A(x, y) = u + a(x, y)$. Since $\mathbf{D}^2(f \circ A)(x, y) \cdot (e_1, e_2) = \mathbf{D}^2 f(u) \cdot (v, w)$ (Exercise **2.3E**), it suffices to prove this formula: $\mathbf{D}^2(f \circ A) \cdot (x,y) \cdot (e_1, e_2) = \mathbf{D}^2(f \circ A)(x, y) \cdot (e_2, e_1)$; i.e.,

$$\frac{\partial^2 (f \circ A)}{\partial x \partial y} = \frac{\partial^2 (f \circ A)}{\partial y \partial x}$$

(see **2.4.13**). Let $g = f \circ A : V = A^{-1}(U) \subset \mathbb{R}^2 \to F$. Since for any $\lambda \in F^*$, $\partial^2(\lambda \circ g)/\partial x \partial y = \lambda(\partial^2 g/\partial x \partial y)$, using the Hahn-Banach theorem **2.2.13**, it suffices to prove that

$$\frac{\partial^2 \varphi}{\partial x \partial y} = \frac{\partial^2 \varphi}{\partial y \partial x}$$

where $\varphi = \lambda \circ g : V \subset \mathbb{R}^2 \to \mathbb{R}$, which is a standard result from calculus. For the sake of completeness we recall the proof. Applying the mean value theorem twice, we get

$$\begin{aligned}
S_{h,k} &= [\varphi(x+h, y+k) - \varphi(x, y+k)] - [\varphi(x+h, y) - \varphi(x, y)] \\
&= \left(\frac{\partial \varphi}{\partial x}(c_{h,k}, y+k) - \frac{\partial \varphi}{\partial x}(c_{h,k}, y)\right)k \\
&= \frac{\partial^2 \varphi}{\partial x \partial y}(c_{h,k}, d_{h,k})hk
\end{aligned}$$

for some $c_{h,k}, d_{h,k}$ lying between x and $x+h$ and y and $y+k$, respectively. By interchanging the two middle terms in $S_{h,k}$ we can derive in the same way that

$$S_{h,k} = \frac{\partial^2 \varphi}{\partial y \partial x}(\gamma_{h,k}, \delta_{h,k})\, hk.$$

Equating these two formulas for $S_{h,k}$, cancelling h, k, and letting $h \to 0, k \to 0$, the continuity of $D^2\varphi$ gives the result.

For general r, proceed by induction:

$$\begin{aligned}
D^r f(u) \cdot (v_1, v_2, \ldots, v_n) &= D^2(D^{r-2} f)(u) \cdot (v_1, v_2) \cdot (v_3, \ldots, v_n) \\
&= D^2(D^{r-2} f)(u) \cdot (v_2, v_1) \cdot (v_3, \ldots, v_n) \\
&= D^r f(u) \cdot (v_2, v_1, v_3, \ldots, v_n).
\end{aligned}$$

Let σ be any permutation of $\{2, \ldots, n\}$, so by the inductive hypothesis

$$D^{r-1} f(u)(v_2, \ldots, v_n) = D^{r-1} f(u)(v_{\sigma(2)}, \ldots, v_{\sigma(n)}).$$

Take the derivative of this relation with respect to $u \in U$ keeping $v_2, \ldots, v_n$ fixed and get (Exercise **2.4F**):

$$D^r f(u)(v_1, \ldots, v_n) = D^r f(u)(v_1, v_{\sigma(2)}, \ldots, v_{\sigma(n)}).$$

Since any permutation can be written as a product of the transposition $\{1, 2, 3, \ldots, n\} \to \{2, 1, 3, \ldots, n\}$ (if necessary) and a permutation of $\{2, \ldots, n\}$, the result follows. ■

Suppose $U \subset E$ is an open set. Then as $+ : E \times E \to E$ is continuous, there exists an open set $\tilde{U} \subset E \times E$ with these three properties: (i) $U \times \{0\} \subset \tilde{U}$, (ii) $u + \xi h \in U$ for all $(u, h) \in \tilde{U}$ and $0 \le \xi \le 1$, and (iii) $(u, h) \in \tilde{U}$ implies $u \in U$. For example let

$$\tilde{U} = \{(+)^{-1}(U)\} \cap (U \times E)$$

Let us call such a set $\tilde{U}$, temporarily, a ***thickening*** of U. See Figure 2.4.2.

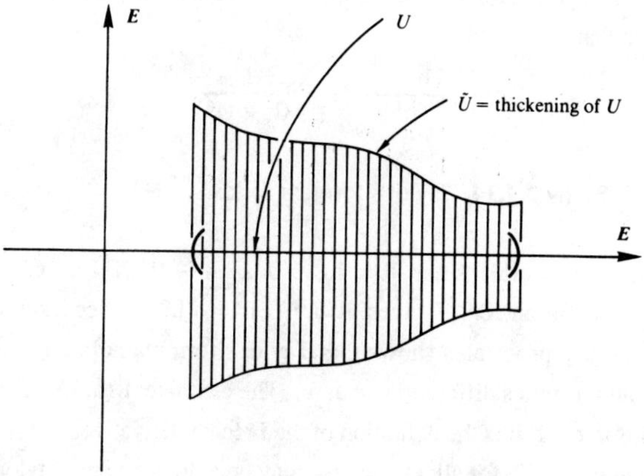

Figure 2.4.2

2.4.15 Taylor's Theorem *A map* $f : U \subset E \to F$ *is of class* C^r *iff there are continuous mappings*

$$\varphi_p : U \subset E \to L^p_s(E, F), \quad p = 1, \ldots, r \quad \textit{and} \quad R : \tilde{U} \to L^r_s(E, F),$$

where $\tilde{U}$ *is some thickening of* U, *such that for all* $(u, h) \in \tilde{U}$,

$$f(u + h) = f(u) + \frac{\varphi_1(u)}{1!} \cdot h + \frac{\varphi_2(u)}{2!} \cdot h^2 + \cdots + \frac{\varphi_r(u)}{r!} \cdot h^r + R(u, h) \cdot h^r,$$

where $h^p = (h, \ldots, h)$ (p *times*) *and* $R(u, 0) = 0$. *If* f *is* C^r *then necessarily* $\varphi_p = \mathbf{D}^p f$ *and*

$$R(u, h) = \int_0^1 \frac{(1 - t)^{r-1}}{(r-1)!} (\mathbf{D}^r f(u + th) - \mathbf{D}^r f(u))\, dt .$$

Proof We shall prove the "only if" part. The converse is proved in Supplement **2.4B**. Leibniz' rule gives the following ***integration by parts*** formula. If $[a, b] \subset U \subset \mathbb{R}$ and $\psi_i : U \subset \mathbb{R} \to E_i$, $i = 1, 2$ are C^1 mappings and $B \in L(E_1, E_2; F)$ is a bilinear map of $E_1 \times E_2$ to F, then

$$\int_a^b B(\psi_1'(1), \psi_2(t))\, dt = B(\psi_1(b), \psi_2(b)) - B(\psi_1(a), \psi_2(a)) - \int_a^b B(\psi_1(t), \psi_2'(t))\, dt .$$

Assume f is a C^r mapping. If $r = 1$, then by **2.4.7**

$$f(u + h) = f(u) + \left(\int_0^1 \mathbf{D}f(u + th)\, dt \right) \cdot h = f(u) + \mathbf{D}f(u) \cdot h + \left(\int_0^1 (\mathbf{D}f(u + th) - \mathbf{D}f(u))\, dt \right) \cdot h$$

and the formula is proved. For general $k \le r$ proceed by induction choosing in the integration by parts formula $E_1 = \mathbb{R}$, $E_2 = E$, $B(s, e) = se$, $\psi_2(t) = D^kf(u + th)\cdot h^k$, and $\psi_1(t) = -(1-t)^k/k!$, and taking into account that

$$\int_0^1 \frac{(1-t)^k}{k!}\,dt = \frac{1}{(k+1)!}.$$

Since $D^kf(u) \in L^k_s(E, F)$ by **2.4.14**, Taylor's formula follows. ■

Note that $R(u, h)\cdot h^r = o(h^r)$ since $R(u, h) \to 0$ as $h \to 0$. If f is C^{r+1} then the mean value inequality and a bound on $D^{r+1}f$ gives $R(u, h)\cdot h^r = o(h^{r+1})$. See Exercise **2.4M** for the differentiability of R. The proof also shows that Taylor's formula holds if f is $(r-1)$ times differentiable on U and r times differentiable at u. The estimate $R(u, h)\cdot h^r = o(h^r)$ is proved directly by induction; for $r = 1$ it is the definition of the Fréchet derivative.

If f is C^∞ (that is, is C^r for all r) then we may be able to extend Taylor's formula into a convergent power series. If we can, we say f is of class C^ω, or ***analytic***. A standard example of a C^∞ function that is not analytic is the following function from $\mathbb{R}$ to $\mathbb{R}$ (Fig. 2.4.3)

$$\theta(x) = \begin{cases} \exp\{-1/(1-x^2)\}, & |x| < 1 \\ 0 & , |x| \geq 1. \end{cases}$$

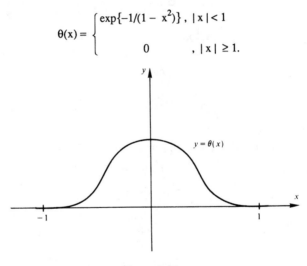

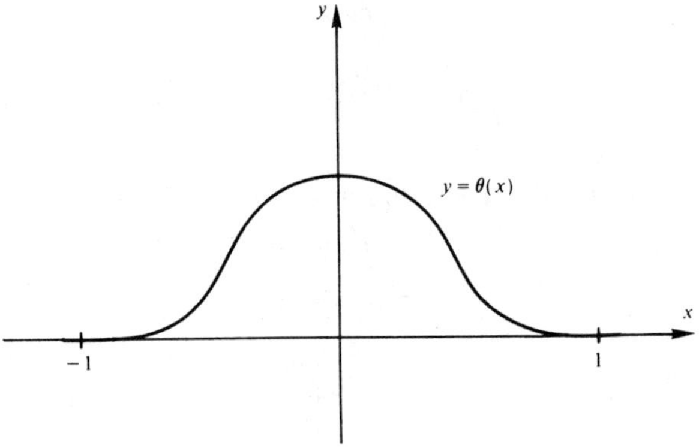

Figure 2.4.3

This function is C^∞, and all derivatives are 0 at $x = \pm 1$. (To see this note that for $|x| < 1$, $f^{(n)}(x) = Q_n(x)(1-x^2)^{-2n}\exp(-1/(1-x^2))$, where $Q_n(x)$ are polynomials given recursively by $Q_0(x) = 1$, $Q_{n+1}(x) = (1-x^2)^2Q_n'(x) + 2x(2n-1-2nx^2)Q_n(x)$.) Hence all coefficients of the Taylor series around these points vanish. Since the function is not identically 0 in any neighborhood of ± 1, it cannot be analytic there.

2.4.16 Example Differentiating Under the Integral Let $U \subset E$ be open and $f: [a, b] \times U \to F$. For $t \in [a, b]$, define $g(t): U \to F$ by $g(t)(u) = f(t, u)$. If, for each t, $g(t)$ is of class C^r

and if the maps $(t, u) \in [a, b] \times U \mapsto \mathbf{D}^j(g(t))(u) \in L^j_s(E, F)$ are continuous, then $h : U \to F$, defined by

$$h(u) = \int_a^b f(t, u)\, dt = \int_a^b g(t)(u)\, dt$$

is C^r and

$$\mathbf{D}^j h(u) = \int_a^b \mathbf{D}^j_u f(t, u) dt, \quad j = 1, \dots, r,$$

where $\mathbf{D}_u$ means the partial derivative in u. For $r = 1$, write

$$\left\| h(u + e) - h(u) - \int_a^b \mathbf{D}\,(g(t))(u)\cdot e\, dt \right\| = \left\| \int_a^b \left(\int_0^1 (\mathbf{D}(g(t))(u + se)\cdot e - \mathbf{D}(g(t))(u)\cdot e) ds \right) dt \right\|$$

$$\leq (b - a)\| e \| \sup_{a \leq t \leq b,\ 0 \leq t \leq 1} \| \mathbf{D}(g(t))(u + se) - \mathbf{D}(g(t))(u) \| = o(e) .$$

For $r > 1$ one can also use an argument like this, but the converse to Taylor's theorem also yields the result rather easily. Indeed, if $R(t, u, e)$ denotes the remainder for the C^r Taylor expansion of $g(t)$, then with $\varphi_p = \mathbf{D}^p h = \int_a^b \mathbf{D}^p[g(t)]\, dt$, the remainder for h is clearly $R(u, e) = \int_a^b R(t, u, e)\, dt$. But $R(t, u, e)\, dt \mapsto 0$ as $e \mapsto 0$ uniformly in t, so $R(u, e)$ is continuous and $R(u, 0) = 0$. Thus h is C^r. ♦

☞ SUPPLEMENT 2.4A
The Leibniz and Chain Rules

Here the explicit formulas are given for the k-th order derivatives of products and compositions. The proofs are straightforward but quite messy induction arguments, which will be left to the interested reader.

1 The Higher Order Leibniz Rule Let E, F_1, F_2, and G be Banach spaces, $U \subset E$ an open set, $f : U \to F_1$, and $g : U \to F_2$ of class C^k and $B \in L(F_1, F_2; G)$. Let $f \times g : U \to F_1 \times F_2$ denote the mapping $(f \times g)(e) = (f(e), g(e))$ and let $B(f, g) = B \circ (f \times g)$. Thus $B(f, g)$ is of class C^k and by Leibniz' rule,

$$\mathbf{D}B(f, g)(p) \cdot e = B(\mathbf{D}f(p) \cdot e, g(p)) + B(f(p), \mathbf{D}g(p) \cdot e).$$

Higher derivatives of f and g are maps

$$\mathbf{D}^i f : U \to L^i(E; F_1), \quad \mathbf{D}^{k-i} g : U \to L^{k-i}(E; F_2),$$

where

$$\mathbf{D}^0 f = f, \qquad \mathbf{D}^0 g = g, \qquad L^0(E; F_1) = F_1, \qquad L^0(E; F_2) = F_2.$$

Denote by

$$\lambda^{i,k-i} \in L(L^i(E; F_1), L^{k-i}(E, F_2); L^k(E; G)),$$

the bilinear mapping defined by

$$[\lambda^{i,k-i}(A_1, A_2)](e_1,\ldots,e_k) = B(A_1(e_1,\ldots,e_i), A_2(e_{i+1},\ldots, e_k))$$

for $A_1 \in L^i(E; F_1)$, $A_2 \in L^{k-i}(E; F_2)$, and $e_1,\ldots,e_k \in E$. Then

$$\lambda^{i,k-i}(\mathbf{D}^i f, \mathbf{D}^{k-i} g) : U \to L^k(E; G)$$

is defined by

$$\lambda^{i,k-i}(\mathbf{D}^i f, \mathbf{D}^{k-i} g)(p) = \lambda^{i,k-i}(\mathbf{D}^i f(p), \mathbf{D}^{k-i} g(p))$$

for $p \in U$. Leibniz' rule for k-th derivatives is

$$\mathbf{D}^k B(f, g) = \mathrm{Sym}^k \circ \sum_{i=0}^{k} \binom{k}{i} \lambda^{i,k-i}(\mathbf{D}^i f, \mathbf{D}^{k-i} g),$$

where $\mathrm{Sym}^k : L^k(E; G) \to L^k_s(E; G)$ is the symmetrization operator, given by (see Exercise **2.2I**):

$$(\mathrm{Sym}^k A)(e_1,\ldots, e_k) = \frac{1}{k!} \sum_{\sigma \in S_k} A(e_{\sigma(1)},\ldots,e_{\sigma(k)}),$$

where S_k is the group of permutations of $\{1, \ldots, k\}$. Explicitly, taking advantage of the symmetry of higher order derivaties, this formula is

$$\mathbf{D}^k B(f, g)(p) \cdot (e_1,\ldots, e_k)$$

$$= \sum_{\substack{\sigma \in S_k \\ \sigma(1)<\cdots<\sigma(i) \\ \sigma(i+1)<\cdots<\sigma(k)}} \sum_{i=0}^{k} \binom{k}{i} B(\mathbf{D}^i f(p)\cdot(e_{\sigma(1)},\ldots, e_{\sigma(i)}),\ \mathbf{D}^{k-i} g(p)(e_{\sigma(i+1)},\ldots, e_{\sigma(k)})).$$

2 The Higher Order Chain Rule Let E, F, and G be Banach spaces and $U \subset F$ and $V \subset F$ be open sets. Let $f : U \to V$, and $g : V \to G$ be maps of class C^k. By the usual chain rule, $g \circ f : U \to G$ is of class C^k and

$$\mathbf{D}(g \circ f)(p) = \mathbf{D}g(f(p)) \circ \mathbf{D}f(p)$$

for $p \in U$. For every tuple $(i, j_1,\ldots, j_i)$, where $i > 1$, and $j_1 + \ldots + j_i = k$, define the continuous multilinear map

$$\lambda^{i,j_1,\ldots,j_i} : L^i(F; G) \times L^{j_1}(E; F) \times \cdots \times L^{j_i}(E; F) \to L^k(E; G)$$

by $$\lambda^{i,j_1,\ldots,j_i}(A, B_1,\ldots, B_i)\cdot(e_1,\ldots,e_k) = A(B_1(e_i,\ldots, e_{j_1}),\ldots, B_i(e_{j_1+\cdots+j_{i-1}+1},\ldots,e_k))$$

for $A \in L^i(F; G)$, $B_\ell \in L^{j_\ell}(E; F)$, $\ell = 1,\ldots, i$, and $e_1,\ldots, e_k \in E$. Since $\mathbf{D}^{j_\ell}f : U \to L^{j_\ell}(E; F)$, we can define

$$\lambda^{i,j_1,\ldots,j_i} \circ (\mathbf{D}^i g \circ f \times \mathbf{D}^{j_1} f \times \cdots \times \mathbf{D}^{j_i}) : U \to L^k(E; G)$$

by

$$p \mapsto \lambda^{i,j_1,\ldots,j_i}(\mathbf{D}^i g(f(p)),\ \mathbf{D}^{j_1}f(p),\ldots, \mathbf{D}^{j_i}f(p)) .$$

With these notations, the k-th order chain rule is

$$\mathbf{D}^k(g \circ f) = \mathrm{Sym}^k \circ \sum_{i=1}^{k} \sum_{j_1+\cdots+j_1=k} \frac{k!}{j_1!\cdots j_i!} \lambda^{i,j_1,\ldots,j_i} \circ (\mathbf{D}^i g \circ f \times \mathbf{D}^{j_1}f \times \cdots \times \mathbf{D}^{j_i}f) ,$$

where $\mathrm{Sym}^k : L^k(E; G) \to L^k_s(E; G)$ is the symmetrization opertor. Taking into account the symmetry of higher order derivatives, the explicit formula at $p \in U$ and $e_1, \ldots, e_k \in E$, is

$$\mathbf{D}^k(g \circ f)(p) \cdot (e_1,\ldots, e_k) =$$

$$\sum_{i=1}^{k} \sum_{j_1+\cdots+j_i=k} \sum \mathbf{D}^i g(f(p))(\mathbf{D}^{j_1}f(p) \cdot (e_{\ell_1},\ldots, e_{\ell_{j_1}}),\ldots, \mathbf{D}^{j_i}f(p) \cdot (e_{\ell_{j_1+\cdots+j_{i-1}+1}},\ldots, e_{\ell_k}))$$

where the third sum is taken for $\ell_1 < \cdots \ell_{j_1},\ldots, \ell_{j_1+\cdots+j_{i-1}+1} < \cdots < \ell_k$.

✍

☞ SUPPLEMENT 2.4B
The Converse to Taylor's Theorem

This theorem goes back to Marcinkiewicz and Zygmund [1936], Whitney [1943a], and Glaeser [1958]. The proof of the converse that we shall follow is due to Nelson [1969]. Assume the formula in the theorem holds where $\varphi_p = \mathbf{D}^p f$, $1 \le p \le r$, and that $R(u, h)$ has the desired expression. If $r = 1$, the formula reduces to the definition of the derivative. Hence $\varphi_1 = \mathbf{D}f$, f is C^1, and thus $R(u, h)$ has the desired form, using **2.4.7**. Inductively assume the theorem is true for $r = p - 1$. Thus $\varphi_j = \mathbf{D}^j f$, for $1 \le j \le p - 1$. Let $h, k \in E$ be small in norm such that $u + h + k \in U$. Write the formula in the theorem for $f(u + h + k)$ in two different ways:

$$f(u+h+k) = f(u+h) + \mathbf{D}f(u+h)\cdot k + \cdots + \frac{1}{(p-1)!}\mathbf{D}^{p-1}f(u+h)\cdot k^{p-1} + \frac{1}{p!}\varphi_p(u+h)\cdot k^p + R_1(u+k,k)\cdot k^p;$$

$$f(u+h+k) = f(u) + \mathbf{D}f(u)\cdot(h+k) + \cdots + \frac{1}{(p-1)!}\mathbf{D}^{p-1}f(u)\cdot(h+k)^{p-1} + \frac{1}{p!}\varphi_p(u)\cdot(h+k)^p + R_2(u,h+k)\cdot(h+k)^p.$$

Subtracting them and collecting terms homogeneous in k^j we get:

$$g_0(h) + g_1(h)\cdot k + \cdots + g_{p-1}(h)\cdot k^{p-1} + g_p(h)\cdot k^p = R_1(u+h,k)\cdot k^p - R_2(u,h+k)\cdot(h+k)^p,$$

where $g_j(h) \in L^j(E;F)$, $g_j(0) = 0$ is given by

$$g_j(h) = \frac{1}{j!}\left[\mathbf{D}^j f(u+h) - \mathbf{D}^j f(u) - \sum_{i=1}^{p-1-j}\frac{1}{i!}\mathbf{D}^{j+i}f(u)\cdot h^i - \frac{1}{(p-j)!}\varphi_p(u)\cdot h^{p-j}\right], \qquad 0 \le j \le p-2;$$

$$g_{p-1}(h) = \frac{1}{(p-1)!}[\mathbf{D}^{p-1}f(u+h) - \mathbf{D}^{p-1}f(u) - \varphi_p(u)\cdot h];$$

and

$$g_p(h) = \frac{1}{p!}[\varphi_p(u+h) - \varphi_p(u)].$$

Let $\|k\|$ satisfy $\frac{1}{4}\|h\| \le \|k\| \le \frac{1}{2}\|h\|$. Since

$$\begin{aligned}\|R_1(u+h,k)\cdot k^p - R_2(u,h+k)\cdot(h+k)^p - g_p(h)\cdot k^p\| &\le (\|R_1(u+h,k)\| + \|g_p(h)\|)\|k\|^p + \|R_2(u,h+k)\|(\|h\|+\|k\|)^p \\ &\le \{\|R_1(u+h,k)\| + \|g_p(h)\| + \|R_2(u,h+k)\|\}(1+3^p)\|h\|^p/2^p\end{aligned}$$

and the quantity in braces $\{\ \} \to 0$ as $h \to 0$, it follows that

$$R_1(u+h,k)\cdot k^p - R_2(u,h+k)\cdot(h+k)^p - g_p(h)\cdot k^p = o(h^p).$$

Hence

$$g_0(h) + g_1(h)\cdot k + \cdots + g_{k-1}(h)\cdot k^{p-1} = o(h^p).$$

We claim that subject to the condition $\frac{1}{4}\|h\| \le \|k\| \le \frac{1}{2}\|h\|$, each term of this sum is $o(h^p)$. If $\lambda_1, \dots, \lambda_p$ are *distinct* numbers, replace k by $\lambda_j k$ in the foregoing, and get a $p \times p$ linear system in the unknowns $g_0(h), \dots, g_{p-1}(h)\cdot k^{p-1}$ with Vandermonde determinant $\Pi_{i<j}(\lambda_i - \lambda_j) \neq 0$ and right-hand side a column vector all of whose entries are $o(h^p)$. Solving this system we get the result claimed. In particular,

$$(\mathbf{D}^{p-1}f(u+k) - \mathbf{D}^{p-1}f(u) - \varphi_p(u)\cdot h)\cdot k^{p-1} = g_{p-1}(h)\cdot k^{p-1} = o(h^p).$$

Using polarization (see Supplement **2.2B**) we get

$$\| D^{p-1}f(u+h) - D^{p-1}f(u) - \varphi_p(u)\cdot h \|$$

$$\leq \frac{(p-1)^{p-1}}{(p-1)!} \left\| (D^{p-1}f(u+h) - D^{p-1}f(u) - \varphi_p(u)\cdot h)' \right\|$$

$$= \frac{(p-1)^{p-1}}{(p-1)!} \sup_{\|e\| \leq 1} \| (D^{p-1}f(u+h) - D^{p-1}f(u) - \varphi_p(u)\cdot h)\cdot e^{p-1} \|$$

$$= \frac{(p-1)^{p-1}}{(p-1)!} \sup_{\|k\| \leq \|h\|/2} \left\| (D^{p-1}f(u+h) - D^{p-1}f(u) - \varphi_p(u)\cdot h)\cdot \left(\frac{2k}{\|h\|}\right)^{p-1} \right\|$$

$$= \frac{(2(p-1))^{p-1}}{(p-1)!\|h\|^{p-1}} \sup_{\|k\| \leq \|h\|/2} \left\| (D^{p-1}f(u+h) - D^{p-1}f(u) - \varphi_p(u)\cdot h)\cdot k^{p-1} \right\|$$

$$= \frac{(2(p-1))^{p-1}}{(p-1)!\|h\|^{p-1}} o(h^p).$$

Since $o(h^p)/\|h\|^p \to 0$ as $h \to 0$, this relation proves that $D^{p-1}f$ is differentiable and $D^pf(u) = \varphi_p(u)$. Thus f is of class C^p, φ_p being continuous, and the formula for R follows by subtracting the given formula for $f(u+h)$ from Taylor's expansion. ■

The converse of Taylor's theorem provides an alternative proof that $D^rf(u) \in L^r_s(E; F)$. Observe first that in the proof of Taylor's expansion for a C^r map f the symmetry of $D^jf(u)$ was never used, so if one symmetrizes the $D^jf(u)$ and calls them φ_j, the same expansion holds. But then the converse of **2.4.12** says that $\varphi_j = D^jf$.

We shall consider here simple versions of two theorems from global analysis, which shall be used in Supplement **4.1C**, namely the smoothness of the evaluation mapping and the "omega-lemma."

Let $I = [0, 1]$ and E be a Banach space. The vector space $C^r(I; E)$ of C^r-maps $(r > 0)$ of I into E is a Banach space with respect to the norm

$$\|f\|_k = \max_{1 \leq i \leq k} \sup_{t \in I} \|D^if(t)\|$$

(see Exercise **2.4H**). If U is open in E, then the set $C^r(I; U) = \{f \in C^r(I; E) \mid f(I) \subset U\}$ is checked to be open in $C^r(I; E)$.

2.4.17 Proposition *The evaluation map*

$$\mathrm{ev} : C^r(I; U) \times]0, 1[\to U$$

defined by

$$\mathrm{ev}(f, t) = f(t)$$

is C^r *and its kth derivative is given by*

$$D^k ev(f, t)\cdot((g_1, s_1), \ldots, (g_k, s_k)) = D^k f(t)\cdot(s_1, \ldots, s_k) + \sum_{i=1}^{k} D^{i-1} g_i(t)\cdot(s_1, \ldots, s_{i-1}, s_{i+1}, \ldots, s_k)$$

where $(g_i, s_i) \in C^r(I; E) \times \mathbb{R}, \qquad i = 1, \ldots, k.$

Proof For $(g, s) \in C^r(I; E) \times \mathbb{R}$, define the norm $\|(g, s)\| = \max(\| g \|_k, | s |)$. Note that the right-hand side of the formula in the statement is symmetric in the arguments (g_i, s_i), $i = 1, \ldots, k$. We shall let this right-hand side be denoted $\varphi_k : C^r(I; U) \times]0, 1[\to L^k{}_s(C^r(I; E) \times \mathbb{R}; E)$. Note that $\varphi_0(f, t) = f(t)$ and that the proposition holds for $r = 0$ by uniform continuity of f on I since $\| f(t) - g(s) \| \le \| f(t) - f(s) \| + \| f - g \|_0$. Since

$$\lim_{(g,\, s) \to (0,\, 0)} \frac{D^r g(t)\cdot s^r}{\| (g, s) \|^r} = 0$$

for all $t \in]0, 1[$, by Taylor's theorem for g we get

$$\begin{aligned} ev(f + g, t + s) &= f(t + s) + g(t + s) \\ &= \sum_{i=0}^{r} \frac{1}{i!} (D^i f(t) \cdot s^i + D^i g(t) \cdot s^i) + R(t, s) \cdot s^r \\ &= f(t) + \sum_{i=1}^{r} \frac{1}{i!} \varphi_i(f, t) \cdot (g, s)^i + R((f, t), (g, s)) \cdot (g, s)^r, \end{aligned}$$

where

$$R((f, t), (g, s)) \cdot ((g_1, s_1), \ldots, (g_r, s_r)) = R(t, s) \cdot (s_1, \ldots, s_r) + \sum_{i=1}^{r} D^r g_i(t) \cdot (s_1, \ldots, s_r),$$

which is symmetric in its arguments and $R((f, t), (0, 0)) = 0$. By the converse to Taylor's theorem, the proposition is proved if we show that every φ_i, $1 \le i \le r$, is continuous. Since

$$\| D^{k-1} g_i(t) - D^{k-1} g_i(s) \| \le | t - s | \sup_{u \in I} \| D^k g_i(u) \| \le | t - s | \, \| g_i \|_r$$

by the mean value theorem, the inequality

$$\begin{aligned} \| (\varphi_k(f, t) - \varphi_k(g, s))\cdot((g_1, s_1), \ldots, (g_k, s_k)) \| &\le \| D^k f(t) - D^k g(s) \| \, | s_1 | \cdots | s_k | + \\ &+ \sum_{i=1}^{k} \| D^{k-1} g_i(t) - D^{k-1} g_i(s) \| \, | s_1 | \cdots | s_{i-1} | \, | s_{i+1} | \cdots | s_k | \end{aligned}$$

implies

$$\| \varphi_k(f, t) - \varphi_k(g, s) \| \le \| D^k f(t) - D^k g(s) \| + k \, | t - s |$$

$$\leq \| \mathbf{D}^k f(t) - \mathbf{D}^k f(s) \| + \| \mathbf{D}^k f(s) - \mathbf{D}^k g(s) \| + k \mid t - s \mid$$
$$\leq \| \mathbf{D}^k f(t) - \mathbf{D}^k f(s) \| + 2k \, \|(f, t) - (g, s)\|.$$

Thus the uniform continuity of $\mathbf{D}^k f$ on I implies the continuity of φ_k at (f, t). ■

Let us consider now the omega lemma. (This is terminology of Abraham [1963]. Various results of this type can be traced back to earlier works of Sobolev [1939] and Eells [1958].)

Let M be a compact topological space and E, F be Banach spaces. With respect to the norm

$$\| f \| = \sup_{m \in M} \| f(m) \|,$$

the vector space $C^0(M, E)$ of continuous E-valued maps on M, is a Banach space. If U is open in E, it is easy to see that $C^0(M, U) = \{ f \in C^0(M, E) \mid f(M) \subset U \}$ is open.

2.4.18 Omega Lemma *Let* $g : U \to F$ *be a* C^r *map,* $r > 0$. *The map*

$$\Omega_g : C^0(M, U) \to C^0(M, F) \quad \textit{defined by} \quad \Omega_g(f) = g \circ f$$

is also of class C^r. *The derivative of* Ω_g *is*

$$\mathbf{D}\Omega_g(f) \cdot h = [(\mathbf{D}g) \circ f] \cdot h$$

i.e.,

$$[\mathbf{D}\Omega_g(f) \cdot h](x) = \mathbf{D}g(f(x)) \cdot h(x).$$

The formula for $\mathbf{D}\Omega_g$ is quite plausible. Indeed, we have

$$[\mathbf{D}\Omega_g(f) \cdot h](x) = \frac{d}{d\varepsilon} \Omega_g(f + \varepsilon h)(x) \Big|_{\varepsilon=0} = \frac{d}{d\varepsilon} g(f(x) + \varepsilon h(x)) \Big|_{\varepsilon=0}.$$

By the chain rule this is $\mathbf{D}g(f(x)) \cdot h(x)$. This shows that if Ω_g is differentiable, then $\mathbf{D}\Omega_g$ must be as stated in the proposition.

Proof Let $f \in C^0(M, U)$. By continuity of g and compactness of M,

$$\| \Omega_g(f) - \Omega_g(f') \| = \sup_{m \in M} \| g(f(m)) - g(f'(m)) \|$$

is small as soon as $\| f - f' \|$ is small; i.e., Ω_g is continuous at each point f. Let

$$A_i : C^0(M, L^i_s(E; F)) \to L^i_s(C^0(M, E); C^0(M, F))$$

be given by

$$A_i(H)(h_1, \ldots, h_i)(m) = H(m)(h_1(m), \ldots, h_i(m))$$

for $H \in C^0(M, L^i(E; F))$, $h_1,\ldots, h_i \in C^0(M, E)$ and $m \in M$. The maps A_i are clearly linear and are continuous with $\| A_i \| \le 1$. Since $\mathbf{D}^i g : U \to L^i_s(E; F)$ is continuous, the preceding argument shows that the maps

$$\Omega_{\mathbf{D}^i g} : C^0(M, U) \to C^0(M, L^i_s(E; F))$$

are continuous and hence

$$A_i \circ \Omega_{\mathbf{D}^i g} : C^0(M, U) \to L^i_s(C^0(M, E); C^0(M, F))$$

is continuous. The Taylor theorem applied to g yields

$$g(f(m) + h(m)) = g(f(m)) + \sum_{i=1}^{r} \frac{1}{i!} \mathbf{D}^i g(f(m))\cdot h(m)^i + R(f(m), h(m))\cdot h(m)^i$$

so that defining

$$[(\mathbf{D}^i g \circ f)\cdot h^i](m) = \mathbf{D}^i g(f(m))\cdot h(m)^i,$$

and

$$[R(f, h)\cdot(h_1,\ldots, h_r)](m) = R(f(m), h(m))\cdot(h_1(m),\ldots, h_r(m))$$

we see that R is continous, $R(f, 0) = 0$, and

$$\Omega_g(f + g) = g \circ (f + h) = g \circ f + \sum_{i=1}^{r} \frac{1}{i!} (\mathbf{D}^i g \circ f) \cdot h^i + R(f, h) \cdot h^i$$

$$= \Omega_g(f) + \sum_{i=1}^{r} \frac{1}{i!} (A_i \circ \Omega_{\mathbf{D}^i g})(f) \cdot h^i + R(f, h) \cdot h^i.$$

Thus by the converse of Taylor's theorem $\mathbf{D}^i\Omega_g = A_i \circ \Omega_{\mathbf{D}^i g}$ and Ω_g is of class C^r. ■

This proposition can be generalized to the Banach space $C^r(I, E)$, $I = [a, b]$, equipped with the norm $\| \cdot \|_r$ given by the maximum of the norms of the first r derivatives; i.e., $\| f \|_r = \max_{0 \le i \le r} \sup_{t \in I} \| f^{(i)}(t) \|$. If g is C^{r+q}, then $\Omega_g : C^r(I, E) \to C^{r-k}(I, F)$ is C^{q+k}. Readers are invited to convince themselves that the foregoing proof works with only trivial modifications in this case. This version of the omega lemma will be used in Supplement **4.1C**.

For applications to partial differential equations, the most important generalizations of the two previous propositions is to the case of Sobolev maps of class H^s; see for example Palais [1968], Marsden [1974b] and Marsden and Hughes [1982] for proofs and applications.

✍

☞ Supplement 2.4C
The Functional Derivative and the Calculus of Variations

Differential calculus in infinite dimensions has many applications, one of which is to the calculus of variations. We give some of the elementary aspects here. We shall begin with some notation and a generalization of the notion of the dual space.

Let E and F be Banach spaces. A continuous bilinear functional $\langle\, ,\rangle : E \times F \to \mathbb{R}$ is called E-***non-degenerate*** if $\langle x, y\rangle = 0$ for all $y \in F$ implies $x = 0$. Similarly, it is F-***non-degenerate*** if $\langle x, y\rangle = 0$ for all $x \in E$ implies $y = 0$. If it is both, we just say $\langle\, ,\rangle$ is ***non-degenerate.*** Equivalently, the two linear maps of E to F^* and F to E^* defined by $x \mapsto \langle x, \cdot\rangle$ and $y \mapsto \langle \cdot, y\rangle$, respectively, are one to one. If they are isomorphisms, $\langle\, ,\rangle$ is called E- or F-***strongly non-degenerate.*** A non-degenerate bilinear form $\langle\, ,\rangle$ thus represents *certain* linear functionals on F in terms of elements in E. We say E and F are in ***duality*** if there is a non-degenerate bilinear functional $\langle\, ,\rangle : E \times F \to \mathbb{R}$, also called a ***pairing*** of E with F. If the functional is strongly non-degenerate, we say the duality is ***strong.***

2.4.19 Examples

A Let $E = F^*$. Let $\langle\, ,\rangle : F^* \times F \to \mathbb{R}$ be given by $\langle \varphi, y\rangle = \varphi(y)$ so the map $E \to F^*$ is the identity. Thus, $\langle\, ,\rangle$ is E-strongly non-degenerate by the Hahn-Banach theorem. It is easily checked that $\langle\, ,\rangle$ is F-non-degenerate. (If it is F^* strongly non-degenerate, F is called ***reflexive.***)

B Let $E = F$ and $\langle\, ,\rangle : E \times E \to \mathbb{R}$ be an inner product on E. Then $\langle\, ,\rangle$ is non-degenerate since $\langle\, ,\rangle$ is positive definite. If E is a Hilbert space, then $\langle\, ,\rangle$ is a strongly non-degenerate pairing by the Riesz representation theorem. ♦

2.4.20 Definition *Let* E *and* F *be normed spaces and* $\langle\, ,\rangle$ *be an* E*-weakly non-degenerate pairing. Let* $f : F \to \mathbb{R}$ *be differentiable at the point* $\alpha \in F$. *The* ***functional derivative*** $\delta f/\delta\alpha$ *of* f *with respect to* α *is the unique element in* E, *if it exists, such that*

$$Df(\alpha) \cdot \beta = \left\langle \frac{\delta f}{\delta \alpha}, \beta \right\rangle \quad \textit{for all} \quad \beta \in F. \tag{1}$$

Likewise, if $g : E \to \mathbb{R}$ *and* $\langle\, ,\rangle$ *is* F*–weakly degenerate, we define* $\frac{\delta g}{\delta v} \in F$, *if it exists, by*

$$Dg(v) \cdot v' = \left\langle v', \frac{\delta f}{\delta v} \right\rangle \quad \textit{for all} \quad v' \in E. \tag{1)'$$

Often E and F are spaces of mappings, as in the following example.

2.4.21 Example Let $\Omega \subset \mathbb{R}^n$ be an open bounded set and consider the space $E = C^0(D)$, of continuous real valued functions on D where $D = \mathrm{cl}(\Omega)$. Take $F = C^0(D) = E$. The L^2-pairing on $E \times F$ is the bilinear map given by

$$\langle\,,\rangle : C^0(D) \times C^0(D) \to \mathbb{R}, \ \langle f, g\rangle = \int_\Omega f(x)g(x)\, d^n x.$$

Let r be a positive integer and define $f : E \to \mathbb{R}$ by $f(\varphi) = \frac{1}{2}\int_\Omega [\varphi(x)]^r\, d^n x$. Then using the calculus rules from this section, we find

$$Df(\varphi) \cdot \psi = \int_\Omega r[\varphi(x)]^{r-1}\psi(s)\, d^n x.$$

Thus, $\dfrac{\delta f}{\delta\varphi} = r\varphi^{r-1}$. ♦

Suppose, more generally, that f is defined on a Banach space E of functions φ on a region Ω in $\mathbb{R}^n$. The functional derivative $\dfrac{\delta f}{\delta\varphi}$ of f with respect to φ is the unique element $\dfrac{\delta f}{\delta\varphi} \in E$, if it exists, such that

$$Df(\varphi) \cdot \psi = \left\langle \frac{\delta f}{\delta\varphi}, \psi \right\rangle = \int_\Omega \left(\frac{\delta f}{\delta\varphi}\right)(x)\psi(x)\, d^n x \quad \text{for all } \in E.$$

The functional derivative may be determined in examples by

$$\int_\Omega \frac{\delta f}{\delta\varphi}(x)\psi(x)\, d^n x = \left.\frac{d}{d\varepsilon}\right|_{\varepsilon=0} f(\varphi + \varepsilon\psi). \tag{2}$$

A basic result in the calculus of variations is the following.

2.4.22 Proposition *Let* E *be a space of functions, as above. A necessary condition for a differentiable function* $f : E \to \mathbb{R}$ *to have an extremum at* φ *is that*

$$\frac{\delta f}{\delta\varphi} = 0.$$

Proof If f has an extremum at φ, then for each ψ, the function $h(\varepsilon) = f(\varphi + \varepsilon\psi)$ has an extremum at $\varepsilon = 0$. Thus, by elementary calculus, $h'(0) = 0$. Since ψ is arbitrary, the result follows from (2). ■

Sufficient conditions for extrema in the calculus of variations are more delicate. See, for example, Bolza [1904] and Morrey [1966].

2.4.23 Examples

A Suppose that $\Omega \subset \mathbb{R}$ is an interval and that f, as a functional of $\varphi \in C^k(\Omega)$, $k \geq 1$, is of the form

$$f(\varphi) = \int_\Omega F\left(x, \varphi(x), \frac{d\varphi}{dx}\right) dx \tag{3}$$

for some smooth function $F : \Omega \times \mathbb{R} \times \mathbb{R} \to \mathbb{R}$, so that the right hand side of (3) is defined. We call F the ***density*** associated with f. It can be shown by using the results of the preceding supplement that f is smooth. Using the chain rule,

$$\int_\Omega \frac{\delta f}{\delta \varphi}(x)\psi(x)\, dx = \frac{d}{d\varepsilon}\bigg|_{\varepsilon=0} \int_\Omega F\left(x, \varphi + \varepsilon\psi, \frac{d(\varphi + \varepsilon\psi)}{dx}\right) dx$$

$$= \int_\Omega D_2F\left(x, \varphi(x), \frac{d\varphi}{dx}\right)\psi(x)\, dx + \int_\Omega D_3F\left(x, \varphi(x), \frac{d\varphi}{dx}\right)\frac{d\psi}{dx}\, dx,$$

where $D_2F = \dfrac{\partial F}{\partial \varphi}$ and $D_3F = \dfrac{\partial F}{\partial\left(\dfrac{\partial \varphi}{\partial x}\right)}$ denote the partial derivatives of F with respect to its second and third arguments. Integrating by parts, this becomes

$$\int_\Omega D_2F\left(x, \varphi(x), \frac{d\varphi}{dx}\right)\psi(x)\, dx - \int_\Omega \left(\frac{d}{dx} D_3F\left(x, \varphi(x), \frac{d\varphi}{dx}\right)\right)\psi(x)\, dx$$

$$+ \int_{\partial\Omega} D_3F\left(x, \varphi(x), \frac{d\varphi}{dx}\right)\psi(x)\, dx.$$

Let us now restrict our attention to the space of ψ's which vanish on the boundary $\partial\Omega$ of Ω. In that case we get

$$\frac{\delta f}{\delta \varphi} = D_2F - \frac{d}{dx} D_3F\,.$$

Rewriting this according to the designation of the second and third arguments of F as φ and $\dfrac{d\varphi}{dx}$, respectively, we obtain

$$\frac{\delta f}{\delta \varphi} = \frac{\partial F}{\partial \varphi} - \frac{d}{dx}\frac{\partial F}{\partial\left(\dfrac{d\varphi}{dx}\right)}\,. \tag{4}$$

By a similar argument, if $\Omega \subset \mathbb{R}^n$, (4) generalizes to

$$\frac{\delta f}{\delta \varphi} = \frac{\partial F}{\partial \varphi} - \frac{d}{dx}\frac{\partial F}{\partial\left(\dfrac{d\varphi}{dx}\right)}\,. \tag{4}$$

(Here, a sum on repeated indices is assumed.) Thus, f has an extremum at φ only if

$$\frac{\partial F}{\partial \varphi} - \frac{d}{dx^k} \frac{\partial F}{\partial\left(\frac{\partial \varphi}{\partial x^k}\right)} = 0 .$$

This is called the ***Euler-Lagrange equation*** in the calculus of variations.

B Assume that in example **A**, the density F associated with f depends also on higher derivatives, i.e., $F = F(x, \varphi(x), \varphi_x, \varphi_{xx}, \dots)$, where $\varphi_x = \frac{d\varphi}{dx}$, $\varphi_{xx} = \frac{d^2\varphi}{dx^2}$, etc. Therefore

$$f(\varphi) = \int_\Omega F(x, \varphi(x), \varphi_x, \varphi_{xx}, \dots)\, dx .$$

By an analogous argument, formula (4) generalizes to

$$\frac{\delta f}{\delta \varphi} = \frac{\partial F}{\partial \varphi} - \frac{d}{dx}\left(\frac{\partial F}{\partial \varphi_x}\right) + \frac{d^2}{dx^2}\left(\frac{\partial F}{\partial \varphi_{xx}}\right) - \cdots \tag{6}$$

C Consider a closed curve γ in $\mathbb{R}^3$ such that γ lies above the boundary $\partial\Omega$ of a region Ω in the xy-plane, as in Figure 2.4.4.

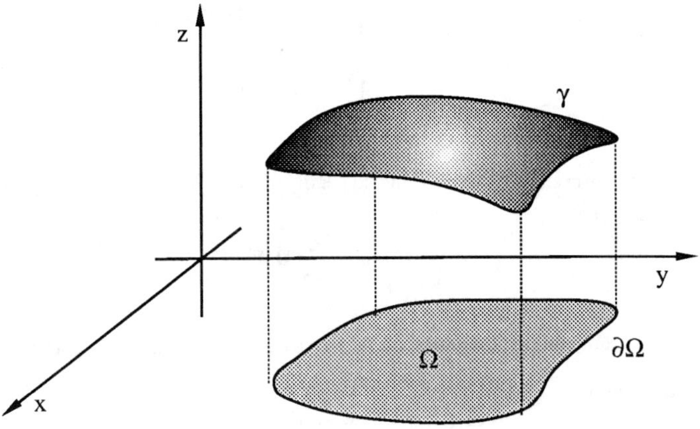

Figure 2.4.4

Consider differentiable surfaces in $\mathbb{R}^3$ (i.e., two dimensional manifolds of $\mathbb{R}^3$) which are graphs of C^k functions $\varphi : \Omega \subset \mathbb{R}^2 \to \mathbb{R}$, so that $(x, y, \varphi(x,y))$ are coordinates on the surface. What is the surface of least area whose boundary is γ? From elementary calculus we know that the area as a function of φ is given by

$$A(\varphi) = \int_\Omega \sqrt{1 + \varphi_x^2 + \varphi_y^2}\ dx\, dy .$$

From equation (5), a necessary condition for φ to minimize A is that

$$\frac{\delta A}{\delta \varphi} = -\frac{\varphi_{xx}(1+\varphi_y^2) - 2\varphi_x\varphi_y\varphi_{xy} + \varphi_{yy}(1+\varphi_x^2)}{(1+\varphi_x^2+\varphi_y^2)^{3/2}} = 0, \tag{7}$$

for $(x, y) \in \Omega$. We relate this to the classical theory of surfaces as follows. A surface has two principal curvatures κ_1 and κ_2; ***the mean curvature*** κ is defined to be their average: that is, $\kappa = (\kappa_1 + \kappa_2)/2$. An elementary theorem of geometry asserts that κ is given by the formula

$$\kappa = \frac{\varphi_{xx}(1+\varphi_y^2) - 2\varphi_x\varphi_y\varphi_{xy} + \varphi_{yy}(1+\varphi_x^2)}{(1+\varphi_x^2+\varphi_y^2)^{1/2}}. \tag{8}$$

If the surface represents a sheet of rubber, the mean curvature represents the net force due to internal stretching. Comparing (7) and (8) we find the well-known result that a minimal surface, i.e., a surface with minimal area, has zero mean curvature. ♦

Now consider the case in which f is a differentiable function of n variables, i.e., f is defined on a product of n function spaces F_i, $i = 1, \dots, n$; $f : F_1 \times \cdots \times F_n \to \mathbb{R}$ and we have pairings $\langle\, , \rangle_i : E_i \times F_i \to \mathbb{R}$.

2.4.24 Definition *The* i–th *partial functional derivative* $\dfrac{\delta f}{\delta \varphi_i}$ *of* f *with respect to* $\varphi_i \in F_i$ *is defined by*

$$\left\langle \frac{\delta f}{\delta \varphi_i}, \psi_i \right\rangle_i = \left.\frac{d}{d\varepsilon}\right|_{\varepsilon=0} f(\varphi_1, \dots, \varphi_i + \varepsilon\psi_i, \dots, \varphi_n)$$

$$= D_i f(\varphi_1, \dots, \varphi_n)\cdot\psi_i = Df(\varphi_1, \dots, \varphi_n)(0, \dots, \psi_i, \dots, 0). \tag{9}$$

*The **total functional derivative** is given by*

$$\left\langle \frac{\delta f}{\delta(\varphi_1, \dots, \varphi_n)}, (\psi_1, \dots, \psi_n) \right\rangle = Df(\varphi_1, \dots, \varphi_n)\cdot(\psi_1, \dots, \psi_n)$$

$$= \sum_{i=1}^{n} D_i f(\varphi_1, \dots, \varphi_n)(0, \dots, \psi_i \dots, 0) = \sum_{i=1}^{n} \left\langle \frac{\delta f}{\delta \varphi_i}, \psi_i \right\rangle_i.$$

2.4.25 Examples

A Suppose that f is a function of n functions $\varphi_i \in C^k(\Omega)$, where $\Omega \subset \mathbb{R}^n$, and their first partial derivatives, and is of the form

$$f(\varphi_1, \ldots, \varphi_n) = \int_\Omega F\left(x, \varphi_i, \frac{\partial \varphi_i}{\partial x^i}\right) d^n x.$$

It follows that

$$\frac{\delta f}{\delta \varphi_i} = \frac{\partial F}{\partial \varphi_i} - \frac{\partial}{\partial x^k} \frac{\partial F}{\partial\left(\dfrac{\partial \varphi_i}{\partial x^k}\right)} \qquad \text{(sum on k).} \tag{10}$$

B Classical Field Theory As discussed in Goldstein [1980], §12, Lagrange's equations for a field $\eta = \eta(x, t)$ with components η^a follow from Hamilton's variational principle. When the Lagrangian L is given by a Lagrangian density $\mathcal{L}$, i.e., L is of the form

$$L(\eta) = \iint_{\Omega \subset \mathbb{R}^3} \mathcal{L}\left(x^j, \eta^a, \frac{\partial \eta^a}{\partial x^j}, \frac{\partial \eta^a}{\partial t}\right) d^n x \, dt, \tag{11}$$

the variational principle states that η should be a critical point of L. Assuming appropriate boundary conditions, this results in the ***equations of motion***

$$0 = \frac{\delta L}{\delta \eta^a} = \frac{d}{dt} \frac{\partial \mathcal{L}}{\partial\left(\dfrac{\partial \eta^a}{\partial t}\right)} - \frac{\partial \mathcal{L}}{\partial \eta^a} + \frac{\partial}{\partial x^k}\left(\frac{\partial \mathcal{L}}{\partial\left(\dfrac{\partial \eta^a}{\partial x^k}\right)}\right) \tag{12}$$

(sum on k is understood). Regarding L as a function of η^a and $\dot{\eta}^a = \dfrac{\partial \eta^a}{\partial t}$, the equations of motion take the form:

$$\frac{d}{dt} \frac{\delta L}{\delta \dot{\eta}^a} = \frac{\delta L}{\delta \eta^a}. \tag{13}$$

C Let $\Omega \subset \mathbb{R}^n$ and let $C^k_\partial(\Omega)$ stand for the C^k functions vanishing on $\partial\Omega$. Let $f : C^k_\partial(\Omega) \to \mathbb{R}$ be given by the Dirichlet integral

$$f(\varphi) = \frac{1}{2} \int_\Omega \sum_{i=1}^n \left(\frac{\partial \varphi}{\partial x^i}\right)^2 d^n x.$$

Using the standard inner product $\langle\, , \rangle$ in $\mathbb{R}^n$, we may write

$$f(\varphi) = \frac{1}{2} \int_\Omega \langle \nabla\varphi, \nabla\varphi \rangle \, d^n x.$$

Differentiating with respect to φ:

$$\begin{aligned} Df(\varphi)\cdot\psi &= \frac{d}{d\varepsilon}\Big|_{\varepsilon=0} \frac{1}{2} \int_\Omega \langle \nabla(\varphi + \varepsilon\psi), \nabla(\varphi + \varepsilon\psi) \rangle \, d^n x \\ &= \int_\Omega \langle \nabla\varphi, \nabla\psi \rangle \, d^n x \end{aligned}$$

$$= -\int_\Omega \nabla^2\varphi(x)\cdot\psi(x)\ d^n x \qquad \text{(integrating by parts)} .$$

Thus $\dfrac{\delta f}{\delta\varphi} = -\nabla^2\varphi$, the Laplacian of φ.

D The Stretched String Consider a string of length ℓ and mass density σ, stretched horizontally under a tension τ, with ends fastened at $x = 0$ and $x = \ell$. Let $u(x, t)$ denote the vertical displacement of the string at x, at time t. We have $u(0, t) = u(\ell, t) = 0$. The potential energy V due to small vertical displacements is shown in elementary mechanics texts to be

$$V = \int_0^\ell \frac{1}{2}\tau\left(\frac{\partial u}{\partial x}\right)^2 dx ,$$

and the kinetic energy T is

$$T = \int_0^\ell \frac{1}{2}\sigma\left(\frac{\partial u}{\partial t}\right)^2 dx .$$

From the definitions, we get

$$\frac{\delta V}{\delta u} = -\tau\frac{\partial^2 u}{\partial x^2}, \quad \text{and} \quad \frac{\delta T}{\delta \dot{u}} = \sigma\dot{u}.$$

Then with the Lagrangian $L = T - V$, the equations of motion (13) become the ***wave equation***

$$\sigma\frac{\partial^2 u}{\partial t^2} - \tau\frac{\partial^2 u}{\partial x^2} = 0 . \quad \blacklozenge$$

Next we formulate a chain rule for functional derivatives. Let $\langle\, ,\rangle : E \times F \to \mathbb{R}$ be a weakly nondegenerate pairing between E and F. If $A \in L(F, F)$, its ***adjoint*** $A^* \in L(E, E)$, if it exists, is defined by $\langle A^* v, \alpha\rangle = \langle v, A\alpha\rangle$ for all $v \in E$ and $\alpha \in F$.

Let $\varphi : F \to F$ be a differentiable map and $f : F \to \mathbb{R}$ be differentiable at $\alpha \in F$. From the chain rule,

$$D(f \circ \varphi)(\alpha)\cdot\beta = Df(\varphi(\alpha))\cdot(D\varphi(\alpha)\cdot\beta), \quad \text{for} \quad \beta \in F.$$

Hence assuming that all functional derivatives and adjoints exist, the preceding relation implies

$$\left\langle \frac{\delta(f\circ\varphi)}{\delta\alpha}, \beta \right\rangle = \left\langle \frac{\delta f}{\delta\gamma}, D\varphi(\alpha)\cdot\beta \right\rangle = \left\langle D\varphi(\alpha)^*\cdot\frac{\delta f}{\delta\gamma}, \beta \right\rangle$$

where $\gamma = \varphi(\alpha)$, i.e.,

$$\frac{\delta(f\circ\varphi)}{\delta\alpha} = D\varphi(\alpha)^*\cdot\frac{\delta f}{\delta\gamma}. \tag{14}$$

Similarly if $\psi : \mathbb{R} \to \mathbb{R}$ is differentiable then for $\alpha, \beta \in F$,

$$\mathbf{D}(\psi \circ f)(\alpha) \cdot \beta = \mathbf{D}\psi(f(\alpha)) \cdot (\mathbf{D}f(\alpha) \cdot \beta)$$

where the first dot on the right hand side is ordinary multiplication by $\mathbf{D}\psi(f(\alpha)) \in \mathbb{R}$. Hence

$$\left\langle \frac{\delta(\psi \circ f)}{\delta \alpha}, \beta \right\rangle = \mathbf{D}\psi(f(\alpha)) \cdot \left\langle \frac{\delta f}{\delta \alpha}, \beta \right\rangle = \left\langle \psi'(f(\alpha)) \frac{\delta f}{\delta \alpha}, \beta \right\rangle$$

i.e.,

$$\frac{\delta(\psi \circ f)}{\delta \alpha} = \psi'(f(\alpha)) \frac{\delta f}{\delta \alpha}. \tag{15}$$

We conclude with a discussion of extrema for real valued functions on Banach spaces.

2.4.26 Definition *Let* $f : U \subset E \to \mathbb{R}$ *be a continuous function,* U *open in* E. *We say* f *has **a local minimum (resp. maximum)** at* $u_0 \in U$, *if there is a neighborhood* V *of* u_0, $V \subset U$ *such that* $f(u_0) \le f(u)$ $(f(u_0) \ge f(u))$ *for all* $u \in V$. *If the inequality is strict,* u_0 *is called a **strict local minimum (resp. maximum)**. The point* u_0 *is called a **global minimum (resp. maximum)** if* $f(u_0) \le f(u)$ (resp. $f(u_0) \ge f(u)$) *for all* $u \in U$. *Local maxima and minima are called **local extrema**.*

2.4.27 Proposition *Let* $f : U \subset E \to \mathbb{R}$ *be a continuous function differentiable at* $u_0 \in U$. *If* f *has a local extremum at* u_0, *then* $\mathbf{D}f(u_0) = 0$.

Proof If u_0 is a local minimum, then there is a neighborhood V of U such that $f(u_0 + th) - f(u_0) \ge 0$ for all $h \in V$. Therefore, the limit of $[f(u_0 + th) - f(u_0)]/t$ as $t \to 0$, $t \ge 0$ is ≥ 0 and as $t \to 0$, $t \le 0$ is ≤ 0. Since both limits equal $\mathbf{D}f(u_0)$, it must vanish. ■

This criterion is not sufficient as the elementary calculus example $f : \mathbb{R} \to \mathbb{R}$, $f(x) = x^3$ shows. Also, if U is not open, the values of f on the boundary of U must be examined separately.

2.4.28 Proposition *Let* $f : U \subset E \to \mathbb{R}$ *be twice differentiable at* $u_0 \in U$.

(i) *If* u_0 *is a local minimum (maximum), then* $\mathbf{D}^2f(u_0) \cdot (e, e) \ge 0$ (≤ 0) *for all* $e \in E$.

(ii) *If* u_0 *is a non-degenerate critical point* f, *i.e.,* $\mathbf{D}f(u_0) = 0$ *and* $\mathbf{D}^2f(u_0)$ *defines an isomorphism of* E *with* E^*, *and if* $\mathbf{D}^2f(u_0) \cdot (e, e) > 0$ (< 0) *for all* $e \ne 0$, $e \in E$, *then* u_0 *is a strict local minimum (maximum) of* f.

Proof (i) By Taylor's formula, in a neighborhood V of U_0, $0 \le f(u_0 + h) - f(u_0) = \frac{1}{2}\mathbf{D}f(u_0)(h, h) + o(h^2)$ for all $h \in V$. If $e \in E$ is arbitrary, for small $t \in \mathbb{R}$, $te \in V$, so that $0 \le \frac{1}{2}\mathbf{D}^2f(u_0)(te, te) + o(t^2e^2)$ implies $\mathbf{D}^2f(u_0)(e, e) + 2o(t^2e^2)/t^2 \ge 0$. Now let $t \to 0$.

(ii) Denote by $T : E \to E^*$ the isomorphism defined by $e \mapsto \mathbf{D}^2f(u_0) \cdot (e, \cdot)$, so that there exists $a > 0$ such that

$$a\,\|e\| \le \|Te\| = \sup_{\|e'\|=1} |\langle Te, e'\rangle| = \sup_{\|e'\|=1} |\mathbf{D}^2 f(u_0)\cdot(e, e')|.$$

By hypothesis and symmetry of the second derivative,

$$0 < \mathbf{D}^2 f(u_0)\cdot(e + se', e + se') = s^2\mathbf{D}^2 f(u_0)\cdot(e', e') + 2s\mathbf{D}^2 f(u_0)\cdot(e, e') + \mathbf{D}^2 f(u_0)\cdot(e, e)$$

which is a quadratic form in s. Therefore its discriminant must be negative, i.e.,

$$|\mathbf{D}^2 f(u_0)\cdot(e, e')|^2 < \mathbf{D}^2 f(u_0)\cdot(e', e')\mathbf{D}^2 f(u_0)\cdot(e, e) \le \|\mathbf{D}^2 f(u_0)\|\,\mathbf{D}^2 f(u_0)\cdot(e, e),$$

and we get

$$a\,\|e\| \le \sup_{\|e'\|=1} |\mathbf{D}^2 f(u_0)\cdot(e, e')| \le \|\mathbf{D}^2 f(u_0)\|^{1/2}[\mathbf{D}^2 f(u_0)\cdot(e, e)]^{1/2}.$$

Therefore, letting $m = a^2/\|\mathbf{D}^2 f(u_0)\|$, the following inequality holds for any $e \in E$:

$$\mathbf{D}^2 f(u_0)\cdot(e, e) \ge m\,\|e\|^2.$$

Thus, by Taylor's theorem we have

$$f(u_0 + h) - f(u_0) = \frac{1}{2}\mathbf{D}^2 f(u_0)\cdot(h, h) + o(h^2) \ge m\,\|h\|^2/2 + o(h^2).$$

Let $\varepsilon > 0$ be such that if $\|h\| < \varepsilon$, then $|o(h^2)| \le m\,\|h\|^2/4$, which implies $f(u_0 + h) - f(u_0) \ge m\,\|h\|^2/4 > 0$ for $h \ne 0$, and thus u_0 is a strict local minimum of f. ■

The condition in (i) is not sufficient for f to have a local minimum at u_0. For example, $f : \mathbb{R}^2 \to \mathbb{R}$, $f(x, y) = x^2 - y^4$ has $f(0, 0) = 0$, $\mathbf{D}f(0, 0) = 0$, $\mathbf{D}^2 f(0, 0)\cdot(x, y)^2 = 2x^2 \ge 0$ and in any neighborhood of the origin, f changes sign. The conditions in (ii) are not necessary for f to have a strict local minimum at u_0. For example $f : \mathbb{R} \to \mathbb{R}$, $f(x) = x^4$ has $f(0) = f'(0) = f''(0) = f'''(0) = 0$, $f^{(4)}(0) > 0$ and 0 is a strict global minimum for f. On the other hand, if the conditions in (ii) hold and u_0 is the only critical point of a differentiable function $f : U \to \mathbb{R}$, then u_0 is a strict global minimum of f. For if there was another point $u_1 \in U$ with $f(u_1) \le f(u_0)$ on the straight line segment $(1 - t)u_0 + tu_1$, $t \in [0, 1]$ there exists a point u_2 such that $f(u_2) > f(u_0) \ge f(u_1)$ since by (ii) u_0 is a strict local minimum. Therefore, there exists u_3 on this segment, $u_3 \ne u_0, u_1$ such that $f(u_3) = f(u_0)$. But then by the mean value theorem **2.4.8** there exists $u_4 \ne u_0, u_3$ such that $\mathbf{D}f(u_4) = 0$ which contradicts uniqueness of the critical point. Finally, care has to be taken with the statement in (ii): non-degeneracy holds in the topology of E. If E is continuously embedded in another Banach space F and $\mathbf{D}^2 f(u_0)$ is non-degenerate in F only, u_0 need not even be a minumum. For example, consider the smooth map

$$f : L^4([0, 1]) \to \mathbb{R} , \; f(u) = \frac{1}{2}\int_0^1 (u(x)^2 - u(x)^4) \, dx$$

and note

$$f(0) = 0 , \; \mathbf{D}f(0) = 0 , \text{ and } \mathbf{D}^2 f(0)(v, v) = \int_0^1 v(x)^2 \, dx > 0 \text{ for } v \neq 0,$$

and that $\mathbf{D}^2 f(0)$ defines an isomorphism of $L^4([0, 1])$ with $L^{4/3}([0, 1])$. Alternatively, $\mathbf{D}^2 f(0)$ is non-degenerate on $L^2([0, 1])$ *not* on $L^4([0, 1])$. Also note that in any neighborhood of 0 in $L^4([0, 1])$, f changes sign: $f(1/n) = (n^2 - 1)/2n^4 \geq 0$ for $n \geq 2$, but $f(u_n) = -12/n < 0$ for $n \geq 1$ if

$$u_n = \begin{cases} 2, & \text{on } [0, 1/n] \\ 0, & \text{elsewhere} \end{cases}$$

and both $1/n$, u_n converge to 0 in $L^4([0, 1])$. Thus, even though $\mathbf{D}^2 f(0)$ is positive, 0 is not a minimum of f. (See Ball and Marsden [1984] for more sophisticated examples of this sort.) ✍

Exercises

2.4A Show that if $g : U \subset E \to L(F, G)$ is C^r, then $f : U \times F \to G$, defined by $f(u, v) = (g(u))(v)$, $u \in U$, $v \in F$ is also C^r. (*Hint:* Apply the Leibniz rule with $L(F, G) \times F \to G$ the evaluation map.)

2.4B Show that if $f : U \subset E \to L(F, G)$, $g : U \subset E \to L(G, H)$ are C^r mappings then so is $h : U \subset E \to L(F, H)$, defined by $h(u) = g(u) \circ f(u)$.

2.4C Extend Leibniz' rule to multilinear mappings and find a formula for the derivative.

2.4D Define a map $f : U \subset E \to F$ to be of ***class*** T^1 if it is differentiable, its tangent map $Tf : U \times E \to F \times F$ is continuous and $\| \mathbf{D}f(x) \|$ is locally bounded.

(i) For E and F finite dimensional, show that this is equivalent to C^1.

(ii) (Project) Investigate the validity of the chain rule and Taylor's theorem for T^r maps .

(iii) (Project) Show that the function developed in Smale [1964] is T^2 but is *not* C^2.

2.4E Suppose that $f : E \to F$ (where E, F are real Banach spaces) is ***homogeneous of degree*** k; i.e., $f(te) = t^k f(e)$ for all $t \in \mathbb{R}$, and $e \in E$.

(i) Show that if f is differentiable, then $\mathbf{D}f(u) \cdot u = kf(u)$. (*Hint:* Let $g(t) = f(tu)$ and compute dg/dt.)

(ii) If $E = \mathbb{R}^n$ and $F = \mathbb{R}$, show that this relation is equivalent to

$$\sum_{i=1}^{n} x_i \frac{\partial f}{\partial x_i} = kf(x).$$

Show that maps multilinear in k variables are homogeneous of degree k. Give other examples.

(iii) If f is C^k show that $f(e) = (1/k!)\ \mathbf{D}^k f(0) \cdot e^k$, i.e., $f \in S^k(E, F)$ and thus it is C^∞. (*Hint:* $f(0) = 0$; inductively applying Taylor's theorem and replacing at each step h by th, show that

$$f(h) = \frac{1}{k!}\mathbf{D}^k f(0) \cdot h^k + o(t^k h)/t^k.)$$

2.4F Let $e_1, \ldots, e_{n-1} \in E$ be fixed and $f : U \subset E \to F$ be n times differentiable. Show that the map $g : U \subset E \to F$ defined by $g(u) = \mathbf{D}^{n-1} f(u) \cdot (e_1, \ldots, e_{n-1})$ is differentiable and $\mathbf{D}g(u) \cdot e = \mathbf{D}^n f(u) \cdot (e, e_1, \ldots, e_{n-1})$.

2.4G (i) Prove the following refinement of **2.4.14.** If f is C^1 and $\mathbf{D}_1\mathbf{D}_2 f(u)$ exists and is continuous in u, then $\mathbf{D}_2\mathbf{D}_1 f(u)$ exists and these are equal.

(ii) The hypothesis in (i) cannot be weakened: show that the function

$$f(x, y) = \begin{cases} \dfrac{xy(x^2 - y^2)}{x^2 + y^2}, & \text{if } (x, y) \neq (0, 0) \\ 0, & \text{if } (x, y) = (0, 0) \end{cases}$$

is C^1, has $\partial^2 f/\partial x \partial y$, $\partial^2 f/\partial y \partial x$ continuous on $\mathbb{R}^2 \setminus \{0, 0)\}$, but that $\partial^2 f(0, 0)/\partial x \partial y \neq \partial^2 f(0, 0)/\partial y \partial x$.

2.4H For $f : U \subset E \to F$, show that the second tangent map is given as follows:

$$T^2 f : (U \times E) \times (E \times E) \to (F \times F) \times (F \times F)$$

$$(u, e_1, e_2, e_3) \mapsto \big(f(u), \mathbf{D}f(u) \cdot e_1, \mathbf{D}f(u) \cdot e_2,\ \mathbf{D}^2 f(u) \cdot (e_1, e_2) + \mathbf{D}f(u) \cdot e_3\big).$$

2.4I Let $f : \mathbb{R}^2 \to \mathbb{R}$ be defined by $f(x, y) = 2x^2y/(x^4 + y^2)$ if $(x, y) \neq (0, 0)$ and 0 if $(x, y) = (0, 0)$. Show that

(i) f is discontinuous at (0, 0), hence is not differentiable at (0, 0);

(ii) all directional derivatives exist at (0, 0); i.e., f is Gâteaux differentiable.

2.4J Differentiating sequences Let $f_n : U \subset E \to F$ be a sequence of C^r maps, where E and F are Banach spaces. If $\{f_n\}$ converges pointwise to $f : U \to F$ and if $\{\mathbf{D}^j f_n\}$, $0 \le j \le r$, converges locally uniformly to a map $g^j : U \to L^j_s(E, F)$, then show that f is C^r, $\mathbf{D}^j f = g^j$ and $\{f_n\}$ converges locally uniformly to f. (*Hint:* For $r = 1$ use the mean value inequality and continuity of g^1 to conclude that

$$\| f(u+h) - f(u) - g^1(u) \cdot h \| \le \| f(u+h) - f_n(u+h) - [f(u) - f_n(u)] \|$$
$$+ \| f_n(u+h) - f_n(u) - \mathbf{D}f_n(u) \cdot h \| + \| \mathbf{D}f_n(u) \cdot h - g^1(u) \cdot h \| \le \varepsilon \| h \|.$$

For general r use the converse to Taylor's theorem.)

2.4K α **lemma** In the context of **2.4.18** let $\alpha(g) = g \circ f$. Show that α is continuous linear and hence is C^∞.

2.4L Consider the map $\Phi : C^1([0,1]) \to C^0([0,1])$ given by $\Phi(f)(x) = \exp[f'(x)]$. Show that Φ is C^∞ and compute $\mathbf{D}\Phi$.

2.4M (H. Whitney [1943a]) Let $f : U \subset E \to F$ be of class C^{k+p} with Taylor expansion

$$f(b) = f(a) + \mathbf{D}f(a) \cdot (b-a) + \cdots + \frac{1}{k!}\mathbf{D}^k f(a) \cdot (b-a)^k$$
$$+ \left\{ \int_0^1 \frac{(1-t)^{k-1}}{(k-1)!} [\mathbf{D}^k f((1-t)a + tb) - \mathbf{D}^k f(a)]\, dt \right\} \cdot (b-a)^k.$$

(i) Show that the remainder $R_k(a,b)$ is C^{k+p} for $b \ne a$ and C^p for $a, b \in E$. If $E = F = \mathbb{R}$, $R_k(a,a) = 0$, and

$$\lim_{b \to a} (| b - a |^i \, \mathbf{D}^{i+p} R_k(a,b)) = 0, \ 1 \le i \le k\,.$$

(For generalizations to Banach spaces, see Tuan and Ang [1979].)

(ii) Show that the conclusion in (i) cannot be improved by considering $f(x) = | x |^{k+p+1/2}$.

2.4N (H. Whitney [1943b]). Let $f : \mathbb{R} \to \mathbb{R}$ be an even (resp. odd) function; i.e., $f(x) = f(-x)$ (resp. $f(x)) = -f(-x)$).

(i) Show that $f(x) = g(x^2)$ (resp. $f(x) = xg(x^2)$) for some g.

(ii) Show that if f is C^{2k} (C^{2k+1}) then g is C^k (*Hint:* Use the converse to Taylor's theorem).

(iii) Show that (ii) is still true if $k = \infty$.

(iv) Let $f(x) = | x |^{2k+1+1/2}$ to show that the conclusion in (ii) cannot be sharpened.

2.4O (M. Buchner, J. Marsden, and S. Schecter [1983]). Let $E = L^4([0,1])$ and let $\varphi : \mathbb{R} \to \mathbb{R}$ be a C^∞ function such that $\varphi'(\lambda) = 1$, if $-1 \le \lambda \le 1$ and $\varphi'(\lambda) = 0$ if $|\lambda| \ge 2$. Assume φ is monotone increasing with $\varphi = -M$ for $\lambda \le -2$ and $\varphi = M$ for $\lambda \ge 2$. Define the map $h : E \to \mathbb{R}$ by

$$h(u) = \frac{1}{3}\int_0^1 \varphi([u(x)]^3)\,dx\,.$$

(i) Show that h is C^3 using the converse to Taylor's theorem. (*Hint:* Let $\psi(\lambda) = \varphi(\lambda^3)$, write out Taylor's theorem for $r = 3$ for $\psi(\lambda)$, and plug in $u(x)$ for λ.)

(ii) The formal L^2 gradient of h (i.e., the functional derivative $\delta h/\delta u$) is given by

$$\nabla h(u) = \frac{1}{3}\,\psi'(u)$$

where $\psi(\lambda) = \varphi(\lambda^3)$. Show that $\nabla h : E \to E$ is C^0 but is not C^1. (*Hint:* Its derivative would be $v \mapsto \psi''(u)v/3$. Let $a \in [0, 1]$ be such that $\varphi''(a)/3 \neq 0$ and let $u_n = a$ on $[0, 1/n]$, $u_n = 0$ elsewhere; $v_n = n^{1/4}$ on $[0, 1/n]$, $v_n = 0$ elsewhere. Show that in $L^4([0, 1])$, $u_n \to 0$, $\| v_n \| = 1$, $\psi''(u_n) \cdot v_n$ does not converge to 0, but $\psi''(0) = 0$.) Using the same method, show h is not C^4 on $L^4([0, 1])$.

(iii) Show: if q is a positive integer and $E = L^q([0, 1])$, then h is C^{q-1} but is not C^q.

(iv) Let

$$f(u) = \frac{1}{2}\int_0^1 |\, u(x)\,|^2\,dx + h(u).$$

Show that on $L^4([0, 1])$, f has a formally non-degenerate critical point at 0 (i.e., $D^2f(0)$ defines an isomorphism of $L^2([0, 1])$, yet this critical point is *not isolated.* (*Hint:* Consider the function $u_n = -1$ on $[0, 1/n]$; 0 on $]1/n, 1]$.) This exercise is continued in **5.4H**.

2.4P Let E be the space of maps $\mathbf{A} : \mathbb{R}^3 \to \mathbb{R}^3$ with $\mathbf{A}(x) \to \mathbf{0}$ as $x \to 0$ sufficiently rapidly. Let $f : E \to \mathbb{R}$ and show

$$\frac{\delta}{\delta \mathbf{A}}\, f(\text{curl } \mathbf{A}) = \text{curl}\, \frac{\delta f}{\delta \mathbf{A}}\,.$$

(*Hint:* Specify whatever smoothness and fall-off hypotheses you need; use $\mathbf{A} \cdot \text{curl } \mathbf{B} - \mathbf{B} \cdot \text{curl } \mathbf{A} = \text{div}(\mathbf{B} \times \mathbf{A})$, the divergence theorem, and the chain rule.)

2.4Q (i) Let $E = \{\mathbf{B} \mid \mathbf{B}$ is a vector field on $\mathbb{R}^3$ vanishing at ∞ and such that $\text{div } \mathbf{B} = 0\}$ and pair E with itself via $\langle \mathbf{B}, \mathbf{B}' \rangle = \int \mathbf{B}(x) \cdot \mathbf{B}'(x)\,dx$. Compute $\delta F/\delta \mathbf{B}$, where F is defined by $F = (1/2)\int \| \mathbf{B} \|^2\, d^3x$.

(ii) Let

$E = \{\mathbf{B} \mid \mathbf{B}$ is a vector field on $\mathbb{R}^3$ vanishing at ∞ such that $\mathbf{B} = \nabla \times \mathbf{A}$ for some $\mathbf{A}\}$

and let $\quad F = \{\mathbf{A}' \mid \mathbf{A}'$ is a vector field on $\mathbb{R}^3$, $\text{div } \mathbf{A}' = 0\}$

with the pairing $\langle \mathbf{B}, \mathbf{A}' \rangle = \int \mathbf{A} \cdot \mathbf{A}'\, d^3x$. Show that this pairing is well defined. Compute $\delta F/\delta \mathbf{B}$, where F is as in (i). Why is your answer different?

§2.5 *The Inverse and Implicit Function Theorems*

The Inverse and Implicit Function Theorems are pillars of nonlinear analysis and geometry, so we give them special attentionin this section. Throughout, E, F, ..., are assumed to be Banach spaces. In the finite-dimensional case these theorems have a long and complex history; the infinite-dimensional version is apparently due to Hildebrandt and Graves [1927].

We first consider the Inverse Function Theorem. This states that if the linearization of the equation $f(x) = y$ is uniquely invertible, then locally so is f; i.e., we can uniquely solve $f(x) = y$ for x as a function of y. To formulate the theorem, the following terminology is useful.

2.5.1 Definition *A map* $f : U \subset E \to V \subset F$ *(U, V open) is a* C^r ***diffeomorphism*** *if* f *is of class* C^r*, is a bijection (that is, one-to-one and onto* V*), and* f^{-1} *is also of class* C^r.

2.5.2 Inverse Mapping Theorem *Let* $f : U \subset E \to F$ *be of class* C^r, $r \geq 1$, $x_0 \in U$, *and suppose* $\mathbf{D}f(x_0)$ *is a linear isomorphism. Then* f *is a* C^r *diffeomorphism of some neighborhood of* x_0 *onto some neighborhood of* $f(x_0)$ *and, moreover,* $\mathbf{D}f^{-1}(y) = [\mathbf{D}f(f^{-1}(y))]^{-1}$ *for* y *in this neighborhood of* $f(x_0)$.

Although our immediate interest is the finite-dimensional case, for Banach spaces it is good to keep in mind the Banach Isomorphism Theorem: If $T : E \to F$ is linear, bijective, and continuous, then T^{-1} is continuous. (See **2.2.19**.)

To prove the theorem, we assemble a few lemmas. First recall the contraction mapping principle from §**1.2**.

2.5.3 Lemma *Let* M *be a complete metric space with distance function* $d : M \times M \to \mathbb{R}$. *Let* $F : M \to M$ *and assume there is a constant* λ, $0 \leq \lambda < 1$, *such that for all* $x, y \in M$,

$$d(F(x), F(y)) \leq \lambda d(x, y) .$$

Then F *has a unique fixed point* $x_0 \in M$; *that is,* $F(x_0) = x_0$.

This result is the basis of many important existence theorems in analysis. The other fundamental fixed point theorem in analysis is the Schauder Fixed Point Theorem, which states that a continuous map of a compact convex set (in a Banach space, say) to itself, has a fixed point -- not necessarily unique, however.

2.5.4 Lemma *The set* $\mathbf{GL}(E, F)$ *of linear isomorphisms from* E *to* F *is open in* $L(E, F)$.

Proof We can assume $E = F$. Indeed, if $\varphi_0 \in \mathbf{GL}(E, F)$, the linear map $\psi \mapsto \varphi_0^{-1} \circ \psi$ from $L(E, F)$ to $L(E, E)$ is continuous and $\mathbf{GL}(E, F)$ is the inverse image of $\mathbf{GL}(E, E)$. Let

$$\| \alpha \| = \sup_{\substack{e \in E \\ \|e\|=1}} \| \alpha(e) \|$$

be the norm on $L(E, F)$ relative to given norms on E and F. For $\varphi \in \mathbf{GL}(E, E)$, we need to prove that ψ sufficiently near φ is also invertible. We will show that $\| \psi - \varphi \| < \| \varphi^{-1} \|^{-1}$ implies $\psi \in \mathbf{GL}(E, E)$. The key is that $\| \cdot \|$ is an algebra norm. That is, $\| \beta \circ \alpha \| \leq \| \beta \| \| \alpha \|$ for $\alpha \in L(E, E)$ and $\beta \in L(E, E)$ (see Section **2.2**). Since $\psi = \varphi \circ (I - \varphi^{-1} \circ (\varphi - \psi))$, φ is invertible, and our norm assumption shows that $\| \varphi^{-1} \circ (\varphi - \psi)\| < 1$, it is sufficient to show that $I - \xi$ is invertible whenever $\| \xi \| < 1$. (I is the identity operator.) Consider the following sequence called the ***Neumann series***):

$$\begin{aligned}
\xi_0 &= I \,, \\
\xi_1 &= I + \xi \,, \\
\xi_2 &= I + \xi + \xi \circ \xi \,, \\
&\cdots \\
\xi_n &= I + \xi + \xi \circ \xi + \cdots + (\xi \circ \xi \circ \cdots \circ \xi) \,.
\end{aligned}$$

Using the triangle inequality and the foregoing norm inequality, we can compare this sequence to the sequence of real numbers, $1, 1 + \| \xi \|, 1 + \| \xi \| + \| \xi \|^2, \ldots,$ which we know is a Cauchy sequence since $\| \xi \| < 1$. Because $L(E, E)$ is complete, ξ_n must converge. The limit, say ρ, is the inverse of $I - \xi$. Indeed $(I - \xi)\xi_n = I - (\xi \circ \xi \circ \cdots \circ \xi)$, so the result follows. ■

2.5.5 Lemma *Let* $I : \mathbf{GL}(E, F) \to \mathbf{GL}(F, E)$ *be given by* $\varphi \mapsto \varphi^{-1}$. *Then* I *is of class* C^∞ *and* $\mathbf{D}I(\varphi) \cdot \psi = -\varphi^{-1} \circ \psi \circ \varphi^{-1}$. (For $\mathbf{D}^r I$, see Supplement **2.5E**.)

Proof We may assume $\mathbf{GL}(E, F) \neq \varnothing$. If we can show that $\mathbf{D}I(\varphi)\cdot\psi = -\varphi^{-1} \circ \psi \circ \varphi^{-1}$, then it will follow from Leibniz' rule that I is of class C^∞. Indeed $\mathbf{D}I = B(I, I)$ where $B \in L^2(L(F,E); L(L(E, F), L(F, E)))$ is defined by $B(\psi_1, \psi_2)(A) = -\psi_1 \circ A \circ \psi_2$, where $\psi_1, \psi_2 \in L(F, E)$ and $A \in L(E, F)$, which shows inductively that if I is C^k then it is C^{k+1}. Since the mapping $\psi \mapsto -\varphi^{-1} \circ \psi \circ \varphi^{-1}$ is linear ($\psi \in L(E, F)$), we must show that

$$\lim_{\psi \to \varphi} \frac{\| \psi^{-1} - (\varphi^{-1} - \varphi^{-1} \circ \psi \circ \varphi^{-1} + \varphi^{-1} \circ \varphi \circ \varphi^{-1}) \|}{\| \psi - \varphi \|} = 0.$$

Note that

$$\begin{aligned}
\psi^{-1} - (\varphi^{-1} - \varphi^{-1} \circ \psi \circ \varphi^{-1} + \varphi^{-1} \circ \varphi \circ \varphi^{-1}) &= \psi^{-1} - 2\varphi^{-1} + \varphi^{-1} \circ \psi \circ \varphi^{-1} \\
&= \psi^{-1} \circ (\psi - \varphi) \circ \varphi^{-1} \circ (\psi - \varphi) \circ \varphi^{-1} \,.
\end{aligned}$$

Again, using $\| \beta \circ \alpha \| \leq \| \alpha \| \, \| \beta \|$ for $\alpha \in L(E, F)$ and $\beta \in L(F, G)$, we get

$$\| \psi^{-1} \circ (\psi - \varphi) \circ \varphi^{-1} \circ (\psi - \varphi) \circ \varphi^{-1} \| \leq \| \psi^{-1} \| \, \| \psi - \varphi \|^2 \, \| \varphi^{-1} \|^2 .$$

With this inequality, the limit is clearly zero. ■

To prove the Inverse Mapping Theorem, we first note that it is enough to prove it under the simplifying assumptions $x_0 = 0$, $f(x_0) = 0$, $E = F$, and $\mathbf{D}f(0)$ is the identity. Indeed, replace f by $h(x) = \mathbf{D}f(x_0)^{-1} \circ [f(x + x_0) - f(x_0)]$.

Now let $g(x) = x - f(x)$ so $\mathbf{D}g(0) = 0$. Choose $r > 0$ so that $\| x \| \leq r$ implies $\| \mathbf{D}g(x) \| \leq 1/2$, which is possible by continuity of $\mathbf{D}g$. Thus by the Mean Value Inequality, $\| x \| \leq r$ implies $\| g(x) \| \leq r/2$. Let $B_\varepsilon(0) = \{x \in E \mid \| x \| \leq \varepsilon\}$. For $y \in B_{r/2}(0)$, let $g_y(x) = y + x$. If $y \in B_{r/2}(0)$ and $x_1, x_2 \in B_r(0)$, then $\| y \| \leq r/2$ and $\| g(x) \| \leq r/2$, so

(i) $\| g_y(x) \| \leq \| y \| + \| g(x) \| \leq r$,

and, by the Mean Value Inequality,

(ii) $\| g_y(x_1) - g_y(x_2) \| \leq \| x_1 - x_2 \|/2$.

Thus by Lemma **2.5.3**, g_y has a unique fixed point x in $B_r(0)$. This point x is the unique solution of $f(x) = y$. Thus f has an inverse

$$f^{-1} : V_0 = D_{r/2}(0) \to U_0 = f^{-1}(D_{r/2}(0)) \subset D_r(0) .$$

From (ii) with $y = 0$, we have $\| (x_1 - f(x_1)) - (x_2 - f(x_2)) \| \leq \| x_1 - x_2 \|/2$, and so

$$\| x_1 - x_2 \| - \| f(x_1) - f(x_2) \| \leq \frac{\| x_1 - x_2 \|}{2} ,$$

i.e.,

$$\| x_1 - x_2 \| \leq 2\| f(x_1) - f(x_2) \| .$$

Thus we have

(iii) $\| f^{-1}(y_1) - f^{-1}(y_2) \| \leq 2 \| y_1 - y_2 \|$,

so f^{-1} is continuous.

From Lemma **2.5.4** we can choose r small enough so that $\mathbf{D}f(x)^{-1}$ exists for $x \in D_r(0)$. Moreover, by continuity, $\| \mathbf{D}f(x)^{-1} \| \leq M$ for some M and all $x \in D_r(0)$ can be assumed as well. If $y_1, y_2 \in D_{r/2}(0)$, $x_1 = f^{-1}(y_1)$, and $x_2 = f^{-1}(y_2)$, then

$$\begin{aligned} \| f^{-1}(y_1) - f^{-1}(y_2) - \mathbf{D}f(x_2)^{-1} \cdot (y_1 - y_2) \| &= \| x_1 - x_2 - \mathbf{D}f(x_2)^{-1} \cdot [f(x_1) - f(x_2)] \| \\ &= \| \mathbf{D}f(x_2)^{-1} \cdot \{\mathbf{D}f(x_2) \cdot (x_1 - x_2) - f(x_1) + f(x_2)\} \| \\ &\leq M\| f(x_1) - f(x_2) - \mathbf{D}f(x_2) \cdot (x_1 - x_2) \| . \end{aligned}$$

This, together with (iii), shows that f^{-1} is differentiable with derivative $\mathbf{D}f(x)^{-1}$ at $f(x)$; i.e.,

$D(f^{-1}) = I \circ Df \circ f^{-1}$ on $V_0 = D_{r/2}(0)$. This formula, the chain rule, and Lemma **2.5.5** show inductively that if f^{-1} is C^{k-1} then f^{-1} is C^k for $1 \le k \le r$. ■

This argument also proves the following: *if* $f: U \to V$ *is a* C^r *homeomorphism where* $U \subset E$ *and* $V \subset F$ *are open sets, and* $Df(u) \in GL(E, F)$ *for* $u \in U$, *then* f *is a* C^r *diffeomorphism.*

For a Lipschitz Inverse Function Theorem see Exercise **2.5K.**

☞ SUPPLEMENT 2.5A
The Size of the Neighborhoods in the Inverse Mapping Theorem

An analysis of the preceeding proof also gives *explicit estimates* on the size of the ball on which $f(x) = y$ is solvable. Such estimates are sometimes useful in applications. The easiest one to use in examples involves estimates on the second derivative (we thank M. Buchner for suggestions about this).

2.5.6 Proposition *Suppose* $f: U \subset E \to F$ *is of class* C^r, $r \ge 2$, $x_0 \in U$ *and* $Df(x_0)$ *is an isomorphism. Let*

$$L = \| Df(x_0) \| \quad \text{and} \quad M = \| Df(x_0)^{-1} \| .$$

Assume

$$\| D^2f(x) \| \le K \quad \text{for} \quad \| x - x_0 \| \le R \quad \text{and} \quad B_R(x_0) \subset U .$$

Let

$$P = \min\left(\frac{1}{2KM}, R\right), \quad Q = \min\left(\frac{1}{2NL}, \frac{P}{M}, P\right), \quad S = \min\left(\frac{1}{2KM}, \frac{Q}{2L}, Q\right).$$

Here $N = 8M^3K$ *Then* f *maps an open set* $G \subset D_P(x_0)$ *diffeomorphically onto* $D_{P/2M}(y_0)$ *and* f^{-1} *maps an open set* $H \subset D_Q(y_0)$ *diffeomorphically onto* $D_{Q/2L}(x_0)$. *Moreover,* $B_{Q/2L}(x_0) \subset G \subset D_P(x_0)$ *and* $B_{S/2M}(y_0) \subset H \subset D_Q(y_0) \subset D_{P/2M}(y_0)$. *See Figure 2.5.1.*

Proof We can assume $x_0 = 0$ and $f(x_0) = 0$. From

$$Df(x) = Df(0) + \int_0^1 D(Df(tx)) \cdot x \, dt = Df(0) \cdot \left\{ I + [Df(0)]^{-1} \cdot \int_0^1 D^2f(tx) \cdot x \, dt \right\}$$

and the fact that

$$\|(I + A)^{-1}\| \le 1 + \| A \| + \| A \|^2 + \cdots = \frac{1}{1 - \| A \|}$$

for $\| A \| < 1$ (see the proof of **2.5.5**), we get

$$\| \mathbf{D}f(x)^{-1} \| \leq 2M \quad \text{if} \quad \| x \| \leq R \quad \text{and} \quad \| x \| \leq \frac{1}{2MK} \, ,$$

i.e., if $\| x \| \leq P$.

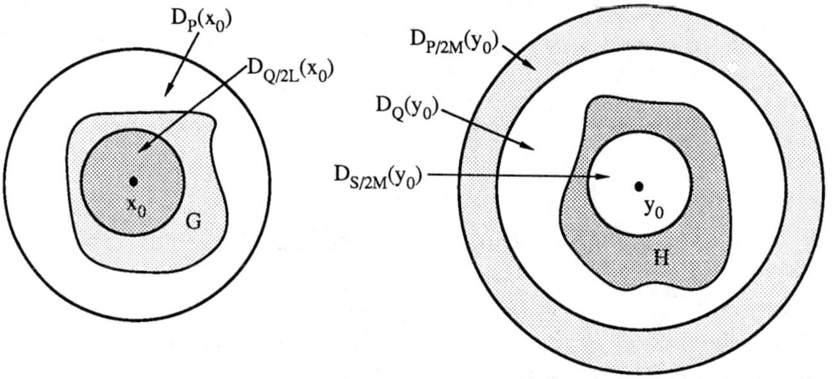

Figure 2.5.1

As in the proof of the Inverse Function Theorem , let $g_y(x) = [\mathbf{D}f(0)]^{-1} \cdot (y + \mathbf{D}f(0)x - f(x))$. Write

$$\varphi(x) = \mathbf{D}f(0) \cdot x - f(x) = \int_0^1 \mathbf{D}\varphi(sx) \cdot x \, ds = -\int_0^1 \int_0^1 \mathbf{D}^2 f(tsx) \cdot (sx, x) \, dt \, ds$$

to obtain $g_y(x) = [\mathbf{D}f(0)^{-1}] \cdot (y + \varphi(x))$, $\| \varphi(x) \| \leq K \| x \|^2$ if $\| x \| \leq P$, and

$$\| g_y(x) \| \leq M(\| y \| + K \| x \|^2) \, .$$

Hence for $\| y \| \leq P/2M$, g_y maps $B_P(0)$ to $B_S(0)$. Similarly we get $\| g_y(x_1) - g_y(x_2) \| \leq \| x_1 - x_2 \| /2$ from the Mean Value Inequality and the estimate

$$\| \mathbf{D}g_y(x) \| = \| \mathbf{D}f(0)^{-1} \| \left(\left\| \int_0^1 \mathbf{D}^2 f(tx)x \, dt \right\| \right) \leq M(K \| x \|) \leq \frac{1}{2},$$

if $\| x \| \leq P$. Thus, as in the previous proof, $f^{-1} : B_{P/2M}(0) \to B_P(0)$ is defined and there exists an open set $G \subset B_P(0)$ diffeomorphic via f to the open ball $D_{P/2M}(0)$.

Taking the second derivative of the relation $f^{-1} \circ f = \text{identity}$ on G, we get

$$\mathbf{D}^2 f^{-1}(f(x))(\mathbf{D}f(x) \cdot u_1, \mathbf{D}f(x) \cdot u_2) + \mathbf{D}f^{-1}(f(x)) \cdot \mathbf{D}^2 f(x)(u_1, u_2) = 0$$

for any $u_1, u_2 \in E$. Let $v_i = \mathbf{D}f(x) \cdot u_i$, $i = 1, 2$, so that

$$\mathbf{D}^2 f^{-1}(f(x)) \cdot (v_1, v_2) = -\mathbf{D}f^{-1}(f(x)) \cdot \mathbf{D}^2 f(x)(\mathbf{D}f(x)^{-1} \cdot v_1, \mathbf{D}f(x)^{-1} \cdot v_2)$$

and hence

$$\| D^2f^{-1}(f(x))\,(v_1, v_2) \| \le \| Df^{-1}(f(x)) \|^3 \| D^2f(x) \| \| v_1 \| \| v_2 \| \le 8M^3K \| v_1 \| \| v_2 \|$$

since $x \in G \subset D_P(0)$ and on $B_P(0)$ we have the inequality $\| Df(x)^{-1} \| \le 2M$. Thus on $B_{P/2M}(0)$ the following estimate holds:

$$\| D^2f^{-1}(y) \| \le 8M^3K\ .$$

By the previous argument with f replaced by f^{-1}, R by P/2M, L by M, and K by $N = 8M^3K$, it follows that there is an open set $H \subset D_Q(0)$, $Q = \min\{1/2KM, Q/2L, Q\}$ such that $f^{-1} : H \to D_{Q/2L}(0)$ is a diffeomorphism. Since f^{-1} is a diffeomorphism on $B_Q(0)$ and H is one of its open subsets, it follows that $B_{Q/2L}(0) \subset G$.

Finally, replacing R by Q/2L, we conclude the existence of a ball $B_{S/2M}(0)$, where $S = \min\{1/2KM, Q/2L, Q\}$, on which f^{-1} is a diffeomorphism. Therefore $B_{S/2M}(0) \subset H$. ■ ✍

In the study of manifolds and submanifolds, the argument used in the following is of central importance.

2.5.7 Implicit Function Theorem *Let* $U \subset E$, $V \subset F$ *be open and* $f : U \times V \to G$ *be* C^r, $r \ge 1$. *For some* $x_0 \in U$, $y_0 \in V$ *assume* $D_2f(x_0, y_0) : F \to G$ *is an isomorphism. Then there are neighborhoods* U_0 *of* x_0 *and* W_0 *of* $f(x_0, y_0)$ *and a unique* C^r *map* $g : U_0 \times W_0 \to V$ *such that for all* $(x, w) \in U_0 \times W_0$,

$$f(x, g(x, w)) = w\ .$$

Proof Define the map $\Phi : U \times V \to E \times G$ by $(x, y) \mapsto (x, f(x, y))$. Then $D\Phi(x_0, y_0)$ is given by

$$D\Phi(x_0, y_0) \cdot (x_1, y_1) = \begin{pmatrix} I & 0 \\ D_1f(x_0, y_0) & D_2f(x_0, y_0) \end{pmatrix} \begin{pmatrix} x_1 \\ y_1 \end{pmatrix}$$

which is an isomorphism of $E \times F$ with $E \times G$. Thus Φ has a unique C^r local inverse, say $\Phi^{-1} : U_0 \times W_0 \to U \times V$, $(x, w) \mapsto (x, g(x, w))$. The g so defined is the desired map. ■

Applying the chain rule to the relation $f(x, g(x, w))$, one can compute the derivatives of g:

$$D_1g(x, w) = -[D^2f(x, g(x, w))]^{-1} \circ D_1f(x, g(x, w))\ ,$$

$$D_2g(x, w) = [D_2f(x, g(x, w))]^{-1}\ .$$

2.5.8 Corollary *Let* $U \subset E$ *be open and* $f : U \to F$ be C^r, $r \ge 1$. *Suppose* $Df(x_0)$ *is surjective and* ker $Df(x_0)$ *is complemented. Then* $f(U)$ *contains a neighborhood of* $f(x_0)$.

Proof Let $E_1 = \ker Df(x_0)$ and $E = E_1 \oplus E_2$. Then $D_2f(x_0) : E_2 \to F$ is an isomorphism, so the hypotheses of **2.5.7** are satisfied and thus $f(U)$ contains W_0 provided by that theorem. ■

Since in finite-dimensional spaces every subspace splits, the foregoing corollary implies that if $f : U \subset \mathbb{R}^n \to \mathbb{R}^m$, $n \geq m$, and the Jacobian of f at every point of U has rank m, then f is an open mapping. This statement generalizes directly to Banach spaces, but it is *not* a consequence of the Implicit Function Theorem anymore, since not every subspace is split. This result goes back to Graves [1950]. The proof given in Supplement **2.5B** follows Luenberger [1969].

2.5.9 Local Surjectivity Theorem *If* $f : U \subset E \to F$ *is* C^1 *and* $\mathbf{D}f(u_0)$ *is onto for some* $u_0 \in U$, *then* f *is locally onto; i.e., there exist open neighborhoods* U_1 *of* u_0 *and* V_1 *of* $f(u_0)$ *such that* $f \mid U_1 : U_1 \to V_1$ *is onto. In particular, if* $\mathbf{D}f(u)$ *is onto for all* $u \in U$, *then* f *is an open mapping.*

☞ SUPPLEMENT 2.5B
Proof of the Local Surjectivity Theorem

Proof Recall from Section **2.1** that $E/\ker \mathbf{D}f(u_0) = E_0$ is a Banach space with norm $\|[x]\| = \inf\{\|x + u\| \mid u \in \ker \mathbf{D}f(u_0)\}$, where $[x]$ is the equivalence class of x. To solve $f(x) = y$ we set up an iteration scheme in E_0 and E simultaneously. Now $\mathbf{D}f(u_0)$ induces an isomorphism $T : E_0 \to F$, so $T^{-1} \in L(F, E_0)$ exists by the Banach Isomorphism Theorem. Let $x = u_0 + h$ and write $f(x) = y$ as

$$T^{-1}(y - f(u_0 + h)) = 0 .$$

To solve this equation, define a sequence $L_n \in E/\ker \mathbf{D}f(u_0)$ (so the element L_n is a coset of $\ker \mathbf{D}f(u_0)$) and $h_n \in L_n \subset E$ inductively by $L_0 = \ker \mathbf{D}f(u_0)$, $h_0 \in L_0$ small, and

$$L_n = L_{n-1} + T^{-1}(y - f(u_0 + h_{n-1})) , \tag{1}$$

and selecting $h_n \in L_n$ such that

$$\| h_n - h_{n-1} \| \leq 2\| L_n - L_{n-1} \| . \tag{2}$$

The latter is possible since $\| L_n - L_{n-1} \| = \inf\{\| h - h_{n-1} \| \mid h \in L_n\}$. Since $h_{n-1} \in L_{n-1}$, $L_{n-1} = T^{-1}(\mathbf{D}f(u_0) \cdot h_{n-1})$, so

$$L_n = T^{-1}(y - f(u_0 + h_{n-1}) + \mathbf{D}f(u_0) \cdot h_{n-1}) .$$

Subtracting this from the expression for L_{n-1} gives

$$L_n - L_{n-1} = -T^{-1}(f(u_0 + h_{n-1}) - f(u_0 + h_{n-2}) - \mathbf{D}f(u_0) \cdot (h_{n-1} - h_{n-2})) . \tag{3}$$

For $\varepsilon > 0$ given, there is a neighborhood U of u_0 such that $\| \mathbf{D}f(u) - \mathbf{D}f(u_0) \| < \varepsilon$ for $u \in U$, since f is C^1. Assume inductively that $u_0 + h_{n-1} \in U$ and $u_0 + h_{n-2} \in U$. Then from the Mean Value Inequality,

$$\| L_n - L_{n-1} \| \leq \varepsilon \| T^{-1} \| \| h_{n-1} - h_{n-2} \| . \tag{4}$$

By (2),

$$\| h_n - h_{n-1} \| \le 2\| L_n - L_{n-1} \| \le 2\varepsilon \| T^{-1} \| \| h_{n-1} - h_{n-2} \| .$$

Thus if ε is small,

$$\| h_n - h_{n-1} \| \le \tfrac{1}{2} \| h_{n-1} - h_{n-2} \|.$$

Starting with h_0 small and $\| h_1 - h_0 \| < \frac{1}{2} \| h_0 \|$, $u_0 + h_n$ remains inductively in U since

$$\begin{aligned} \| h_n \| &\le \| h_0 \| + \| h_1 - h_0 \| + \| h_2 - h_1 \| + \cdots + \| h_n - h_{n-1} \| \\ &\le \left(1 + \frac{1}{2} + \cdots + \frac{1}{2^{n-1}} \right) \| h_0 \| \le 2 \| h_0 \| . \end{aligned}$$

It follows that h_n is a Cauchy sequence, so it converges to some point, say h. Correspondingly, L_n converges to L and $h \in L$. Thus from (1), $0 = T^{-1}(y - f(u_0 + h))$ and so $y = f(u_0 + h)$. ■

This proves that for y near $y_0 = f(u_0)$, $f(x) = y$ has a solution. If there is a solution $g(y) = x$ which is C^1, then $\mathbf{D}f(x_0) \circ \mathbf{D}g(y_0) = I$ and so range $\mathbf{D}g(y_0)$ is an algebraic complement to ker $\mathbf{D}f(x_0)$. It follows that if range $\mathbf{D}g(y_0)$ is closed, then ker $\mathbf{D}f(x_0)$ is split.

In many applications to nonlinear partial differential equations, methods of functional analysis and elliptic operators can be used to show that ker $\mathbf{D}f(x_0)$ does split, even in Banach spaces. Such a splitting theorem is called the ***Fredholm alternative***. For illustrations of this idea in geometry and relativity, see Fischer and Marsden [1975], [1979], and in elasticity, see Marsden and Hughes [1983, Chapter 6]. For such applications, **2.5.8** usually sufficies.

The locally injective counterpart of this theorem is the following.

2.5.10 Local Injectivity Theorem *Let* $f : U \subset E \to F$ *be a* C^1 *map,* $\mathbf{D}f(u_0)(E)$ *be closed in* F, *and* $\mathbf{D}f(u_0) \in GL(E, \mathbf{D}f(u_0)(E))$. *Then there exists a neighborhood* V *of* u_0, $V \subset U$, *on which* f *is injective. The inverse* $f^{-1} : f(V) \to U$ *is Lipschitz continuous.*

Proof Since $(\mathbf{D}f(u_0))^{-1} \in L(\mathbf{D}f(u_0)(E), E)$, there is a constant $M > 0$ such that $\| \mathbf{D}f(u_0) \cdot e \| \ge M\| e \|$ for all $e \in E$. By continuity of $\mathbf{D}f$, there exists $r > 0$ such that $\| \mathbf{D}f(u) - \mathbf{D}f(u_0) \| < M/2$ whenever $\| u - u_0 \| < 3r$. By the Mean Value Inequality, for $e_1, e_2 \in D_r(u_0)$

$$\begin{aligned} \| f(e_1) - f(e_2) - \mathbf{D}f(u_0)(e_1 - e_2) \| &\le \sup_{t \in [0, 1]} \| \mathbf{D}f(e_1 + t(e_2 - e_1)) - \mathbf{D}f(u_0) \| \| e_1 - e_2 \| \\ &\le \frac{M\| e_1 - e_2 \|}{2} \end{aligned}$$

since $\| u_0 - e_1 - t(e_2 - e_1) \| < 3r$. Thus

$$M\| e_1 - e_2 \| \le \| \mathbf{D}f(u_0) \cdot (e_1 - e_2) \| \le \| f(e_1) - f(e_2) \| + \frac{M}{2} \| e_1 - e_2 \| ;$$

i.e.,

$$\frac{M}{2} \| e_1 - e_2 \| \le \| f(e_1) - f(e_2) \|,$$

which proves that f is injective on $D_r(u_0)$ and that $f^{-1} : f(D_r(u_0)) \to U$ is Lipschitz continuous. ■

We now give an example of the use of the Implicit Function Theorem to prove an existence theorem for differential equations. For this and related examples, we choose the spaces to be infinite dimensional. In fact, E, F, G, ··· will be suitable spaces of functions. The map f will often be a nonlinear differential operator. The linear map $\mathbf{D}f(x_0)$ is called the ***linearization*** of f about x_0. (Phrases like "first variation," "first-order deformation," and so forth are also used.)

2.5.11 Example Let E be the space of all C^1-functions $f : [0, 1] \to \mathbb{R}$ with the norm

$$\| f \|_1 = \sup_{x \in [0,1]} | f(x) | + \sup_{x \in [0,1]} \left| \frac{df(x)}{dx} \right|$$

and F the space of all C^0-functions with the norm $\| f \|_0 = \sup_{x \in [0, 1]} | f(x) |$. These are Banach spaces (see Exercise **2.1C**). Let $\Phi : E \to F$ be defined by $\Phi(f) = df/dx + f^3$. It is easy to check that Φ is C^∞ and $\mathbf{D}\Phi(0) = d/dx : E \to F$. Clearly $\mathbf{D}\Phi(0)$ is surjective (Fundamental Theorem of Calculus). Also ker $\mathbf{D}\Phi(0)$ consists of E_1 = all constant functions. This is complemented because it is finite dimensional; explicitly, a complement consists of functions with zero integral. Thus Corollary **2.5.8** yields the following:

There is an $\varepsilon > 0$ *such that if* $g : [0, 1] \to \mathbb{R}$ *is a continuous function with* $| g(x) | < \varepsilon$, *then there is a* C^1 *function* $f: [0, 1] \to \mathbb{R}$ *such that*

$$\frac{df}{dx} + f^3(x) = g(x). \quad ♦$$

☞ SUPPLEMENT 2.5C

An Application of the Inverse Function Theorem to a Nonlinear Partial Differential Equation

Let $\Omega \subset \mathbb{R}^n$ be a bounded open set with smooth boundary. Consider the problem

$$\nabla^2\varphi + \varphi^3 = f \text{ in } \Omega , \quad \varphi + \varphi^7 = g \text{ on } \partial\Omega$$

for given f and g. We claim that for f and g small, this problem has a unique small solution. For partial differential equations of this sort one can use the Sobolev spaces $H^s(\Omega, \mathbb{R})$ consisting of maps $\varphi : \Omega \to \mathbb{R}$ whose first s distributional derivatives lie in L^2. (One uses Fourier transforms to define this space if s is not an integer.) In the Sobolev spaces $E = H^s(\Omega, \mathbb{R})$, $F = H^{s-2}(\Omega, \mathbb{R}) \times H^{s-1/2}(\partial\Omega, \mathbb{R})$, if $s > n/2$ the map

$$\Phi : E \to F, \ \varphi \mapsto (\nabla^2\varphi + \varphi^3, (\varphi + \varphi^7)|\partial\Omega)$$

is C^∞ (use Supplement **2.4B**) and the linear operator

$$\mathbf{D}\Phi(0) \cdot \varphi = (\nabla^2\varphi, \varphi|\partial\Omega)$$

is an isomorphism. The fact that $\mathbf{D}\Phi(0)$ is an isomorphism is a result on the solvability of the Dirichlet problem from the theory of elliptic linear partial differential equations. See, for example, Friedman [1969]. (In the C^k spaces, $\mathbf{D}\Phi(0)$ is *not* an isomorphism.) The result claimed above now follows from the Inverse Function Theorem. ✍

The following series of consequences of the Inverse Function Theorem are important technical tools in the study of manifolds. The first two results give, roughly speaking, sufficient conditions to "straighten out" the range (respectively, the domain) of f in a neighborhood of a point, thus making f look like an inclusion (respectively, a projection).

2.5.12 Local Immersion Theorem *Let* $f : U \subset E \to F$ *be of class* C^r, $r \geq 1$, $u_0 \in U$ *and suppose that* $\mathbf{D}f(u_0)$ *is one to one and has a closed split image* F_1 *with closed complement* F_2. *(If* $E = \mathbb{R}^m$ and $F = \mathbb{R}^n$, *assume only that* $\mathbf{D}f(u_0)$ *has trivial kernel.) Then there are two open sets* $U' \subset F$ *and* $V \subset E \oplus F_2$ *where* $f(u_0) \in U'$ *and a* C^r *diffeomorphism* $\varphi : U' \to V$ *such that* $(\varphi \circ f)(e) = (e, 0)$ *for all* $e \in V \cap (E \times \{0\}) \subset E$.

The intuition for $E = F_1 = \mathbb{R}^2$, $F_2 = \mathbb{R}$ (i.e., $m = 2$, $n = 3$) is given in Figure 2.5.2. The function φ flattens out the image of f. Notice that this is intuitively correct; we expect the range of f to be an m-dimensional "surface" so it should be possible to flatten it to a piece of $\mathbb{R}^m$. Note that the range of a linear map of rank m is a linear subspace of dimension exactly m, so this result expresses, in a sense, a generalization of the linear case. Also note that **2.5.10**, the Local Injectivity Theorem, follows from the more restrictive hypotheses of **2.5.12**

Proof Define $g : U \times F_2 \subset E \times F_2 \to F = F_1 \oplus F_2$ by $g(u, v) = f(u) + (0, v)$ and note that $g(u, 0) = f(u)$. Now $\mathbf{D}g(u_0, 0) = (\mathbf{D}f(u_0), I_{F_2}) \in GL(E \oplus F_2, F)$ by the Banach Isomorphism Theorem (I_{F_2} denotes the identity mapping of F_2 and $(A, B) \in L(E \oplus E', F \oplus F')$ for $A \in L(E, F)$, and $B \in L(E', F')$ is defined by $(A, B)(e, e') = (Ae, Be')$.) By the Inverse Function Theorem there exist open sets U' and V such that $(u_0, 0) \in V \subset E \oplus F_2$, and $g(u_0, 0) = f(u_0) \in U' \subset F$

and a C^r diffeomorphism $\varphi : U' \to V$ such that $\varphi^{-1} = g \mid V$. Hence for $(e, 0) \in V$, $(\varphi \circ f)(e) = (\varphi \circ g)(e, 0) = (e, 0)$. ■

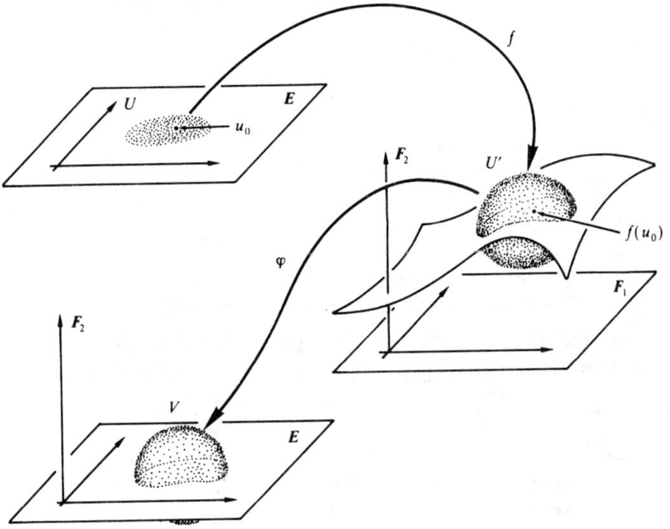

Figure 2.5.2

2.5.13 Local Submersion Theorem *Let* $f : U \subset E \to F$ *be of class* C^r, $r \geq 1$, $u_0 \in U$ *and suppose* $\mathbf{D}f(u_0)$ *is surjective and has split kernel* E_2 *with closed complement* E_1. *(If* $E = \mathbb{R}^m$ *and* $F = \mathbb{R}^n$, *assume only that* rank $(\mathbf{D}f(u_0)) = n$.) *Then there are open sets* U' *and* V *such that* $u_0 \in U' \subset U \subset E$ *and* $V \subset F \oplus E_2$ *and a* C^r *diffeomorphism* $\psi : V \to U'$ *with the property that* $(f \circ \psi)(u, v) = u$ *for all* $(u, v) \in V$.

The intuition for $E_1 = E_2 = F = \mathbb{R}$ is given in Figure 2.5.3, which should be compared to Figure 2.5.2. Note that this theorem implies the results of **2.5.9**, the Local Surjectivity Theorem, but the hypotheses are more stringent.

Proof By the Banach Isomorphism Theorem (Section **2.2**), $\mathbf{D}_1 f(u_0) \in GL(E_1, F)$. Define the map $g : U \subset E_1 \oplus E_2 \to F \oplus E_2$ by $g(u_1, u_2) = (f(u_1, u_2), u_2)$ and note that

$$\mathbf{D}g(u_0)\cdot(e_1, e_2) = \begin{bmatrix} \mathbf{D}_1 f(u_0) & \mathbf{D}_2 f(u_0) \\ 0 & I_{E_2} \end{bmatrix} \begin{bmatrix} e_1 \\ e_2 \end{bmatrix}$$

so that $\mathbf{D}g(u_0) \in GL(E, F \oplus E_2)$. By the Inverse Function Theorem there are open sets U' and V such that $u_0 \in U' \subset U \subset E$, $V \subset F \oplus E_2$ and a C^r diffeomorphism $\psi : V \to U'$ such that $\psi^{-1} = g \mid U'$. Hence if $(u, v) \in V$, $(u, v) = (g \circ \psi)(u, v) = (f(\psi(u, v)), \psi_2(u, v))$, where

$\psi = \psi_1 \times \psi_2$; i.e., $\psi_2(u, v) = v$ and $(f \circ \psi)(u, v) = u$. ■

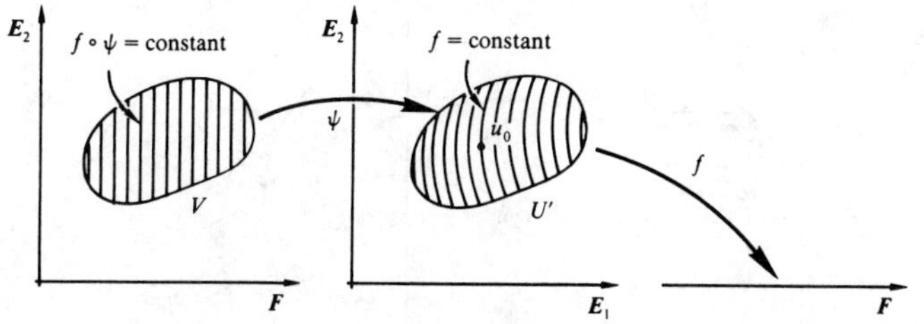

Figure 2.5.3

2.5.14 Local Representation Theorem *Let* $f : U \subset E \to F$ *be of class* C^r, $r > 1$, $u_0 \in U$ *and suppose* $\mathbf{D}f(u_0)$ *has closed split image* F_1 *with closed complement* F_2 *and split kernel* E_2 *with closed complement* E_1. *(If* $E = \mathbb{R}^m$, $F = \mathbb{R}^n$, *assume that* rank $(\mathbf{D}f(u_0)) = k$, $k \le n$, $k \le m$, *so that* $F_2 = \mathbb{R}^{n-k}$, $F_1 = \mathbb{R}^k$, $E_1 = \mathbb{R}^k$, $E_2 = \mathbb{R}^{m-k}$.) *Then there are open sets* U' *and* V *with the property that* $u_0 \in U' \subset U \subset E$, $V \subset F_1 \oplus E_2$ *and a* C^r *diffeomorphism* $\psi : V \to U'$ *such that* $(f \circ \psi)(u, v) = (u, \eta(u, v))$, *where* $\eta : V \to F_2$ *is a* C^r *map satisfying* $\mathbf{D}\eta(u_0) = 0$.

Proof Write $f = f_1 \times f_2$, where $f_i : U \to F_i$, $i = 1, 2$. Then f satisfies the conditions of **2.5.13**, and thus there exists a C^r diffeomorphism $\psi : V \subset F_1 \oplus E_2 \to U' \subset E$ such that the composition $f_1 \circ \psi$ is given by $(f_1 \circ \psi)(u, v) = u$. Let $\eta = f_2 \circ \psi$. ■

To use **2.5.12** (or **2.5.13**) in finite dimensions, we must have the rank of $\mathbf{D}f$ equal to the dimension of its domain space (or the range space). However, we can also use the Inverse Function Theorem to tell us that if $\mathbf{D}f(x)$ has *constant rank* k in a neighborhood of x_0, then we can straighten out the domain of f with some invertible function ψ such that $f \circ \psi$ depends only on k variables. Then we can apply Proposition **2.5.12**. This is the essence of the following theorem. Roughly speaking, the theorem says that if $\mathbf{D}f$ has rank k on $\mathbb{R}^m$, then $m - k$ variables are redundant and can be eliminated. As a trivial example, if $f : \mathbb{R}^2 \to \mathbb{R}$ is defined by setting $f(x, y) = x - y$, $\mathbf{D}f$ has rank 1, and so we can express f using just one variable, namely, let $\psi(x, y) = (x + y, y)$ so that $(f \circ \psi)(x, y) = x$, which depends only on x.

2.5.15 Rank Theorem *Let* $f : U \subset E \to F$ *be of class* C^r, $r \ge 1$, $u_0 \in U$ *and suppose* $\mathbf{D}f(u_0)$ *has closed split image* F_1 *with closed complement* F_2 *and split kernel* E_2 *with closed complement* E_1. *In addition, assume that for all* u *in a neighborhood of* $u_0 \in U$, $\mathbf{D}f(u)(E)$ *is a closed subspace of* F *and* $\mathbf{D}f(u) \mid E_1 : E_1 \to \mathbf{D}f(u)(E)$ *is a Banach space isomorphism. (In case* $E = \mathbb{R}^m$ *and* $F = \mathbb{R}^n$, *assume only that* rank $(\mathbf{D}f(u)) = k$ *for* u *in a neighborhood of* u_0.) *Then there exist open sets* $U_1 \subset F_1 \oplus E_2$, $U_2 \subset E$, $V_1 \subset F$, *and* $V_2 \subset F$ *and there are* C^r

diffeomorphisms $\varphi : V_1 \to V_2$ *and* $\psi : U_1 \to U_2$ *such that* $(\varphi \circ f \circ \psi)(x, e) = (x, 0)$.

The intuition is given by Figure 2.5.4 for $E = \mathbb{R}^2$, $F = \mathbb{R}^2$ and $k = 1$.

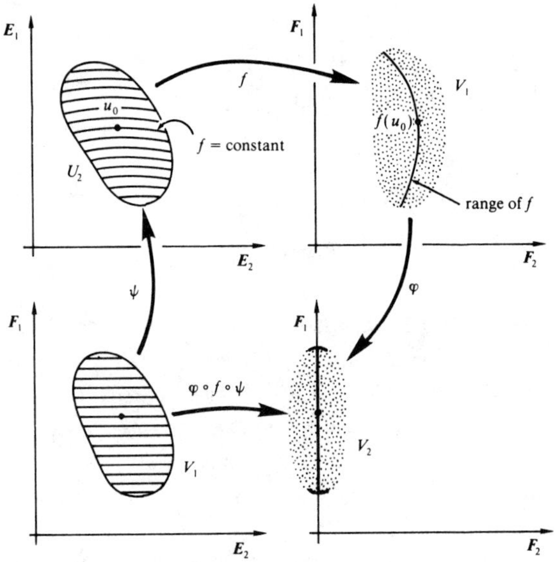

Figure 2.5.4

Remark It is clear that the theorem implies $E_1 \oplus \ker(\mathbf{D}f(u)) = E$ and $\mathbf{D}f(u)(E) \oplus F_2 = F$ for u in a neighborhood of u_0 in U, because $\varphi \circ f \circ \psi$ has these properties. These seemingly stronger conditions can in fact be shown directly to be equivalent to the hypotheses in the theorem by the use of the openness of $GL(E, E)$ in $L(E, E)$.

Proof By the Local Representation Theorem there is a C^r diffeomorphism $\psi : U_1 \subset F_1 \oplus E_2 \to U_2 \subset E$ such that $f(x, y) := (f \circ \psi)(x, y) = (x, \eta(x, y))$. Let $P_1 : F \to F_1$ be the projection. Since $\mathbf{D}f(x, y) \cdot (w, e) = (w, \mathbf{D}\eta(x, y) \cdot (w, e))$, it follows that $P_1 \circ \mathbf{D}f(x, y))(w, e) = (w, 0)$, for $w \in F_1$ and $e \in E_2$. In particular $P_1 \circ \mathbf{D}f(x, y) \mid F_1 \times \{0\} = I_1$, the identity on F_1, which shows that $\mathbf{D}f(x, y) \mid F_1 \times \{0\} : F_1 \times \{0\} \to \mathbf{D}f(x, y)(F_1 \oplus E_2)$ is injective. In finite dimensions this implies that it is an isomorphism, since $\dim(F_1) = \dim(\mathbf{D}f(x, y)(F_1 \oplus E_2))$. In infinite dimensions this is our hypothesis. Thus we get $\mathbf{D}f(x, y) \circ P_1 \mid \mathbf{D}f(x, y)(F_1 \oplus E_2) = \text{identity}$. Let $(w, \mathbf{D}\eta(x, y)(w, e)) \in \mathbf{D}f(x, y)(F_1 \oplus E_2)$. Since

$$
\begin{aligned}
(\mathbf{D}f(x, y) \circ P_1)(w, \mathbf{D}\eta(x, y) \cdot (w, e)) &= \mathbf{D}f(x, y) \cdot (w, 0) \\
&= (w, \mathbf{D}\eta(x, y) \cdot (w, 0)) = (w, \mathbf{D}_1\eta(x, y) \cdot w) \ ,
\end{aligned}
$$

we must have $\mathbf{D}\eta(x, y) \cdot e = 0$ for all $e \in E_2$; i.e., $\mathbf{D}_2\eta(x, y) = 0$. However, $\mathbf{D}_2 f(x, y) \cdot e = (0, \mathbf{D}_2\eta(x, y) \cdot e)$, which says that $\mathbf{D}_2 f(x, y) = 0$; i.e., f does not depend on the variable $y \in E_2$. Define $\mathcal{f}(x) = f(x, y) = (f \circ \psi)(x, y)$, so $\mathcal{f} : P_1'(V) \subset F_1 \to F$ where $P_1' : F_1 \oplus E_2 \to F_1$ is the projection. Now $\mathcal{f}$ satisfies the conditions of **2.5.12** at $P_1'(\psi^{-1}(u_0))$ and hence there exists a C^r diffeomorphism $\varphi : V_1 \to V_2$, where $V_1, V_2 \subset F$, such that $(\varphi \circ \mathcal{f})(z) = (z, 0)$; that is, we have $(\varphi \circ f \circ \psi)(x, y) = (x, 0)$. ■

2.5.16 Example Functional Dependence Let $U \subset \mathbb{R}^n$ be an open set and let the functions $f_1, ..., f_n : U \to \mathbb{R}$ be smooth. The functions $f_1, ..., f_n$ are said to be ***functionally dependent*** at $x_0 \in U$ if there is a neighborhood V of the point $(f_1(x_0), ..., f_n(x_0)) \in \mathbb{R}^n$ and a smooth function $F : V \to \mathbb{R}$ such that $\mathbf{D}F \neq 0$ on a neighborhood of $(f_1(x_0), ..., f_n(x_0))$, and

$$F(f_1(x), ..., f_n(x)) = 0$$

for all x in some neighborhood of x_0. Show:

(i) *If* $f_1, ..., f_n$ *are functionally dependent at* x_0, *then the determinant of* $\mathbf{D}f$, *denoted*

$$JF = \frac{\partial(f_1, \ldots, f_n)}{\partial(x_1, \ldots, x_n)} ,$$

vanishes at x_0.

(ii) *If*

$$\frac{\partial(f_1, ..., f_{n-1})}{\partial(x_1, ..., x_{n-1})} \neq 0 \quad \textit{and} \quad \frac{\partial(f_1, ..., f_n)}{\partial(x_1, ..., x_n)} = 0$$

on a neighborhood of x_0, *then* $f_1, ..., f_n$ *are functionally dependent, and* $f_n = G(f_1, ..., f_{n-1})$ *for some* G.

Solution Let $f = (f_1, ..., f_n)$.

(i) We have $F \circ f = 0$, so $\mathbf{D}F(f(x)) \circ \mathbf{D}f(x) = 0$. Now if $Jf(x_0) \neq 0$, $\mathbf{D}f(x)$ would be invertible in a neighborhood of x_0, implying $\mathbf{D}F(f(x)) = 0$. By the Inverse Function Theorem, this implies $\mathbf{D}F(y) = 0$ on a whole neighborhood of $f(x_0)$.

(ii) The conditions of (ii) imply that $\mathbf{D}f$ has rank $n - 1$. Hence by the Rank Theorem, there are mappings φ and ψ such that

$$(\varphi \circ f \circ \psi)(x_1, ..., x_n) = (x_1, ..., x_{n-1}, 0) .$$

Let F be the last component of φ. Then $F(f_1, ..., f_n) = 0$. Since φ is invertible, $\mathbf{D}F \neq 0$.

It follows from the Implicit Function Theorem that we can locally solve $F(f_1, ..., f_n) = 0$ for $f_n = G(f_1, ..., f_{n-1})$, provided we can show $\Delta = \partial F/\partial y_n \neq 0$. As we saw before, $\mathbf{D}F(f(x)) \circ \mathbf{D}f(x) = 0$, or, in components with $y = f(x)$,

$$\left(\frac{\partial F}{\partial y_1} \cdots \frac{\partial F}{\partial y_n}\right)\begin{bmatrix} \frac{\partial f_1}{\partial x_1} & \cdots & \frac{\partial f_1}{\partial x_n} \\ \vdots & & \vdots \\ \frac{\partial f_n}{\partial x_1} & \cdots & \frac{\partial f_n}{\partial x_n} \end{bmatrix} = (0, 0, \ldots, 0).$$

If $\partial F/\partial y_n = 0$, we would have

$$\left(\frac{\partial F}{\partial y_1}, \ldots, \frac{\partial F}{\partial y_{n-1}}\right)\begin{bmatrix} \frac{\partial f_1}{\partial x_1} & \cdots & \frac{\partial f_1}{\partial x_{n-1}} \\ \vdots & & \vdots \\ \frac{\partial f_{n-1}}{\partial x_1} & \cdots & \frac{\partial f_{n-1}}{\partial x_{n-1}} \end{bmatrix} = (0, 0, \ldots, 0)$$

or

$$\left(\frac{\partial F}{\partial y_1}, \ldots, \frac{\partial F}{\partial y_{n-1}}\right) = (0, 0, \ldots, 0)$$

since the square matrix is invertible by the assumption that

$$\frac{\partial(f_1, \ldots, f_{n-1})}{\partial(x_1, \ldots, x_{n-1})} \neq 0.$$

This implies $\mathbf{D}f = 0$, which is not true. Hence $\partial F/\partial y_n \neq 0$, and we have the desired result.

The reader should note the analogy between linear dependence and functional dependence, where rank or determinant conditions are replaced by the analogous conditions on the Jacobian matrix. ♦

☞ SUPPLEMENT 2.5D
The Hadamard-Levy Theorem

This supplement gives sufficient conditions which together with the hypotheses of the Inverse Function Theorem guarantee that a C^k map f between Banach spaces is a *global* diffeomorphism. To get a feel for these supplementary conditions, consider a C^k function $f : \mathbb{R} \to \mathbb{R}$, $k \geq 1$, satisfying $1/|f'(x)| < M$ for all $x \in \mathbb{R}$. Then f is a local diffeomorphism at every point of $\mathbb{R}$ and thus is an open map. In particular, $f(\mathbb{R})$ is an open interval $]a, b[$. The condition $|f'(x)| > 1/M$ implies that f is either strictly increasing or strictly decreasing. Let us assume that f is strictly increasing. If $b < +\infty$, then the line $y = b$ is a horizontal asymptote of the graph of f and therefore we should have $\lim_{x \to \infty} f'(x) = 0$ contradicting $|f'(x)| > 1/M$. One similarly shows that $a = -\infty$ and the same proof works if $f'(x) < -1/M$. The theorem below generalizes this result to the case of Banach spaces.

2.5.17 The Hadamard-Levy Theorem *Let* $f : E \to F$ *be a* C^k *map of Banach spaces,* $k \geq 1$. *If* $Df(x)$ *is an isomorphism of* E *with* F *for every* $x \in E$ *and if there is a constant* $M > 0$ *such that* $\| Df(x)^{-1} \| < M$ *for all* $x \in E$, *then* f *is a diffeomorphism.*

The key to the proof of the theorem consists of a homotopy lifting argument. If X is a topological space, a continuous map $\varphi : X \to F$ is said to ***lift to*** E ***through*** f, if there is a continuous map $\psi : X \to E$ satisfying $f \circ \psi = \varphi$.

2.5.18 Lemma *Let* X *be a connected topological space,* $\varphi : X \to F$ *a continuous map and let* $f : E \to F$ *be a* C^1 *map with* $Df(x)$ *an isomorphism for every* $x \in E$. *Fix* $u_0 \in E$, $v_0 \in F$, *and* $x_0 \in X$ *satisfying* $f(u_0) = v_0$ *and* $\varphi(x_0) = v_0$. *Then if a lift* ψ *of* φ *through* f *with* $\varphi(x_0) = u_0$ *exists, it is unique.*

Proof Let ψ' be another lift and define the sets $X_1 = \{x \in X \mid \varphi(x) = \psi'(x)\}$ and $X_2 = \{x \in X \mid \psi(x) \neq \psi'(x)\}$, so that $X = X_1 \cup X_2$ and $X_1 \cap X_2 = \varnothing$. We shall prove that both X_1, X_2 are open. Since $x_0 \in X_1$, connectedness of X implies $X_2 = \varnothing$ and the lemma will be proved.

If $x \in X_1$, let U be an open neighborhood of $\psi(x) = \psi'(x)$ on which f is a diffeomorphism. Then $\psi^{-1}(U) \cap \psi'^{-1}(U)$ is an open neighborhood of x contained in X_1.

If $x \in X_2$, let U (resp. U′) be an open neighborhood of the point $\psi(x)$ (resp. of $\psi'(x)$) on which f is a diffeomorphism and such that $U \cap U' = \varnothing$. Then the set $\psi^{-1}(U) \cap \psi'^{-1}(U')$ is an open neighborhood of x contained in X_2. ■

A path $\gamma : [0, 1] \to G$, where G is a Banach space, is called C^1 if $\gamma \mid]0, 1[$ is uniformly C^1 and the extension by continuity of γ' to [0, 1] has the values $\gamma'(0)$, $\gamma'(1)$ equal to

$$\gamma'(0) = \lim_{h \downarrow 0} \frac{\gamma(h) - \gamma(0)}{h}, \quad \gamma'(1) = \lim_{h \downarrow 0} \frac{\gamma(1) - \gamma(1-h)}{h}.$$

2.5.19 Homotopy Lifting Lemma *Under the hypotheses of Theorem* **2.5.17,** *let* H(t, s) *be a continuous map of* $[0, 1] \times [0, 1]$ *into* F *such that for each fixed* $s \in [0, 1]$ *the path* $t \mapsto H(t, s)$ *is* C^1. *In addition, assume that* H *fixes endpoints, i.e.,* $H(0, s) = y_0$ *and* $H(1, s) = y_1$, *for all* $s \in [0, 1]$. *If* $y_0 = f(x_0)$ *for some* $x_0 \in E$, *there exists a unique lift* K *of* H *through* f *which is* C^1 *in* t *for every* s. See Figure 2.5.5.

Proof Uniqueness follows by **2.5.18.** By the Inverse Function Theorem, there are open neighborhoods U of x_0 and V of y_0 such that $f \mid U : U \to V$ is a diffeomorphism. Since the open set $H^{-1}(U)$ contains the closed set $\{0\} \times [0, 1]$, there exists $\varepsilon > 0$ such that $[0, \varepsilon[\times [0, 1] \subset H^{-1}(U)$. Let $K : [0, \varepsilon[\times [0, 1] \to E$ be given by $K = f^{-1} \circ H$. Consider the set $A = \{\delta \in [0, 1] \mid H : [0, \delta[\times [0, 1] \to F$ can be lifted through f to $E\}$ which contains the interval $[0, \varepsilon[$. If $\alpha = \sup A$ we shall show first that $\alpha \in A$ and second that $\alpha = 1$. This will prove the existence of the lifting K.

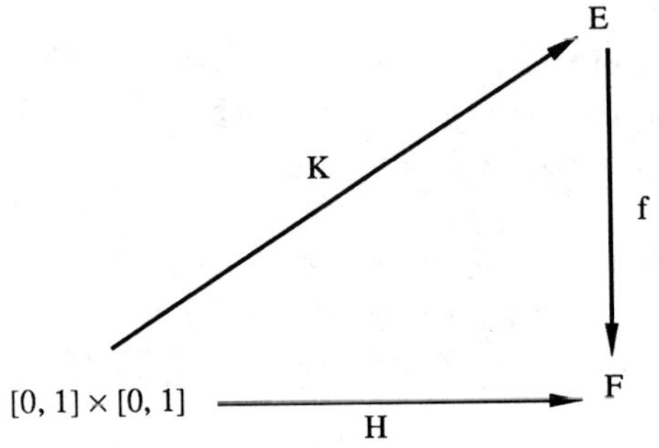

Figure 2.5.5

To show that $\alpha \in A$, note that for $0 \le t < \alpha$ we have $f \circ K = H$ and thus $\mathbf{D}f(K(1, s)) \circ \partial K/t = \partial H/\partial t$, which implies that

$$\left\| \frac{\partial K}{\partial t} \right\| \le M \sup_{t,\, s \in [0,\, 1]} \left\| \frac{\partial H}{\partial t} \right\| = N .$$

Thus by the Mean Value Inequality, if $\{t_n\}$ is an increasing sequence in A converging to α,

$$\| K(t_n, s) - K(t_m, s) \| \le N | t_n - t_m | ,$$

which shows that $\{K(t_n, s)\}$ is a Cauchy sequence in E, uniformly in $s \in [0, 1]$. Let

$$K(\alpha, s) = \lim_{t_n \uparrow \alpha} K(t_n, s) .$$

By continuity of f and H we have

$$f(K(\alpha, s)) = \lim_{t_n \uparrow \alpha} f(K(t_n, s)) = \lim_{t_n \uparrow \alpha} H(t_n, s) = H(\alpha, s) ,$$

which proves that $\alpha \in A$.

Next we show that $\alpha = 1$. If $\alpha < 1$ consider the curves $s \mapsto K(\alpha, s)$ and $s \mapsto H(\alpha, s) = f(K(\alpha, s))$. For each $s \in [0, 1]$ choose open neighborhoods U_s of $K(\alpha, s)$ and V_s of $H(\alpha, s)$ such that $f | U_s : U_s \to V_s$ is a diffeomorphism. By compactness of the path $K(\alpha, s)$ in s, i.e. of the set $\{K(\alpha, s) \mid s \in [0, 1]\}$, finitely many of the U_s, say $U_1, \ldots, U_n$, cover it. Therefore the corresponding $V_1, \ldots, V_n$ cover $\{H(\alpha, s) \mid s \in [0, 1]\}$. Since $H^{-1}(V_i)$ contains the point (α, s_i), there exists $\varepsilon > 0$ such that

$$]\alpha - \varepsilon_i, \alpha + \varepsilon_i[\times]s_i - \delta_i, s_i + \delta_i[\subset H^{-1}(V_i) ,$$

where $]s_i - \delta_i, s_i + \delta_i[= H(\alpha, \cdot)^{-1}(V_i)$ and in particular $]s_i - \delta_i, s_i + \delta_i[$, $i = 1, ..., n$ cover $[0, 1]$. Let $\varepsilon = \min\{\varepsilon_1, ..., e_n\}$ and define $K : [0, \alpha + \varepsilon[\times [0, 1] \to E$ by

$$K(t, s) = \begin{cases} K(t, s), & \text{if } (t, s) \in [0, \alpha[\times \{(0, 1)\} \\ (f \mid U_i)^{-1}(H(t, s)), & \text{if } (t, s) \in [\alpha, \alpha + \varepsilon[\times]s_i - \delta_i, s_i + \delta_i[, \ i = 1, \ldots, n. \end{cases}$$

By **2.5.18**, K is a lifting of H, contradicting the definition of α.

Finally, K is C^1 in t for each s by the chain rule:

$$\frac{\partial K}{\partial t} = Df(K(t, s))^{-1} \circ \frac{\partial H}{\partial t}. \quad \blacksquare$$

Proof of 2.5.17 Let $y_0, y \in F$ and consider the path $\gamma(t) = (1 - t)y_0 + ty$. Regarding γ as defined on $[0, 1] \times [0, 1]$, independent of the second variable, the Homotopy Lifting Lemma guarantees the existence of a C^1 path $\delta : [0, 1] \to E$ lifting γ, i.e., $f \circ \delta = \gamma$. In particular, $f(\delta(1)) = \gamma(1) = y$ and thus f is surjective.

To show that f is injective, assume $x_1 \neq x_2$, $f(x_1) = f(x_2)$, and consider the path $\delta(t) = (1 - t)x_1 + tx_2$. Then $\gamma(t) = f(\delta(t))$. By the Homotopy Lifting Lemma, there exists a lift K of H through f. From $f \circ K = H$ it follows that the continuous curve $s \mapsto K(0, s)$ is mapped by f to the point $f(x_1)$, thus contradicting the Inverse Function Theorem.

Therefore f is a bijective map which is a local diffeomorphism around every point, i.e., f is a diffeomorphism of E with F. $\blacksquare$

Remarks (i) The uniform bound on $\| Df(x)^{-1} \|$ can be replaced by properness of the map, i.e., if $f(x_n) \to y$ there exists a convergent subsequence $\{x_m\}$, $x_m \to x$ with $f(x) = y$ (see Exercise **1.5J**). Indeed, the only place where the uniform bound on $\| Df(x)^{-1} \|$ was used is in the Homotopy Lifting Lemma in the argument that $\alpha = \sup A \in A$. If f is proper, this is shown in the following way. Let $\{t(n)\}$ be an increasing sequence in A converging to α. Then $H(t(n), s) \to H(\alpha, s)$ and from $f \circ K = H$ on $[0, \alpha[\times [0, 1]$, it follows that $f(K(t(m), s)) \to H(\alpha, s)$ uniformly in $s \in [0, 1]$. Thus, by properness of f, there is a subsequence $\{t(m)\}$ such that $K(t(m), s)$ is convergent for every s. Put $K(\alpha, s) = \lim_{t(m) \uparrow \alpha} K(t(n), s)$ and proceed as before.

(ii) If E and F are finite dimensional, properness of f is equivalent to: the inverse image of every compact set in F is compact in E (see Exercise **1.5J**).

(iii) Conditions on f like the one in (ii) or in the theorem are necessary as the following counterexample shows. Let $f : \mathbb{R}^2 \to \mathbb{R}^2$ be given by (e^x, ye^{-x}) so that $f(\mathbb{R}^2)$ is the right open half plane and in particular f is not onto. However

$$\mathbf{D}f(x, y) = \begin{bmatrix} e^x & 0 \\ -ye^{-x} & e^{-x} \end{bmatrix}$$

is clearly an isomorphism for every $(x, y) \in \mathbb{R}^2$. But f is neither proper nor does the norm $\| \mathbf{D}f(x, y)^{-1} \|$ have a uniform bound on $\mathbb{R}^2$. For example, the inverse image of the compact set $[0, 1] \times \{0\}$ is $]-\infty, 0] \times \{0\}$ and $\| \mathbf{D}f(x, y)^{-1} \| = C[e^{-2x} + e^{2x} + y^2e^{-2x}]^{1/2}$, which is unbounded $x \to +\infty$.

(iv) See Wu and Desoer [1972] and Ichiraku [1985] for useful references to the theorem and applications. ♦

If $E = F = H$ is a Hilbert space, then the Hadamard-Levy Theorem has an important consequence. We have seen that in the case of $f : \mathbb{R} \to \mathbb{R}$ with a uniform bound on $1/| f'(x) |$, the strong monotonicity of f played a key role in the proof that f is a diffeomorphism.

2.5.20 Definition *Let* H *be a Hilbert space. A map* $f : H \to H$ *is* ***strongly monotone*** *if there exists* $a > 0$ *such that* $\langle f(x) - f(y), x-y \rangle \geq a\| x - y \|^2$.

As in calculus, for differentiable maps strong monotonicity takes on a familiar form.

2.5.21 Lemma *Let* $f : H \to H$ *be a differentiable map of the Hilbert space* H *onto itself. Then* f *is strongly monotone if and only if* $\langle \mathbf{D}f(x) \cdot u, u \rangle \geq a\| u \|^2$ *for some* $a > 0$.

Proof If f is strongly monotone, $\langle f(x + tu) - f(x), tu \rangle \geq at^2\| u \|^2$ for any $x, u \in H$, $t \in \mathbb{R}$. Dividing by t^2 and taking the limit as $t \to 0$ yields the result.

Conversely, integrating both sides of $\langle \mathbf{D}f(x + tu) \cdot u, u \rangle \geq a\| u \|^2$ from 0 to 1 gives the strong monotonicity condition. ■

2.5.22 Lax-Milgram Lemma *Let* H *be a real Hilbert space and* $A \in L(H, H)$ *satisfy the estimate* $\langle Ae, e \rangle \geq a\| e \|^2$ *for all* $e \in H$. *Then* A *is an isomorphism and* $\| A^{-1} \| \leq 1/a$.

Proof The condition clearly implies injectivity of A. To prove A is surjective, we show first that A(H) is closed and then that the orthogonal complement $A(H)^{\perp}$ is $\{0\}$. Let $f_n = A(e_n)$ be a sequence which converges to $f \in H$. Since $\| Ae \| \geq a\| e \|$ by the Schwarz inequality, we have

$$\| f_n - f_m \| = \| A(e_n - e_m) \| \geq a\| e_n - e_m \| ,$$

and thus $\{e_n\}$ is a Cauchy sequence in H. If e is its limit we have $Ae = f$ and thus $f \in A(H)$.

To prove $A(H)^{\perp} = \{0\}$, let $u \in A(H)^{\perp}$ so that $0 = \langle Au, u \rangle \geq a\| u \|^2$ whence $u = 0$.

By Banach's Isomorphism Theorem **2.2.16**, A is a Banach space isomorphism of H with itself. Finally, replacing e by $A^{-1}f$ in $\| Ae \| \geq a\| e \|$ yields $\| A^{-1}f \| \leq \| f \|/a$, i.e., $\| A^{-1} \| \leq 1/a$. ■

Lemmas **2.5.21**, **2.5.22**, and the Hadamard-Levy Theorem imply the following global Inverse Function Theorem on the real Hilbert space.

2.5.23 Theorem *Let* H *be a real Hilbert space and* $f : H \to H$ *be a strongly monotone* C^k *mapping* $k \geq 1$. *Then* f *is a* C^k *diffeomorphism.*

SUPPLEMENT 2.5E
The Inversion Map

Let E and F be isomorphic Banach spaces and consider the inversion map $I : GL(E, F) \to GL(F, E)$; $I(\varphi) = \varphi^{-1}$. We have shown that I is C^∞ and

$$\mathbf{D}I(\varphi) \cdot \psi = -\varphi^{-1} \circ \psi \circ \varphi^{-1}$$

for $\varphi \in GL(E, F)$ and $\psi \in L(E, F)$. We shall give below the formula for $\mathbf{D}^k I$. The proof is straightforward and done by a simple induction argument that will be left to the reader. Define the map $\alpha^{k+1} : L(F, E) \times \ldots \times L(F, E)$ {there are $k + 1$ factors} $\to L^k(L(E, F); L(F, E))$ by

$$\alpha^{k+1}(\chi_1, \ldots, \chi_{k+1}) \cdot (\psi_1, \ldots, \psi_k) = (-1)^k \chi_1 \circ \psi_1 \circ \chi_2 \circ \psi_2 \circ \cdots \circ \chi_k \circ \psi_k \circ \chi_{k+1} ,$$

where $\chi_i \in L(F, E)$, $i = 1, \ldots, k + 1$ and $\psi_j \in L(E, F)$, $j = 1, \ldots, k$. Let $I \times \ldots \times I$ {with $k +$ 1 factors} be the mapping of $GL(E, F)$ to $GL(F, E) \times \ldots \times GL(F, E)$ {with $k + 1$ factors} defined by $(I \times \cdots \times I)(\varphi) = (\varphi^{-1}, \ldots, \varphi^{-1})$. Then

$$\mathbf{D}^k I = k!\, \mathrm{Sym}^k \circ \alpha^{k+1} \circ (I \times \cdots \times I) ,$$

where Sym^k denotes the symmetrization operator. Explicitly, for $\varphi \in GL(E, E)$, $\psi_1, \ldots, \psi_k \in L(E, F)$, this formula becomes

$$\mathbf{D}^k I(\varphi) \cdot (\psi_1, \ldots, \psi_k) = (-1)^k \sum_{\sigma \in S_k} \varphi^{-1} \circ \psi_{\sigma(1)} \circ \varphi^{-1} \circ \cdots \circ \varphi^{-1} \circ \psi_{\sigma(k)} \circ \varphi^{-1},$$

where S_k is the group of permutations of $\{1, \ldots, k\}$ (see Supplements **2.2B** and **2.4A**).

Exercises

2.5A Let $f : \mathbb{R}^4 \to \mathbb{R}^2$ be defined by $f(x, y, u, v) = (u^3 + vx + y, uy + v^3 - x)$. At what points can we solve $f(x, y, u, v) = (0, 0)$ for (u, v) in terms of (x, y)? Compute $\partial u/\partial x$.

2.5B (i) Let E be a Banach space. Using the Inverse Function Theorem, show that each A in a neighborhood of the identy map in $\mathcal{GL}(E, E)$ has a unique square root.

(ii) Show that for $A \in L(E, E)$ the series

$$B = I - \frac{1}{2}(I - A) - \frac{1}{2^2 \cdot 2!}(I - A)^2 - \cdots - \frac{1 \cdot 3 \cdot 5 \cdots (2n-3)}{2^n n!}(I - A)^n - \cdots$$

is absolutely convergent for $\| I - A \| < 1$. Check directly that $B^2 = A$.

2.5C (i) Let $A \in L(E, E)$ and let

$$e^A = \sum_{n=0}^{\infty} \frac{A^n}{n!} .$$

Show this series is absolutely convergent and find an estimate for $\| e^A \|$, $A \in L(E, E)$.

(ii) Show that if $AB = BA$, then $e^{A+B} = e^A e^B = e^B e^A$. Conclude that $(e^A)^{-1} = e^{-A}$; i.e., $e^A \in \mathcal{GL}(E, E)$.

(iii) Show that $e^{(\cdot)} : L(E, E) \to \mathcal{GL}(E, E)$ is analytic.

(iv) Use the Inverse Function Theorem to conclude that $A \mapsto e^A$ has a unique inverse around the origin. Call this inverse $A \mapsto \log A$ and note that $\log I = 0$.

(v) Show that if $\| I - A \| < 1$, the function $\log A$ is given by the absolutely convergent power series

$$\log A = \sum_{n=1}^{\infty} \frac{(-1)^{n-1}}{n} (A - I)^n.$$

(vi) If $\| I - A \| < 1$, $\| I - B \| < 1$, and $AB = BA$, conclude that $\log(AB) = \log A + \log B$. In particular, $\log A^{-1} = -\log A$.

2.5D Show that the Implicit Function Theorem implies the Inverse Function Theorem. (*Hint:* Apply the Implicit Function Theorem to $g : U \times F \to F$, $g(u, v) = f(u) - v$, for $f : U \subset E \to F$.)

2.5E Let $f : \mathbb{R}^2 \to \mathbb{R}^2$ be C^∞ and satisfy the Cauchy-Riemann equations (see Exercise **2.3F**):

$$\frac{\partial f_1}{\partial x} = \frac{\partial f_2}{\partial y}, \quad \frac{\partial f_1}{\partial y} = -\frac{\partial f_2}{\partial x}.$$

Show that $\mathbf{D}f(x, y) = 0$ iff $\det(\mathbf{D}f(x, y)) = 0$. Show that the local inverse (where it exists) also satisfies the Cauchy-Riemann equations. Give a counterexample for the first statement, if f does not satisfy Cauchy-Riemann.

2.5F Let $f : \mathbb{R} \to \mathbb{R}$ be given by $f(x) = x + x^2 \cos(1/x)$ if $x \neq 0$, $f(0) = 0$. Show that

(i) f is continuous;

(ii) f is differentiable at all points;

(iii) the derivative is discontinuous at $x = 0$;

(iv) $f'(0) \neq 0$;

(v) f has no inverse in any neighborhood of $x = 0$. (This shows that in the Inverse Function Theorem the continuity hypothesis on the derivative cannot be dropped.)

2.5G It is essential to have Banach spaces in the Inverse Function Theorem rather than more general spaces such as topological vector spaces or Fréchet spaces. (The following example of the failure of Theorem **2.5.2** in Fréchet spaces is due to M. McCracken.)

Let $\mathcal{H}(\Delta)$ denote the set of all analytic functions on the open unit disk in $\mathbb{C}$, with the topology of uniform convergence on compact subsets. Let $F : \mathcal{H}(\Delta) \to \mathcal{H}(\Delta)$ be defined by

$$\sum_{n=0}^{\infty} a_n z^n \mapsto \sum_{n=0}^{\infty} a_n^2 z^n.$$

Show that F is C^∞ and that

$$DF\left(\sum_{n=0}^{\infty} a_n z^n\right) \cdot \left(\sum_{n=0}^{\infty} b_n z^n\right) = \sum_{n=0}^{\infty} 2\, a_n b_n z^n .$$

(Define the Fréchet derivative in $\mathcal{H}(\Delta)$ as part of your answer.) If $a_0 = 1$ and $a_n = 1/n$, $n \neq 1$, then

$$DF\left(\sum_{n=1}^{\infty} \frac{z^n}{n}\right)$$

is a bounded linear isomorphism. However, since

$$F\left(z + \frac{z^2}{2} + \cdots + \frac{z^{k-1}}{k-1} - \frac{z^k}{k} + \frac{z^{k+1}}{k+1} + \cdots\right) = F\left(\sum_{n=1}^{\infty} \frac{z^n}{n}\right)$$

conclude that F is not locally injective. (Consult Schwartz [1967], Sternberg [1969], and Hamilton [1982] for more sophisticated versions of the Inverse Function Theorem valid in

Fréchet spaces.)

2.5H Generalized Lagrange Multiplier Theorem (Luenberger [1969]). Let $f : U \subset E \to$ and $g : U \subset E \to G$ be C^1 and suppose $\mathbf{D}g(u_0)$ is surjective. Suppose f has a local extremum (maximum or minimum) at u_0 subject to the constraint $g(u) = 0$. Then prove

(i) $\mathbf{D}f(u_0) \cdot h = 0$ for all $h \in \ker \mathbf{D}g(u_0)$, and

(ii) there is a $\lambda \in G^*$ such that $\mathbf{D}f(u_0) = \lambda \mathbf{D}g(u_0)$.

(See Supplement **3.5A** for the geometry behind this result).

2.5I Let $f : U \subset \mathbb{R}^m \to \mathbb{R}^n$ be a C^1 map.

(i) Show that the set $G_r = \{x \in U \mid \text{rank } \mathbf{D}f(x) \geq r\}$ is open in U. (*Hint:* If $x_0 \in G_r$, let $M(x_0)$ be a square block of the matrix of $\mathbf{D}f(x_0)$ in given bases of $\mathbb{R}^m$ and $\mathbb{R}^n$ of size $\geq r$ such that $\det M(x_0) \neq 0$. Using continuity of the determinant function, what can you say about $\det M(x)$ for x near x_0?)

(ii) We say R is the maximal rank of $\mathbf{D}f(x)$ on U if $R = \sup_{x \in U} (\text{rank } \mathbf{D}f(x))$. Show that $V_R = \{x \in U \mid \text{rank } \mathbf{D}f(x) = R\}$ is open in U. Conclude that if $\text{rank } \mathbf{D}f(x_0)$ is maximal then $\text{rank } \mathbf{D}f(x)$ stays maximal in a neighborhood of x_0.

(iii) Define $O_i = \text{int}\{x \in U \mid \text{rank } \mathbf{D}f(x) = i\}$ and let r_0 be the maximal rank of $\mathbf{D}f(x)$, $x \in U$. Show that $O_1 \cup \ldots \cup O_R$ is dense in U. (*Hint:* Let $x \in U$ and let V be an arbitrary neighborhood of x. If Q denotes the maximal rank of $\mathbf{D}f(x)$ on $x \in V$, use (ii) to argue that $V \cap O_Q = \{x \in V \mid \text{rank } \mathbf{D}f(x) = Q\}$ is open and nonempty in V.)

(iv) Show that if a C^1 map $f : U \subset \mathbb{R}^m \to \mathbb{R}^n$ is injective (surjective onto an open set), then $m \leq n$ $(m \geq n)$. (*Hint:* Use the Rank Theorem and (ii).)

2.5J Uniform Contraction Principle (Hale [1969] and Chow and Hale [1982]).

(i) Let $T : \text{cl}(U) \times V \to E$ be a C^k map, where $U \subset E$ and $V \subset F$ are open sets. Suppose that for fixed $y \in V$, $T(x, y)$ is a contraction in x, uniformly in y. If g(y) denotes the unique fixed point of $T(x, y)$, show that g is C^k. (*Hint:* Proceed directly as in the proof of the Inverse Mapping Theorem.)

(ii) Use (i) to prove the Inverse Mapping Theorem.

2.5K Lipschitz Inverse Function Theorem (Hirsch and Pugh [1970]).

(i) Let (X_i, d_i) be metric spaces and $f : X_1 \to X_2$. The map f is called ***Lipschitz*** if there exists a constant L such that $d_2(f(x), f(y)) \leq L d_1(x, y)$ for all $x, y \in X_1$. The smallest such L is the ***Lipschitz constant*** L(f). Thus, if $X_1 = X_2$ and $L(f) < 1$, then f is a contraction. If f is not Lipschitz, set $L(f) = \infty$. Show that if $g : (X_2, d_2) \to (X_3, d_3)$, then $L(g \circ f) \leq L(g)L(f)$. Show that if X_1, X_2 are normed vector spaces and $f, g : X_1 \to X_2$, then $L(f + g) \leq L(f) + L(g)$, $L(f) - L(g) \leq L(f - g)$.

(ii) Let E be a Banach space, U an open set in E such that the closed ball $B_r(0) \subset U$. Let $f : U \to E$ be given by $f(x) = x + \varphi(x)$, where $\varphi(0) = 0$ and φ is a contraction. Show that $f(D_r(0)) \supset D_{r(1-L(\varphi))}(0)$, that f is invertible on $f^{-1}(D_{r(1-L(\varphi))}(0))$, and that f^{-1} is Lipschitz with constant $L(f^{-1}) \le 1/(1-L(\varphi))$. (*Hint:* If $\|y\| < r(1 - L(\varphi))$, define $F : U \to E$ by $F(x) = y - \varphi(x)$. Apply the Contraction Mapping Principle in $B_r(0)$ and show that the fixed point is in $D_r(0)$. Finally, note that $(1 - L(\varphi))\| x_1 - x_2 \| \le \| x_1 - x_2 \| - \| \varphi(x_1) - \varphi(x_2) \| \le \| f(x_1) - f(x_2) \|$.)

(iii) Let U be an open set in the Banach space E, V be an open set in the Banach space F, $x_0 \in U$, $B_r(x_0) \subset U$. Let $\alpha : U \to V$ be a homeomorphism. Assume that $\alpha^{-1} : V \to U$ is Lipschitz and let $\psi : U \to F$ be another Lipschitz map. Assume $L(\psi)L(\alpha^{-1}) < 1$ and define $f = \alpha + \psi : U \to F$. Denote $y_0 = f(x_0)$. Show that $f(\alpha^{-1}(D_r(x_0))) \supset D_{r(1-L(\psi)L(\alpha^{-1}))}(y_0)$, that f is invertible on $f^{-1}(D_{r(1-L(\psi)L(\alpha^{-1}))}(y_0)$, and that f^{-1} is Lipschitz with constant $L(f^{-1}) \le 1/(L(\alpha^{-1})^{-1} - L(\psi))$. (*Hint:* Replacing ψ by the map $x \mapsto \psi(x) - \psi(x_0)$ and V by $V + \{\psi(x_0)\}$, we can assume that $\psi(x_0) = 0$ and $f(x_0) = \alpha(x_0) = y_0$. Next, replace this new f by $x \mapsto f(x + x_0) - f(x_0)$, U by $U - \{x_0\}$, and the new V by $V + \{y_0\}$; thus we can assume that $x_0 = 0$, $y_0 = 0$, $\psi(0) = 0$, $\alpha(0) = 0$. Then $f \circ \alpha^{-1} = I + \psi \circ \alpha^{-1}$, $(\psi \circ \alpha^{-1})(0) = 0$, $L(\psi \circ \alpha^{-1}) \le L(\psi)L(\alpha^{-1}) < 1$, so (ii) is applicable.)

(iv) Show that $| L(f^{-1}) - L(\alpha^{-1}) | \to 0$ as $L(\psi) \to 0$. Let $\alpha : \mathbb{R} \to \mathbb{R}$ be the homeomorphism defined by $\alpha(x) = x$ if $x \le 0$ and $\alpha(x) = 2x$ if $x \ge 0$. Show that both α and α^{-1} are Lipschitz. Let $\psi(x) = c =$ constant. Show that $L(\psi) = 0$ and if $c \ne 0$, then $L(f^{-1} - \alpha^{-1}) \ge 1/2$. Prove, however, that if α, f are diffeomorphisms, then $L(f^{-1} - \alpha^{-1}) \to 0$ as $L(\psi) \to 0$.

2.5L Use the Inverse Function Theorem to show that simple roots of polynomials are smooth functions of their coefficients. Conclude that simple eigenvalues of operators of $\mathbb{R}^n$ are smooth functions of the operator. (*Hint:* If $p(t) = a_n t^n + a_{n-1}t^{n-1} + \ldots + a_0$, define a smooth map $F : \mathbb{R}^{n+2} \to \mathbb{R}$ by $F(a_n, \ldots, a_0, \lambda) = p(\lambda)$ and note that if λ_0 is a simple eigenvalue, $\partial F(\lambda_0)/\partial\lambda \ne 0$.)

2.5M (i) Let E, F be Banach spaces, $f : U \to V$ a C^r bijective map, $r \ge 1$, between two open sets $U \subset E$, $V \subset F$. Assume that for each $x \in U$, $\mathbf{D}f(x)$ has closed split image and is one-to-one. Use the Local Immersion Theorem to show that f is a C^r diffeomorphism.

(ii) What fails for $y = x^3$?

2.5N Let E be a Banach space, $U \subset E$ open and $f : U \to \mathbb{R}$ a C^r map, $r \ge 2$. We say that $u \in U$ is a ***critical point*** of f, if $\mathbf{D}f(u) = 0$. The critical point u is called ***strongly non-degenerate*** if $\mathbf{D}^2 f(u)$ induces a Banach space isomorphism of E with its dual E^*. Use the Inverse Function Theorem on $\mathbf{D}f$ to show that strongly non-degenerate points are isolated, i.e., each strongly non-degenerate point is unique in one of its neighborhoods. (A counter-example, if $\mathbf{D}^2 f$ is only injective, is given in Exercise **2.4O**.)

2.5O For $u : S^1 \to \mathbb{R}$, consider the equation

$$\frac{du}{d\theta} + u^2 - \frac{1}{2\pi}\int_0^{2\pi} u^2\, d\theta = \varepsilon \sin\theta$$

where θ is a 2π-periodic angular variable and ε is a constant. Show that if ε is sufficiently small, this equation has a solution.

2.5P Use the Implicit Function Theorem to study solvability of

$$\nabla^2\varphi + \varphi^3 = f \text{ in } \Omega \text{ and } \frac{\partial\varphi}{\partial n} = g \text{ on } \partial\Omega$$

where Ω is a region in $\mathbb{R}^n$ with smooth boundary, as in Supplement **2.5C**.

2.5Q Let E be a finite dimensional vector space.

(i) Show that $\det(\exp A) = e^{\operatorname{trace} A}$. (*Hint:* Show it for A diagonalizable and then use Exercise **2.2L**(i).)

(ii) If E is real, show that $\exp(L(E, F)) \cap \{A \in GL(E) \mid \det A < 0\} = \varnothing$. This shows that the exponential map is not onto.

(iii) If E is complex, show that the exponential map is onto. For this you will need to recall the following facts from linear algebra. Let p be the characteristic polynomial of $A \in L(E, E)$, i.e., $p(\lambda) = \det(A - \lambda I)$. Assume that p has m distinct roots $\lambda_1, \ldots, \lambda_m$ such that the multiplicity of λ_i is k_i. Then

$$E = \bigoplus_{i=1}^{m} \ker(A - \lambda_i I)^{k_i} \text{ and } \dim(\ker(A - \lambda_i I)^{k_i}) = k_i.$$

Thus, to prove the exponential is onto, it suffices to prove it for operators $S \in GL(E)$ for which the characteristic polynomial is $(\lambda - \lambda_0)^k$. (*Hint:* Since S is invertible, $\lambda_0 \neq 0$, so write $\lambda_0 = e^z$, $z \in \mathbb{C}$. Let $N = \lambda_0^{-1}S - I$ and

$$A = \sum_{i=1}^{k-1} \frac{(-1)^{i-1}N^i}{i} .$$

By the Cayley-Hamilton Theorem (see Exercise **2.2.L**(ii)), $N^k = 0$, and from the fact that $\exp(\log(1 + w)) = 1 + w$ for all $w \in \mathbb{C}$, it follows that $\exp(A + zI) = \lambda_0 \exp A = \lambda_0(I + N) = S$.)

Chapter 3
MANIFOLDS AND VECTOR BUNDLES

We are now ready to study manifolds and the differential calculus of maps between manifolds. Manifolds are an abstraction of the idea of a smooth surface in Euclidean space. This abstraction has proved useful because many sets that are smooth in some sense are not presented to us as subsets of Euclidean space. The abstraction strips away the containing space and makes constructions intrinsic to the manifold itself. This point of view is well worth the geometric insight it provides.

§3.1 *Manifolds*

The basic idea of a manifold is to introduce a local object that will support differentiation processes and then to patch these local objects together smoothly. Before giving the formal definitions it is good to have an example in mind. In $\mathbb{R}^{n+1}$ consider the n-sphere S^n; that is, the set of $x \in \mathbb{R}^{n+1}$ such that $\| x \| = 1$ ($\| \cdot \|$ denotes the usual Euclidean norm). We can construct bijections from subsets of S^n to $\mathbb{R}^n$ in several ways. One way is to project stereographically from the south pole onto a hyperplane tangent to the north pole. This is a bijection from S^n, with the south pole removed, onto $\mathbb{R}^n$. Similarly, we can interchange the roles of the poles to obtain another bijection. (See Figure 3.1.1.)

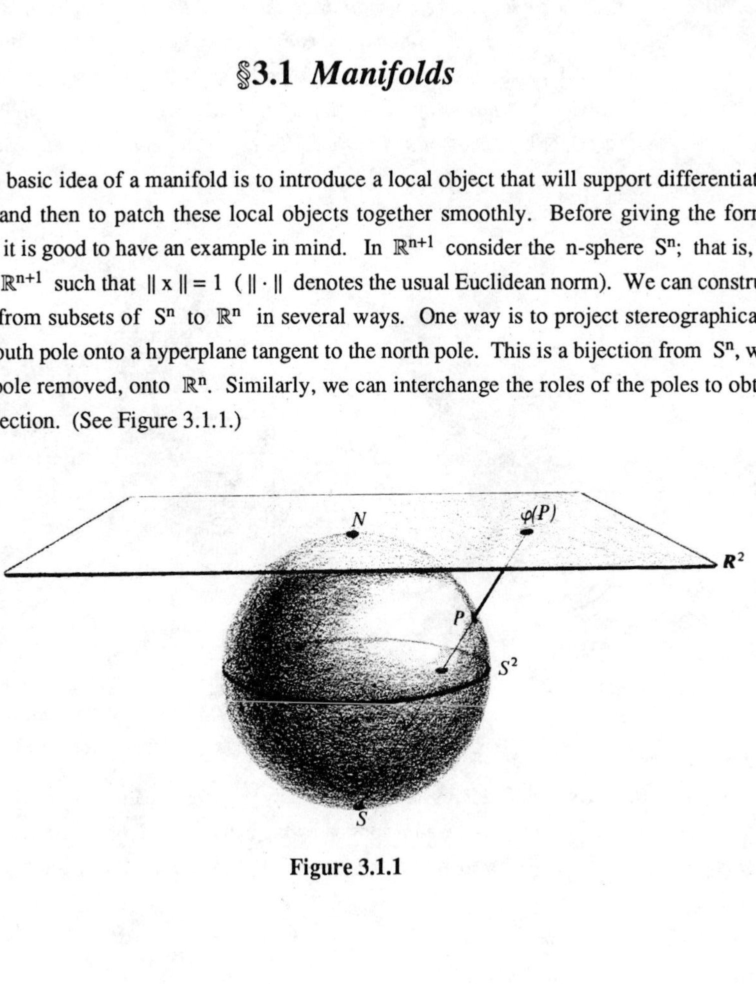

Figure 3.1.1

With the usual relative topology on S^n as a subset of $\mathbb{R}^{n+1}$, these maps are homeomorphisms from their domain to $\mathbb{R}^n$. Each map takes the sphere minus the two poles to an open subset of $\mathbb{R}^n$. If we go from $\mathbb{R}^n$ to the sphere by one map, then back to $\mathbb{R}^n$ by the other, we get a smooth map from an open subset of $\mathbb{R}^n$ to $\mathbb{R}^n$. Each map assigns a coordinate system to S^n minus a pole. The union of the two domains is S^n, but no single homeomorphism can be used between S^n and $\mathbb{R}^n$; however, we can cover S^n using two of them. In this case they are ***compatible***; that is, in the region covered by both coordinate systems, the change of coordinates is smooth. For some studies of the sphere, and for other manifolds, two coordinate systems will not suffice. We thus allow all other coordinate systems compatible with these. For example, on S^2 we want to allow spherical coordinates (θ, φ) since they are convenient for many computations.

3.1.1 Definition *Let* S *be a set. A* ***chart*** *on* S *is a bijection* φ *from a subset* U *to* S *to an open subset of a Banach space. We sometimes denote* φ *by* (U, φ), *to indicate the domain* U *of* φ. *A* C^k ***atlas*** *on* S *is a family of charts* $\mathcal{A} = \{(U_i, \varphi_i) \mid i \in I\}$ *such that*

MA1 $S = \cup \{U_i \mid i \in I\}$.

MA2 *Any two charts in* $\mathcal{A}$ *are compatible in the sense that the overlap maps between members of* $\mathcal{A}$ *are* C^k *diffeomorphisms: for two charts* (U_i, φ_i) *and* (U_j, φ_j) *with* $U_i \cap U_j \neq \varnothing$, *we form the* ***overlap map***: $\varphi_{ji} = \varphi_j \circ \varphi_i^{-1} \mid \varphi_i(U_i \cap U_j)$, *where* $\varphi_i^{-1} \mid \varphi_i(U_i \cap U_j)$ *means the restriction of* φ_i^{-1} *to the set* $\varphi_i(U_i \cap U_j)$. *We require that* $\varphi_i(U_i \cap U_j)$ *is open and that* φ_{ji} *be a* C^k *diffeomorphism.* (See Figure 3.1.2).

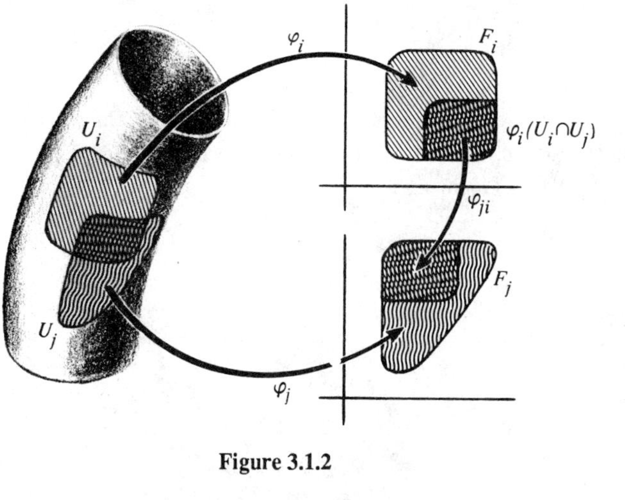

Figure 3.1.2

3.1.2 Examples

A Any Banach space F admits an atlas formed by the single chart (F, identity).

B A less trivial example is the atlas formed by the two charts of S^n discussed previously. More explicitly, if $N = (1, 0, ..., 0)$ and $S = (-1, ..., 0, 0)$ are the north and south poles of S^n, the stereographic projections from N and S are

$$\varphi_1 : S^n \backslash \{N\} \to \mathbb{R}^n,\ \varphi_1(x^1, ..., x^{n+1}) = (x^2/(1 - x^1), ..., x^{n+1}/(1 - x^1)) \ ,$$

and

$$\varphi_2 : S^n \backslash \{S\} \to \mathbb{R}^n,\ \varphi^2(x^1, ..., x^{n+1}) = (x^2/(1 + x^1), ..., x^{n+1}/(1 + x^1))$$

and the overlap map $\varphi_2 \circ \varphi_1^{-1} : \mathbb{R}^n \backslash \{0\} \to \mathbb{R}^n \backslash \{0\}$ is given by $(\varphi_2 \circ \varphi_1^{-1})(z) = z/\| z \|^2$, $z \in \mathbb{R}^n \backslash \{0\}$, which is clearly a C^∞ diffeomorphism of $\mathbb{R}^n \backslash \{0\}$ to itself. ♦

3.1.3 Definition *Two* C^k *atlases* $\mathcal{A}_1$ *and* $\mathcal{A}_2$ *are* ***equivalent*** *if* $\mathcal{A}_1 \cup \mathcal{A}_2$ *is a* C^k *atlas. A* C^k ***differentiable structure*** $\mathcal{D}$ *on* S *is an equivalence class of atlases on* S. *The union of the atlases in* $\mathcal{D}$, $\mathcal{A}_\mathcal{D} = \bigcup \{\mathcal{A} \mid \mathcal{A} \in \mathcal{D}\}$ *is the* ***maximal atlas*** *of* $\mathcal{D}$, *and a chart* $(U, \varphi) \in \mathcal{A}_\mathcal{D}$ *is an* ***admissible local chart***. *If* $\mathcal{A}$ *is a* C^k *atlas on* S, *the union of all atlases equivalent to* $\mathcal{A}$ *is called the* C^k *differentiable structure* ***generated by*** $\mathcal{A}$. *A* ***differentiable manifold*** M *is a pair* $(S, \mathcal{D})$, *where* S *is a set and* $\mathcal{D}$ *is a* C^k *differentiable structure on* s. *We shall often identify* M *with the underlying set* S *for notational convenience. If a covering by charts takes their values in a Banach space* E, *then* E *is called the* ***model space*** *and we say that* M *is a* C^k ***Banach manifold modeled on*** E.

If we make a choice of a C^k atlas $\mathcal{A}$ on S then we obtain a maximal atlas by including all charts whose overlap maps with those in $\mathcal{A}$ are C^k. In practice it is sufficient to specify a particular atlas on S to determine a manifold structure for S.

3.1.4 Example An alternative atlas for S^n has the following $2(n + 1)$ charts: $(U_i^\pm, \psi_i^\pm)$, $i = 1, ..., n + 1$, where $U_i^\pm = \{x \in S^n \mid \pm x^i > 0\}$ and $\psi_i^\pm : U_i^\pm \to \{y \in \mathbb{R}^n \mid \| y \| < 1\}$ is defined by

$$\psi_i^\pm(x^1, ..., x^{n+1}) = (x^1, ..., x^{i-1}, x^{i+1}, ..., x^{n+1});$$

ψ_i projects the hemisphere containing the pole $(0, ..., \pm 1, ..., 0)$ onto the open unit ball in the tangent space to the sphere at that pole. It is verified that this atlas and the one in **3.1.2B** with two charts are equivalent. The overlap maps of this atlas are given by

$$(\psi_j^\pm \circ (\psi_i^\pm)^{-1})(y^1, ..., y^n) = \left(y^1, ..., y^{j-1}, y^{j+1}, ..., y^{i-1}, \pm\sqrt{1 - \| y \|^2}, y^i, ..., y^n\right),$$

where $j < 1$. ♦

3.1.5 Definition *Let* M *be a differentiable manifold. A subset* $A \subset M$ *is called* ***open*** *if for each* $a \in A$ *there is an admissible local chart* (U, φ) *such that* $a \in U$ *and* $U \subset A$.

3.1.6 Proposition *The open sets in* M *define a topology.*

Proof Take as basis of the topology the family of finite intersections of chart domains. ■

3.1.7 Definition *A differentiable manifold* M *is an* ***n-manifold*** *when every chart has values in an* n*-dimensional vector space. Thus for every point* $a \in M$ *there is an admissible local chart* (U, φ) *with* $a \in U$ *and* $\varphi(U) \subset \mathbb{R}^n$. *We write* $n = \dim M$. *An* n*-manifold will mean a Hausdorff, differentiable* n*-manifold in this book. A differentiable manifold is called a* ***finite-dimensional manifold*** *if its connected components are all* n*-manifolds* (n *can vary with the component). A differentiable manifold is called a* ***Hilbert manifold*** *if the model space is a Hilbert space.* (One can similarly form a manifold modeled on any linear space in which one has a theory of differential calculus. For example mathematicians often speak of a "Fréchet manifolds," a "LCTVS manifold," etc. We have chosen to stick with Banach manifolds here primarily to avail ourselves of the inverse function theorem. See Exercise **2.5G**.)

No assumption on the connectedness of a manifold has been made. In fact, in many applications the manifolds are disconnected (see Exercise **3.1C**). Since manifolds are locally arcwise connected, their components are both open and closed.

3.1.8 Examples

A Every discrete topological space S is a 0-manifold, the charts being given by the pairs $(\{s\}, \varphi_s)$, where $\varphi_s : s \mapsto 0$ and $s \in S$.

B Every Banach space is a manifold; its differentiable structure is given by the atlas with the single identity chart.

C S^n with a maximal atlas generated by the atlas with two charts described in Example **3.1.2B** or **3.1.4** makes S^n into an n-manifold. The resulting topology is the same as that induced on S^n as a subset of $\mathbb{R}^{n+1}$.

D A set can have more than one differentiable structure. For example, $\mathbb{R}$ has the following incompatible charts:

$$(U_1, \varphi_1) : U_1 = \mathbb{R}, \quad \varphi_1(r) = r^3 \in \mathbb{R}; \quad \text{and} \quad (U_2, \varphi_2) : U_2 = \mathbb{R}, \quad \varphi_2(r) = r \in \mathbb{R}.$$

They are not compatible since $\varphi_2 \circ \varphi_1^{-1}$ is not differentiable at the origin. Nevertheless, these two structures are "diffeomorphic" (Exercise **3.2H**), but structures can be "essentially different" on more complicated sets (e.g., S^7). That S^7 has two nondiffeomorphic differentiable structures is a famous result of J. Milnor [1956]. Recently similar phenomena have been found on $\mathbb{R}^4$ by Donaldson [1983]; see also Fried and Uhlenbeck [1984].

E Essentially the only one-dimensional paracompact connected manifolds are $\mathbb{R}$ and S^1. This means that all others are diffeomorphic to $\mathbb{R}$ or S^1 (diffeomorphic will be precisely defined later). For example, the circle with a knot is diffeomorphic to S^1. (See Fig. 3.1.3.) See Milnor [1965] or Guillemin and Pollack [1974] for proofs.

F A general two-dimensional compact connected manifold is the sphere with "handles" (see Fig. 3.1.4). This includes, for example, the torus, whose precise definition will be given in the next section. This classification of two-manifolds is described in Massey [1976] and Hirsch [1976].

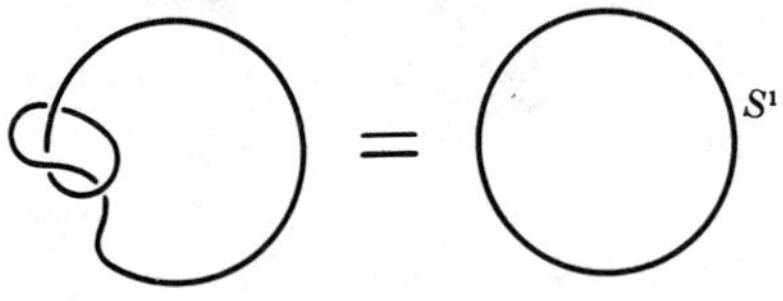

Figure 3.1.3

Figure 3.1.4

G Grassmann Manifolds Let E be a Banach space and consider the set $\mathbb{G}(E)$ of all split subspaces of E. For $F \in \mathbb{G}(E)$, let G denote one of its complements, i.e., $E = F \oplus G$, let

$$U_G = \{H \in \mathbb{G}(E) \mid E = H \oplus G\}, \text{ and define } \varphi_{F,G} : U_G \to L(F, G)$$

by

$$\varphi_{F,G}(H) = \pi_F(H, G) \circ \pi_G(H, F)^{-1},$$

where $\pi_F(G) : E \to G$, $\pi_G(F) : E \to F$ denote the pojections induced by the direct sum decomposition $E = F \oplus G$, and $\pi_F(H, G) = \pi_F(G) \mid H$, $\pi_G(H, F) = \pi_G(F) \mid H$. The inverse appearing in the definition of $\varphi_{F,G}$ exists as the following argument shows. If $H \in U_G$, i.e., if $E = F \oplus G = H \oplus G$, then the maps $\pi_G(H, F) \in L(H, F)$ and $\pi_G(F, H) \in L(F, H)$ are invertible and one is the inverse of the other, for if $h = f + g$, then $f = h - g$, for $f \in F$, $g \in G$, and $h \in H$, so that $(\pi_G(F, H) \circ \pi_G(H, F))(h) = \pi_G(F, H)(f) = h$, and $(\pi_G(H, F) \circ \pi_G(F, H))(f) = \pi_G(H, F)(h) = f$. In particular, $\varphi_{F,G}$ has the alternative expression

$$\varphi_{F,G} = \pi_F(H, G) \circ \pi_G(F, H).$$

Note that we have shown that $H \in U_G$ implies $\pi_G(H, F) \in L(H, F)$ is an isomorphism. The converse is also true, i.e., if $\pi_G(H, F)$ is an isomorphism for some split subspace H of E then $E = H \oplus G$. Indeed, if $x \in H \cap G$, then $\pi_G(H, F)(x) = 0$ and so $x = 0$, i.e., $H \cap G = \{0\}$. If $e \in E$, then we can write

$$e = (\pi_G(F, H) \circ \pi_G(F))(e) + [e - (\pi_G(F, H) \circ \pi_G(F))(e)]$$

with the first summand an element of H. Since $\pi_G(F) \circ \pi_G(F, H) = \text{identity}$ on F, we have $\pi_G(F)\,[e - (\pi_G(F, H) \circ \pi_G(F))(e)] = 0$, i.e., the second summand is an element of G, and thus $E = H + G$. Therefore $E = H \oplus G$ and we have the alternative definition of U_G as

$$U_G = \{H \in \mathbb{G}(E) \mid \pi_G(H, F) \text{ is an isomorphism of } H \text{ with } F\}.$$

Let us next show that $\varphi_{F,G} : U_G \to L(F, G)$ is bijective. For $\alpha \in L(F, G)$ define the graph of α by $\Gamma_{F,G}(\alpha) = \{f + \alpha(f) \mid f \in F\}$ which is a closed subspace of $E = F \oplus G$. Then $E = \Gamma_{F,G}(\alpha) \oplus G$, i.e., $\Gamma_{F,G}(\alpha) \in U_G$, since any $e \in E$ can be written as $e = f + g = (f + \alpha(f)) + (g - \alpha(f))$ for $f \in F$ and $g \in G$, and also $\Gamma_{F,G}(\alpha) \cap G = \{0\}$ since $f + \alpha(f) \in G$ for $f \in F$ iff $f \in F \cap G = \{0\}$. We have

$$\varphi_{F,G}(\Gamma_{F,G}(\alpha)) = \pi_F(\Gamma_{F,G}(\alpha), G) \circ \pi_G(F, \Gamma_{F,G}(\alpha))$$

$$f = (f + \alpha(f)) - \alpha(f) \quad \mapsto \quad \pi_F(\Gamma_{F,G}(\alpha), G)(f + \alpha(f)) \mapsto \alpha(f)$$

i.e. $\varphi_{F,G} \circ \Gamma_{F,G} =$ identity on $L(F, G)$, and

$$\begin{aligned} \Gamma_{F,G}(\pi_F(H, G) \circ \pi_G(F, H)) &= \{f + (\pi_F(H, G) \circ \pi_G(F, H))(f) \mid f \in F\} \\ &= \{f + \pi_F(H, G)(h) \mid f \in F, f = h + g, h \in H, \text{ and } g \in G\} \\ &= \{f - g \mid f \in F, f = h + g, h \in H, \text{ and } g \in G\} = H , \end{aligned}$$

i.e. $\Gamma_{F,G} \circ \varphi_{F,G} =$ identity on U_G. Thus $\varphi_{F,G}$ is a bijective map which sends $H \in U_G$ to an element of $L(F, G)$ whose graph in $F \oplus G$ is H. We have thus shown that $(U_G, \varphi_{F,G})$ is a chart on $\mathbb{G}(E)$.

To show that $\{(U_G, \varphi_{F,G}) \mid E = F \oplus G\}$ is an atlas on $\mathbb{G}(E)$, note that $\bigcup_{F \in \mathbb{G}(E)} \bigcup_G U_G = \mathbb{G}(E)$, where the second union is taken over all $G \in \mathbb{G}(E)$ such that $E = H \oplus G = F \oplus G$ for some $H \in \mathbb{G}(E)$. Thus **MA1** is satisfied. To prove **MA2**, let $(U_{G'}, \varphi_{F',G'})$ be another chart on $\mathbb{G}(E)$ with $U_G \cap U_{G'} \neq \varnothing$. We need to show that $\varphi_{F,G}(U_G \cap U_{G'})$ is open in $L(F, G)$ and that $\varphi_{F,G} \circ \varphi_{F',G'}{}^{-1}$ is a C^∞ diffeomorphism of $L(F', G')$ to $L(F, G)$.

Step 1. Proof of the openness of $\varphi_{F,G}(U_G \cap U_{G'})$. Let $\alpha \in \varphi_{F,G}(U_G \cap U_{G'}) \subset L(F, G)$ and let $H = \Gamma_{F,G}(\alpha)$. Then $E = H \oplus G = H \oplus G'$. Assume for the moment that we can show the existence of an $\varepsilon > 0$ such that if $\beta \in L(H, G)$ and $\|\beta\| < \varepsilon$, then $\Gamma_{H,G}(\beta) \oplus G' = E$. Then if $\alpha' \in L(F, G)$ is such that $\|\alpha'\| < \varepsilon/\|\pi_G(H, F)\|$, we get $\Gamma_{H,G}(\alpha' \circ \pi_G(H, F)) \oplus G' = E$. We prove that $\Gamma_{F,G}(\alpha + \alpha') = \Gamma_{H,G}(\alpha' \circ \pi_G(H, F))$. Indeed, since the inverse of $\pi_G(H, F) \in GL(H, F)$ is $I + \alpha$, where I is the identity mapping on F, for any $h \in H$, $\pi_G(H, F)(h) + ((\alpha + \alpha') \circ \pi_G(H, F))(h) = [(I + \alpha) \circ \pi_G(H, F)](h) + (\alpha' \circ \pi_G(H, F))(h) = h + (\alpha' \circ \pi_G(H, F))(h)$, whence the desired equality between the graphs of $\alpha + \alpha'$ in $F \oplus G$ and $\alpha' \circ \pi_G(H, F)$ in $H \oplus G$. Thus we have shown that $\Gamma_{F,G}(\alpha + \alpha') \oplus G' = E$. Since we always have $\Gamma_{F,G}(\alpha + \alpha') \oplus G = E$ (since $\Gamma_{F,G}$ is bijective with range U_G), we conclude that $\alpha + \alpha' \in \varphi_{F,G}(U_G \cap U_{G'})$ thereby proving openness of $\varphi_{F,G}(U_G \cap U_{G'})$.

To complete the proof of Step 1 we therefore have to show that *if* $E = H \oplus G = H \oplus G'$ *then there is an* $\varepsilon > 0$ *such that for all* $\beta \in L(H, G)$ *satisfying* $\|\beta\| < \varepsilon$, *we have* $\Gamma_{H,G}(\beta) \oplus G' = E$. This in turn is a consequence of the following statement: if $E = H \oplus G = H \oplus G'$ then there is an $\varepsilon > 0$ such for all $\beta \in L(H, G)$ satisfying $\|\beta\| < \varepsilon$, we have $\pi_{G'}(\Gamma_{H,G}(\beta), H) \in$

$GL(\Gamma_{H,G}(\beta), H)$. Indeed, granted this last statement, write $e \in E$ as $e = h + g'$, for some $h \in H$ and $g' \in G'$, use the bijectivity of $\pi_{G'}(\Gamma_{H,G}(\beta), H)$ to find an $x \in \Gamma_{H,G}(\beta)$ such that $h = \pi_{G'}(\Gamma_{H,G}(\beta), H)(x)$, and note that $\pi_{G'}(H)(h - x) = 0$, i.e. $h - x = g_1' \in G'$; see Figure 3.1.5. Therefore $e = x + (g_1' + g') \in \Gamma_{H,G}(\beta) + G'$. In addition, we also have $\Gamma_{H,G}(\beta) \cap G' = \{0\}$, for if $z \in \Gamma_{H,G}(\beta) \cap G'$, then $\pi_{G'}(\Gamma_{H,G}(\beta), H)(z) = 0$, whence $z = 0$ by injectivity of the mapping $\pi_{G'}(\Gamma_{H,G}(\beta), H)$; thus we have shown $E = \Gamma_{H,G}(\beta) \oplus G'$.

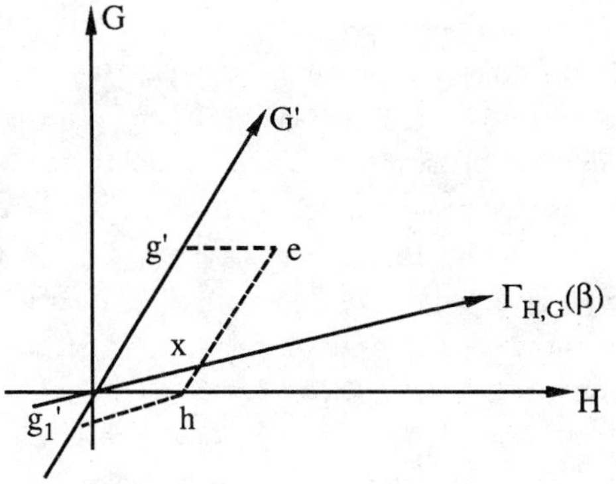

Figure 3.1.5

Finally, assume that $E = H \oplus G = H \oplus G'$. Let us prove *that there is an* $\varepsilon > 0$ *such that if* $\beta \in L(H, G)$, satisfies $\| \beta \| < \varepsilon$, *then* $\pi_{G'}(\Gamma_{H,G}(\beta), H) \in GL(\Gamma_{H,G}(\beta), H)$. Because of the identity $\pi_G(H, \Gamma_{H,G}(\beta)) = I + \beta$, where I is the identity mapping on H, we have

$$\begin{aligned} \| I - \pi_{G'}(H) \circ \pi_G(H, \Gamma_{H,G}(\beta)) \| &= \| I - \pi_{G'}(H) \circ (I + \beta) \| \\ &= \| \pi_G(H') \circ (I - (I + \beta)) \| \leq \| \pi_{G'}(H) \| \, \| \beta \| < 1 \end{aligned}$$

provided that $\| \beta \| < \varepsilon = 1 / \| \pi_{G'}(H) \|$. Therefore we get $I - (I - \pi_{G'}(H) \circ \pi_G(H, \Gamma_{H,G}(\beta)) = \pi_{G'}(H) \circ \pi_G(H, \Gamma_{H,G}(\beta)) \in GL(H, H)$. Therefore, since $\pi_G(H, \Gamma_{H,G}(\beta)) \in GL(H, \Gamma_{H,G}(\beta))$ has inverse $\pi_G(\Gamma_{H,G}(\beta), H)$, we obtain

$$\begin{aligned} \pi_{G'}(\Gamma_{H,G}(\beta), H) &= \pi_{G'}(H) \mid \Gamma_{H,G}(\beta) \\ &= [\pi_{G'}(H) \circ \pi_G(H, (\Gamma_{H,G}(\beta))] \circ \pi_G(\Gamma_{H,G}(\beta), H) \in GL(\Gamma_{H,G}(\beta), H). \end{aligned}$$

Step 2. Proof that the overlap maps are C^∞. Let $(U_G, \varphi_{F,G})$, $(U_{G'}, \varphi_{F',G'})$ be two charts at the points $F, F' \in \mathbb{G}(E)$ such that $U_G \cap U_{G'} \neq \varnothing$. If $\alpha \in \varphi_{F,G}(U_G \cap U_{G'})$, then $I + \alpha \in GL(F, \Gamma_{F,G}(\alpha))$, where I is the identity mapping on F, and $\pi_{G'}(\Gamma_{F,G}(\alpha), F') \in GL(\Gamma_{F,G}(\alpha), F')$ since $\Gamma_{F,G}(\alpha) \in U_G \cap U_{G'}$. Therefore $\pi_{G'}(F') \circ (I + \alpha) \in GL(F, F')$ and we get

$$(\varphi_{F',G'} \circ \varphi_{F,G}{}^{-1})(\alpha) = \varphi_{F',G'}(\Gamma_{F,G}(\alpha)) = \pi_{F'}(\Gamma_{F,G}(\alpha), G') \circ \pi_{G'}(F', \Gamma_{F,G}(\alpha))$$

$$= \pi_{F'}(\Gamma_{F,G}(\alpha), G') \circ \pi_{G'}(F', \Gamma_{F,G}(\alpha)) \circ \pi_{G'}(F') \circ (I + \alpha) \circ [\pi_{G'}(F') \circ (I + \alpha)]^{-1}$$

$$= \pi_{F'}(G') \circ (I + \alpha) \circ [\, \pi_{G'}(F') \circ (I + \alpha)]^{-1}$$

which is a C^∞ map from $\varphi_{F,G}(U_G \cap U_{G'}) \subset L(F, G)$ to $\varphi_{F',G'}(U_G \cap U_{G'}) \subset L(F', G')$. Since its inverse is $\beta \in L(F', G') \mapsto \pi_F(G) \circ (I' + \beta) \circ [\pi_G(F) \circ (I' + \beta)]^{-1} \in L(F, G)$, where I' is the identity mapping on F', it follows that the maps $\varphi_{F',G'} \circ \varphi_{F,G}{}^{-1}$ are diffeomorphisms.

Thus $\mathbb{G}(E)$ *is a* C^∞ *Banach manifold, locally modeled on* $L(F, G)$.

Let $\mathbb{G}_n(E)$ (resp. $\mathbb{G}^n(E)$) denote the space of n-dimensional (n-codimensional) subspaces of E. From the preceding proof we see that $\mathbb{G}_n(E)$ and $\mathbb{G}^n(E)$ are connected components of $\mathbb{G}(E)$ and so are also manifolds. The classical Grassmann manifolds are $\mathbb{G}_n(\mathbb{R}^m)$, where $m \geq n$ (n-planes in m space). They are connected $n(m - n)$-manifolds. Furthermore, $\mathbb{G}_n(\mathbb{R}^m)$ *is compact.* To see this, consider the set $F_{n,m}$ of orthogonal sets of n unit vectors in $\mathbb{R}^m$. Since $F_{n,m}$ is closed and bounded in $\mathbb{R}^m \times \cdots \times \mathbb{R}^m$ (n times), $F_{n,m}$ is compact. Thus $\mathbb{G}_n(\mathbb{R}^m)$ is compact, as the continuous image of $F_{n,m}$ by the map $\{e_1, ..., e_n\} \mapsto$ the span of $e_1, ..., e_n$.

H Projective spaces Let $\mathbb{RP}^n = \mathbb{G}_1(\mathbb{R}^{n+1})$ = the set of lines in $\mathbb{R}^{n+1}$. Thus from the previous example, $\mathbb{RP}^n$ is a compact connected real n-manifold. Similarly $\mathbb{CP}^n$, the set of complex lines in $\mathbb{C}^n$, is a compact connected (complex) n-manifold. There is a projection $\pi : S^n \to \mathbb{RP}^n$ defined by $\pi(x) = \text{span}(x)$, which is a diffeomorphism restricted to any open hemisphere. Thus any chart for S^n produces one for $\mathbb{RP}^n$ as well. ♦

Exercises

3.1A Let $S = \{(x, y) \in \mathbb{R}^2 \mid xy = 0\}$. Construct two "charts" by mapping each axis to the real line by $(x, 0) \mapsto x$ and $(0, y) \mapsto y$. What fails in the definition of a manifold?

3.1B Let $S =]0, 1[\times]0, 1[\subset \mathbb{R}^2$ and for each $s, 0 \leq s \leq 1$ let $\mathcal{V}_s = \{s\} \times]0, 1[$ and $\varphi_s : \mathcal{V}_s \to \mathbb{R}$, $(s, t) \mapsto t$. Does this make S into a one-manifold?

3.1C Let $S = \{(x, y) \in \mathbb{R}^2 \mid x^2 - y^2 = 1\}$. Show that the two charts $\varphi_1 : \{(x, y) \in S \mid \pm x > 0\} \to \mathbb{R}$, $\varphi_\pm(x, y) = y$ define a manifold structure on the disconnected set S.

3.1D On the topological space $\mathbb{M}$ obtained from $[0, 2\pi] \times \mathbb{R}$ by identifying the point $(0, x)$ with $(2\pi, -x)$, $x \in \mathbb{R}$, consider the following two charts:

(i) $(]0, 2\pi[\times \mathbb{R}$, identity), and

(ii) $(([0, \pi[\cup]\pi, 2\pi[) \times \mathbb{R}, \varphi)$, where φ is defined by $\varphi(\theta, x) = (\theta, x)$ if $0 \leq \theta < \pi$ and $\varphi(\theta, x) = (\theta - 2\pi, -x)$ if $\pi < \theta < 2\pi$. Show that these two charts define a manifold structure on $\mathbb{M}$. This manifold is called the ***Möbius band*** (see Fig. 3.4.3 and Example **3.4.8C** for an alternative description). Note that the chart (ii) joins 2π to 0 and twists the second factor $\mathbb{R}$, as required by the topological structure of $\mathbb{M}$.

(iii) Repeat a construction like (ii) for $\mathbb{K}$, the ***Klein bottle*** .

3.1E Compactification of $\mathbb{R}^n$. Let $\{\infty\}$ be a one point set and let $\mathbb{R}^n{}_c = \mathbb{R}^n \cup \{\infty\}$. Define the charts (U, φ) and $(U_\infty, \varphi_\infty)$ by $U = \mathbb{R}^n$, φ = identity on $\mathbb{R}^n$, $U_\infty = \mathbb{R}^n{}_c \setminus \{0\}$, $\varphi_\infty(x) = x / \| x \|^2$, if $x \neq \infty$ and $\varphi_\infty(x) = 0$, if $x = \infty$.

(i) Show that the atlas $\mathcal{A}_c = \{(U, \varphi), (U_\infty, \varphi_\infty)\}$ defines a smooth manifold structure on $\mathbb{R}^n{}_c$.

(ii) Show that with the topology induced by $\mathcal{A}_c$, $\mathbb{R}^n{}_c$ becomes a compact topological space. It is called the ***one-point compactification*** of $\mathbb{R}^n$.

(iii) Show that if $n = 2$, the differentiable structure of $\mathbb{R}^2{}_c = \mathbb{C}_c$ can be alternatively given by the chart (U, φ) and the chart (U_∞, ψ_∞), where $\psi_\infty(z) = z^{-1}$, if $z \neq \infty$ and $\psi_\infty(z) = 0$, if $z = \infty$.

(iv) Show that stereographic projection induces a homeomorphism of $\mathbb{R}^n{}_c$ with S^n.

3.1F (i) Define an equivalence relation $\sim$ on S^n by $x \sim y$ if $x = \pm y$. Show that $S^n/\sim$ is homeomorphic with $\mathbb{RP}^n$.

(ii) Show that

(a) $e^{i\theta} \in S^1 \mapsto e^{2i\theta} \in S^1$, and

(b) $(x, y) \in S^1 \mapsto (xy^{-1}$, if $y \neq 0$ and ∞, if $y = 0) \in \mathbb{R}_c \cong S^2$ (see Exercise **3.1E**) induce homeomorphisms of S^1 with $\mathbb{RP}^1$.

(iii) Show that neither S^n nor $\mathbb{RP}^n$ can be covered by a single chart.

3.1G (i) Define an equivalence relation on $S^{2n+1} \subset \mathbb{C}^{2(n+1)}$ by $x \sim y$ if $y = e^{i\theta}x$ for some $\theta \in \mathbb{R}$. Show $S^{2n+1}/\sim$ is homeomorphic to $\mathbb{CP}^n$.

(ii) Show that

(a) $(u, v) \in S^3 \subset \mathbb{C}^2 \mapsto 4(-u\bar{v}, | v |^2 - | u |^2) \in S^2$, and

(b) $(u, v) \in S^3 \subset \mathbb{C}^2 \mapsto (uv^{-1}$, if $v \neq 0$, and ∞, if $v = 0) \in \mathbb{R}^2{}_c \cong S^2$ (see Exercise **3.1E**) induce homeomorphisms of S^2 with $\mathbb{CP}^1$. The map in (a) is called the ***classical Hopf fibration***; it will be studied further in Section **3.4.**

3.1H Flag manifolds Let F_n denote the set of sequences of nested linear subspaces $V_1 \subset V_2 \subset \dots \subset V_{n-1}$ in $\mathbb{R}^n$ (or $\mathbb{C}^n$), where $\dim V_i = i$. Show that F_n is a compact manifold and compute its dimension. (Flag manifolds are typified by F_n and come up in the study of symplectic geometry and representations of Lie groups.) (*Hint:* Show that F_n is in bijective correspondence with the quotient $\mathbb{GL}(n)$/upper triangular matrices.)

§3.2 *Submanifolds, Products, and Mappings*

A submanifold is the nonlinear analogue of a subspace in linear algebra. Likewise, the product of two manifolds, producing a new manifold, is the analogue of a product vector space. The analogue of linear transformations are the C^r maps between manifolds, also introduced in this section. We are not yet ready to differentiate these mappings; this will be possible after we introduce the tangent bundle in §3.3.

If M is a manifold and $A \subset M$ is an open subset of M, the differentiable structure of M naturally induces one on A. We call A an ***open submanifold*** of M. For example, $\mathbb{G}_n(E)$, $\mathbb{G}^n(E)$ are open submanifolds of $\mathbb{G}(E)$ (see Example **3.1.8G**). We would also like to say that S^n is a submanifold of $\mathbb{R}^{n+1}$, although it is a closed subset. To motivate the general definition we notice that there are charts in $\mathbb{R}^{n+1}$ in which a neighborhood of S^n becomes part of the subspace $\mathbb{R}^n$. Figure 3.2.1 illustrates this for $n = 1$.

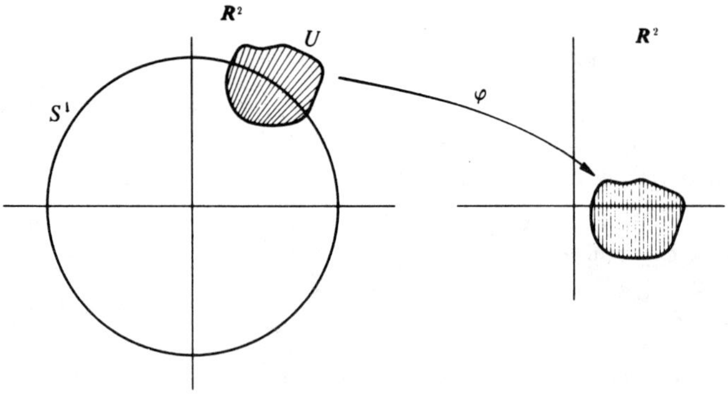

Figure 3.2.1

3.2.1 Definition *A* ***submanifold*** *of a manifold* M *is a subset* $B \subset M$ *with the property that for each* $b \in B$ *there is an admissible chart* (U, φ) *in* M *with* $b \in U$ *which has the* ***submanifold property****, namely,*

SM $\varphi : U \to E \times F$, *and* $\varphi(U \cap B) = \varphi(U) \cap (E \times \{0\})$.

An open subset V of M is a submanifold in this sense. Here we merely take $F = \{0\}$, and for $x \in V$ use any chart (U, φ) of M for which $x \in U$.

3.2.2 Proposition *Let* B *be a submanifold of a manifold* M. *Then* B *itself is a manifold with differentiable structure generated by the atlas:*

$$\{(U \cap B, \varphi \mid U \cap B) \mid (U, \varphi) \textit{ is an admissible chart in } M \textit{ having property } \mathbf{SM} \textit{ for } B\} .$$

Furthermore, the topology on B *is the relative topology.*

Proof If $U_i \cap U_j \cap B \neq \varnothing$, and (U_i, φ_i) and (U_j, φ_j) both have the submanifold property, and if we write $\varphi_i = (\alpha_i, \beta_i)$ and $\varphi_j = (\alpha_j, \beta_j)$, where $\alpha_i : U_i \to E$, $\alpha_j : U_j \to E$, $\beta_i : U_i \to F$, and $\beta_j : U_j \to F$, then the maps

$$\alpha_i \mid U_i \cap B : U_i \cap B \to \varphi_i(U_i) \cap (E \times \{0\}) \quad \text{and} \quad \alpha_j \mid U_j \cap B : U_j \cap B \to \varphi_j(U_j) \cap (E \times \{0\})$$

are bijective. The overlap map $(\varphi_j \mid U_j \cap B) \circ (\varphi_i \mid U_i \cap B)^{-1}$ is given by $(e, 0) \mapsto ((\alpha_j \circ \alpha_i^{-1})(e), 0) = \varphi_{ji}(e, 0)$ and is C^∞, being the restriction of a C^∞ map. The last statement is a direct consequence of the definition of relative topology and **3.2.1.** ■

If M is an n-manifold and B a submanifold of M, the ***codimension*** of B in M is defined by codim B = dim M − dim B. Note that open submanifolds are characterized by having codimension zero.

In §**3.5** methods are developed for proving that various subsets are actually submanifolds, based on the Implicit Function Theorem. For now we do a case "by hand."

3.2.3 Example To show that $S^n \subset \mathbb{R}^{n+1}$ is a submanifold, it is enough to observe that the charts in the atlas $\{(U_i^\pm, \psi_i^\pm)\}$, $i = 1, \ldots, n + 1$ of S^n come from charts of $\mathbb{R}^{n+1}$ with the submanifold property (see **3.1.4**): the $2(n + 1)$ maps

$$\chi_i^\pm : \{x \in \mathbb{R}^{n+1} \mid \pm x^i > 0\} \to \{y \in \mathbb{R}^{n+1} \mid (y^{n+1} + 1)^2 > (y^1)^2 + \ldots + (y^n)^2\}$$

given by

$$\chi_i^\pm(x^1, \ldots, x^{n+1}) = (x^1, \ldots, x^{i-1}, x^{i+1}, \ldots, x^{n+1}, \|x\| - 1)$$

are C^∞ diffeomorphisms, and charts in an atlas of $\mathbb{R}^{n+1}$. Since $(\chi_i^\pm \mid U_i^\pm)(x^1, \ldots, x^{n+1}) = (x^1, \ldots, x^{i-1}, x^{i+1}, \ldots, x^{n+1}, 0)$, they have the submanifold property for S^n. ♦

3.2.4 Definition *Let* $(S_1, \mathcal{D}_1)$ *and* $(S_2, \mathcal{D}_2)$ *be two manifolds. The* ***product manifold*** $(S_1 \times S_2, \mathcal{D}_1 \times \mathcal{D}_2)$ *consists of the set* $S_1 \times S_2$ *together with the differentiable structure* $\mathcal{D}_1 \times \mathcal{D}_2$ *generated by the atlas* $\{(U_1 \times U_2, \varphi_1 \times \varphi_2) \mid (U_i, \varphi_i) \textit{ is a chart of } (S_i, \mathcal{D}_i), i = 1, 2\}$.

That the set in **3.2.4** is an atlas follows from the fact that if $\psi_1 : U_1 \subset E_1 \to V_1 \subset F_1$ and $\psi_2 : U_2 \subset E_2 \to V_2 \subset F_2$, then $\psi_1 \times \psi_2$ is a diffeomorphism iff ψ_1 and ψ_2 are, and in this case $(\psi_1 \times \psi_2)^{-1} = \psi_1^{-1} \times \psi_2^{-1}$. It is clear that the topology on the product manifold is the product topology. Also, if S_1, S_2 are finite dimensional, $\dim(S_1 \times S_2) = \dim S_1 + \dim S_2$. Inductively one defines the product of a finite number of manifolds. A simple example of a product manifold is the n-***torus*** $\mathbb{T}^n = S^1 \times \cdots \times S^1$ (n times).

The final topic of this section is that of maps between manifolds. The following definition introduces two important ideas: the local representative of a map and the concept of a C^r map between manifolds.

3.2.5 Definition *Suppose* $f : M \to N$, *where* M *and* N *are* C^k *manifolds (that is,* f *maps the underlying set of* M *into that of* N*). We say* f *is of* ***class*** $C^r, 0 \le r \le k$, *if for each* x *in* M *and an admissible chart* (V, ψ) *of* N *with* $f(x) \in V$, *there is a chart* (U, φ) *of* M *satisfying* $x \subset U$, *and* $f(U) \subset V$, *and such that the local representative of* f, $f_{\varphi\psi} = \psi \circ f \circ \varphi^{-1}$, *is of class* C^r. (See Figure 3.2.2.)

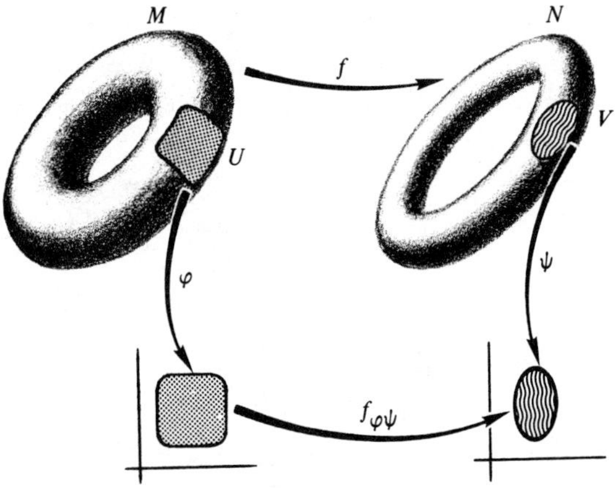

Figure 3.2.2

For $r = 0$, this is consistent with the definition of continuity of f, regarded as a map between topological spaces (with the manifold topologies).

3.2.6 Proposition *Let* $f : M \to N$ *be a continuous map of manifolds. Then* f *is* C^r *iff the local representatives of* f *relative to a collection of charts which cover* M *and* N *are* C^r.

Proof Assume that the local representatives of f relative to a collection of charts covering M and N are C^r. If (U, φ) and (U, φ') are charts in M and (V, ψ), (V, ψ') are charts in N such that $f_{\varphi\psi}$ is C^r, then the Composite Mapping Theorem and condition **MA2** of **3.1.1** show that $f_{\varphi'\psi'} = (\psi' \circ \psi^{-1}) \circ f_{\varphi\psi} \circ (\varphi' \circ \varphi^{-1})^{-1}$ is also C^r. Moreover, if φ'' and ψ'' are restrictions of φ and

ψ to open subsets of U and V, then $f_{\varphi''\psi''}$ is also C^r. Finally, note that if f is C^r on open submanifolds of M, then it is C^r on their union. That f is C^r now follows from the fact that any chart of M can be obtained from the given collection by change of diffeomorphism, restrictions, and/or unions of domains, all three operations preserving the C^r character of f. This argument also demonstrates the converse. ■

Any map from (open subsets of) E to F which is C^r in the Banach space sense is C^r in the sense of Definition **3.2.5**. Other examples of C^∞ maps are the antipodal map $x \mapsto -x$ of S^n and the transation map by $(\theta_1, \ldots, \theta_n)$ on $\mathbb{T}^n$ given by $(\exp(ir_1), \ldots, \exp(ir_n)) \mapsto \exp(i(r_1 + \theta_1)), \ldots, \exp(i(r_n + \theta_n))$. From the previous proposition and the Composite Mapping Theorem, we get the following.

3.2.7 Proposition *If* $f: M \to N$ *and* $g: N \to P$ *are* C^r *maps, then so is* $g \circ f$.

3.2.8 Definition *A map* $f: M \to N$, *where* M *and* N *are manifolds, is called a* C^r ***diffeomorphism*** *if* f *is of class* C^r, *is a bijection, and* $f^{-1}: N \to M$ *is of class* C^r. *If a diffeomorphism exists between two manifolds, they are called* ***diffeomorphic***.

It follows from **3.2.7** that the set $\mathfrak{Diff}^r(M)$ of C^r diffeomorphisms of M forms a group under composition. This large and intricate group will be encountered again several times in the book.

Exercises

3.2A Show that

(i) if (U, φ) is a chart of M and $\psi: \varphi(U) \to V \subset F$ is a diffeomorphism, then $(U, \psi \circ \varphi)$ is an admissible chart of M, and

(ii) admissible local charts are diffeomorphisms.

3.2B A C^1 diffeomorphism that is also a C^r map is a C^r diffeomorphism. (*Hint:* Use the comments after the proof of **2.5.2**.)

3.2C Show that if $N_i \subset M_i$ are submanifolds, $i = 1, \ldots, n$, then $N_1 \times \cdots \times N_n$ is a submanifold of $M_1 \times \cdots \times M_n$.

3.2D Show that every submanifold N of a manifold M is locally closed in M; i.e., every point $n \in N$ has a neighborhood U in M such that $N \cap U$ is closed in U.

3.2E Show that $f_i: M_i \to N_i$, $i = 1, \ldots, n$ are all C^r iff $f_1 \times \cdots \times f_n: M_1 \times \cdots \times M_n \to N_1 \times \cdots \times N_n$ is C^r.

3.2F Let M be a set and $\{M_i\}_{i \in I}$ a covering of M, each M_i being a manifold. Assume that for every pair of indices (i, j), $M_i \cap M_j$ is an open submanifold in both M_i and M_j. Show that there is a unique manifold structure on M for which the M_i are open submanifolds. The differentiable structure on M is said to be obtained by the ***collation*** of the differentiable structures of M_i.

3.2G Show that the map $F \mapsto F^0 = \{u \in F^* \mid u \mid F = 0\}$ of $\mathbb{G}(E)$ into $\mathbb{G}(E^*)$ is a C^∞ map. If $E = E^{**}$ (i.e., E is reflexive) it restricts to a C^∞ diffeomorphism of $\mathbb{G}^n(E)$ onto $\mathbb{G}_n(E^*)$ for all $n = 1, 2, \ldots$. Conclude that $\mathbb{RP}^n$ is diffeomorphic to $\mathbb{G}^n(\mathbb{R}^{n+1})$.

3.2H Show that the two differentiable structures of $\mathbb{R}$ defined in **3.1.8D** are diffeomorphic. (*Hint:* Consider the map $x \mapsto x^{1/3}$.)

3.2I (i) Show that S^1 and $\mathbb{RP}^1$ are diffeomorphic manifolds (see Exercise **3.1F**(b)).
(ii) Show that $\mathbb{CP}^1$ is diffeomorphic to S^2 (see Exercise **3.1G**(b)).

3.2J Let $M_\lambda = \{(x, |x|^\lambda) \mid x \in \mathbb{R}\}$, where $\lambda \in \mathbb{R}$. Show that
(i) if $\lambda \le 0$, M_λ is a C^∞ submanifold of $\mathbb{R}^2$;
(ii) if $\lambda > 0$ is an even integer, M_λ is a C^∞ submanifold of $\mathbb{R}^2$;
(iii) if $\lambda > 0$ is an odd integer or not an integer, then M_λ is a $C^{[\lambda]}$ submanifold of $\mathbb{R}^2$ which is not $C^{[\lambda]+1}$, where $[\lambda]$ denotes the smallest integer $\ge \lambda$, i.e., $[\lambda] \le \lambda < [\lambda] + 1$;
(iv) in case (iii), show that M_λ is the union of three disjoint C^∞ submanifolds of $\mathbb{R}^2$.

3.2K Let M be a C^k submanifold. Show that the diagonal $\Delta = \{(m, m) \mid m \in M\}$ is a closed C^k submanifold of $M \times M$.

3.2L Let E be a Banach space. Show that the map $x \mapsto Rx(R^2 - \| x \|^2)^{-1/2}$ is a diffeomorphism of the open ball of radius R with E. Conclude that any manifold M modeled on E has an atlas $\{(U_i, \varphi_i)\}$ for which $\varphi_i(U_i) = E$.

3.2M If $f : M \to N$ is of class C^k and S is a submanifold of M, show that $f \mid S$ is of class C^k.

3.2N Let M and N be C^r manifolds and $f : M \to N$ be a continuous map. Show that f is of class C^k, $1 \le k \le r$ if and only if for any open set U in N and any C^k map $g : U \to E$, E a Banach space, the map $g \circ f : f^{-1}(U) \to E$ is C^k.

3.2O Let $\pi : S^n \to \mathbb{RP}^n$ denote the projection. Show that $f : \mathbb{RP}^n \to M$ is smooth iff the map $f \circ \pi : S^n \to M$ is smooth; here M denotes another smooth manifold.

3.2P Covering Manifolds Let M and N be smooth manifolds and let $p : M \to N$ be a smooth map. The map p is called a ***covering***, or equivalently, M is said to ***cover*** N, if p is surjective and each point $n \in N$ admits an open neighborhood V such that $p^{-1}(V)$ is a union of disjoint open sets, each diffeomorphic via p to V.

(i) **Path lifting property** Suppose $p : M \to N$ is a covering and $p(m_0) = n_0$, where $n_0 \in N$ and $m_0 \in M$. Let $c : [0, 1] \to N$ be a C^k path, $k \geq 0$, starting at $n_0 = c(0)$. Show that there is a unique C^k path $d : [0, 1] \to M$, such that $d(0) = m_0$ and $p \circ d = c$. (*Hint:* Partition [0, 1] into a finite set of closed intervals $[t_i, t_{i+1}]$, $i = 0, ..., n-1$, where $t_0 = 0$ and $t_n = 1$, such that each of the sets $c([t_i, t_{i+1}])$ lies entirely in a neighborhood V_i guaranteed by the covering property of p. Let U_0 be the open set in the union $p^{-1}(V_0)$ containing m_0. Define $d_0 : [0, t_1] \to U_0$ by $d_0 = p^{-1} \circ c \mid [0, t_1]$. Let V_1 be the open set containing $c([t_1, t_2])$ and U_1 be the open set in the union $p^{-1}(V_1)$ containing $d(t_1)$. Define the map $d_1 : [t_1, t_2] \to U_1$ by $d_1 = p^{-1} \circ c \mid [t_1, t_2]$. Now proceed inductively. Show that d so obtained is C^k if c is and prove the construction is independent of the partition of [0, 1].)

(ii) **Homotopy lifting property** In the hypotheses and notations of (i), let $H : [0, 1] \times [0, 1] \to N$ be a C^k map, $k \geq 0$ and assume that $H(0, 0) = n_0$. Show that there is a unique C^k-map $K : [0, 1] \times [0, 1] \to M$ such that $K(0, 0) = m_0$ and $p \circ K = H$. (*Hint:* Apply the reasoning in (i) to the square $[0, 1] \times [0, 1]$.)

(iii) Show that if two curves in N are homotopic via a homotopy keeping the endpoints fixed, then the lifted curves are also homotopic via a homotopy keeping the endpoints fixed.

(iv) Assume that $p_i : M_i \to N$ are coverings of N with M_i connected, $i = 1, 2$. Show that if M_1 is simply connected, then M_1 is also a covering of M_2. (*Hint:* Choose points $n_0 \in N$, $m_1 \in M_1$, $m_2 \in M_2$ such that $p_i(m_i) = n_0$, $i = 1, 2$. Let $x \in M_1$ and let $c_1(t)$ be a C^k-curve (k is the differentiability class of M_1, M_2, and N) in M_1 such that $c_1(0) = m_1$, $c_1(1) = x$. Then $c(t) = (p \circ c_1)(t)$ is a curve in N connecting n_0 to $p_1(x)$. Lift this curve to a curve $c_2(t)$ in M_2 connecting m_2 to $y = c_2(1)$ and define $q : M_1 \to M_2$ by $q(x) = y$. Show by (iii) that q is well defined and C^k. Then show that q is a covering.)

(v) Show that if $p_i : M_i \to N$, $i = 1, 2$ are coverings with M_1 and M_2 simply connected, then M_1 and M_2 are C^k-diffeomorphic. This is why a simply connected covering of N is called ***the universal covering manifold*** of N.

3.2Q Construction of the universal covering manifold Let N be a connected (hence arcwise connected) manifold and fix $n_0 \in N$. Let M denote the set of homotopy classes of paths $c : [0, 1] \to N$, $c(0) = n_0$, keeping the endpoints fixed. Define $p : M \to N$ by $p([c]) = c(1)$, where [c] is the homotopy class of c.

(i) Show that p is onto since N is arcwise connected.

(ii) For an open set U in N define $U_{[c]} = \{[c * d] \mid d \text{ is a path in } U \text{ starting at } c(1)\}$. (See Exercise **1.6F** for the definition of $c * d$.) Show that $\mathcal{B} = \{\varnothing, U_{[c]} \mid$ c is a path in N starting at n_0 and U is open in N$\}$ is a basis for a topology on M. Show that if N is Hausdorff, so is M. Show that p is continuous.

(iii) Show that M is arcwise connected. (*Hint:* A continuous path $\varphi : [0, 1] \to M$, $\varphi(0) = [c]$, $\varphi(1) = [d]$ is given by $\varphi(s) = [c_s]$, for $s \in [0, 1/2]$, and $\varphi(s) = [d_s]$, for $s \in [1/2, 1]$, where $c_s(t) = c((1 - 2s)t)$, $d_s(t) = d((2s - 1)t)$.)

(iv) Show that p is an open map. (*Hint:* If $n \in p(U_{[c]})$ then the set of points in U that can be joined to n by paths in U is open in N and included in $p(U_{[c]})$.)

(v) Use (iv) to show that $p : M \to N$ is a covering. (*Hint:* Let U be a contractible chart domain of N and show that $p^{-1}(U) = \bigcup U_{[c]}$, where the union is over all paths c with $p([c]) = n$, n a fixed point in U.)

(vi) Show that M is simply connected. (*Hint:* If $\psi : [0, 1] \to M$ is a loop based at [c], i.e., ψ is continuous and $\psi(0) = \psi(1) = [c]$, then $H : [0, 1] \times [0, 1] \to M$ given by $H(\cdot, s) = [c_s]$, $c_s(t) = c(ts)$ is a homotopy of [c] with the constant path [c(0)].)

(vii) If (U, φ) is a chart on N whose domain is such that $p^{-1}(U)$ is a disjoint union of open sets in M each diffeomorphic to U (see (v)), define $\psi : V \to E$ by $\psi = \varphi \circ p \mid V$. Show that the atlas defined in this way defines a manifold structure on M. Show that M is locally diffeomorphic to N.

§3.3 *The Tangent Bundle*

Recall that for $f : U \subset E \to V \subset F$ of class C^{r+1} we define the tangent of f, $Tf : TU \to TV$ by setting $TU = U \times E$, $TV = V \times F$, and $Tf(u, e) = (f(u), \mathbf{D}f(u) \cdot e)$ and that the chain rule reads $T(g \circ f) = Tg \circ Tg$. If for each open set U in some vector space E, $\tau_U : TU \to U$ denotes the projection, the diagram

$$\begin{array}{ccc} TU & \xrightarrow{\;Tf\;} & TV \\ {\scriptstyle \tau_U}\downarrow & & \downarrow{\scriptstyle \tau_V} \\ U & \xrightarrow[\;f\;]{} & V \end{array}$$

is commutative, that is, $f \circ \tau_U = \tau_V \circ Tf$.

The tangent operation T can now be extended from this local context to the context of differentiable manifolds and mappings. During the definitions it may be helpful to keep in mind the example of the family of tangent spaces to the sphere $S^n \subset \mathbb{R}^{n+1}$.

A major advance in differential geometry occurred when it was realized how to define the tangent space to an abstract manifold independent of any embedding in $\mathbb{R}^n$. (The history is not completely clear to us, but this idea seems to be primarily due to Riemann, Weyl, and Levi-Cività and was "well known" by 1920.) Several alternative ways to do this can be used according to taste. (See below and Spivak [1979] for further information.)

1 ***Coordinate approach*** Using transformation properties of vectors under coordinate changes, one defines a tangent vector at $m \in M$ to be an equivalence class of triples (U, φ, e), where $\varphi : U \to E$ is a chart and $e \in E$, with two triples identified if they are related by the tangent of the corresponding overlap map evaluated at the point corresponding to $m \in M$.

2 ***Derivation approach*** This approach characterizes a vector by specifying a map that gives the derivative of a general function in the direction of that vector.

3 ***The ideal approach*** This is a variation of alternative 2. Here T_mM is defined to be the dual of $I^{(0)}_m / I^{(1)}_m$, where $I^{(j)}_m$ is the ideal of functions on M vanishing up to order j at m.

4 ***The curves approach*** This is the method followed here. We abstract the idea that a tangent vector to a surface is the velocity vector of a curve in the surface.

If [a, b] is a closed interval, a continuous map $c : [a, b] \to M$ is said to be ***differentiable*** at the endpoint a if there is a chart (U, φ) at c(a) such that

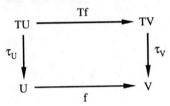

$$\lim_{t \downarrow a} \frac{(\varphi \circ c)(t) - (\varphi \circ c)(a)}{t - a}$$

exists and is finite; this limit is denoted by $(\varphi \circ c)'(a)$. If (V, ψ) is another chart at $c(a)$ and we let $v = (\varphi \circ c)(t) - (\varphi \circ c)(a)$, then in $U \cap V$ we have

$$(\psi \circ \varphi^{-1})((\varphi \circ c)(t)) - (\psi \circ \varphi^{-1})((\varphi \circ c)(a)) = \mathbf{D}(\psi \circ \varphi^{-1})((\varphi \circ c)(a)) \cdot v + o(\| v \|)$$

whence

$$\frac{(\psi \circ c)(t) - (\psi \circ c)(a)}{t - a} = \frac{\mathbf{D}(\psi \circ \varphi^{-1})(\varphi \circ c)(a) \cdot v}{t - a} + \frac{o(\| v \|)}{t - a} .$$

Since

$$\lim_{t \downarrow a} \frac{v}{t - a} = (\varphi \circ c)'(a) \quad \text{and} \quad \lim_{t \downarrow a} \frac{o(\| v \|)}{t - a} = 0 ,$$

it follows that

$$\lim_{t \downarrow a} \frac{[(\psi \circ c)(t) - (\psi \circ c)(a)]}{t - a} = \mathbf{D}(\psi \circ \varphi^{-1})((\varphi \circ c)(a)) \cdot (\varphi \circ c)'(a)$$

and therefore the map $c : [a, b] \to M$ is differentiable at a in the chart (U, φ) iff it is differentiable at a in the chart (V, ψ). In summary, *it makes sense to speak of differentiability of curves at an endpoint of a closed interval.* The map $c : [a, b] \to M$ is said to be ***differentiable*** if $c \mid]a, b[$ is differentiable and if c is differentiable at the endpoints a and b. The map $c : [a, b] \to M$ is said to be ***of class*** C^1 if it is differentiable and if $(\varphi \circ c)' : [a, b] \to E$ is continuous for any chart (U, φ) satisfying $U \cap c([a, b]) \neq \varnothing$, where E is the model space of M.

3.3.1 Definition *Let* M *be a manifold and* $m \in M$. *A* ***curve at*** m *is a* C^1 *map* $c : I \to M$ *from an interval* $I \subset \mathbb{R}$ *into* M *with* $0 \in I$ *and* $c(0) = m$. *Let* c_1 *and* c_2 *be curves at* m *and* (U, φ) *an admissible chart with* $m \in U$. *Then we say* c_1 *and* c_2 *are* ***tangent at*** m ***with respect to*** φ *if and only if* $(\varphi \circ c_1)'(0) = (\varphi \circ c_2)'(0)$.

Thus two curves are tangent with respect to φ if they have identical tangent vectors (same direction and speed) in the chart φ; see Figure 3.3.1. The reader can safely assume in what follows that I is an open interval; the use of closed intervals becomes essential when defining tangent vectors to a manifold with boundary at a boundary point; this will be discussed in Chapter 7.

3.3.2 Proposition *Let* c_1 *and* c_2 *be two curves at* $m \in M$. *Suppose* (U_β, φ_β) *are admissible charts with* $m \in U_\beta$, $\beta = 1, 2$. *Then* c_1 *and* c_2 *are tangent at* m *with respect to* φ_1 *if and only if they are tangent at* m *with respect to* φ_2.

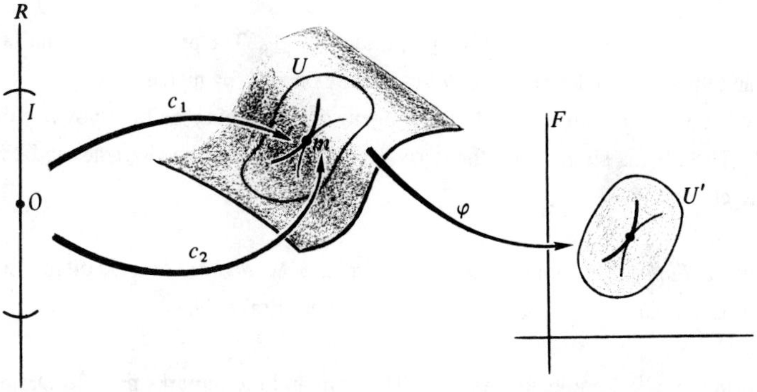

Figure 3.3.1

Proof By taking restrictions if necessary we may suppose that $U_1 = U_2$. Since we have the identity $\varphi_2 \circ c_i = (\varphi_2 \circ \varphi_1^{-1}) \circ (\varphi_1 \circ c_i)$, the C^1 Composite Mapping Theorem in Banach spaces implies that $(\varphi_2 \circ c_1)'(0) = (\varphi_2 \circ c_2)'(0)$ iff $(\varphi_1 \circ c_1)'(0) = (\varphi_1 \circ c_2)'(0)$. ∎

This proposition guarantees that the *tangency of curves* at $m \in M$ *is a notion that is independent of the chart used.* Thus we say c_1, c_2 are ***tangent at*** $m \in M$ if c_1, c_2 are tangent at m with respect to φ, for any local chart φ at m. It is evident that tangency at $m \in M$ is an equivalence relation among curves at m. An equivalence class of such curves is denoted $[c]_m$, where c is a representative of the class.

3.3.3 Definition *For a manifold* M *and* $m \in M$ *the* ***tangent space to*** M ***at*** m *is the set of equivalence classes of curves at* m:

$$T_mM = \{[c]_m \mid c \textit{ is a curve at } m\} .$$

For a subset $A \subset M$, *let* $TM \mid A = \bigcup_{m \in A} T_mM$ *(disjoint union). We call* $TM = TM \mid M$ *the* ***tangent bundle of*** M. *The mapping* $\tau_M : TM \to M$ *defined by* $\tau_M([c]_m) = m$ *is the* ***tangent bundle projection*** *of* M.

Let us show that if $M = U$, an open set in a Banach space E, TU as defined here can be identified with $U \times E$. This will establish consistency with our usage of T in §**2.3**.

3.3.4 Lemma *Let* U *be an open subset of* E, *and* c *be a curve at* $u \in U$. *Then there is a unique* $e \in E$ *such that the curve* $c_{u,e}$ *defined by* $c_{u,e}(t) = u + te$ (*with* t *belonging to an interval* I *such that* $c_{u,e}(I) \subset U$) *is tangent to* c *at* u.

Proof By definition, $\mathbf{D}c(0)$ is the unique linear map in $L(\mathbb{R}, E)$ such that the curve $g : \mathbb{R} \to E$ given by $g(t) = u + \mathbf{D}c(0) \cdot t$ is tangent to c at $t = 0$. If $e = \mathbf{D}c(0) \cdot 1$, then $g = c_{u,e}$. ∎

Define a map $i : U \times E \to T(U)$ by $i(u, e) = [c_{u,e}]_u$. The preceding lemma says that i is a bijection and thus we can define a manifold structure on TU by means of i.

It will be convenient to define the tangent of a mapping before showing that TM is a manifold. The idea is simply that the derivative of a map can be characterized by its effect on tangents to curves.

3.3.5 Lemma *Suppose* c_1 *and* c_2 *are curves at* $m \in M$ *and are tangent at* m. Let $f : M \to N$ *be of class* C^1. *Then* $f \circ c_1$ *and* $f \circ c_2$ *are tangent at* $f(m) \in N$.

Proof From the C^1 Composite Mapping Theorem and the remarks prior to Definition **3.3.1**, it follows that $f \circ c_1$ and $f \circ c_2$ are of class C^1. For tangency, let (V, ψ) be a chart on N with $f(m) \in V$. We must show that $(\psi \circ f \circ c_1)'(0) = (\psi \circ f \circ c_2)'(0)$. But $\psi \circ f \circ c_\alpha = (\psi \circ f \circ \varphi^{-1}) \circ (\varphi \circ c_\alpha)$, where (U, φ) is a chart on M with $f(U) \subset V$. Hence the result follows from the C^1 Composite Mapping Theorem. ■

3.3.6 Definition *If* $f : M \to N$ *is of class* C^1, *we define* $Tf : TM \to TN$ *by*

$$Tf([c]_m) = [f \circ c]_{f(m)} .$$

We call Tf *the* ***tangent*** *of* f.

Tf is well defined, for if we choose any other representative from $[c]_m$, say c_1, then c and c_1 are tangent at m and hence $f \circ c$ and $f \circ c_1$ are tangent at $f(m)$, i.e., $[f \circ c]_{f(m)} = [f \circ c_1]_{f(m)}$. By construction the following diagram commutes.

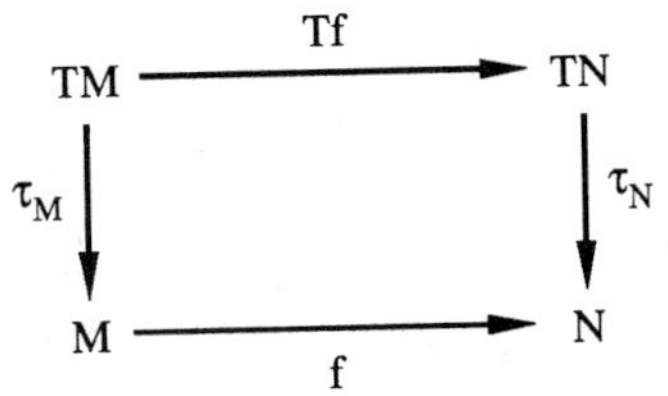

The basic properties of T are summarized in the following.

3.3.7 Composite Mapping Theorem

(i) *Suppose* $f : M \to N$ *and* $g : N \to K$ *are* C^r *maps of manifolds. Then* $g \circ f : M \to K$ *is of class* C^r *and*

$$T(g \circ f) = Tg \circ Tf .$$

(ii) *If* $h : M \to M$ *is the identity map, then* $Th : TM \to TM$ *is the identity map.*

(iii) *If* $f : M \to N$ *is a diffeomorphism, then* $Tf : TM \to TN$ *is a bijection and* $(Tf)^{-1} = T(f^{-1})$.

Proof (i) Let (U, φ), (V, ψ), (W, ρ) be charts of M, N, K, with $f(U) \subset V$ and $g(V) \subset W$. Then the local representatives are

$$(g \circ f)_{\varphi\rho} = \rho \circ g \circ f \circ \varphi^{-1} = \rho \circ g \circ \psi^{-1} \circ \psi \circ f \circ \varphi^{-1} = g_{\psi\rho} \circ f_{\varphi\psi}.$$

By the Composite Mapping Theorem in Banach spaces, this, and hence $g \circ f$, is of class C^r. Moreover,

$$T(g \circ f)[c]_m = [g \circ f \circ c]_{(g \circ f)(m)}$$

and

$$(Tg \circ Tf)[c]_m = Tg([f \circ c]_{f(m)}) = [g \circ f \circ c]_{(g \circ f)(m)}.$$

Hence $T(g \circ f) = Tg \circ Tf$.

Part (ii) follows from the definition of T. For (iii), f and f^{-1} are diffeomorphisms with $f \circ f^{-1}$ the identity of N, while $f^{-1} \circ f$ is the identity on M. Using (i) and (ii), $Tf \circ Tf^{-1}$ is the identity on TN while $Tf^{-1} \circ Tf$ is the identity on TM. Thus (iii) follows. ∎

Next, let us show that in the case of local manifolds, Tf as defined in §**2.4**, which we temporarily denote f', coincides with Tf as defined here.

3.3.8 Lemma *Let* $U \subset E$ *and* $V \subset F$ *be local manifolds (open subsets) and* $f : U \to V$ *be of class* C^1. *Let* $i : U \times E \to TU$ *be the map defined following* **3.3.4**. *Then the diagram*

$$\begin{array}{ccc} U \times E & \xrightarrow{\quad f' \quad} & V \times F \\ {\scriptstyle i}\downarrow & & \downarrow{\scriptstyle i} \\ TU & \xrightarrow[\quad Tf \quad]{} & TV \end{array}$$

commutes; that is, $Tf \circ i = i \circ f'$.

Proof For $(u, e) \in U \times E$, we have $(Tf \circ i)(u, e) = Tf \cdot [c_{u,e}]_u = [f \circ c_{u,e}]_{f(u)}$. Also, we have the identities $(i \circ f')(u, e) = i(f(u), \mathbf{D}f(u) \cdot e) = [c_{f(u),\, \mathbf{D}f(u)\cdot e}]_{f(u)}$. These will be equal provided the curves $t \mapsto f(u + te)$ and $t \mapsto f(u) + t(\mathbf{D}f(u) \cdot e)$ are tangent at $t = 0$. But this is clear from the definition of the derivative and the Composite Mapping Theorem. ∎

This lemma states that if we identify $U \times E$ and TU by means of i then we should correspondingly identify f' and Tf. Thus we will just write Tf and suppress the identification. Theorem **3.3.7** implies the following.

3.3.9 Lemma *If* $f : U \subset E \to V \subset F$ *is a* C^r *diffeomorphism, then* $Tf : U \times E \to V \times F$ *is a* C^r *diffeomorphism.*

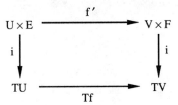

For a chart (U, φ) on a manifold M, we define $T\varphi : TU \to T(\varphi(U))$ by $T\varphi([c]_u) = (\varphi(u), (\varphi \circ c)'(0))$. Then $T\varphi$ is a bijection, since φ is a diffeomorphism. Hence, on TM we can regard $(TU, T\varphi)$ as a local chart.

3.3.10 Theorem *Let* M *be a* C^{r+1} *manifold and* $\mathcal{A}$ *an atlas of admissible charts. Then* $T\mathcal{A} = \{(TU, T\varphi) \mid (U, \varphi) \in \mathcal{A}\}$ *is a* C^r *atlas of* TM *called the* ***natural atlas.***

Proof Since the union of chart domains of $\mathcal{A}$ is M, the union of the corresponding TU is TM. To verify **MA2**, suppose we have $TU_i \cap TU_j \neq \varnothing$. Then $U_i \cap U_j \neq \varnothing$ and therefore the overlap map $\varphi_i \circ \varphi_j^{-1}$ can be formed by restriction of $\varphi_i \circ \varphi_j^{-1}$ to $\varphi_j(U_i \cap U_j)$. The chart overlap map $T\varphi_i \circ (T\varphi_j)^{-1} = T(\varphi_i \circ \varphi_j^{-1})$ is a C^r diffeomorphism by Lemma **3.3.9**. ■

Hence TM has a natural C^r manifold structure induced by the differentiable structure of M. If M is n-dimensional, Hausdorff, and second countable, TM will be 2n-dimensional, Hausdorff, and second countable. Since the local representative of τ_M is $(\varphi \circ \tau_M \circ T\varphi^{-1})(u, e) = u$, the tangent bundle projection is a C^r map.

Let us next develop some of the simplest properties of tangent maps. First of all, let us check that tangent maps are smooth.

3.3.11 Proposition *Let* M *and* N *be* C^{r+1} *manifolds, and let* $f : M \to N$ *be a map of class* C^{r+1}. *Then* $Tf : TM \to TN$ *is a map of class* C^r.

Proof It is enough to check that Tf is a C^r map using the natural atlas. For $m \in M$ choose charts (U, φ) and (V, ψ) on M and N so that $m \in U$, $f(m) \in V$ and $f_{\varphi\psi} = \psi \circ f \circ \varphi^{-1}$ is of class C^{r+1}. Using $(TU, T\varphi)$ for TM and $(TV, T\psi)$ for TN, the local representative $(Tf)_{T\varphi, T\psi} = T\psi \circ Tf \circ T\varphi^{-1} = Tf_{\varphi\psi}$ is given by $Tf_{\varphi\psi}(u, e) = (u, \mathbf{D}f_{\varphi\psi}(u) \cdot e)$, which is a C^r map. ■

Now that TM has a manifold structure we can form higher tangents. For mappings $f : M \to N$ of class C^r, define $T^r f : T^r M \to T^r N$ inductively to be the tangent of $T^{r-1} f : T^{r-1} M \to T^{r-1} N$. Induction shows: *If* $f : M \to N$ *and* $g : N \to K$ *are* C^r *mappings of manifolds, then* $g \circ f$ *is of class* C^r *and* $T^r(g \circ f) = T^r g \circ T^r f$.

Let us apply the tangent construction to the manifold TM and its projection. This gives the tangent bundle of TM, namely $\tau_{TM} : T(TM) \to TM$. In coordinates, if (U, φ) is a chart in M, then $(TU, T\varphi)$ is a chart of TM, $(T(TU), T(T\varphi))$ is a chart of $T(TM)$, and thus the local representative of τ_{TM} is $(T\varphi \circ \tau_{TM} \circ T(T\varphi^{-1})) : (u, e, e_1, e_2) \mapsto (u, e)$. On the other hand, taking the tangent of the map $\tau_M : TM \to M$, we get $T\tau_M : T(TM) \to TM$. The local representative of $T\tau_M$ is

$$(T\varphi \circ T\tau_M \circ T(T\varphi^{-1}))(u, e, e_1, e_2) = T(\varphi \circ \tau_M \circ T\varphi^{-1})(u, e, e_1, e_2) = (u, e_1) .$$

Applying the commutative diagram for Tf following **3.3.6** to the case $f = \tau_M$, we get what is commonly known as the ***dual tangent rhombic***:

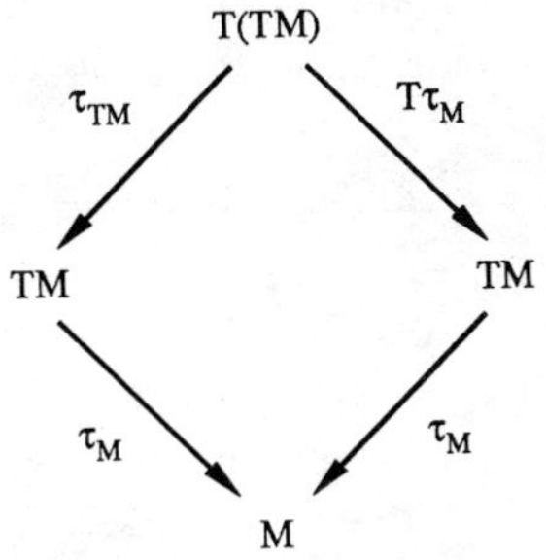

Let us now turn to a discussion of tangent bundles of product manifolds. Here and in what follows, tangent vectors will often be denoted by single letters such as $v \in T_mM$.

3.3.12 Proposition *Let* M_1 *and* M_2 *be manifolds and* $p_i : M_1 \times M_2 \to M_i$, $i = 1, 2$, *the two canonical projections. The map* $(Tp_1, Tp_2) : T(M_1 \times M_2) \to TM_1 \times TM_2$ *which is defined by* $(Tp_1, Tp_2)(v) = (Tp_1(v), Tp_2(v))$ *is a diffeomorphism of the tangent bundle* $T(M_1 \times M_2)$ *with the product manifold* $TM_1 \times TM_2$.

Proof The local representative of this map is $(u_1, u_2, e_1, e_2) \in U_1 \times U_2 \times E_1 \times E_2 \mapsto ((u_1, e_1), (u_2, e_2)) \in (U_1 \times E_1) \times (U_2 \times M_2)$, which clearly is a local diffeomorphism. ■

Since the tangent is just a global version of the derivative, statements concerning partial derivatives might be expected to have analogues on manifolds. To effect these analogies, we globalize the definition of partial derivatives.

Let M_1, M_2, and N be manifolds, and $f : M_1 \times M_2 \to N$ be a C^r map. For $(p, q) \in M_1 \times M_2$, let $i_p : M_2 \to M_1 \times M_2$ and $i_q : M_1 \to M_1 \times M_2$ be given by

$$i_p(y) = (p, y) \ , \quad i_q(x) = (x, q) \ ,$$

and define $T_1f(p, q) : T_p M_1 \to T_{f(p, q)}N$ and $T_2f(p, q) : T_qM_2 \to T_{f(p, q)}N$ by

$$T_1f(p, q) = T_p(f \circ i_q) \ , \quad T_2f(p, q) = T_q(f \circ i_p) \ .$$

With these notations the following proposition giving the behavior of T under products is a straightforward verification using the definition and local differential calculus.

3.3.13 Proposition *Let* M_1, M_2, N, *and* P *be manifolds,* $g_i : P \to M_i$, $i = 1, 2$, *and* $f : M_1 \times M_2 \to N$ *be* C^r *maps,* $r \geq 1$. *Identify* $T(M_1 \times M_2)$ *with* $TM_1 \times TM_2$. *Then the following statements hold.*

(i) $T(g_1 \times g_2) = Tg_1 \times Tg_2$.

(ii) $Tf(u_p, v_q) = T_1f(p, q)(u_p) + T_2f(p, q)(v_q)$, *for* $u_p \in T_pM_1$ *and* $v_q \in T_qM_2$.

(iii) **Implicit Function Theorem** *If* $T_2f(p, q)$ *is an isomorphism, then there exist open neighborhoods* U *of* p *in* M_1, W *of* $f(p, q)$ *in* N, *and a unique* C^r *map* $g : U \times W \to M_2$ *such that for all* $(x, w) \in U \times W$,

$$f(x, g(x, w)) = w .$$

In addition,

$$T_1g(x, w) = -(T_2f(x, g(x, w)))^{-1} \circ (T_1f(x, g(x, w)))$$

and

$$T_2g(x, w) = (T_2f(x, g(x, w)))^{-1} .$$

3.3.14 Examples

A The tangent bundle TS^1 of the circle Consider the atlas with the four charts $\{(U_i^{\pm}, \psi_i^{\pm}) \mid i = 1, 2\}$ of $S^1 = \{(x, y) \in \mathbb{R}^2 \mid x^2 + y^2 = 1\}$ from **3.1.4**. Let us construct the natural atlas for $TS^1 = \{((x, y), (u, v)) \in \mathbb{R}^2 \times \mathbb{R}^2 \mid x^2 + y^2 = 1, \langle (x, y), (u, v)\rangle = 0\}$. Since the map $\psi_1^+ : U_1^+ = \{(x, y) \in S^1 \mid x > 0\} \to]-1, 1[$ is given by $\psi_1^+(x, y) = y$, by definition of the tangent we have

$$T_{(x, y)}\psi_1^+(u, v) = (y, v) \ , \quad T\psi_1^+ : TU_1^+ \to]-1, 1[\times \mathbb{R} .$$

Proceed in the same way with the other three charts. Thus, for example, $T_{(x, y)}\psi_2^{-1}(u, v) = (x, u)$ and hence for $x \in]-1, 0[$,

$$(T\psi_2^- \circ T(\psi_1^+)^{-1})(y, v) = \left(\sqrt{1 - y^2}, -\frac{yv}{\sqrt{1 - y^2}}\right) .$$

This gives a complete description of the tangent bundle. But more can be said. Thinking of S^1 as the multiplicative group of complex numbers with modulus 1, we shall show that the group operations are C^∞: the inversion $I : s \mapsto s^{-1}$ has local representative $(\psi_1^{\pm} \circ I \circ (\psi_1^{\pm})^{-1})(x) = -x$ and the composition $C : (s_1, s_2) \mapsto s_1s_2$ has local representative

$$(\psi_1 \circ C \circ (\psi_1^{\pm} \times \psi_1^{\pm})^{-1})(x_1, x_2) = x_1\sqrt{1 - x_2^2} + x_2\sqrt{1 - x_1^2}$$

(here $\pm$ can be taken in any order). Thus for each $s \in S^1$, the map $L_s : S^1 \to S^1$ defined by $L_s(s') = ss'$, is a diffeomorphism. This enables us to define a map $\lambda : TS^1 \to S^1 \times \mathbb{R}$ by $\lambda(v_s) = (s, T_sL_s^{-1}(v_s))$, which is easily seen to be a diffeomorphism. *Thus* TS^1 *is diffeomorphic to* $S^1 \times \mathbb{R}$. See Figure 3.3.2.

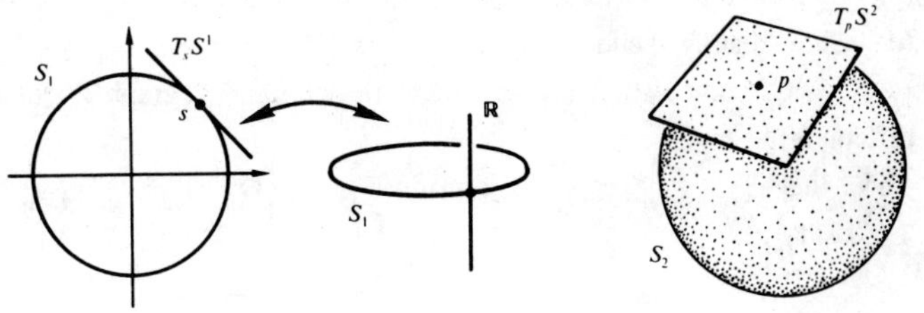

Trivial tangent bundle   Non trivial tangent bundle

Figure 3.3.2

B The tangent bundle $T\mathbb{T}^n$ to the n-torus Since $\mathbb{T}^n = S^1 \times \cdots \times S^1$ (n times) and $TS^1 \cong S^1 \times \mathbb{R}$, it follows that $T\mathbb{T}^n \cong \mathbb{T}^n \times \mathbb{R}^n$.

C The tangent bundle TS^2 to the sphere The previous examples yielded trivial tangent bundles. In general this is not the case, the tangent bundle to the two-sphere being a case in point, which we now describe. Choose the atlas with six charts $\{(U_i^{\pm}, \psi_i^{\pm}) \mid i = 1, 2, 3\}$ of S^2 that were given in **3.1.4.** Since

$$\psi_1^{\pm} : U_1^{+} = \{(x^1, x^2, x^3) \in S^2 \mid x_1 > 0\} \to D_1(0) = \{(x, y) \in \mathbb{R}^2 \mid x^2 + y^2 < 1\} ,$$

$$\psi_1^{+}(x^1, x^2, x^3) = (x^2, x^3) ,$$

we have

$$T_{(x^1, x^2, x^3)}\psi_1^{+}(v^1, v^2, v^3) = (x^2, x^3, v^2, v^3) ,$$

where $x^1v^1 + x^2v^2 + x^3v^3 = 0$. Similarly, construct the other five charts. For example, one of the twelve overlap maps for $x^2 + y^2 < 1$, and $y < 0$, is

$$(T\psi_3^{-} \circ (T\psi_1^{+})^{-1})\,(x, y, u, v) = \left(\sqrt{1 - x^2 - y^2},\ x,\ \frac{-ux}{\sqrt{1 - x^2 - y^2}} - \frac{vy}{\sqrt{1 - x^2 - y^2}},\ u\right).$$

One way to see that TS^2 is not trivial is to use the topological fact that any vector field on S^2 must vanish somewhere. We shall prove this in **§7.5.** ♦

Exercises

3.3A Let M and N be manifolds and $f : M \to N$.

(i) Show that

(a) if f is C^∞, then $\text{graph}(f) = \{(m, f(m)) \in M \times N \mid m \in M\}$ is a C^∞ submanifold of $M \times N$ and

(b) $T_{(m, f(m))}(M \times N) \cong T_{(m, f(m))}(\text{graph}(f)) \oplus T_{f(m)}N$ for all $m \in M$.

(c) Show that the converse of (a) is false. (*Hint:* $x \in \mathbb{R} \mapsto x^{1/3} \in \mathbb{R}$.)

(d) Show that if (a) and (b) hold, then f is C^∞.

(ii) If f is C^∞ show that the canonical projection of graph(f) onto M is a diffeomorphism.

(iii) Show that $T_{(m, f(m))}(\text{graph}(f)) \cong \text{graph}(T_m f) = \{(v_m, T_m f(v_m)) \mid v_m \in T_m M\} \subset T_m M \times T_{f(m)} N$.

3.3B (i) Show that there is a map $s_M : T(TM) \to T(TM)$ such that $s_M \circ s_M$ = identity and the diagram

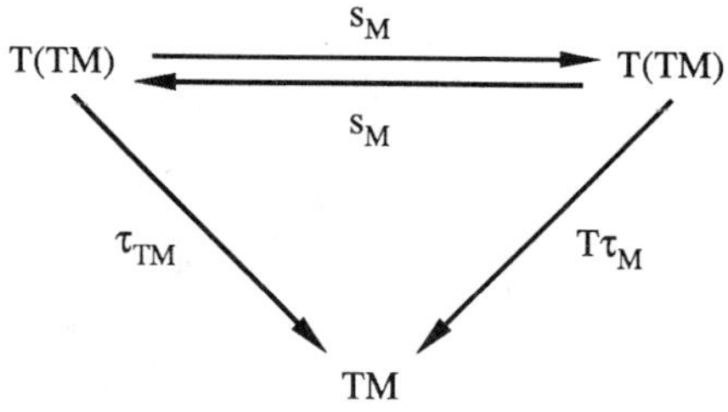

commutes. (*Hint:* In local natural coordinate charts $s_M(u, e, e_1, e_2) = (u, e_1, e, e_2)$.) One calls s_M the ***canonical involution*** on M and says that T(TM) is a ***symmetric rhombic.***

(ii) Verify that for $f : M \to N$ of class C^2, $T^2 f \circ s_M = s_N \circ T^2 f$.

(iii) If X is a vector field on M, that is, a section of $\tau_M : TM \to M$, show that TX is a section of $T\tau_M : T^2M \to TM$ and $X^1 = s_M \circ TX$ is a section of $\tau_{TM} : T^2M \to TM$. (A ***section*** σ of a map $f : A \to B$ is a map $\sigma : B \to A$ such that $f \circ \sigma$ = identity on B.)

3.3C (i) Let $\mathbb{S}(S^2) = \{v \in TS^2 \mid \|v\| = 1\}$ be the ***circle bundle*** of S^2. Prove that $\mathbb{S}(S^2)$ is a submanifold of TS^2 of dimension three.

(ii) Define $f : \mathbb{S}(S^2) \to \mathbb{RP}^3$ by f(x, y, v) = the line through the origin in $\mathbb{R}^4$ determined by the vector with components (x, y, v^1, v^2). Show that f is a diffeomorphism.

3.3D Let M be an n-dimensional submanifold of $\mathbb{R}^N$. Define the ***Gauss map*** $\Gamma : M \to \mathbb{G}_{n, N-n}$ by $\Gamma(m) = T_m M - m$, i.e., $\Gamma(m)$ is the n dimensional subspace of $\mathbb{R}^N$ through the origin, which, when translated by m, equals $T_m M$. Show that Γ is a smooth map.

3.3E Let $f : \mathbb{T}^2 \to \mathbb{R}$ be a smooth map. Show that f has at least four critical points (points where Tf vanishes). (*Hint:* Parametrize $\mathbb{T}^2$ using angles θ, φ and locate the maximum and minimum points of $f(\theta, \varphi)$ for φ fixed, say $(\theta_{max}(\varphi), \varphi)$ and $(\theta_{min}(\varphi), \varphi)$; now maximize and minimize f as φ varies. How many critical points must $f : S^2 \to \mathbb{R}$ have?)

§3.4 *Vector Bundles*

Roughly speaking, a vector bundle is a manifold with a vector space attached to each point. During the formal definitions we may keep in mind the example of the tangent bundle to a manifold, such as the n-sphere S^n. Similarly, the collection of normal lines to S^n form a vector bundle.

The definitions will follow the pattern of those for a manifold. Namely, we obtain a vector bundle by smoothly patching together local vector bundles. The following terminology for vector space products and maps will be useful.

3.4.1 Definition *Let* E *and* F *be vector spaces with* U *an open subset of* E. *We call the Cartesian product* $U \times F$ ***a local vector bundle****. We call* U *the* ***base space****, which can be identified with* $U \times \{0\}$*, the* ***zero section****. For* $u \in U$, $\{u\} \times F$ *is called the* ***fiber*** *of* u, *which we endow with the vector space structure of* F. *The map* $\pi : U \times F \to U$ *given by* $\pi(u, f) = u$ *is called the* ***projection*** *of* $U \times F$. *(Thus, the fiber over* $u \in U$ *is* $\pi^{-1}(u)$. *Also note that* $U \times F$ *is an open subset of* $E \times F$ *and so is a local manifold.)*

Next, we introduce the idea of a local vector bundle map. The main idea is that such a map must map a fiber linearly to a fiber.

3.4.2 Definition *Let* $U \times F$ *and* $U' \times F'$ *be local vector bundles. A map* $\varphi : U \times F \to U' \times F'$ *is called a* C^r ***local vector bundle map*** *if it has the form* $\varphi(u, f) = (\varphi_1(u), \varphi_2(u) \cdot f)$ *where* $f_1 : U \to U'$ *and* $\varphi_2 : U \to L(F, F')$ *are* C^r. *A local vector bundle map that has an inverse which is also a local vector bundle map is called a* ***local vector bundle isomorphism.*** (See Figure 3.4.1.)

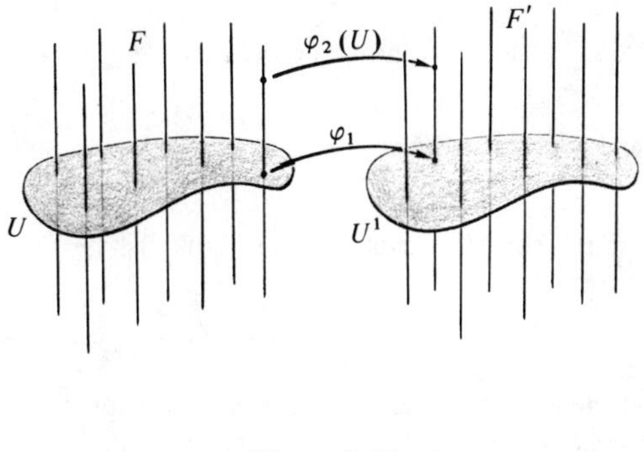

Figure 3.4.1

A local vector bundle map $\varphi : U \times F \to U' \times F'$ maps the fiber $\{u\} \times F$ into the fiber $\{\varphi_1(u)\} \times F'$ and so restricted is linear. By Banach's Isomorphism Theorem it follows that a local vector bundle map φ is a local vector bundle isomorphism iff $\varphi_2(u)$ is a Banach space isomorphism for every $u \in U$.

☞ SUPPLEMENT 3.4A
Smoothness of Local Vector Bundle Maps

In some examples, to check whether a map φ is a C^∞ local vector bundle map, one is faced with the rather unpleasant task of verifying that $\varphi_2 : U \to L(F, F')$ is C^∞. It would be nice to know that the smoothness of φ as a function of two variables suffices. This is the context of the next proposition. We state the result for C^∞, but the proof contains a C^r result (with an interesting derivative loss) which is discussed in the ensuing Remark **A**.

3.4.3 Proposition *A map* $\varphi : U \times F \to U' \times F'$ *is a* C^∞ *local vector bundle map if* φ *is* C^∞ *and is of the form* $\varphi(u, f) = (\varphi_1(u), \varphi_2(u) \cdot f)$, *where* $f_1 : U \to U'$ *and* $\varphi_2 : U \to L(F, F')$ *are* C^∞.

Proof (M. Craioveanu and T. Ratiu [1976]) The evaluation map $ev : L(F, F') \times F \to F'$; $ev(T, f) = T(f)$ is clearly bilinear and continuous. First assume φ is a C^r local vector map, so $\varphi_2 : U \to L(F, F')$ is C^r. Now write $\varphi_2(u) \cdot f = (ev \circ (\varphi_2 \times I))(u, f)$. By the composite mapping theorem, it follows that φ_2 is C^r as a function of two variables. Thus φ is C^r by **2.4.12** (iii).

Conversely, assume $\varphi(u, f) = (\varphi_1(u), \varphi_2(u) \cdot f)$ is C^∞. Then again by **2.4.12** (iii), $\varphi_1(u)$ and $\varphi_2(u) \cdot f$ are C^∞ as functions of two variables. To show that $\varphi_2 : U \to L(E, F')$ is C^∞, it suffices to prove the following: *if* $h : U \times F \to F'$ is C^r, $r \geq 1$, *and such that* $h(u, \cdot) \in L(F, F')$ *for all* $u \in U$, *then the map* $h' : U \to L(F, F')$, *defined by* $h'(u) = h(u, \cdot)$ *is* C^{r-1}. This will be shown by induction on r.

If $r = 1$ we prove continuity of h' in a disk around $u_0 \in U$ in the following way. By continuity of **D**h, there exists $\varepsilon > 0$ such that for all $u \in D_\varepsilon(u_0)$ and $v \in D_\varepsilon(0)$, $\| \mathbf{D}_1 h(u, v) \| \leq N$ for some $N > 0$. The mean value inequality yields

$$\| h(u, v) - h(u', v) \| \leq N \| u - u' \|$$

for all $u, u' \in D_\varepsilon(u_0)$ and $v \in D_\varepsilon(0)$. Thus

$$\| h'(u) - h'(u') \| = \sup_{\|v\| \leq 1} \| h(u, v) - h(u', v) \| < \frac{N}{\varepsilon} \| u - u' \|,$$

proving that h' is continuous.

Let $r > 1$ and inductively assume that the statement is true for $r - 1$. Let $S : L(F, L(E, F')) \to L(E, L(F, F'))$ be the canonical isometry: $S(T)(e) \cdot f = T(f) \cdot e$. We shall prove that

$$\mathbf{D}h' = S \circ (\mathbf{D}_1 h)', \tag{1}$$

where $(\mathbf{D}_1h)'(u) \cdot v = \mathbf{D}_1h(u, v)$. Thus, if h is C^r, $\mathbf{D}_1h$ is C^{r-1}, by induction $(\mathbf{D}_1h)'$ is C^{r-2}, and hence by (1), $\mathbf{D}h'$ will be C^{r-2}. This will show that h' is C^{r-1}.

For relation (1) to make sense, we first show that $\mathbf{D}_1h(u, \cdot) \in L(F, L(E, F'))$. Since

$$\mathbf{D}_1h(u, v) \cdot w = \frac{\lim_{t \to 0} [h'(u + tw) - h'(u)] \cdot v}{t} = \lim_{n \to \infty} A_n v ,$$

for all $v \in F$, where

$$A_n = n\left(h'\left(u + \frac{1}{n}w\right) - h'(u)\right) \in L(F, F'),$$

it follows by the Uniform Boundedness Principle (or rather its corollary **2.2.21**) that $\mathbf{D}_1h(u, \cdot) \cdot w \in L(F, F')$. Thus $(v, w) \mapsto \mathbf{D}_1h(u, v) \cdot w$ is linear continuous in each argument and hence is bilinear continuous (Exercise **2.2J**), and consequently $v \mapsto \mathbf{D}_1h(u, v) \in L(E, F')$ is linear and continuous.

Relation (1) is proved in the following way. Fix $u_0 \in U$ and let ε and N be positive constants such that

$$\| \mathbf{D}_1h(u, v) - \mathbf{D}_1h(u', v) \| \leq N \| u - u' \| \tag{2}$$

for all $u, u' \in D_{2\varepsilon}(u_0)$ and $v \in D_\varepsilon(0)$. Apply the mean value inequality to the C^{r-1} map $g(u) = h(u, v) - \mathbf{D}_1h(u', v) \cdot u$ for fixed $u' \in D_{2\varepsilon}(u_0)$ and $v \in D_\varepsilon(0)$ to get

$$\| h(u + w, v) - h(u, v) - \mathbf{D}_1h(u', v) \cdot w \| = \| g(u + w) - g(u) \| \leq \| w \| \sup_{t \in [0,1]} \| \mathbf{D}g(u + tw) \|$$

$$= \| w \| \sup_{t \in [0,1]} \| \mathbf{D}_1h(u + tw, v) - \mathbf{D}_1h(u', v) \|$$

for $w \in D_\varepsilon(u_0)$. Letting $u' \to u$ and taking into account (2) we get

$$\| h(u + w, v) - h(u, v) - \mathbf{D}_1h(u, v) \cdot w \| \leq N \| w \|^2;$$

i.e.,

$$\| h'(u + w) \cdot v - h'(u) \cdot v - [(S \circ (\mathbf{D}_1h)')(u) \cdot w](v) \| \leq N \| w \|^2$$

for all $v \in D_\varepsilon(0)$, and hence

$$\| h'(u + w) - h'(u) - (S \circ (\mathbf{D}_1h)') \cdot w \| \leq \frac{N}{\varepsilon} \| w \|^2$$

thus proving (1). ■

Remarks

A If F is finite dimensional and if $h : U \times F \to F'$ is C^r, $r \geq 1$, and is such that $h(u, \cdot) \in L(F, F')$ for all $u \in U$, then $h' : U \to L(F, F')$ given by $h'(u) = h(u, \cdot)$ is also C^r. In other words, Proposition **3.4.3** holds for C^r-maps. Indeed, since $F = \mathbb{R}^n$ for some n, $L(F, F') \cong F' \times \cdots \times F'$ (n times) so it suffices to prove the statement for $F = \mathbb{R}$. Thus we want to show that if $h : U \times \mathbb{R} \to F'$ is C^r and $h(u, 1) = g(u) \in F'$, then $g : U \to F'$ is also C^r. Since $h(u, x) = xg(u)$ for all $(u, x) \in U \times \mathbb{R}$ by linearity of h in the second argument, it follows that $h' = g$ is a C^r map.

B If F is infinite dimensional the result in the proof of Proposition **3.4.3** cannot be improved even if $r = 0$. The following counterexample is due to A.J. Trunba. Let $h : [0, 1] \times L^2[0, 1] \to L^2[0, 1]$ be given by

$$h(x, \varphi) = \int_0^1 \sin \frac{2\pi t}{x} \varphi(t)\, dt$$

if $x \neq 0$, and $h(0, \varphi) = 0$. Continuity at each $x \neq 0$ is obvious and at $x = 0$ it follows by the Riemann Lebesque Lemma (the Fourier coefficients of a uniformly bounded sequence in L^2 relative to an orthonormal set converge to zero). Thus h is C^0. However, since $h(x, \sin(2\pi t/x)) = {}^1/_2 - (x/4\pi) \sin(4\pi/x)$, we have $h(1/n, \sin 2\pi nt) = {}^1/_2$ and therefore its L^2-norm is $1/\sqrt{2}$; this says that $\| h'(1/n) \| \geq 1/\sqrt{2}$ and thus h' is not continuous. ✍

Any linear map $A \in L(E, F)$ defines a local vector bundle map $\varphi_A : E \times E \to E \times F$ by $\varphi(u, e) = (u, Ae)$. Another example of a local vector bundle map was encountered in §**2.4**: if the map $f : U \subset E \to V \subset F$ is C^{r+1}, then $Tf : U \times E \to V \times F$ is a C^r local vector bundle map and $Tf(u, e) = (f(u), \mathbf{D}f(u) \cdot e)$.

Using these local notions, we are now ready to define a vector bundle.

3.4.4 Definition *Let* S *be a set. A* ***local bundle chart*** *of* S *is a pair* (W, φ) *where* $W \subset S$ *and* $\varphi : W \subset S \to U \times F$ *is a bijection onto a local bundle* $U \times F$; U *and* F *may depend on* φ. *A* ***vector bundle atlas*** *on* S *is a family* $\mathcal{B} = \{(W_i, \varphi_i)\}$ *of local bundle charts satisfying:*

VB1 = **MA1** *of* **3.1.1**: $\mathcal{B}$ *covers* S; *and*

VB2 *for any two local bundle charts* (W_i, φ_i) *and* (W_j, φ_j) *in* $\mathcal{B}$ *with* $W_i \cap W_j \neq \varnothing$, $\varphi_i(W_i \cap W_j)$ *is a local vector bundle, and the overlap map* $\psi_{ji} = \varphi_j \circ \varphi_i^{-1}$ *restricted to* $\varphi_i(W_i \cap W_j)$ *is a* C^∞ *local vector bundle isomorphism.*

If $\mathcal{B}_1$ *and* $\mathcal{B}_2$ *are two vector bundle atlases on* S, *we say that they are* ***VB-equivalent*** *if* $\mathcal{B}_1 \cup \mathcal{B}_2$ *is a vector bundle atlas. A* ***vector bundle structure*** *on* S *is an equivalence class of vector bundle atlases. A* ***vector bundle*** E *is a pair* $(S, \mathcal{V})$, *where* S *is a set and* $\mathcal{V}$ *is a vector bundle structure on* S. *A chart in an atlas of* $\mathcal{V}$ *is called an* ***admissible vector bundle chart*** *of* E. *As with manifolds, we often identify* E *with the underlying set* S.

The intuition behind this definition is depicted in Figure 3.4.2.

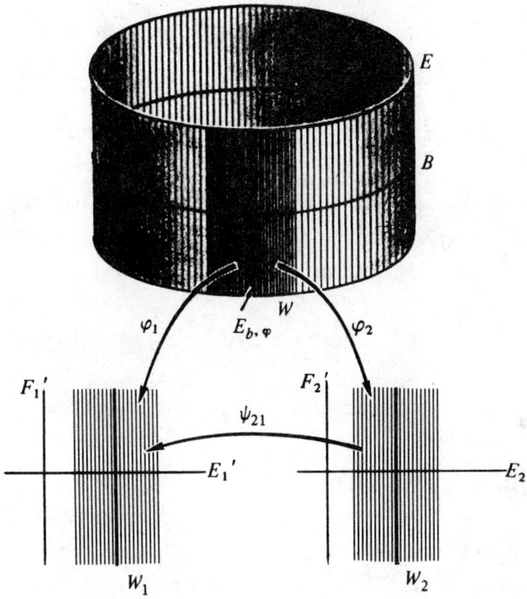

Figure 3.4.2

As in the case of manifolds, if we make a choice of vector bundle atlas $\mathcal{B}$ on S then we obtain a maximal vector bundle atlas by including all charts whose overlap maps with those in $\mathcal{B}$ are C^∞ local vector bundle isomorphisms. Hence a particular vector bundle atlas suffices to specify a vector bundle structure on S. Vector bundles are special types of manifolds. Indeed **VB1** and **VB2** give **MA1** and **MA2** in particular, so $\mathcal{V}$ induces a differentiable structure on S.

3.4.5 Definition *For a vector bundle* $E = (S, \mathcal{V})$ *we define the* ***zero section*** *(or* ***base****) by*

$$B = \left\{ e \in E \,\middle|\, \textit{there exists } (W, \varphi) \in \mathcal{V} \textit{ and } u \in U \textit{ with } e = \varphi^{-1}(u, 0) \right\} ,$$

that is, B *is the union of all the zero sections of the local vector bundles (identifying* W *with a local vector bundle via* $\varphi : W \to U \times F$*).*

If $(U, \varphi) \in \mathcal{V}$ *is a vector bundle chart, and* $b \in U$ *with* $\varphi(b) = (u, 0)$*, let* $E_{b,\varphi}$ *denote the subset* $\varphi^{-1}(\{u\} \times F)$ *of* S *together with the structure of a vector space induced by the bijection* φ.

The next few propositions derive basic properties of vector bundles that are sometimes included in the definition.

3.4.6 Proposition

(i) *If* b *lies in the domain of two local bundle charts* φ_1 *and* φ_2, *then*

$$E_{b,\varphi_1} = E_{b,\varphi_2},$$

where the equality means equality as topological spaces and as vector spaces.

(ii) *For* $v \in E$, *there is a unique* $b \in B$ *such that* $v \in E_{b,\varphi}$, *for some (and so all)* (U, φ).

(iii) B *is a submanifold of* E.

(iv) *The map* π, *defined by* $\pi : E \to B$, $\pi(e) = b$ [*in*(ii)] *is surjective and* C^∞.

Proof (i) Suppose $\varphi_1(b) = (u_1, 0)$ and $\varphi_2(b) = (u_2, 0)$. We may assume that the domains of φ_1 and φ_2 are identical, for $E_{b,\varphi}$ is unchanged if we restrict φ to any local bundle chart containing b. Then $\alpha = \varphi_1 \circ \varphi_2^{-1}$ is a local vector bundle isomorphism. But we have

$$E_{b,\varphi_1} = \varphi_1^{-1}(\{u_1\} \times F_1) = (\varphi_2^{-1} \circ \alpha^{-1})\{u_1\} \times F_1) = \varphi_2^{-1}(\{u_2\} \times F_2) = E_{b,\varphi_2}.$$

Hence $E_{b,\varphi_1} = E_{b,\varphi_2}$ as sets, and it is easily seen that addition and scalar multiplication in E_{b,φ_1} and E_{b,φ_2} are identical as are the topologies.

For (ii) note that if $v \in E$, $\varphi_1(v) = (u_1, f_1)$, $\varphi_2(v) = (u_2, f_2)$, $b_1 = \varphi_1^{-1}(u_1, 0)$, and $b_2 = \varphi_2^{-1}(u_2, 0)$, then $\psi_{21}(u_2, f_2) = (u_1, f_1)$, so ψ_{21} gives a linear isomorphism $\{u_2\} \times F_2 \to \{u_1\} \times F_1$, and therefore $\varphi_1(b_2) = \psi_{21}(u_2, 0) = (u_1, 0) = \varphi_1(b_1)$, or $b_2 = b_1$.

To prove (iii) we verify that for $b \in B$ there is an admissible chart with the submanifold property. To get such a manifold chart, we choose an admissible vector bundle chart (W, φ), $b \in W$. Then $\varphi(W \cap B) = U \times \{0\} = \varphi(W) \cap (E \times \{0\})$.

Finally, for (iv), it is enough to check that π is C^∞ using local bundle charts. But this is clear, for such a representative is of the form $(u, f) \mapsto (u, 0)$. That π is onto is clear. ■

The fibers of a vector bundle inherit an intrinsic vector space structure and a topology independent of the charts, but there is no norm that is chart independent. Putting particular norms on fibers is extra structure to be considered later in the book. Sometimes the phrase ***Banachable space*** is used to indicate that the topology comes from a complete norm but we are not nailing down a particular one.

The following summarizes the basic properties of a vector bundle.

3.4.7 Theorem *Let* E *be a vector bundle. The* ***zero section*** *(or* ***base****)* B *of* E *is a submanifold of* E *and there is a map* $\pi : E \to B$ *(sometimes denoted* $\pi_{BE} : E \to B$*) called the* ***projection*** *that is of class* C^∞, *and is surjective (onto). Moreover, for each* $b \in B$, $\pi^{-1}(b)$, *called the* ***fiber*** *over* b, *has a Banachable vector space structure induced by any admissible vector bundle chart, with* b *the zero element.*

Because of these properties we sometimes write "the vector bundle $\pi : E \to B$" instead of "the vector bundle $(E, \mathcal{V})$." Fibers are often denoted by $E_b = \pi^{-1}(b)$. If the base B and the map π are understood, we just say "the vector bundle E."

A most commonly encountered vector bundle is the tangent bundle $\tau_M : TM \to M$ of a manifold M. To see that the tangent bundle, as we defined it in the previous section, is a vector bundle in the sense of this section, we use the following lemma.

3.4.8 Lemma *If* $f : U \subset E \to V \subset F$ *is a diffeomorphism of open sets in Banach spaces, then* $Tf : U \times E \to V \times F$ *is a local vector bundle isomorphism.*

Proof Since $Tf(u, e) = (f(u), \mathbf{D}f(u) \cdot e)$, Tf is a local vector bundle mapping. But as f is a diffeomorphism, $(Tf)^{-1} = T(f^{-1})$ is also a local vector bundle mapping, and hence Tf is a vector bundle isomorphism. ■

Let $\mathcal{A} = \{(U, \varphi)\}$ be an atlas of admissible charts on a manifold M that is modeled on a Banach space E. In the previous section we constructed the atlas $T\mathcal{A} = \{(TU, T\varphi)\}$ of the manifold TM. If $U_i \cap U_j \neq \varnothing$, then the overlap map $T\varphi_i \circ T\varphi_j^{-1} = T(\varphi_i \circ \varphi_j^{-1}) : \varphi_j(U_i \cap U_j) \times E \to \varphi_i(U_i \cap U_j) \times E$ has the expression $(u, e) \mapsto ((\varphi_i \circ \varphi_j^{-1})(u), \mathbf{D}(\varphi_i \circ \varphi_j^{-1})(u) \cdot e)$. By lemma **3.4.8**, $T(\varphi_i \circ \varphi_j^{-1})$ is a local vector bundle isomorphism. This proves the first part of the following theorem.

3.4.9 Theorem *Let* M *be a manifold and* $\mathcal{A} = \{(U, \varphi)\}$ *be an atlas of admissible charts.*

(i) *Then* $T\mathcal{A} = \{(TU, T\varphi)\}$ *is a vector bundle atlas of* TM, *called the natural atlas.*

(ii) *If* $m \in M$, *then* $\tau_M^{-1}(m) = T_mM$ *is a fiber of* TM *and its base* B *is diffeomorphic to* M *by the map* $\tau_M | B : B \to M$.

Proof (ii) Let (U, φ) be a local chart at $m \in M$, with $\varphi : U \to \varphi(U) \subset E$ and $\varphi(m) = u$. Then $T\varphi : TM | U \to \varphi(u) \times E$ is a natural chart of TM, so that

$$T\varphi^{-1}(\{u\} \times E) = T\varphi^{-1}\{[c_{u,e}]_u \mid e \in E\}$$

by definition of $T\varphi$, and this is exactly T_mM. For the second assertion, $\tau_M | B$ is obviously a bijection, and its local representative with respect to $T\varphi$ and φ is the natural identification determined by $\varphi(U) \times \{0\} \to \varphi(U)$, a diffeomorphism. ■

Thus, T_mM is isomorphic to the Banach space E, the model space of M, M is identified with the zero section of TM, and τ_M is identified with the bundle projection onto the zero section. It is also worth recalling that the local representative τ_M is $(\varphi \circ \tau_M \circ T\varphi^{-1})(u, e) = u$, i.e., just the projection of $\varphi(U) \times E$ to $\varphi(U)$.

3.4.10 Examples

A Any manifold M is a vector bundle with zero-dimensional fiber, namely $M \times \{0\}$.

B The ***cylinder*** $E = S^1 \times \mathbb{R}$ is a vector bundle with $\pi : E \to B = S^1$ the projection on the first factor (Figure 3.4.3). This is a ***trivial vector bundle*** in the sense that it is a product. The cylinder is diffeomorphic to TS^1 by Example **3.3.14A.**

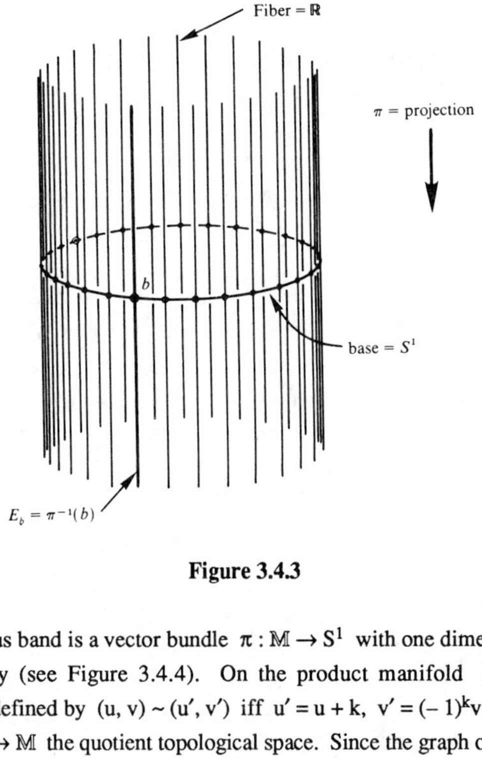

Figure 3.4.3

C The Möbius band is a vector bundle $\pi : M \to S^1$ with one dimensional fiber obtained in the following way (see Figure 3.4.4). On the product manifold $\mathbb{R} \times \mathbb{R}$, consider the equivalence relation defined by $(u, v) \sim (u', v')$ iff $u' = u + k$, $v' = (-1)^k v$ for some $k \in \mathbb{Z}$ and denote by $p : \mathbb{R} \times \mathbb{R} \to M$ the quotient topological space. Since the graph of this relation is closed and p is an open map, M is a Hausdorff space. Let $[u, v] = p(u, v)$ and define the projection $\pi : M \to S^1$ by $\pi[u, v] = e^{2\pi i u}$. Let $V_1 =]\,0, 1[\, \times \mathbb{R}$, $V_2 =]-{}^1\!/_2, {}^1\!/_2[\, \times \mathbb{R}$, $U_1 = S^1 \backslash \{1\}$ and $U_2 = S^1 \backslash \{-1\}$ and then note that $p \mid V_1 : V_1 \to \pi^{-1}(U_1)$ and $p \mid V_2 : V_2 \to \pi^{-1}(U_2)$ are homeomorphisms and that $M = \pi^{-1}(U_1) \cup \pi^{-1}(U_2)$. Let $\{(U_1, \varphi_1), (U_2, \varphi_2)\}$ be an atlas with two charts for S^1 (see **3.1.2**). Define $\psi_j : \pi^{-1}(U_j) \to \mathbb{R} \times \mathbb{R}$ by $\psi_j = \chi_j \circ (p \mid V_j)^{-1}$ and $\chi_j : V_j \to \mathbb{R} \times \mathbb{R}$ by $\chi_j(u, v) = (\varphi_j(e^{2\pi i u}), (-1)^{j+1} v)$, $j = 1, 2$ and observe that χ_j and ψ_j are homeomorphisms. Since the composition $\psi_2 \circ \psi_1^{-1} : (\mathbb{R} \times \mathbb{R}) \backslash (\{0\} \times \mathbb{R}) \to (\mathbb{R} \times \mathbb{R}) \backslash (\{0\} \times \mathbb{R})$ is given by the formula $(\psi_2 \circ \psi_1^{-1})(x, y) = ((\varphi_2 \circ \varphi_1^{-1})(x), -y)$, we see that $\{(\pi^{-1}(U_1), \psi_1), (\pi^{-1}(U_2), \psi_2)\}$ forms a vector bundle atlas of M.

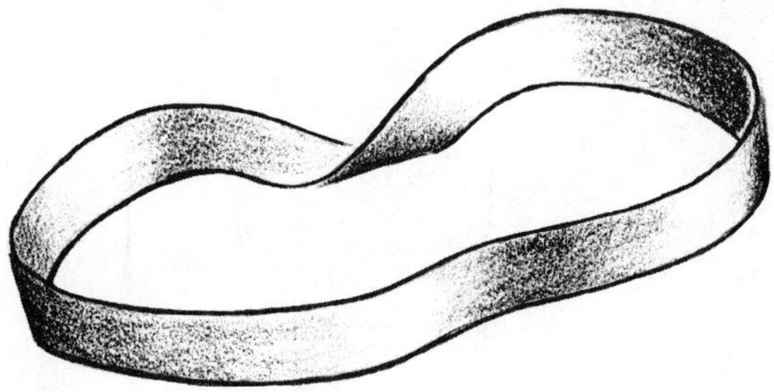

Figure 3.4.4

D The Grassmann bundles (universal bundles) We now define vector bundles $\gamma_n(E) \to \mathbb{G}_n(E)$, $\gamma^n(E) \to \mathbb{G}^n(E)$, and $\gamma(E) \to \mathbb{G}(E)$ which play an important role in the classification of isomorphism classes of vector bundles (see for example Hirsch [1976]). The definition of the projection $\rho : \gamma_n(E) \to \mathbb{G}_n(E)$ is the following (see Example **3.1.8G** for notations): recalling $\gamma_n(E) = \{(F, v) \mid F$ is an n-dimensional subspace of E and $v \in F\}$, we set $\rho(F, v) = F$. The charts $(\rho^{-1}(U_G), \psi_{FG})$, where $E = F \oplus G$, $\psi_{FG}(H, v) = (\varphi_{FG}(H), \pi_G(H, F)(v))$, and $\psi_{FG} : \rho^{-1}(U_G) \to L(F, G) \times F$, define a vector bundle structure on $\gamma_n(E)$ since the overlap maps are

$$(\psi_{F'G'} \circ \psi_{FG}^{-1})(T, f) = ((\varphi_{F'G'} \circ \varphi_{FG}^{-1})(T), (\pi_{G'}(\text{graph}(T), F') \circ \pi_G(\text{graph}(T), F)^{-1})(f)).$$

where $T \in L(F, G)$, $f \in F$, and graph(T) denotes the graph of T in $E \times F$; smoothness in T is shown as in **3.1.8G**. The fiber dimension of this bundle is n. A similar construction holds for $\mathbb{G}^n(E)$ yielding $\gamma^n(E)$; the fiber codimension in this case is also n. Similarly $\gamma(E) \to \mathbb{G}(E)$ is obtained with not necessarily isomorphic fibers at different points of $\mathbb{G}(E)$. ♦

3.4.11 Definition *Let E and E′ be two vector bundles. A map $f : E \to E'$ is called a C^r* ***vector bundle mapping (local isomorphism)*** *when for each $v \in E$ and each admissible local bundle chart (V, ψ) of E′ for which $f(v) \in V$, there is an admissible local bundle chart (W, φ) with $f(W) \subset W'$ such that the local representative $f_{\varphi\psi} = \psi \circ f \circ \varphi^{-1}$ is a C^r local vector bundle mapping (local isomorphism). A bijective local vector bundle isomorphism is called a* ***vector bundle isomorphism.***

This definition makes sense only for local vector bundle charts and not for all manifold charts. Also, such a W is not guaranteed by the continuity of f, nor does it imply it. However, if we first check that f is ***fiber preserving*** (which it must be) and is continuous, then such an open set W is guaranteed. This fiber-preserving character is made more explicit in the following.

3.4.12 Proposition *Suppose $f : E \to E'$ is a C^r vector bundle map, $r \geq 0$. Then:*

(i) f *preserves the zero section:* $f(B) \subset B'$;

(ii) f *induces a unique mapping* $f_B : B \to B'$ *such that the following diagram commutes:*

$$\begin{array}{ccc} E & \xrightarrow{\ f\ } & E' \\ {\scriptstyle \pi}\downarrow & & \downarrow{\scriptstyle \pi'} \\ B & \xrightarrow[\ f_B\ ]{} & B' \end{array}$$

that is, $\pi' \circ f_B = f_B \circ \pi$. *(Here,* π and π' *are the projection maps.) Such a map* f *is called a* ***vector bundle map over*** f_B.

(iii) *A* C^∞ *map* $g : E \to E'$ *is a vector bundle map iff there is a* C^∞ *map* $g_B : B \to B'$ *such that* $\pi' \circ g = g_B \circ \pi$ *and* g *restricted to each fiber is a linear continuous map into a fiber.*

Proof (i) Suppose $b \in B$. We must show $f(b) \in B'$. That is, for a vector bundle chart (V, ψ) with $f(b) \in V$ we must show $\psi(f(b)) = (v, 0)$. Since we have a chart (W, φ) such that $b \in W$, $f(W) \subset V$, and $\varphi(b) = (u, 0)$, it follows that $\psi(f(b)) = (\psi \circ f \circ \varphi^{-1})(u, 0)$ which is of the form $(v, 0)$ by linearity of $f_{\varphi\psi}$ on each fiber.

For (ii), let $f_B = f \,|\, B : B \to B'$. With the notations above,

$$\psi \,|\, B' \circ \pi' \circ f \circ \varphi^{-1} = \pi'_{\psi,\psi|B'} \circ f_{\varphi\psi}$$

and

$$\psi \,|\, B' \circ f_B \circ \pi \circ \varphi^{-1} = (f_B)_{\varphi|B,\psi|B'} \circ \pi_{\varphi,\varphi|B}$$

which are equal by (i) and because the local representatives of π and π' are projections onto the first factor. Also, if $f_{\varphi\psi} = (\alpha_1, \alpha_2)$, then $(f_B)_{\varphi\psi} = \alpha_1$, so f_B is C^r.

One half of (iii) is clear from (i) and (ii). For the converse we see that in local representation, g has the form

$$g_{\varphi\psi}(u, f) = (\psi \circ g \circ \varphi^{-1})(u, f) = (\alpha_1(u), \alpha_2(u) \cdot f),$$

which defines α_1 and α_2. Since g is linear on fibers, $\alpha_2(u)$ is linear. Thus, the local representatives of g with respect to admissible local bundle charts are local bundle mappings by Proposition **3.4.3**. ■

We also note that the composition of two vector bundle mappings is again a vector bundle mapping.

3.4.13 Examples

A Let M and N be C^{r+1} manifolds and $f : M \to N$ a C^{r+1} map. Then $Tf : TM \to TN$ is a C^r vector bundle map of class C^r. Indeed the local representative of Tf, $(Tf)_{T\varphi,T\psi} = T(f_{\varphi\psi})$ is a local vector bundle map, so the result follows from Proposition **3.3.11**.

B The proof of Proposition **3.3.13** shows that $T(M_1 \times M_2)$ and $TM_1 \times TM_2$ are isomorphic as vector bundles over the identity of $M_1 \times M_2$. They are usually identified.

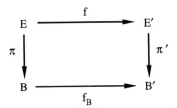

C To get an impression of how vector bundle maps work, let us show that the cylinder $S^1 \times \mathbb{R}$ and the Möbius band $\mathbb{M}$ are *not* vector bundle isomorphic. If $\varphi : \mathbb{M} \to S^1 \times \mathbb{R}$ were such an isomorphism, then the image of the curve $c : [0, 1] \to \mathbb{M}$, $c(t) = [t, 1]$ by φ would never cross the zero section in $S^1 \times \mathbb{R}$, since $[s, 1]$ is never zero in all fibers of $\mathbb{M}$; i.e., the second component of $(\varphi \circ c)(t) \neq 0$ for all $t \in [0, 1]$. But $c(1) = [1, 1] = [0, -1] = -[0, 1] = -c(0)$ so that the second components of $\varphi \circ c$ at $t = 0$ and $t = 1$ are in absolute value equal and of opposite sign, which, by the intermediate value theorem, implies that the second component of $\varphi \circ c$ vanishes somewhere.

D It is shown in differential topology that for any vector bundle E with an n-dimensional base B and k-dimensional fiber there exists a vector bundle map $\varphi : E \to B \times \mathbb{R}^p$, where $p \geq k + n$, with $\varphi_B = I_B$ and which, when restricted to each fiber, is injective (Hirsch [1976]). Write $\varphi(v) = (\pi(v), F(v))$ so $F : E \to \mathbb{R}^p$ is linear on fibers. With the aid of this theorem, analogous in spirit to the Whitney embedding theorem, we can construct a vector bundle map $\Phi : E \to \gamma_k(\mathbb{R}^p)$ by $\Phi(v) = (F(E_b), F(v))$ where $v \in E_b$. Note that $\Phi_B : B \to \mathbb{G}_k(\mathbb{R}^p)$ maps $b \in B$ to the k-plane $F(E_b)$ in $\mathbb{R}^p$. Furthermore, note that E is vector bundle isomorphic to the pull-back bundles $\Phi^*(\gamma_k(\mathbb{R}^p))$ (see Exercise **3.4.O** for the definition of pull-back bundles). It is easy to check that $\varphi \mapsto \Phi$ is a bijection. Mappings $f : B \to \mathbb{G}_k(\mathbb{R}^p)$ such that $f^*(\gamma_k(\mathbb{R}^p))$ is isomorphic to E are called ***classifying maps for*** E; they play a central role in differential topology since they convert the study of vector bundles to homotopy theory (see Hirsch [1976] and Husemoller [1978]). ♦

A second generalization of a local C^r mapping, $f : U \subset E \to F$, globalizes not f but rather its graph mapping $\lambda_f : U \to U \times F$; $u \mapsto (u, f(u))$.

3.4.14 Definition *Let* $\pi : E \to B$ *be a vector bundle. A* C^r ***local section*** *of* π *is a* C^r *map* $\xi : U \to E$, *where* U *is open in* B, *such that for each* $b \in U$, $\pi(\xi(b)) = b$. *If* $U = B$, ξ *is called a* C^r ***global section***, *or simply a* C^r ***section*** *of* π. *Let* $\Gamma^r(\pi)$ *denote the set of all* C^r *sections of* π, *together with the obvious real (infinite-dimensional) vector space structure.*

The condition on ξ merely says that $\xi(b)$ lies in the fiber over b. The C^r sections form a linear function space suitable for global linear analysis. As will be shown in later chapters, this general construction includes spaces of vector and tensor fields on manifolds. The space of section of a vector bundle differs from the more general class of global C^r maps from one manifold to another, which is a ***nonlinear*** function space. (See, for example, Eells [1958], Palais [1968], Eliasson [1967], or Ebin and Marsden [1970] for further details.)

Submanifolds were defined in the preceding section. There are two analogies for vector bundles.

3.4.15 Definition *If* $\pi : E \to B$ *is a vector bundle and* $M \subset B$ *a submanifold, the* ***restricted bundle*** $\pi_M : E_M = E \mid M \to M$ *is defined by* $E_M = \bigcup_{m \in M} E_m$, $\pi_M = \pi \mid E_M$.

The restriction $\pi_M : E_M \to M$ is a vector bundle whose charts are induced by the charts of E in the following way. Let (V, ψ_1), $\psi_1 : V \to V' \subset E' \times \{0\}$, be a chart of M induced by the chart (U, φ_1) of B with the submanifold property, where $(\pi^{-1}(U), \varphi)$ (with $\varphi(e) = (\varphi_1(\pi(e)), \varphi_2(e))$, $\varphi : \pi^{-1}(U) \to U' \times F$, and $U' \subset E' \times E'' = E$) is a vector bundle chart of E. Then $\psi : \pi_M{}^{-1}(V) \to V' \times F$, $\psi(e) = (\psi_1(\pi(e)), \varphi_2(e))$ defines a vector bundle chart of E_M. It can be easily verified that the overlap maps satisfy **VB2**.
For example the restriction of any vector bundle to a chart domain of the base defined by a vector bundle chart gives a bundle isomorphic to a local vector bundle.

3.4.16 Definition *Let* $\pi : E \to B$ *be a vector bundle. A subset* $F \subset E$ *is called a* ***subbundle*** *if for each* $b \in B$ *there is a vector bundle chart* $(\pi^{-1}(U), \varphi)$ *of* E *where* $b \in U \subset B$ *and* $\varphi : \pi^{-1}(U) \to U' \times F$, *and a split subspace* G *of* F *such that* $\varphi(\pi^{-1}(U) \cap F) = U' \times (G \times \{0\})$.

These induced charts are verified to form a vector bundle atlas for $\pi \mid F : F \to B$. Note that subbundles have the ***same*** base as the original vector bundle. Intuitively, the restriction cuts the base keeping the fibers intact, while a subbundle has the same base but smaller fiber, namely $F_b = F \cap E_b$. Note that a subbundle F is a closed submanifold of E.

For example $\gamma_k(\mathbb{R}^n)$ is a subbundle of both $\gamma_k(\mathbb{R}^{n+1})$ and $\gamma_{k+1}(\mathbb{R}^{n+1})$, the canonical inclusions being given by $(F, x) \mapsto (F \times \{0\}, (x, 0))$ and $(F, x) \mapsto (F \times \mathbb{R}, (x, 0))$, respectively.

3.4.17 Proposition *Let* $\pi : E \to B$ *be a vector bundle and* $F \subset E$ *a subbundle. Consider the following equivalence relation on* $E : v \sim v'$ *if there is a* $b \in B$ *such that* $v, v' \in E_b$ *and* $v - v' \in F_b$. *The quotient set* $E/\sim$ *has a unique vector bundle structure for which the canonical projection* $p : E \to E/\sim$ *is a vector bundle map over the identity. This vector bundle is called the* ***quotient*** E/F *and has fibers* $(E/F)_b = E_b/F_b$.

Proof Since $F \subset E$ is a subbundle there is a vector bundle chart $\varphi : \pi^{-1}(U) \to U' \times F$ and split subspaces $F_1, F_2, F_1 \oplus F_2 = F$, such that $\varphi \mid \pi^{-1}(U) \cap *F : (\pi \mid F)^{-1}(U) \to U' \times (F_1 \times \{0\})$ is a vector bundle chart for F. The map π induces a unique map $\Pi : E/\sim \to B$ such that $\Pi \circ p = \pi$. Similarly φ induces a unique map $\Phi : \Pi^{-1}(U) \to U' \times (\{0\} \times F_2)$ by the condition $\Phi \circ p = \varphi \mid \varphi^{-1}(U' \times (\{0\} \times F_2))$, which is seen to be a homeomorphism. One verifies that the overlap map of two such Φ is a local vector bundle isomorphism, thus giving a vector bundle structure to $E/\sim$, with fiber E_b/F_b, for which $p : E \to E/\sim$ is a vector bundle map. From the definition of Φ it follows that the structure is unique if p is to be a vector bundle map over the identity. ■

3.4.18 Proposition *Let* $\pi : E \to B$ *and* $\rho : F \to B$ *be vector bundles and* $f : E \to F$ *a vector bundle map over the identity. Let* $f_b : E_b \to F_b$ *be the restriction of* f *to the fiber over* $b \in B$ *and define the* ***kernel*** *of* f *by* $\ker(f) = \bigcup_{b \in B} \ker(f_b)$ *and the* ***range*** *of* f *by* $\mathrm{range}(f) = \bigcup_{b \in B} \mathrm{range}(f_b)$.

(i) $\ker(f)$ *and* $\mathrm{range}(f)$ *are subbundles of* E *and* F *respectively iff for every* $b \in B$ *there are vector bundle charts* $(\pi^{-1}(U), \varphi)$ *of* E *and* $(\rho^{-1}(U), \psi)$ *of* F *such that the local representative of* f *has the form*

$$f_{\varphi\psi} : U' \times (F_1 \times F_2) \to U' \times (G_1 \times G_2),$$

where

$$f_{\varphi\psi}((u, (f_1, f_2))) = (u, (\chi(u) \cdot f_2, 0)),$$

and $\chi(u) : F_2 \to G_1$ *is a continuous linear isomorphism.*

(ii) *If* E *has finite-dimensional fiber, the condition in* (i) *is equivalent to the local constancy of the rank of the linear map* $f_b : E_b \to F_b$.

Proof (i) It is enough to prove the result for local vector bundles. But there it is trivial since $\ker(f_{\varphi\psi})_u = F_1$ and $\mathrm{range}(f_{\varphi\psi})_u = G_1$.

(ii) Fix $u \in U'$ and put $(f_{\varphi\psi})_u(F) = G_1$. Then since G_1 is closed and finite dimensional in G, it splits; let $G = G_1 \oplus G_2$. Let $F_1 = \ker(f_{\varphi\psi})_u$; F is finite dimensional and hence $F = F_1 \oplus F_2$. Then $(f_{\varphi\psi})_u : F_2 \to G_1$ is an isomorphism. Write

$$(f_{\varphi\psi})_{u'} = \begin{bmatrix} a(u') & b(u') \\ c(u') & d(u') \end{bmatrix} : \begin{bmatrix} F_1 \\ F_2 \end{bmatrix} \to \begin{bmatrix} G_1 \\ G_2 \end{bmatrix}$$

for $u' \in U'$ and note that $b(u')$ is an isomorphism. Therefore $b(u')$ is an isomorphism for all u' in a neighborhood of u by **2.5.4**. We can assume that this neighborhood is U', by shrinking U' if necessary. Note also that $a(u) = 0$, $c(u) = 0$, $d(u) = 0$. The rank of $(f_{\varphi\psi})_{u'}$ is constant in a neighborhood of u, so shrink U' further, if necessary, so that $(f_{\varphi\psi})_{u'}$ has constant rank for all $u' \in U'$. Since $b(u')$ is an isomorphism, $a(u')(F_1) + b(u')(F_2) = G_1$ and since the rank of $(f_{\varphi\psi})_{u'}$ equals the dimension of G_1, it follows that $c(u') = 0$ and $d(u') = 0$ for all $u' \in U'$. Then

$$\lambda_{u'} = \begin{bmatrix} I & 0 \\ -b(u')^{-1}a(u') & I \end{bmatrix} \in GL(F_1 \oplus F_2, F_1 \oplus F_2)$$

and

$$(f_{\varphi\psi})_{u'} \circ \lambda_{u'} = \begin{bmatrix} 0 & b(u') \\ 0 & 0 \end{bmatrix}$$

which yields the form of the local representative in (i) after fiberwise composing $\varphi_{u'}$ with $\lambda_{u'}{}^{-1}$. ■

3.4.19 Definition *A sequence of vector bundle maps over the identity* $E \xrightarrow{f} F \xrightarrow{g} G$ *is* ***exact at*** F *if* range(f) = ker (g). *It is* ***split fiber exact*** *if* ker(f), range(g), *and* range(f) = ker(g) *split in each fiber. It is* ***bundle exact*** *if it is split fiber exact and* ker(f), range(g), *and* range(f) = ker(g) *are subbundles.*

3.4.20 Proposition *Let E, F, and G be vector bundles over a manifold* B *and let*

$$E \xrightarrow{g} F \xrightarrow{f} G$$

be a split fiber exact sequence of smooth bundle maps. Then the sequence is bundle exact; i.e., ker(f), range(f) = ker(g), *and* range(g) *are subbundles of* E, F, *and* G *respectively.*

Proof Fixing $b \in B$, set $A = \ker(f_b)$, $B = \ker(g_b) = \operatorname{range}(f_b)$, $C = \operatorname{range}(g_b)$, and let D be a complement for C in G_b, so $E_b = A \times B$, $F_b = B \times C$, and $G_b = C \times D$. Let $\varphi : U \to U'$ be a chart on B at b, $\varphi(b) = 0$, defining vector bundle charts on E, F, and G. Then the local representatives $f' : U' \times A \times B \to U' \times B \times C$, $g' : U' \times B \times C \to U' \times C \times D$ of f and g respectively are the identity mappings on U' and can be written as matrices of operators

$$f'_{u'} = \begin{bmatrix} z & w \\ x & y \end{bmatrix} : \begin{bmatrix} A \\ B \end{bmatrix} \to \begin{bmatrix} B \\ C \end{bmatrix} \text{ and } g'_{u'} = \begin{bmatrix} \beta & \gamma \\ \alpha & \delta \end{bmatrix} : \begin{bmatrix} B \\ C \end{bmatrix} \to \begin{bmatrix} C \\ D \end{bmatrix}$$

depending smoothly on $u' \in U'$. Now since w_0 and γ_0 are isomorphisms by Banach's isomorphism theorem, shrink U and U' such that $w_{u'}$ and $\gamma_{u'}$ are isomorphisms for all $u' \in U'$. By exactness, $g'_{u'} \circ f'_{u'} = 0$, which in terms of the matrix representations becomes

$$x = -\gamma^{-1} \circ \beta \circ z, \quad y = -\gamma^{-1} \circ \beta \circ w, \quad \alpha = -\delta \circ y \circ w^{-1}, \text{ i.e.,}$$

$$x = y \circ w^{-1} \circ z \text{ and } \alpha = \delta \circ \gamma^{-1} \circ \beta.$$

Extend $f'_{u'}$ to the map $h_{u'} : A \times B \times C \to A \times B \times C$ depending smoothly on $u' \in U'$ by

$$h_{u'} = \begin{bmatrix} I & 0 & 0 \\ z & w & 0 \\ x & y & I \end{bmatrix}.$$

We find maps a, b, c, d, k, m, n, p such that

$$\begin{bmatrix} I & 0 & 0 \\ 0 & a & b \\ 0 & c & d \end{bmatrix} h_{u'} \begin{bmatrix} k & p & 0 \\ m & n & 0 \\ 0 & 0 & I \end{bmatrix} = I$$

which can be accomplished by choosing $k = I$, $d = I$, $p = 0$, $b = 0$, $a = w^{-1}$, $n = I$, $m = -w^{-1} \circ z$, $c = -y \circ w^{-1}$ and taking into account that $x = y \circ w^{-1} \circ z$. This procedure gives isomorphisms

$$\lambda = \begin{bmatrix} a & b \\ c & d \end{bmatrix} = \begin{bmatrix} w^{-1} & 0 \\ -y \circ w^{-1} & I \end{bmatrix} : \begin{bmatrix} B \\ C \end{bmatrix} \to \begin{bmatrix} B \\ C \end{bmatrix}$$

$$\mu = \begin{bmatrix} k & p \\ m & n \end{bmatrix} = \begin{bmatrix} I & 0 \\ -w^{-1} \circ z & I \end{bmatrix} : \begin{bmatrix} A \\ B \end{bmatrix} \to \begin{bmatrix} A \\ B \end{bmatrix}$$

depending smoothly on $u' \in U$ such that

$$\lambda \circ f'_{u'} \circ \mu = \begin{bmatrix} 0 & I \\ 0 & 0 \end{bmatrix} : \begin{bmatrix} A \\ B \end{bmatrix} \to \begin{bmatrix} B \\ C \end{bmatrix}.$$

Proposition **3.4.18**(ii) shows that ker(f) and range(f) are subbundles. The same procedure applied to $g'_{u'}$ proves that ker(g) and range(g) are subbundles and thus the fiber split exact sequence $E \xrightarrow{f} F \xrightarrow{g} G$ is bundle exact. ∎

As a special case note that $0 \longrightarrow F \xrightarrow{g} G$ is split fiber exact when g_b is injective and has split range. Here 0 is the trivial bundle over B with zero-dimensional fiber and the first arrow is injection to the zero section. Similarly, taking $G = 0$ and g the zero map, the sequence $E \xrightarrow{f} E$ $F \longrightarrow 0$ is split fiber exact when f_b is surjective with split kernel. In both cases range(g) and ker(f) are subbundles by **3.4.20**. In **3.4.20,** and these cases in particular, we note that if the sequences are split fiber exact at b, then they are also split fiber exact in a neighborhood of b by the openness of $GL(E, E)$ in $L(E, E)$.

A split fiber exact sequence of the form

$$0 \longrightarrow E \xrightarrow{f} F \xrightarrow{g} G \longrightarrow 0$$

is called a ***short exact sequence***. By **3.4.20** and **3.4.17,** any split fiber exact sequence

$$E \xrightarrow{f} F \xrightarrow{g} G$$

induces a short exact sequence

$$0 \longrightarrow E/\ker(f) \xrightarrow{[f]} F \xrightarrow{g} \mathrm{range}(g) \longrightarrow 0$$

where $[f]([e]) = f(e)$ for $e \in E$.

3.4.21 Definition *A short exact sequence*

$$0 \longrightarrow E \xrightarrow{f} F \xrightarrow{g} G \longrightarrow 0$$

is said to be ***split exact*** *if there is a split fiber exact sequence* $0 \longrightarrow G \xrightarrow{h} F$ *such that* $g \circ h$ *is the identity on* G

The geometric meaning of this concept will become clear after we introduce a few additional constructions with vector bundles.

3.4.22 Definition *If* $\pi : E \to B$ *and* $\pi' : E' \to B'$ *are two vector bundles, the* **product bundle** $\pi \times \pi' : E \times E' \to B \times B'$ *is defined by the vector bundle atlas:*

$$\{(\pi^{-1}(U) \times \pi'^{-1}(U'), \varphi \times \psi \mid (\pi^{-1}(U), \varphi), U \subset B \text{ and } (\pi'^{-1}(U'), \psi), U' \subset B' \text{ are vector bundle charts of } E \text{ and } E', \text{ respectively}\}.$$

It is straightforward to check that the product atlas verifies conditions **VB1** and **VB2** of **3.4.4**.

Below we present a general construction, special cases of which are used repeatedly in the rest of the book. It allows the transfer of vector space constructions into vector bundle constructions. The abstract procedure will become natural in the context of examples given below in **3.4.24** and later in the book.

3.4.23 Definition *Let* I *and* J *be finite sets and consider two families* $\mathcal{E} = (E_k)_{k \in I \cup J}$, and $\mathcal{E}' = (E'_k)_{k \in I \cup J}$ *of Banachable spaces. Let*

$$L(\mathcal{E}, \mathcal{E}') = \prod_{i \in I} L(E_i, E'_i) \times \prod_{j \in J} L(E'_j, E_j)$$

and let

$$(A_k) \in L(\mathcal{E}, \mathcal{E}');$$

i.e., $A_i \in L(E_i, E'_i)$, $i \in I$, *and* $A_j \in L(E'_j, E_j)$, $j \in J$. *An assignment* Ω *taking any family* $\mathcal{E}$ *to a Banach space* $\Omega\mathcal{E}$ *and any sequence of linear maps* (A_k) *to a linear continuous map* $\Omega(A_k) \in L(\Omega\mathcal{E}, \Omega\mathcal{E}')$ *satisfying* $\Omega(I_{E_k}) = I_{\Omega\mathcal{E}}$, $\Omega((B_k) \circ (A_k)) = \Omega((B_k)) \circ \Omega((A_k))$ *(composition is taken componentwise) and is such that the induced map* $\Omega : L(\mathcal{E}, \mathcal{E}') \to L(\Omega\mathcal{E}, \Omega\mathcal{E}')$ *is* C^∞, *will be called a* **tensorial construction of type** (I, J).

3.4.24 Proposition *Let* Ω *be a tensorial construction of type* (I, J) *and* $\mathcal{E} = (E^k)_{k \in I \cup J}$ *be a family of vector bundles with the same base* B. *Let*

$$\Omega\mathcal{E} = \bigcup_{b \in B} \Omega\mathcal{E}_b, \text{ where } \mathcal{E}_b = (E_b{}^k)_{k \in I \cup J}.$$

Then $\Omega\mathcal{E}$ *has a unique vector bundle structure over* B *with* $(\Omega\mathcal{E})_b = \Omega\mathcal{E}_b$ *and* $\pi : \Omega\mathcal{E} \to B$ *sending* $\Omega\mathcal{E}_b$ *to* $b \in B$, *whose atlas is given by the charts* $(\pi^{-1}(U), \psi)$, *where* $\psi : \pi^{-1}(U) \to U' \times \Omega((F^k))$ *is defined as follows. Let*

$$(\pi_k{}^{-1}(U), \varphi^k), \ \varphi^k : \pi_k{}^{-1}(U) \to U' \times F^k, \ \varphi^k(e^k) = (\varphi_1(\pi_k(e^k)), \varphi_2{}^k(e^k))$$

be vector bundle charts on E^k *inducing the same manifold chart on* B. *Define* $\psi(x) = (\varphi_1(\pi(x)), \Omega(\psi_{\pi(x)})(x))$ *by* $\psi_{\pi(x)} = (\psi^k{}_{\pi(x)})$ *where* $\psi^i{}_{\pi(x)} = (\varphi^i{}_2)^{-1}$, *for* $i \in I$ *and* $\psi^j{}_{\pi(x)} = (\varphi^j{}_2)$ *for* $j \in J$.

Proof We need to show that the overlap maps are local vector bundle isomorphisms. We have

$$(\psi' \circ \psi^{-1})(u, e) = ((\varphi'_1 \circ \varphi_1^{-1})(u),\ \ \Omega((\varphi'^k_2 \circ (\varphi_2^k)^{-1}(u)) \cdot e)),$$

the first component of which is trivially C^∞. The second component is also C^∞ since each φ^k is a vector bundle chart by the Composite Mapping Theorem, and by the fact that Ω is smooth. ■

3.4.25 Examples

A Whitney sum Choose for the tensorial construction the following: $J = \varnothing$, $I = \{1, ..., n\}$, and $\Omega\mathcal{E}$ is the single Banach space $E_1 \times \cdots \times E_n$. Let $\Omega((A_i)) = A_1 \times \cdots \times A_n$. The resulting vector bundle is denoted by $E_1 \oplus \cdots \oplus E_n$ and is called the ***Whitney sum***. The fiber over $b \in B$ is just the sum of the component fibers.

B Vector bundles of bundle maps Let E_1, E_2 be two vector bundles. Choose for the tensorial construction the following: I, J are one-point sets $I = \{1\}$, $J = \{2\}$, $\Omega(E_1, E_2) = L(E_2, E_1)$, $\Omega(A_1, A_2) \cdot S = A_1 \circ S \circ A_2$ for $S \in L(E_1, E_1)$. The resulting bundle is denoted by $L(E_2, E_1)$. The fiber over $b \in B$ consists of the linear maps of $(E_2)_b$ to $(E_1)_b$.

C Dual bundle This is a particular case of Example **B** for which $E = E_2$ and $E_1 = B \times \mathbb{R}$. The resulting bundle is denoted E^*; the fiber over $b \in B$ is the dual $E_b{}^*$. If $E = TM$, then E^* is called the ***cotangent bundle*** of M and is denoted by T^*M.

D Vector bundle of multilinear maps Let $E_0, E_1, ..., E_n$ be vector bundles over the same base. The space of n-multilinear maps (in each fiber) $L(E_1, ..., E_n; E_0)$ is a vector bundle over B by the choice of the following tensorial construction : $I = \{0\}$, $J = \{1,..., n\}$, $\Omega(E_0, ..., E_n) = L^n(E_1, ..., E_n; E_0)$, $\Omega(A_0, A_1, ..., A_n) \cdot S = A_0 \circ S \circ (A_1 \times \cdots \times A_n)$ for $S \in L^n(E_1, ..., E_n; E_0)$. One similarly considers $L^k{}_s(E; E_0)$ and $L^k{}_a(E; E_0)$, the vector bundle of symmetric and antisymmetric k-linear vector bundle maps of $E \times E \times \cdots \times E$ to E_0. ♦

3.4.26 Proposition *A short exact sequence of vector bundles*

$$0 \longrightarrow E \overset{f}{\longrightarrow} F \overset{g}{\longrightarrow} G \longrightarrow 0$$

is split if and only if there is a vector bundle isomorphism $\varphi : F \to E \oplus G$ *such that* $\varphi \circ f = i$ *and* $p \circ \varphi = g$, *where* $i : E \to E \oplus G$ *is the inclusion* $u \mapsto (u, 0)$ *and* $p : E \oplus G \to G$ *is the projection* $(u, w) \mapsto w$.

Proof Note that $0 \longrightarrow E \overset{i}{\longrightarrow} E \oplus G \overset{p}{\longrightarrow} G \longrightarrow 0$ is a split exact sequence; the splitting is given by $w \in E \mapsto (0, w) \in E \oplus G$. If there is an isomorphism $\varphi : F \to E \oplus G$ as in the statement of the proposition, define $h : G \to F$ by $h(w) = \varphi^{-1}(0, w)$. Since φ is an isomorphism and G is a subbundle of $E \oplus G$, it follows that $0 \longrightarrow G \overset{h}{\longrightarrow} F$ is split fiber exact. Moreover $(g \circ h)(w) = (g \circ \varphi^{-1})(0, w) = p(0, w) = w$.

Conversely, assume that

$$0 \longrightarrow G \overset{h}{\longrightarrow} F \text{ is a splitting of } 0 \longrightarrow E \overset{f}{\longrightarrow} F \overset{g}{\longrightarrow} G \longrightarrow 0,$$

i.e., $g \circ h =$ identity on G. Then range(h) is a subbundle of F (by definition **3.4.19**) which is isomorphic to G by h. Since $g \circ h =$ identity, it follows that range(h) $\cap$ ker(g) $= 0$. Moreover, since any $v \in F$ can be written in the form $v = (v - h(g(v))) + h(g(v))$, with $h(g(v)) \in$ range(h) and $v - h(g(v)) \in$ ker(g), it follows that $F = \ker(g) \oplus \text{range}(h)$. Since the inverse of φ is given by $(u, v) \mapsto (f(u), h(v))$, it follows that the map φ is a smooth vector bundle isomorphism and that the identities $\varphi \circ f = i$, $p \circ \varphi = g$ hold. ■

We next give a brief account of a useful generalization of vector bundles, the locally trivial fiber bundles.

3.4.27 Definition *A* C^k ***fiber bundle,*** *where* $k \geq 0$, *with* ***typical fiber*** F *(a given manifold) is a* C^k *surjective map of* C^k *manifolds* $\pi : E \to B$ *which is locally a product, i.e., the* C^k *manifold* B *has an open atlas* $\{(U_\alpha, \varphi_\alpha)\}_{\alpha \in A}$ *such that for each* $\alpha \in A$ *there is a* C^k *diffeomorphism* χ_α : $\pi^{-1}(U_\alpha) \to U_\alpha \times F$ *such that* $p_\alpha \circ \chi_\alpha = \pi$, *where* $p_\alpha : U_\alpha \times F \to U_\alpha$ *is the projection. The* C^k *manifolds* E *and* B *are called the* ***total space*** *and* ***base*** *of the fiber bundle respectively. For each* $b \in B$, $\pi^{-1}(b) = E_b$ *is called the* ***fiber*** *over* b. *The* C^k *diffeomorphisms* χ_α *are called* ***fiber bundle charts.*** *If* $k = 0$, E, B, F *are required to be only topological spaces and* $\{U_\alpha\}$ *an open covering of* B.

Each fiber $E_b = \pi^{-1}(b)$, for $b \in B$, is a closed C^k submanifold of E, which is C^k diffeomorphic to F via $\chi_\alpha \mid E_b$. The total space E is the disjoint union of all of its fibers. By the local product property, the C^k manifold structure of E is given by an atlas whose charts are products, i.e., any chart on E contains a chart of the form $\rho_{\alpha\beta} = (\varphi_\alpha \times \psi_\beta) \circ \chi_\alpha : \chi_\alpha^{-1}(U_\alpha \times V_\beta) \to \varphi_\alpha(U_\alpha) \times \psi_\beta(V_\beta)$, where $(U_\alpha, \varphi_\alpha)$ is a chart on B satisfying the property of the definition and thus giving rise to χ_α, and (V_β, ψ_β) is any chart on F. Note that the maps $\chi_{\alpha b} = \chi_\alpha \mid E_b$: $E_b \to F$ are C^k diffeomorphisms. If $(U_{\alpha'}, \varphi_{\alpha'})$ and $\chi_{\alpha'}$ are as in **3.4.27** with $U_\alpha \cap U_{\alpha'} \neq \varnothing$, then the diffeomorphism $\chi_{\alpha'} \circ \chi_\alpha^{-1} : (U_\alpha \cap U_{\alpha'}) \times F \to (U_\alpha \cap U_{\alpha'}) \times F$ is given by $(\chi_{\alpha'} \circ \chi_\alpha^{-1})(u, f) = (u, (\chi_{\alpha' u} \circ \chi_{\alpha u}^{-1})(f))$ and therefore $\chi_{\alpha' u} \circ \chi_{\alpha u}^{-1} : F \to F$ is a C^k diffeomorphism. This proves the uniqueness part in the following proposition.

3.4.28 Proposition *Let* E *be a set,* B *and* F *be* C^k *manifolds, and let* $\pi : E \to B$ *be a surjective map. Assume that*

(i) *there is a* C^k *atlas* $\{(U_\alpha, \varphi_\alpha)\}$ *of* B *and a family of bijective maps* $\chi_\alpha : \pi^{-1}(U_\alpha) \to U_\alpha \times F$ *satisfying* $p_\alpha \circ \chi_\alpha = \pi$, *where* $p_\alpha : U_\alpha \times F \to U_\alpha$ *is the projection, and that*

(ii) *the maps* $\chi_{\alpha'} \circ \chi_\alpha^{-1} : U_\alpha \times F \to U_{\alpha'} \times F$ *are* C^k *diffeomorphisms whenever* $U_{\alpha'} \cap U_\alpha \neq \varnothing$.

Then there is a unique C^k *manifold structure on* E *for which* $\pi : E \to B$ *is a* C^k *locally trivial fiber bundle with typical fiber* F.

Proof Define the atlas of E by $(\chi_\alpha^{-1}(U_\alpha \times V_\beta), \rho_{\alpha\beta})$, where $(U_\alpha, \varphi_\alpha)$ is a chart in the atlas of B given in (i), $\chi_\alpha : \pi^{-1}(U_\alpha) \to U_\alpha \times F$ is the bijective map given in (i), (V_β, ψ_β) is any chart on F, and $\rho_{\alpha\beta} = (\varphi_\alpha \times \psi_\beta) \circ \chi_\alpha$. If $(U_{\alpha'}, \varphi_{\alpha'})$ is another chart of of the atlas of B in (i) and $(V_{\beta'}, \psi_{\beta'})$ is another chart on F such that $U_\alpha \cap U_{\alpha'} \neq \varnothing$ and $V_\beta \cap V_{\beta'} \neq \varnothing$, then the overlap map $\rho_{\alpha'\beta'} \circ \rho_{\alpha\beta}^{-1} = (\varphi_{\alpha'} \times \psi_{\beta'}) \circ \chi_{\alpha'} \circ \chi_\alpha^{-1} \circ (\varphi_\alpha^{-1} \times \psi_\beta^{-1})$ is C^k by (i). Thus $\{(\chi_\alpha^{-1}(U_\alpha \times V_\beta), \rho_{\alpha\beta})\}$ is a C^k atlas on E relative to which $\pi : E \to B$ is a C^k locally trivial fiber bundle by (i). The differentiable structure on E is unique by the remarks preceding this proposition. ■

Many of the concepts introduced for vector bundles have generalizations to fiber bundles. For instance, ***local*** and ***global sections*** are defined as in **3.4.14**. Given a fiber bundle $\pi : E \to B$, the restricted bundle $\pi_M : E_M = E \mid M \to M$, for M a submanifold of B is defined as in **3.4.15.** A ***locally trivial subbundle*** of $\pi : E \to F$ with typical fiber G, a submanifold of F, is a submanifold E' of E such that the map $\pi' = \pi \mid E' : E' \to B$ is onto and satisfies the following property: if $\chi_\alpha : \pi^{-1}(U_\alpha) \to U_\alpha \times F$ is a local trivialization of E, then $\chi'_\alpha = \chi_\alpha \mid \pi'^{-1}(U_\alpha) \to U_\alpha \times G$ are local trivializations. Thus $\pi' : E' \to B$ is a locally trivial fiber bundle in its own right. Finally, ***locally trivial fiber bundle maps***, or ***fiber bundle morphisms*** are defined in the following way. If $\pi' : E' \to B'$ is another locally trivial fiber bundle with typical fiber F', then a smooth map $f : E \to E'$ is called ***fiber preserving*** if $\pi(e_1) = \pi(e_2)$ implies $(\pi' \circ f)(e_1) = (\pi' \circ f)(e_2)$, for $e_1, e_2 \in E$. Thus f determines a map $f_B : B \to B'$ satisfying $\pi' \circ f = \pi \circ f_B$. The map f_B is smooth since for any chart $(U_\alpha, \varphi_\alpha)$ of B inducing a local trivialization $\chi_\alpha : \pi^{-1}(U_\alpha) \to U_\alpha \times F$, the map f_B can be written as $f_B(b) = (\pi \circ f \circ \chi_\alpha^{-1})(b, n)$, for any fixed $n \in F$. The pair (f, f_B) is called a ***locally trivial fiber bundle map*** or ***fiber bundle morphism***. An invertible fiber bundle morphism is called a ***fiber bundle isomorphism.***

3.4.29 Examples

A Any manifold is a locally trivial fiber bundle with typical fiber a point.

B Any vector bundle $\pi : E \to B$ is a locally trivial fiber bundle whose typical fiber is the model of the fiber E_b. Indeed, if $\varphi : W \to U' \times F$, where U' open in E, is a local vector bundle chart, by Proposition **3.4.6**, $\varphi \mid \varphi^{-1}(U \times \{0\}) : U \to U' \subset E$, $U = W \cap B$, is a chart on the base B and $\chi : \pi^{-1}(U) \to U \times F$ defined by $\chi(e) = (\pi(e), (p_2 \circ \varphi)(e))$, where $p_2 : U' \times F \to F$ is the projection, is a local trivialization of E. In fact, *any locally trivial fiber bundle* $\pi : E \to B$ *whose typical fiber* F *is a Banach space is a vector bundle, iff the maps* $\chi_{\alpha b} : E_b \to F$ *induced by the local trivializations* $\chi_\alpha : \pi^{-1}(U_\alpha) \to U_\alpha \times F$, *are linear and continuous*. Indeed, under these hypotheses, the vector bundle charts are given by $(\varphi_\alpha \times \mathrm{id}_F) \circ \chi_\alpha : \pi^{-1}(U_\alpha) \to U_\alpha \times F$, where id_F is the identity mapping on F.

C Many of the topological properties of a vector bundle are determined by its fiber bundle structure. For example, a *vector bundle* $\pi : E \to B$ *is trivial if and only if it is trivial as a fiber bundle*. Clearly, if E is a trivial vector bundle, then it is also a trivial fiber bundle. The converse is also true, but requires topological ideas beyond the scope of this book. (See, for instance,

Steenrod [1952].)

D The *Klein bottle* $\mathbb{K}$ (see Fig. 1.4.2) is a locally trivial fiber bundle $\pi : \mathbb{K} \to S^1$ with typical fiber S^1. The space $\mathbb{K}$ is defined as the quotient topological space of $\mathbb{R}^2$ by the relation $(a, b) \sim (a + k, (-1)^k b + n)$ for all $k, n \in \mathbb{Z}$. Let $p : \mathbb{R}^2 \to \mathbb{K}$ be the projection $p(a, b) = [a, b]$ and define the surjective map $\pi : \mathbb{K} \to S^1$ by $\pi([a, b]) = e^{2\pi i a}$. Let $\{(U_j, \varphi_j) \mid j = 1, 2\}$ be the atlas of S^1 given in Example **3.1.2**, i.e., $\varphi_j : S^1 \setminus \{(0, (-1)^{j+1})\} \to \mathbb{R}$, $\varphi_j(x, y) = y/(1 - (-1)^j x)$, which satisfy $(\varphi_2 \circ \varphi_1^{-1})(z) = 1/z$, for $z \in \mathbb{R} \setminus \{0\}$. Define $\chi_j : \pi^{-1}(U_j) \to U_j \times S^1$ by $\chi_j([a, b]) = (e^{2\pi i a}, e^{2\pi i b})$ and note that $p_j \circ \chi_j = \pi$, where $p_j : U_j \times S^1 \to U_j$ is the projection. Since $\chi_2 \circ \chi_1^{-1} : (S^1 \setminus \{(0, 1)\}) \times S^1 \to (S^1 \setminus \{(0, -1)\}) \times S^1$ is the identity, Proposition **3.4.28** implies that $\mathbb{K}$ is a locally trivial fiber bundle with typical fiber S^1. Further topological results show that this bundle is nontrivial; see Exercise 4.1P. (Later we will prove that $\mathbb{K}$ is non - orientable - see Chapter 7.)

E Consider the smooth map $\pi_n : S^n \to \mathbb{RP}^n$ which associates to each point of S^n the line through the origin it determines. Then $\pi_n : S^n \to \mathbb{RP}^n$ is a locally trivial fiber bundle whose typical fiber is a two-point set. This is easily seen by taking for each pair of antipodal points two small antipodal disks and projecting them to an open set U in $\mathbb{RP}^n$; thus $\pi_n^{-1}(U)$ consists of the disjoint union of these disks and the fiber bundle charts simply send this disjoint union to itself. This bundle is not trivial since S^n is connected and two disjoint copies of $\mathbb{RP}^n$ are disconnected. These fiber bundles are also called the ***real Hopf fibrations***.

F This example introduces the ***classical Hopf fibration*** $h : S^3 \to S^2$ which is the fibration with the lowest dimensional total space and base among the series of ***complex Hopf fibrations*** $\kappa_n : S^{2n+1} \to \mathbb{CP}^n$ with typical fiber S^1 (see Exercise **3.4U**). To describe $h : S^3 \to S^2$ it is convenient to introduce the division algebra of quaternions $\mathbb{H}$.

For $x \in \mathbb{R}^4$ write $x = (x^0, \mathbf{x}) \in \mathbb{R} \times \mathbb{R}^3$ and introduce the product

$$(x^0, \mathbf{x})(y^0, \mathbf{y}) = (x^0 y^0 - \mathbf{x} \cdot \mathbf{y}, x^0 \mathbf{y} + y^0 \mathbf{x} + \mathbf{x} \times \mathbf{y}).$$

Relative to this product and the usual vector space structure, $\mathbb{R}^4$ becomes a non-commutative field denoted by $\mathbb{H}$ and whose elements are called ***quaternions***. The identity element in $\mathbb{H}$ is $(1, \mathbf{0})$, the inverse of $(x^0, \mathbf{x})$ is $(x^0, \mathbf{x})^{-1} = (x^0, -\mathbf{x})/\| x \|^2$, where $\| x \|^2 = (x^0)^2 + (x^1)^2 + (x^2)^2 + (x^3)^2$. Associativity of the product comes down to the vector identity $\mathbf{a} \times (\mathbf{b} \times \mathbf{c}) = \mathbf{b}(\mathbf{a} \cdot \mathbf{c}) - \mathbf{c}(\mathbf{a} \cdot \mathbf{b})$. Alternatively, the quaternions written as linear combinations of the form $x^0 + ix^1 + jx^2 + kx^3$, where $i = (0, \mathbf{i})$, $j = (0, \mathbf{j})$, $k = (0, \mathbf{k})$ obey the multiplication rules $ij = k$, $jk = i$, $ki = j$, $i^2 = j^2 = k^2 = -1$. Quaternions with $x^0 = 0$ are called ***pure quaternions*** and the ***conjugation*** $x \mapsto x^*$ given by $i^* = -i$, $j^* = -j$, $k^* = -k$ is an automorphism of the $\mathbb{R}$-algebra $\mathbb{H}$. Then $\| x \|^2 = xx^*$ and $\| xy \| = \| x \| \| y \|$ for all $x, y \in \mathbb{H}$. Finally, the dot product in $\mathbb{R}^4$ and the product of $\mathbb{H}$ are connected by the relation $xz \cdot yz = (x \cdot y)\| z \|^2$, for all $x, y, z \in \mathbb{H}$.

Fix $y \in \mathbb{H}$. The conjugation map $c_y : \mathbb{H} \to \mathbb{H}$ defined by $c_y(x) = yxy^{-1}$ is norm preserving and hence orthogonal. Since it leaves the vector $(x^0, \mathbf{0})$ invariant, it defines an orthogonal transformation of $\mathbb{R}^3$. A simple computation shows that this orthogonal transformation of $\mathbb{R}^3$ is given by

$$\mathbf{x} \mapsto \mathbf{x} + \frac{2}{\| \mathbf{y} \|} [(\mathbf{x} \cdot \mathbf{y})\mathbf{y} - y^0(\mathbf{x} \times \mathbf{y}) - (\mathbf{y} \cdot \mathbf{y})\mathbf{x}]$$

from which one can verify that its determinant equals one, i.e., it is an element of $\mathbf{SO}(3)$. Let $\pi : S^3 \to \mathbf{SO}(3)$ denote its restriction to the unit sphere in $\mathbb{R}^4$. Choosing $\mathbf{x} \in \mathbb{R}^3$, define $\rho_\mathbf{x} : \mathbf{SO}(3) \to S^2$ by $\rho_\mathbf{x}(A) = A\mathbf{x}$ so that by composition we get $h_\mathbf{x} = \rho_\mathbf{x} \circ \pi : S^3 \to S^2$. It is easily verified that the inverse image of any point under $h_\mathbf{x}$ is a circle. Taking for $\mathbf{x} = -\mathbf{k}$, minus the third standard basis vector in $\mathbb{R}^3$, $h_\mathbf{x}$ becomes the ***standard Hopf fibration*** $h : S^3 \to S^2$,

$$h(y^0, y^1, y^1, y^3) = (-2y^1y^3 - 2y^0y^2, 2y^0y^1 - 2y^2y^3, (y^1)^2 + (y^2)^2 - (y^0)^2 - (y^3)^2)$$

which, by substituting $w^1 = y^0 + iy^3$, $w^2 = y^2 + iy^1 \in \mathbb{C}$ takes the classical form

$$h(w^1, w^2) = (-2w^1\bar{w}^2, | w^2 |^2 - | w^1 |^2).$$

Interestingly enough, the Hopf fibration enters into a number of problems in classical mechanics from rigid body dynamics to the dynamics of coupled oscillators (see Marsden and Ratiu [1989] for instance - in fact, the map h above is an example of the important notion of what is called a *momentum map*).

The Hopf fibration is nontrivial. A rigorous proof of this fact is not so elementary and historically was what led to the introduction of the Hopf invariant, a precursor of characteristic classes (Hopf [1931] and Hilton and Wylie [1956]). We shall limit ourselves to a geometric description of this bundle which exhibits its non-triviality. In fact we shall describe how each pair of fibers are linked. Cut S^2 along an equator to obtain the closed northern and southern hemispheres, each of which is diffeomorphic to two closed disks D_N and D_S. Their inverse images in S^3 are two solid tori $S^1 \times D_N$ and $S^1 \times D_S$. We think of S^3 as the compactification of $\mathbb{R}^3$ and as the union of two solid tori glued along their common boundary by a diffeomorphism which identifies the parallels of one with meridians of the other and vice-versa. The Hopf fibration on S^3 is then obtained in the following way. Cut each of these two solid tori along a meridian, twist them by 2π and glue each one back together. The result is still two solid tori but whose embedding in $\mathbb{R}^3$ is changed: they have the same parallels but twisted meridians; each two meridians are now linked (see Figure 3.4.5). Now glue the two twisted solid tori back together along their common boundary by the diffeomorphism identifying the twisted meridians of one with the parallels of the other and vice-versa, thereby obtaining the total space S^3 of the Hopf fibration. ♦

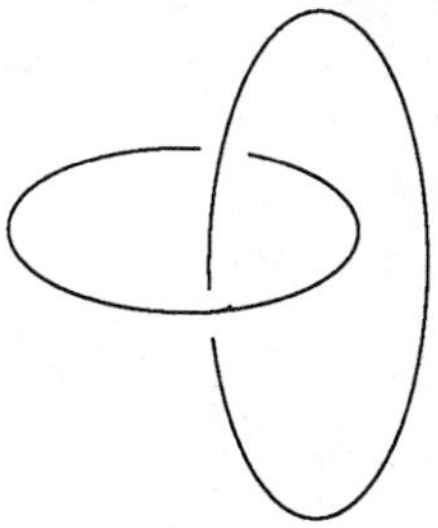

Figure 3.4.5

Topological properties of the total space E of a locally trivial fiber bundle are to a great extent determined by the topological properties of the base B and the typical fibers F. We present here only some elementary connectivity properties; other results can be found in Supplement **5.5C** and §**7.5**.

3.4.30 Path Lifting Theorem *Let* $\pi : E \to B$ *be a locally trivial* C^0 *fiber bundle and let* $c : [0, 1] \to B$ *be a continuous path starting at* $c(0) = b$. *Then for each* $c_0 \in \pi^{-1}(b_0)$, *there is a unique continuous path* $\tilde{c} : [0, 1] \to E$ *such that* $\tilde{c}(0) = c_0$ *and* $\pi \circ \tilde{c} = c$.

Proof Cover the compact set $c([0, 1])$ by a finite number of open sets U_i, $i = 0, 1, \ldots, n-1$ such that each $\chi_i : \pi^{-1}(U_i) \to U_i \times F$ is a fiber bundle chart. Let $0 = t_0 < t_1 < \cdots < t_n = 1$ be a partition of $[0, 1]$ such that $c([t_i, t_{i+1}]) \subset U_i$, $i = 0, \ldots, n-1$. Let $\chi_0(e_0) = (b_0, f_0)$ and define $c_0(t) = \chi_0^{-1}(c(t), f_0)$ for $t \in [0, t_0]$. Then $\tilde{c}_0$ is continuous and $\pi \circ \tilde{c}_0 = c \mid [0, t_0]$. Let $\chi_1(\tilde{c}_0(t_1)) = (c(t_1), f_1)$ and define $\tilde{c}_1(t) = \chi_1^{-1}(c(t), f_1)$ for $t \in [t_1, t_2]$. Then $\tilde{c}_1$ is continuous and $\pi \circ \tilde{c}_1 = c \mid [t_1, t_2]$. In addition, if $e_1 = \chi_0^{-1}(c(t_1), f_0)$ then

$$\lim_{t \uparrow t_1} \tilde{c}_0(t) = e_1 \quad \text{and} \quad \lim_{t \downarrow t_1} \tilde{c}_1(t) = \chi_1^{-1}(c(t_1), f_1) = \tilde{c}_0(t_1) = e_1,$$

i.e., the map $[0, t_2] \to E$ which equals $\tilde{c}_0$ on $[0, t_1]$ and $\tilde{c}_1$ on $[t_1, t_2]$ is continuous. Now proceed similarly on $U_2, \ldots, U_{n-1}$. ■

Note that if $\pi : E \to B$ is a C^k locally trivial fiber bundle and $c : [0, 1] \to B$ is a piecewise C^k-map, the above construction yields a C^k piecewise ***lift*** $\tilde{c} : [0, 1] \to E$.

3.4.31 Corollary *Let* $\pi : E \to B$ *be a* C^k *locally trivial fiber bundle* $k \geq 0$, *with base* B *and typical fiber* F *pathwise connected. Then* E *is pathwise connected. If* $k \geq 1$, *only connectivity of* B *and* F *must be assumed.*

Proof Let $e_0, e_1 \in E$, $b_0 = \pi(e_0)$, $b_1 = \pi(e_1) \in B$. Since B is pathwise connected, there is a continuous path $c : [0, 1] \to B$, $c(0) = b_0$, $c(1) = b_1$. By Theorem **3.4.30**, there is a continuous path $\tilde{c} : [0, 1] \to E$ with $\tilde{c}(0) = e_0$. Let $\tilde{c}(1) = e'_1$. Since the fiber $\pi^{-1}(b_1)$ is connected there is a continous path $d : [1, 2] \to \pi^{-1}(b_1)$ with $d(1) = e'_1$ and $d(2) = e_1$. Thus γ defined by

$$\gamma(t) = \tilde{c}(t), \text{ if } t \in [0, 1] \quad \text{and} \quad \gamma(t) = d(t), \text{ if } t \in [1, 2],$$

is a continuous path with $\gamma(0) = e_0$, $\gamma(2) = e_1$. Thus E is pathwise connected. ■

In Supplement **5.5C** we shall prove that if $\pi : E \to B$ is a C^0 locally trivial fiber bundle over a paracompact simply connected base with simply connected typical fiber F, then E is simply connected.

☞ SUPPLEMENT 3.4B
Fiber Bundles over Contractible Spaces

This supplement proves that any C^0 fiber bundle $\pi : E \to B$ over a contractible base B is trivial.

3.4.32 Lemma *Let $\pi : E \to B \times [0, 1]$ be a C^0 fiber bundle. If $\{V_i \mid i = 1,..., n\}$ is a finite cover of $[0, 1]$ by open intervals such that $E \mid B \times V_i$ is a trivial C^0 fiber bundle, then E is trivial.*

Proof By induction it suffices to prove the result for $n = 2$, i.e., prove that if $E \mid B \times [0, t]$ and $E \mid B \times [t, 1]$ are trivial, then E is trivial. If F denotes the typical fiber of E, by hypothesis there are C^0 trivializations over the identity $\varphi_1 : E \mid B \times [0, t] \to B \times [0, t] \times F$ and $\varphi_2 : E \mid B \times [t, 1] \to B \times [t, 1] \times F$. The map $\varphi_2 \circ \varphi_1^{-1} : B \times \{t\} \times F \to B \times \{t\} \times F$ is a homeomorphism of the form $(b, t, f) \mapsto (b, t, \alpha_b(f))$, where $\alpha_b : F \to F$ is a homeomorphism depending continuously on b. Define the homeomorphism $\chi : (b, s, f) \in B \times [t, 1] \times F \mapsto (b, s, \alpha_b^{-1}(f)) \in B \times [t, 1] \times F$. Then the trivialization $\chi \circ \varphi_2 : E \mid B \times [t, 1] \to B \times [t, 1] \times F$ sends any $e \in \pi^{-1}(B \times \{t\})$ to $\varphi_1(e)$. Therefore, the map that sends e to the element of $B \times [0, 1] \times F$ given by $\varphi_1(e)$, if $\pi(e) \in B \times [0, t]$ and $(\chi \circ \varphi_2)(e)$, if $\pi(e) \in B \times [t, 1]$ is a continuous trivialization of E. ▼

3.4.33 Lemma *Let $\pi : E \to B \times [0, 1]$ be a C^0 fiber bundle. Then there is an open covering $\{U_i\}$ of B such that $E \mid U_i \times [0, 1]$ is trivial.*

Proof There is a covering of $B \times [0, 1]$ by sets of the form $W \times V$ where W is open in B and V is open in $[0, 1]$, such that $E \mid W \times V$ is trivial. For each $b \in B$ consider the family Φ_b of sets $W \times V$ for which $b \in W$. By compactness of $[0, 1]$, there is a finite subcollection $V_1, ..., V_n$ of the V's which cover $[0, 1]$. Let $W_1, ..., W_n$ be the corresponding W's in the family Φ_b and let $U_b = W_1 \cap ... \cap W_n$. But then $E \mid U_b \times W_i$, $i= 1,.., n$ are all trivial and thus by Lemma **3.4.32**, $E \mid U_b \times [0, 1]$ is trivial. Then $\{U_b \mid b \in B\}$ is the desired open covering of B. ▼

3.4.34 Lemma *Let $\pi : E \to B \times [0, 1]$ be a C^0 fiber bundle such that $E \mid B \times \{0\}$ is trivial. Then E is trivial.*

Proof By Lemma **3.4.33**, there is an open cover $\{U_i\}$ of B such that $E \mid U_i \times [0, 1]$ is trivial; let φ_i be the corresponding trivializations. Denote by $\varphi : E \mid B \times \{0\} \to B \times \{0\} \times F$ the trivialization guaranteed in the hypothesis of the lemma, where F is the typical fiber of E. We modify all φ_i in such a way that $\varphi_i : E \mid U_i \times \{0\} \to U_i \times \{0\} \times F$ coincides with $\varphi : E \mid U_i \times \{0\} \to U_i \times \{0\} \times F$ in the following way.

The homeomorphism $\varphi_i \circ \varphi^{-1} : U_i \times \{0\} \times F \to U_i \times \{0\} \times F$ is of the form $(b, 0, f) \mapsto (b, 0, \alpha_b^i(f))$ for $\alpha_b^i : F \to F$ a homeomorphism depending continuously on $b \in B$. Define χ_i :

$U_i \times [0, 1] \times F \to U_i \times [0, 1] \times F$ by $\chi_i(b, s, f) = (b, s, (\alpha_b{}^i)^{-1}(f))$. Then $\psi_i = \chi_i \circ \varphi_i : E \mid U_i \times [0, 1] \to U_i \times [0, 1] \times F$ maps any $e \in \pi^{-1}(B \times \{0\})$ to $\varphi(e)$.

Assume each φ_i on $E \mid U_i \times \{0\}$ equals φ on $E \mid U_i \times \{0\}$. Define $\lambda_i : E \mid U_i \times [0, 1] \to U_i \times \{0\} \times F$ to be the composition of the map $(b, s, f) \in U_i \times [0, 1] \times F \mapsto (b, 0, f) \in U_i \times \{0\} \times F$ with φ_i. Since each φ_i coincides with φ on $E \mid U_i \times \{0\}$, it follows that whenever $U_i \cap U_j \neq \varnothing$, λ_i and λ_j coincide on $E \mid (U_i \cap U_j) \times [0, 1]$, so that the collection of all $\{\lambda_i\}$ define a fiber bundle map $\lambda : E \to B \times \{0\} \times F$ over the map $\chi : (b, s) \in B \times [0, 1] \mapsto (b, 0) \in B \times \{0\}$. By the fiber bundle version **3.4W** of exercise **3.4O**(i) and (iii), E equals the pull-back $\chi^*(B \times \{0\} \times F)$. Since the bundle $B \times \{0\} \times F \to B \times \{0\}$ is trivial, so is its pull-back E. ▼

3.4.35 Theorem *Let* $\pi : E \to B$ *be any* C^0 *fiber bundle over a contractible space* B. *Then* E *is trivial.*

Proof By hypothesis, there is a homotopy $h : B \times [0, 1] \to B$ such that $h(b, 0) = b_0$ and $h(b, 1) = b$ for any $b \in B$, where $b_0 \in B$ is a fixed element of B. Then the pull-back bundle h^*E is a fiber bundle over $B \times [0, 1]$ whose restrictions to $B \times \{0\}$ and $B \times \{1\}$ equals the trivial fiber bundle over $\{b_0\}$ and E over $B \times \{1\}$, respectively. By lemma **3.4.34**, h^*E is trivial over $B \times [0, 1]$ and thus E, which is isomorphic to $E \mid B \times \{1\}$, is also trivial. ■

All previous proofs go through without any modifications to the C^k-case, once manifolds with boundary are defined (see §**7.1**). ✍

Exercises

3.4A Let $N \subset M$ be a submanifold. Show that TN is a subbundle of $TM \mid N$ and thus is a submanifold of TM.

3.4B Find an explicit example of a fiber-preserving diffeomorphism between vector bundles that is not a vector bundle isomorphism.

3.4C Let $\rho : \mathbb{R} \times S^n \to S^n$ and $\sigma : \mathbb{R}^{n+1} \times S^n \to S^n$ be trivial vector bundles. Show that

$$TS^n \oplus (\mathbb{R} \times S^n) \cong (\mathbb{R}^{n+1} \times S^n).$$

(*Hint:* Realize ρ as the vector bundle whose one-dimensional fiber is the normal to the sphere.)

3.4D (i) Let $\pi : E \to B$ be a vector bundle. Show that $TE \mid B$ is vector bundle isomorphic to $E \oplus TB$. Conclude that E is isomorphic to a subbundle of TE. (*Hint:* The short exact sequence $0 \longrightarrow E \longrightarrow TE \mid B \overset{T\pi}{\longrightarrow} TB \longrightarrow 0$ splits via Ti, where $i : B \longrightarrow E$ is the inclusion of B as the zero section of E; apply **3.4.26**.)

(ii) Show that the isomorphism φ_E found in (i) is natural, i.e., if $\pi' : E' \to B'$ is another vector bundle, $f : E \to E'$ is a vector bundle map over $f_B : B \to B'$, and $\varphi_{E'} : TE' \mid B \to E' \oplus TB'$ is the isomorphism in (i) for $\pi' : E' \to B'$, then

$$\varphi_{E'} \circ Tf = (f \oplus Tf_B) \circ \varphi_E.$$

3.4E Show that the mapping $s : E \oplus E \to E$, $s(e, e') = e + e'$ (fiberwise addition) is a vector bundle mapping over the identity.

3.4F Write down explicitly the charts in examples **3.4.25** given by Theorem **3.4.24**.

3.4G (i) A vector bundle $\pi : E \to B$ is called ***stable*** if its Whitney sum with a trival bundle over B is trivial. Show that TS^n is stable, but the Möbius band $\mathbb{M}$ is not.

(ii) Two vector bundles $\pi : E \to B$, $\rho : E \to B$ are called ***stably isomorphic*** if the Whitney sum of E with some trivial bundle over B is isomorphic with the Whitney sum of F with (possibly another) trivial vector bundle over B. Let KB be the set of stable isomorphism classes of vector bundles with finite dimensional fiber over B. Show that the operations of Whitney sum and of tensor product induce on KB a ring structure. Find a surjective ring homomorphism of KB onto $\mathbb{Z}$.

3.4H A vector bundle with one dimensional fibers is called a ***line bundle.*** Show that any line bundle which admits a global nowhere vanishing section is trivial.

3.4I Generalize example **3.4.25B** to vector bundles with different bases. If $\pi : E \to M$ and $\rho : F \to N$ are vector bundles, show that the set $\bigcup_{(m,n) \in M \times N} L(E_m, F_n)$ is a vector bundle with base $M \times N$. Describe the fiber and compute the relevant dimensions in the finite dimensional case.

3.4J Let N be a submanifold of M. The ***normal bundle*** $\nu(N)$ of N is defined to be $\nu(N) = (TM \mid N) / TN$. Assume that N has finite codimension k. Show that $\nu(N)$ is trivial iff there are smooth maps $X_i : N \to TM$, $i = 1, \ldots, k$ such that $X_i(n) \in T_nM$ and $\{X_i(n) \mid i = 1, \ldots, k\}$ span a subspace V_n satisfying $T_nM = T_nM \oplus V_n$ for all $n \in N$. Show that $\nu(S^n)$ is trivial.

3.4K Let N be a submanifold of M. Prove that the ***conormal bundle*** defined by $\mu(N) = \{\alpha \in T_n{}^*M \mid \langle \alpha, u \rangle = 0 \text{ for all } u \in T_nN \text{ and all } n \in N\}$ in a subbundle of $T^*M \mid N$ which is isomorphic to the normal bundle $\nu(N)$ defined in Exercise **3.4J**. Generalize the constructions and statements of **3.4J** and the current Exercise to an arbitrary vector subbundle F of a vector bundle E.

3.4L (i) Use the fact that S^3 is the unit sphere in the associative division algebra $\mathbb{H}$ to show that TS^3 is trivial.

(ii) **Cayley numbers** Consider on $\mathbb{R}^8 = \mathbb{H} \oplus \mathbb{H}$ the usual Euclidean inner product $\langle\, , \rangle$ and define a multiplication in $\mathbb{R}^8$ by $(a_1, b_1)(a_2, b_2) = (a_1a_2 - b_2{}^*b_1,\ b_2a_1 + b_1a_2{}^*)$ where $a_i, b_i \in \mathbb{H}$, $i = 1, 2$, and the multiplication on the right hand side is in $\mathbb{H}$. Prove the relation $\langle\alpha_1\beta_1, \alpha_2\beta_2\rangle + \langle\alpha_2\beta_1, \alpha_1\beta_2\rangle = 2\langle\alpha_1, \alpha_2\rangle\langle\beta_1, \beta_2\rangle$ where $\langle\alpha, \beta\rangle$ denotes the dot product in $\mathbb{R}^8$. Show that if one defines the conjugate of (a, b) by $(a, b)^* = (a^*, -b)$, then $\|(a, b)\|^2 = (a, b)[(a, b)^*]$. Prove that $\|\alpha\beta\| = \|\alpha\|\,\|\beta\|$ for all $\alpha, \beta \in \mathbb{R}^8$. Use this relation to show that $\mathbb{R}^8$ is a nonassociative division algebra over $\mathbb{R}$, $\mathbb{C}$, and $\mathbb{H}$. $\mathbb{R}^8$ with this algebraic structure is called the algebra of ***Cayley numbers*** or ***algebra of octaves***; it is denoted by $\mathbb{O}$.

(iii) Show that $\mathbb{O}$ is generated by 1 and seven symbols $e_1, \ldots, e_7$ satisfying the relations $e_i{}^2 = -1$, $e_ie_j = -e_je_i$, $e_1e_2 = e_3$, $e_1e_4 = e_5$, $e_1e_6 = -e_7$, $e_2e_5 = e_7$, $e_2e_4 = e_6$, $e_3e_4 = e_7$, $e_3e_5 = -e_6$, together with 14 additional relations obtained by cyclic permutations of the indices in the last 7 relations. (*Hint:* The isomorphism is given by associating 1 to the element $(1, 0, \ldots, 0) \in \mathbb{R}^8$ and to e_i the vector in $\mathbb{R}^8$ having all entries zero with the exception of the (i+1)-st which is 1.)

(iv) Show that any two elements of $\mathbb{O}$ generate an associative algebra isomorphic to a subalgebra of $\mathbb{H}$. (*Hint:* Show that any element of $\mathbb{O}$ is of the form $a + be_4$ for $a, b \in \mathbb{H}$.)

(v) Since S^7 is the unit sphere in $\mathbb{O}$, show that TS^7 is trivial.

3.4M (i) Let $\pi : E \to B$ be a locally trivial fiber bundle. Show that $V = \ker(T\pi)$ is a vector subbundle of TE, called the ***vertical bundle***. A vector subbundle H of TE such that $V \oplus H = TE$ is called a ***horizontal subbundle***. Show that $T\pi$ induces a vector bundle map $H \to TE$ over π which is an isomorphism on each fiber.

(ii) If $\pi : E \to M$ is a vector bundle, show that each fiber V_v of V, $v \in E$ is naturally identified with E_b, where $b = \pi(v)$. Show that there is a natural isomorphism of T_0E with $T_bB \oplus E_b$, where 0 is the zero vector in E_b. Argue that there is in general no such natural isomorphism of T_vE for $v \neq 0$.

3.4N Let E_n be the trivial vector bundle $\mathbb{RP}^n \times \mathbb{R}^{n+1}$.

(i) Show that $F_n = \{([x], \lambda x) \mid x \in \mathbb{R}^{n+1}, \lambda \in \mathbb{R}\}$ is a line subbundle of E_n. (*Hint:* Define $f : E_n \to \mathbb{RP}^n \times \mathbb{R}^{n+1}$ by $f([x], y) = ([x], y - (x\cdot y)x/\|x\|^2)$ and show that f is a vector bundle map having the restriction to each fiber a linear map of rank n. Apply **3.3.16**(ii).)

(ii) Show that F_n is isomorphic to $\gamma_1(\mathbb{R}^{n+1})$.

(iii) Show that F_1 is isomorphic to the Möbius band $\mathbb{M}$.

(iv) Show that F_n is the quotient bundle of the normal bundle $\nu(S^n)$ to S^n by the equivalence relation which identifies antipodal points and takes the outward normal to the inward normal. Show that the projection map $\nu(S^n) \to F_n$ is a 2 to 1 covering map.

(v) Show that F_n is nontrivial for all $n \geq 1$. (*Hint:* Use (iv) to show that any section σ of F_n vanishes somewhere; do this by considering the associated section σ^* of the trivial normal bundle to S^n and using the intermediate value theorem.)

(vi) Show that any line bundle over S^1 is either isomorphic to the cylinder $S^1 \times \mathbb{R}$ or the Möbius band $\mathbb{M}$.

3.4O (i) Let $\pi : E \to B$ be a vector bundle and $f : B' \to B$ a smooth map. Define the ***pull-back bundle*** $f^*\pi : f^*E \to B'$ by

$$f^*E = \{(v, b') \mid \pi(v) = f(b')\}, \qquad f^*\pi(v, b') = b'$$

and show that it is a vector bundle over B', whose fibers over b' equal $E_{f(b')}$. Show that $h : f^*E \to E$, $h(e, b) = e$, is a vector bundle map which is the identity on every fiber. Show that the pull-back bundle of a trivial bundle is trivial.

(ii) If $g : B'' \to B'$ show that $(f \circ g)^*\pi : (f \circ g)^*E \to B''$ is isomorphic to the bundles $g^*f^*\pi : g^*f^*E \to B''$. Show that isomoprhic vector bundles have isomorphic pull-backs.

(iii) If $\rho : E' \to B'$ is a vector bundle and $g : E' \to E$ is a vector bundle map inducing the map $f : B' \to B$ on the zero sections, then prove there exists a unique vector bundle map $g^* : E' \to f^*E$ inducing the identity on B' and is such that $h \circ g^* = g$.

(iv) Let $\sigma : F \to B$ be a vector bundle and $u : F \to E$ be a vector bundle map inducing the identity on B. Show that there exists a unique vector bundle map $f^*u : f^*F \to f^*E$ inducing the identity on B' and making the diagram

$$\begin{array}{ccc} F & \xrightarrow{\;f^*u\;} & f^*E \\ \downarrow & & \downarrow \\ F & \xrightarrow[\;u\;]{} & E \end{array}$$

commutative, where vertical arrows are projections.

(v) If $\pi : E \to B$, $\pi' : E' \to B$ are vector bundles and if $\Delta : B \to B \times B$ is the diagonal map $b \mapsto (b, b)$, show that $E \oplus E' \cong \Delta^*(E \times E')$.

(vi) Let $\pi : E \to B$ and $\pi' : E' \to B$ be vector bundles and denote by $p_i : B \times B \to B$, $i =$ 1, 2 the projections. Show that $E \times E' \cong p_1{}^*(E) \oplus p_2{}^*(E')$ and that the following sequences are split exact:

$$0 \to E \to E \oplus E' \to E' \to 0.$$
$$0 \to E' \to E \oplus E' \to E \to 0.$$

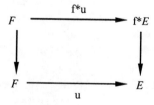

$$0 \to p_1^*(E) \to E \times E' \to p_2^*(E') \to 0.$$
$$0 \to p_2^*(E') \to E \times E' \to p_1^*(E) \to 0.$$

3.4P (i) Show that $\mathbb{G}_k(\mathbb{R}^n)$ is a submanifold of $\mathbb{G}_{k+1}(\mathbb{R}^{n+1})$. Denote by $i : \mathbb{G}_k(\mathbb{R}^n) \to \mathbb{G}_{k+1}(\mathbb{R}^{n+1})$, $i(F) = F \times \mathbb{R}$ the canonical inclusion map.

(ii) If $\rho : \mathbb{R} \times \mathbb{G}_k(\mathbb{R}^n) \to \mathbb{G}_k(\mathbb{R}^n)$ is the trivial bundle show that $i^*(\gamma_{k+1}(\mathbb{R}^{n+1}))$ is isomorphic to $\gamma_k(\mathbb{R}^n) \oplus (\mathbb{R} \times \mathbb{G}_k(\mathbb{R}^n))$.

3.4Q Show that $T(M_1 \times M_2) \cong p_1^*(TM_1) \oplus p_2^*(TM_2)$ where $p_i : M_1 \times M_2 \to M_i$, $i = 1, 2$ are the canonical projections and $p_i^*(TM_i)$ denotes the pull-back bundle defined in Exercise 3.4O.

3.4R (i) Let $\pi : E \to B$ be a vector bundle. Show that there is a short exact sequence

$$0 \to \pi^* E \to T \xrightarrow{f} E \xrightarrow{g} \pi^*(TB) \to 0$$

where $f(v, v') = \left.\frac{d}{dt}\right|_{t=0} (v + tv')$ and $g(u_v) = (T_v\pi(u_v), \pi(v))$.

(iii) Show that $\ker(T\pi)$ is a subbundle of TE, called the ***vertical subbundle*** of TE. Any subbundle $H \subset TE$ such that $TE = \ker(T\pi) \oplus H$, is called a ***horizontal subbundle*** of TE. Show that $T\pi$ induces an isomorphism of H with $\pi^*(TB)$.

(iv) Show that if TE admits a horizontal subbundle then the sequence in (i) splits.

3.4S Let $\pi : E \to B$, $\rho : F \to C$ be vector bundles and let $f : B \to C$ be a smooth map. Define $L_f(E, F) = \Gamma_f^* L(E, F)$, where $\Gamma_f : b \in B \mapsto (b, f(b)) \in B \times C$ is the graph map defined by f. Show that sections of $L_f(E, F)$ coincide vector bundle maps $E \to F$ over f.

3.4T Let M be an n-manifold. A ***frame*** at $m \in M$ is an isomorphism $\alpha : T_mM \to \mathbb{R}^n$. Let $\mathbb{F}(M) = \{(m, \alpha) \mid \alpha \text{ is a frame at } m\}$. Define $\pi : \mathbb{F}(M) \to M$ by $\pi(m, \alpha) = m$.

(i) Let (U, φ) be a chart on M. Show that $(m, \alpha) \in \pi^{-1}(U) \mapsto (m, T_m\varphi \circ \alpha^{-1}) \in U \times GL(\mathbb{R}^n)$ is a diffeomorphism. Prove that these diffeomorphisms as (U, φ) vary over a maximal atlas of M define by collation a manifold structure on $\mathbb{F}(M)$. Prove that $\pi : \mathbb{F}(M) \to M$ is a locally trivial fiber bundle with typical fiber $GL(n)$.

(ii) Prove that the sequence

$$0 \to \ker(T\pi) \xrightarrow{i} T\mathbb{F}(M) \xrightarrow{\pi^*\tau} \pi^*(TM) \to 0$$

is short exact, where i is the inclusion and $\pi^*\tau$ is the vector bundle projection $\pi^*(TM) \to \mathbb{F}(M)$ induced by the tangent bundle projection $\tau : TM \to M$.

(iii) Show that $\ker(T\pi)$ and $\pi^*(TM)$ are trivial vector bundles.

(iv) A splitting $0 \to \pi^*(TM) \xrightarrow{h} T\mathbb{F}(M)$ of the short sequence in (ii) is called a ***connection*** on M. Show that if M has a connection, then $T\mathbb{F}(M) = \ker(T\pi) \oplus H$, where H is a subbundle of $T\mathbb{F}(M)$ whose fibers are isomorphic by $T\pi$ to the fiber of TM.

3.4U (i) Generalize the Hopf fibration to the ***complex Hopf fibrations*** $\kappa_n : S^{2n+1} \to \mathbb{CP}^n$ with fiber S^1.

(ii) Replace in (i) $\mathbb{C}$ by the division algebra of quaternions $\mathbb{H}$. Generalize (i) to the ***quaternionic Hopf fibrations*** $\chi_n : S^{4n+3} \to \mathbb{HP}^n$ with fiber S^3. $\mathbb{HP}^n$ is the quaternionic space defined as the set of one dimensional vector subspaces over $\mathbb{H}$ in $\mathbb{H}^{n+1}$. Is anything special happening when $n = 1$? Describe.

3.4V (i) Try to define $\mathbb{OP}^n$, where $\mathbb{O}$ are the Cayley numbers. Show that the proof of transitivity of the equivalence relation in $\mathbb{O}^{n+1}$ requires associativity.

(ii) Define $p'(a, b) = ab^{-1}$ if $b \neq 0$ and $p'(a, b) = \infty$ if $b = 0$, where S^8 is thought of as the one-point compactification of $\mathbb{R}^8 = \mathbb{O}$ (see Exercise **3.1E**). Show p' is smooth and prove that $p = p' \mid S^{15}$ is onto. Proceed as in Example **3.4.29D** and show that $p : S^{15} \to S^8$ is a fiber bundle with typical fiber S^7 whose bundle structure is given by an atlas with two fiber bundle charts.

3.4W Define the pull-back of fiber bundles and prove properties analogous to those in Exercise 3.4O.

§3.5 *Submersions, Immersions, and Transversality*

The notions of submersion, immersion, and transversality are geometric ways of stating various hypotheses needed for the inverse function theorem, and are central to large portions of calculus on manifolds. One immediate benefit is easy proofs that various subsets of manifolds are actually submanifolds.

3.5.1 Local Diffeomorphisms Theorem *Let* M *and* N *be manifolds,* $f : M \to N$ *be of class* C^r, $r \geq 1$ *and* $m \in M$. *Suppose* Tf *restricted to the fiber over* $m \in M$ *is an isomorphism. Then* f *is a* C^r *diffeomorphism from some neighborhood of* m *onto some neighborhood of* f(m).

Proof In local charts, the hypothesis reads: $(\mathbf{D}f_{\varphi\psi})(u)$ is an isomorphism, where $\varphi(m) = u$. Then the inverse function theorem guarantees that $f_{\varphi\psi}$ restricted to a neighborhood of u is a C^r diffeomorphism. Composing with chart maps gives the result. ■

The local results **2.5.9** and **2.5.13** give the following:

3.5.2 Local Onto Theorem *Let* M *and* N *be manifolds and* $f : M \to N$ *be of class* C^r, *where* $r \geq 1$. *Suppose* Tf *restricted to the fiber* T_mM *is surjective to* $T_{f(m)}N$. *Then*

(i) f *is locally onto at* m; *i.e., there are neighborhoods* U *of* m *and* V *of* f(m) *such that* $f \mid U : U \to V$ *is onto; in particular, if* Tf *is surjective on each tangent space, then* f *is an open mapping;*

(ii) *if, in addition, the kernel* $\ker(T_mf)$ *is split in* T_mM *there are charts* (U, φ) *and* (V, ψ) *with* $m \in U$, $f(U) \subset V$, $\varphi : U \to U' \times V'$, $\varphi(m) = (0, 0)$, $\psi : V \to V'$ *and* $f_{\varphi\psi} : U' \times V \to V'$ *is the projection onto the second factor.*

Proof It suffices to prove the results locally, and these follow from **2.5.9** and **2.5.13**. ■

The notions of submersion and immersion correspond to the local surjectivity and injectivity theorems from §**2.5**. Let us first examine submersions, building on the preceding theorem.

3.5.3 Definition *Suppose* M *and* N *are manifolds with* $f : M \to N$ *of class* C^r, $r \geq 1$. *A point* $n \in N$ *is called a* ***regular value*** *of* f *if for each* $m \in f^{-1}(\{n\})$, T_mf *is surjective with split kernel. Let* $\mathcal{R}_f$ *denote the set of regular values of* $f : M \to N$; *note* $N \backslash f(M) \subset \mathcal{R}_f \subset N$. *If, for each* m *in a set* S, T_mf *is surjective with split kernel, we say* f *is a* ***submersion*** *on* S. *Thus* $n \in \mathcal{R}_f$ *iff* f *is a submersion on* $f^{-1}(\{n\})$. *If* T_mf *is not surjective,* $m \in M$ *is called a* ***critical point*** *and* $n = f(m) \in N$ *a* ***critical value*** *of* f.

3.5.4 Submersion Theorem *Let* $f: M \to N$ *be of class* C^∞ *and* $n \in \mathcal{R}_f$. *Then the level set* $f^{-1}(n) = \{m \mid m \in M, f(m) = n\}$ *is a closed submanifold of* M *with tangent space given by* $T_m f^{-1}(n) = \ker T_m f$.

Proof If $f^{-1}(n) = \varnothing$ the theorem is satisfied. Otherwise, for $m \in f^{-1}(n)$ we find charts (U, φ), (V, ψ) as described in **3.5.2**. Because $\varphi(U \cap f^{-1}(n)) = f_{\varphi\psi}{}^{-1}(0) = U' \times \{0\}$, we get the submanifold property. (See Fig. 3.5.1.) Since $f_{\varphi\psi} : U' \times V' \to V'$ is the projection onto the second factor, where $U' \subset E$ and $V' \subset F$, we have $T_u(f_{\varphi\psi}{}^{-1}(0)) = T_u U' = E = \ker(T_u f_{\varphi\psi})$ for $u \in U'$, which is the local version of the second statement. ■

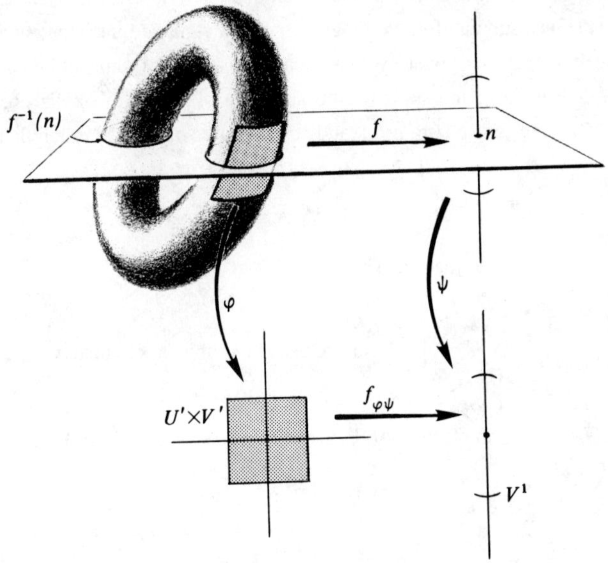

Figure 3.5.1

If N is finite dimensional and $n \in \mathcal{R}_f$, observe that codim $(f^{-1}(n)) = \dim N$, from the second statement of **3.5.4**. (This makes sense even if M is infinite dimensional.) Sard's theorem, discussed in the next section, implies that $\mathcal{R}_f$ is dense in N.

3.5.5 Examples

A We shall use the preceding theorem to show that $S^n \subset \mathbb{R}^{n+1}$ is a submanifold. Indeed, let $f : \mathbb{R}^{n+1} \to \mathbb{R}$ be defined by $f(\mathbf{x}) = \| \mathbf{x} \|^2$, so $S^n = f^{-1}(1)$. To show that S^n is a submanifold, it suffices to show that 1 is a regular value of f. Suppose $f(\mathbf{x}) = 1$. Identifying $T\mathbb{R}^{n+1} = \mathbb{R}^{n+1} \times \mathbb{R}^{n+1}$, and the fiber over $\mathbf{x}$ with elements of the second factor, we get

$$(T_{\mathbf{x}}f)(\mathbf{v}) = \mathbf{D}f(\mathbf{x}) \cdot \mathbf{v} = 2\langle \mathbf{x}, \mathbf{v} \rangle.$$

Since $\mathbf{x} \neq \mathbf{0}$, this linear map is not zero, so as the range is one-dimensional, it is surjective. The same argument shows that the unit sphere in Hilbert space is a submanifold.

B Stiefel Manifolds Define

$$\mathrm{St}(m, n; k) = \{A \in L(\mathbb{R}^m, \mathbb{R}^n) \mid \text{rank } A = k\}, \quad \text{where } k \leq \min(m, n).$$

Using the preceding theorem we shall prove that $\mathrm{St}(m, n; k)$ *is a submanifold of* $L(\mathbb{R}^m, \mathbb{R}^n)$ *of codimension* $(m - k)(n - k)$; this manifold is called the ***Stiefel manifold*** and plays an important role in the study of principal fiber bundles. To show that $\mathrm{St}(m, n; k)$ is a submanifold, we will prove that every point $A \in \mathrm{St}(m, n; k)$ has an open neighborhood U in $L(\mathbb{R}^m, \mathbb{R}^n)$ such that $\mathrm{St}(m, n; k) \cap U$ is a submanifold in $L(\mathbb{R}^m, \mathbb{R}^n)$ of the right codimension; since the differentiable structures on intersections given by two such U coincide (being induced from the manifold structure of $L(\mathbb{R}^m, \mathbb{R}^n)$), the submanifold structure of $\mathrm{St}(m, n; k)$ is obtained by collation (Exercise **3.2F**). Let $A \in \mathrm{St}(m, n; k)$ and choose bases of $\mathbb{R}^m, \mathbb{R}^n$ such that

$$A = \begin{bmatrix} \mathbf{a} & \mathbf{b} \\ \mathbf{c} & \mathbf{d} \end{bmatrix}$$

with $\mathbf{a}$ an invertible $k \times k$ matrix. The set

$$U = \left\{ \begin{bmatrix} \mathbf{x} & \mathbf{y} \\ \mathbf{z} & \mathbf{v} \end{bmatrix} \middle| \ \mathbf{x} \text{ is an invertible } k \times k \text{ matrix} \right\}$$

is open in $L(\mathbb{R}^m, \mathbb{R}^n)$. An element of U has rank k iff $\mathbf{v} - \mathbf{z}\mathbf{x}^{-1}\mathbf{y} = \mathbf{0}$. Indeed

$$\begin{bmatrix} \mathbf{I} & \mathbf{0} \\ -\mathbf{z}\mathbf{x}^{-1} & \mathbf{I} \end{bmatrix}$$

is invertible and

$$\begin{bmatrix} \mathbf{I} & \mathbf{0} \\ -\mathbf{z}\mathbf{x}^{-1} & \mathbf{I} \end{bmatrix} \begin{bmatrix} \mathbf{x} & \mathbf{y} \\ \mathbf{z} & \mathbf{v} \end{bmatrix} = \begin{bmatrix} \mathbf{x} & \mathbf{y} \\ \mathbf{0} & \mathbf{v} - \mathbf{z}\mathbf{x}^{-1}\mathbf{y} \end{bmatrix},$$

so

$$\text{rank} \begin{bmatrix} \mathbf{x} & \mathbf{y} \\ \mathbf{z} & \mathbf{v} \end{bmatrix} = \text{rank} \begin{bmatrix} \mathbf{x} & \mathbf{y} \\ \mathbf{0} & \mathbf{v} - \mathbf{z}\mathbf{x}^{-1}\mathbf{y} \end{bmatrix}$$

equals k iff $\mathbf{v} - \mathbf{z}\mathbf{x}^{-1}\mathbf{y} = \mathbf{0}$. Define $f : U \to L(\mathbb{R}^{m-k}, \mathbb{R}^{n-k})$ by

$$f\left(\begin{bmatrix} \mathbf{x} & \mathbf{y} \\ \mathbf{z} & \mathbf{v} \end{bmatrix} \right) = \mathbf{v} - \mathbf{z}\mathbf{x}^{-1}\mathbf{y}.$$

The preceding remark shows that $f^{-1}(0) = \mathrm{St}(m, n; k) \cap U$ and thus if f is a submersion, $f^{-1}(0)$

is a submanifold of $L(\mathbb{R}^m, \mathbb{R}^n)$ of codimension equal to

$$\dim L(\mathbb{R}^{m-k}, \mathbb{R}^{n-k}) = (m-k)(n-k).$$

To see that f is a submersion, note that for $\mathbf{x}, \mathbf{y}, \mathbf{z}$ fixed, the map $\mathbf{v} \mapsto \mathbf{v} - \mathbf{z}\mathbf{x}^{-1}\mathbf{y}$ is a diffeomorphism of $L(\mathbb{R}^{m-k}, \mathbb{R}^{n-k})$ to itself.

C Orthogonal Group Let $\mathbf{O}(n)$ be the set of elements $\mathbf{Q}$ of $L(\mathbb{R}^n, \mathbb{R}^n)$ that are orthogonal, i.e., $\mathbf{Q}\mathbf{Q}^T = \text{Identity}$. We shall prove that $\mathbf{O}(n)$ *is a compact submanifold of dimension* $n(n-1)/2$. This manifold is called the ***orthogonal group*** of $\mathbb{R}^n$; the group operations (composition of linear operators and inversion) being smooth in $L(\mathbb{R}^n, \mathbb{R}^n)$ are therefore smooth in $\mathbf{O}(n)$, i.e., $\mathbf{O}(n)$ is an example of a ***Lie group***. To show that $\mathbf{O}(n)$ is a submanifold, let $\mathfrak{sym}(n)$ denote the vector space of symmetric linear operators $\mathbf{S}$ of $\mathbb{R}^n$, i.e., $\mathbf{S}^T = \mathbf{S}$; its dimension equals $n(n+1)/2$. The map $f : L(\mathbb{R}^n, \mathbb{R}^n) \to \mathfrak{sym}(n)$, $f(\mathbf{Q}) = \mathbf{Q}\mathbf{Q}^T$ is smooth and has derivative $T_{\mathbf{Q}}f(\mathbf{A}) = \mathbf{A}\mathbf{Q}^T + \mathbf{Q}\mathbf{A}^T = \mathbf{A}\mathbf{Q}^{-1} + \mathbf{Q}\mathbf{A}^T$ at $\mathbf{Q} \in \mathbf{O}(n)$. This linear map from $L(\mathbb{R}^n, \mathbb{R}^n)$ to $\mathfrak{sym}(n)$ is onto since for any $\mathbf{S} \in \mathfrak{sym}(n)$, $T_{\mathbf{Q}}f(\mathbf{S}\mathbf{Q}/2) = \mathbf{S}$. Therefore, by Theorem **3.5.4**, $f^{-1}(\text{Identity}) = \mathbf{O}(n)$ is a closed submanifold of $L(\mathbb{R}^n, \mathbb{R}^n)$ of dimension equal to $n^2 - n(n+1)/2 = n(n-1)/2$. Finally, $\mathbf{O}(n)$ is compact since it lies on the unit sphere of $L(\mathbb{R}^n, \mathbb{R}^n)$.

D Orthogonal Stiefel Manifold Let $k \le n$ and

$$F_{k,n} = \mathbf{OSt}(n, n; k) = \{\text{orthonormal } k\text{-tuple of vectors in } \mathbb{R}^n\}.$$

We shall prove that $\mathbf{OSt}(n, n; k)$ *is a compact submanifold of* $\mathbf{O}(n)$ of *dimension* $nk - k(k+1)/2$; it is called the ***orthogonal Stiefel manifold.*** Any n-tuple of orthonormal vectors in $\mathbb{R}^n$ is obtained from the standard basis $e_1, \ldots, e_n$ of $\mathbb{R}^n$ by an orthogonal transformation. Since any k-tuple of orthonormal vectors can be completed via the Gram-Schmidt procedure to an orthonormal basis, the set $\mathbf{OSt}(n, n; k)$ equals $f^{-1}(0)$, where $f : \mathbf{O}(n) \to \mathbf{O}(n-k)$ is given by letting $f(\mathbf{Q}) = \mathbf{Q}'$, where

$\mathbf{Q}'$ = the $(n-k) \times (n-k)$ matrix obtained from $\mathbf{Q}$ by removing its first k rows and columns.

Since $T_{\mathbf{Q}}f(\mathbf{A}) = \mathbf{A}'$ is onto, it follows that f is a submersion. Therefore, $f^{-1}(\mathbf{0})$ is a closed submanifold of $\mathbf{O}(n)$ of dimension equal to $n(n-1)/2 - (n-k)(n-k-1)/2 = nk - k(k+1)/2$. ♦

3.5.6 Definition *A* C^r *map* $f : M \to N$, $r \ge 1$, is *called an* ***immersion at*** m *if* $T_m f$ *is injective with closed split range in* $T_{f(m)}N$. *If* f *is an immersion at each* m, *we just say* f *is an* ***immersion.***

3.5.7 Immersion Theorem *For a* C^r *map* $f : M \to N$, *where* $r \ge 1$, *the following are equivalent:*

(i) f *is an immersion at* m;

(ii) *there are charts* (U, φ) *and* (V, ψ) *with* $m \in U$, $f(U) \subset V$, $\varphi : U \to U'$, $\psi : V \to U' \times V'$ *and* $\varphi(m) = 0$ *such that* $f_{\varphi\psi} : U' \to U' \times V'$ *is the inclusion* $u \mapsto (u, 0)$;

(iii) *there is a neighborhood* U *of* m *such that* f(U) *is a submanifold in* N *and* f *restricted to* U *is a diffeomorphism of* U *onto* f(U).

Proof The equivalence of (i) and (ii) is guaranteed by the local immersion Theorem **2.5.12**. Assuming (ii), choose U and V given by that theorem to conclude that f(U) is a submanifold in V. But V is open in N and hence f(U) is a submanifold in N proving (iii). The converse is a direct application of the definition of a submanifold. ■

It should be noted that the theorem does *not* imply that f(M) is a submanifold in N. For example $f : S^1 \to \mathbb{R}^2$, given in polar coordinates by $r = \cos 2\theta$, is easily seen to be an immersion (by computing Tf using the curve $c(\theta) = \cos 2\theta$) on S^1 but $f(S^1)$ is not a submanifold of $\mathbb{R}^2$: any neighborhood of 0 in $\mathbb{R}^2$ intersects $f(S^1)$ in a set with "corners" which is not diffeomorphic to an open interval. In such cases we say f is an ***immersion with self-intersections***. See Fig. 3.5.2.

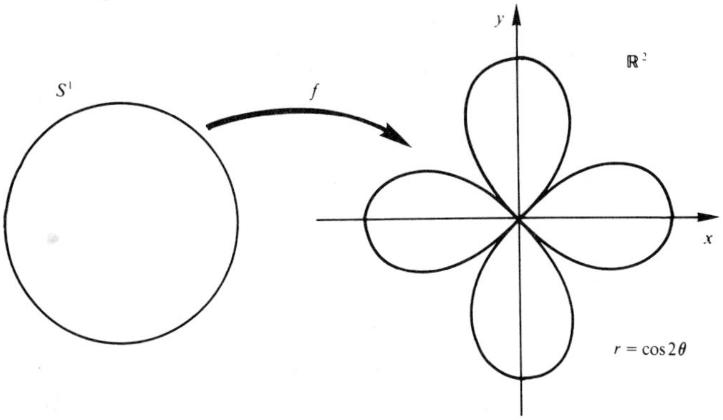

Figure 3.5.2

In the preceding example f is not injective. But even if f is an injective immersion, f(M) need not be a submanifold of N, as the following example shows. Let f be a curve whose image is as shown in Fig. 3.5.3. Again the problem is at the origin: any neighborhood of zero does *not* have the relative topology given by N.

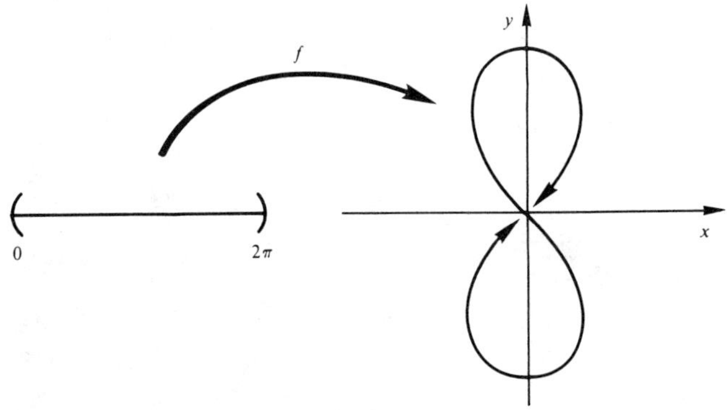

Figure 3.5.3

If $f : M \to N$ is an injective immersion, $f(M)$ is called an ***immersed submanifold*** of N.

3.5.9 Definition *An immersion* $f : M \to N$ *that is a homeomorphism onto* $f(M)$ *with the relative topology induced from* N *is called an* ***embedding****.*

Thus, if $f : M \to N$ is an embedding, then $f(M)$ is a submanifold of N.

The following is an important situation in which an immersion is guaranteed to be an embedding; the proof is a straightforward application of the definition of relative topology.

3.5.9 Embedding Theorem *An injective immersion which is an open or closed map onto its image is an embedding.*

The condition "$f : M \to N$ is closed" is implied by "f is proper", i.e. *each sequence* $x_n \in M$ *with* $f(x_n)$ *convergent to* y N *has a convergent subsequence* $x_{n(i)}$ *in* M *such that* $f(x_{n(i)})$ *converges to* y. Indeed, if this hypothesis holds, and A is a closed subset of M, then $f(A)$ is shown to be closed in N in the following way. Let $x_n \in A$, and suppose $f(x_n) = y_n$ converges to $y \in N$. Then there is a subsequence $\{z_m\}$ of $\{x_n\}$, such that $z_m \to x$. Since $A = \mathrm{cl}(A)$, $x \in A$ and by continuity of f, $y = f(x) \in f(A)$; i.e., $f(A)$ is closed. If N is infinite dimensional, this hypothesis is assured by the condition "the inverse image of every compact set in N is compact in M". This is clear since in the preceding hypothesis one can choose a compact neighborhood V of the limit of $f(x_n)$ in N so that for n large enough, all x_n belong to the compact neighborhood $f^{-1}(V)$ in M. The reader should note that while both hypotheses in the proposition are necessary, properness of f is only sufficient. An injective nonproper immersion whose image is a submanifold is, for example, the map $f :]0, \infty[\to \mathbb{R}^2$ given by

$$f(t) = \left(t \cos \frac{1}{t},\ t \sin \frac{1}{t}\right).$$

This is an open map onto its image so **3.5.9** applies; the submanifold $f(]0, \infty[)$ is a spiral around the origin.

3.5.10 Definition *A* C^r *map* $f : M \to N$, $r \geq 1$, *is said to be* ***transversal*** *to the submanifold* P *of* N *(denoted* $f \pitchfork P$*) if either* $f^{-1}(P) = \varnothing$, *or if for every* $m \in f^{-1}(P)$,

T1 $(T_m f)(T_m M) + T_{f(m)}P = T_{f(m)}N$ *and*

T2 *the inverse image* $(T_m f)^{-1}(T_{f(m)}P)$ of $T_{f(m)}P$ *splits in* $T_m M$.

The first condition **T1** is purely algebraic; no splitting assumptions are made on $(T_m f)(T_m M)$, nor need the sum be direct. If M is a Hilbert manifold, or if M is finite dimensional, then the splitting condition **T2** in the definition is automatically satisfied.

3.5.11 Examples

A If each point of P is a regular value of f, then $f \pitchfork P$ since $(T_m f)(T_m M) = T_{f(m)}N$ in this case.

B Assume that M and N are finite-dimensional manifolds with dim(P) + dim(M) < dim(N). Then $f \pitchfork P$ implies $f(M) \cap P = \varnothing$. This is seen by a dimension count: if there were a point $m \in f^{-1}(P) \cap M$, then $\dim(N) = \dim((T_m f)(T_m M) + T_{f(m)}P) \leq \dim(M) + \dim(P) < \dim(N)$ which is absurd.

C Let $M = \mathbb{R}^2$, $N = \mathbb{R}^3$, P = the (x, y) plane in $\mathbb{R}^3$, $a \in \mathbb{R}$ and define $f_a : M \to N$, by $f_a(x, y) = (x, y, x^2 + y^2 + a)$. Then $f \pitchfork P$ if $a \neq 0$; see Fig. 3.5.4. This example also shows intuitively that if a map is not transversal to a submanifold it can be perturbed very slightly to a transversal map; for a discussion of this phenomenon we refer to the Supplement **3.6B**. ♦

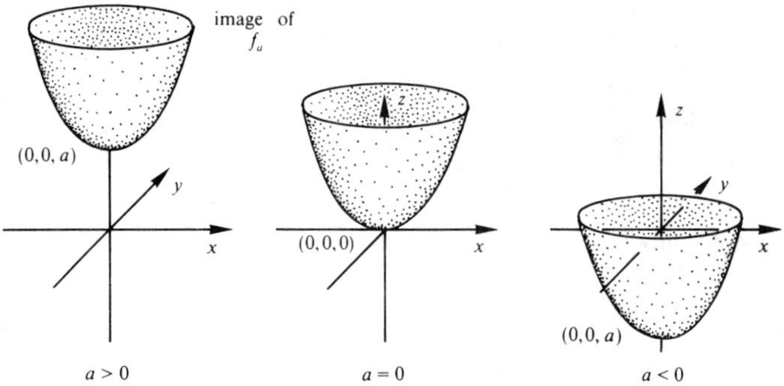

Figure 3.5.4

3.5.12 Transversal Mapping Theorem *Let* $f : M \to N$ *be a* C^∞ *map and* P *a submanifold of* N. *If* $f \pitchfork P$, *then* $f^{-1}(P)$ *is a submanifold of* M *and* $T_m(f^{-1}(P)) = (T_m f)^{-1}(T_{f(m)}P)$ *for all* $m \in f^{-1}(P)$. *If* P *has finite codimension in* N, *then* $\operatorname{codim}(f^{-1}(P)) = \operatorname{codim}(P)$.

Proof Let (V, ψ) be a chart at $f(m_0) \in P$ in N with the submanifold property for P; let

$$\psi(V) = V_1 \times V_2 \subset F_1 \oplus F_2, \quad \psi(V \cap P) = V_1 \times \{0\}, \quad \psi(f(m_0)) = (0, 0)$$

and denote by $p_2 : V_1 \times V_2 \to V_2$ the canonical projection. Let (U, φ) be a chart at m_0 in M, such that $\varphi(m_0) = 0$, $\varphi : U \to \varphi(U) \subset E$ and $f(U) \subset V$. For $m \in U \cap f^{-1}(P)$,

$$T_m(p_2 \circ \psi \circ f \mid U) = p_2 \circ T_{f(m)}\psi \circ T_m f \quad \text{and} \quad T_m(\psi \circ f)\,(T_m M) + F_1 = F_1 \oplus F_2$$

(by transversality of f on P). Hence $T_m(p_2 \circ \psi \circ f \mid U) : T_m U = T_m M \to F_2$ is onto. Its kernel is $(T_m f)^{-1}(T_{f(m)}P)$ since $\ker p_2 = F_1$ and $(T_{f(m)}\psi)^{-1}(F_1) = T_{f(m)}P$, and thus it is split in $T_m M$. In other words, 0 is a regular value of $p_2 \circ \psi \circ f \mid U : U \to F_2$ and thus $(p_2 \circ \psi \circ f \mid U)^{-1}(0) = f^{-1}(P \cap V)$ is a submanifold of U, and hence of M whose trangent space at $m \in U$ equals

$\ker(T_m(p_2 \circ \psi \circ f \mid U)) = (T_m f)^{-1}(T_{f(m)}P)$ by the Submersion Theorem **3.5.4**. Thus $f^{-1}(P \cap V)$ is a submanifold of M for any chart domain V with the submanifold property; i.e., $f^{-1}(P)$ is a submanifold of M. If P has finite codimension then F_2 is finite dimensional and thus again by the Submersion Theorem,

$$\operatorname{codim}(f^{-1}(P)) = \operatorname{codim} f^{-1}(P \cap V) = \dim(F_2) = \operatorname{codim}(P). \quad \blacksquare$$

Notice that this theorem reduces to the Submersion Theorem if P is a point.

3.5.13 Corollary *Suppose that* M_1 *and* M_2 *are submanifolds of* M, $m \in M_1 \cap M_2$, $T_mM_1 + T_mM_2 = T_mM$, *and that* $T_mM_1 \cap T_mM_2$ *splits in* T_mM *for all* $m \in M_1 \cap M_2$; *this condition is denoted* $M_1 \pitchfork M_2$ *and we say* M_1 *and* M_2 *are* ***transversal***. *Then* $M_1 \cap M_2$ *is a submanifold of* M *and* $T_m(M_1 \cap M_2) = T_mM_1 \cap T_mM_2$. M_1 *and* M_2 *are said to* ***intersect cleanly*** *when this conclusion holds. (Transversaltiy thus implies clean intersection.) If both* M_1 *and* M_2 *have finite codimension in* M, *then* $\operatorname{codim}(M_1 \cap M_2) = \operatorname{codim}(M_1) + \operatorname{codim}(M_2)$.

Proof The inclusion $i_1 : M_1 \to M$, satifies $i_1 \pitchfork M_2$, and $i_1^{-1}(M_2) = M_1 \cap M_2$. Now apply the previous theorem. $\blacksquare$

3.5.14 Examples

A In $\mathbb{R}^3$, the unit sphere $M_1 = \{(x, y, z) \mid x^2 + y^2 + z^2 = 1\}$ intersects the cylinder $M_2 = \{(x, y, z) \mid x^2 + y^2 = a\}$ transversally if $0 < a \neq 1$; $M_1 \cap M_2 = \varnothing$ if $a > 1$ and $M_1 \cap M_2$ is the union of two circles if $0 < a < 1$ (Fig. 3.5.5).

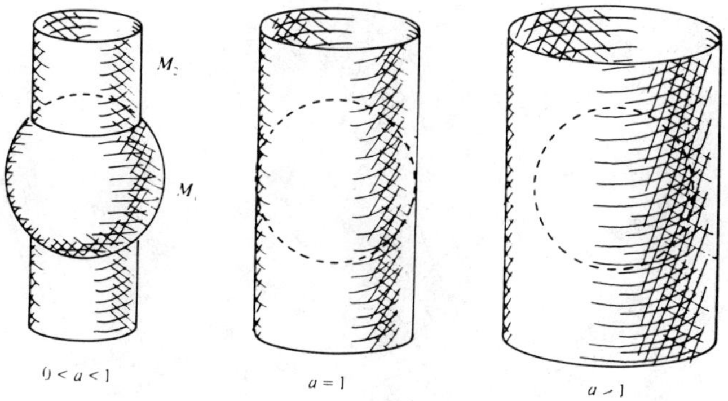

Figure 3.5.5

B The twisted ribbon M_1 in Fig. 3.5.6 does not meet M_2, the xy plane, in a manifold, so M_1 is not transversal to M_2.

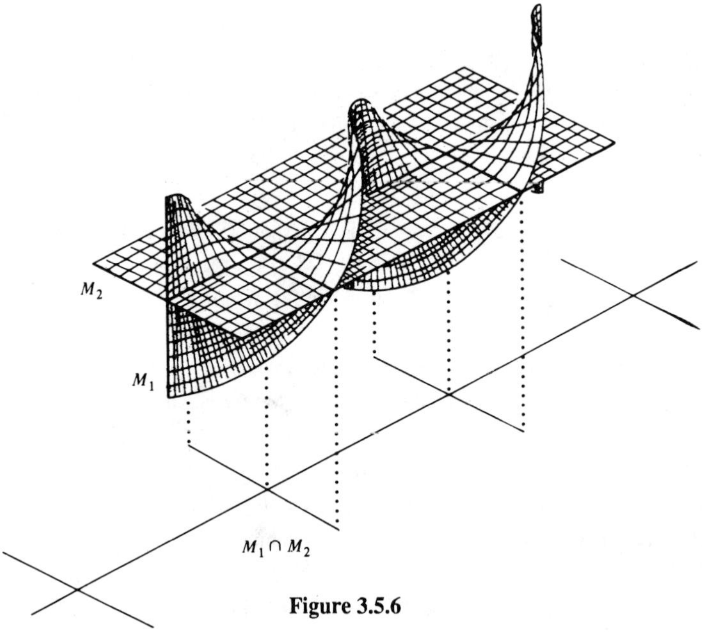

Figure 3.5.6

C Let M be the xy-plane in $\mathbb{R}^3$ and N be the graph of $f(x, y) = (xy)^2$. Even though TN $\cap$ TM has constant dimension (equal to 2), N $\cap$ M is not a manifold (Fig. 3.5.7). ♦

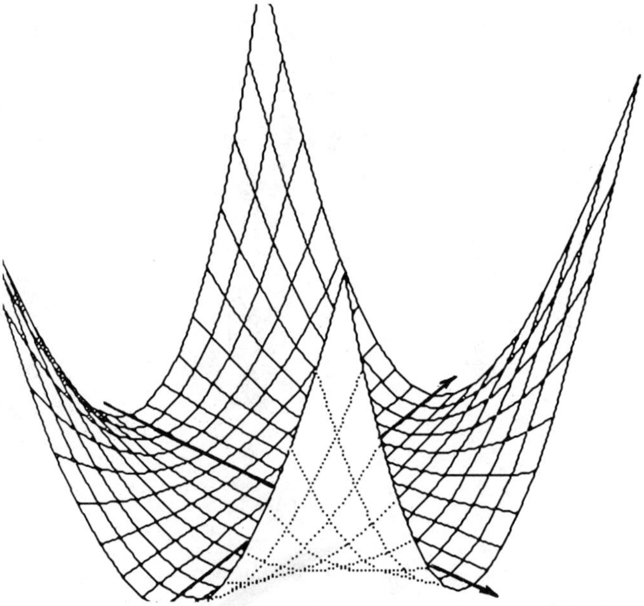

Figure 3.5.7

There is one more notion connected with geometric ways to state the implict function theorem that generalizes submersions in a different way than transversality. Roughly speaking, instead of requiring that a map $f : M \to N$ have onto tangent map at each point, one asks that its rank be constant.

3.5.15 Definition *A* C^r *map* $f : M \to N$, $r \geq 1$ *is called a* ***subimmersion*** *if for each point* $m \in M$ *there is an open neighborhood* U *of* m, *a manifold* P, *a submersion* $s : U \to P$, *and an immersion* $j : P \to N$ *such that* $f \mid U = j \circ s$.

Note that submersions and immersions are subimmersions. The order submersion followed by immersion in this definition is important, because the opposite order would yield nothing. Indeed, if $f : M \to N$ is any C^∞ mapping, then we can write $f = p_2 \circ j$, where $j : M \to M \times N$, is given by $j(m) = (m, f(m))$ and $p_2 : M \times N \to N$ the canonical projection. Clearly j is an immersion and p_2 a submersion, so any mapping can be written as an immersion followed by a submersion.

The following will connect the notion of submersion to "constant rank."

3.5.16 Proposition

(i) *A* C^∞ *map* $f : M \to N$ *is a subimmersion iff for every* $m \in M$ *there is a chart* (U, φ) *at* m, *where* $\varphi : U \to V_1 \times U_2 \subset F_1 \oplus E_2$, $\varphi(m) = (0, 0)$, *and a chart* (V, ψ) *at* $f(m)$ *where* $f(U) \subset V$, $\psi : V \to V_1 \times V_2 \subset F_1 \oplus F_2$, $\psi(f(m)) = (0, 0)$ *such that* $f_{\varphi\psi}(x, y) = (x, 0)$.

(ii) *If* M *or* N *are finite dimensional,* f *is a subimmersion iff the rank of the linear map* $T_m f : T_m M \to T_{f(m)} N$ *is constant for* m *in each connected component of* M.

Proof (i) follows from **3.5.2**, **3.5.7**(ii) and the composite mapping theorem; alternatively, one can use **2.5.15**. If M or N are finite dimensional, then necessarily $\operatorname{rank}(T_m f)$ is finite and thus by (i) the local representative $T_m f$ has constant rank in a chart at m; i.e., the rank of $T_m f$ is constant on connected components of M, thus proving (ii). The converse follows by **2.5.15** and (i). ■

3.5.17 Subimmersion Theorem *Suppose* $f : M \to N$ *is* C^∞, $n_0 \in N$ *and* f *is a subimmersion in an open neighborhood of* $f^{-1}(n_0)$. *(If* M *or* N *are finite dimensional this is equivalent to* $T_m f$ *having constant rank in a neighborhood of each* $m \in f^{-1}(n_0)$.*) Then* $f^{-1}(n_0)$ *is a submanifold of* M *with* $T_m f^{-1}(n_0) = \ker(T_m f)$.

Proof If $f^{-1}(n_0) = \varnothing$ there is nothing to prove. If $m \in f^{-1}(n_0)$ find charts (U, φ) at m and (V, ψ) at $f(m) = n_0$ given by **3.5.16**(i). Since $\varphi(U \cap f^{-1}(n_0)) = f_{\varphi\psi}{}^{-1}(0) = \{0\} \times U_2$ we see that (U, φ) has the submanifold property for $f^{-1}(n_0)$. In addition, if $u \in U_2$, then

$$T_{(0,u)}(f_{\varphi\psi}^{-1}(0)) = T_u U_2 = E_2 = \ker(T_{(0,u)} f_{\varphi\psi}),$$

which is the local version of the second statement. ■

Notice that

$$\operatorname{codim}(f^{-1}(n_0)) = \operatorname{rank}(T_m f) \quad \text{for} \quad m \in f^{-1}(n_0)$$

if $\operatorname{rank}(T_m f)$ is finite. The subimmersion theorem reduces to **3.5.4** when f is a submersion. The immersion part of the subimmersion f implies a version of **3.5.7**(iii).

3.5.18 Fibration Theorem *The following are equivalent for a* C^∞ *map* $f : M \to N$:

(i) *the map* f *is a subimmersion (if* M *or* N *are finite dimensional, this is equivalent to the rank of* $T_m f$ *being locally constant);*

(ii) *for each* $m \in M$ *there is a neighborhood* U *of* m, *a neighborhood* V *of* $f(m)$, *and a submanifold* Z *of* M *with* $m \in Z$ *such that* $f(U)$ *is a submanifold of* N *and* f *induces a diffeomorphism of* $f^{-1}(V) \cap Z \cap U$ *onto* $f(U) \cap V$;

(iii) $\ker Tf$ *is a subbundle of* TM *(called the* ***tangent bundle to the fibers****) and for each* $m \in M$, *the image of* $T_m f$ *is closed and splits in* $T_{f(m)}N$.

If one of these hold and if f *is open (or closed) onto its image, then* $f(M)$ *is a submanifold of* N.

Proof To show that (i) implies (ii), choose for U and V the chart domains given by **3.5.16**(i) and let $Z = \varphi^{-1}(V_1 \times \{0\})$. Then

$$\psi(f(U) \cap V) = f_{\varphi\psi}(V_1 \times U_2) = V_1 \times \{0\},$$

i.e., $f(U)$ is a submanifold of N. In addition, the local expression of $f : f^{-1}(V) \cap Z \cap U \to f(U) \cap V$ is $(x, 0) \mapsto (x, 0)$, thus proving that so restricted, f is a diffeomorphism. Reading this argument backward shows that (ii) implies (i).

We have $T\varphi(\ker Tf \cap TU) = V_1 \times U_2 \times \{0\} \times E_2$, since $f_{\varphi\psi}(x, y) = (x, 0)$, thus showing that $(TU, T\varphi)$ has the subbundle property for $\ker Tf$. The same local expression shows that $(T_m f)(T_m M)$ is closed and splits in $T_{f(m)}M$ and thus (i) implies (iii). The reverse is proved along the same lines.

If f is open (closed) onto its image, then $f(U) \cap V$ [resp. $\operatorname{int}(f(\operatorname{cl}(U))) \cap V$] can serve as a chart domain for $f(M)$ with the relative topology induced from N. Thus $f(M)$ is a submanifold of N. ■

The relationship between transversality and subimmersivity is given by the following.

3.5.19 Proposition *Let* $f : M \to N$ *be smooth,* P *a submanifold of* N, *and assume that* $f \pitchfork P$. *Then there is an open neighborhood* W *of* P *in* N *such that* $f \mid f^{-1}(W) : f^{-1}(W) \to W$ *is a subimmersion.*

Proof Let $m \in f^{-1}(P)$ and write $T_m(f^{-1}(P)) = \ker(T_m f) \oplus C_m$ so that $T_m f : C_m \to T_n P$ is an isormorphism, where $n = f(m)$. As in the proof of **3.5.12**, this situation can be locally straightened out, so we can assume that $M \subset E_1 \times E_2$, where P is open in E_1 and $E_1 = \ker T_m f \oplus C$ for a complementary space C. The map f restricted to C is a local immersion, so projection to C followed by the restriction of f to C writes f as the composition of a submersion followed by an immersion. We leave it to the reader to expand the details of this argument. ■

In some applications, the closedness of f follows from properness of f; see the discussion following **3.5.9**.

If M or N is finite dimensional, then f being a subimmersion is equivalent to ker Tf being a subbundle of TM. Indeed, $\mathrm{range}(T_m f) = T_m M / \ker(T_m f)$ and thus $\dim(\mathrm{range}\,(T_m f)) = \mathrm{codim}(\ker\,(T_m f)) = \text{constant}$ and f is hence a subimmersion by **3.5.16**(ii). Theorem **3.5.19** is an infinite dimensional version of this.

We have already encountered subimmersions in the study of vector bundles. Namely, the condition in **3.4.16**(i) (which insures that for a vector bundle map f over the identity ker f and range f are subbundles) is nothing other than f being a subimmersion.

We conclude this section with a study of quotient manifolds.

3.5.20 Definition *An equivalence relation* R *on a manifold* M *is called* **regular** *if the quotient space* M/R *carries a manifold structure such that the canonical projection* $\pi : M \to M/R$ *is a submersion. If* R *is a regular equivalence relation, then* M/R *is called the* **quotient manifold** *of* M *by* R.

Since submersions are open mappings, π and hence the regular equivalence relations R are open.

Quotient manifolds are characterized by their effect on mappings.

3.5.21 Proposition *Let* R *be a regular equivalent relation on* M.

(i) *A map* $f : M/R \to N$ *is* C^r, $r \geq 1$ *iff* $f \circ \pi : M \to N$ *is* C^r.

(ii) *Any* C^r *map* $g : M \to N$ *compatible with* R, *i.e.,* xRy *implies* $g(x) = g(y)$, *defines a unique* C^r *map* $\hat{g} : M/R \to N$ *such that* $\hat{g} \circ \pi = g$.

(iii) *The manifold structure of* M/R *is unique.*

Proof (i) If f is C^r, then so is $f \circ \pi$ by the composite mapping theorem. Conversely, let $f \circ \pi$ be C^r. Since π is a submersion it can be locally expressed as a projection and thus there exist charts (U, φ) at $m \in M$ and (V, ψ) at $\pi(m) \in M/R$ such that $\varphi(U) = U_1 \times U_2 \subset E_1 \oplus E_2$, $\psi(V) = U_2 \subset E_2$, and $\pi_{\varphi\psi}(x, y) = y$. Hence if (W, χ) is a chart at $(f \circ \pi)(m)$ in N satisfying $(f \circ \pi)(U) \subset W$, then $f_{\psi\chi} = (f \circ \pi)_{\varphi\chi} \mid \{0\} \times U_2$ and thus $f_{\psi\chi}$ is C^r.

(ii) The mapping $\hat{g}$ is uniquely determined by $\hat{g} \circ \pi = g$. It is C^r by (i).

(iii) Let $(M/R)_1$ and $(M/R)_2$ be two manifold structures on M/R having π as a submersion. Apply (ii) for $(M/R)_1$ with $N = (M/R)_2$ and $g = \pi$ to get a unique C^∞ map h :

$(M/R)_1 \to (M/R)_2$ such that $h \circ \pi = \pi$. Since π is surjective, h = identity. Interchanging the roles of the indices 1 and 2, shows that the identity mapping induces a C^∞ map of $(M/R)_2$ to $(M/R)_1$. Thus, the identity induces a diffeomorphism. ■

3.5.22 Corollary *Let* M *and* N *be manifolds,* R *and* Q *regular equivalence relations on* M *and* N, *respectively, and* $f: M \to N$ *a* C^r *map,* $r \geq 1$, *compatible with* R *and* Q; *i.e., if* xRy *then* f(x)Qf(y). *Then* f *induces a unique* C^r *map* $\varphi: M/R \to N/Q$ *and the diagram*

$$\begin{array}{ccc} TM & \xrightarrow{\;Tf\;} & TN \\ {\scriptstyle T\pi_M}\downarrow & & \downarrow{\scriptstyle T\pi_N} \\ T(M/R) & \xrightarrow[\;T\phi\;]{} & T(N/Q) \end{array}$$

commutes.

Proof The map φ is uniquely determined by $\pi_N \circ f = \varphi \circ \pi_M$. Since $\pi_N \circ f$ is C^r, φ is C^r by **3.5.21**. The diagram is obtained by applying the chain rule to $\pi_N \circ f = \varphi \circ \pi_M$. ■

The manifold M/R might *not* be Hausdorff. By **1.4.10** it is Hausdorff iff the graph of R is closed in $M \times M$; R is open since it is regular. For an example of a non-Hausdorff quotient manifold see Exercise **3.5H**.

There is, in fact, a bijective correspondence between surjective submersions and quotient manifolds. More precisely, we have the following.

3.5.23 Proposition *Let* $f: M \to N$ *be a submersion and let* R *be the equivalence relation defined by* f; *i.e.,* xRy *iff* f(x) = f(y). *Then* R *is regular,* M / R *is diffeomorphic to* f(M), *and* f(M) *is open in* N.

Proof As f is a submersion, it is an open mapping, so f(M) is open in N. Moreover, since f is open, so is the equivalence R and thus f induces a homeomorphism of M / R onto f(M) (see the comments following **1.4.9**). Put the differentiable structure on M / R that makes the homeomorphism into a diffeomorphism. Then M / R is a manifold and the projection is clearly a submersion, since f is. ■

This construction provides a number of examples of quotient manifolds.

3.5.24 Examples

A The base space of any vector bundle is a quotient manifold. Take the submersion to be the vector bundle projection.

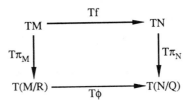

B The circle S^1 is a quotient manifold of $\mathbb{R}$ defined by the submersion $\theta \mapsto e^{i\theta}$; we can then write $S^1 = \mathbb{R}/2\pi\mathbb{Z}$, including the differentiable structure.

C The Grassmannian $\mathbb{G}_k(E)$ is a quotient manifold in the following way. Define the set D by $D = \{(x_1, ..., x_k) \mid x_i \in E,\ x_1, ..., x_k$ are linearly dependent$\}$. Since D is open in the product $E \times \cdots \times E$ (k times), one can define the map $\pi : D \to \mathbb{G}_k(E)$ by $\pi(x_1, ..., x_k) = \operatorname{span}\{x_1, ..., x_k\}$. Using the charts described in Example **3.1.8G**, one finds that π is a submersion. In particular, the projective spaces $\mathbb{RP}^n$ and $\mathbb{CP}^n$ are quotient manifolds.

D The Möbius band as explained in **3.4.10C** is a quotient manifold.

E Quotient bundles are quotient manifolds (see **3.4.17**). ♦

We close with an important characterization of regular equivalence relations due to Godement, as presented by Serre [1965].

3.5.25 Theorem *An equivalence relation* R *on a manifold* M *is regular iff*

(i) graph(R) *is a submanifold of* $M \times M$, *and*

(ii) $p_1 : \text{graph}(R) \to M,\ p_1(x, y) = x$ *is a submersion.*

Proof First assume that R is regular. Since $\pi : M \to M/R$ is a submersion, so is the product $\pi \times \pi : M \times M \to (M/R) \times (M/R)$ so that $\text{graph}(R) = (\pi \times \pi)^{-1}(\Delta_{M/R})$, where the set $\Delta_{M/R} = \{([x], [x]) \mid [x] \in M/R\}$ is the diagonal of $M/R \times M/R$, is a submanifold of $M \times M$ (**3.5.4**). This proves (i). To verify (ii), let $(x,y) \in \text{graph}(R)$ and $v_x \in T_xM$. Since π is a submersion and $\pi(x) = \pi(y)$, there exists $v_y \in T_yM$ such that $T_y\pi(v_y) = T_x\pi(v_x)$; i.e., $(v_x, v_y) \in T_{(x,y)}(\text{graph}(R))$ by **3.5.4**. But then $T_{(x,y)}p_1(v_x, v_y) = v_x$, showing that p_1 is a submersion.

To prove the converse, we note that the equivalence relation is open, i.e., that $\pi^{-1}(\pi(U))$ is an open subset of M whenever U is open in M. Indeed $\pi^{-1}(\pi(U)) = p_1((M \times U) \cap \text{graph}(R))$ which is open since p_1 is an open map being a submersion by (ii). Second, the diffeomorphism $s : (x, y) \mapsto (y, x)$ of graph(R) shows that p_1 is a submersion iff $p_2 : \text{graph}(R) \to M,\ p_2(x, y) = y$ is a submersion since $p_2 = p_1 \circ s$. The rest of the proof consists of two major steps: a reduction to a local problem and the solution of the local problem.

Step 1 *If* $M = \bigcup_i U_i$, *where* U_i *are open subsets of* M *such that* $R_i = R \cap (U_i \times U_i)$ *is a regular equivalence relation in* U_i, *then* R *is regular.*

Openness of R implies openness of R_i and of $U_i^* = \pi^{-1}(\pi(U_i))$. Let us first show that $R_i^* = R \cap (U_i^* \times U_i^*)$ is regular on U_i^*. Let $\pi_i : U_i \to U_i/R_i$ and $\pi_i^* : U_i^* \to U_i^*/R_i^*$ denote the canonical projections. We prove that the existence of a manifold structure on U_i/R_i and submersivity of π_i imply that U_i^*/R_i^* has a manifold structure and that π_i^* is a submersion. For this purpose let $\lambda_i : U_i/R_i \to U_i^*/R_i^*$ be the bijective map induced by the inclusion $j_i : U_i \to U_i^*$ and endow U_i^*/R_i^* with the manifold structure making λ_i into a diffeomorphism. Thus π_i^* is a submersion iff $\rho_i = \lambda_i^{-1} \circ \pi_i^* : U_i^* \to U_i/R_i$ is a submersion. Since $\lambda_i \circ \pi_i = \pi_i^* \circ j_i$, it follows that $\rho_i \mid U_i = \pi_i$ is a submersion and therefore the composition $(\rho_i \mid U_i) \circ p_2 : (U_i^* \times U_i) \cap \text{graph}(R) \to U_i/R_i$ is a submersion. The relations $(\rho_i \mid U_i) \circ p_2 = \rho_i \circ p_1$ show that $\rho_i \circ p_1$ is a submersion and since p_1 is a surjective submersion this implies that ρ_i is a submersion (see exercise **3.5F(iv)**).

Thus, in the statement of Step 1, we can assume that all open sets U_i are such that $U_i = \pi^{-1}(\pi(U_i))$. Let R_{ij} be the equivalence relation induced by R on $U_i \cap U_j$. Since $U_i \cap U_j / R_{ij}$ is open in both U_i / R_i and U_j / R_j, it follows that it has two manifold structures. Since π_i and π_j are submersions, they will remain submersions when restricted to $U_i \cap U_j / R_{ij}$. Therefore R_{ij} is regular and by proposition **3.5.21**(iii) the manifold structures on $U_i \cap U_j / R_{ij}$ induced by the equivalence relations U_i / R_i and U_j / R_j coincide. Therefore there is a unique manifold structure on M / R such that U_i / R_i are open submanifolds; this structure is obtained by collation (see exercise **3.2F**). The projection π is a submersion since $\pi_i = \pi \mid U_i$ is a submersion for all i.

Step 2 *For each* $m \in M$ *there is an open neighborhood* U *of* m *such that* $R \cap (U \times U) = R_U$ *is regular.*

The main technical work is contained in the following.

3.5.26 Lemma *For each* $m \in M$ *there is an open neighborhood* U *of* M, *a submanifold* S *of* U *and a smooth map* $s : U \to S$ *such that* $[u] \cap S = \{s(u)\}$; S *is called a* ***local slice*** *of* R.

Let us assume the lemma and use it to prove Step 2. The inclusion of S into U is a right inverse of s and thus s is a submersion. Now define $\varphi : S \to U / R_U$ by $\varphi(u) = [u]$. By the lemma, φ is a bijective map. Put the manifold structure on U / R_U making φ into a diffeomorphism. The relation $\varphi \circ s = \pi \mid U$ shows that $\pi \mid U$ is submersive and thus R_U is regular.

Proof of Lemma 3.5.26 In the entire proof, $m \in M$ is fixed. Define the space F by $F = \{v \in T_mM \mid (0, v) \in T_{(m,m)}(\text{graph}(R))\}$, then $\{0\} \times F = \ker T_{(m,m)}p_1$ and thus by hypothesis (ii) in the theorem, $\{0\} \times F$ splits in $T_{(m,m)}(\text{graph}(R))$. The latter splits in $T_mM \times T_mM$ by hypothesis (i) and thus $\{0\} \times F$ splits in $T_mM \times T_mM$. Since $\{0\} \times F$ is a closed subspace of $\{0\} \times T_mM$, it follows that F splits in T_mM (see Exercise **2.1G**). Let G be a closed complement of F in T_mM and choose locally a submanifold P of M, $m \in P$, such that $T_mP = G$. Define the set Q by $Q = (M \times P) \cap \text{graph}(R)$. Since $Q = p_2^{-1}(P)$ and $p_2 : \text{graph}(R) \to M$ is a submersion, Q is a submanifold of graph(R).

We claim that $T_{(m,m)}p_1 : T_{(m,m)}Q \to T_mM$ is an isomorphism. Since $T_{(m,m)}Q = (T_{(m,m)}p_2)^{-1}(T_mP)$, it follows that $\ker(T_{(m,m)}p_1 \mid T_{(m,m)}Q) = \ker T_{(m,m)}p_1 \cap (T_{(m,m)}p_2)^{-1}(T_mP) = (\{0\} \times F) \cap (T_mM \times G) = \{0\} \times (F \cap G) = \{(0, 0)\}$, i.e., $T_{(m,m)}p_1 \mid T_{(m,m)}Q$ is injective. Now let $u \in T_mM$ and choose $v \in T_mM$ such that $(u, v) \in T_{(m,m)}(\text{graph}(R))$. If $v = v_1 + v_2$, $v_1 \in F$, $v_2 \in G$, then $(u, v_2) = (u, v) - (0, v_1) \in T_{(m,m)}(\text{graph}(R)) + (\{0\} \times F) \subset T_{(m,m)}(\text{graph}(R))$ and $T_{(m,m)}p_2(u, v_2) = v_2 \in T_mP$, i.e., $(u, v_2) \in (T_{(m,m)}p_2)^{-1}(T_mP) = T_{(m,m)}Q$. Then $T_{(m,m)}p_1(u, v_2) = u$ and hence $T_{(m,m)}p_1 \mid T_{(m,m)}Q$ is onto.

Thus $p_1 : Q \to M$ is a local diffeomorphism at (m, m), that is, there are open neighborhoods U_1 and U_2 of m, $U_1 \subset U_2$ such that $p_1 : Q \cap (U_1 \times U_2) \to U_1$ is a diffeomorphism. Let σ be the inverse of p_1 on U_1. Since σ is of the form $\sigma(x) = (x, s(x))$, this defines a smooth map $s : U_1 \to P$. Note that if $x \in U_1 \cap P$, then (x, x) and (x, s(x)) are two points in $Q \cap (U_1 \times U_2)$ with the same image in U_1 and hence are equal. This shows that $s(x) = x$ for $x \in U_1 \cap P$.

Set $U = \{x \in U_1 \mid s(x) \in U_1 \cap P\}$ and let $S = U \cap P$. Since s is smooth and $U_1 \cap P$ is open in P it follows that U is open in U_1 hence in M. Also, $m \in U$ since $m \in U_1 \cap P$ and so $m = s(m)$. Let us show that $s(U) \subset S$, i.e., that if $x \in U$, then $s(x) \in U$ and $s(x) \in P$. The last relation is obvious from the definition of s. To show that $s(x) \in U$ is equivalent to proving that $s(x) \in U_1$, which is clear, and that $s(s(x)) \in U_1 \cap P$. However, since $s(s(x)) = s(x)$, because $s(x) \in P$, it follows from $x \in U$ that $s(x) \in U_1 \cap P$. Thus we have found an open neighborhood U of m, a submanifold S of U, and a smooth map $s : U \to S$ which is the identity in $U \cap S$.

Finally, we show that s(x) is the only element of S equivalent to $x \in U$. But this is clear since there is exactly one point in $(U \times S) \cap \text{graph}(R)$, namely $(x, s(x))$ mapped by p_1 into x, since $p_1 \mid (U \times S) \cap \text{graph}(R)$ is a diffeomorphism m. ■

The above proof shows that in condition (ii) of the theorem, p_1 can be replaced by p_2. Also recall that M / R is Hausdorff iff R is closed.

☞ SUPPLEMENT 3.5A
Lagrange Multipliers

Let M be a smooth manifold and $i : N \to M$ a submanifold of M, i denoting the inclusion mapping. If $f : M \to \mathbb{R}$, we want to determine necessary and sufficient conditions for $n \in N$ to be a critical point of $f \mid N$, the restriction of f to N. Since $f \mid N = f \circ i$, the chain rule gives $T_n(f \mid N) = T_n f \circ T_n i$; thus $n \in N$ is a critical point of $f \mid N$ iff $T_n f \mid T_n N = 0$. This condition takes a simple form if N happens to be the inverse image of a point under submersion.

3.5.27 Lagrange Multiplier Theorem *Let* $g : M \to P$ *be a smooth submersion,* $p \in P$, $N = g^{-1}(p)$ *and let* $f : M \to \mathbb{R}$ *be* C^r, $r \geq 1$. *A point* $n \in N$ *is a critical point of* $f \mid N$ *if there exists* $\lambda \in T_p{}^*P$, *called a* ***Lagrange multiplier***, *such that* $T_n f = \lambda \circ T_n g$.

Proof First assume such a λ exists. Since $T_n N = \ker T_n g$,

$$(\lambda \circ T_n g \circ T_n i)(v_n) = \lambda(T_n g(v_n)) = 0 \quad \text{for all } v_n \in T_n N;$$

i.e., $0 = (\lambda \circ T_n g) \mid T_n N = T_n f \mid T_n N$.

Conversely, assume $T_n f \mid T_n N = 0$. By the local normal form for submersions, there is a chart (U, φ) at n, $\varphi : U \to U_1 \times V_1 \subset E \times F$ such that $\varphi(U \cap N) = \{0\} \times V_1$ satisfying $\varphi(n) = (0, 0)$, and a chart (V, ψ) at p, $\psi : V \to U_1 \subset E$ where $g(U) \subset V$, $\psi(p) = 0$, and such that $g_{\varphi\psi}(x, y) = (\psi \circ g \circ \varphi^{-1})(x, y) = x$ for all $(x, y) \in U_1 \times V_1$. If $f_\varphi = f \circ \varphi^{-1} : U_1 \times V_1 \to \mathbb{R}$, we have for all $v \in F$, $\mathbf{D}_2 f_\varphi(0, 0) \cdot v = 0$ since $T_n f \mid T_n N = 0$. Thus, letting $\mu = \mathbf{D}_1 f_\varphi(0, 0) \in E^*$,

$u \in E$ and $v \in F$, we get

$$\mathbf{D}f_\varphi(0,0)\cdot(u,v) = \mu(u) = (\mu \circ \mathbf{D}g_{\varphi\psi})(0,0)\cdot(u,v); \text{ i.e., } \mathbf{D}f_\varphi(0,0) = (\mu \circ \mathbf{D}g_{\varphi\psi})(0,0).$$

To pull this local calculation back to M and P, let $\lambda = \mu \circ T_p\psi \in T_p{}^*P$, so composing the foregoing relation with $T_n\varphi$ on the right we get $T_nf = \lambda \circ T_ng$. ■

3.5.28 Corollary *Let* $g : M \to P$ *be transversal to the submanifold* W *of* P, $N = g^{-1}(W)$, *and let* $f : M \to \mathbb{R}$ *be* C^r, $r \geq 1$. *Let* $E_{g(n)}$ *be a closed complement to* $T_{g(n)}W$ *in* $T_{g(n)}P$ *so* $T_{g(n)}P = T_{g(n)}W \oplus E_{g(n)}$ *and let* $\pi : T_{g(n)}P \to E_{g(n)}$ *be the projection. A point* $n \in N$ *is a critical point of* $f \mid N$ *iff there exists* $\lambda \in E_{g(n)}{}^*$ *called a* ***Lagrange multiplier*** *such that* $T_nf = \lambda \circ \pi \circ T_ng$.

Proof By **3.5.12**, there is a chart (U, φ) at n, with $\varphi(U) = U_1 \times U_2 \subset E_1 \times E_2$, $\varphi(U \cap N) = \{0\} \times U_2$, and $\varphi(n) = (0, 0)$, and a chart (V, ψ) at $g(n)$ satisfying $\psi(V) = U_1 \times V_1 \subset F_1 \times F$, $\psi(V \cap W) = \{0\} \times V_1$, $\psi(g(n)) = (0, 0)$, and $g(U) \subset V$, such that the local representative satisfies $g_{\varphi\psi}(x, y) = (\psi \circ g \circ \varphi^{-1})(x, y) = (x, \eta(x, y))$ for all $(x, y) \in U_1 \times V_1$. Let $\rho : E_1 \times F \to E_1$ be the canonical projection. By the Lagrange multiplier theorem applied to the composition $\rho \circ g_{\varphi\psi} : U_1 \times U_2 \to U_1$, $(0, 0) \in U_1 \times U_2$ is a critical point of $f \mid \{0\} \times U_2$ iff there is a point $\mu \in E_1{}^*$ such that $\mathbf{D}f_\varphi(0, 0) = \mu \circ \rho \circ \mathbf{D}g_{\varphi\psi}(0, 0)$. Composing this relation on the right with $T_n\varphi$ and letting $\lambda = \mu \circ T_{g(n)}\psi$ and $\pi = (T_{g(n)}\psi)^{-1} \mid E_{g(n)} \circ \rho \circ T_{g(n)}\psi : T_{g(n)}P \to E_{g(n)}$, we get the required identity $T_nf = \lambda \circ \pi \circ T_ng$. ■

If P is a Banach space F, then **3.5.27** can be formulated in the following way.

3.5.29 Corollary *Let* F *be a Banach space,* $g : M \to F$ *a smooth submersion,* $N = g^{-1}(0)$, *and* $f : M \to \mathbb{R}$ *be* C^r, $r \geq 1$. *The point* $n \in N$ *is a critical point of* $f \mid N$ *iff there exists* $\lambda \in F^*$, *called a* ***Lagrange multiplier***, *such that* n *is a critical point of* $f - \lambda \circ g$.

Notes

1. λ depends not just on $f \mid N$ but also on how f is extended off N.
2. This form of the Lagrange multiplier theorem is extensively used in the calculus of variations to study critical points of functions with constraints; cf. Caratheodory [1965].
3. We leave it to the reader to generalize **3.5.28** in the same spirit.
4. There are generalizations to $f : M \to \mathbb{R}^k$, which we invite the reader to formulate.

The name ***Lagrange multiplier*** is commonly used in conjuction with the previous corollary in Euclidean spaces. Let U be an open set in $\mathbb{R}^n$, $F = \mathbb{R}^p$, $g = (g^1, \ldots, g^p) : U \to \mathbb{R}^p$ a submersion and $f : U \to \mathbb{R}$ smooth. Then $x \in N = g^{-1}(0)$ is a critical point of $f \mid N$, iff there exists

$$\lambda = \sum_{i=1}^{p} \lambda_i e^i \in (\mathbb{R}^p)^*,$$

where $e^1, ..., e^p$ is the standard dual basis in $\mathbb{R}^p$, such that n is a critical point of

$$f - \lambda \circ g = f - \sum_{i=1}^{p} \lambda_i g^i.$$

In calculus, the real numbers λ_i are referred to as ***Lagrange multipliers***. Thus, to find a critical point $x = (x^1, ..., x^m) \in N \subset \mathbb{R}^m$ of $f \mid N$ one solves the system of $m + p$ equations

$$\frac{\partial f}{\partial x^j}(x) - \sum_{i=1}^{p} \lambda_i \frac{\partial g^i}{\partial x^j}(x) = 0, \qquad j = 1, ..., m$$

$$g^i(x) = 0, \qquad i = 1, ..., p$$

for the $m + p$ unknowns $x^1, ..., x^m, \lambda_1, ..., \lambda_p$.

For example, let $N = S^2 \subset \mathbb{R}^3$ and $f : \mathbb{R}^3 \to \mathbb{R}$; $f(x, y, z) = z$. Then $f \mid S^2$ is the height function on the sphere and we would expect $(0, 0, \pm 1)$ to be the only critical poins of $f \mid S^2$; note that f itself has *no* critical points. The method of Lagrange multipliers, with $g(x, y, z) = x^2 + y^2 + z^2 - 1$, gives

$$0 - 2x\lambda = 0, \quad 0 - 2y\lambda = 0, \quad 0 - 2z\lambda = 0, \text{ and } x^2 + y^2 + z^2 = 1.$$

The only solutions are $\lambda = \pm 1/2$, $x = 0$, $y = 0$, $z = \pm 1$, and indeed these correspond to the maximum and minimum points for f on S^2. See an elementary text such as Marsden and Tromba [1988] for additional examples. For more advanced applications, see Luenberger [1969].

The reader will recall from advanced calculus that maximum and minimum tests for a critical point can be given in terms of the Hessian, i.e., matrix of second derivtives. For constrained problems there is a similar test involving ***bordered Hessians***. Bordered Hessians are simply the Hessians of $h = f - \lambda g + c(\lambda - \lambda_0)^2$ in (x, λ)-space. Then the Hessian test for maxima and minima apply; a maximum or minimum of h clearly implies the same for f on a level set of g. See Marsden and Tromba [1988, pp. 224-30] for an elementary treatment and applications. ✍

Exercises

3.5A (i) Show that the set $\mathbf{SL}(n, \mathbb{R})$ of elements of $L(\mathbb{R}^n, \mathbb{R}^n)$ with determinant 1 is a closed submanifold of dimension n^2-1; $\mathbf{SL}(n, \mathbb{R})$ is called the ***special linear group***. Generalize to the complex case.

(ii) Show that $\mathbf{O}(n)$ has two connected components. The component of the identity $\mathbf{SO}(n)$

is called the ***special orthogonal group.***

(iii) Let $\mathcal{U}(n) = \{U \in L(\mathbb{C}^n, \mathbb{C}^n) \mid UU^* = \text{Identity}\}$ be the ***unitary group.*** Show that $\mathcal{U}(n)$ is a compact submanifold of dimension n^2 of $L(\mathbb{C}^n, \mathbb{C}^n)$ and $\mathcal{O}(2n)$.

(iv) Show that the special unitary group $\mathcal{SU}(n) = \mathcal{U}(n) \cap \mathcal{SL}(n, \mathbb{C})$ is a compact n^2-1 dimensional manifold.

(v) Define $\mathbb{J} = \begin{bmatrix} 0 & I \\ -I & 0 \end{bmatrix} \in L(\mathbb{R}^{2n}, \mathbb{R}^{2n})$, where I is the identity of $\mathbb{R}^n$. Show that $\mathcal{Sp}(2n, \mathbb{R}) = \{Q \in L(\mathbb{R}^{2n}, \mathbb{R}^{2n}) \mid Q\mathbb{J}Q^T = \mathbb{J}\}$ is a compact submanifold of dimension $2n^2 + n$; it is called the ***symplectic group.***

3.5B Define $\mathcal{USt}(n, n; k)$ and $\mathcal{SpSt}(2n, 2n; 2k)$ analogous to the definition of $\mathcal{OSt}(n, n; k)$ in example **3.5.5D**. Show that they are compact manifolds and compute their dimensions.

3.5C (i) Let $P \subset \mathcal{O}(3)$ be defined by

$$P = \{Q \in \mathcal{O}(3) \mid \det Q = +1, Q = Q^T\} \setminus \{I\}.$$

Show that P is a two-dimensional compact submanifold of $\mathcal{O}(3)$.

(ii) Define $f: \mathbb{RP}^2 \to \mathcal{O}(3)$, $f(\ell) =$ the rotation through π about the line ℓ. Show that f is a diffeomorphism of $\mathbb{RP}^2$ onto P.

3.5D (i) If N is a submanifold of dimension n in an m-manifold M, show that for each $x \in N$ there is an open neighborhood $U \subset M$ with $x \in U$ and a submersion $f : U \subset M \to \mathbb{R}^{m-n}$, such that $N \cap U = f^{-1}(0)$.

(ii) Show that $\mathbb{RP}^1$ is a submanifold of $\mathbb{RP}^2$, which is not the level set of any submersion of $\mathbb{RP}^2$ into $\mathbb{RP}^1$; in fact, there are no such submersions. (*Hint:* $\mathbb{RP}^1$ is one-sided in $\mathbb{RP}^2$).

3.5E (i) Show that if $f : M \to N$ is a subimmersion, $g : N \to P$ an immersion and $h : Z \to M$ a submersion, then $g \circ f \circ h$ is a subimmersion.

(ii) Show that if $f_i : M_i \to N_i$, $i = 1, 2$ are immersions (submersions, subimmersions), then so is $f_1 \times f_2 : M_1 \times M_2 \to N_1 \times N_2$.

(iii) Show that the composition of two immersions (submersions) is again an immersion (submersion). Show that this fails for subimmersions.

(iv) Let $f : M \to N$ and $g : N \to P$ be C^r, $r \geq 1$. If $g \circ f$ is an immersion, show that f is an immersion. If $g \circ f$ is a submersion and if f is onto, show that g is a submersion.

(v) Show that if f is an immersion (resp. embedding, submersion, subimmersion) then so is Tf.

3.5F (i) Let M be a manifold, R a regular equivalence relation and S another equivalence relation implied by R; i.e., graph $R \subset$ graph S. Denote by S / R the equivalence relation induced by S on M / R. Show that S is regular iff S / R is and in this case establish a diffeomorphism $(M / R) / (S / R) \to M / S$.

(ii) Let M_i, $i = 1, 2$ be manifolds and R_i be regular equivalences on M_i. Denote by R the equivalence on $M_1 \times M_2$ defined by $R_1 \times R_2$. Show that M / R is diffeomorphic to $(M_1 / R_1) \times (M_2 / R_2)$.

3.5G The line with two origins Let M be the quotient topological space obtained by starting with $(\mathbb{R} \times \{0\}) \cap (\mathbb{R} \times \{1\})$ and identifying $(t, 0)$ with $(t, 1)$ for $t \neq 0$. Show that this is a one-dimensional non-Hausdorff manifold. Find an immersion $\mathbb{R} \to M$.

3.5H Let $f : M \to N$ be C^∞ and denote by $h : TM \to f^*(TN)$ the vector bundle map over the identity uniquely defined by the pull-back. Prove the following:

(i) f is an immersion iff $0 \longrightarrow TM \overset{h}{\longrightarrow} f^*(TN)$ is fiber split exact;

(ii) f is a submersion iff $TM \overset{h}{\longrightarrow} f^*(TN) \longrightarrow 0$ is fiber split exact;

(iii) f is a subimmersion iff $\ker(h)$ and $\operatorname{range}(h)$ are subbundles.

3.5I Let A be a real nonsingular symmetric $n \times n$ matrix and c a nonzero real number. Show that the quadratic surface $\{x \in \mathbb{R}^n \mid \langle Ax, x\rangle = c\}$ is an $(n-1)$-submanifold of $\mathbb{R}^n$.

3.5J Steiner's Roman Surface Let $f: S^2 \to \mathbb{R}^4$ be defined by

$$f(x, y, z) = (yz, xz, xy, x^2 + 2y^2 + 3z^2).$$

(i) Show that $f(p) = f(q)$ if and only if $p = \pm q$.

(ii) Show that f induces an immersion $f' : \mathbb{RP}^2 \to \mathbb{R}^4$.

(iii) Let $g : \mathbb{RP}^2 \to \mathbb{R}^3$ be the first three components of f'. Show that g is a "topological" immersion and try to draw the surface $g(\mathbb{RP}^2)$ (see Spivak [1979] for the solution).

3.5K Covering maps Let $f : M \to N$ be smooth and M compact, $\dim(M) = \dim(N) < \infty$. If n is a regular value of f, show that $f^{-1}(n)$ is a finite set $\{m_1, \ldots, m_k\}$ and that there exists an open neighborhood V of n in N and disjoint open neighborhoods $U_1, \ldots, U_k$ of $m_1, \ldots, m_k$ such that $f^{-1}(V) = U_1 \cup \cdots \cup U_k$, and $f \mid U_k : U_i \to V$, $i = 1, \ldots, k$ are all diffeomorphisms. Show k is constant if M is connected and f is a submersion.

3.5L Let $f : M \to N$ and $g : N \to P$ be smooth maps, such that $g \pitchfork V$ where V is a submanifold of P. Show that $f \pitchfork g^{-1}(V)$ iff $g \circ f \pitchfork V$.

3.5M Show that an injective immersion $f : M \to N$ is an embedding iff $f(M)$ is a closed submanifold of an open submanifold of N. Show that if $f : M \to N$ is an embedding, f is a diffeomorphism of M onto $f(M)$. (*Hint:* See Exercise **2.5L**.)

3.5N Show that the map $p : \mathbf{St}(n, n; k) = \mathbb{G}_k(\mathbb{R}^n)$ defined by $p(A) = \text{range } A$ is a surjective submersion.

3.5O Show that $f : \mathbb{RP}^n \times \mathbb{RP}^m \to \mathbb{RP}^{mn+m+n}$ given by $(x, y) \mapsto [x_0y_0, x_0y_1, \ldots, x_iy_j, \ldots, x_ny_m]$ is an embedding. (This embedding is used in algebraic geometry to define the product of quasiprojective varities; it is called the ***Segre embedding***.)

3.5P Show that $\{(x, y) \in \mathbb{RP}^n \times \mathbb{RP}^m \mid n \le m, \ \Sigma_{i=0, \ldots, n}\, x_iy_i = 0\}$ is an $(m + n - 1)$-manifold. It is usually called a ***Milnor manifold***.

3.5Q Fiber product of manifolds Let $f : M \to P$ and $g : N \to P$ be C^∞ mappings such that $(f, g) : M \times N \to P \times P$ is transversal to the diagonal of $P \times P$. Show that the set defined by $M \times_P N = \{(m, n) \in M \times N \mid f(m) = g(n)\}$ is a submanifold of $M \times N$. If M and N are finite dimensional, show that $\operatorname{codim}(M \times_P N) = \dim P$.

3.5R (i) Let H be a Hilbert space and $\mathcal{GL}(H)$ the group of all isomorphisms $A : H \to H$ that are continuous. As we saw earlier, $\mathcal{GL}(H)$ is open in $L(H, H)$ and multiplication and inversion are C^∞ maps, so $\mathcal{GL}(H)$ is a Lie group. Show that $\mathbf{O}(H) \subset \mathcal{GL}(H)$ defined by $\mathbf{O}(H) = \{A \in \mathcal{GL}(H) \mid \langle Ax, Ay\rangle = \langle x, y\rangle \text{ for all } x, y \in H\}$ is a smooth submanifold and hence also a Lie group.

(ii) Show that the tangent space at the identity of $\mathcal{GL}(H)$ (the Lie algebra) consists of all bounded skew adjoint operators, as follows. Let $S(H) = \{A \in L(H, H) \mid A^* = A\}$, where A^* is the adjoint of A. Define $f : \mathcal{GL}(H) \to S(H)$, by $f(A) = A^*A$. Show f is C^∞, $f^{-1}(I) = \mathbf{O}(H)$, and $\mathbf{D}f(A) \cdot B = B^*A + A^*B$. Show that f is a submersion, and $\ker \mathbf{D}f(A) = \{B \in L(H, H) \mid B^*A + A^*B = 0\}$, which splits; a complement is the space $\{T \in L(H, H) \mid T^*A = A^*T\}$ since any U splits as

$$U = \frac{1}{2}(U - A^{*-1}U^*A) + \frac{1}{2}(U + A^{*-1}U^*A).$$

3.5S (i) If $f : M \to N$ is a smooth map of finite dimensional manifolds and $m \in M$, show that there is an open neighborhood U of m such that $\operatorname{rank}(T_xf) \ge \operatorname{rank}(T_mf)$ for all $x \in U$. (*Hint:* Use the local expression of T_mf as a Jacobian matrix.)

(ii) Let M be a finite dimensional connected manifold and $f : M \to M$ a smooth map satisfying $f \circ f = f$. Show that $f(M)$ is a closed connected submanifold of M. What is its dimension? (*Hint:* Show f is a closed map. For $m \in f(M)$ show that $\operatorname{range}(T_mf) = \ker(\text{Identity} - T_mf)$ and thus $\operatorname{rank}(T_mf) + \operatorname{rank}(\text{Identity} - T_mf) = \dim M$. Both ranks can only increase in a neighborhood of m by (i), so $\operatorname{rank}(T_mf)$ is locally constant on f(M). Thus there is a neighborhood U of f(M) such that the rank of f on U is bigger than or equal to the rank of f on f(M). Use rank(AB) $\leq$ rank A and the fact that $f \circ f = f$ to show that the rank of f on U is smaller than or equal to the rank of f in f(M). Therefore rank of f on U is constant. Apply **3.5.18**(iii).)

3.5T (i) Let $\alpha, \beta : E \to \mathbb{R}$ be continuous linear maps on a Banach space E such that $\ker \alpha = \ker \beta$. Show that α and β are proportional. (*Hint:* Split $E = \ker \alpha \oplus \mathbb{R}$.)

(ii) Let $f, g : M \to \mathbb{R}$ be smooth functions with 0 a regular value of both f and g and $N = f^{-1}(0) = g^{-1}(0)$. Show that for all $x \in N$, $\mathbf{d}f(x) = \lambda(x)\, \mathbf{d}g(x)$ for a smooth function $\lambda : N \to \mathbb{R}$.

3.5U Let $f : M \to N$ be a smooth map, $P \subset N$ a submanifold, and assume $f \pitchfork P$. Use **3.5.10** and Exercise **3.4J** to show the vector bundle isomorphism $\nu(f^{-1}(P)) \cong f^*(\nu(P))$. (*Hint:* Look at $Tf \,|\, f^{-1}(P)$ and compute the kernel of the induced map $TM \,|\, f^{-1}(P) \to \nu(P)$. Obtain a vector bundle map $\nu(f^{-1}(P)) \to \nu(P)$ which is an isomorphism on each fiber. Then invoke the universal property of the pull-back of vector bundles; see Exercise **3.4M**.)

3.5V (i) Recall (Exercise **3.2A**) that S^n with antipodal points identified is diffeomorphic to $\mathbb{RP}^n$. Conclude that any closed hemisphere of S^n with antipodal points on the great circle identified is also diffeomorphic to $\mathbb{RP}^n$.

(ii) Let B^n be the closed unit ball in $\mathbb{R}^n$. Map B^n to the upper hemisphere of S^n by mapping $x \mapsto (x, (1 - \|x\|^2)^{1/2})$. Show that this map is a diffeomorphism of an open neighborhood of B^n to an open neighborhood of the upper hemisphere $S^n_+ = \{x \in S^n \mid x^{n+1} \geq 0\}$, mapping B^n homeomorphically to S^n_+ and homeomorphically to the great circle $\{x \in S^n \mid x^{n+1} = 0\}$. Use (i) to show that B^n with antipodal points on the boundary identified is diffeomorphic to $\mathbb{RP}^n$.

(iii) Show that $\mathbf{SO}(3)$ is diffeomorphic to $\mathbb{RP}^3$; the diffeomorphism is induced by the map sending the closed unit ball B^3 in $\mathbb{R}^3$ to $\mathbf{SO}(3)$ via $(x, y, z) \mapsto$ (the rotation about (x, y, z) by the right hand rule in the plane perpendicular to (x, y, z) through the angle $\pi(x^2 + y^2 + z^2)^{1/2}$.

3.5W (i) Show that $S^n \times \mathbb{R}$ embeds in $\mathbb{R}^{n+1}$. (*Hint:* The image is a "fat" sphere.)

(ii) Describe explicitly in terms of trigonometric functions the embedding of $\mathbb{T}^2$ into $\mathbb{R}^3$.

(iii) Show that $S^{a(1)} \times \cdots \times S^{a(k)}$, where $a(1) + \cdots a(k) = n$ embeds in $\mathbb{R}^{n+1}$. (*Hint:* Show that its product with $\mathbb{R}$ embeds in $\mathbb{R}^{n+1}$ by (i).)

3.7X Let $f : M \to M$ be an involution without fixed points, i.e., $f \circ f =$ identity and $f(m) \neq m$ for all m. Let R be the equivalence relation determined by f, i.e., $m_1 R m_2$ iff $f(m_1) = f(m_2)$.

(i) Show R is a regular equivalence relation.

(ii) Show that the differentiable structure of M/ R is uniquely determined by the property: the projection $\pi : M \to M / R$ is a local diffeomorphism.

3.5Y Connected sum of manifolds Let M and N to be two Hausdorff manifolds modeled on the same Banach space E. Let $m \in M$, $n \in N$ and let (U_0, φ_0) be a chart at m and let (V_0, ψ_0) be a chart at n such that $\varphi_0(m) = \psi_0(n) = 0$ and $\varphi_0(U)$, $\psi_0(V)$ contain the closed unit ball in E. Thus, if B denotes the open unit ball in E, $\varphi_0(U)\setminus B$ and $\psi_0(V)\setminus B$ are nonempty. If $\mathcal{A}$ and $\mathcal{B}$ are atlases of M and N respectively, let $\mathcal{A}_m$, $\mathcal{B}_n$ be the induced atlases on $M\setminus\{m\}$ and $N\setminus\{n\}$ respectively. Define $\sigma : B\setminus\{0\} \to B\setminus\{0\}$ by $\sigma(\|x\|, x/\|x\|) = (1 - \|x\|, x/\|x\|)$ and observe that $\sigma^2 =$ identity. Let W be the disjoint union of $M\setminus\{m\}$ with $N\setminus\{n\}$ and define an equivalence relation R in W by

$$\begin{aligned} v_1 R v_2 \text{ iff } & \big(w_1 = w_2\big) \text{ or } \big(w_1 \in M\setminus\{m\},\ \varphi_0(w_1) \in B\setminus\{0\} \\ & \text{and } w_2 \in N\setminus\{n\},\ \psi_0(w_2) \subset B\setminus\{0\} \text{ and } \varphi_0(w_1) = (\sigma \circ \psi_0)(w_2)\big) \\ & \text{or } \big(\text{same condition with } w_1 \text{ and } w_2 \text{ interchanged}\big). \end{aligned}$$

(i) Show that R is an equivalence relation on W.

(ii) Show R is regular.

(iii) If $\pi : W \to W/R$ denotes the projection, show that for any open set O in the atlas of W, $\pi : O \mapsto \pi(O)$ is a diffeomorphism.

(iv) Show that $\{(\pi(U), \varphi \circ (\pi \mid U)^{-1}), (\pi(V), \psi \circ (\pi \mid V)^{-1}) \mid (U, \varphi) \in \mathcal{A}_m, (V, \psi) \in \mathcal{B}_n\}$ is an atlas defining the differentiable structure of W/R.

(v) W/R is denoted by $M \# N$. Draw $\mathbb{T}^2 \# \mathbb{T}^2$ and identify $T^2 \# \mathbb{RP}^2$ and $\mathbb{RP}^2 \# \mathbb{RP}^2$.

(vi) Prove that $M \# (N \# P) \approx (M \# N) \# P$, $M \# N \approx N \# M$, $M \# S_E \approx M$, where M, N, and P are all modeled on E, S_E is the unit sphere in E, and $\approx$ denotes "diffeomorphic."

(vii) Compute $\mathbb{R}^n \# \mathbb{R}^n \# \cdots \# \mathbb{R}^n$ (k times) for all positive integers n and k and show that it embeds in $\mathbb{R}^n$.

3.5Z (i) Let $a > 0$ and define $\chi_a : \mathbb{R} \to \mathbb{R}$ by $\chi_a(x) = \exp(-x^2/(a^2 - x^2))$, if $x \in \,]-a, a[$ and $\chi_a(x) = 0$ if $x \in \mathbb{R} \setminus \,]-a, a[$. Show that this is a C^∞ function and satisfies the inequalities $0 \le \chi_a(x) \le 1$, $|\chi_a'(x)| < 1$ for all $x \in \mathbb{R}$, and $\chi_a(0) = 1$.

(ii) Fix $a > 0$ and $\lambda \in E^*$, where E is a Banach space whose norm is of class C^r away from the origin and $r \ge 1$. Write $E = \ker \lambda \oplus \mathbb{R}$; this is always possible since any closed finite codimensional space splits (see §**2.2**). Define, for any $t \in \mathbb{R}$, $f_{\lambda,a,t} : E \to E$ by $f_{\lambda,a,t}(u,x) = (u, x + t\chi_a(\|u\|)$ where $u \in \ker \lambda$ and $x \in \mathbb{R}$. Show that $f_{\lambda,a,t}$ satisfies $f_{\lambda,a,t}(0, 0) = (0, 1)$ and $f_{\lambda,a,t} \mid (E\setminus \mathrm{cl}B_a(0)) =$ identity. (*Hint:* Show that $f_{\lambda,a,t}$ is a bijective local diffeomorphism.)

(iii) Let M be a C^r Hausdorff manifold modeled on a Banach space E whose norm is C^r on $E\setminus\{0\}$, $r \ge 1$. Assume $\dim M \ge 2$. Let C be a closed set in M and assume that $M \setminus C$ is connected. Let $\{p_1,\ldots, p_k\}$, $\{q_1,\ldots, q_k\}$ be two finite subsets of $M \setminus C$. Show that there exists a C^r diffeomorphism $\varphi : M \to M$ such that $\varphi(p_i) = q_i$, $i = 1,$

..., k and $\varphi \mid C$ = identity. Show that if $k = 1$, the result holds even if $\dim M = 1$. (*Hint:* For $k = 1$, define an equivalence relation on $M \setminus C : m \sim n$ iff there is a diffeomorphism $\psi : M \to M$ homotopic to the identity such that $\varphi(m) = n$ and $\psi \mid C$ = identity. Show that the equivalence classes are open in $M \setminus C$ in the following way. Let $\varphi : U \to E$ be a chart at m, $\varphi(m) = 0$, $U \subset M \setminus C$, and let $n \in U$, $n \neq m$. Use the Hahn-Banach theorem to show that there is $\lambda \in E^*$ such that φ can be modified to satisfy $\varphi(m) = (0, 0)$, $\varphi(n) = (0, 1)$, where $E = \ker \lambda \oplus \mathbb{R}$. Use (ii) to find a diffeomorphism $h : U \to U$ homotopic to the identity on U, satisfying $h(m) = n$ and $h \mid (U \setminus A)$ = identity, where A is a closed neighborhood of n. Then $f : M \to M$ which equals h on U and the identity on $M \setminus U$ establishes $m \sim n$. For general k proceed by induction, using the connectedness of $M \setminus C \setminus \{q_1, ..., q_{k-1}\}$ and finding by the case $k = 1$ a diffeomorphism g homotopic to the identity on M sending $h(p_k)$ to q_k and keeping $C \cup \{q_1, ..., q_k\}$ fixed; $h : M \to M$ is the diffeomorphism given by induction which keeps C fixed and sends p_i to q_i for $r = 1, ..., k-1$. Then $f = g \circ h$ is the desired diffeomorphism.)

§3.6 *The Sard and Smale Theorems*

This section is devoted to the classical Sard theorem and its infinite-dimensional genralization due to Smale [1965]. We first develop a few properties of sets of measure zero in $\mathbb{R}^n$.

A subset $A \subset \mathbb{R}^m$ is said to have ***measure zero*** if, for every $\varepsilon > 0$, there is a countable covering of A by open sets U_i, such that the sum of the volumes of U_i is less than ε. Clearly a countable union of sets of measure zero has measure zero.

3.6.1 Lemma *Let* $U \subset \mathbb{R}^m$ *be open and* $A \subset U$ *be of measure zero. If* $f : U \to \mathbb{R}^m$ *is a* C^1 *map, then* $f(A)$ *has measure zero.*

Proof Let A be contained in a countable union of relatively compact sets C_n. If we show that $A \cap C_n$ has measure zero, then A has measure zero since it will be a countable union of sets of measure zero. But C_n is relatively compact and thus there exists $M > 0$ such that $\| \mathbf{D}f(x) \| \leq M$ for all $x \in C_n$. By the mean value theorem, the image of a cube of edge length d is contained in a cube of edge length $d\sqrt{m}\,M$. ■

3.6.2 Fubini Lemma *Let* A *be a countable union of compact sets in* $\mathbb{R}^n$, *fix an integer* r *satisfying* $0 \leq r \leq n-1$ *and assume that* $A_c = A \cap (\{c\} \times \mathbb{R}^{n-r})$ *has measure zero in* $\mathbb{R}^{n-r}$ *for all* $c \in \mathbb{R}^r$. *Then* A *has measure zero.*

Proof By induction we reduce to the case $r = n-1$. It is enough to work with one element of the union, so we may assume A itself is compact and hence there exists and interval $[a, b]$ such that $A \subset [a, b] \times \mathbb{R}^{n-1}$. Since A_c is compact and has measure zero, for each $c \in [a, b]$ there is a finite number of closed cubes $K_{c,1}, \ldots, K_{c,N(c)}$ in $\mathbb{R}^{n-1}$ the sum of whose volumes is less than ε and such that $\{c\} \times K_{c,i}$ cover A_c, $i = 1, \ldots, N(c)$. Find a closed interval I_c with c in its interior such that $I_c \times K_{c,i} \subset A_c \times \mathbb{R}^{n-1}$. Thus the family $\{I_c \times K_{c,i} \mid i = 1, \ldots, N(c), c \in [a, b]\}$ covers $A \cap ([a, b] \times \mathbb{R}^{n-1}) = A$. Since $\{\operatorname{int}(I_c) \mid c \in [a, b]\}$ covers $[a, b]$, we can choose a finite subcovering $I_{c(1)}, \ldots, I_{c(M)}$. Now find another covering $J_{c(1)}, \ldots, J_{c(K)}$ such that each $J_{c(i)}$ is contained in some $I_{c(j)}$ and such that the sum of the lengths of all $J_{c(i)}$ is less than $2(b-a)$. Consequently $\{J_{c(j)} \times K_{c(j),i} \mid j = 1, \ldots, K, i = 1, \ldots, N_{c(j)}\}$ cover A and the sum of their volumes is less than $2(b-a)\varepsilon$. ■

Let us recall the following notations from **§3.5**. If M and N are C^1 manifolds and $f : M \to N$ is a C^1 map, a point $x \in M$ is a ***regular point*** of f iff $T_x f$ is surjective, otherwise x is a ***critical point*** of f. If $C \subset M$ is the set of critical points of f, then $f(C) \subset N$ is the set of ***critical values*** of f and $N \backslash f(C)$ is the set of ***regular values*** of f, which is denoted by $\mathcal{R}_f$ or $\mathcal{R}(f)$. In addition, for $A \subset M$ we define $\mathcal{R}_f \mid A$ by $\mathcal{R}_f \mid A = N \setminus f(A \cap C)$. In particular, if $U \subset M$ is open, $\mathcal{R}_f \mid U = \mathcal{R}(f \mid U)$.

3.6.3 Sard's Theorem in $\mathbb{R}^n$. *Let* $U \subset \mathbb{R}^m$ *be open and* $f : U \to \mathbb{R}^n$ *be of class* C^k, *where* $k > \max(0, m - n)$. *Then the set of critical values of* f *has measure zero in* $\mathbb{R}^n$.

Note that if $m < n$, then f is only required to be at least C^1.

Proof (Complete only for $k = \infty$) Denote by $C = \{x \in U \mid \text{rank } \mathbf{D}f(x) < n\}$ the set of critical points of f. We shall show that f(C) has measure zero in $\mathbb{R}^n$. If $m = 0$, then $\mathbb{R}^m$ is one point and the theorem is trivially true. Suppose inductively the theorem holds for $m - 1$.

Let $C_i = \{x \in U \mid \mathbf{D}^j f(x) = 0 \text{ for } j = 1, ..., i\}$, and write C as the following union of disjoint sets:

$$C = (C \setminus C_1) \cup (C_1 \setminus C_2) \cup \cdots \cup (C_{k-1} \setminus C_k) \cup C_k.$$

The proof that f(C) has measure zero is divided in three steps.

1 $f(C_k)$ has measure zero.

2 $f(C \setminus C_1)$ has measure zero.

3 $f(C_s \setminus C_{s+1})$ has mesure zero, where $1 \le s \le k - 1$.

Proof of Step 1 Since $k \ge 1$, $kn \ge n + k - 1$. But $k \ge m - n + 1$, so that $kn \ge m$.

Let $K \subset U$ be a closed cube with edges parallel to the coordinate axes. We will show that $f(C_k \cap K)$ has measure zero. Since C_k can be covered by countably many such cubes, this will prove that $f(C_k)$ has measure zero. By Taylor's theorem, the compactness of K, and the definition of C_k, we have

$$f(y) = f(x) + R(x, y) \quad \text{where} \quad \| R(x, y) \| \le M \| y - x \|^{k+1} \tag{1}$$

for $x \in C_k \cap K$ and $y \in K$. Here M is a constant depending only on $\mathbf{D}^k f$ and K. Let e be the length of the edge of K. Choose an integer ℓ, subdivide K into ℓ^m cubes with edge e/2, and choose any cube K' of this subdivision which intersects C_k. For $x \in C_k \cap K'$ and $y \in K'$, we have $\| x - y \| \le \sqrt{m}\,(e/\ell)$. By (1), $f(K') \subset L$ where L is the cube of edge $N\ell^{k-1}$ with center f(x); $N = 2M((m)^{1/2}\ell)^{k+1}$. The volume of L is $N^n \ell^{-n(k+1)}$. There are at most ℓ^m such cubes; hence, $f(C_k \cap K)$ is contained in a union of cubes whose total volume V satisfies

$$V \le N^n \ell^{m-n(k+1)}.$$

Since $m \le kn$, $m - n(k + 1) < 0$, so $V \to 0$ as $\ell \to \infty$, and thus $f(C_k \cap K)$ has measure zero.

Proof of Step 2 Write $C \setminus C_1 = \{x \in U \mid 1 \le \text{rank } \mathbf{D}f(x) < n\} = K_1 \cup \cdots \cup K_{n-1}$, where

$$K_q = \{x \in U \mid \text{rank } \mathbf{D}f(x) = q\}$$

and it suffices to show that $f(K_q)$ has measure zero for $q = 1, \ldots, n-1$. Since K_q is empty for $q > m$, we may assume $q \le m$. As before it will suffice to show that each point K_q has a neighborhood V such that $f(V \cap K_q)$ has measure zero.

Choose $x_0 \in K_q$. By the local representation theorem **2.5.14** we may assume that x_0 has a neighborhood $V = V_1 \times V_2$, where $V_1 \subset \mathbb{R}^q$ and $V_2 \subset \mathbb{R}^{m-q}$ are open balls, such that for $t \in V_1$ and $x \in V_2$, $f(t, x) = (t, \eta(t, x))$. Hence $\eta : V_1 \times V_2 \to \mathbb{R}^{n-q}$ is a C^k map. For $t \in V_1$ define $\eta_t : V_2 \to \mathbb{R}^{n-q}$ by $\eta_t(x) = \eta(t, x)$ for $x \in V_2$. Then for every $t \in V_1$,

$$K_q \cap (\{t\} \times V_2) = \{t\} \times \{x \in V_2 \mid \mathbf{D}\eta_t(x) = 0\}.$$

This is because, for $(t, x) \in V_1 \times V_2$, $\mathbf{D}f(t, x)$ is given by the matrix

$$\mathbf{D}f(t, x) = \begin{bmatrix} I_q & 0 \\ * & \mathbf{D}\eta_t(x) \end{bmatrix}.$$

Hence rank $\mathbf{D}f(t, x) = q$ iff $\mathbf{D}\eta_t(x) = 0$.

Now η_t is C^k and $k \ge m - n = (m - q) - (n - q)$. Since $q \ge 1$, by induction we find that the critical values of η_t, and in particular $\eta_t(\{x \in V_2 \mid \mathbf{D}\eta_t(x) = 0\})$, has measure zero for each $t \in V_2$. By Fubini's lemma, $f(K_q \cap V)$ has measure zero. Since K_q is covered by countably many such V, this shows that $f(K_q)$ has measure zero.

Proof of Step 3 To show $f(C_s \setminus C_{s+1})$ has measure zero, it suffices to show that every $x \in C_s \setminus C_{s+1}$ has a neighborhood V such that $f(C_s \cap V)$ has measure zero; then since $C_s \setminus C_{s+1}$ is covered by countably many such neighborhoods V, it follows that $f(C_s \setminus C_{s+1})$ has measure zero.

Choose $x_0 \in C_s \setminus C_{s+1}$. All the partial derivatives of f at x_0 of order less than or equal to s are zero, but some partial derivative of order $s + 1$ is not zero. Hence we may assume that $\mathbf{D}_1 w(x_0) \ne 0$ and $w(x_0) = 0$, where $\mathbf{D}_1$ is the partial derivative with respect to x_1 and that w has the form

$$w(x) = \mathbf{D}_{i(1)} \cdots \mathbf{D}_{i(s)} f(x).$$

Define $h : U \to \mathbb{R}^m$ by

$$h(x) = (w(x), x_2, \ldots, x_m),$$

where $x = (x_1, x_2, \ldots, x_m) \in U \subset \mathbb{R}^m$. Clearly h is C^{k-s} and $\mathbf{D}h(x_0)$ is nonsingular; hence there is an open neighborhood V of x_0 and an open set $W \subset \mathbb{R}^m$ such that $h : V \to W$ is a C^{k-s} diffeomorphism. Let $A = C_s \cap V$, $A' = h(A)$ and $g = h^{-1}$. We would like to consider the function $f \circ g$ and then arrange things such that we can apply the inductive hypothesis to it. If $k = \infty$, there is no trouble. But if $k < \infty$, then $f \circ g$ is only C^{k-s} and the inductive hypothesis would not apply anymore. However, all we are really interested in is that some C^k function $F : W \to \mathbb{R}^n$ exists such that $F(x) = (f \circ g)(x)$ for all $x \in A'$ and $\mathbf{D}F(x) = 0$ for all $x \in A'$. The existence of such a function is guaranteed by the Kneser-Glaeser rough composition theorem (Abraham and Robbin [1969]). For $k = \infty$, we take $F = f \circ g$. In any case, define the open set $W_0 \subset \mathbb{R}^{m-1}$ by

$$W_0 = \{(x_2, ..., x_m) \in \mathbb{R}^{m-1} \mid (0, x_2, ..., x_n) \in W\}$$

and $F_0 : W_0 \to \mathbb{R}^m$ by

$$F_0(x_2, ..., x_m) = F(0, x_2, ..., x_m)$$

Let $S = \{(x_2, ..., x_m) \in W_0 \mid DF_0(x_2, ..., x_m) = 0\}$.

By the induction hypothesis, $F_0(S)$ has measure zero. But $A' = h(C_s \cap V) \subset 0 \times S$ since for $x \in A'$, $DF(x) = 0$ and since for $x \in C_s \cap V$,

$$h(x) = (w(x), x_2, ..., x_m) = (0, x_2, ..., x_m)$$

because w is an s-th deriviative of f. Hence

$$f(C_s \cap V) = F(h(C_s \cap V)) \subset F(0 \times S) = F_0(S),$$

and so $f(C_s \cap V)$ has measure zero. As $C_s \setminus C_{s+1}$ is covered by countably many such V, the sets $f(C_s \setminus C_{s+1})$ have measure zero $(s = 1, ..., k - 1)$. ■

The smoothness assumption $k \geq 1 + \max(0, m-n)$ cannot be weakened as the following counterexample shows.

3.6.7 Example Devil's Staircase Phenomenon The ***Cantor set*** C is defined by the following construction. Remove the open interval]–1/3, 2/3[from the closed interval [0, 1]. Then remove the middle thirds]1/9, 2/9[and]7/9, 8/9[from the closed intervals [0, 1/3] and [2/3, 1] respectively and continue this process of removing the middle third of each remaining closed interval indefinitely. The set C is the remaining set. Since we have removed a (countable) union of open intervals, C *is closed.* The total length of the removed intervals equals $(1/3)\sum_{n\geq 0}(2/3)^n = 1$ and thus C has measure zero in [0, 1]. On the other hand each point of C is approached arbitrarily closely by a sequence of endpoints of the intervals removed, i.e., each point of C is an accumulation point of $[0, 1] \setminus C$. Each open subinterval of [0, 1] has points in common with at least one of the deleted intervals which means that the union of all these deleted intervals is dense in [0, 1]. Therefore C *is nowhere dense.* Expand each number x in [0, 1] in a ternary expansion $0.a_1a_2 ...$ i.e., $x = \sum_{n\geq 0} 3^{-n}a_n$, where $a_n = 0, 1$, or 2. Then it is easy to see that C *consists of all numbers whose ternary expansion involves only* 0 *and* 2. (The number 1 equals 0.222... .) Thus C is in bijective correspondence with all sequences valued in a two-point set, i.e., the cardinality of C is that of the continuum; i.e., C is uncountable.

We shall construct a C^1 function $f : \mathbb{R}^2 \to \mathbb{R}$ which is not C^2 and which contains [0, 2] among its critical values. Since the measure of this set equals 2 this contradicts the conclusion of

Sard's theorem. Note, however, that there is no contradiction with the statement of Sard's theorem since f is only C^1. We start the construction by noting that the set $C + C = \{x + y \mid x, y \in C\}$ equals $[0, 2]$. The reader can easily be convinced of this fact by expanding every number in $[0, 2]$ in a ternary expansion and solving the resulting undetermined system of infinitely many equations. (The number 2 equals 1.222.... .) Assume that we have constructed a C^1-function $g : \mathbb{R} \to \mathbb{R}$ which is not C^2 and which contains C among its critical values. The function $f(x, y) = g(x) + g(y)$ is C^1, is not C^2 and if $c_1, c_2 \in C$, then there are critical points $x_1, x_2 \in [0, 1]$ such that $g(x_i) = c_i$, $i = 1, 2$; i.e., (x_1, x_2) is a critical point of f and its critical value is $c_1 + c_2$. Since $C + C = [0, 2]$, the set of critical values of f contains $[0, 2]$.

We proceed to the construction of a function $g : \mathbb{R} \to \mathbb{R}$ having all points of C as critical values. At the k-th step in the construction of C, we delete 2^{k-1} open intervals, each of length 3^{-k}. On these 2^{k-1} intervals, construct (smooth) congruent bump functions of height 2^{-k} and area (const.) $2^{-k}3^{-k}$ (Fig. 3.6.1).

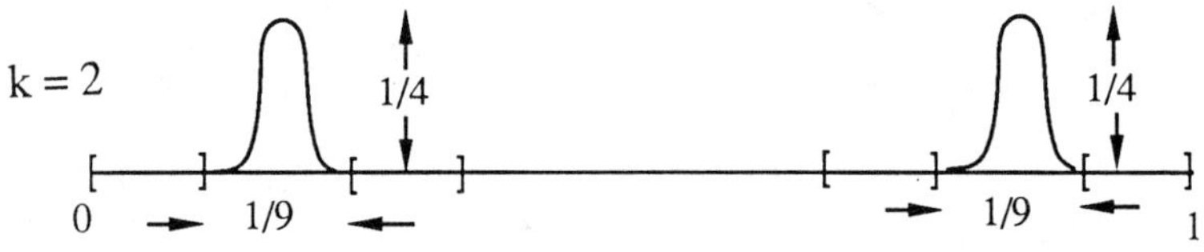

Figure 3.6.1

These define a smooth function h_k; let g_k be the integral from $-\infty$ to x of h_k, so $g_k' = h_k$ and g_k is smooth. At each endpoint of the intervals, h_k vanishes, i.e., the finite set of endpoints occurring in the k-th step of the construction of C is among the critical values of g_k. It is easy to see that $h = \Sigma_{k\geq 1} h_k$ is a uniformly Cauchy series and that $g = \Sigma_{k\geq 1} g_k$ is pointwise Cauchy; note that g_k is monotone and $g_k(1) - g_k(0) = (\text{const.})\, 3^{-k}$. Therefore g defines a C^1 function with $g' = h$. The reader can convince himself that h has arbitrarily steep slopes so that g is not C^2. The above example was given by Grinberg [1981]. Other examples of this sort are due to Whitney [1935] and Kaufman [1979]. ♦

We proceed to the global version of Sard's theorem on finite-dimensional manifolds. Recall that a subset of a topological space is ***residual*** if it is the countable intersection of open dense sets. The Baire category theorem **1.7.3** asserts that a residual subset of a a locally compact space or of a complete pseudometric space is dense. A topological space is called ***Lindelöf*** if every open covering has a countable subcovering. In particular, second countable topological spaces are Lindelöf. (See **1.1.6**.)

3.6.8 Sard's Theorem for Manifolds *Let* M *and* N *be finite-dimensional* C^k *manifolds,* $\dim(M) = m$, $\dim(N) = n$ *and* $f : M \to N$ *a* C^k *mapping,* $k \geq 1$. *Assume* M *is Lindelöf and* $k > \max(0, m - n)$. *Then* $\mathcal{R}_f$ *is residual and hence dense in* N.

Proof Denote by C the set of critical points of f. We will show that every $x \in M$ has a neighborhood Z such that $\mathcal{R}_f \mid \mathrm{cl}(Z)$ is open and dense. Then, since M is Lindelöf we can find a countable cover $\{Z_i\}$ of X with $\mathcal{R}_f \mid \mathrm{cl}(Z_i)$ open and dense. Since $\mathcal{R}_f = \bigcap_i \mathcal{R}_f \mid \mathrm{cl}(Z_i)$, it will follow that $\mathcal{R}_f$ is residual.

Choose $x \in M$. We want a neighborhood Z of x with $\mathcal{R}_f \mid \mathrm{cl}(Z)$ open and dense. By taking local charts we may assume that M is an open subset of $\mathbb{R}^m$ and $N = \mathbb{R}^n$. Choose an open neighborhood Z of x such that cl(Z) is compact. Then $C = \{x \in M \mid \operatorname{rank} \mathbf{D}f(x) < n\}$ is closed, so $\mathrm{cl}(Z) \cap C$ is compact, and hence $f(\mathrm{cl}(Z) \cap C)$ is compact. But $f(\mathrm{cl}(Z) \cap C)$ is a subset of the set of critical values of f and hence, by Sard's theorem in $\mathbb{R}^n$, has measure zero. A closed set of measure zero is nowhere dense; hence $\mathcal{R}_f \mid \mathrm{cl}(Z) = \mathbb{R}^n \setminus f(\mathrm{cl}(Z) \cap C)$ is open and dense. ■

We leave it to the reader to show that the concept of measure zero makes sense on an n-manifold and to deduce that the set of critical values of f has measure zero in N.

To consider the infinite-dimensional version of Sard's theorem, we first analyze the regular points of a map.

3.6.9 Lemma *The set* SL(E, F) *of linear continuous split surjective maps is open in* L(E, F).

Proof Choose $A \in SL(E, F)$, write $E = F \oplus K$ where K is the kernel of A, and define $A' : E \to F \times K$ by $A'(e) = (A(e), p(e))$ where $p : E = F \oplus K \to K$ is the projection. By the closed graph theorem, p is continuous; hence $A' \in GL(E, F \times K)$. Consider the map $T : L(E, F \times K) \to L(E, F)$ given by $T(B) = \pi \circ B \in L(E, F \times K)$, where $\pi : F \times K \to F$ is the projection. Then T is linear, continuous $(\|\pi \circ B\| \leq \|\pi\| \, \|B\|)$, and surjective; hence, by the open mapping theorem, T is an open mapping. Since $GL(E, F \times K)$ is open in $L(E, F \times K)$, it follows that $T(GL(E, F \times K))$ is open in L(E, F). But $A = T(A')$ and $T(GL(E, F \times K)) \subset SL(E, F)$. This shows that SL(E, F) is open. ■

3.6.10 Proposition *Let* $f : M \to N$ *be a* C^1 *mapping of manifolds. Then the set of regular points is open in* M. *Consequently the set of critical points of* f *is closed in* M.

Proof It suffices to prove the proposition locally. Thus, if E, F are the model spaces for M and N, respectively, and $x \in U \subset E$ is a regular point of f, then $\mathbf{D}f(x) \in SL(E, F)$. Since $\mathbf{D}f : U \to L(E, F)$ is continuous, $(\mathbf{D}f)^{-1}(SL(E, F))$ is open in U by lemma **3.6.9**. ■

3.6.11 Corollary *Let* $f : M \to N$ *be* C^1 *and* P *a submanifold of* N. *The set* $\{m \in M \mid f$ *is transversal to* P *at* $m\}$ *is open in* M.

Proof Assume f is transversal to P at $m \in M$. Choose a submanifold chart (V, φ) at $f(m) \in P$, $\varphi : V \to F_1 \times F_2$, $\varphi(V \cap P) = F_1 \times \{0\}$. Hence if $\pi : F_1 \times F_2 \to F_2$ is the canonical projection, $V \cap P = \varphi^{-1}(F_1 \times \{0\}) = (\pi \circ \varphi)^{-1}\{0\}$. Clearly, $\pi \circ \varphi : V \cap P \to F_2$ is a submersion so that by **3.5.4**, $\ker T_{f(m)}(\pi \circ \varphi) = T_{f(m)}P$. Thus f is transversal to P at the poing $f(m)$ iff $T_{f(p)}N = \ker T_{f(m)}(\pi \circ \varphi) + T_{f(p)}P$ and $(T_m f)^{-1}(T_{f(m)}P) = \ker T_m(\pi \circ \varphi \circ f)$ splits in T_mM. Since $\varphi \circ \pi$ is a submersion this is equivalent to $\pi \circ \varphi \circ f$ being submersive at $m \in M$ (see Exercise **2.2E**). From Proposition **3.6.10**, the set where $\pi \circ \varphi \circ f$ is submersive is open in U, hence in M, where U is a chart domain such that $f(U) \subset V$. ■

3.6.12 Example If M and N are Banach manifolds, the Sard theorem is false without further assumptions. The following couterexample is, so far as we know, due to Bonic, Douady, and Kupka. Let $E = \{x = (x_1, x_2, ...) \mid x_i \in \mathbb{R}, \| x \|^2 = \sum_{j\geq 1}(x_j / j)^2 < \infty\}$, which is a Hilbert space with respect to the usual algebraic operations on components and the inner product $\langle x, y\rangle = \sum_{j\geq 1} x_j y_j / j^2$. Consider the map $f : E \to \mathbb{R}$ given by $f(x) = \sum_{j\geq 1}(-2x_j^3 + 3x_j^2)/2$, which is defined since $x \in E$ implies $| x_i | < c$ for some $c > 0$ and thus

$$\left| \frac{-2x_j^3 + 3x_j^2}{2^j} \right| \leq \frac{2c^3j^3 + 3c^2j^2}{2^j} < \frac{c'j^3}{2^j};$$

i.e., the series $f(x)$ is majorized by the convergent series $c'\sum_{j\geq 1} j^3/2^j$. We have $\mathbf{D}f(x) \cdot v = \sum_{j\geq 1} 6(-x_j^2 + x_j)v_j / 2^j$; i.e., f is C^1. In fact f is C^∞. Moreover, $\mathbf{D}f(x) = 0$ iff all coefficients of v_j are zero, i.e., iff $x_j = 0$ or $x_j = 1$. Hence the set of critical points is $\{x \in E \mid x_j = 0 \text{ or } 1\}$ so that the set of critical values is

$$\{f(x) \mid x_j = 0 \text{ or } x_j = 1\} = \left\{ \sum_{j=1}^{\infty} \frac{x_j}{2^j} \,\middle|\, x_j = 0 \text{ or } x_j = 1 \right\} = [0, 1].$$

But clearly $[0, 1]$ has measure one. ♦

Sard's theorem holds, however, if enough restrictions are imposed on f. The generalization we consider is due to Smale [1965]. The class of linear maps allowed are Fredholm operators which have splitting properties similar to those in the Fredholm alternative theorem.

3.6.13 Definition *Let* E *and* F *be Banach spaces and* $A \in L(E, F)$. *Then* A *is called a* ***Fredholm operator*** *if:*

(i) A *is double splitting; i.e., both the kernel and the image of* A *are closed and have closed complement;*

(ii) *the kernel of* A *is finite dimensional;*

(iii) *the range of* A *has finite codimension.*

In this case, if $n = \dim(\ker A)$ *and* $p = \operatorname{codim}(\operatorname{range}(A))$, $\operatorname{index}(A) := n - p$ *is the* ***index*** *of* A. *If*

*M and N are C^1 manifolds and $f : M \to N$ is a C^1 map, we say f is a **Fredholm map** if for every $x \in M$, $T_x f$ is a Fredholm operator.*

Condition (i) follows from (ii) and (iii); see exercises **2.2H** and **2.2N**. A map g between topological spaces is called ***locally closed*** if every point in the domain of definition of g has an open neighborhood U such that $g \mid cl(U)$ is a closed map (i.e., maps closed sets to closed sets).

3.6.14 Lemma *A Fredholm map is locally closed.*

Proof By the local representative theorem we may suppose our Fredholm map has the form $f(e, x) = (e, \eta(e, x))$, for $e \in D_1$, and $x \in D_2$, where $f : D_1 \times D_2 \to E \times \mathbb{R}^p$ and $D_1 \subset E$, $D_2 \subset \mathbb{R}^n$ are open unit balls. Let U_1 and U_2 be open balls with $cl(U_1) \subset D_1$ and $cl(U_2) \subset D_2$. Let $U = U_1 \times U_2$ so that $cl(U) = cl(U_1) \times cl(U_2)$. Then $f \mid cl(U)$ is closed. To see this, suppose $A \subset cl(U)$ is closed; to show $f(A)$ is closed, choose a sequence $\{(e_i, y_i)\}$ such that $(e_i, y_i) \to (e, y)$ as $i \to \infty$ and $(e_i, y_i) \in f(A)$, say $(e_i, y_i) = f(e_i, x_i)$, where $(e_i, x_i) \in A$. Since $x_i \in cl(U_2)$ and $cl(U_2)$ is compact, we may assume $x_i \to x \in cl(U_2)$. Then $(e_i, x_i) \to (e, x)$. Since A is closed, $(e, x) \in A$, and $f(e, x) = (e, y)$, so $(e, y) \in f(A)$. Thus $f(A)$ is closed. ■

3.6.15 The Smale-Sard Theorem *Let M and N be C^k manifolds with M Lindelöf and assume that $f : M \to N$ is a C^k Fredholm map, $k \geq 1$. Suppose that $k > \max(0, \text{index}(T_x f))$ for every $x \in M$. Then $\mathcal{R}_f$ is a residual subset of N.*

Proof It suffices to show that every $x_0 \in M$ has a neighborhood Z such that $\mathcal{R}(f \mid Z)$ is open and dense in Z.

Choose $z \in M$. We shall construct a neighborhood Z of z so that $\mathcal{R}(f \mid Z)$ is open and dense. By the local representation theorem we may choose charts (U, α) at z and (V, β) at $f(z)$ such that $\alpha(U) \subset E \times \mathbb{R}^n$, $\beta(V) \subset E \times \mathbb{R}^p$ and the local representative $f_{\alpha\beta} = \beta \circ f \circ \alpha^{-1}$ of f has the form $f_{\alpha\beta}(e, x) = (e, \eta(e, x))$ for $(e, x) \in \alpha(U)$. (Here $x \in \mathbb{R}^n$, $e \in E$, and $\eta : \alpha(U) \to \mathbb{R}^p$.) The index of $T_z f$ is $n - p$ and so $k > \max(0, n - p)$ by hypothesis.

We now show that $\mathcal{R}(f \mid U)$ is dense in N. Indeed it suffices to show that $\mathcal{R}(f_{\alpha\beta})$ is dense in $E \times \mathbb{R}^p$. For $e \in E$, $(e, x) \in \alpha(U)$, define $\eta_e(x) = \eta(e, x)$. Then for each e, η_e is a C^k map defined on an open set of $\mathbb{R}^n$. By Sard's theorem, $\mathcal{R}(\eta_e)$ is dense in $\mathbb{R}^n$ for each $e \in E$. But for $(e, x) \in \alpha(U) \subset E \times \mathbb{R}^n$, we have

$$\mathbf{D}f_{\alpha\beta}(e, x) = \begin{bmatrix} I & 0 \\ * & \mathbf{D}\eta_e(x) \end{bmatrix}$$

so $\mathbf{D}f_{\alpha\beta}(e, x)$ is surjective iff $\mathbf{D}\eta_e(x)$ is surjective. Thus for $e \in E$

$$\{e\} \times \mathcal{R}(\eta_e) = \mathcal{R}(f_{\alpha\beta}) \cap (\{e\} \times \mathbb{R}^p)$$

and so $\mathcal{R}(f_{\alpha\beta})$ intersects every plane $\{e\} \times \mathbb{R}^p$ in a dense set and is, therefore, dense in $E \times \mathbb{R}^p$, by **3.6.2**. Thus $\mathcal{R}(f \mid U)$ is dense as claimed.

By lemma **3.6.14** we can choose an open neighborhood Z of z such that $\mathrm{cl}(Z) \subset U$ and $f \mid \mathrm{cl}(Z)$ is closed. By Proposition **3.6.10** the set C of critical points of f is closed in M. Hence, $f(\mathrm{cl}(Z) \cap C)$ is closed in N and so $\mathcal{R}(f \mid \mathrm{cl}(Z)) = N \setminus f(\mathrm{cl}(Z) \cap C)$ is open in N. Since $\mathcal{R}(f \mid U) \subset \mathcal{R}(f \mid \mathrm{cl}(Z))$, this latter set is also dense.

We have shown that every point z has an open neighborhood Z such that $\mathcal{R}(f \mid \mathrm{cl}(Z))$ is open and dense in Z. Repeating the argument of **3.6.8** shows that $\mathcal{R}_f$ is residual (recall that M is Lindelöf). ■

Sard's theorem deals with the genericity of the surjectivity of the derivative of a map. We now address the dual question of genericity of the injectivity of the derivative of a map.

3.6.16 Lemma *The set* $\mathbb{IL}(E, F)$ *of linear continuous split injective maps is open in* $L(E, F)$.

Proof Let $A \in \mathbb{IL}(E, F)$. Then $A(E)$ is closed and $F = A(E) \oplus G$ for G a closed subspace of F. The map $\Gamma : E \times G \to F$; defined by $\Gamma(e, g) = A(e) + g$ is clearly linear, bijective, and continuous, so by Banach's isomorphism theorem $\Gamma \in GL(E \times G, F)$. The map $P : L(E \times G, F) \to L(E, F)$ given by $P(B) = B \mid E$ is linear, continuous, and onto, so by the open mapping theorem it is also open. Moreover $P(\Gamma) = A$ and $P(GL(E \times G, F)) \subset \mathbb{IL}(E, F)$ for if $B \in GL(E \times G, F)$ then $F = B(E) \oplus B(G)$ where both $B(E)$ and $B(G)$ are closed in F. Thus A has an open neighborhood $P(GL(E \times G, F))$ contained in $\mathbb{IL}(E, F)$. ■

3.6.17 Proposition *Let* $f : M \to N$ *be a* C^1*-map of manifolds. The set* $P = \{x \in M \mid f$ *is an immersion at* $x\}$ *is open in* M.

Proof It suffices to prove the proposition locally. If E and F are the models of M and N respectively and if $f : U \to E$ is immersive at $x \in U \subset E$, then $\mathbf{D}f(x) \in \mathbb{IL}(E, F)$. By lemma **3.6.16**, $(\mathbf{D}f)^{-1}(\mathbb{IL}(E, F))$ is open in U since $\mathbf{D}f : U \to L(E, F)$ is continuous. ■

The analog of the openness statements in proposition **3.6.10** and **3.6.17** for subimmersions follows from definition **3.5.15**. Indeed, if $f : M \to N$ is a C^1 map which is a subimmersion at $x \in M$, then there is an open neighborhood U of x, a manifold P, a submersion $s : U \to P$, and an immersion $j : P \to N$ such that $f \mid U = j \circ s$. But this says that f is subimmersive at every point of U. Thus we have the following.

3.6.18 Proposition *Let* $f : M \to N$ *be a* C^1*-map of manifolds. Then the set* $P = \{x \in M \mid f$ *is a subimmersion at* $x\}$ *is open in* M.

If M or N are finite dimensional then $P = \{x \in M \mid \text{rank } T_x f \text{ is locally constant}\}$ by proposition **3.5.16**. Lower semicontinuity of the rank (i.e., each point $x \in M$ admits an open neighborhood of U such that rank $T_yf \geq$ rank T_xf for all $y \in U$; see exercise **2.5I**(i)) implies that P is dense. Indeed, if V is any open subset of M, by lower semicontinuity $\{x \in V \mid \text{rank } T_xf \text{ is maximal}\}$ is open in V and obviously contained in P. Thus we have proved the following.

3.6.19 Proposition *Let* $f : M \to N$ *be a* C^1*-map of manifolds where at least one of* M *or* N *are finite dimensional. Then the set* $P = \{x \in M \mid f \text{ is a subimmersion at } x\}$ *is dense in* M.

3.6.20 Corollary *Let* $f : M \to N$ *be a* C^1 *injective map of manifolds and let* $\dim(M) = m$. *Then the set* $P = \{x \in M \mid f \text{ is immersive at } x\}$ *is open and dense in* M. *In particular, if* $\dim(N) = n$, *then* $m \leq n$.

Proof By **3.6.18** and **3.6.19**, it suffices to show that if $f : M \to N$ is a C^1-injective map which is subimmersive at x, then it is immersive at x. Indeed, if $f \mid U = j \circ s$ where U is an open neighborhood of x on which j is injective, then the submersion s must also be injective. Since submersions are locally onto, this implies that s is a diffeomorphism in a neighborhood of x, i.e., f restricted to a sufficiently small neighborhood of x is an immersion. ■

There is a second proof of this corollary that is independent of proposition **3.6.19**. It relies ultimately on the existence and uniqueness of integral curves of C^1 vector fields. This material will be treated in Chapter 4, but we include this proof here for completeness.

Alternative Proof of 3.6.20 (*D. Burghelea*) It suffices to work in a local chart V. We shall use induction on k to show that $\mathrm{cl}(U_{i(1), \ldots, i(k)}) \supset V$, where

$$U_{i(1), \ldots, i(k)} = \left\{ x \in V \;\middle|\; T_xf\left(\frac{\partial}{\partial x^{i(1)}}\right), \ldots, T_xf\left(\frac{\partial}{\partial x^{i(k)}}\right) \text{ linearly independent} \right\}.$$

The case $k = n$ gives then the statement of the theorem. Note that by the preceding proposition $U_{i(1), \ldots, i(k)}$ is open in V.

The statement is obvious for $k = 1$ since if it fails T_xf would vanish on an open subset of V and thus f would be constant on V, contradicting the injectivity of f. Assume inductively that the statement for k holds; i.e., $U_{i(1), \ldots, i(k)}$ is open in V and $\mathrm{cl}(U_{i(1), \ldots, i(k)}) \supset V$. Define

$$U'_{i(k+1)} = \left\{ x \in U_{i(1), \ldots, i(k)} \;\middle|\; T_xf\left(\frac{\partial}{\partial x^{i(k+1)}}\right) \neq 0 \right\}$$

and notice that it is open in $U_{i(1), \ldots, i(k)}$ and thus in V. It is also dense in $U_{i(1), \ldots, i(k)}$ (by the case $k = 1$) and hence in V by induction. Define the following subset of $U_{i(1), \ldots, i(k+1)}$

$$U'_{i(1), \ldots, i(k+1)} = \left\{ x \in U'_{i(k+1)} \;\middle|\; T_x f\left(\frac{\partial}{\partial x^{i(1)}}\right), \ldots, T_x f\left(\frac{\partial}{\partial x^{i(k+1)}}\right) \text{ linearly indepenent} \right\}.$$

We prove that $U'_{i(1), \ldots, i(k+1)}$ is dense in $U'_{i(k+1)}$, which then shows that $\mathrm{cl}(U_{i(1), \ldots, i(k+1)}) \supset V$. If this were not the case, there would exist an open set $W \subset U'_{i(k+1)}$ such that

$$a^1(x) T_x f\left(\frac{\partial}{\partial x^{i(1)}}\right) + \cdots + a^k(x) T_x f\left(\frac{\partial}{\partial x^{i(k)}}\right) + T_x f\left(\frac{\partial}{\partial x^{i(k+1)}}\right) = 0$$

for some C^1–functions $a^1, \ldots, a^k$ nowhere zero on W. Let $c :]-\varepsilon, \varepsilon[\to W$ be an integral curve of the vector field

$$a^1 \frac{\partial}{\partial x^{i(1)}} + \cdots + a^k \frac{\partial}{\partial x^{i(k)}} + \frac{\partial}{\partial x^{i(k+1)}}.$$

Then $(f \circ c)'(t) = T_{c(t)} f(c'(t)) = 0$, so $f \circ c$ is constant on $]-\varepsilon, \varepsilon[$ contradicting injectivity of f. ■

There is no analogous result for surjective maps known to us; an example of a surjective function $\mathbb{R} \to \mathbb{R}$ which has zero derivative on an open set is given in Figure 3.6.2. However surjectivity of f can be replaced by a topological condition which then yields a result similar to the one in Corollary **3.6.20**.

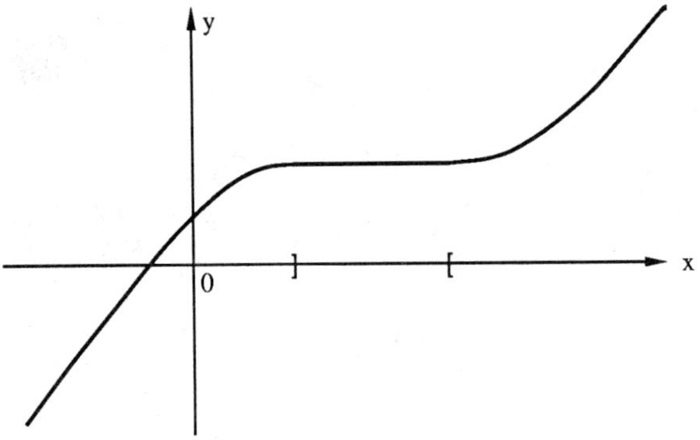

Figure 3.6.2

Corollary 3.6.21 *Let* $f : M \to N$ *be a* C^1*-map of manifolds where* $\dim(N) = n$. *If* f *is an open map, then the set* $\{x \in M \mid f$ *is a submersion at* $x\}$ *is dense in* M. *In particular, if* $\dim(M) = m$, *then* $m \geq n$.

Proof It suffices to prove that if f is a C^1-open map which is subimmersive at x, then it is submersive at x. This follows from the relation $f \mid U = j \circ s$ and the openness of f and s, for then j is necessarily open and hence a diffeomorphism by theorem **3.5.7**(iii). ■

☞ SUPPLEMENT 3.6A
An Application of Sard's Theorem to Fluid Mechanics

The Navier-Stokes equations governing homogeneous incompressible flow in a region Ω in $\mathbb{R}^3$ for a velocity field $\mathbf{u}$ are

$$\frac{\partial \mathbf{u}}{\partial t} + (\mathbf{u} \cdot \nabla)\mathbf{u} - \nu\Delta\mathbf{u} = -\nabla p + \mathbf{f} \quad \text{in } \Omega \tag{1a}$$

$$\operatorname{div} \mathbf{u} = 0 \tag{1b}$$

where $\mathbf{u}$ is parallel to $\partial\Omega$ (so fluid does not escape) and

$$\mathbf{u} = \varphi \text{ on } \partial\Omega. \tag{1c}$$

Here $\mathbf{f}$ is a given external forcing function assumed divergence free, p is the pressure (unknown), φ is a given boundary condition and ν is the viscosity. ***Stationary solutions*** are defined by setting $\partial\mathbf{u}/\partial t = 0$. Given $\mathbf{f}, \varphi$ and ν the set of possible stationary solutions $\mathbf{u}$ is denoted $S(\mathbf{f}, \varphi, \nu)$. A theorem of Foias and Temam [1977] states (amongst other things) that there is an open dense set O in the Banach space of all $(\mathbf{f}, \varphi)$'s such that $S(\mathbf{f}, \varphi, \nu)$ is finite for each $(\mathbf{f}, \varphi) \in O$.

We refer the reader to the cited paper for the precise (Sobolev) spaces needed for $\mathbf{f}, \varphi, \mathbf{u}$, and rather give the essential idea behind the proof. Let E be the space of possible $\mathbf{u}$'s (of class H^2, $\operatorname{div} \mathbf{u} = 0$ and $\mathbf{u}$ parallel to $\partial\Omega$), F the product of the space H of (L^2) divergence free vector fields with the space of vector fields on $\partial\Omega$ (of class $H^{3/2}$). We can rewrite the equation

$$(\mathbf{u} \cdot \nabla)\mathbf{u} - \nu\Delta\mathbf{u} = -\nabla p + \mathbf{f} \tag{2}$$

as

$$\nu A\mathbf{u} + B(\mathbf{u}) = \mathbf{f} \tag{3}$$

where $A\mathbf{u} = -P_H\Delta\mathbf{u}$, P_H being the orthogonal projection to H (this is a special instance of the Hodge decomposition; see **§7.5** for details) and $B(\mathbf{u}) = P_H((\mathbf{u} \cdot \nabla)\mathbf{u})$. The orthogonal projection operator really encodes the pressure term. Effectively, p is solved for by taking the divergence of (2) to give Δp in terms of $\mathbf{u}$ and the normal component of (2) gives the normal derivative of p. The resulting Neuman problem is solved, thereby eliminating p from the problem. Define the map

$$\Phi_\nu : E \to F \text{ by } \Phi_\nu(u) = (\nu Au + B(u),\ u \,|\, \partial\Omega).$$

One shows that Φ_ν is a C^∞ map by using the fact that A is a bounded linear operator and B is obtained from a continuous bilinear operator; theorems about Sobolev spaces are also required here. Moreover, elliptic theory shows that the derivative of Φ_ν is a Fredholm operator, so Φ_ν is a Fredholm map. In fact, from self adjointness of A, one sees that Φ_v has index zero.

The Sard-Smale theorem shows that the set of regular values of Φ_v forms a residual set $\mathcal{O}_v$. It is easy to see that $\mathcal{O}_v = \mathcal{O}$ is independent of v. Now since Φ_v has index zero, at a regular point, $\mathbf{D}\Phi_v$ is an isomorphism, so Φ_v is a local diffeomorphism. Thus we conclude that $S(\mathbf{f}, \varphi, v)$ is discrete and that $\mathcal{O}$ is open (Foias and Temam [1977] give a direct proof of openness of $\mathcal{O}$ rather than using the implicit function theorem). One knows, also from elliptic theory that $S(\mathbf{f}, \varphi, v)$ is compact, so being discrete, it is finite.

One can also prove a similar generic finiteness result for an open dense set of boundaries $\partial\Omega$ using a transversality analogue of the Smale-Sard theorem (see Supplement **3.6B**), as was pointed out by A. J. Tromba. We leave the precise formulation as a project for the reader.

SUPPLEMENT 3.6B
The Parametric Transversality Theorem

3.6.22 Density of Transversal Intersection *Let* P, M, N *be* C^k *manifolds,* $S \subset N$ *a submanifold (not necessarily closed) and* $F : P \times M \to N$ *a* C^k *map,* $k \geq 1$. *Assume*

(i) M *is finite dimensional* (dim M = m) *and that* S *has finite codimension* q *in* N.

(ii) $P \times M$ *is Lindelöf.*

(iii) $k > \max(0, n-q)$

(iv) $F \pitchfork S$

Then $\pitchfork(F, S) := \{p \in P \mid F_p : M \to N$ *is transversal to* S *at all points of* $S\}$ *is residual in* P.

The idea is this. Since $F \pitchfork S$, $F^{-1}(S) \subset P \times M$ is a submanifold. The projection $\pi : F^{-1}(S) \to P$ has the property that a value of π is a regular value iff F_p is transverse to S. We then apply Sard's theorem to π.

A main application is this: consider a family of perturbations $f :]-1, 1[\times M \to N$ of a given map $f_0 : M \to N$, where $f(0, x) = f_0(x)$. Suppose $f \pitchfork S$. Then there exist t's arbitrarily close to zero such that $f_t \pitchfork S$; i.e., slight perturbations of f_0 are transversal to S.

For the proof we need two lemmas.

3.6.23 Lemma *Let* E *and* F *be Banach spaces,* dim F = n, $pr_1 : E \times F \to E$ *the projection onto the first factor, and* $G \subset E \times F$ *a closed subspace of codimension* q. *Denote by* p *the restriction of* pr_1 *to* G. *Then* p *is a Fredholm operator of index* $n - q$.

Proof Let $H = G + (\{0\} \times F)$ and $K = G \cap (\{0\}) \times F)$. Since F is finite dimensional and G is closed, it follows that H is closed in $E \times F$ (see Exercise **2.2M**(ii)). Moreover, H has finite codimension since it contains the finite codimensional subspace G. Therefore H is split (see Exercise **2.2N**) and thus there exists a finite dimensional subspace $S \subset E \times \{0\}$ such that $E \times F = H \oplus S$. Since $K \subset F$, choose closed subspaces $G_0 \subset G$ and $F_0 \subset \{0\} \times F$ such that $G = G_0 \oplus K$ and $\{0\} \times F = K \oplus F_0$. Thus $H = G_0 \oplus K \oplus F_0$ and $E \times F = G_0 \oplus K \oplus F_0 \oplus S$. Note that $pr_1 \mid G_0 \oplus S : G_0 \oplus S \to E$ is an isomorphism, $K = \ker p$, and $pr_1(S)$ is a finite dimensional complement to $p(G)$ in F. Thus p is a Fredholm operator and its index equals $\dim(K) - \dim(S) = \dim(K \oplus F_0) - \dim(S \oplus F_0)$. Since $K \oplus F_0 = \{0\} \times F$ and $F_0 \oplus S$ is a complement to G in $E \times F$ (having therefore dimension q by hypothesis), the index of p equals $n - q$. ▼

3.6.24 Lemma *In the hypothesis of Theorem* **3.6.22**, *let* $V = F^{-1}(S)$. *Let* $\pi' : P \times M \to P$ *be the projection onto the first factor and let* $\pi = \pi' \mid V$. *Then* π *is a* C^k *Fredholm map of constant index* $n - q$.

Proof By (iv), V is a C^k submanifold of $P \times M$ so that π is a C^k map. The map $T_{(p,m)}\pi : T_{(p,m)}V \to T_pP$ is Fredholm of index $n - q$ by lemma **3.6.23**: E is the model of P, F the model of M, and G the model of V. ▼

Proof of 3.6.22 We shall prove below that p *is a regular value of* π *if and only if* $F_p \pitchfork S$. If this is shown, since $\pi : V \to P$ is a C^k Fredholm map of index $n - q$ by lemma **3.6.24**, the codimension of V in $E \times F$ equals the codimension of S in N which is q, $k > \max(0, n - q)$, and V is Lindelöf as a closed subspace of the Lindelöf space $P \times M$, the Smale-Sard theorem **3.6.15** implies that $\pitchfork(F, S)$ is residual in P.

By definition, (iv) is equivalent to the following statement:

(a) *for every* $(p, m) \in P \times M$ *satisfying* $F(p, m) \in S$, $T_{(m,p)}F(T_pP \times T_mM) + T_{F(p,m)}S = T_{F(p,m)}N$ *and* $(T_{(m,p)}F)^{-1}(T_{F(p,m)}S)$ *splits in* $T_pP \times T_mM$.

Since M is finite dimensional, the map $m \in M \mapsto F(p, m) \in N$ for fixed $p \in P$ is transversal to S if and only if

(b) *for every* $m \in M$ *satisfying* $F(p, m) \in S$, $T_mF_p(T_mM) + T_{F(p, m)}S = T_{F(p, m)}S$.

Since π is a Fredholm map, the kernel of $T\pi$ at any point splits being finite dimensional (see Exercise **2.2N**). Therefore p is a regular value of π if and only if

(c) *for every* $m \in M$ *satisfying* $F(p, m) \in S$ *and every* $v \in T_pP$, *there exists* $u \in T_mM$ *such that* $T_{(m,p)}F(v, u) \in T_{F(p, m)}S$.

We prove the equivalence of (b) and (c). First assume (c), take $m \in M$, $p \in P$ such that $F(p, m) \in S$ and let $w \in T_{F(p,m)}S$. By (a) there exists $v \in T_pP$, $u_1 \in T_mM$, $z \in T_{F(p,m)}S$ such that $T_{(m,p)}F(v, u_1) + z = w$. By (c) there exists $u_2 \in T_mM$ such that $T_{(m,p)}F(v, u_2) \in T_{F(p,m)}S$. Therefore,

$$\begin{aligned} w &= T_{(m,p)}F(v, u_1) - T_{(m,p)}F(v, u_2) + T_{(m,p)}F(v, u_2) + z \\ &= T_{(m,p)}F(0, u_1 - u_2) + T_{(m,p)}F(v, u_2) + z \\ &= T_{(m,p)}F(0, u) + z' \in T_mF_p(T_mM) + T_{F(p,m)}S, \end{aligned}$$

where $u = u_1 - u_2$ and $z' = T_{(m,p)}F(v, u_2) + z \in T_{F(p,m)}S$. Thus (b) holds.

Conversely, let (b) hold, take $p \in P$, $m \in M$ such that $F(p, m) \in S$ and let $v \in T_pP$. Pick $u_1 \in T_mM$, $z_1 \in T_{F(p,m)}S$ and define $w = T_{(m,p)}F(v, u_1) + z_1$. By (b), there exist $u_2 \in T_mM$ and $z_2 \in T_{F(p,m)}S$ such that $w = T_mF_p(u_2) + z_2$. Subtract these two relations to get

$$0 = T_{(m,\,p)}F(v, u_1) - T_mF_p(u_2) + z_1 - z_2 = T_{(m,\,p)}F(v, u_1 - u_2) + z_1 - z_2,$$

i.e., $T_{(m,p)}F(v, u_1 - u_2) = z_2 - z_1 \in T_{F(m,p)}S$ and therefore (c) holds. ■

There are many other very useful theorems about genericity of transversal intersection. We refer the reader to Golubitsky and Guillemin [1974], Hirsch [1976], and Kahn [1985] for the finite dimensional results and to Abraham [1963b] and Abraham-Robbin [1968] for the infinite dimensional case and the situation when P is a manifold of maps.

Exercises

3.6A Construct a C^∞ function $f : \mathbb{R} \to \mathbb{R}$ whose set of critical points equals $[0, 1]$. This shows that the set of regular points is not dense in general.

3.6B Construct a C^∞ function $f : \mathbb{R} \to \mathbb{R}$ which has each rational number as a critical value. (*Hint:* Since $\mathbb{Q}$ is countable, write it as a sequence $\{q_n \mid n = 0, 1, \ldots\}$. Construct on the closed interval $[n, n+1]$ a C^∞ function which is zero near n and $n+1$ and equal to q_n on an open interval. Define f to equal f_n on $[n, n+1]$.)

3.6C Show that if $m < n$ there is no C^1 map of an open set of $\mathbb{R}^m$ *onto* an open set of $\mathbb{R}^n$.

3.6D A manifold M is called ***C^k-simply connected***, if it is connected and if every C^k map $f : S^1 \to M$ is C^k-homotopic to a constant, i.e., there exist a C^k-map $H :]-\varepsilon, 1 + \varepsilon[\times S^1 \to M$ such that for all $s \in S^1$, $H(0, s) = f(s)$ and $H(1, s) = m_0$, where $m_0 \in M$.

(i) Show that the sphere S^n, $n \geq 2$, is C^k-simply connected for any $k \geq 1$. (*Hint:* By Sard, there exists a point $x \in S^n \setminus f(S^1)$. Then use the stereographic projection defined by x.)

(ii) Show that S^n, $n \geq 2$, is C^0-simply connected. (*Hint:* Approximate the continuous map $g : S^1 \to S^n$ by a C^1-map $f : S^1 \to S^n$. Show that one can choose f to be homotopic to g.)

(iii) Show that S^1 is not simply connected.

3.6E Let M and N be submanifolds of $\mathbb{R}^n$. Show that the set $\{x \in \mathbb{R}^n \mid M$ intersects $N + x$ transversally$\}$ is dense in $\mathbb{R}^n$. Find an example when it is not open.

3.6F Let $f : \mathbb{R}^n \to \mathbb{R}$ be C^2 and consider for each $\mathbf{a} \in \mathbb{R}^n$ the map $f_{\mathbf{a}}(\mathbf{x}) = f(\mathbf{x}) + \mathbf{a} \cdot \mathbf{x}$. Prove that the set $\{\mathbf{a} \in \mathbb{R}^n \mid$ the matrix $[\partial^2 f_{\mathbf{a}}(x_0)/\partial x^i \partial x^j]$ is nonsingular for every critical point x_0 of $f_{\mathbf{a}}\}$ is a dense set in $\mathbb{R}^n$ which is a countable intersection of open sets. (*Hint:* Use Supplement **3.6B**; when is the map $(\mathbf{a}, \mathbf{x}) \mapsto \nabla f(\mathbf{x}) + \mathbf{a}$ transversal to $\{0\}$?)

3.6G Let M be a C^2 manifold and $f: M \to \mathbb{R}$ a C^2 map. A critical point m_0 of f is called ***non-degenerate***, if in a local chart (U, φ) at m_0, $\varphi(m_0) = 0$, $\varphi : U \to E$, the bilinear continuous map $D^2(f \circ \varphi)^{-1}(0) : E \times E \to \mathbb{R}$ is strongly non-degenerate, i.e., it induces an isomorphism of E with E^*.

(i) Show that the notion of non-degeneracy is chart independent. Functions all of whose critical poins are nondegenerate are called ***Morse functions***.

(ii) Assume M is a C^2 submanifold of $\mathbb{R}^n$ and $f: M \to \mathbb{R}$ is a C^2 function. For $\mathbf{a} \in \mathbb{R}^n$ define $f_\mathbf{a} : M \to \mathbb{R}$ by $f_\mathbf{a}(\mathbf{x}) = f(\mathbf{x}) + \mathbf{a} \cdot \mathbf{x}$. Show that the set $\{\mathbf{a} \in \mathbb{R}^n \mid f_\mathbf{a}$ is a Morse function$\}$ is a dense subset of $\mathbb{R}^n$ which is a countable intersection of open sets. Show that if M is compact, this set is open in $\mathbb{R}^n$. (*Hint:* Show first that if $\dim M = m$ and $(x^1, \ldots, x^n)$ are the coordinates of a point $x \in M$ in $\mathbb{R}^n$, there is a neighborhood of $\mathbf{x}$ in $\mathbb{R}^n$ such that m of these coordinates define a chart on M. Cover M with countably many such neighborhoods. In such a neighborhood U, consider the function $g : U \subset \mathbb{R}^n \to \mathbb{R}$ defined by $g(\mathbf{x}) = f(\mathbf{x}) + a^{m+1}x^{m+1} + \cdots + a^n x^n$. Apply Exercise **3.6F** to the map $f_{\mathbf{a}'}(\mathbf{x}) = g(\mathbf{x}) + a^1x^1 + \cdots + a^m x^m$, $\mathbf{a}' = (a^1, \ldots, a^m)$ and look at the set $S = \{\mathbf{a} \in \mathbb{R}^n \mid f_\mathbf{a}$ is not Morse on $U\}$. Consider $S \cap (\mathbb{R}^m \times \{a^{m+1}, \ldots, a^n\})$ and apply Lemma **3.6.2**,)

(iii) Assume M is a C^2–submanifold of $\mathbb{R}^n$. Show that there is a linear map $\alpha : \mathbb{R}^n \to \mathbb{R}$ such that $\alpha \mid M$ is a Morse function.

(iv) Show that the "height functions" on S^n and $\mathbb{T}^n$ are Morse functions.

3.6H Let E and F be Banach spaces. A linear map $T: E \to F$ is called *compact* if it maps bounded sets into relatively compact sets.

(i) Show that a compact map is continuous.

(ii) Show that the set $K(E, F)$ of compact linear operators from E to F is a closed subspace of $L(E, F)$.

(iii) If G is another Banach space, show that $L(F, G) \circ K(E, F) \subset K(E, G)$, and that $K(E, F) \circ L(G, E) \subset K(G, F)$.

(iv) Show that if $T \in K(E, F)$, then $T^* \in K(F^*, E^*)$.

3.6I (F. Riesz) Show that if $K \in K(E, F)$ where E and F are Banach spaces and a is a scalar (real or complex), then $T = \text{Identity} + aK$ is a Fredholm operator. (*Hint:* It suffices to prove the result for $a = -1$. Show $\ker T$ is a locally compact space by proving that $K(D) = D$, where D is the open unit ball in $\ker T$. To prove that $T(E)$ is closed and finite dimensional show that $\dim(E / T(E)) = \dim(E / T(E))^* = \dim(\ker T^*) = \dim(\ker(\text{Identity} - K^*)) < \infty$ and use Exercise **2.2H**.)

3.6J Show that there exist Fredholm operators of any index. (*Hint:* Consider the shifts $(x_1, x_2, \ldots) \mapsto (0, \ldots, 0, x_1, x_2)$ and $(x_1, x_2 \ldots) \mapsto (x_n, x_{n+1}, \ldots)$ in $\ell^2(\mathbb{R})$.)

3.6K Show that if $T \in L(E, F)$ is a Fredholm operator, then $T^* \in L(F^*, E^*)$ is a Fredholm operator and $\text{index}(T^*) = -\text{index}(T)$.

3.6L (i) Let E, F, G be Banach spaces and $T \in L(E, F)$. Assume that there are $S, S' \in L(F, E)$ such that $S \circ T - \text{Identity} \in K(E, E)$ and $T \circ S' - \text{Identity} \in K(F, F)$. Show that T is Fredholm. (*Hint:* Use Exercise **3.6I**.)

(ii) Use (i) to prove that $T \in L(E, F)$ is Fredholm if and only if there exists an operator $S \in L(F, E)$ such that $(S \circ T - \text{Identity})$ and $(T \circ S - \text{Identity})$ have finite dimensional range. (*Hint:* If T is Fredholm, write $E = \ker T \oplus F_0$, $F = T(E) \oplus F_0$ and show that $T_0 = T \mid E_0 : E_0 \to T(E)$ is a Banach space isomorphism. Define $S \in L(F, E)$ by $S \mid T(E) = T_0^{-1}$, $S \mid F_0 = 0$.)

(iii) Show that if $T \in L(E, F)$, $K \in K(E, F)$ then $T + K$ is Fredholm.

(iv) Show that if $T \in L(E, F)$, $S \in L(F, G)$ are Fredholm, then so is $S \circ T$ and that $\text{index}(S \circ T) = \text{index}(S) + \text{index}(T)$.

3.6M Let E, F be Banach spaces.

(i) Show that the set $\text{Fred}_q(E, F) = \{T \in L(E, F) \mid T \text{ is Fredholm, index}(T) = q\}$ is open in $L(E, F)$. (*Hint:* Write $E = \ker T \oplus E_0$, $F = T(E) \oplus F_0$ and define $\tilde{T} : E \oplus F_0 \to F \oplus \ker T$ by $\tilde{T}(z \oplus x, y) = (T(x) \oplus y, z)$, for $x \in E_0$, $z \in \ker T$, $y \in F_0$. Show that $\tilde{T} \in GL(E \oplus F_0, F \oplus \ker T)$. Define $\rho : L(E \oplus F_0, F \oplus \ker T) \to L(E, F)$ by $\rho(S) = \pi \circ S \circ i$, where $\pi : F \oplus \ker T \to F$ is the projection and $i : e \in E \mapsto (e, 0) \in E \oplus F_0$ is the inclusion. Show that ρ is a continuous linear surjective map and hence open. Prove $\rho(GL(E \oplus F_0, F \oplus \ker T) \subset \text{Fred}_q(E, F)$, $\rho(\tilde{T}) = T$.)

(ii) Conclude from (i) that the index map from $\text{Fred}(E, F)$ to $\mathbb{Z}$ is constant on each connected component of $\text{Fred}(E, F) = \{T \in L(E, F) \mid T \text{ is Fredholm}\}$. Show that if $E = F = \ell^2(\mathbb{R})$ and $T(t)(x_1, x_2, \ldots) = (0, tx_2, x_3, \ldots)$ then $\text{index}(T(t))$ equals 1, but $\dim(\ker(T(t))$ and $\dim(\ell^2(\mathbb{R}) / T(t)(\ell^2(\mathbb{R})))$ jump at $t = 0$.

(iii) **Homotopy invariance of the index** Show that if $\varphi : [0, 1] \to \text{Fred}(E, F)$ is continuous, then $\text{index}(\varphi(0)) = \text{index}(\varphi(1))$. (*Hint:* Let $a = \sup\{t \in [0, 1] \mid s < t \text{ implies index}(f(s)) = \text{index}(f(0))\}$. By (i) we can find $\varepsilon > 0$ such that $|b - a| < \varepsilon$ implies $\text{index}(f(b)) = \text{index}(f(a))$. Let $b = a - \varepsilon/2$ and thus $\text{index}(f(0)) = \text{index}(f(b)) = \text{index}(f(a))$. Show by contradiction that $a = 1$.)

(iv) If $T \in \text{Fred}(E, F)$, $K \in K(E, F)$, show that $\text{index}(T + K) = \text{index}(T)$. (*Hint:* $T + K(E, F)$ is connected; use (ii).)

(v) **The Fredholm alternative** Let $K \in K(E, F)$ and $a \neq 0$. Show that the equation $K(e) = ae$ has only the trivial solution iff for any $v \in E$, there exists $u \in E$ such that $K(u) = au + v$. (*Hint:* $K - a(\text{Identity})$ is injective iff $(1/a)K - (\text{Identity})$ is injective. By (iv) this happens iff $(1/a)K - (\text{Indentity})$ is onto.)

3.6N Using Exercise **3.5B**, show that the map $\pi : SL(\mathbb{R}^m, \mathbb{R}^k) \times IL(\mathbb{R}^k, \mathbb{R}^n) \to \mathbf{St}(m, n; k)$, where $k \le \min(m, n)$, defined by $\pi(A, B) = B \circ A$, is a smooth locally trivial fiber bundle with typtical fiber $GL(\mathbb{R}^k)$.

3.6O (i) Let M and N be smooth finite dimensional manifolds and let $f : M \to N$ be a C^1 bijective immersion. Show that f is a C^1 diffeomorphism. (*Hint:* If $\dim M < \dim N$, then f(M) has measure zero in N, so f could not be bijective.)

(ii) Formulate an infinite dimensional version of (i).

Chapter 4
Vector Fields and Dynamical Systems

This chapter studies vector fields and the dynamical systems they determine. The ensuing chapters will study the related topics of tensors and differential forms. A basic operation introduced in this chapter is the Lie derivative of a function or a vector field. It is introduced in two different ways, algebraically as a type of directional derivative and dynamically as a rate of change along a flow. The *Lie derivative formula* asserts the equivalence of these two definitions. The Lie derivative is a basic operation used extensively in differential geometry, general relativity, Hamiltonian mechanics, and continuum mechanics.

§4.1 *Vector Fields and Flows*

This section introduces vector fields and the flows they determine. This topic puts together and globalizes two basic ideas we learn in undergraduate calculus: the study of vector fields on the one hand and differential equations on the other.

4.1.1 Definition *Let* M *be a manifold. A* ***vector field*** *on* M *is a section of the tangent bundle* TM *of* M. *The set of all* C^r *vector fields on* M *is denoted by* $\mathfrak{X}^r(M)$ *and the* C^∞ *vector fields by* $\mathfrak{X}^\infty(M)$ *or* $\mathfrak{X}(M)$.

Thus a vector field X on a manifold M is a mapping $X : M \to TM$ such that $X(m) \in T_mM$ for all $m \in M$. In other words, a vector field assigns to each point of M a vector based (i.e., bound) at that point.

4.1.2 Example Consider the force field determined by Newton's law of gravitation. Here the manifold is $\mathbb{R}^3$ minus the origin and the vector field is

$$\mathbf{F}(x, y, x) = -\frac{mMG}{r^3}\,\mathbf{r},$$

where m is the mass of a test body, M is the mass of the central body, G is the constant of gravitation, **r** is the vector from the origin to (x, y, z), and $r = (x^2 + y^2 + z^2)^{1/2}$; see Fig. 4.1.1.♦

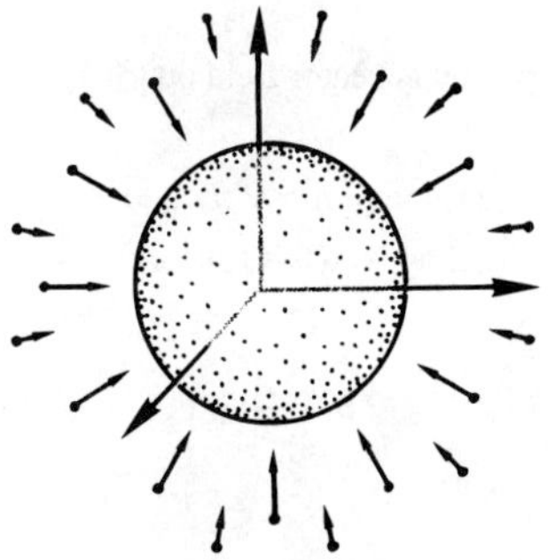

Figure 4.1.1

The study of dynamical systems, also called flows, may be motivated as follows. Consider a physical system that is capable of assuming various "states" described by points in a set S. For example, S might be $\mathbb{R}^3 \times \mathbb{R}^3$ and a state might be the position and momentum (**q**, **p**) of a particle. As time passes, the state evolves. If the state is $s_0 \in S$ at time λ and this changes to s at a later time t, we set

$$F_{t,\lambda}(s_0) = s$$

and call $F_{t,\lambda}$ the ***evolution operator***; it maps a state at time λ to what the state would be at time t; i.e., after time $t - \lambda$ has elapsed. "Determinism" is expressed by the law

$$F_{\tau,t} \circ F_{t,\lambda} = F_{\tau,\lambda} \qquad F_{t,t} = \text{identity},$$

sometimes called the ***Chapman-Kolmogorov law***.

The evolution laws are called ***time independent*** when $F_{t,\lambda}$ depends only on $t - \lambda$; i.e.,

$$F_{t,\lambda} = F_{s,\mu} \qquad \text{if} \qquad t - \lambda = s - \mu.$$

Setting $F_t = F_{t,0}$, the preceding law becomes the ***group property***:

$$F_t \circ F_\tau = F_{t+\tau}, \quad F_0 = \text{identity}.$$

We call such an F_t a ***flow*** and $F_{t,\lambda}$ a ***time-dependent flow***, or as before, an evolution operator. If the system is nonreversible, that is, defined only for $t \geq \lambda$, we speak of a ***semi-flow***.

It is usually not $F_{t,\lambda}$ that is given, but rather the ***laws of motion***. In other words, differential equations are given that we must solve to find the flow. These equations of motion have the form

$$\frac{ds}{dt} = X(s), \quad s(0) = s_0$$

where X is a (possibly time-dependent) vector field on S.

4.1.3 Example The motion of a particle of mass m under the influence of the gravitational forcc field in Example **4.1.2** is determined by Newton's second law:

$$m\frac{d^2\mathbf{r}}{dt^2} = \mathbf{F};$$

i.e., by the ordinary differentatial equations

$$m\frac{d^2x}{dt^2} = -\frac{mMGx}{r^3};$$

$$m\frac{d^2y}{dt^2} = -\frac{mMGy}{r^3};$$

$$m\frac{d^2z}{dt^2} = -\frac{mMGz}{r^3}.$$

Letting $\mathbf{q} = (x, y, z)$ denote the position and $\mathbf{p} = m(d\mathbf{r}/dt)$ the momentum, these equations become

$$\frac{d\mathbf{q}}{dt} = \frac{\mathbf{p}}{m}; \quad \frac{d\mathbf{p}}{dt} = \mathbf{F}(\mathbf{q}).$$

The phase space here is the manifold $(\mathbb{R}^3\backslash\{\mathbf{0}\}) \times \mathbb{R}^3$, i.e., the cotangent manifold of $\mathbb{R}^3\backslash\{\mathbf{0}\}$. The right-hand side of the preceding equations define a vector field on this six-dimensional manifold by $X(\mathbf{q}, \mathbf{p}) = ((\mathbf{q}, \mathbf{p}), (\mathbf{p}/m, \mathbf{F}(\mathbf{q})))$. In courses on mechanics or differential equations, it is shown how to integrate these equations explicitly, producing trajectories, which are planar conic sections. These trajectories comprise the flow of the vector field. ♦

Let us now turn to the elaboration of these ideas when a vector field X is given on a manifold M. If M = U is an open subset of a Banach space E, then a vector field on U is a map $X : U \to U \times E$ of the form $X(x) = (x, V(x))$. We call V the ***principal part*** of X. However, having a separate notation for the principal part turns out to be an unnecessary burden. By abuse of notation, in linear spaces we shall write a vector field simply as a map $X : U \to E$ and shall mean the vector field $x \mapsto (x, X(x))$. When it is necessary to be careful with the distinction, we shall be.

If M is a manifold and $\varphi : U \subset M \to V \subset E$ is a local coordinate chart for M, then a vector field X on M induces a vector field X on E called the ***local representative*** of X by the formula $\mathsf{X}(x) = T\varphi \cdot X(\varphi^{-1}(x))$. If $E = \mathbb{R}^n$ we can identify the principal part of the vector field X with an n-component vector function $(X^1(x), ..., X^n(x))$. Thus we sometimes just say "the vector field X whose local representative is $(X^i) = (X^1, ..., X^n)$."

Recall that a *curve* c at a point m of a manifold M is a C^1–map from an open interval I of $\mathbb{R}$ into M such that $0 \in I$ and $c(0) = m$. For such a curve we may assign a tangent vector at each point $c(t)$, $t \in I$, by $c'(t) = T_t c(1)$.

4.1.4 Definition *Let* M *be a manifold and* $X \in \mathfrak{X}(M)$. *An* ***integral curve*** *of* X *at* $m \in M$ *is a curve* c *at* m *such that* $c'(t) = X(c(t))$ *for each* $t \in I$.

In case $M = U \subset E$, a curve $c(t)$ is an integral curve of $X : U \to E$ when

$$c'(t) = X(c(t)),$$

where $c' = dc/dt$. If X is a vector field on a manifold M and X denotes the principal part of its local representative in a chart φ, a curve c on M is an integral curve of X when

$$\frac{d\mathsf{c}}{dt}(t) = \mathsf{X}(\mathsf{c}(t)),$$

where $\mathsf{c} = \varphi \circ c$ is the ***local representative*** of the curve c. If M is an n-manifold and the local representatives of X and c are $(X^1, ..., X^n)$ and $(c^1, ..., c^n)$ respectively, then c is an integral curve of X when the following system of ordinary differential equations is satisfied

$$\begin{aligned} \frac{dc^1}{dt}(t) &= X^1(c^1(t), \ \ldots, \ c^n(t)) \\ &\vdots \\ \frac{dc^n}{dt}(t) &= X^n(c^1(t), \ \ldots, c^n(t)). \end{aligned}$$

The reader should chase through the definitions to verify this assertion.

These equations are autonomous, corresponding to the fact that X is time independent. If X were time dependent, time t would appear explicitly on the right-hand side. As we saw in Example **4.1.3**, the preceding system of equations includes equations of higher order (by their usual reduction to first-order systems) and the Hamilton equations of motion as special cases.

4.1.5 Local Existence, Uniqueness, and Smoothness Theorem *Let* E *be a Banach space and suppose* $X : U \subset E \to E$ *is of class* C^∞. *For each* $x_0 \in U$, *there is a curve* $c : I \to U$ *at* x_0 *such that* $c'(t) = X(c(t))$ *for all* $t \in I$. *Any two such curves are equal on the intersection of their domains. Furthurmore, there is a neighborhood* U_0 *of the point* $x_0 \in U$, *a real number* $a > 0$, *and a* C^∞ *mapping* $F : U_0 \times I \to E$, *where* I *is the open interval* $]-a, a[$, *such that the curve* $c_u : I \to E$, *defined by* $c_u(t) = F(u, t)$ *is a curve at* $u \in E$ *satisfying the differential equations* $c'_u(t) = X(c_u(t))$ *for all* $t \in I$.

4.1.6 Lemma *Let* E *be a Banach space,* $U \subset E$ *an open set, and* $X : U \subset E \to E$ *a Lipschitz map; i.e. there is a constant* $K > 0$ *such that* $\| X(x) - X(y) \| \le K \| x - y \|$ *for all* $x, y \in U$. *Let* $x_0 \in U$ *and suppose the closed ball of radius* b, $B_b(x_0) = \{x \in E \mid \| x - x_0 \| \le b\}$ *lies in* U, *and* $\| X(x) \| \le M$ *for all* $x \in B_b(x_0)$. *Let* $t_0 \in \mathbb{R}$ *and let* $\alpha = \min(1/K, b/M)$. *Then there is a unique* C^1 *curve* $x(t)$, $t \in [t_0 - \alpha, t_0 + \alpha]$ *such that* $x(t) \in B_b(x_0)$ *and*

$$\begin{cases} x'(t) = X(x(t)) \\ x(t_0) = x_0 \ . \end{cases}$$

Proof The conditions $x'(t) = X(x(t))$, $x(t_0) = x_0$ are equivalent to the integral equation

$$x(t) = x_0 + \int_{t_0}^{t} X(x(s))\, ds \ .$$

Put $x_0(t) = x_0$ and define inductively

$$x_{n+1}(t) = x_0 + \int_{t_0}^{t} X(x_n(s))\, ds \ .$$

Clearly $x_n(t) \in B_b(x_0)$ for all n and $t \in [t_0 - \alpha, t_0 + \alpha]$ by definition of α. We also find by induction that

$$\| x_{n+1}(t) - x_n(t) \| \le MK^n \, | t - t_0 |^{n+1}/(n + 1)! \ .$$

Thus $x_n(t)$ converges uniformly to a continuous curve $x(t)$. Clearly $x(t)$ satisfies the integral equation and thus is the solution we sought.

For uniqueness, let $y(t)$ be another solution. By induction we find that $\| x_n(t) - y(t) \| \le MK^n \, | t - t_0 |^{n+1}/(n + 1)!$; thus, letting $n \to \infty$ gives $x(t) = y(t)$. ▼

The same argument holds if X depends explicitly on t and/or on a parameter ρ, is jointly continuous in (t, ρ, x), and is Lipschitz in x uniformly in t and ρ. Since $x_n(t)$ is continuous in (x_0, t_0, ρ) so is $x(t)$, being a uniform limit of continuous functions; thus the integral curve is jointly continuous in (x_0, t_0, ρ).

4.1.7 Gronwall's Inequality *Let* $f, g : [a, b[\to \mathbb{R}$ *be continuous and nonnegative. Suppose that for all* t *satisfying* $a \le t \le b$,

$$f(t) \le A + \int_a^t f(s)g(s)\,ds, \textit{ for a constant } A \ge 0.$$

Then

$$f(t) \le A \exp\left(\int_a^t g(s)\,ds\right) \textit{ for all } t \in [a, b[\,.$$

Proof First suppose $A > 0$. Let

$$h(t) = A + \int_a^t f(s)g(s)\,ds;$$

thus $h(t) > 0$. Then $h'(t) = f(t)g(t) \le h(t)g(t)$. Thus $h'(t)/h(t) \le g(t)$. Integration gives

$$h(t) \le A \exp\left(\int_a^t g(s)\,ds\right).$$

This gives the result for $A > 0$. If $A = 0$, then we get the result by replacing A by $\varepsilon > 0$ for every $\varepsilon > 0$; thus h and hence f is zero. ▼

4.1.8 Lemma *Let* X *be as in Lemma* **4.1.6**. *Let* $F_t(x_0)$ *denote the solution* (= *integral curve*) *of* $x'(t) = X(x(t))$, $x(0) = x_0$. *Then there is a neighborhood* V *of* x_0 *and a number* $\varepsilon > 0$ *such that for every* $y \in V$ *there is a unique integral curve* $x(t) = F_t(y)$ *satisfying* $x'(t) = X(x(t))$ *for all* $t \in [-\varepsilon, \varepsilon]$, *and* $x(0) = y$. *Moreover,*

$$\| F_t(x) - F_t(y) \| \le e^{K|t|} \| x - y \|.$$

Proof Choose $V = B_{b/2}(x_0)$ and $\varepsilon = \min(1/K, b/2M)$. Fix an arbitrary $y \in V$. Then $B_{b/2}(y) \subset B_b(x_0)$ and hence $\| X(z) \| \le M$ for all $z \in B_{b/2}(y)$. By lemma **4.1.5** with x_0 replaced by y, b by $b/2$, and t_0 by 0, there exists an integral curve $x(t)$ of $x'(t) = X(x(t))$ for $t \in [-\varepsilon, \varepsilon]$ and satisfying $x(0) = y$. This proves the first part. For the second, let $f(t) = \| F_t(x) - F_t(y) \|$. Clearly

$$f(t) = \left\| \int_0^t [X(F_s(x)) - X(F_s(y))]\,ds + x - y \right\| \le \|x - y\| + K\int_0^t f(s)\,ds,$$

so the result follows from Gronwall's inequality. ▼

This result shows that $F_t(x)$ depends in a continuous, indeed Lipschitz, manner on the initial condition x and is jointly continuous in (t, x). Again, the same result holds if X depends explicitly on t and on a parameter ρ is jointly continuous in (t, ρ, x), and is Lipschitz in x uniformly in t and ρ; $(F_{t,\lambda})^{\rho}(x)$ is the unique integral curve $x(t)$ satisfying $x'(t) = X(x(t), t, \rho)$ and $x(\lambda) = x$. By the remark following lemma **4.1.6**, $(F_{t,\lambda})^{\rho}(x)$ is jointly continuous in the variables (λ, t, ρ, x), and is Lipschitz in x, uniformly in (λ, t, ρ). The next result shows that

F_t is C^k if X is, and completes the proof of **4.1.5**. For the next lemma, recall that a C^1-function is locally Lipschitz.

4.1.9 Lemma *Let* X *in Lemma* **4.1.6** *be of class* C^k, $1 \le k \le \infty$, *and let* $F_t(x)$ *be defined as before. Then locally in* (t, x), $F_t(x)$ *is of class* C^k *in* x *and is* C^{k+1} *in the* t*-variable.*

Proof We define $\psi(t, x) \in L(E, E)$, the set of continuous linear maps of E to E, to be the solution of the "linearized" or "first variation" equations:

$$\frac{d}{dt}\psi(t, x) = DX(F_t(x)) \circ \psi(t, x), \quad \text{with} \quad \psi(0, x) = \text{identity},$$

where $DX(y) : E \to E$ is the derivative of X taken at the point y. Since the vector field $\psi \mapsto DX(F_t(x)) \circ \psi$ on L(E, E) (depending explicitly on t and on the parameter x) is Lipschitz in ψ, uniformly in (t, x) in a neighborhood of every (t_0, x_0), by the remark following **4.1.8** it follows that $\psi(t, x)$ is continuous in (t, x) (using the norm topology on L(E, E)).

We claim that $DF_t(x) = \psi(t, x)$. To show this, fix t, set $\theta(s, h) = F_s(x + h) - F_s(x)$, and write

$$\begin{aligned}\theta(t, h) - \psi(t, x) \cdot h &= \int_0^t \{X(F_s(x + h)) - X(F_s(x))\}\, ds - \int_0^t [DX(F_s(x) \circ \psi(s, x)] \cdot h\, ds \\ &= \int_0^t DX(F_s(x)) \cdot [\theta(s, h) - \psi(s, x) \cdot h]\, ds \\ &\quad + \int_0^t \{X(F_s(x + h)) - X(F_s(x)) - DX(F_s(x)) \cdot [F_s(x + h) - F_s(x)]\}\, ds\,.\end{aligned}$$

Since X is of class C^1, given $\varepsilon > 0$, there is a $\delta > 0$ such that $\| h \| < \delta$ implies the second term is dominated in norm by

$$\int_0^t \varepsilon \| F_s(x + h) - F_s(x) \|\, ds\,,$$

which is, in turn, smaller than $A\varepsilon \| h \|$ for a positive constant A by lemma **4.1.8.** By Gronwall's inequality we obtain $\| \theta(t, h) - \psi(t, x) \cdot h \| \le (\text{constant})\ \varepsilon \| h \|$. It follows that $DF_t(x) \cdot h = \psi(t, x) \cdot h$. Thus both partial derivatives of $F_t(x)$ exist and are continuous; therefore $F_t(x)$ is of class C^1.

We prove $F_t(x)$ is C^k by induction on k. Begin with the equation defining F_t:

$$\frac{d}{dt} F_t(x) = X(F_t(x))$$

so

$$\frac{d}{dt}\frac{d}{dt}F_t(x) = DX(F_t(x)) \cdot X(F_t(x))$$

and

$$\frac{d}{dt}DF_t(x) = DX(F_t(x)) \cdot DF_t(x)) \ .$$

Since the right-hand sides are C^{k-1}, so are the solutions by induction. Thus F itself is C^k. ■

Again there is an analogous result for the evolution operator $(F_{t,\lambda})^\rho(x)$ for a time-dependent vector field $X(x, t, \rho)$, which depends on extra parameters ρ in a Banach space P. If X is C^k, then $(F_{t,\lambda})^\rho(x)$ is C^k in all variables and is C^{k+1} in t and λ. The variable ρ can be easily dealt with by suspending X to a new vector field obtained by appending the trivial differential equation $\rho' = 0$; this defines a vector field on $E \times P$ and **4.1.5** may be applied to it. The flow on $E \times P$ is just $F_t(x, \rho) = (F_t^\rho(x), \rho)$.

For another more "modern" proof of **4.1.5** see Supplement **4.1C**. That alternative proof has a technical advantage: it works easily for other types of differentiability assumptions on X or on F_t, such as Hölder or Sobolev differentiability; this result is due to Ebin and Marsden [1970].

The mapping F gives a locally unique integral curve c_u for each $u \in U_0$, and for each $t \in I$, $F_t = F \mid (U_0 \times \{t\})$ maps U_0 to some other set. It is convenient to think of each point u being allowed to "flow for time t" along the integral curve c_u (see Fig. 4.1.2 and our opening motivation). This is a picture of a U_0 "flowing," and the system (U_0, a, F) is a local flow of X, or ***flow box***. The analogous situation on a manifold is given by the following.

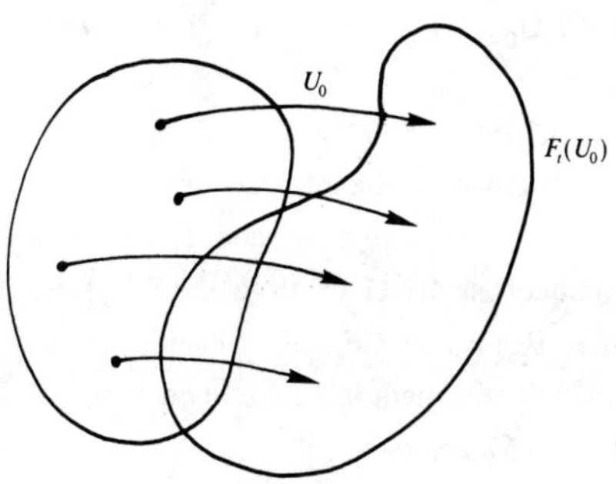

Figure 4.1.2

4.1.10 Definition *Let* M *be a manifold and* X *a* C^r *vector field on* M, $r \geq 1$. *A **flow box** of* X *at* $m \in M$ *is a triple* (U_0, a, F), *where*

(i) $U_0 \subset M$ *is open,* $m \in U_0$, *and* $a \in \mathbb{R}$, *where* $a > 0$ *or* $a = +\infty$;

(ii) $F : U_0 \times I_a \to M$ *is of class* C^r, *where* $I_a =]-a, a[$;

(iii) *for each* $u \in U_0$, $c_u : I_a \to M$ *defined by* $c_u(t) = F(u, t)$ *is an integral curve of* X *at the point* u;

(iv) *if* $F_t : U_0 \to M$ *is defined by* $F_t(u) = F(u, t)$, *then for* $t \in I_a$, $F_t(U_0)$ *is open, and* F_t *is a* C^r *diffeomorphism onto its image.*

Before proving the existence of a flow box, it is convenient first to establish the following, which concerns uniqueness.

4.1.11 Global Uniqueness *Suppose* c_1 *and* c_2 *are two integral curves of* X *at* $m \in M$. *Then* $c_1 = c_2$ *on the intersection of their domains.*

Proof This does not follow at once from **4.1.5** for c_1 and c_2 may lie in different charts. (Indeed, if the manifold is not Hausdorff, Exercise **4.1M** shows that this proposition is false.) Suppose $c_1 : I_1 \to M$ and $c_2 : I_2 \to M$. Let $I = I_1 \cap I_2$, and let $K = \{t \in I \mid c_1(t) = c_2(t)\}$; K is closed since M is Hausdorff. We will now show that K is open. From **4.1.5**, K contains some neighborhood of 0. For $t \in K$ consider $c_1{}^t$ and $c_2{}^t$, where $c^t(s) = c(t + s)$. Then $c_1{}^t$ and $c_2{}^t$ are integral curves at $c_1(t) = c_2(t)$. Again, by **4.1.5** they agree on some neighborhood of 0. Thus some neigborhood of t lies in K, and so K is open. Since I is connnected, K = I. ■

4.1.12 Proposition *Suppose* (U_0, a, F) *is a triple satisfying* (i), (ii), *and* (iii) *of* **4.1.10**. *Then for* t, s *and* $t + s \in I_a$,

$$F_{t+s} = F_t \circ F_s = F_s \circ F_t \quad \textit{and } F_0 \textit{ is the identity map,}$$

whenever the compositions above are defined. Moreover, if $U_t = F_t(U_0)$ *and* $U_t \cap U_0 \neq \varnothing$, *then* $F_t \mid U_{-t} \cap U_0 : U_{-t} \cap U_0 \to U_0 \cap U_t$ *is a diffeomorphism and its inverse is* $F_{-t} \mid U_0 \cap U_t$.

Proof $F_{t+s}(u) = c_u(t + s)$, where c_u is the integral curve defined by F at u. But $d(t) = F_t(F_s(u)) = F_t(c_u(s))$ is the integral curve through $c_u(s)$ and $f(t) = c_u(t + s)$ is also an integral curve at $c_u(s)$. Hence by global uniqueness **4.1.11** we have $F_t(F_s(u)) = c_u(t + s) = F_{t+s}(u)$. To show that $F_{t+s} = F_s \circ F_t$, observe that $F_{t+s} = F_{s+t} = F_s \circ F_t$. Since $c_u(t)$ is a curve at u, $c_u(0) = u$, so F_0 is the identity. Finally, the last statement is a consequence of $F_t \circ F_{-t} = F_{-t} \circ F_t =$ identity. Note, however, that $F_t(U_0) \cap U_0 = \varnothing$ can occur. ■

4.1.13 Existence and Uniqueness of Flow Boxes *Let* X *be a* C^r *vector field on a manifold* M. *For each* $m \in M$ *there is a flow box of* X *at* m. *Suppose* (U_0, a, F), (U_0', a', F') *are two flow boxes at* $m \in M$. *Then* F *and* F′ *are equal on* $(U_0 \cap U_0') \times (I_a \cap I_{a'})$.

Proof (Uniqueness). Again we emphasize that this does not follow at once from **4.1.5**, since U_0 and U_0' need not be chart domains. However, for each point $u \in U_0 \cap U_0'$ we have $F \mid \{u\} \times I = F' \mid \{u\} \times I$, where $I = I_a \cap I_{a'}$. This follows from **4.1.11** and **4.1.10**(iii). Hence F = F′ on the set $(U_0 \cap U_0') \times I$.

(Existence). Let (U, φ) be a chart in M with $m \in U$. It is enough to establish the result in $\varphi(U)$ by means of the local representation. Thus let (U_0', a, F') be a flow box of X, the local representative of X, at $\varphi(m)$ as given by **4.1.5**, with

$$U_0' \subset U' = \varphi(U) \quad \text{and} \quad F'(U_0' \times I_a) \subset U', \quad U_0 = \varphi^{-1}(U_0')$$

and let

$$F : U_0 \times I_a \to M; \quad (u, t) \mapsto \varphi^{-1}(F'(\varphi(u), t)).$$

Since F is continuous, there is a $b \in\,]0, a[\, \subset \mathbb{R}$ and $V_0 \subset U_0$ open, with $m \in V_0$, such that $F(V_0 \times I_b) \subset U_0$. We contend that (V_0, b, F) is a flow box at m (where F is understood as the restriction of F to $V_0 \times I_b$). Parts (i) and (ii) of **4.1.10** follow by construction and (iii) is a consequence of the remarks following **4.1.4** on the local representation. To prove (iv), note that for $t \in I_b$, F_t has a C^r inverse, namely, F_{-t} as $V_t \cap U_0 = V_t$. It follows that $F_t(V_0)$ is open. And, since F_t and F_{-t} are both of class C^r, F_t is a C^r diffeomorphism. ■

As usual, there is an analogous result for time- (or parameter-) dependent vector fields. The following result shows that near a point m satisfying $X(m) \neq 0$, the flow can be transformed by a change of variables so that the integral curves become straight lines.

4.1.14 Straightening Out Theorem *Let* X *be a vector field on a manifold* M *and suppose at* $m \in M$, $X(m) \neq 0$. *Then there is a local chart* (U, φ) *with* $m \in U$ *such that*

(i) $\varphi(U) = V \times I \subset G \times \mathbb{R} = E$, $V \subset G$, *open, and* $I =]-a, a[\, \subset \mathbb{R}$, $a > 0$;

(ii) $\varphi^{-1} | \{v\} \times I : I \to M$ *is an integral curve of* X at $\varphi^{-1}(v, 0)$, *for all* $v \in V$;

(iii) *the local representative* X *has the form* $X(y, t) = (y, t; 0, 1)$.

Proof Since the result is local, by taking any initial coordinate chart, it suffices to prove the result in E. We can arrange things so that we are working near $0 \in E$ and $X(0) = (0, 1) \in E = G \oplus \mathbb{R}$ where G is a complement to the span of $X(0)$. Letting (U_0, b, F) be a flow box for X at 0 where $U_0 = V_0 \times\,]-\varepsilon, \varepsilon[$ and V_0 is open in G, define

$$f_0 : V_0 \times I_b \to E \quad \text{by} \quad f_0(y, t) = F_t(y, 0).$$

But

$$Df_0(0, 0) = \text{Identity}$$

since

$$\left.\frac{\partial F_t(0, 0)}{\partial t}\right|_{t=0} = X(0) = (0,1) \quad \text{and} \quad F_0 = \text{Identity}.$$

By the inverse mapping theorem there are open neighborhoods $V \times I_a \subset V_0 \times I_b$ and $U = f_0(V \times I_a)$ of $(0, 0)$ such that $f = f_0 | V \times I_a : V \times I_a \to U$ is a diffeomorphism. Then $f^{-1} : U$

$\to V \times I_a$ can serve as chart for (i). Notice that $c = f \mid \{y\} \times I : I \to U$ is the integral curve of X through $(y, 0)$ for all $y \in V$, thus proving (ii). Finally, the expression of the vector field X in this local chart given by f^{-1} is $\mathbf{D}f^{-1}(y, t) \cdot X(f(y, t)) = \mathbf{D}f^{-1}(c(t)) \cdot c'(t) = (f^{-1} \circ c)'(t) = (0, 1)$, since $(f^{-1} \circ c)(t) = (y, t)$, thus proving (iii). ■

In §4.3 we shall see that singular points, where the vector field vanishes, are of great interest in dynamics. The straightening out theorem does not claim anything about these points. Instead, one needs to appeal to more sophisticated normal form theorems; see Guckenheimer and Holmes [1983].

Now we turn our attention from local flows to global considerations. These ideas center on considering the flow of a vector field as a whole, extended as far as possible in the t-variable.

4.1.15 Definition *Given a manifold* M *and a vector field* X *on* M, *let* $\mathcal{D}_X \subset M \times \mathbb{R}$ *be the set of* $(m, t) \in M \times \mathbb{R}$ *such that there is an integral curve* $c : I \to M$ *of* X *at* m *with* $t \in I$. *The vector field* X *is* ***complete*** *if* $\mathcal{D}_X = M \times \mathbb{R}$. *A point* $m \in M$ *is called* ***σ–complete***, *where* $\sigma = +, -,$ *or* $\pm$, if $\mathcal{D}_X \cap (\{m\} \times \mathbb{R})$ *contains all* (m, t) *for* $t > 0, < 0,$ *or* $t \in \mathbb{R}$, *respectively. Let* $T^+(m)$ (resp. $T^-(m)$) *denote the sup (resp. inf) of the times of existence of the integral curves through* m; $T^+(m)$ *resp.* $T^-(m)$ *is called the* ***positive (negative) lifetime of*** m.

Thus, X is complete iff each integral curve can be extended so that its domain becomes $]-\infty, \infty[$; i.e., $T^+(m) = \infty$ and $T^-(m) = -\infty$ for all $m \in M$.

4.1.16 Examples

A For $M = \mathbb{R}^2$, let X be the constant vector field, whose principal part is $(0, 1)$. Then X is complete since the integral curve of X through (x, y) is $t \mapsto (x, y + t)$.

B On $M = \mathbb{R}^2 \setminus \{0\}$, the same vector field is not complete since the inegral curve of X through $(0, -1)$ cannot be extended beyond $t = 1$; in fact as $t \to 1$ this integral curve tends to the point $(0, 0)$. Thus $T^+(0, -1) = 1$, while $T^-(0, -1) = -\infty$.

C On $\mathbb{R}$ consider the vector field $X(x) = 1 + x^2$. This is not complete since the integral curve c with $c(0) = 0$ is $c(\theta) = \tan\theta$ and thus it cannot be continuously extended beyond $-\pi/2$ and $\pi/2$; i.e., $T^{\pm}(0) = \pm\,\pi/2$. ♦

4.1.17 Proposition *Let* M *be a manifold and* $X \in \mathfrak{X}^r(M)$, $r \geq 1$. *Then*

(i) $\mathcal{D}_X \supset M \times \{0\}$;

(ii) $\mathcal{D}_X$ *is open in* $M \times \mathbb{R}$;

(iii) *there is a unique* C^r *mapping* $F_X : \mathcal{D}_X \to M$ *such that the mapping* $t \mapsto F_X(m, t)$ *is an integral curve at* m *for all* $m \in M$;

(iv) *for* $(m, t) \in \mathcal{D}_X$, $(F_X(m, t), s) \in \mathcal{D}_X$ *iff* $(m, t + s) \in \mathcal{D}_X$; *in this case*

$$F_X(m, t + s) = F_X(F_X(m, t), s).$$

Proof. Parts (i) and (ii) follow from the flow box existence theorem. In (iii), we get a unique map $F_X : \mathcal{D}_X \to M$ by the global uniqueness and local existence of integral curves: $(m, t) \in \mathcal{D}_X$ when the integral curve $m(s)$ through m exists for $s \in [0, t]$. We set $F_X(m, t) = m(t)$. To show F_X is C^r, note that in a neighborhood of a fixed m_0 and for small t, it is C^r by local smoothness. To show F_X is globally C^r, first note that (iv) holds by global uniqueness. Then in a neighborhood of the compact set $\{m(s) \mid s \in [0, t]\}$ we can write F_X as a composition of finitely many C^r maps by taking short enough time steps so the local flows are smooth. ■

4.1.18 Definition *Let* M *be a manifold and* $X \in \mathfrak{X}^r(M)$, $r \geq 1$. *Then the mapping* F_X *is called the* ***integral*** *of* X, *and the curve* $t \mapsto F_X(m, t)$ *is called the* ***maximal integral curve*** *of* X *at* m. *In case* X *is complete,* F_X *is called the* ***flow*** *of* X.

Thus, if X is complete with flow F, then the set $\{F_t \mid t \in \mathbb{R}\}$ is a group of diffeomorphisms on M, sometimes called a ***one-parameter group of diffeomorphisms.*** Since $F_n = (F_1)^n$ (the n–th power), the notation F^t is sometimes convenient and is used where we use F_t. For incomplete flows, (iv) says that $F_t \circ F_s = F_{t+s}$ wherever it is defined. Note that $F_t(m)$ is defined for $t \in\,]T^-(m), T^+(m)[$. The reader should write out similar definitions for the time-dependent case and note that the lifetimes depend on the starting time t_0.

4.1.19 Proposition *Let* X be C^r, *where* $r \geq 1$. *Let* c(t) *be a maximal integral curve of* X *such that for every finite open interval*]a, b[*in the domain* $]T^-(c(0)), T^+(c(0))[$ *of* c, c(]a, b[) *lies in a compact subset of* M. *Then* c *is defined for all* $t \in \mathbb{R}$.

Proof It suffices to show that $a \in I$, $b \in I$, where I is the inteval of definition of c. Let $T_n \in\,]a, b[$, $t_n \to b$. By compactness we can assume some subsequence $c(t_{n(k)})$ converges, say, to a point x in M. Since the domain of the flow is open, it contains a neighborhood of (x, 0). Thus, there are $\varepsilon > 0$ and $\tau > 0$ such that integral curves starting at points (such as $c(t_{n(k)})$) for large k) closer than ε to x persist for a time longer than τ. This serves to extend c to a time greater than b, so $b \in I$ since c is maximal. Similarly, $a \in I$. ■

The ***support*** of a vector field X defined on a manifold M is defined to be the closure of the set $\{m \in M \mid X(m) \neq 0\}$.

4.1.20 Corollary. *A* C^r *vector field with compact support on a manifold* M *is complete. In particular, a* C^r *vector field on a compact manifold is complete.*

Completeness corresponds to well-defined dynamics persisting eternally. In some circumstances (shock waves in fluids and solids, singularities in general relativity, etc.) one has to live with incompleteness or overcome it in some other way. Because of its importance we give two additional criteria. In the first result we use the notation $X[f] = \mathbf{d}f \cdot X$ for the derivative of f in the direction X. Here $f: E \to \mathbb{R}$ and $\mathbf{d}f$ stands for the derivative map. In standard coordinates on $\mathbb{R}^n$,

$$\mathbf{d}f(x) = \left(\frac{\partial f}{\partial x^1}, \ldots, \frac{\partial f}{\partial x^n}\right) \text{ and } X[f] = \sum_{i=1}^{n} X^i \frac{\partial f}{\partial x^i}.$$

4.1.21 Proposition *Suppose* X *is a* C^k *vector field on the Banach space* E, $k \geq 1$, *and* $f: E \to \mathbb{R}$ *is a* C^1 ***proper** map; that is, if* $\{x_n\}$ *is any sequence in* E *such that* $f(x_n) \to a$, *then there is a convergent subseqence* $\{x_{n(i)}\}$. *Suppose there are constants* $K, L \geq 0$ *such that*

$$|X[f](m)| \leq K |f(m)| + L \quad \textit{for all} \quad m \in E.$$

Then the flow of X *is complete.*

Proof From the chain rule we have $(\partial/\partial t)f(F_t(m)) = X[f](F_t(m))$, so that

$$f(F_t(m)) - f(m) = \int_0^t X[f](F_\tau(m))\, d\tau .$$

Applying the hypothesis and Gronwall's inequality we see that $|f(F_t(m))|$ is bounded and hence relatively compact on any finite t-interval, so as f is proper, a repetition of the argument in the proof of **4.1.19** applies. ■

Note that the same result holds if we replace "properness" by "inverse images of compact sets are bounded" and assume X has a uniform existence time on each bounded set. This version is useful in some infinite dimensional examples.

4.1.22 Proposition *Let* E *be a Banach space and* X *a* C^r *vector field on the Banach space* E, $r \geq 1$. *Let* σ *be any integral curve of* X. *Assume* $\|X(\sigma(t))\|$ *is bounded on finite t-intervals. Then* $\sigma(t)$ *exists for all* $t \in \mathbb{R}$.

Proof. Suppose $\|X(\sigma(t))\| < A$ for $t \in]a, b[$ and let $t_n \to b$. For $t_n < t_m$ we have

$$\|\sigma(t_n) - \sigma(t_m)\| \leq \int_{t_n}^{t_m} \|\sigma'(t)\|\, dt = \int_{t_n}^{t_m} \|X(\sigma(t))\|\, dt \leq A|t_m - t_n| .$$

Hence $\sigma(t_n)$ is a Cauchy sequence and therefore, converges. Now argue as in **4.1.19.** ■

4.1.23 Examples

A Let X be a C^r vector field, $r \geq 1$, on the manifold M admitting a ***first integral,*** i.e. a function $f: M \to \mathbb{R}$ such that $X[f] = 0$. If all level sets $f^{-1}(r)$, $r \in \mathbb{R}$ are compact, X is complete. Indeed, each integral curve lies on a level set of f so that the result follows by **4.1.19.**

B Newton's equations for a moving particle of mass m in a potential field in $\mathbb{R}^n$ are given by $\ddot{\mathbf{q}}(t) = -(1/m)\,\nabla V(\mathbf{q}(t))$, for $V : \mathbb{R}^n \to \mathbb{R}$ a smooth function. We shall prove that *if there are constants* $a, b \in \mathbb{R}$, $b \geq 0$ *such that* $(1/m)V(\mathbf{q}) \geq a - b\|\,\mathbf{q}\,\|^2$, *then every solution exists for all time*. To show this, rewrite the second order equations as a first order system $\dot{\mathbf{q}} = (1/m)\mathbf{p}$, $\dot{\mathbf{p}} = -\nabla V(\mathbf{q})$ and note that the energy $E(\mathbf{q}, \mathbf{p}) = (1/2m)\,\|\,\mathbf{p}\,\|^2 + V(\mathbf{q})$ is a first integral. Thus, for any solution $(\mathbf{q}(t), \mathbf{p}(t))$ we have $\beta = E(\mathbf{q}(t), \mathbf{p}(t)) = E(\mathbf{q}(0), \mathbf{p}(0)) \geq V(\mathbf{q}(0))$. We can assume $\beta > V(\mathbf{q}(0))$, i.e. $\mathbf{p}(0) \neq 0$, for if $\mathbf{p}(t) \equiv 0$, then the conclusion is trivially satisifed; thus there exists a t_0 for which $\mathbf{p}(t_0) \neq 0$ and by time translation we can assume that $t_0 = 0$. Thus we have

$$\|\,\mathbf{q}(t)\,\| \leq \|\,\mathbf{q}(t) - \mathbf{q}(0)\,\| + \|\,\mathbf{q}(0)\,\| \leq \|\,\mathbf{q}(0)\,\| + \int_0^t \|\,\dot{\mathbf{q}}(s)\,\|\;ds$$

$$= \|\,\mathbf{q}(0)\,\| + \int_0^t \sqrt{2\left[\beta - \frac{1}{m}V(\mathbf{q}(s))\right]}\;ds$$

$$\leq \|\,\mathbf{q}(0)\,\| + \int_0^t \sqrt{2(\beta - a + b\,\|\,\mathbf{q}(s)\,\|^2)}\;\;ds$$

or in differential form

$$\frac{d}{dt}\|\,\mathbf{q}(t)\,\| \leq \sqrt{2(\beta - a + b\,\|\,\mathbf{q}(t)\|^2)}$$

whence

$$t \geq \int_{\|\mathbf{q}(0)\|}^{\|\mathbf{q}(t)\|} \frac{du}{\sqrt{2(\beta - a + bu^2)}}\,. \tag{1}$$

Now let $r(t)$ be the solution of the differential equation

$$\frac{d^2r(t)}{dt^2} = -\frac{d}{dr}(a - br^2)(t) = 2br(t)\,,$$

which, as a second order equation with constant coefficients, has solutions for all time for any initial conditions. Choose

$$r(0) = \|\,\mathbf{q}(0)\,\|,\;\; [\,\dot{r}(0\,)]^2 = 2(\beta - a + b\|\,\mathbf{q}(0)\,\|^2)$$

and let $r(t)$ be the corresponding solution. Since

$$\frac{d}{dt}\left(\frac{1}{2}\,\dot{r}(t)^2 + a - br(t)^2\right) = 0\,,$$

it follows that $(1/2)\,\dot{r}(t)^2 + a - br(t)^2 = (1/2)\,\dot{r}(0)^2 + a - br(0)^2 = \beta$, i.e.

$$\frac{dr(t)}{dt} = \sqrt{2(\beta - a + br(t)^2)}$$

whence
$$t = \int_{\|q(0)\|}^{r(t)} \frac{du}{\sqrt{2(\beta - \alpha + \beta u^2)}} \ . \qquad (2)$$

Comparing the two expressions (1) and (2) and taking into account that the integrand is > 0, it follows that for any finite time interval for which $\mathbf{q}(t)$ is defined, we have $\| \mathbf{q}(t) \| \leq r(t)$, i.e., $\mathbf{q}(t)$ remains in a compact set for finite t-intervals. But then $\dot{\mathbf{q}}(t)$ also lies in a compact set since $\|\dot{\mathbf{q}}(t)\| \leq 2(\beta - a + b\| \mathbf{q}(s) \|^2)$. Thus by **4.1.19,** the solution curve $(\mathbf{q}(t), \mathbf{p}(t))$ is defined for any $t \geq 0$. However, since $(\mathbf{q}(-t), \mathbf{p}(-t))$ is the value at t of the integral curve with initial conditions $(-\mathbf{q}(0), -\mathbf{p}(0))$, it follows that the solution also exists for all $t \leq 0$. (This example will be generalized in §**8.1** to any Lagrangian system on a complete Riemannian manifold whose energy function is kinetic energy of the metric plus a potential, with the potential obeying an inequality of the sort here).

The following counterexample shows that the condition $V(\mathbf{q}) \geq a - b\| \mathbf{q} \|^2$ cannot be relaxed much further. Take $n = 1$ and $V(q) = -\varepsilon^2 q^{2+(4/\varepsilon)}/8$, $\varepsilon > 0$. Then the equation $\ddot{q} = \varepsilon(\varepsilon + 2)q^{1+(4/\varepsilon)}/4$ has the solution $q(t) = 1/(t-1)^{\varepsilon/2}$, which cannot be extended beyond $t = 1$.

C Let E be a Banach space. Suppose

$$A(x) = A \cdot x + B(x) ,$$

where A is a bounded linear operator of E to E and B is ***sublinear***; i.e., $B : E \to E$ is C^r with $r \geq 1$ and satisfies $\| B(x) \| \leq K\| x \| + L$ for constants K and L. We shall show that X is complete by using **4.1.22.** (In $\mathbb{R}^n$, **4.1.21** can also be used with $f(x) = \| x \|^2$). Let x(t) be an integral curve of X on the bounded interval [0, T]. Then

$$x(t) = x(0) + \int_0^t (A \cdot x(s) + B(x(s)))\, ds$$

Hence

$$\|x(t)\| \leq \| x(0) \| \int_0^t (\| A \| + K) \| x(s) \| \, ds + Lt.$$

By Gronwall's inequality,

$$\| x(t) \| \leq (LT + \| x(0) \|)e^{(\|A\|+K)t}$$

Hence x(t) and so X(x(t)) remain bounded on bounded t-intervals. ♦

4.1.24 Proposition *Let* X *be a* C^r *vector field on the manifold* M, $r \geq 1$, $m_0 \in M$, *and* $T^+(m_0)(T^-(m_0))$ *the positive (negative) lifetime of* m_0. *Then for each* $\varepsilon > 0$, *there exists a neighborhood* V *of* m_0 *such that for all* $m \in V$, $T^+(m) > T^+(m_0) - \varepsilon$ *(respectively,* $T^-(m_0) <$

$T^-(m_0) + \varepsilon$). [*One says* $T^+(m_0)$ *is a lower semi-continuous function of* m.]

Proof Cover the segment $\{F_t(m_0) \mid t \in [0, T^+(m_0) - \varepsilon]\}$ with a finite number of neighborhoods $U_0, \ldots, U_n$, each in a chart domain and such that $\varphi_i(U_i)$ is diffeomorphic to the open ball $B_{b(i)/2}(0)$ in E given in the proof of lemma **4.1.8**, where φ_i is the chart map. Let $m_i \in U_i$ be such that $\varphi_i(m_i) = 0$ and t(i) such that $F_{t(i)}(m_0) = m_i$, $i = 0, \ldots, n$, $t(0) = 0$, $t(n) = T^+(m_0) - \varepsilon$. By lemma **4.1.8** the time of existence of all integral curves starting in U_i is uniformly at least $\alpha(i) > 0$. Pick points $p_i \in U_i \cap U_{i+1}$ and let s(i) be such that $F_{s(i)}(m_0) = p_i$, $i = 0, \ldots, n-1$, $s(0) = 0$, $p_0 = m_0$, $s(i) < s(i+1)$, $s(i+1) - t(i) < \alpha(i)$, $t(i+1) - s(i+1) < \alpha(i+1)$, $s(i+1) - s(i) < \min(\alpha(i), \alpha(i+1))$; see Figure 4.1.3.

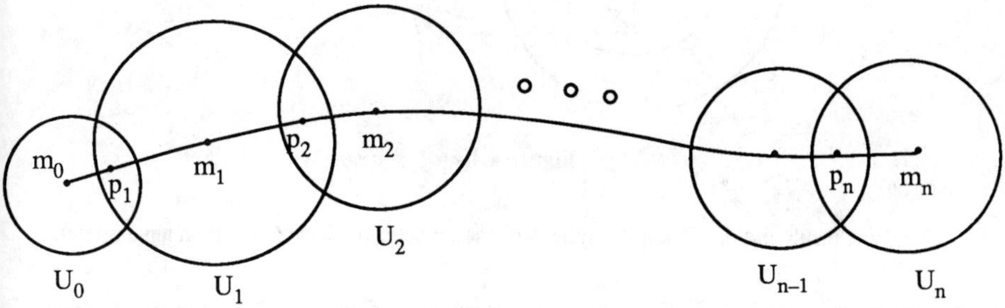

Figure 4.1.3

Let $W_1 = U_0 \cap U_1$. Since $s(2) - s(1) < \alpha(1)$ and any integral curve starting in $W_1 \subset U_1$ exists for time at least $\alpha(1)$, the domain of $F_{s(2)-s(1)}$ includes W_1 and hence $F_{s(2)-s(1)}(W_1)$ makes sense. Define the open set $W_2 = F_{s(2)-s(1)}(W_1) \cap U_1 \cap U_2$ and use $s(3) - s(2) < \alpha(2)$, $W_2 \subset U_2$ to conclude that the domain of $F_{s(3)-s(2)}$ contains U_2. Define the open set $W_3 = F_{s(3)-s(2)}(W_2) \cap U_2 \cap U_3$ and inductively define $W_i = F_{s(i)-s(i-1)}(W_{i-1}) \cap U_{i-1} \cap U_i$, $i = 1, \ldots, n$, which are open sets. Since $s(1) < \alpha(0)$ and $W_1 \subset U_0$, the domain of $F_{-s(1)}$ includes W_1 and thus $V_1 = F_{-s(1)}(W_1) \cap U_0$ is an open neighborhood of m_0. Since $s(2) - s(1) < \alpha(1)$ and $W_2 \subset U_1$, the domain of $F_{-s(2)+s(1)}$ contains W_2 and so $F_{-s(2)+s(1)}(W_2) \subset W_1$ makes sense. Therefore $F_{-s(2)}(W_2) = F_{-s(1)}(F_{-s(2)+s(1)}(W_2)) \subset F_{-s(1)}(W_1)$ exists and is an open neighborhood of m_0. Put $V_2 = F_{-s(2)}(W_2) \cap U_0$. Now proceed inductively to show that $F_{-s(i)}(W_i) \subset F_{-s(i-1)}(W_{i-1})$ makes sense and is an open neighborhood of m_0, $i = 1, 2, \ldots, n$; see Figure 4.1.4. Let $V_i = F_{-s(i)}(W_i) \cap U_0$, $i = 1, \ldots, n$, open neighborhoods containing m_0. Any integral curve c(t) starting in V_n exists thus for time at least s(n) and $F_{s(n)}(V_n) \subset W_n \subset U_n$. Now consider the integral curve starting at c(s(n)) whose time of existence is at least $\alpha(n)$. By uniqueness, c(t) can be smoothly extended to an integral curve which exists for time at least $s(n) + \alpha(n) > t_n = T^+(m_0) - \varepsilon$. ■

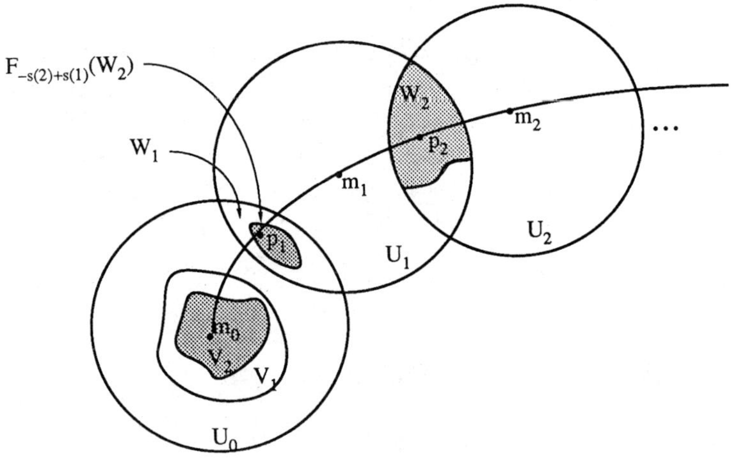

Figure 4.1.4

The same result and proof hold for time dependent vector fields depending on a parameter.

4.1.25 Corollary *Let* X_t *be a* C^r *time-dependent vector field on* M, $r \geq 1$, *and let* m_0 *be an* ***equilibrium*** *of* X_t, *i.e.* $X_t(m_0) = 0$, *for all* t. *Then for any* T *there exists a neighborhood* V *of* m_0 *such that any* $m \in V$ *has integral curve existing for time* $t \in [-T, T]$.

Proof Since $T^+(m_0) = +\infty$, $T^-(m_0) = -\infty$, the previous proposition gives the result. ∎

☞ Supplement 4.1A
Product Formulas

A result of some importance in both theoretical and numerical work concerns writing a flow in terms of iterates of a known mapping. Let $X \in \mathfrak{X}(M)$ with flow F_t (maximally extended). Let $K_\varepsilon(x)$ be a given map defined in some open set of $[0, \infty[\times M$ containing $\{0\} \times M$ and taking values in M, and assume that

(i) $K_0(x) = x$

and (ii) $K_\varepsilon(x)$ is C^1 in ε with derivative continuous in (ε, x).

We call K the ***algorithm***.

4.1.26 Theorem *Let* X *be a* C^r *vector field,* $r \geq 1$. *Assume that the algorithm* $K_\varepsilon(x)$ *is consistent with* X *in the sense that*

$$X(x) = \frac{\partial}{\partial \varepsilon} K_\varepsilon(x)\Big|_{\varepsilon=0}$$

Then, if (t, x) *is in the domain of* $F_t(x)$, $(K_{t/n})^n(x)$ *is defined for* n *sufficiently large and converges to* $F_t(x)$ *as* $n \to \infty$. *Conversely, if* $(K_{t/n})^n(x)$ *is defined and converges for* $0 \le t \le T$, *then* (T, x) *is in the domain of* F *and the limit is* $K_t(x)$.

In the following proof the notation $O(x)$, $x \in \mathbb{R}$ is used for any continuous function in a neighborhood of the origin such that $O(x)/x$ is bounded. Recall from **§2.1** that $o(x)$ denotes a continuous function in a neighborhood of the origin satisfying $\lim_{x \to 0} o(x)/x = 0$.

Proof. First, we prove that convergence holds locally. We begin by showing that for any x_0, the iterates $(K_{t/n})^n(x_0)$ are defined if t is sufficiently small. Indeed, on a neighborhood of x_0, $K_\varepsilon(x) = x + O(\varepsilon)$, so if $(K_{t/j})^j(x)$ is defined for x in a neighborhood of x_0, for $j = 1, \ldots, n-1$, then

$$\begin{aligned}(K_{t/n})^n(x) - x &= ((K_{t/n})^n x - (K_{t/n})^{n-1} x) + ((K_{t/n})^{n-1} - (K_{t/n})^{n-2} x) + \ldots + (K_{t/n}(x) - x) \\ &= O(t/n) + \ldots + O(t/n) = O(t).\end{aligned}$$

This is small, independent of n for t sufficiently small; so, inductively, $(K_{t/n})^n(x)$ is defined and remains in a neighborhood of x_0 for x near x_0.

Let β be a local Lipschitz constant for X so that $\| F_t(x) - F_t(y) \| \le e^{\beta|t|}\| x - y \|$. Now write

$$\begin{aligned}F_t(x) - (K_{t/n})^n(x) &= (F_{t/n})^n(x) - (K_{t/n})^n(x) \\ &= (F_{t/n})^{n-1}F_{t/n}(x) - (F_{t/n})^{n-1}K_{t/n}(x) \\ &\quad + (F_{t/n})^{n-2}F_{t/n}(y_1) - (F_{t/n})^{n-2}K_{t/n}(y_1) \\ &\quad + \ldots + (F_{t/n})^{n-k}F_{t/n}(y_{k-1}) - (F_{t/n})^{n-k}K_{t/n}(y_{k-1}) \\ &\quad + \ldots + F_{t/n}(y_{n-1}) - K_{t/n}(y_{n-1})\end{aligned}$$

where $y_k = (K_{t/n})^k(x)$. Thus

$$\begin{aligned}\| F_t(x) - (K_{t/n})^n(x) \| &\le \sum_{k=1}^{n} e^{\beta| t |(n-k)/n} \| F_{t/n}(y_{k-1}) - K_{t/n}(y_{k-1}) \| \\ &\le n e^{\beta| t |} o(t/n) \to 0 \quad \text{as } n \to \infty,\end{aligned}$$

since $F_\varepsilon(y) - K_\varepsilon(y) = o(\varepsilon)$ by the consistency hypothesis.

Now suppose $F_t(x)$ is defined for $0 \le t \le T$. We shall show $(K_{t/n})^n(x)$ converges to $F_t(x)$. By the foregoing proof and compactness, if N is large enough, $F_{t/N} = \lim_{n \to \infty} (K_{t/nN})^n$ uniformly on a neighborhood of the curve $t \mapsto F_t(x)$. Thus, for $0 \le t \le T$,

$$F_t(x) = (F_{t/N})^N(x) = \lim_{n\to\infty} (K_{t/nN})^N(x).$$

By uniformity in t,

$$F_T(x) = \lim_{j\to\infty} (K_{T/j})^j(x).$$

Conversely, let $(K_{t/n})^n(x)$ converge to a curve $c(t)$, $0 \le t \le T$. Let $S = \{t \mid F_t(x)$ is defined and $c(t) = F_t(x)\}$. From the local result, S is a nonempty open set. Let $t(k) \in S$, $t(k) \to t$. Thus $F_{t(k)}(x)$ converges to $c(t)$, so by local existence theory, $F_t(x)$ is defined, and by continuity, $F_t(x) = c(t)$. Hence $S = [0, T]$ and the proof is complete. ■

4.1.27 Corollary *Let* $X, Y \in \mathfrak{X}(M)$ *with flows* F_t *and* G_t. *Let* S_t *be the flow of* $X + Y$. *Then for* $x \in M$,

$$S_t(x) = \lim_{n\to\infty} (F_{t/n} \circ G_{t/n})^n(x).$$

The left-hand side is defined iff the right-hand side is. This follows from **4.1.26** by setting $K_\varepsilon(x) = (F_\varepsilon \circ G_\varepsilon)(x)$. For example, for $n \times n$ matrices A and B, **4.1.27** yields the classical formula

$$e^{(A+B)} = \lim_{n\to\infty} (e^{A/n}e^{B/n})^n .$$

To see this, define for any $n \times n$ matrix C a vector field $X_C \in \mathfrak{X}(\mathbb{R}^n)$ by $X_C(x) = Cx$. Since X_C is linear in C and has flow $F_t(x) = e^{tC}x$, the formula follows from **4.1.27** by letting $t = 1$.

The topic of this supplement will continue in Supplement **4.2A**. The foregoing proofs were inspired by Nelson [1969] and Chorin et al. [1978].

Supplement 4.1B
Invariant Sets

If X is a smooth vector field on a manifold M and $N \subset M$ is a submanifold, the flow of X will leave N invariant (as a set) iff X is tangent to N. If N is not a submanifold (e.g., N is an open subset together with a non-smooth boundary) the situation is not so simple; however, for this there is a nice criterion going back to Nagumo [1942]. Our proof follows Brezis [1970].

4.1.28 Theorem *Let* X *be a locally Lipschitz vector field on an open set* $U \subset E$, *where* E *is a Banach space. Let* $G \subset U$ *be relatively closed and set* $d(x, G) = \inf\{\|x - y\| \mid y \in G\}$. *The following are equivalent:*

(i) $\lim_{h\downarrow 0}(d(x + hX(x), G)/h) = 0$ *locally uniformly in* $x \in G$ *(or pointwise if* $E = \mathbb{R}^n$*)*;

(ii) *if* $x(t)$ *is the integral curve of* X *starting in* G, *then* $x(t) \in G$ *for all* $t \ge 0$ *in the domain of* $x(\cdot)$.

Note that $x(t)$ need not lie in G for $t \le 0$; so G is only + invariant. (We remark that if X is only continuous the theorem fails.) We give the proof assuming $E = \mathbb{R}^n$ for simplicity.

Proof Assume (ii) holds. Setting $x(t) = F_t(x)$, where F_t is the flow of X and $x \in G$, for small h we get

$$d(x + hX(x), G) \le \| x(h) - x - hX(x) \| = |h| \left\| \frac{x(h) - x}{h} - X(x) \right\| ,$$

from which (i) follows.

Now assume (i). It suffices to show $x(t) \in G$ for small t. Near $x = x(0) \in G$, say on a ball of radius r, we have

$$\| X(x_1) - X(x_2) \| \le K \| x_1 - x_2 \|$$

and

$$\| F_t(x_1) - F_t(x_2) \| \le e^{Kt} \| x_1 - x_2 \| .$$

We can assume $\| F_t(x) - x \| < r/2$. Set $\varphi(t) = d(F_t(x), G)$ and note that $\varphi(0) = 0$, so that for small t, $\varphi(t) < r/2$. Since G is relatively closed, and $E = \mathbb{R}^n$, $d(F_t(x), G) = \| F_t(x) - y_t \|$ for some $y_t \in G$. (In the general Banach space case an approximation argument is needed here.) Thus, $\| y_t - x \| < r$. For small h, $\| F_h(y_t) - x \| < r$, so that

$$\begin{aligned}
\varphi(t + h) &= \inf_{z \in G} \| F_{t+h}(x) - z \| \\
&\le \inf_{z \in G} \{ \| F_{t+h}(x) - F_h(y_t) \| + \| F_h(y_t) - y_t - hX(y_t) \| + \| y_t + hX(y_t) - z \| \} \\
&= \| F_{t+h}(x) - F_h(y_t) \| + \| F_h(y_t) - y_t - hX(y_t) \| + d(y_t + hX(y_t), G) \\
&\le e^{Kh} \| y_t - F_t(x) \| + \| F_h(y_t) - y_t - hX(y_t) \| + d(y_t + hX(y_t), G)
\end{aligned}$$

or

$$\frac{\varphi(t + h) - \varphi(t)}{h} \le \left(\frac{e^{Kh} - 1}{h} \right) \varphi(t) + \left\| \frac{F_h(y_t) - y_t}{h} - X(y_t) \right\| + d(y_t + hX(y_t), G)/h.$$

Hence

$$\limsup_{h \downarrow 0} \frac{\varphi(t + h) - \varphi(t)}{h} \le K\varphi(t).$$

As in Gronwall's inequality, we may conclude that

$$\varphi(t) \le e^{Kt} \varphi(0)$$

so $\varphi(t) = 0$. ∎

4.1.29 Example Let X be a C^∞ vector field on $\mathbb{R}^n$, let $g : \mathbb{R}^n \to \mathbb{R}$ be smooth, and let $\lambda \in \mathbb{R}$ be a regular value for g, so $g^{-1}(\lambda)$ is a submanifold; see Figure 4.1.5. Let $G = g^{-1}(]-\infty, \lambda])$ and suppose that on $g^{-1}(\lambda)$,

$$\langle X, \operatorname{grad} g \rangle \le 0.$$

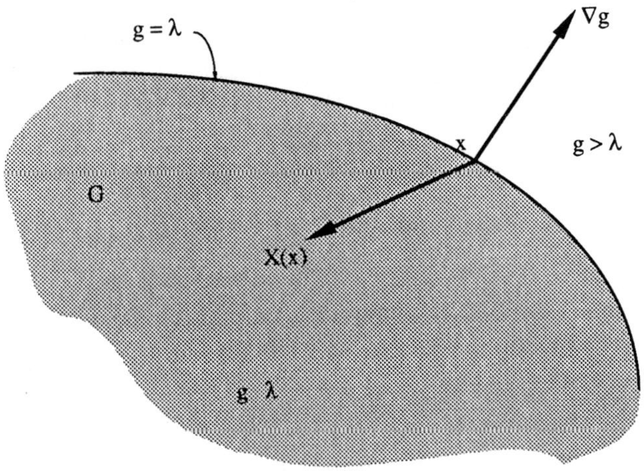

Figure 4.1.5 The set $G = g^{-1}(]-\infty, \lambda[)$ is invariant if X does not point strictly outwards at ∂G.

Then G is + invariant under F_t as may be seen by using **4.1.28**. This result has been generalized to the case where ∂G might not be smooth by Bony [1969]. See also Redheffer [1972] and Martin [1973]. Related references are Yorke [1967], Hartman [1972], and Crandall [1972]. ✍

☞ Supplement 4.1C
A second Proof of the Existence and Uniqueness of Flow Boxes

We shall now give an alternative "modern" proof of Theorems **4.1.5** and **4.1.13**, namely, if $X \in \mathfrak{X}^k(M)$, $k \geq 1$, then for each $m \in M$ there exists a unique C^k flow box at m. The basic idea is due to Robbin [1968] although similar alternative proofs were simultaneously discovered by Abraham and Pugh [unpublished] and Marsden [1968b, p. 368]. The present exposition follows Robbin [1968] and Ebin and Marsden [1970].

Step 1 *Existence and uniqueness of integral curves for* C^1 *vector fields.*

Proof Working on a local chart, we may assume that $X : D_r(0) \to E$, where $D_r(0)$ is the open disk at the origin of radius r in the Banach space E. Let $U = D_{r/2}(0)$, $I = [-1, 1]$ and define

$$\Phi : \mathbb{R} \times C^1{}_0(I, U) \to C^0(I, E)$$

by

$$\Phi(s, \gamma)(t) = \frac{d\gamma}{dt}(t) - sX(\gamma(t)),$$

where $C^i(I, E)$ is the Banach space of C^i-maps of I into E, endowed with the $\|\cdot\|_i$-norm (see Supplement **2.4B**), $C^i_0(I, E) = \{f \in C^i(I, E) \mid f(0) = 0\}$ is a closed subspace of $C^i(I, E)$ and $C^i_0(I, U) = \{f \in C^i_0(I, E) \mid f(I) \subset U\}$ is open in $C^i_0(I, E)$. We first show that Φ is a C^1-map.

The map $d/dt : C^1_0(I, E) \to C^0(I, E)$ is clearly linear and is continuous since $\| d/dt \| \le 1$. Moreover, if $d\gamma/dt = 0$ on I, then γ is constant and since $\gamma(0) = 0$, it follows $\gamma = 0$; i.e., d/dt is injective. Given $\delta \in C^0(I, E)$,

$$\gamma(t) = \int_0^t \delta(s)\, ds$$

defines an element of $C^i_0(I, E)$ with $d\gamma/dt = \delta$, i.e., d/dt is a Banach space isomorphism from $C^1_0(I, E)$ to $C^0(I, E)$.

From these remarks and the Ω-lemma (**2.4.18**), it follows that Φ is a C^1–map. Moreover, $\mathbf{D}_\gamma\Phi(0, 0) = d/dt$ is an isommorphism of $C^1_0(I, E)$ with $C^0(I, E)$. Since $\Phi(0, 0) = 0$, by the implicit function theorem there is an $\varepsilon > 0$ such that $\Phi(\varepsilon, \gamma) = 0$ has a unique solution $\gamma_\varepsilon(t)$ in $C^1_0(I, U)$. The unique integral curve sought is $\gamma(t) = \gamma_\varepsilon(t/\varepsilon)$, $-\varepsilon \le t \le \varepsilon$. ▼

The same argument also works in the time-dependent case. It also shows that γ varies continuously with X.

Step 2 *The local flow of a* C^k *vector field* X *is* C^k.

Proof First, suppose $k = 1$. Modify the definition of Φ in Step 1 by setting $\Psi : \mathbb{R} \times U \times C^1_0(I, U) \to C^0(I, E)$,

$$\Psi(s, x, \gamma)(t) = \gamma'(t) - sX(x + \gamma(t)).$$

As in Step 1, Ψ is a C^1-map and $\mathbf{D}_\gamma\Psi(0, 0, 0)$ is an isomorphism, so $\Psi(\varepsilon, x, \gamma) = 0$ can be locally solved for γ giving a map $H_\varepsilon : U \to C^1_0(I, U)$, $\varepsilon > 0$. The local flow is $F(x, t) = x + H_\varepsilon(x)(t/\varepsilon)$, as in Step 1. By **2.4.17** (differentiability of the evaluation map), F is C^1. Moreover, if $v \in E$, we have $\mathbf{D}F_t(x) \cdot v = v + (\mathbf{D}H_\varepsilon(x) \cdot v)(t/\varepsilon)$, so that the mixed partial derivative

$$\frac{d}{dt}\, \mathbf{D}F_t(x) \cdot v = \frac{1}{\varepsilon}\, (\mathbf{D}H_\varepsilon(x) \cdot v) \left(\frac{t}{\varepsilon}\right)$$

exists and is jointly continuous in (t, x). By exercise **2.4G**, $\mathbf{D}(dF_t(x)/dt)$ exists and equals $(d/dt)(\mathbf{D}F_t(x))$.

Next we prove the result for $k \ge 2$. Consider the Banach space $F = C^{k-1}(\mathrm{cl}(U), E)$ and the map $\omega_X : F \to F; \eta \mapsto X \circ \eta$. This map is C^1 by the Ω-lemma (remarks following **2.4.18**). Regarding ω_X as a vector field on F, it has a unique C^1 integral curve η_t with η_0 = identity, by Step 1. This integral curve is the local flow of X and is C^{k-1} since it lies in F. Since $k \ge 2$, η_t is at least C^1 and so one sees that $\mathbf{D}\eta_t = u_t$ satisfies $du_t/dt = \mathbf{D}X(\eta_t) \cdot u_t$, so by Step 1 again, u_t lies in C^{k-1}. Hence η_t is C^k. ■

The following is a useful alternative argument for proving the result for $k = 1$ from that for $k \geq 2$. For $k = 1$, let $X^n \to X$ in C^1, where X^n are C^2. By the above, the flows of X^n are C^2 and by Step 1, converge uniformly i.e. in C^0, to the flow of X. From the equations for $D\eta_t^n$, we likewise see that $D\eta_t^n$ converges uniformly to the solution of $du_t/dt = DX(\eta_t) \cdot u_t$, u_0 = identity. It follows by elementary analysis (see Exercise **2.4J** or Marsden [1974a, p. 109]) that η_t is C^1 and $D\eta_t = u_t$. ■

This proof works with minor modifications on manifolds with vector fields and flows of Sobolev class H^s or Holder class $C^{k+\alpha}$; see Ebin and Marsden [1970] and Bourguignon and Brezis [1974]. In fact the foregoing proof works in any function spaces for which the Ω–lemma can be proved. Abstract axioms guaranteeing this are given in Palais [1968]. ✍

Exercises

4.1A Find an explicit formula for the flow $F_t : \mathbb{R}^2 \to \mathbb{R}^2$ of the harmonic oscillator equation $\ddot{x} + \omega^2 x = 0$, $\omega \in \mathbb{R}$ a constant.

4.1B Show that if (U_0, a, F) is a flow box for X, then (U_0, a, F_-) is a flow box for $-X$, where $F_-(u, t) = F(u, -t)$ and $(-X)(m) = -(X(m))$.

4.1C Show that the integral curves of a C^r vector field X on an n-manifold can be defined locally in the neighborhood of a point where X is nonzero by n equations $\psi_i(m, t) = c_i$ = constant, $i = 1, ..., n$ in the $n+1$ unknowns (m, t). Such a system of equations is called a ***local complete system of integrals***. (*Hint:* Use the straightening-out theorem.)

4.1D Prove the following generalization of Gronwall's inequality. *Supppose* $v(t) \geq 0$ *satisfies*

$$v(t) \leq C + \left| \int_0^t p(s)\, v(s)\, ds \right|,$$

where $C \geq 0$ *and* $p \in L^1$. *Then*

$$v(t) \leq C \exp\left(\left| \int_0^t p(s)\, ds \right| \right).$$

Use this to generalize **4.1.23**(iii) to allow A to be a time-dependent matrix.

4.1.E Let $F_t = e^{tX}$ be the flow of a linear vector field X on E. Show that the solution of the equation

$$\dot{x} = X(x) + f(x)$$

with initial conditions x_0 satisfies the ***variation of constants formula***

$$x(t) = e^{tX}x_0 + \int_0^t e^{(t-s)X} f(x(s))\, ds.$$

4.1F Let $F(m, t)$ be a C^∞ mapping of $M \times \mathbb{R}$ to M such that $F_{t+s} = F_t \circ F_s$ and F_0 = identity (where $F_t(m) = F(m, t)$). Show that there is a unique C^∞ vector field X whose flow is F.

4.1G Let $\sigma(t)$ be an integral curve of a vector field X and let $g : M \to \mathbb{R}$. Let $\tau(t)$ satisfy $\tau'(t) = g(\sigma(\tau(t)))$. Then show $t \mapsto \sigma(\tau(t))$ is an integral curve of gX. Show by example that even if X is complete, gX need not be.

4.1H (i) **Gradient Flows** Let $f : \mathbb{R}^n \to \mathbb{R}$ be C^1 and let $X = (\partial f/\partial x^1, \ldots, \partial f/\partial x^n)$ be the gradient of f. Let F be the flow of X. Show that $f(F_t(x) \geq f(F_s(x))$ if $t \geq s$.

(ii) Use (i) to find a vector field X on $\mathbb{R}^n$ such that $X(0) = 0$, $X'(0) = 0$, yet 0 is globally attracting; i.e., every integral curve converges to 0 as $t \to \infty$. This exercise continues in **4.3K**.

4.1I Let c be a locally Lipschitz increasing function, $c(t) > 0$ for $t \geq 0$ and assume that the differential equation $r'(t) = c(r(t))$ has the solution with $r(0) = r_0 \geq 0$ existing for time $t \in [0, T]$. Conclude that $r(t) \geq 0$ for $t \in [0, T]$. Prove the following ***comparison lemmas.***

(i) If $h(t)$ is a continuous function on $[0, T]$, $h(t) \geq 0$, satisfying $h'(t) \leq c(h(t))$ on $[0, T]$, $h(0) = r_0$, then show that $h(t) \leq r(t)$. (*Hint:* Prove that

$$\int_{r_0}^{h(t)} \frac{dx}{c(x)} \leq t = \int_{r_0}^{r(t)} \frac{dx}{c(x)}$$

and use strict positivity of the integrand.)

(ii) Generalize (i) to the case $h(0) \leq r_0$. (*Hint:* The function $\mathsf{h}(t) = h(t) + r_0 - h(0) \leq h(t)$ satisfies the hypotheses in (i).)

(iii) If $f(t)$ is a continuous function on $[0, T]$, $f(t) \geq 0$, satisfying

$$f(t) \leq r_0 + \int_0^t c(f(s))\, ds$$

on $[0, T]$, then show that $f(t) \le r(t)$. (*Hint:*

$$h(t) = r_0 + \int_0^t c(f(s))\, ds \ge f(t)$$

satisfies the hypothesis in (i) since $h'(t) = c(f(t)) \le c(h(t))$.)

(iv) If in addition

$$\int_0^\infty \frac{dx}{c(x)} = +\infty ,$$

show that the solution $f(t) \ge 0$ exists for all $t \ge 0$.

(v) If $h(t)$ is only continuous on $[0, T]$ and

$$h(t) \le r_0 + \int_0^t c(h(s))\, ds ,$$

show that $h(t) \le r(t)$ on $[0, T]$. (*Hint:* Approximate $h(t)$ by a C^1-function $g(t)$ and show that $g(t)$ satisfies the same inequality as $h(t)$.).

4.1J (i) Let $X = y^2\partial/\partial x$ and $Y = x^2\partial/\partial y$. Show that X and Y are complete on $\mathbb{R}^2$ but $X + Y$ is not. (*Hint:* Note that $x^3 - y^3 =$ constant and consider an integral curve with $x(0) = y(0)$.)

(ii) Prove the following theorem:

Let H be a Hilbert space and X and Y be locally Lipschitz vector fields that satisfy the following:

(a) X and Y are bounded and Lipschitz on bounded sets;

(b) there is a constant $\beta \ge 0$ such that

$$\langle Y(x), x \rangle \le \beta \| x \|^2 \quad \text{for all } x \in H;$$

(c) there is a locally Lipschitz monotone increasing function $c(t) > 0$, $t \ge 0$, such that

$$\int_0^\infty \frac{dx}{c(x)} = +\infty ,$$

and if $x(t)$ is an integral curve of X,

$$\frac{d}{dt} \| x(t) \| \le c(\| x(t) \|) .$$

Then X, Y and $X + Y$ are positively complete.

Note: One may assume $\| X(x_0) \| \leq c(\| x_0 \|)$ in (c) instead of $(d/dt) \| x(t) \| \leq c(\| x(t) \|)$. (*Hint:* Find a differential inequality for $(1/2)(d/dt) \| u(t) \|^2$, where $u(t)$ is an integral curve of $X + Y$ and then use Exercise **4.I**(iii).)

4.1K Prove the following result on the ***convergence of flows:***

Let X_α be locally Lipschitz vector fields on M for α in some topological space. Suppose the Lipschitz constants of X_α are locally bounded as $\alpha \to \alpha(0)$ and $X_\alpha \to X_{\alpha(0)}$ locally uniformly. Let $c(t)$ be an integral curve of $X_{\alpha(0)}$, $0 \leq t \leq T$ and $\varepsilon > 0$. Then the integral curves $c_\alpha(t)$ of X_α with $c_\alpha(0) = c(0)$ are defined for the interval $t \in [0, T - \varepsilon]$ for α sufficiently close to $\alpha(0)$ and $c_\alpha(t) \to c(t)$ uniformly in $t \in [0, T - \varepsilon]$ as $\alpha \to \alpha(0)$. If the flows are complete $F_t^\alpha \to F_t$ locally uniformly. (The vector fields may be time dependent if the estimates are locally t-uniform.) (*Hint:* Show that

$$\| c_\alpha(t) - c(t) \| \leq k \int_0^t \| c_\alpha(\tau) - c(\tau) \| \, d\tau + \int_0^t \| X_\alpha(c(\tau)) - X_{\alpha(0)}(c(\tau)) \| \, d\tau$$

and conclude from Gronwall's inequality that $c_\alpha(t) \to c(t)$ for $\alpha \to \alpha(0)$ since the second term $\to 0$. This estimate shows that $c_\alpha(t)$ exists as long as $c(t)$ does on any compact subinterval of $[0, T[$.)

4.1L Prove that the C^r flow of a C^{r+1} vector field is a C^1 function of the vector field by utilizing Supplement **4.1C**. (*Caution.* It is known that the C^k flow of a C^k vector field cannot be a C^1 function of the vector field; see Ebin and Marsden [1970] for the explanation and further references).

4.1.M **Nonunique integral curves on non-Hausdorff manifolds** Let M be the line with two origins (see Exercise **3.5H**) and consider the vector field $X : M \to TM$ which is defined by $X([x, i]) = x$, $i = 1, 2$; here $[x, i]$ denotes a point of the quotient manifold M. Show that through every point other than $[0, 0]$ and $[0, 1]$, there are exactly two integral curves of X. Show that X is complete. (*Hint:* The two distinct integral curves pass respectively through $[0, 0]$ and $[0, 1]$.)

4.1N Give another proof of Theorem **4.1.5** using Exercise **2.5J**.

4.1O Give examples of vector fields satisfying the following conditions:

(i) on $\mathbb{R}$ and S^1 with no critical points; generalize to $\mathbb{R}^n$ and $\mathbb{T}^n$;

(ii) on S^1 with exactly k critical points; generalize to $\mathbb{T}^n$;

(iii) on $\mathbb{R}^2$ and $\mathbb{RP}^2$ with exactly one critical point and all other orbits closed;

(iv) on the Möbius band with no critical points and such that the only integral curve intersecting the zero section is the zero section itself;

(v) on S^2 with precisely two critical points and one closed orbit;

(vi) on S^2 with precisely one critical point and no closed orbit;

(vii) on S^2 with no critical points on a great circle and nowhere tangent to it; show that any such vector field has its integral curves intersecting this great circle at most once;

(viii) on $\mathbb{T}^2$ with no critical points, all orbits closed and winding exactly k times around $\mathbb{T}^2$.

4.1P Let $\pi : M \to N$ be a surjective submersion. A vector field $X \in \mathfrak{X}(M)$ is called ***π-vertical*** if $T\pi \circ X = 0$. If $\mathbb{K}$ is the Klein bottle, show that $\pi : \mathbb{K} \to S^1$ given by $\pi([a, b]) = e^{2\pi ia}$ is a surjective submersion; see Figure 4.1.6. Prove that $\mathbb{K}$ is a non trivial S^1-bundle. (*Hint:* If it were trivial, there would exist a nowhere zero vertical vector field on $\mathbb{K}$. In Figure 4.1.6, this means that arrows go up on the left and down on the right hand side. Follow a path from left to right and argue by the intermediate value theorem that the vector field must vanish somewhere.)

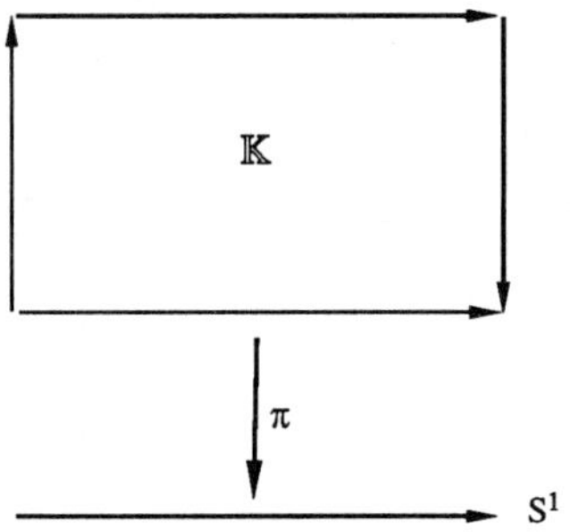

Figure 4.1.6 The Klein bottle as an S^1-bundle.

§4.2 *Vector Fields as Differential Operators*

In the previous section vector fields were studied from the point of view of dynamics; that is in terms of the flows they generate. Before continuing the development of dynamics, we shall treat some of the algebraic aspects of vector fields. The specific goal of the section is the development of the Lie derivative of functions and vector fields and its relationship with flows. One important feature is the behavior of the constructions under mappings. The operations should be as natural or covariant as possible when subjected to a mapping.

We begin with a discussion of the action of mappings on functions and vector fields. First, recall some notation. Let $C^r(M, F)$ denote the space of C^r maps $f : M \to F$, where F is a Banach space, and let $\mathfrak{X}^r(M)$ denote the space of C^r vector fields on M. Both are vector spaces with the obvious operations of addition and scalar multiplication. For brevity we write $\mathcal{F}(M) = C^\infty(M, \mathbb{R})$, $\mathcal{F}^r(M) = C^r(M, \mathbb{R})$ and $\mathfrak{X}(M) = \mathfrak{X}^\infty(M)$. Note that $\mathcal{F}^r(M)$ has an *algebra structure;* that is, for $f, g \in \mathcal{F}^r(M)$ the product fg defined by $(fg)(m) = f(m)g(m)$ obeys the usual algebraic properties of a product such as $fg = gf$ and $f(g + h) = fg + fh$.

4.2.1 Definitions

(i) *Let* $\varphi : M \to N$ *be a* C^r *mapping of manifolds and* $f \in \mathcal{F}^r(N)$. *Define the* ***pull-back*** *of* f *by* φ *by*

$$\varphi^* f = f \circ \varphi \in \mathcal{F}^r(M).$$

(ii) *If* f *is a* C^r *diffeomorphism and* $X \in \mathfrak{X}^r(M)$, *the* ***push-forward*** *of* X *by* φ *is defined by*

$$\varphi_* X = T\varphi \circ X \circ \varphi^{-1} \in \mathfrak{X}^r(N).$$

Consider local charts (U, χ), $\chi : U \to U' \subset E$ on M and (V, ψ), $\psi : V \to V' \subset F$ on N, and let $(T\chi \circ X \circ \chi^{-1})(u) = (u, \mathbf{X}(u))$, where $\mathbf{X} : U' \to E$ is the local representative of X. Then from the chain rule and the definition of push-forward, the local representative of $\varphi_* X$ is

$$(T\psi \circ (\varphi_* X) \circ \psi^{-1})(v) = (v, \mathbf{D}(\psi \circ \varphi \circ \chi^{-1})(u) \cdot \mathbf{X}(u)),$$

where $v = (\psi \circ \varphi \circ \chi^{-1})(u)$. The different point of evaluation on each side of the equation corresponds to the necessity of having φ^{-1} in the definition. If M and N are finite dimensional, x^i are local coordinates on M and y^j local coordinates on N, the preceding formula gives the components of $\varphi_* X$ by

$$(\varphi_* X)^j(y) = \frac{\partial \varphi^j}{\partial x^i}(x)\, X^i(x)$$

where $y = \varphi(x)$.

We can interchange "pull-back" and "push-forward" by changing φ to φ^{-1}, i.e., defining φ_* (resp. φ^*) by $\varphi_* = (\varphi^{-1})^*$ (resp. $\varphi^* = (\varphi^{-1})_*$). Thus the ***push-forward*** of a function f on M is $\varphi_* f = f \circ \varphi^{-1}$ and the ***pull-back*** of a vector field Y on N is $\varphi^* Y = (T\varphi)^{-1} \circ Y \circ \varphi$ (Fig. 4.2.1). Notice that φ must be a diffeomorphism in order that the pull-back and push-forward operations make sense, the only exception being pull-back of functions. Thus vector fields can only be pulled back and pushed forward by diffeomorphisms. However, even when φ is not a diffeomorphism we can talk about φ-related vector fields as follows.

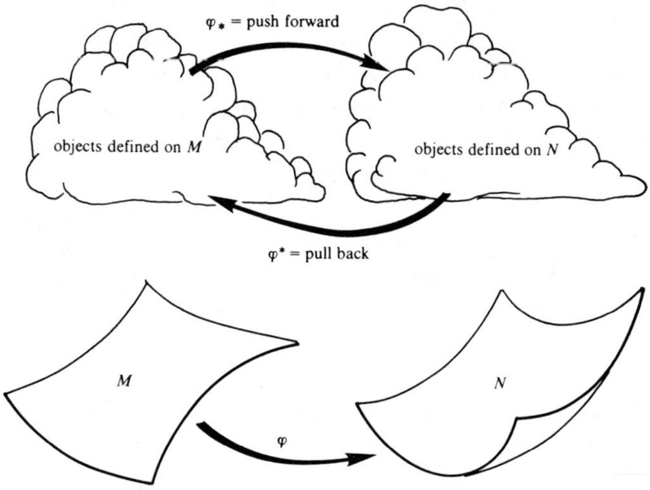

Figure 4.2.1

4.2.2 Definition *Let* $\varphi : M \to N$ *be a* C^r *mapping of manifolds. The vector fields* $X \in \mathfrak{X}^{r-1}(M)$ *and* $Y \in \mathfrak{X}^{r-1}(N)$ *are called* ***φ-related****, denoted* $X \sim_\varphi Y$, *if* $T\varphi \circ X = Y \circ \varphi$.

Note that if φ is diffeomorphism and X and Y are φ-related, then $Y = \varphi_* X$. In general however, X can be φ-related to more than one vector field on N. φ-relatedness means that the following diagram commutes:

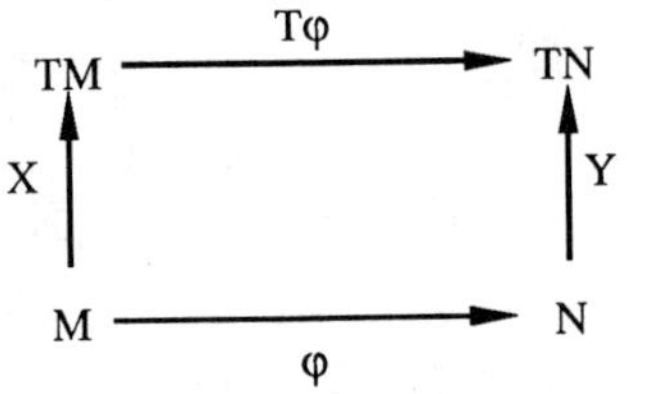

4.2.3 Proposition

(i) *Pull-back and push-forward are linear maps, and* $\varphi^*(fg) = (\varphi^*f)(\varphi^*g)$, $\varphi_*(fg) = (\varphi_*f)(\varphi_*g)$.

(ii) *If* $X_i \sim_\varphi Y_i$, $i = 1, 2$, *and* $a, b \in \mathbb{R}$, *then* $aX_1 + bX_2 \sim_\varphi aY_1 + bY_2$.

(iii) *For* $\varphi : M \to N$ *and* $\psi : N \to P$, *we have*

$$(\psi \circ \varphi)^* = \varphi^* \circ \psi^* \quad \textit{and} \quad (\psi \circ \varphi)_* = \psi_* \circ \varphi_* .$$

(iv) *If* $X \in \mathfrak{X}(M)$, $Y \in \mathfrak{X}(N)$, $Z \in \mathfrak{X}(P)$, $X \sim_\varphi Y$ *and* $Y \sim_\psi Z$, *then* $X \sim_{\psi \circ \varphi} Z$.

In this proposition it is understood that all maps are diffeomorphims with the exception of the pull-back of functions and the relatedness of vector fields.

Proof (i) This consists of straightforward verifications. For example, if $X_i \sim_\varphi Y_i$, $i = 1, 2$, then $T\varphi \circ (aX_1 + bX_2) = aT\varphi \circ X_1 + bT\varphi \circ X_2 = aY_1 \circ \varphi + bY_2 \circ \varphi$, i.e., $aX_1 + bX_2 \sim_\varphi aY_1 + bY_2$.

(ii) These relations on functions are simple consequences of the definition, and the ones on $\mathfrak{X}(P)$ and $\mathfrak{X}(M)$ are proved in the following way using the chain rule:

$$T(\psi \circ \varphi) \circ X = T\psi \circ T\varphi \circ X = T\psi \circ Y \circ \varphi = Z \circ \psi \circ \varphi. \blacksquare$$

In this development we can replace $\mathcal{F}(M)$ by $C^r(M, F)$ with little change; i.e., we can replace real-valued functions by F-valued functions.

The behavior of flows under these operations is as follows:

4.2.4 Proposition *Let* $\varphi : M \to N$ *be a* C^r*-mapping of manifolds,* $X \in \mathfrak{X}^r(M)$ and $Y \in \mathfrak{X}^r(N)$. *Let* F^X_t *and* F^Y_t *denote the flows of* X *and* Y *respectively. Then* $X \sim_\varphi Y$ *iff* $\varphi \circ F^X_t = F^Y_t \circ \varphi$. *In particular, if* φ *is a diffeomorphism, then the equality* $Y = \varphi_*X$ *holds iff the flow of* Y *is* $\varphi \circ F^X_t \circ \varphi^{-1}$. *In particular,* $(F^X_s)_*X = X$.

Proof Taking the time derivative of the relation $(\varphi \circ F^X_t)(m) = (F^Y_t \circ \varphi)(m)$, for $m \in M$, using the chain rule and definition of the flow, we get

$$T\varphi\left(\frac{\partial F^X_t(m)}{\partial t}\right) = \frac{\partial F^Y_t}{\partial t}(\varphi(m)), \text{ i.e.}$$

$$(T\varphi \circ X \circ F^X_t)(m) = (Y \circ F^Y_t \circ \varphi)(m) = (Y \circ \varphi \circ F^X_t)(m),$$

which is equivalent to $T\varphi \circ X = Y \circ \varphi$. Conversely, if this relation is satisfied, let $c(t) = F^X_t(m)$ denote the integral curve of X through $m \in M$. Then

$$\frac{d(\varphi \circ c)(t)}{dt} = T\varphi\left(\frac{dc(t)}{dt}\right) = T\varphi(X(c(t))) = Y((\varphi \circ c)(t))$$

says that $\varphi \circ c$ is the integral curve of Y through $\varphi(c(0)) = \varphi(m)$. By uniqueness of integral curves, we get $(\varphi \circ F^X_t)(m) = (\varphi \circ c)(t) = F^Y_t(\varphi(m))$. The last statement is obtained by taking $\varphi = F^X_s$ for fixed s. ■

We call $\varphi \circ F_t \circ \varphi^{-1}$ the ***push-forward*** of F_t by φ since it is the natural way to construct a diffeomorphism on N out of one on M. See Fig. 4.2.2. Thus, **4.2.4** says that *the flow of the push-forward of a vector field is the push forward of its flow.*

Next we define how vector fields operate on functions. This is done by means of the directional derivative. Let $f : M \to \mathbb{R}$, so $Tf : TM \to T\mathbb{R} = \mathbb{R} \times \mathbb{R}$. Recall that a tangent vector to $\mathbb{R}$ at a base point $\lambda \in \mathbb{R}$ is a pair (λ, μ), the number μ being the principal part. Thus we can write Tf acting on a vector $v \in T_mM$ in the form

$$Tf \cdot v = (f(m), \mathbf{d}f(m) \cdot v).$$

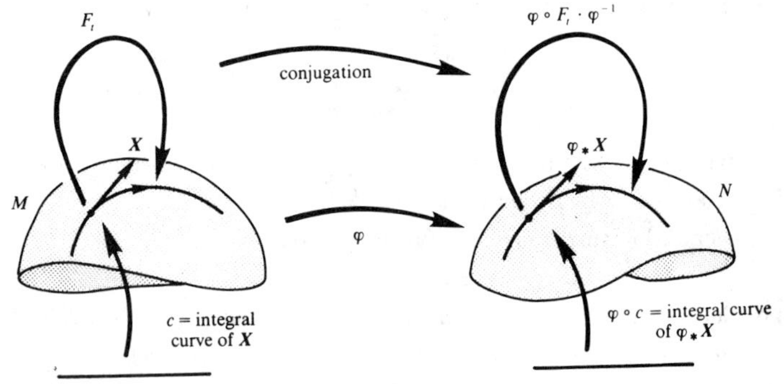

Figure 4.2.2

This defines, for each $m \in M$, the element $\mathbf{d}f(m) \in T_m^*M$. Thus $\mathbf{d}f$ is a section of T^*M, a ***covector field***, or ***one-form.***

4.2.5 Definition *The covector field* $\mathbf{d}f : M \to T^*M$ *defined this way is called the* ***differential*** *of* f.

For F-valued functions, $f : M \to F$, where F is a Banach space, a similar definition gives $\mathbf{d}f(m) \in L(T_mM, F)$ and we speak of $\mathbf{d}f$ as an F-***valued one-form.***

Clearly if f is C^r, then $\mathbf{d}f$ is C^{r-1}. Let us now work out $\mathbf{d}f$ in local charts for $f \in \mathcal{F}(M)$. If $\varphi : U \subset M \to V \subset E$ is a local chart for M, then the local representative of f is the map $\mathsf{f} : V \to \mathbb{R}$ defined by $\mathsf{f} = f \circ \varphi^{-1}$. The local representative of Tf is the tangent map for local manifolds:

$$Tf(x, v) = (f(x), \mathbf{D}f(x) \cdot v).$$

Thus the local representative of $\mathbf{d}f$ is the derivative of the local representative of f. In particular, if M is finite dimensional and local coordinates are denoted $(x^1, ..., x^n)$, then the local components of $\mathbf{d}f$ are

$$(\mathbf{d}f)_i = \frac{\partial f}{\partial x^i}.$$

The introduction of $\mathbf{d}f$ leads to the following.

4.2.6 Definition *Let* $f \in \mathcal{F}^r(M)$ *and* $X \in \mathfrak{X}^{r-1}(M)$, $r \geq 1$. *Define the* ***directional*** *or* ***Lie derivative*** *of* f *along* X *by*

$$\mathbf{L}_X f(m) \equiv X[f](m) = \mathbf{d}f(m) \cdot X(m),$$

for any $m \in M$. *Denote by* $X[f] = \mathbf{d}f(X)$ *the map* $m \in M \mapsto X[f](m) \in \mathbb{R}$. *If* f *is* F*-valued, the same definition is used, but now* X[f] *is* F*-valued.*

The local representative of X[f] in a chart is given by the function $x \mapsto \mathbf{D}f(x) \cdot X(x)$, where f and X are the local representatives of f and X. In particular, if M is finite dimensional then we have

$$X[f] \equiv \mathbf{L}_X f = \sum_{i=1}^{n} \frac{\partial f}{\partial x^i} X^i.$$

Evidently if f is C^r and X is C^{r-1} then X[f] is C^{r-1}.

Let us observe that from the chain rule, $\mathbf{d}(f \circ \varphi) = \mathbf{d}f \circ T\varphi$ where $\varphi : N \to M$ is a C^r map of manifolds, $r \geq 1$. For real-valued functions, Leibniz' rule gives

$$\mathbf{d}(fg) = f\mathbf{d}g + g\mathbf{d}f.$$

(If f is F-valued, g is G-valued and $B : F \times G \to H$ is a continuous bilinear map of Banach spaces, this generalizes to $\mathbf{d}(B(f, g)) = B(\mathbf{d}f, g) + B(f, \mathbf{d}g)$.

4.2.7 Proposition

(i) *Suppose* $\varphi : M \to N$ *is a diffeomorphism. Then* $\mathbf{L}_X$ *is natural with respect to* ***push-forward*** *by* φ. *That is, for each* $f \in \mathcal{F}(M)$,

$$\mathbf{L}_{\varphi_* X}(\varphi_* f) = \varphi_* \mathbf{L}_X \varphi$$

or the following diagram commutes:

$$\begin{array}{ccc} \mathcal{F}(M) & \xrightarrow{\varphi_*} & \mathcal{F}(N) \\ \Big\downarrow \mathcal{L}_X & & \Big\downarrow \mathcal{L}_{\varphi_* X} \\ \mathcal{F}(M) & \xrightarrow[\varphi_*]{} & \mathcal{F}(N) \end{array}$$

(ii) $\mathbf{L}_X$ ***is natural with respect to restrictions.*** *That is, for* U *open in* M *and* $f \in \mathcal{F}(M)$, $\mathbf{L}_{X|U}(f \mid U) = (\mathbf{L}_X f) \mid U$; *or if* $\mid U : \mathcal{F}(M) \to \mathcal{F}(U)$ *denotes restriction to* U, *the following diagram commutes:*

$$\begin{array}{ccc} \mathcal{F}(M) & \xrightarrow{|U} & \mathcal{F}(U) \\ \Big\downarrow \mathcal{L}_X & & \Big\downarrow \mathcal{L}_{X|U} \\ \mathcal{F}(M) & \xrightarrow[|U]{} & \mathcal{F}(U) \end{array}$$

Proof For (i), if $n \in N$ then

$$\mathbf{L}_{\varphi_* X}(\varphi_* f)(n) = \mathbf{d}(f \circ \varphi^{-1}) \cdot (\varphi_* X)(n) = \mathbf{d}(f \circ \varphi^{-1})(n) \cdot (T\varphi \circ X \circ \varphi^{-1})(n)$$

$$= \mathbf{d}f(\varphi^{-1}(n)) \cdot (X \circ \varphi^{-1})(n) = \varphi_*(\mathbf{L}_X f)(n).$$

(ii) follows from $\mathbf{d}(f \mid U) = (\mathbf{d}f) \mid U$, which itself is clear from the definition of **d**. ■

This proposition is readily generalized to F-valued C^r functions.

Since $\varphi^* = (\varphi^{-1})_*$, the Lie derivative is also natural with respect to pull-back by φ. This has a generalization to φ-related vector fields as follows.

4.2.8 Proposition *Let* $\varphi : M \to N$ *be a* C^r *map,* $X \in \mathfrak{X}^{r-1}(M)$ *and* $Y \in \mathfrak{X}^{r-1}(N)$. *If* $X \sim_\varphi Y$, *then* $\mathbf{L}_X(\varphi^* f) = \varphi^* \mathbf{L}_Y f$ *for all* $f \in C^r(N, F)$; *i.e., the following diagram commutes:*

$$\begin{array}{ccc} C^r(N, F) & \xrightarrow{\varphi^*} & C^r(M, F) \\ \Big\downarrow \mathcal{L}_Y & & \Big\downarrow \mathcal{L}_X \\ C^{r-1}(N, F) & \xrightarrow[\varphi^*]{} & C^{r-1}(M, F) \end{array}$$

Proof For $m \in M$,

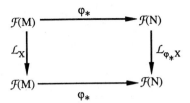

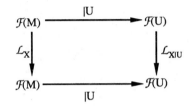

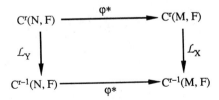

$$\mathbf{L}_X(\varphi^* f)(m) = \mathbf{d}(f \circ \varphi)(m) \cdot X(m) = \mathbf{d}f(\varphi(m)) \cdot (T_m\varphi(X(m)))$$
$$= \mathbf{d}f(\varphi(m)) \cdot Y(\varphi(m)) = \mathbf{d}f(Y)(\varphi(m)) = (\varphi^* \mathbf{L}_Y f)(m). \blacksquare$$

Next we show that $\mathbf{L}_X$ satisfies the Leibniz rule.

4.2.9 Proposition

(i) *The mapping* $\mathbf{L}_X : C^r(M, F) \to C^{r-1}(M, F)$ *is a* ***derivation.*** *That is* $\mathbf{L}_X$ *is* $\mathbb{R}$*-linear and for* $f \in C^r(M, F)$, $g \in C^r(M, G)$ *and* $B : F \times G \to H$ *a bilinear map*

$$\mathbf{L}_X(B(f, g)) = B(\mathbf{L}_X f, g) + B(f, \mathbf{L}_X g).$$

In particular, for real-valued functions, $\mathbf{L}_X(fg) = g\mathbf{L}_X f + f\mathbf{L}_X g$.

(ii) *If* c *is a constant function,* $\mathbf{L}_X c = 0$.

Proof (i) This follows from the Leibniz rule for $\mathbf{d}$ and the definition $\mathbf{L}_X f$. Part (ii) results from the definition. ■

The connection between $\mathbf{L}_X f$ and the flow of X is as follows.

4.2.10 Lie Derivative Formula for Functions *Let* $f \in C^r(M, F)$, $X \in \mathfrak{X}^{r-1}(M)$ *and suppose* X *has a flow* F_t. *Then*

$$\boxed{\frac{d}{dt} F_t^* f = F_t^* \mathbf{L}_X f}$$

Proof By the chain rule, the definition of the differential of a function and the flow of a vector field,

$$\frac{d}{dt}(F_t^* f)(m) = \frac{d}{dt}(f \circ F_t)(m) = \mathbf{d}f(F_t(m)) \cdot \frac{dF_t(m)}{dt}$$
$$= \mathbf{d}f(F_t(m)) \cdot X(F_t(m)) = \mathbf{d}f(X)(F_t(m))$$
$$= (\mathbf{L}_X f)(F_t(m)) = (F_t^* \mathbf{L}_X f)(m). \blacksquare$$

As an application of the Lie derivative formula, we consider the problem of solving a partial differential equation on $\mathbb{R}^{n+1}$ of the form

$$\frac{\partial f}{\partial t}(x, t) = \sum_{i=1}^{n} X^i(x) \frac{\partial f}{\partial x^i}(x, t) \tag{P}$$

with initial condition $f(x, 0) = g(x)$ for given smooth functions $X^i(x)$, $i = 1, \ldots, n$, $g(x)$ and a scalar unknown $f(x, t)$.

4.2.11 Proposition *Suppose* $X = (X^1, \ldots, X^n)$ *has a complete flow* F_t. *Then* $f(x, t) = g(F_t(x))$ *is a solution of the foregoing problem* (P). *(See Exercise* **4.2C** *for uniqueness.)*

Proof

$$\frac{\partial f}{\partial t} = \frac{d}{dt} F_t^* g = F_t^* \mathcal{L}_X g = \mathcal{L}_X (F_t^* g) = X[f]. \quad \blacksquare$$

Thus one can solve this *scalar* equation by computing the orbits of X and pushing (or "dragging along") the graph of g by the flow of X; see Fig. 4.2.3. These trajectories of X are called *characteristics* of (P). (As we shall see below, the vector field X in (P) can be time dependent.)

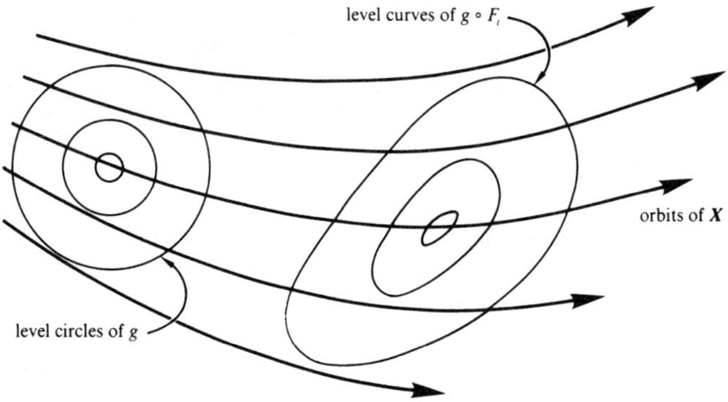

Figure 4.2.3

4.2.13 Example *Solve the partial differential equation*

$$\frac{\partial f}{\partial t} = (x + y)\left(\frac{\partial f}{\partial x} - \frac{\partial f}{\partial y}\right),$$

with initial condition

$$f(x, y, 0) = x^2 + y^2.$$

Solution The vector field $X(x, y) = (x + y, -x - y)$ has a complete flow $F_t(x, y) = ((x + y)t + x, -(x + y)t + y)$, so that the solution of the previous partial differential equation is given by

$$f(x, y, t) = 2(x + y)^2 t^2 + x^2 + y^2 + 2(x^2 - y^2)t. \ \blacklozenge$$

Now we turn to the question of using the Lie derivative to characterize vector fields. We will prove that any derivation on functions uniquely defines a vector field. Because of this, derivations can be (and often are) used to *define* vector fields. (See the introduction to §**3.3**.) In the proof we shall need to localize things in a smooth way, hence the following lemma of general utility is proved first.

4.2.13 Lemma *Let* E *be a* C^r *Banach space, i.e., one whose norm is* C^r *on* $E \setminus \{0\}, r \geq 1$. *Let* U_1 *be an open ball of radius* r_1 *about* x_0 *and* U_2 *an open ball of radius* r_2, $r_1 < r_2$. *Then there is a* C^r *function* $h : E \to \mathbb{R}$ *such that* h *is one on* U_1 *and zero outside* U_2.

We call h a ***bump function.*** Later we will prove more generally that on a manifold M, if U_1 and U_2 are two open sets with $\mathrm{cl}(U_1) \subset U_2$, there is an $h \in \mathcal{F}^r(M)$ such that h is one on U_1 and is zero outside U_2.

Proof By a scaling and translation, we can assume that U_1 and U_2 are balls of radii 1 and 3 and centered at the origin. Let $\theta : \mathbb{R} \to \mathbb{R}$ be given by

$$\theta(x) = \exp(-1/(1 - |x|^2)) \quad \text{if } |x| < 1$$

and set

$$\theta(x) = 0, \quad \text{if } |x| \geq 1.$$

(See the remarks following **2.4.15**.) Now set

$$\theta_1(s) = \frac{\int_{-\infty}^{s} \theta(t)\, dt}{\int_{-\infty}^{\infty} \theta(t)\, dt}$$

so θ_1 is a C^∞ function and is 0 if $s < -1$, and 1 if $s > 1$. Let $\theta_2(s) = \theta_1(2 - s)$, so θ_2 is a C^∞ function that is 1 if $s < 1$ and 0 if $s > 3$. Finally, let $h(x) = \theta_2(\|x\|)$. ∎

The norm on a real Hilbert space is C^∞ away from the origin. The order of differentiability of the norms of some concrete Banach spaces is also known; see Bonic and Frampton [1966], Yamamuro [1974], and Supplement **5.5B**.

4.2.14 Corollary *Let* M *be a* C^r *manifold modeled on a* C^r *Banach space. If* $\alpha_m \in T_m^*M$, *then there is an* $f \in \mathcal{F}^r(M)$ *such that* $\mathbf{d}f(m) = \alpha_m$.

Proof If $M = E$, so $T_mE \cong E$, let $f(x) = \alpha_m(x)$, a linear function on E. Then **d**f is constant and equals α_m.

The general case can be reduced to E using a local chart and a bump function as follows. Let $\varphi : U \to U' \subset E$ be a local chart at m with $\varphi(m) = 0$ and such that U' contains the ball of radius 3. Let $\tilde{\alpha}_m$ be the local representative of α_m and let h be a bump function, 1 on the ball of radius 1 and zero outside the ball of radius 2. let $f(x) = \tilde{\alpha}_m(x)$ and let

$$f = \begin{cases} (hf) \circ \varphi & \text{on } U \\ 0, & \text{on } M \setminus U. \end{cases}$$

It is easily verified that f is C^r and $\mathbf{d}f(m) = \alpha_m$. ■

4.2.15 Proposition

(i) *Let* M *be a* C^r *manifold modeled on a* C^r *Banach space. The collection of operators* $\mathbf{L}_X$ *for* $X \in \mathfrak{X}^r(M)$, *defined on* $C^r(M, F)$ *and taking values in* $C^{r-1}(M, F)$ *forms a real vector space and an* $\mathcal{F}^r(M)$*-module with* $(f\mathbf{L}_X)(g) = f(\mathbf{L}_X g)$, *and is isomorphic to* $\mathfrak{X}^r(M)$ *as a real vector space and as an* $\mathcal{F}^r(M)$*-module. In particular,* $\mathbf{L}_X = 0$ *iff* $X = 0$; *and* $\mathbf{L}_{fX} = f\mathbf{L}_X$.

(ii) *Let* M *be any* C^r *manifold. If* $\mathbf{L}_X f = 0$ *for all* $f \in C^r(U, F)$, *for all open subsets* U *of* M, *then* $X = 0$.

Proof (i) Consider the map $\sigma : X \mapsto \mathbf{L}_X$. It is obviously $\mathbb{R}$ and $\mathcal{F}^r(M)$ linear; i.e.

$$\mathbf{L}_{X_1 + fX_2} = \mathbf{L}_{X_1} + f\mathbf{L}_{X_2}.$$

To show that it is one-to-one, we must show that $\mathbf{L}_X = 0$ implies $X = 0$. But if $\mathbf{L}_X f(m) = 0$, then $\mathbf{d}f(m) \cdot X(m) = 0$ for all f. Hence, $\alpha_m(X(m)) = 0$ for all $\alpha_m \in T_m{}^*M$ by **4.2.14**. Thus $X(m) = 0$ by the Hahn-Banach theorem.

(ii) This has an identical proof with the only exception that one works in a local chart, so it is not necessary to extend a linear functional to the entire manifold M as in **4.2.14.** Thus the condition on the differentiability of the norm of the model space of M can be dropped. ■

4.2.16 Derivation Theorem

(i) *If* M *is finite dimensional and* C^∞, *the collection of all derivations on* $\mathcal{F}(M)$ *is a real vector space isomorphic to* $\mathfrak{X}(M)$. *In particular, for each derivation* θ *there is a unique* $X \in \mathfrak{X}(M)$ *such that* $\theta = \mathbf{L}_X$.

(ii) *Let* M *be a* C^∞ *manifold modeled on a* C^∞ *Banach space* E, *i.e.,* E *has a* C^∞ *norm away from the origin. The collection of all* ($\mathbb{R}$*-linear*) *derivations on* $C^\infty(M, F)$ *(for all Banach spaces* F*) forms a real vector space isomorphic to* $\mathfrak{X}(M)$.

Proof We prove (ii) first. Let θ be a derivation. We wish to construct X such that $\theta = \mathbf{L}_X$. First of all, note that θ is a ***local operator***; that is, if $h \in C^\infty(M, F)$ vanishes on a neighborhood V of m, then $\theta(h)(m) = 0$. Indeed, let g be a bump function equal to one on a neighborhood of m and zero outside V. Thus $h = (1 - g)h$ and so

$$\theta(h)(m) = \theta(1 - g)(m) \cdot h(m) + \theta(h)(m)(1 - g(m)) = 0. \qquad (1)$$

If U is an open set in M, and $f \in C^\infty(U, F)$ define $(\theta \mid U)(f)(m) = \theta(gf)(m)$, where g is a bump function equal to one on a neighborhood of m and zero outside U. By the previous remark, $(\theta \mid U)(f)(m)$ is independent of g, so $\theta \mid U$ is well defined. For convenience we write $\theta = \theta \mid U$.

Let (U, φ) be a chart on M, $m \in U$, and $f \in C^\infty(M, F)$ where $\varphi : U \to U' \subset E$; we can write, for $x \in U'$ and $a = \varphi(m)$,

$$\begin{aligned}(\varphi_* f)(x) &= (\varphi_* f)(a) + \int_0^1 \frac{\partial}{\partial t}(\varphi_* f)[a + t(x - a)]\,dt \\ &= (\varphi_* f)(a) + \int_0^1 \mathbf{D}(\varphi_* f)[a + t(x - a)] \cdot (x - a)\,dt.\end{aligned}$$

This formula holds in some neighborhood $\varphi(V)$ of a. Hence for $u \in V$ we have

$$f(u) = f(m) + g(u) \cdot (\varphi(u) - a), \tag{2}$$

where $g \in C^\infty(V, L(E, F))$ is given by $g(u) = \int_0^1 \mathbf{D}(\varphi_* f)[a + t(\varphi(u) - a)]\,dt$. Applying θ to (2) at $u = m$ gives

$$\theta f(m) = g(m) \cdot (\theta\varphi)(m) = \mathbf{D}(\varphi_* f)(a) \cdot (\theta\varphi)(m). \tag{3}$$

Since θ was given globally, the right hand side of (3) is independent of the chart. Now define X on U by its local representative

$$X_\varphi(x) = (x, \theta(\varphi)(u)),$$

where $x = \varphi(u) \in U'$. It follows that $X \mid U$ is independent of the chart φ and hence $X \in \mathfrak{X}(M)$. Then, for $f \in C^\infty(M, F)$, the local representative of $\mathbf{L}_X f$ is

$$\mathbf{D}(f \circ \varphi^{-1})(x) \cdot X_\varphi(x) = \mathbf{D}(f \circ \varphi^{-1})(x) \cdot (\theta\varphi)(u) = \theta f(u).$$

Hence $\mathbf{L}_X = \theta$. Finally, uniqueness follows from **4.2.15.**

The vector derivative property was used only in establishing (1) and (3). Thus, if M is finite dimensional and θ is a derivation on $\mathcal{F}(M)$, we have as before

$$f(u) = f(m) + g(u) \cdot (\varphi(u) - a) = f(m) + \sum_{i=1}^{n} (\varphi^i(u) - a^i) g_i(u),$$

where $g_i \in \mathcal{F}(V)$ and

$$g_i(m) = \left.\frac{\partial(\varphi_* f)(u)}{\partial x^i}\right|_{u=a}, \quad a = (a^1, \ldots, a^n).$$

Hence (3) becomes

$$\theta f(m) = \sum_{i=1}^{n} g_i(m)\, \theta(\varphi^i)(m) = \sum_{i=1}^{n} \frac{\partial}{\partial x^i} (\varphi_* f)(a)\theta(\varphi^i)(m)$$

and this is again independent of the chart. Now define X on U by its local representative

$$(x, \theta(\varphi^1)(u), \ldots, \theta(\varphi^n)(u))$$

and proceed as before. ■

Remark There is a difficulty with this proof for C^r manifolds and derivations mapping C^r to C^{r-1}. Indeed in (2), g is only C^{r-1} if f is C^r, so θ need not be defined on g. Thus, one has to regard θ as defined on C^{r-1}-functions and taking values in C^{r-2}-functions. Therefore $\theta(\varphi)$ is only C^{r-1} and so the vector field it defines is also only C^{r-2}. Then the above proof shows that $\mathcal{L}_X$ and θ coincide on C^r-functions on M. In Supplement **4.2D** we will prove that $\mathcal{L}_X$ and θ are in fact equal on C^{r-1}-functions, but the proof requires a different argument. ♦

For finite-dimensional manifolds, the preceding theorem provides a local basis for vector fields. If (U, φ), $\varphi : U \to V \subset \mathbb{R}^n$ is a chart on M defning the coordinate functions $x^i : U \to \mathbb{R}$, define n derivations $\partial/\partial x^i$ on $\mathcal{F}(U)$ by

$$\frac{\partial f}{\partial x^i} = \frac{\partial (f \circ \varphi^{-1})}{\partial x^i} \circ \varphi.$$

These derivations are linearly independent with coefficients in $\mathcal{F}(U)$, for if

$$\sum_{i=1}^{n} f^i \frac{\partial}{\partial x^i} = 0, \quad \text{then} \quad \left(\sum_{i=1}^{n} f^i \frac{\partial}{\partial x^i} \right)(x^j) = f^j = 0 \quad \text{for all } j = 1, \ldots, n,$$

since $(\partial/\partial x^i)x^j = \delta^j_i$. By **4.2.16**, $(\partial/\partial x^i)$ can be identified with vector fields on U. Moreover, if $X \in \mathfrak{X}(M)$ has components $X^1, \ldots, X^n$ in the chart φ, then

$$\mathcal{L}_X f = X[f] = \sum_{i=1}^{n} X^i \frac{\partial f}{\partial x^i} = \left(\sum_{i=1}^{n} X^i \frac{\partial}{\partial x^i} \right) f, \quad \text{i.e.,} \quad X = \sum_{i=1}^{n} X^i \frac{\partial}{\partial x^i}.$$

Thus the vector fields $(\partial/\partial x^i)$, $i = 1, \ldots, n$ form a ***local basis*** for the vector fields on M. It should be mentioned however that a ***global basis*** of $\mathfrak{X}(M)$, i.e., n vector fields, $X_1, \ldots, X_n \in \mathfrak{X}(M)$ that are linearly independent over $\mathcal{F}(M)$ and span $\mathfrak{X}(M)$, does not exist in general. Manifolds that do admit such a global basis for $\mathfrak{X}(M)$ are called ***parallelizable.*** It is straightforward to show that a finite-dimensional manifold is parallelizable iff its tangent bundle is trivial. For example, it is shown in differential topology that S^3 is parallelizable but S^2 is not (see Supplement **7.5A**).

This completes the discussion of the Lie derivative on functions. Turning to the Lie derivative on vector fields, let us begin with the following.

4.2.17 Proposition *If* X *and* Y *are* C^r *vector fields on* M, *then*

$$[\mathbf{L}_X, \mathbf{L}_Y] = \mathbf{L}_X \circ \mathbf{L}_Y - \mathbf{L}_Y \circ \mathbf{L}_X$$

is a derivation mapping $C^{r+1}(M, F)$ *to* $C^{r-1}(M, F)$.

Proof More generally, let θ_1 and θ_2 be two derivations mapping C^{r+1} to C^r and C^r to C^{r-1}. Clearly $[\theta_1, \theta_2] = \theta_1 \circ \theta_2 - \theta_2 \circ \theta_1$ is linear and maps C^{r+1} to C^{r-1}. Also, if $f \in C^{r+1}(M, F)$, $g \in C^{r+1}(M, G)$, and $B \in L(F, G; H)$, then

$$\begin{aligned}
[\theta_1, \theta_2](B(f, g)) &= (\theta_1 \circ \theta_2)(B(f, g)) - (\theta_2 \circ \theta_1)(B(f, g)) \\
&= \theta_1\{B(\theta_2(f), g) + B(f, \theta_2(g))\} - \theta_2\{B(\theta_1(f), g) + B(f, \theta_1(g))\} \\
&= B(\theta_1(\theta_2(f)), g) + B(\theta_2(f), \theta_1(g)) + B(\theta_1(f), \theta_2(g)) \\
&\quad + B(f, \theta_1(\theta_2(g))) - B(\theta_2(\theta_1(f)), g) - B(\theta_1(f), \theta_2(g)) \\
&\quad - B(\theta_2(f), \theta_1(g)) - B(f, \theta_2(\theta_1(g))) \\
&= B([\theta_1, \theta_2](f), g) + B(f, [\theta_1, \theta_2](g)). \blacksquare
\end{aligned}$$

Because of **4.2.16** the following definition can be given.

4.2.18 Definition *Let* M *be a manifold modeled on a* C^∞ *Banach space and* $X, Y \in \mathfrak{X}^\infty(M)$. *Then* [X, Y] *is the unique vector field such that* $\mathbf{L}_{[X, Y]} = [\mathbf{L}_X, \mathbf{L}_Y]$. *This vector field is also denoted* $\mathbf{L}_X Y$ *and is called the* ***Lie derivative of*** Y ***with respect to*** X, ***or the Jacobi-Lie bracket*** *of* X *and* Y.

Even though this definition is useful for Hilbert manifolds (in particular for finite-dimensional manifolds), it excludes consideration of C^r vector fields on Banach manifolds modeled on nonsmooth Banach spaces, such as L^p function spaces for p not even. We shall, however, establish an equivalent definition, which makes sense on any Banach manifold and works for C^r vector fields. This alternative definition is based on the following result.

4.2.19 Lie Derivative Formula for Vector Fields *Let* M *be as in* **4.2.18**, $X, Y \in \mathfrak{X}(M)$, *and let* X *have (local) flow* F_t. *Then*

$$\boxed{\frac{d}{dt}(F_t^*Y) = F_t^*(\mathcal{L}_X Y)}$$

(at those points where F_t *is defined).*

Proof. If $t = 0$ this formula becomes

$$\left.\frac{d}{dt}\right|_{t=0} F_t^* Y = \mathcal{L}_X Y. \tag{4}$$

Assuming (4) for the moment,

$$\frac{d}{dt}(F_t^*Y) = \left.\frac{d}{ds}\right|_{s=0} F_{t+s}^* Y = F_t^* \left.\frac{d}{ds}\right|_{s=0} F_s^* Y = F_t^* \mathcal{L}_X Y.$$

Thus the formula in the theorem is equivalent to (4), which is proved in the following way. Both sides of (4) are clearly vector derivations. In view of **4.2.16**, it suffices then to prove that both sides are equal when acting on an aribtrary function $f \in C^\infty(M, F)$. Now

$$\left.\frac{d}{dt}\right|_{t=0} (F_t^*Y)[f](m) = \left.\frac{d}{dt}\right|_{t=0} \{\mathcal{L}_{F_t^*Y}[F_t^*(F_{-t}^* f)]\}(m) = \left.\frac{d}{dt}\right|_{t=0} F_t^*(Y[F_{-t}^* f])(m),$$

by **4.2.7**(i). Using **4.2.10** and Leibniz' rule, this becomes

$$X[Y[f]](m) - Y[X[f]](m) = [X, Y][f](m). \blacksquare$$

Since the formula for $\mathcal{L}_X Y$ in Eq. (4) does not use the fact that the norm of E is C^∞ away from the origin, we can state the following definition of the Lie derivative on any Banach manifold M.

4.2.20 Dynamic Definition of Jacobi-Lie bracket *If* $X, Y \in \mathfrak{X}^r(M)$, $r \geq 1$ *and* X *has flow* F_t, *the* C^{r-1} *vector field* $\mathcal{L}_X Y = [X, Y]$ *on* M *defined by*

$$[X, Y] = \left.\frac{d}{dt}\right|_{t=0} (F_t^* Y)$$

*is called the **Lie derivative** of* Y *with respect to* X, *or the **Lie bracket** of* X *and* Y.

From the point of view of this more general definition, **4.2.19** can be rephrased as follows.

4.2.21 Proposition *Let* $X, Y \in \mathfrak{X}^r(M)$, $1 \leq r$. *Then* $[X, Y] = \mathcal{L}_X Y$ *is the unique* C^{r-1} *vector field on* M *satisfying*

$$[X, Y][f] = X[Y[f]] - Y[X[f]]$$

for all $f \in C^{r+1}(U, F)$, *where* U *is open in* M.

The derivation approach suggessts that if $X, Y \in \mathfrak{X}^r(M)$ then $[X, Y]$ might only be C^{r-2}, since $[X, Y]$ maps C^{r+1} functions to C^{r-1} functions, and differentiates them twice. However **4.2.20** (and the coordinate expression (6) below) show that $[X, Y]$ is in fact C^{r-1}.

4.2.22 Proposition *The bracket* $[X, Y]$ *on* $\mathfrak{X}(M)$, *together with the real vector space structure* $\mathfrak{X}(M)$, *form a* ***Lie algebra.*** *That is,*

(i) $[\ ,\ ]$ *is* $\mathbb{R}$ *bilinear;*

(ii) $[X, X] = 0$ *for all* $X \in \mathfrak{X}(M)$;

(iii) $[X, [Y, Z]] + [Y, [Z, X]] + [Z, [X, Y]] = 0$ *for all* $X, Y, Z \in \mathfrak{X}(M)$ *(**Jacobi identity**).*

The proof is straightforward, applying the brackets in question to an arbitrary function. Unlike $\mathfrak{X}(M)$, the space $\mathfrak{X}^r(M)$ is not a Lie algebra since $[X, Y] \in \mathfrak{X}^{r-1}(M)$ for $X, Y \in \mathfrak{X}^r(M)$. (i) and (ii) imply that $[X, Y] = -[Y, X]$, since $[X + Y, X + Y] = 0 = [X, X] + [X, Y] + [Y, X] + [Y, Y]$. We can describe (iii) by writing $\mathcal{L}_X$ as a ***Lie bracket derivation:***

$$\mathcal{L}_X[Y, Z] = [\mathcal{L}_X Y, Z] + [Y, \mathcal{L}_X Z].$$

Strictly speaking we should be careful using the same symbol $\mathcal{L}_X$ for both definitions of $\mathcal{L}_X f$ and $\mathcal{L}_X Y$. However, the meaning is generally clear from the context. The analog of **4.2.7** on the vector field level is the following.

4.2.23 Proposition

(i) *Let* $\varphi : M \to N$ *be a diffeomorphism and* $X \in \mathfrak{X}(M)$. *Then* $\mathcal{L}_X : \mathfrak{X}(M) \to \mathfrak{X}(M)$ *is* ***natural with respect to push-forward*** *by* φ. *That is,*

$$\mathcal{L}_{\varphi_* X} \varphi_* Y = \varphi_* \mathcal{L}_X Y,$$

or $[\varphi_* X, \varphi_* Y] = \varphi_*[X, Y]$, *or the following diagram commutes:*

$$\begin{array}{ccc} \mathfrak{X}(M) & \xrightarrow{\ \varphi_*\ } & \mathfrak{X}(N) \\ \Big\downarrow {\scriptstyle \mathcal{L}_X} & & \Big\downarrow {\scriptstyle \mathcal{L}_{\varphi_* X}} \\ \mathfrak{X}(M) & \xrightarrow[\ \varphi_*\ ]{} & \mathfrak{X}(N) \end{array}$$

(ii) $\mathcal{L}_X$ *is* ***natural with respect to restrictions.*** *That is, for* $U \subset M$ *open,* $[X \mid U, Y \mid U] = [X, Y] \mid U$; *or the following diagram commutes:*

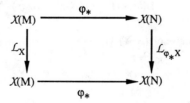

$$\begin{array}{ccc}
\mathfrak{X}(M) & \xrightarrow{\;|U\;} & \mathfrak{X}(U) \\
\Big\downarrow \mathcal{L}_X & & \Big\downarrow \mathcal{L}_{X|U} \\
\mathfrak{X}(M) & \xrightarrow[\;|U\;]{} & \mathfrak{X}(U)
\end{array}$$

Proof For (i), let $f \in \mathcal{F}(V)$, V be open in N, and $\varphi(m) = n \in V$. By **4.2.7(i)**, for any $Z \in \mathfrak{X}(M)$

$$((\varphi_* Z)[f])(n) = Z[f \circ \varphi](m),$$

so we get from **4.2.21**

$$\begin{aligned}
(\varphi_*[X, Y])[f](n) &= [X, Y][f \circ \varphi](m) \\
&= X[Y[f \circ \varphi]](m) - Y[X[f \circ \varphi]](m) \\
&= X[(\varphi_* Y)[f] \circ \varphi](m) - Y[(\varphi_* X)[f] \circ \varphi](m) \\
&= (\varphi_* X)[(\varphi_* Y)[f]](n) - (\varphi_* Y)[(\varphi_* X)[f]](n) \\
&= [\varphi_* X, \varphi_* Y][f](n).
\end{aligned}$$

Thus $\varphi_*[X, Y] = [\varphi_* X, \varphi_* Y]$ by **4.2.15**(ii). (ii) follows from the fact that $\mathbf{d}(f \mid U) = \mathbf{d}f \mid U$. ■

Let us compute the local expression for $[X, Y]$. Let $\varphi : U \to V \subset E$ be a chart on M and let the local representatives of X and Y be X and Y respectively, so $\mathsf{X}, \mathsf{Y} : V \to E$. By **4.2.23**, the local representative of $[X, Y]$ is $[\mathsf{X}, \mathsf{Y}]$. Thus,

$$\begin{aligned}
[\mathsf{X}, \mathsf{Y}][f](x) &= \mathsf{X}[\mathsf{Y}[f]](x) - \mathsf{Y}[\mathsf{X}[f]](x) \\
&= \mathbf{D}(\mathsf{Y}[f])(x) \cdot \mathsf{X}(x) - \mathbf{D}(\mathsf{X}[f])(x) \cdot \mathsf{Y}(x).
\end{aligned}$$

Now $\mathsf{Y}[f](x) = \mathbf{D}f(x) \cdot \mathsf{Y}(x)$ and its derivative may be computed by the product rule. The terms involving the second derivative of f cancel by symmetry of $\mathbf{D}^2 f(x)$ and so we are left with

$$\mathbf{D}f(x) \cdot \{\mathbf{D}\mathsf{Y}(x) \cdot \mathsf{X}(x) - \mathbf{D}\mathsf{X}(x) \cdot \mathsf{Y}(x)\}.$$

Thus the local representative of $[X, Y]$ is

$$[\mathsf{X}, \mathsf{Y}] = \mathbf{D}\mathsf{Y} \cdot \mathsf{X} - \mathbf{D}\mathsf{X} \cdot \mathsf{Y}. \tag{5}$$

If M is n-dimensional and the chart φ gives local coordinates $(x^1, \ldots, x^n)$ then this calculation gives the components of $[X, Y]$ as

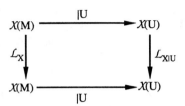

$$[X, Y]^j = \sum_{i=1}^{n} X^i \frac{\partial Y^j}{\partial x^i} - Y^i \frac{\partial X^j}{\partial x^i} \tag{6}$$

i.e. $[X, Y] = (X \cdot \nabla)Y - (Y \cdot \nabla)X$.

Part (i) of **4.2.22** has an important generalization to φ-related vector fields. For this, however, we need first the following preparatory proposition.

4.2.24 Proposition *Let* $\varphi : M \to N$ *be a* C^r *map of* C^r *manifolds,* $X \in \mathfrak{X}^{r-1}(M)$, *and* $X' \in \mathfrak{X}^{r-1}(N)$. *Then* $X \sim_\varphi X'$ *iff* $(X'[f]) \circ \varphi = X[f \circ \varphi]$ *for all* $f \in \mathcal{F}^1(V)$, *where* V *is open in* N.

Proof By definition, $((X'[f]) \circ \varphi)(m) = \mathbf{d}f(\varphi(m)) \cdot X'(\varphi(m))$. By the chain rule,

$$X[f \circ \varphi](m) = \mathbf{d}(f \circ \varphi)(m) \cdot X(m) = \mathbf{d}f(\varphi(m)) \cdot T_m\varphi(X(m)).$$

If $X \sim_\varphi X'$, then $T\varphi \circ X = X' \circ \varphi$ and we have the desired equality. Conversely, if $X[f \circ \varphi] = (X'[f]) \circ \varphi$ for all $f \in \mathcal{F}^1(V)$, and all V open in N, choosing V to be a chart domain and f the pull-back to V of linear functionals on the model space of N, we conclude that $\alpha_n \cdot (X' \circ \varphi)(m) = \alpha_n \cdot (T\varphi \circ X)(m)$, where $n = \varphi(m)$, for all $\alpha_n \in T_n^*N$. Using the Hahn-Banach theorem, we deduce that $(X' \circ \varphi)(m) = (T\varphi \circ X)(m)$, for all $m \in M$. ■

It is to be noted that under differentiability assumptions on the norm on the model space of N (as in **4.2.16**), the condition "for all $f \in \mathcal{F}^1(V)$ and all $V \subset N$" can be replaced by "for all $f \in \mathcal{F}^1(N)$" by using bump functions. This holds in particular for Hilbert (and hence for finite-dimensional) manifolds.

4.2.25 Proposition *Let* $\varphi : M \to N$ *be a* C^r *map of manifolds,* $X, Y \in \mathfrak{X}^{r-1}(M)$, *and* $X', Y' \in \mathfrak{X}^{r-1}(N)$. *If* $X \sim_\varphi X'$ *and* $Y \sim_\varphi Y'$, *then* $[X, Y] \sim_\varphi [X', Y']$.

Proof By **4.2.24** it suffices to show that $([X', Y'][f]) \circ \varphi = [X, Y][f \circ \varphi]$ for all $f \in \mathcal{F}^1(V)$, where V is open in N. We have

$$\begin{aligned}([X', Y'][f]) \circ \varphi &= X'[Y'[f]] \circ \varphi - Y'[X'[f]] \circ \varphi = X[(Y'[f]) \circ \varphi] - Y[(X'[f]) \circ \varphi] \\ &= X[Y[f \circ \varphi]] - Y[X[f \circ \varphi]] = [X, Y][f \circ \varphi]. \ ■\end{aligned}$$

The analog of **4.2.9** is the following.

4.2.26 Proposition *For every* $X \in \mathfrak{X}(M)$, $\mathbf{L}_X$ *is a derivation on* $(\mathcal{F}(M), \mathfrak{X}(M))$. *That is,* $\mathbf{L}_X$ *is* $\mathbb{R}$*-linear and* $\mathbf{L}_X(fY) = (\mathbf{L}_X f)Y + f(\mathbf{L}_X Y)$.

Proof For $g \in C^\infty(U, E)$, where U is open in M, we have

$$[X, fY][g] = \mathcal{L}_X(\mathcal{L}_{fY}g) - \mathcal{L}_{fY}\mathcal{L}_X g = \mathcal{L}_X(f\mathcal{L}_Y g) - f\mathcal{L}_Y\mathcal{L}_X g$$
$$= (\mathcal{L}_X f)\mathcal{L}_Y g + f\mathcal{L}_X \mathcal{L}_Y g - f\mathcal{L}_Y\mathcal{L}_X g,$$

so

$$[X, fY] = (\mathcal{L}_X f)Y + f\,[X, Y] \text{ by } \mathbf{4.2.15}\text{(ii)}. \quad \blacksquare$$

Commutation of vector fields is characterized by their flows in the following way.

4.2.27 Proposition *Let* $X, Y \in \mathfrak{X}^r(M)$, $r \geq 1$, *and let* F_t, G_t *denote their flows. The following are equivalent.*

(i) $[X, Y] = 0$;

(ii) $F_t^*Y = Y$;

(iii) $G_t^*X = X$;

(iv) $F_t \circ G_s = G_s \circ F_t$.

(In (ii) - (iv), *equality is understood, as usual, where the expressions are defined.)*

Proof $F_t \circ G_s = G_s \circ F_t$ iff $G_s = F_t \circ G_s \circ F_t^{-1}$, which by **4.2.4** is equivalent to $Y = F_t^*Y$; i.e., (iv) is equivalent to (ii). Similarly (iv) is equivalent to (iii). If $F_t^*Y = Y$, then

$$[X, Y] = \left.\frac{d}{dt}\right|_{t=0} F_t^*Y = 0.$$

Conversely, if $[X, Y] = \mathcal{L}_X Y = 0$, then

$$\frac{d}{dt} F_t^*Y = \left.\frac{d}{ds}\right|_{s=0} F_{t+s}^*Y = F_t^*[X, Y] = 0$$

so that F_t^*Y is constant in t. For $t = 0$, however, its value is Y, so that $F_t^*Y = Y$ and we have thus showed that (i) and (ii) are equivalent. Similarly (i) and (iii) are equivalent. ■

Just as in **4.2.10**, the formula for the Lie derivative involving the flow can be used to solve special types of first-order linear $n \times n$ systems of partial differential equations. Consider the first-order system:

$$\frac{\partial Y^i}{\partial t}(x, t) = \sum_{j=1}^{n} \left(X^j(x) \frac{\partial Y^i}{\partial x^j}(x, t) - Y^j(x, t) \frac{\partial X^i(x, t)}{\partial x^j} \right) \qquad (P_n)$$

with initial conditions $Y^i(x, 0) = g^i(x)$ for given functions $X^i(x)$, $g^i(x)$ and scalar unknowns $Y^i(x, t)$, $i = 1, \ldots, n$, where $x = (x^1, \ldots, x^n)$.

4.2.28 Proposition *Suppose* $X = (X^1, \ldots, X^n)$ *has a complete flow* F_t. *Then letting* $Y = (Y^1, \ldots, Y^n)$ *and* $G = (g^1, \ldots, g^n)$, $Y = F_t^*G$ *is a solution of the foregoing problem* (P_n). *(See Exercise* **4.2C** *for uniqueness).*

Proof

$$\frac{\partial Y}{\partial t} = \frac{d}{dt} F_t^* G = F_t^*[X, G] = [F_t^* X, F_t^* G] = [X, Y]$$

since $F_t{}^*X = X$ and $Y = F_t{}^*G$. The expression in the problem (P_n) is by (6) the i-th component of $[X, Y]$. ■

4.2.29 Example Solve the system of partial differential equations:

$$\frac{\partial Y^1}{\partial t} = (x+y)\frac{\partial Y^1}{\partial x} - (x+y)\frac{\partial Y^1}{\partial y} - Y^1 - Y^2$$

$$\frac{\partial Y^2}{\partial t} = (x+y)\frac{\partial Y^2}{\partial x} - (x+y)\frac{\partial Y^2}{\partial y} + Y^1 + Y^2$$

with initial conditions $Y^1(x, y, 0) = x$, $Y^2(x, y, 0) = y^2$. The vector field $X(x, y) = (x + y, -x - y)$ has the complete flow $F_t(x, y) = ((x + y)t + x, -(x + y)t + y)$, so that the solution is given by $Y(x, y, t) = F_t{}^*(x, y^2)$; i.e.,

$$Y^1(x, y, t) = ((x+y)t + x)(1-t) - t[y - (x+y)t]^2$$

$$Y^2(x, y, t) = t((x+y)t + x) + (t+1)[y - (x+y)t]^2 . \blacklozenge$$

In later chapters we will need a flow type formula for the Lie derivative of a *time-dependent* vector field, In §**4.1** we discussed the existence and uniqueness of solutions of a time-dependent vector field. Let us formalize and recall the basic facts.

4.2.30 Definition *A* C^r ***time-dependent vector field*** *is a* C^r *map* $X: \mathbb{R} \times M \to TM$ *such that* $X(t, m) \in T_mM$ *for all* $(t, m) \in \mathbb{R} \times M$; i.e., $X_t \in \mathfrak{X}^r(M)$, *where* $X_t(m) = X(t, m)$. *The* ***time-dependent flow*** *or* ***evolution operator*** $F_{t,s}$ *of* X *is defined by the requirement that* $t \mapsto F_{t,s}(m)$ *be the integral curve of* X *starting at* m *at time* $t = s$; *i.e.,*

$$\frac{d}{dt} F_{t,s}(m) = X(t, F_{t,s}(m)) \quad \text{and} \quad F_{s,s}(m) = m.$$

By uniqueness of integral curves we have $F_{t,s} \circ F_{s,r} = F_{t,r}$ (replacing the flow property $F_{t+s} = F_t \circ F_s$), and $F_{t,t} =$ identity. It is customary to write $F_t = F_{t,0}$. If X is time independent, $F_{t,s} = F_{t-s}$. In general $F_t{}^*X_t \neq X_t$. However, the basic Lie derivative formulae still hold.

4.2.31 Theorem *Let* $X_t \in \mathfrak{X}^r(M)$, $r \geq 1$ *for each* t *and suppose* $X(t, m)$ *is continuous in* (t, m). *Then* $F_{t,s}$ *is of class* C^r *and for* $f \in C^{r+1}(M, F)$, *and* $Y \in \mathfrak{X}^r(M)$, *we have*

$$\frac{d}{dt} F_{t,s}^* f = F_{t,s}^*(\mathcal{L}_{X_t} f), \quad \text{and} \tag{i}$$

$$\frac{d}{dt} F_{t,s}^* Y = F_{t,s}^*([X_t, Y]) = F_{t,s}^*(\mathcal{L}_{X_t} Y). \tag{ii}$$

Proof That $F_{t,s}$ is C^r was proved in §4.1. The proof of (i) is a repeat of **4.2.10**:

$$\frac{d}{dt} (F_{t,s}^* f)(m) = \frac{d}{dt} (f \circ F_{t,s})(m) = df(F_{t,s}(m)) \frac{dF_{t,s}(m)}{dt} = df(F_{t,s}(m)) \cdot X_t(F_{t,s}(m))$$

$$= (\mathcal{L}_{X_t} f)(F_{t,s}(m)) = F_{t,s}^* (\mathcal{L}_{X_t} f)(m) .$$

For vector fields, note that by **4.2.7**(i),

$$(F_{t,s}^* Y)[f] = F_{t,s}^*(Y[F_{t,s}^* f]) \tag{7}$$

since $F_{s,t} = F_{t,s}^{-1}$. The result (ii) will be shown to follow from (i), (7), and the next lemma.

4.2.32 Lemma *The following identity holds:*

$$\frac{d}{dt} F_{s,t}^* f = -X_t[F_{s,t}^* f].$$

Proof Differentiating $F_{s,t} \circ F_{t,s} =$ identity in t, we get the *backward differential equation*:

$$\frac{d}{dt} F_{s,t} = -TF_{s,t} \circ X$$

Thus

$$\frac{d}{dt} F_{s,t}^* f(m) = -df(F_{s,t}(m)) \cdot TF_{s,t}(X_t(m))$$

$$= - df(f \circ F_{s,t}) \cdot X_t(m) = -X_t[f \circ F_{s,t}](m). \blacktriangledown$$

Thus from (7) and (i),

$$\frac{d}{dt} (F_{t,s}^* Y)[f] = F_{t,s}^*(X_t[Y[F_{t,s}^* f]]) - F_{t,s}^*(Y[X_t[F_{t,s}^* f]]).$$

By **4.2.21** and (7), this equals $(F_{t,s}^*[X_t, Y])[f]$. ∎

If f and Y are time dependent, then (i) and (ii) read

$$\frac{d}{dt} F_{t,s}^* f = F_{t,s}^* \left(\frac{\partial f}{\partial t} + \mathcal{L}_{X_t} f \right) \tag{8}$$

and

$$\frac{d}{dt} F_{t,s}^* Y = F_{t,s}^*\left(\frac{\partial Y}{\partial t} + [X_t, Y_t]\right). \tag{9}$$

Unlike the corresponding formula for time-independent vector fields, we generally have

$$F_{t,s}^*(\mathcal{L}_{X_t} f) \neq \mathcal{L}_{X_t}(F_{t,s}^* f) \quad \text{and} \quad F_{t,s}^*(\mathcal{L}_{X_t} Y) \neq \mathcal{L}_{F_{t,s}^* X_t} X_t(F_{t,s}^* Y).$$

Time-dependent vector fields on M can be made into time-independent ones on a bigger manifold. Let $t \in \mathfrak{X}(\mathbb{R} \times M)$ denote the vector field which is defined by $t(s, m) = ((s, 1), 0_m) \in T_{(s,m)}(\mathbb{R} \times M) \cong T_s\mathbb{R} \times T_mM$. Let the ***suspension*** of X be the vector field $X' \in \mathfrak{X}(\mathbb{R} \times M)$ where $X'(t, m) = ((t, 1), X(t, m))$ and observe that $X' = t + X$. Since $b : I \to M$ is an integral curve of X at m iff $b'(t) = X(t, b(t))$ and $b(0) = m$, a curve $c : I \to \mathbb{R} \times M$ is an integral curve of X' at $(0, m)$ iff $c(t) = (t, b(t))$. Indeed, if $c(t) = (a(t), b(t))$ then $c(t)$ is an integral curve of X' iff $c'(t) = (a'(t), b'(t)) = X'(c(t))$; that is $a'(t) = 1$ and $b'(t) = X(a(t), b(t))$. Since $a(0) = 0$, we get $a(t) = t$. These observations are summarized in the following (see Fig. **4.2.4**).

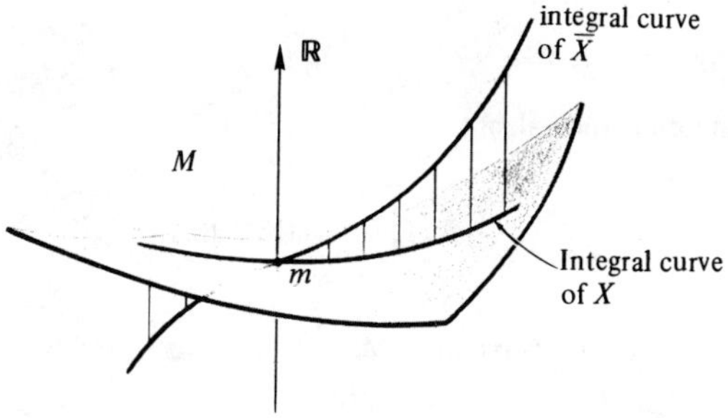

Figure 4.2.4

4.2.33 Proposition *Let* X *be a* C^r*-time-dependent vector field on* M *with evolution operator* $F_{t,s}$. *The flow* F_t *of the suspension* $X' \in \mathfrak{X}^r(\mathbb{R} \times M)$ *is given by* $F_t(s, m) = (t + s, F_{t+s,s}(m))$.

Proof In the preceding notations, $b(t) = F_{t,0}(m)$, $c(t) = F_t(0, m) = (t, F_{t,0}(m))$, and so the statement is proved for $s = 0$. In general, note that $F_t(s, m) = F_{t+s}(0, F_{0,s}(m))$ since $t \mapsto F_{t+s}(0, F_{0,s}(m))$ is the integral curve of X', which at $t = 0$ passes through $F_s(0, F_{0,s}(m)) = (s, (F_{s,0} \circ F_{0,s})m)) = (s, m)$. Thus $F_t(s, m) = F_{t+s}(0, F_{0,s}(m)) = (t + s, (F_{t+s,0} \circ F_{0,s})(m)) = (t + s, F_{t+s,s}(m))$. ■

☞ Supplement 4.2A
Product formulas for the Lie bracket

This box is a continuation of Supplement **4.1A** and gives the flow of the Lie bracket [X, Y] in terms of the flows of the vector fields $X, Y \in \mathfrak{X}(M)$.

4.2.34 Proposition *Let* $X, Y \in \mathfrak{X}(M)$ *have flows* F_t *and* G_t. *If* B_t *denotes the flow of* [X, Y], *then for* $x \in M$,

$$B_t(x) = \lim_{n\to\infty} (G_{-\sqrt{t/n}} \circ F_{-\sqrt{t/n}} \circ G_{\sqrt{t/n}} \circ F_{\sqrt{t/n}})^n(x), \quad t \geq 0.$$

Proof Let

$$K_\varepsilon(x) = (G_{-\sqrt{\varepsilon}} \circ F_{-\sqrt{\varepsilon}} \circ G_{\sqrt{\varepsilon}} \circ F_{\sqrt{\varepsilon}})(x), \quad \varepsilon > 0.$$

The claimed formula follows from **4.1.24** if we show that

$$\frac{\partial}{\partial\varepsilon} K_\varepsilon(x)\bigg|_{\varepsilon=0} = [X, Y](x)$$

for all $x \in M$. This in turn is equivalent to

$$\frac{\partial}{\partial\varepsilon} K_\varepsilon^* f\bigg|_{\varepsilon=0} = [X, Y](x)$$

for any $f \in \mathcal{F}(M)$. By the Lie derivative formula,

$$\frac{\partial}{\partial\varepsilon} K_\varepsilon^* f = \frac{1}{2\sqrt{\varepsilon}} \{F_{\sqrt{\varepsilon}}^* \boldsymbol{L}_X(G_{\sqrt{\varepsilon}}^* F_{-\sqrt{\varepsilon}}^* G_{-\sqrt{\varepsilon}}^* f) + F_{\sqrt{\varepsilon}}^* G_{\sqrt{\varepsilon}}^* \boldsymbol{L}_Y(F_{-\sqrt{\varepsilon}}^* G_{-\sqrt{\varepsilon}}^* f)$$
$$- F_{\sqrt{\varepsilon}}^* G_{\sqrt{\varepsilon}}^* F_{-\sqrt{\varepsilon}}^* \boldsymbol{L}_X(G_{-\sqrt{\varepsilon}}^* f) - F_{\sqrt{\varepsilon}}^* G_{\sqrt{\varepsilon}}^* F_{-\sqrt{\varepsilon}}^* G_{-\sqrt{\varepsilon}}^* \boldsymbol{L}_Y(f)\}.$$

By the chain rule, the limit of this as $\varepsilon \downarrow 0$ is half the $\partial/\partial\sqrt{\varepsilon}$ -derivative of the parenthesis at $\varepsilon = 0$. Again by the Lie derivative formula, this equals a sum of 16 terms, which reduces to the expression [X, Y][f]. ∎

For example, for $n \times n$ matricess A and B, **4.2.34** yields the classical formula

$$e^{[A,B]} = \lim_{n\to\infty} (e^{-A/\sqrt{n}} e^{-B/\sqrt{n}} e^{A/\sqrt{n}} e^{B/\sqrt{n}})^n,$$

where the commutator is given by $[A, B] = AB - BA$. To see this, define for any $n \times n$ matrix C a vector field $X_C \in \mathfrak{X}(\mathbb{R}^n)$ by $X_C(x) = Cx$. Thus X_C has flow $F_t(x) = e^{tC}x$. Note that X_C

is linear in C and satisfies $[X_A, X_B] = -X_{[A,B]}$ as is easily verified. Thus the flow of $[X_B, X_A]$ is $e^{t[A,B]}$ and the formula now follows from **4.2.34.**

The results of **4.1.27** and **4.2.34** had their historical origins in Lie group theory, where they are known by the name of ***exponential formulas.*** The converse of **4.1.25**, namely expressing $e^{tA}e^{tB}$ as an exponential of some matrix for sufficiently small t is the famous Baker-Campbell-Hausdorff formula (see e.g., Varadarajan [1974, §2.15]). The formulas in **4.1.25** and **4.2.34** have certain generalizations to unbounded operators and are called ***Trotter product formulas*** after Trotter [1958]. See Chorin et. al. [1978] for further information. ✍

☞ Supplement 4.2B
The Method of Characteristics

The method used to solve problem (P) also enables one to solve first-order quasi-linear partial differential equations in $\mathbb{R}^n$. Unlike **4.2.11**, the solution will be implicit, not explicit. The equation under consideration in $\mathbb{R}^n$ is

$$\sum_{i=1}^{n} X^i(x^1, \ldots, x^n, f)\frac{\partial f}{\partial x^i} = Y(x^1, \ldots, x^n, f), \tag{Q}$$

where $f = f(x^1, \ldots, x^n)$ is the unknown function and $X^i, Y, i = 1, \ldots, n$ are C^r real-valued functions on $\mathbb{R}^{n+1}$, $r \geq 1$. As initial condition one takes an (n−1)-dimensional submanifold Γ in $\mathbb{R}^{n+1}$ that is nowhere tangent to the vector field

$$\sum_{i=1}^{n} X^i \frac{\partial}{\partial x^i} + Y\frac{\partial}{\partial f}$$

called the ***characteristic vector field*** of (Q). Thus, if Γ is given parametrically by

$$x^i = x^i(t_1, \ldots, t_{n-1}), \quad i = 1, \ldots, n \quad \text{and} \quad f = f(t_1, \ldots, t_{n-1}),$$

this requirement means that the matrix

$$\begin{bmatrix} X^1 & \cdots & X^n & Y \\ \dfrac{\partial x^1}{\partial t_1} & \cdots & \dfrac{\partial x^n}{\partial t_1} & \dfrac{\partial f}{\partial t_1} \\ \vdots & & \vdots & \vdots \\ \dfrac{\partial x^1}{\partial t_{n-1}} & \cdots & \dfrac{\partial x^n}{\partial t_{n-1}} & \dfrac{\partial f}{\partial t_{n-1}} \end{bmatrix}$$

has rank n. It is customary to require that the determinant obtained by deleting the last column be $\neq 0$, for then, as we shall see, the implicit function theorem gives the solution. The function f is found as follows. Consider F_t, the flow of the vector field $\sum_{i=1,\ldots,n} X^i\partial/\partial x^i + Y\partial/\partial f$ in $\mathbb{R}^{n+1}$ and let S be the manifold obtained by sweeping out Γ by F_t. That is, $S = \bigcup\{F_t(\Gamma) \mid t \in \mathbb{R}\}$. The condition that $\sum_{i=1,\ldots,n} X^i\partial/\partial x^i + Y\partial/\partial f$ never be tangent to Γ insures that the manifold Γ is "dragged along" by the flow F_t to produce a manifold of dimension n. If S is described by $f = f(x^1, \ldots, x^n)$ then f is the solution of the partial differential equation. Indeed, the tangent space to S contains the vector $\sum_{i=1,\ldots,n} X^i\partial/\partial x^i + Y\partial/\partial f$; i.e., this vector is perpendicular to the normal $(\partial f/\partial x^1, \ldots \partial f/\partial x^n, -1)$ to the surface $f = f(x^1, \ldots, x^n)$ and thus (Q) is satisfied.

We work parametrically and write the components of F_t as $x^i = x^i(t_1, \ldots, t_{n-1}, t)$, $i = 1, \ldots, n$ and $f = f(t_1, \ldots, t_{n-1}, t)$. Assuming that

$$0 \neq \begin{vmatrix} X^1 & \cdots & X^n \\ \dfrac{\partial x^1}{\partial t_1} & \cdots & \dfrac{\partial x^n}{\partial t_1} \\ \vdots & \cdots & \vdots \\ \dfrac{\partial x^1}{\partial t_{n-1}} & & \dfrac{\partial x^n}{\partial t_{n-1}} \end{vmatrix} = \begin{vmatrix} \dfrac{\partial x^1}{\partial t} & \cdots & \dfrac{\partial x^n}{\partial t} \\ \dfrac{\partial x^1}{\partial t_1} & \cdots & \dfrac{\partial x^n}{\partial t_1} \\ \vdots & \cdots & \vdots \\ \dfrac{\partial x^1}{\partial t_{n-1}} & & \dfrac{\partial x^n}{\partial t_{n-1}} \end{vmatrix}$$

one can locally invert to give $t = (x^1, \ldots, x^n)$, $t_i = t_i(x^1, \ldots, x^n)$, $i = 1, \ldots, n-1$. Substitution into f yields $f = f(x^1, \ldots, x^n)$.

The fundamental assumption in this construction is that the vector field $\sum_{i=1,\ldots,n} X^i\partial/\partial x^i + Y\partial/\partial f$ is never tangent to the $(n-1)$-manifold Γ. The method breaks down if one uses manifolds Γ, which are tangent to this vector field at some point. The reason is that at such a point, no complete information about the derivative of f in a complementary $(n-1)$-dimensional subspace to the characteristic is known.

4.2.35 Example Consider the equation in $\mathbb{R}^2$ given by

$$\frac{\partial f}{\partial x} + f\frac{\partial f}{\partial y} = 3$$

with initial condition $\Gamma = \{(x, y, f) \mid x = s, y = (1/2)s^2 - s, f = s\}$. On this one-manfold

$$\begin{vmatrix} 1 & f \\ \dfrac{\partial x}{\partial s} & \dfrac{\partial y}{\partial s} \end{vmatrix} = \begin{vmatrix} 1 & s \\ 1 & s-1 \end{vmatrix} = -1 \neq 0$$

so that the vector field $\partial/\partial x + f\partial/\partial y + 3\partial/\partial f$ is never tangent to Γ. Its flow is $F_t(x, y, f) = (t + x, (3/2)t^2 + ft + y, 3t + f)$ so that the manifold swept out by Γ along F_t is given by $x(t, s) = t + s$, $y(t, s) = (3/2)t^2 + st + (1/2)s^2 - s$, $f(t, s) = 3t + s$. Eliminating t, s we get

$$f = x - 1 \pm \sqrt{1 - 2x^2 + 4x + 4y}.$$

The solution is defined only for $1 - 2x^2 + 4x + 4y \geq 0$. ♦

Another interesting phenomenon occurs when S can no longer be described in terms of the graph of f; e.g., S "folds over." This corresponds to the formation of *shock waves*. Further information can be found in Chorin and Marsden [1979], Lax [1973], Guillemin and Sternberg [1977], John [1975], and Smoller [1982].

Supplement 4.2C
Automorphisms of Function Algebras

The property of flows corresponding to the derivation property of vector fields is that they are algebra preserving

$$F_t^*(fg) = (F_t^*f)(F_t^*g).$$

In fact it is obvious that every diffeomorphism induces an algebra automorphism of $\mathcal{F}(M)$. The following theorem shows the converse. (We note that there is an analogous result of Mackey [1962] for measurable functions and mesureable automorphisms.)

4.2.36 Theorem *Let* M *be a paracompact second-countable finite dimensinal manifold. Let* $\alpha : \mathcal{F}(M) \to \mathcal{F}(M)$ *be an invertible linear mapping that satisfies* $\alpha(fg) = \alpha(f)\,\alpha(g)$ *for all* $f, g \in \mathcal{F}(M)$. *Then there is a unique* C^∞ *diffeomorphism* $\varphi : M \to M$ *such that* $\alpha(f) = f \circ \varphi$.

Remarks

A There is a similar theorem for paracompact second-countable Banach manifolds; here we assume that there are invertible linear maps $\alpha_F : C^\infty(M, F) \to C^\infty(M, F)$ for each Banach space F such that for any bilinear continuous map $B : F \times G \to H$ we have $\alpha_H(B(f, g)) = B(\alpha_F(f), \alpha_G(g))$ for $f \in C^\infty(M, F)$ and $g \in C^\infty(M, G)$. The conclusion is the same: there is a C^∞ diffeomorphism $\varphi : M \to M$ such that $\alpha_F(f) = f \circ \varphi$ for all F and all $f \in C^\infty(M, G)$. Alternative to assuming this for all F, one can take $F = \mathbb{R}$ and assume that M is modelled on a Banach space that has a norm that is C^∞ away from the origin. We shall make some additional remarks on the infinite-dimensional case in the course of the proof.

B Some of the ideas about partitions of unity are needed in the proof. Although the present proof is self-contained, the reader may wish to consult **§5.6** simultaneously.

C In Chapter 5 we shall see that finite-dimensional paracompact manifolds are metrizable, so by **1.6.14** they are automatically second countable. ♦

Proof *(Uniqueness).* We shall first construct a C^∞ function $\chi : M \to \mathbb{R}$ which takes on the values 1 and 0 at two given points $m_1, m_2 \in M$, $m_1 \neq m_2$. Choose a chart (U, φ) at m_1, such that $m_2 \notin U$ and such that $\varphi(U)$ is a ball of radius r_1 about the origin in E, $\varphi(m_1) = 0$. Let $V \subset U$ by the inverse image by φ of the ball of radius $r_2 < r_1$ and let $\theta : E \to \mathbb{R}$ be a C^∞-bump function as in Lemma **4.2.13**. Then the function $\chi : M \to \mathbb{R}$ given by

$$\chi = \begin{cases} \theta \circ \varphi, & \text{on } U \\ 0, & \text{on } M \setminus U \end{cases}$$

is clearly C^∞ and $\chi(m_1) = 1$, $\chi(m_2) = 0$.

Now assume that $\varphi^* f = \psi^* f$ for all $f \in \mathcal{F}(M)$ for two different diffeomorphisms φ, ψ of M. Then there is a point $m \in M$ such that $\varphi(m) \neq \psi(m)$ and thus we can find $\chi \in \mathcal{F}(M)$ such that $(\chi \circ \varphi)(m) = 1$, $(\chi \circ \psi)(m) = 0$ contradicting $\varphi^*\chi = \psi^*\chi$. Hence $\varphi = \psi$.

The proof of existence is based on the following key lemma.

4.2.37 Lemma *Let* M *be a (finite-dimensional) paracompact second countable manifold and* $\beta : \mathcal{F}(M) \to \mathbb{R}$ *be a nonzero algebra homomorphism. Then there is a unique point* $m \in M$ *such that* $\beta(f) = f(m)$.

Proof (Following suggestions of H. Bercovici). Uniqueness is clear, as before, since for $m_1 \neq m_2$ there exists a bump function $f \in \mathcal{F}(M)$ satisfying $f(m_1) = 0$, $f(m_2) = 1$.

To show existence, note first that $\beta(1) = 1$. Indeed $\beta(1) = \beta(1^2) = \beta(1)\beta(1)$ so that either $\beta(1) = 0$ or $\beta(1) = 1$. But $\beta(1) = 0$ would imply β is identically zero since $\beta(f) = \beta(1 \cdot f) = \beta(1) \cdot \beta(f)$, contrary to our hypotheses. Therefore we must have $\beta(1) = 1$ and thus $\beta(c) = c$ for $c \in \mathbb{R}$. For $m \in M$, let $\mathrm{Ann}(m) = \{f \in \mathcal{F}(M) \mid f(m) = 0\}$. Second, we claim that it is enough to show that there is an $m \in M$ such that $\ker \beta = \{f \in \mathcal{F}(M) \mid \beta(f) = 0\} = \mathrm{Ann}(m)$. Clearly, if $\beta(f) = f(m)$ for some m, then $\ker \beta = \mathrm{Ann}(m)$. Conversely, if this holds for some $m \in M$ and $f \notin \ker \beta$, let $c = \beta(f)$ and note that $f - c \in \ker \beta = \mathrm{Ann}(m)$, so $f(m) = c$ and thus $\beta(f) = f(m)$ for all $f \in \mathcal{F}(M)$.

To prove that $\ker \beta = \mathrm{Ann}(m)$ for some $m \in M$, note that both are ideals in $\mathcal{F}(M)$; i.e., if $f \in \ker \beta$, (resp. Ann(m)), and $g \in \mathcal{F}(M)$, then $fg \in \ker \beta$ (resp. Ann(m)). Moreover, both of them are maximal ideals; i.e., if I is another ideal of $\mathcal{F}(M)$, with $I \neq \mathcal{F}(M)$, and $\ker \beta \subset I$, (resp. $\mathrm{Ann}(m) \subset I$) then necessarily $\ker \beta = I$ (resp. $\mathrm{Ann}(m) = I$). For $\ker \beta$ this is seen in the following way: since $\mathbb{R}$ is a field, it has no ideals except 0 and itself; but $\beta(I)$ is an ideal in $\mathbb{R}$, so $\beta(I) = 0$, i.e., $I = \ker \beta$, or $\beta(I) = \mathbb{R} = \beta(\mathcal{F}(M))$. If $\beta(I) = \mathbb{R}$, then for every $f \in \mathcal{F}(M)$ there exists $g \in I$ such that $f - g \in \ker \beta \subset I$ and hence $f \in g + I \subset I$; i.e., $I = \mathcal{F}(M)$. Similarly, the ideal Ann(m) is maximal since the quotient $\mathcal{F}(M)/\mathrm{Ann}(m)$ is isomorphic to $\mathbb{R}$.

Assume that $\ker \beta \neq \mathrm{Ann}(m)$ for every $m \in M$. By maximality, neither set can be included in the other, and hence for every $m \in M$ there is a relatively compact open neighborhood U_m of m and $f_m \in \ker \beta$ such that $f_m \mid U_m > 0$. Let V_m be an open neighborhood of the closure, $\mathrm{cl}(U_m)$. Since M is paracompact, we can assume that $\{V_m \mid m \in M\}$ is locally finite. Since M is second countable, M can be covered by $\{V_{m(j)} \mid j \in \mathbb{N}\}$. Let $f_j = f_{m(j)}$ and let χ_j be bump functions which are equal to 1 on $\mathrm{cl}(U_{m(j)})$ and vanishing in $M \setminus V_{m(j)}$. If we have the inequality $a_n < 1/[n^2 \sup\{\chi_n(m) f_n^2(m) \mid m \in M\}]$, then the function $f = \sum_{n\geq 1} a_n \chi_n f_n^2$ is C^∞ (since the sum is finite in a neighborhood of every point), $f > 0$ on M, and the series defining f is uniformly convergent, being majorized by $\sum_{n\geq 1} n^{-2}$. If we can show that β can be taken inside the sum, then $\beta(f) = 0$. This construction then produces $f \in \ker \beta$, $f > 0$ and hence $1 = (1/f)f \in \ker \beta$; i.e., $\ker \beta = \mathcal{F}(M)$, contradicting the hypothesis $\beta \neq 0$.

To show that β can be taken inside the series, it suffices to prove the following "g-estimate:" for any $g \in \mathcal{F}(M)$, $|\beta(g)| \leq \sup\{|g(m)| \mid m \in M\}$. Once this is done, then

$$\left|\sum_{m=1}^{N} \beta(a_n \chi_n f_n^2) - \beta(f)\right| = \left|\beta\left(\sum_{m=1}^{N} a_n \chi_n f_n^2 - f\right)\right| \leq \sup\left|\sum_{m=1}^{N} a_n \chi_n f_n^2 - f\right| \to 0$$

as $n \to \infty$ by uniform convergence and boundedness of all functions involved. Thus $\beta(f) = \sum_{n\geq 1} \beta(a_n \chi_n f_n^2)$. To prove the g-estimate, let $\lambda > \sup\{|g(m)| \mid m \in M\}$ so that $\lambda \pm g \neq 0$ on M; i.e., $\lambda \pm g$ are both invertible functions on M. Since β is an algebra homomorphism, $0 \neq \beta(\lambda \pm g) = \lambda \pm \beta(g)$. Thus $\pm\beta(g) \neq \lambda$ for all $\lambda > \sup\{|g(m)| \mid m \in M\}$. Hence we get the estimate $|\beta(g)| \leq \sup\{|g(m)| \mid m \in M\}$. ▼

Remark For infinite-dimensional manifolds, the proof of the lemma is almost identical, with the following changes: we work with $\beta : C^\infty(M, F) \to F$, absolute values are replaced by norms, second countability is in the hypothesis, and the neighborhoods V_m are chosen in such a way that $f_m \mid V_m$ is a bounded function (which is possible by continuity of f_m). ♦

Proof of existence in 4.2.36 For each $m \in M$, define the algebra homomorphism $\beta_m : \mathcal{F}(M) \to \mathbb{R}$ by $\beta_m(f) = \alpha(f)(m)$. Since α is invertible, $\alpha(1) \neq 0$ and since $\alpha(1) = \alpha(1^2) = \alpha(1)\,\alpha(1)$, we have $\alpha(1) = 1$. Thus $\beta_m \neq 0$ for all $m \in M$. By **4.2.37** there exists a unique point, which we call $\varphi(m) \in M$, such that $\beta_m(f) = f(\varphi(m)) = (\varphi^* f)(m)$. This defines a map $\varphi : M \to M$ such that $\alpha(f) = \varphi^* f$ for all $f \in \mathcal{F}(M)$. Since α is an automorphism, φ is bijective and since $\alpha(f) = \varphi^* f \in \mathcal{F}(M)$, $\alpha^{-1}(f) = \varphi_* f \in \mathcal{F}(M)$ for all $f \in \mathcal{F}(M)$, both φ, φ^{-1} are C^∞ (take for f any coordinate function multiplied by a bump function to show that in every chart the local representatives of φ, φ^{-1} are C^∞). ■

The proof of existence in the infinite-dimensional case proceeds in a similar way.

☞ Supplement 4.2D
Derivations on C^r Functions

This supplement investigates to what extent vector fields and tangent vectors are characterized by their derivation properties on functions, if the underlying manifold is finite dimensional and of a finite differentiability class. We start by studying vector fields. Recall from **4.2.9** that a *derivation* θ is an $\mathbb{R}$-linear map from $\mathcal{F}^{k+1}(M)$ to $\mathcal{F}^k(M)$ satisfying the Leibniz rule, i.e., $\theta(fg) = f\theta(g) + g\theta(f)$ for $f, g \in \mathcal{F}^{k+1}(M)$, if the differentiability class of M is at least $k + 1$.

4.2.38 Theorem (A. Blass) *Let* M *be a* C^{k+2} *finite-dimensional manifold, where* $k \geq 0$. *The collection of all derivations* θ *from* $\mathcal{F}^{k+1}(M)$ *to* $\mathcal{F}^k(M)$ *is isomorphic to* $\mathfrak{X}^k(M)$ *as a real vector space.*

Proof By the remark following **4.2.16**, there is a unique C^k vector field X with the property that $\theta \mid \mathcal{F}^{k+2}(M) = \mathbf{L}_X \mid \mathcal{F}^{k+2}(M)$. Thus, all we have to do is show that θ and $\mathbf{L}_X$ agree on C^{k+1} functions. Replacing θ with $\theta - \mathbf{L}_X$, we can assume that θ annihilates all C^{k+2} functions and we want to show that it also annihilates all C^{k+1} functions. As in the proof of **4.2.16**, it suffices to work in a chart, so we can assume without loss of generality that $M = \mathbb{R}^n$.

Let f be a C^{k+1} function and fix $p \in \mathbb{R}^n$. We need to prove that $(\theta f)(p) = 0$. For simplicity, we will show this for $p = 0$, the proof for general p following by centering the following arguments at p instead of 0. Replacing f by the difference between f and its Taylor polynomial of order $k + 1$ about 0, we can assume $f(0) = 0$, and the first $k + 1$ derivatives vanish at 0, since θ evaluated at the origin annihilates any polynomial. We shall prove that $f = g + h$, where g and h are two C^{k+1} functions, satisfying $g \mid U = 0$ and $h \mid V = 0$, where U, V are open sets such that $0 \in \mathrm{cl}(U) \cap \mathrm{cl}(V)$. Then, since θ is a local operator, $\theta g \mid U = 0$ and $\theta h \mid V = 0$, whence by continuity $\theta g \mid \mathrm{cl}(U) = 0$ and $\theta h \mid \mathrm{cl}(V) = 0$. Hence $(\theta f)(0) = (\theta g)(0) + (\theta h)(0) = 0$ and the theorem will be proved.

4.2.39 Lemma *Let* $\varphi : S^{n-1} \subset \mathbb{R}^n \to \mathbb{R}$ be a C^∞ *function and denote by* $\pi : \mathbb{R}^n \setminus \{0\} \to S^{n-1}$, $\pi(x) = x / \|x\|$ *the radial projection. Then for any positive integer* r,

$$D^r(\varphi \circ \pi)(x) = \frac{(\psi \circ \pi)(x}{\|x\|^r}$$

for some C^∞ *function* $\psi : S^{n-1} \to L^r_s(\mathbb{R}^n; \mathbb{R})$. *In particular* $D^k(\varphi \circ \pi)(x) = O(\|x\|^{-r})$ *as* $\|x\| \to 0$.

Proof For $r = 0$ choose $\varphi = \psi$. For $r = 1$, note that

$$D\pi(x) \cdot v = -\frac{1}{\|x\|^2} D\|\cdot\|(x) \cdot v + \frac{v}{\|x\|},$$

and so

$$D(\varphi \circ \pi)(x) \cdot v = \frac{1}{\|x\|} D\varphi(\pi(x)) \left(I - \frac{1}{\|x\|} D\|\cdot\|(x) \right) \cdot v.$$

But the mapping $\ell'(x) = (1/\|x\|)\mathbf{D}\|\cdot\|(x)$ satisfies $\ell'(tx) = \ell'(x)$ for all $t > 0$ so that it is uniquely determined by $\ell = \ell' \mid S^{n-1}$. Hence $\mathbf{D}(\varphi \circ \pi)(x) = (1/\|x\|)(\psi \circ \pi)(x)$, where $\psi(y) = \mathbf{D}\varphi(y) \cdot (I - \ell(y))$, $y \in S^{n-1}$. Now proceed by induction. ▼

Returning to the proof of the theorem, let f be as before, i.e., of class C^{k+1} and $\mathbf{D}^i f(0) = 0$, $0 \le i \le k+1$, and let φ, π be as in the lemma. From Taylor's formula with remainder, we see that $\mathbf{D}^i f(x) = o(\|x\|^{k+1-i})$, $0 \le i \le k+1$, as $x \to 0$. Hence by the product rule and the lemma,

$$\mathbf{D}^i(f \cdot (\varphi \circ \pi))(x) = \sum_{j+\ell=i} o(\|x\|^{k+1-j})O(\|x\|^{-\ell}) = o(\|x\|)^{k+1-i})$$

so that $\mathbf{D}^i(f \cdot (\varphi \circ \pi))$, $0 \le i \le k+1$, can be continuously extended to 0, by making them vanish at 0. Therefore $f \cdot (g \circ \pi)$ is C^{k+1} for all $\mathbb{R}^n$.

Now choose the C^∞ function φ in the lemma to be zero on an open set O_1 and equal to 1 on an open set O_2 of S^{n-1}, $O_1 \cap O_2 = \varnothing$. Then the continuous extension g of $f \cdot (\varphi \circ \pi)$ to $\mathbb{R}^n$ is zero on $U = \pi^{-1}(O_1)$ and agrees with f on $V = \pi^{-1}(O_2)$. Let $h = f - g$ and thus f is the sum of two C^{k+1} functions, each of which vanishes in an open set having 0 in its closure. This completes the proof. ■

We do not know of an example of a derivation not given by a vector field on a C^1–manifold.

In infinite dimensions, the proof would require the norm of the model space to be C^∞ away from the origin and the function ψ in the lemma bounded with all derivatives bounded on the unit sphere. Unfortunately, this does not seem feasible under realistic hypotheses.

The foregoing proof is related to the method of "blowing-up" a singularity; see for example Takens [1974] and Buchner, Marsden, and Schecter [1982]. There are also difficulties with this method in infinite dimensions in other problems, such as the Morse lemma (see Golubitsky and Marsden [1983] and Buchner, Marsden, and Schecter [1983]).

4.2.40 Corollary *Let* M *be a* C^{k+1} *finite-dimensional manifold. Then the only derivative from* $\mathcal{F}^{k+1}(M)$ *to* $\mathcal{F}^k(M)$, *where* $1 \le k < \infty$, *is zero.*

Proof By the theorem, such a derivation is given by a C^{k-1} vector field X. If $X \neq 0$, then for some $f \in \mathcal{F}^{k+1}(M)$, X[f] is only C^{k-1} but not C^k. ■

Next, we turn to the study of the relationship between tangent vectors and germ derivations. On $\mathcal{F}^k(M)$ consider the following equivalence relation: $f \sim_m g$ iff f and g agree on some neighborhood of $m \in M$. Equivalence classes of the relation $\sim_m$ are called ***germs*** at m; they form a vector space denoted by $\mathcal{F}^k_m(M)$. The differential **d** on functions clearly induces an $\mathbb{R}$-linear map, denoted by $\mathbf{d}_m$ on $\mathcal{F}^k_m(M)$ by $\mathbf{d}_m f = \mathbf{d}f(m)$, where we understand f on the left hand side as a germ. It is straightforward to see that $\mathbf{d}_m : \mathcal{F}^k_m(M) \to T_m{}^*M$ is $\mathbb{R}$-linear and satisfies the Leibniz rule. We say that an $\mathbb{R}$-linear map $\theta_m : \mathcal{F}^k_m(M) \to E$, where E is a

Banach space, is a ***germ derivation*** if θ_m satisfies the Leibniz rule. Thus, $\mathbf{d}_m$ is a $T_m{}^*M$-valued germ derivation.

Any tangent vector $v_m \in T_mM$ defines an $\mathbb{R}$-valued germ derivation by $v_m[f] = \langle \mathbf{d}f(m), v_m \rangle$. Conversely, localizing the statement and proof of Theorem **4.2.16**(i) at m, we see that on a C^∞ finite-dimensional manifold, any $\mathbb{R}$-valued germ derivation at m defines a unique tangent vector, i.e., T_mM *is isomorphic to the vector space of* $\mathbb{R}$*-valued germ derivations on* $\mathcal{F}_m(M)$. The purpose of the rest of this supplement is to show that this result is false if M is a C^k–manifold. This is in sharp contrast to Theorem **4.2.38.**

4.2.41 Theorem (Newns and Walker [1956]) *Let* M *be a finite dimensional* C^k *manifold,* $1 \le k < \infty$. *Then there are* $\mathbb{R}$*-valued germ derivations on* $\mathcal{F}^k{}_m(M)$ *which are not tangent vectors. In fact, the vector space of all* $\mathbb{R}$*-valued germ derivations on* $\mathcal{F}^k{}_m(M)$ *is* card($\mathbb{R}$)*-dimensional, where* card($\mathbb{R}$) *is the cardinality of the continuum.*

For the proof, we start with algebraic characterizations of T_mM and the vector space of all germ derivations.

4.2.42 Lemma *Let* $\mathcal{F}^k{}_{m,0}(M) = \{f \in \mathcal{F}^k{}_m(M) \mid \mathbf{d}f(m) = 0\}$. *Then* $\mathcal{F}^k{}_m(M) / \mathcal{F}^k{}_{m,0}(M)$ *is isomorphic to* $T_m{}^*M$. *Therefore, since* M *is finite dimensional* $(\mathcal{F}^k{}_m(M) / \mathcal{F}^k{}_{m,0}(M))^*$ *is isomorphic to* T_mM.

Proof The isomorphism of $\mathcal{F}^k{}_m(M) / \mathcal{F}^k{}_{m,0}(M)$ with $T_m{}^*M$ is given by class of (f) $\mapsto \mathbf{d}f(m)$; this is a direct consequence of Corollary **2.4.14.** ▼

4.2.43 Lemma *Let* $\mathcal{F}^k{}_{m,d}(M) = \mathrm{span}\{1, fg \mid f, g \in \mathcal{F}^k{}_m(M), f(m) = g(m) = 0\}$. *Then the space of* $\mathbb{R}$*-linear germ derivations on* $\mathcal{F}^k{}_m(M)$ *is isomorphic to* $(\mathcal{F}^k{}_m(M) / \mathcal{F}^k{}_{m,d}(M))^*$.

Proof Clearly, if θ_m is a germ derivation $\theta_m(1) = 0$ and $\theta_m(fg) = 0$ for any $f, g \in \mathcal{F}^k{}_m(M)$ with $f(m) = g(m) = 0$, so that θ_m defines a linear functional on $\mathcal{F}^k{}_m(M)$ which vanishes on the space $\mathcal{F}^k{}_{m,d}(M)$. Conversely, if λ is a linear functional on $\mathcal{F}^k{}_m(M)$ vanishing on $\mathcal{F}^k{}_{m,d}(M)$, then λ is a germ derivation, for if $f, g \in \mathcal{F}^k{}_m(M)$, we have

$$fg = (f - f(m))(g - g(m)) + f(m)g + g(m)f - f(m)g(m)$$

so that

$$\lambda(fg) = f(m)\,\lambda(g) + g(m)\,\lambda(f),$$

i.e., the Leibniz rule holds. ▼

4.2.44 Lemma *All germs in* $\mathcal{F}^k{}_{m,d}(M)$ *have* $k + 1$ *derivatives at* m *(even though* M *is only of class* C^k*).*

Proof Since any element of $\mathcal{F}^k_{m,d}(M)$ is of the form $a + bfg$, $f, g \in \mathcal{F}^k_m(M)$, $f(m) = g(m) = 0$, $a, b \in \mathbb{R}$, it suffices to prove the statement for fg. Passing to local charts, we have by the Leibniz rule

$$\mathbf{D}^k(fg) = (\mathbf{D}^k f)g + f(\mathbf{D}^k g) + \varphi ,$$

for

$$\varphi = \sum_{i=1}^{k-1} (\mathbf{D}^i f)(\mathbf{D}^{k-i} g) .$$

Clearly φ is C^1, since the highest order derivative in the expression of φ is $k-1$ and f, g are C^k. Moreover, since $\mathbf{D}^k f$ is continuous and $g(m) = 0$, using the definition of the derivative it follows that $\mathbf{D}[(\mathbf{D}^k f)g](m) = (\mathbf{D}^k f)(m)(\mathbf{D}g)(m)$. Therefore, fg has $k+1$ derivatives at m. ▼

Proof of 4.2.41 We clearly have $\mathcal{F}^k_{m,d}(M) \subset \mathcal{F}^k_{m,0}(M)$. Choose a chart $(x^1, \ldots, x^n)$ at m, $x^i(m) = 0$, and consider the functions $|x^i|^{k+\varepsilon}$, $0 < \varepsilon < 1$. These functions are clearly in $\mathcal{F}^k_{m,0}(M)$, but are not in $\mathcal{F}^k_{m,d}(M)$ by **4.2.44**, since they cannot be differentiated $k+1$ times at m. Therefore, $\mathcal{F}^k_{m,d}(M)$ is strictly contained in $\mathcal{F}^k_{m,0}(M)$ and thus T_mM is a strict subspace of the vector space of germ differentations on $\mathcal{F}^k_m(M)$ by **4.2.42** and **4.2.43**.

The functions $|x^i|^{k+\varepsilon}$, $0 < \varepsilon < 1$ are linearly independent in $\mathcal{F}^k_m M$ modulo $\mathcal{F}^k_{m,d}(M)$, because only a trivial linear combination of such functions has derivatives of order $k+1$ at m. Therefore, the dimension of $\mathcal{F}^k_m(M) / \mathcal{F}^k_{m,d}(M)$ is at least $\operatorname{card}(\mathbb{R})$. Since $\operatorname{card}(\mathcal{F}^k_m(M)) = \operatorname{card}(\mathbb{R})$, it follows that $\dim(\mathcal{F}^k_m(M) / \mathcal{F}^k_{m,d}(M)) = \operatorname{card}(\mathbb{R})$. Consequently, its dual, which by **4.2.43** coincides with the vector space of germ-derivations at m, also has dimension $\operatorname{card}(\mathbb{R})$. ■

Exercises

4.2A (i) On $\mathbb{R}^2$, let $X(x, y) = (x, y; y, -x)$. Find the flow of X.

(ii) Solve the following for $f(t, x, y)$:

$$\frac{\partial f}{\partial t} = y \frac{\partial f}{\partial x} - x \frac{\partial f}{\partial y}$$

if $f(0, x, y) = y \sin x$.

4.2B (i) Let X and Y be vector fields on M with complete flows F_t and G_t, respectively. If $[X, Y] = 0$, show that $X + Y$ has flow $H_t = F_t \circ G_t$. Is this true if X and Y are time dependent?

(ii) Show that if $[X, Y] = 0$ for all $Y \in \mathfrak{X}(M)$, then $X = 0$. (*Hint:* From the local formula conclude first that X is constant; then take for Y linear vector fields and apply the Hahn-Banach theorem. In infinite dimensions, assume the conditions hold locally or that the model spaces are C^∞).

4.2C Show that, under suitable hypotheses, that the solution $f(x, t) = g(F_t(x))$ of problem (P) given in **4.2.11** is unique. (*Hint:* Consider the function

$$E(t) = \int_{\mathbb{R}^n} | f_1(x, t) - f_2(x,t) |^2 \, dx,$$

where f_1 and f_2 are two solutions. Show that $dE/dt \le \alpha E$ for a suitable constant α and conclude by Gronwall's inequality that $E = 0$. The "suitable hypotheses" are conditions that enable integration by parts to be performed in the computation of dE/dt.) Adapt this proof to get uniqueness of the solution in **4.2.28**.

4.2D Let $X, Y \in \mathfrak{X}(M)$ have flows F_t and G_t, respectively. Show that

$$[X, Y] = \left.\frac{d}{dt}\frac{d}{ds}\right|_{t,\, s=0} (F_{-t} \circ G_s \circ F_t).$$

(*Hint:* The flow of $F_t{}^*Y$ is $s \mapsto F_{-t} \circ G_s \circ F_t$.)

4.2E Show that $\mathbf{SO}(n)$ is parallelizable. See Exercise **3.5S** for a proof that $\mathbf{SO}(n)$ is a manifold. (*Hint:* $\mathbf{SO}(n)$ is a group.)

4.2F Solve the following system of partial differential equations.

$$\frac{\partial Y^1}{\partial t} = (x + y)\frac{\partial Y^1}{\partial x} + (4x - 2y)\frac{\partial Y^1}{\partial y} - Y^1 - Y^2$$

$$\frac{\partial Y^2}{\partial t} = (x + y)\frac{\partial Y^2}{\partial x} + (4x - 2y)\frac{\partial Y^2}{\partial y} - 4Y^1 + 2Y^2$$

with initial conditions $Y^1(x, y, 0) = x + y$, $Y^2(x, y, 0) = x^2$. (*Hint:* The flow of the vector field $(x + y, 4x - 2y)$ is

$$(x, y) \mapsto \left(\frac{1}{5}(x - y)e^{-3t} + \frac{1}{5}(4x + y)\, e^{2t}, -\frac{4}{5}(x - y)\, e^{-3t} + \frac{1}{5}(4x + y)\, e^{2t}\right).$$

4.2G Consider the following equation for $f(x, t)$ in ***divergence form:***

$$\frac{\partial f}{\partial t} + \frac{\partial}{\partial x}(H(f)) = 0$$

where H is a given function of f. Show that the characteristics are given by $\dot{x} = -H'(f)$. What does the transversality condition discussed in Supplement **4.2B** become in this case?

4.2H Let M and N be manifolds with N modeled on a Banach space which has a C^k norm away from the origin. Show that a given mapping $\varphi : M \to N$ is C^k iff $f \circ \varphi : M \to \mathbb{R}$ is C^k for all $f \in \mathcal{F}^k(N)$.

4.2I Develop a product formula like that in Supplement **4.1A** for the flow of X + Y for time-dependent vector fields. (*Hint:* You will have to consider ***time-ordered*** products.)

4.2J (Newns and Walker [1956]). In the terminology of Supplement **4.2D**, consider a C^0-manifold modeled on $\mathbb{R}^n$. Show that any germ derivation is identically zero. (*Hint:* Write any $f \in \mathcal{F}^0{}_m(M)$, $f = f(m) + (f - f(m))^{1/3} (f - f(m))^{2/3}$ and apply the derivation.)

4.2K (More on the Lie bracket as a "commutator".) Let M be a manifold, $m \in M$, $v \in T_mM$. Recall that T_mM is a submanifold of T(TM) and that $T_v(T_mM)$ is canonically isomorphic to T_mM. Also recall from Exercise **3.3B** that on T(TM) there is a canonical involution s_M : $T(TM) \to T(TM)$ satisfying $s_M \circ s_M =$ identity on T(TM), $T\tau_M \circ s_M = \tau_{TM}$, and $\tau_{TM} \circ s_M = T\tau_M$, where $\tau_M : TM \to M$ and $\tau_{TM} : T(TM) \to TM$ are the canonical tangent bundle projections. Let $X, Y \in \mathfrak{X}(M)$ and denote by $TX, TY : TM \to T(TM)$ their tangent maps. Prove the following formulae for the Lie bracket:

$$[X, Y](m) = s_M(T_mY(X(m))) - T_mX(Y(m)) = T_mY(X(m)) - s_M(T_mX(Y(m))) ,$$

where the right hand sides, belonging to $T_{X(m)}(T_mM)$ and $T_{Y(m)}(T_mM)$ respectively, are thought of as elements of T_mM. (*Hint:* Show that $T_m\tau_M$ of the right hand sides is zero which proves that the right hand sides are not just elements of $T_{X(m)}(TM)$ and $T_{Y(m)}(TM)$ respectively, but of the indicated spaces. Then pass to a local chart.)

§4.3 *An Introduction to Dynamical Systems*

We have seen quite a bit of theoretical development concerning the interplay between the two aspects of vector fields, namely as differential operators and as ordinary differential equations. It is appropriate now to look a little more closely at geometric aspects of flows.

Much of the work in this section holds for infinite-dimensional as well as finite-dimensional manifolds. The reader who knows or is willing to learn some spectral theory from functional analysis can make the generalization.

This section is intended to link up the theory of this book with courses in ordinary differential equations that the reader may have taken. The section will be most beneficial if it is read with such a course in mind. We begin by introducing some of the most basic terminology regarding the stability of fixed points.

4.3.1 Definition *Let* X *be a* C^1 *vector field on an n-manifold* M. *A point* m *is called a* ***critical point*** *(also called a* ***singular point*** *or an* ***equilibirum point****) of* X *if* $X(m) = 0$. *The* ***linearization*** *of* X *at a critical point* m *is the linear map*

$$X'(m) : T_mM \to T_mM$$

defined by

$$X'(m) \cdot v = \frac{d}{dt}(TF_t(m) \cdot v)\Big|_{t=0}$$

where F_t *is the flow of* X. *The eigenvalues (points in the spectrum) of* $X'(m)$ *are called* ***characteristic exponents*** *of* X *at* m.

Some remarks will clarify this definition. F_t leaves m fixed: $F_t(m) = m$, since $c(t) \equiv m$ is the unique integral curve through m. Conversely, it is obvious that if $F_t(m) = m$ for all t, then m is a critical point. Thus T_mF_t is a linear map of T_mM to itself and so its t-derivative at 0, producing another linear map of T_mM to itself, makes sense.

4.3.2 Proposition *Let* m *be a critical point of* X *and let* (U, φ) *be a chart on* M *with* $\varphi(m) = x_0 \in \mathbb{R}^n$. *Let* $x = (x^1, ..., x^n)$ *denote coordinates in* $\mathbb{R}^n$ *and* $X^i(x^1, ..., x^n)$, $i = 1, ..., n$, *the components of the local representative of* X. *Then the matrix of* $X'(m)$ *in these coordinates is*

$$\left[\frac{\partial X^i}{\partial x^j}\right]_{x=x_0}.$$

Proof This follows from the equations

$$X^i(F_t(x)) = \frac{d}{dt} F_t{}^i(x)$$

after differentiating in x and setting $x = x_0$, $t = 0$. ■

The name "characteristic exponent" arises as follows. We have the linear differential equation

$$\frac{d}{dt} T_m F_t = X'(m) \circ T_m F_t$$

and so

$$T_m F_t = e^{tX'(m)}.$$

Here the exponential is defined, for example, by a power series. The actual computation of these exponentials is learned in differential equations courses, using the Jordan canonical form. (See Hirsch and Smale [1974], for instance.) In particular, if $\mu_1, \ldots, \mu_n$ are the characterisitc exponents of X at m, the eigenvalues of $T_m F_t$ are

$$e^{t\mu_1}, \ldots, e^{t\mu_n}.$$

The characteristic exponents will be related to the following notion of stability of a critical point.

4.3.3 Definition *Let* m *be a critical point of* X. *Then*

(i) m *is **stable** (or **Liapunov stable**) if for any neighborhood* U *of* m, *there is a neighborhood* V *of* m *such that if* $m' \in V$, *then* m' *is + complete and* $F_t(m') \in U$ *for all* $t \geq 0$. *(See Fig.* 4.3.1(a).*)*

(ii) m *is **asymptotically stable** if there is a neighborhood* V *of* m *such that if* $m' \in V$, *then* m *is + complete,* $F_t(V) \subset F_s(V)$ *if* $t > s$ *and*

$$\lim_{t \to +\infty} F_t(V) = \{m\},$$

i.e., for any neighborhood U *of* m, *there is a* T *such that* $F_t(V) \subset U$ *if* $t \geq T$. *(See Fig.* 4.3.1(b).*)*

It is obvious that asymptotic stability implies stability. The harmonic oscillator $\ddot{x} = -x$ giving a flow in the plane shows that stability need not imply asymptotic stability (Fig. 4.3.1(c)).

4.3.4 Liapunov's Stability Criterion *Suppose* X *is* C^1 *and* m *is a critical point of* X. *Assume the spectrum of* $X'(m)$ *is strictly in the left half plane. (In finite dimensions, the characteristic exponents of* m *have negative real parts.) Then* m *is asymptotically stable. (In a similar way, if* $\mathrm{Re}(\mu_i) > 0$, m *is asymptotically unstable, i.e., asymptotically stable as* $t \to -\infty$.*)*

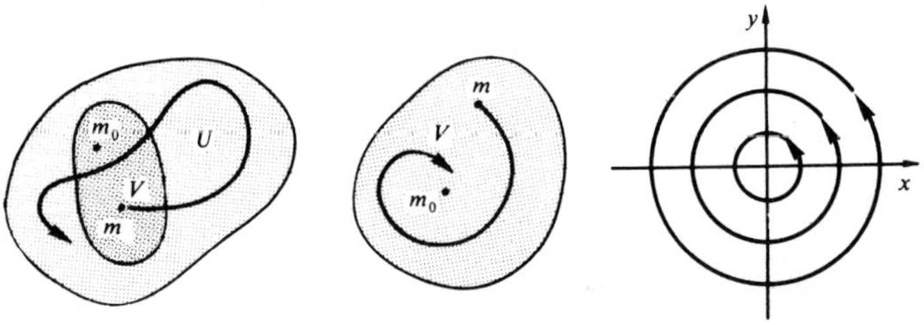

(a) Stable (b) Asymptotically stable (c) Harmonic oscillator

Figure 4.3.1

The proof we give requires some spectral theory that we shall now review. For the finite dimensional case, consult the exercises. This proof in fact can be adapted to work for many partial differential equations (see Marsden and Hughes [1983, Chapters 6, 7, and page 483]).

Let $T : E \to E$ be a bounded linear operator on a Banach space E and let $\sigma(T)$ denote its spectrum; i.e., $\sigma(T) = \{\lambda \in \mathbb{C} \mid T - \lambda I$ is not invertible on the complexification of $E\}$. Then $\sigma(T)$ is non-empty, is compact, and for $\lambda \in \sigma(T)$, $|\lambda| \le \|T\|$. Let $r(T)$ denote its ***spectral radius,*** defined by $r(T) = \sup\{\,|\lambda| \mid \lambda \in \sigma(T)\}$.

4.3.5 Spectral radius formula.

$$r(T) = \lim_{n \to \infty} \|T^n\|^{1/n} .$$

The proof is analogous to the formula for the radius of convergence of a power series and can be supplied without difficulty; cf. Rudin [1973, p. 355]. The following lemma is also not difficult and is proved in Rudin [1973] and Dunford and Schwartz [1963].

4.3.6 Spectral Mapping Theorem. *Let*

$$f(z) = \sum_{n=0}^{\infty} a_n z^n$$

be an entire function and define

$$f(T) = \sum_{n=0}^{\infty} a_n T^n.$$

Then $\sigma(f(T)) = f(\sigma(T))$.

4.3.7 Lemma *Let* $T : E \to E$ *be a bounded linear operator on a Banach space* E. *Let* r *be any number greater than* $r(T)$, *the spectral radius of* T. *Then there is a norm* $|\ |$ *on* E *equivalent to the original norm such that* $|T| \le r$.

Proof From the spectral radius formula, we get $\sup_{n \ge 0}(\|T^n\|/r^n) < \infty$, so if we define $|x| =$

$\sup_{n\geq 0}(\|T^n(x)\|/r^n)$, then $|\ \ |$ is a norm and $\|x\| \leq |x| \leq (\sup_{n\geq 0}(\|T^n\|/r^n))\|x\|$. Hence $|T(x)| = \sup_{n\geq 0}(\|T^{n+1}(x)\|/r^n) = r\sup_{n\geq 0}(\|T^{n+1}(x)\|/r^{n+1}) \leq r|x|$. ■

4.3.8 Lemma *Let* $A : E \to E$ *be a bounded operator on* E *and let* $r > \sigma(A)$ (*i.e., if* $\lambda \in \sigma(A)$, $\mathrm{Re}(\lambda) > r$). *Then there is an equivalent norm* $|\ \ |$ *on* E *such that for* $t \geq 0$, $|e^{tA}| \leq e^{rt}$.

Proof Using Lemma **4.3.6,** e^{rt} is $\geq$ spectral radius of e^{tA}; i.e., $e^{rt} \geq \lim_{n\to\infty}\|e^{ntA}\|^{1/n}$. Set

$$|x| = \sup_{n\geq 0,\, t\geq 0}(\|e^{ntA}(x)\|/e^{rnt})$$

and proceed as in Lemma **4.3.7.** ■

There is an analogous lemma if $r < \sigma(A)$, giving $|e^{tA}| \geq e^{rt}$.

4.3.9 Lemma *Let* $T : E \to E$ *be a bounded linear operator. Let* $\sigma(T) \subset \{z \mid \mathrm{Re}(z) < 0\}$ (*resp.* $\sigma(T) \subset \{z \mid \mathrm{Re}(z) > 0\}$). *Then the origin is an attracting (resp. repelling) fixed point for the flow* $\varphi_t = e^{tT}$ *of* T.

Proof If $\sigma(T) \subset \{z \mid \mathrm{Re}(z) < 0\}$, there is an $r < 0$ with $\sigma(T) < r$, as $\sigma(T)$ is compact. Thus by lemma **4.3.8,** $|e^{tA}| \leq e^{rt} \to 0$ as $t \to +\infty$. ■

Proof of Liapunov's Stability Criterion (4.3.4) We can assume that M is a Banach space E and that $m = 0$. As above, renorm E and find $\varepsilon > 0$ such that $\|e^{tA}\| \leq e^{-\varepsilon t}$, where $A = X'(0)$.

From the local existence theory, there is an r-ball about 0 for which the time of existence is uniform if the initial condition x_0 lies in this ball. Let

$$R(x) = X(x) - \mathbf{D}X(0)\cdot x.$$

Find $r_2 \leq r$ such that $\|x\| \leq r_2$ implies $\|R(x)\| \leq \alpha\|x\|$, where $\alpha = \varepsilon/2$.

Let D be the open $r_2/2$ ball about 0. We shall show that if $x_0 \in D$, then the integral curve starting at x_0 remains in D and $\to 0$ exponentially as $t \to +\infty$. This will prove the result. Let $x(t)$ be the integral curve of X starting at x_0. Suppose $x(t)$ remains in D for $0 \leq t < T$. The equation

$$x(t) = x_0 + \int_0^t X(x(s))\,ds = x_0 + \int_0^t [Ax(s) + R(x(s))]\,ds$$

gives, by the variation of constants formula (Exercise **4.1E**),

$$x(t) = e^{tA}x_0 + \int_0^t e^{(t-s)A}R(x(s))\,ds\,,$$

and so

$$\| x(t) \| \leq e^{-t\varepsilon} \| x_0 \| + \alpha \int_0^t e^{-(t-s)\varepsilon} \| x(s) \| \, ds.$$

Letting $f(t) = e^{t\varepsilon} \| x(t) \|$, the previous inequality becomes

$$f(t) \leq \| x_0 \| + \alpha \int_0^1 f(s) \, ds,$$

and so, by Gronwall's inequality, $f(t) \leq \| x_0 \| e^{\alpha t}$. Thus

$$\| x(t) \| \leq \| x_0 \| e^{(\alpha - \varepsilon)t} = \| x_0 \| e^{-\varepsilon t/2}.$$

Hence $x(t) \in D$, $0 \leq t < T$, so as in **4.1.19**, $x(t)$ may be indefinitely extended in t and the foregoing estimate holds. ■

One can also show that if M is finite dimensional and m is a stable equilibrium, then no eigenvalue of $X'(m)$ has strictly positive real part ; see Hirsch and Smale [1974, pp. 187-190] and the remarks below on invariant manifolds. See Hille and Phillips [1957] and Curtain and Pritchard [1974] for the infinite dimensional linear case.

Another method of proving stability is to use Liapunov functions.

4.3.10 Definition *Let* $X \in \mathfrak{X}^r(M)$, $r \geq 1$, *and let* m *be an equilibrium solution for* X, *i.e.* $X(m) = 0$. *A **Liapunov function** for* X *at* m *is a continuous function* $L : U \to \mathbb{R}$ *defined on a neighborhood* U *of* m, *differentiable on* $U \setminus \{m\}$, *and satisfying the following conditions:*

(i) $L(m) = 0$ *and* $L(m') > 0$ *if* $m' \neq m$;

(ii) $X[L] \leq 0$ *on* $U \setminus \{m\}$;

(iii) *there is a connected chart* $\varphi : V \to E$ *where* $m \in V \subset U$, $\varphi(m) = 0$, *and an* $\varepsilon > 0$ *satisfying* $B_\varepsilon(0) = \{ x \in E \mid \| x \| \leq \varepsilon \} \subset \varphi(V)$, *such that for all* $0 < \varepsilon' \leq \varepsilon$,

$$\inf \{ L(\varphi^{-1}(x)) \mid \| x \| = \varepsilon' \} > 0.$$

The Liapunov function L *s said to be **strict**, if* (ii) *is replaced by* (ii') $X[L] < 0$ *in* $U \setminus \{m\}$.

Conditions (i) and (iii) are called the ***potential well hypothesis***. In finite dimensions, (iii) follows automatically from compactness of the sphere of radius ε' and (i). By the Lie derivative formula, condition (ii) is equivalent to the statement: L is decreasing along integral curves of X.

4.3.11 Theorem *Let* $X \in \mathfrak{X}^r(M)$, $r \geq 1$, *and* m *be an equilibrium of* X. *If there exists a Liapunov function for* X *at* m, *then* m *is stable.*

Proof Since the statement is local, we can assume M is a Banach space E and $m = 0$. By

Lemma **4.1.8**, there is a neighborhood U of 0 in E such that all solutions starting in U exist for time $t \in [-\delta, \delta]$, with δ depending only on X and U, but not on the solution. Now fix $\varepsilon > 0$ as in (iii) such that the open ball $D_\varepsilon(0)$ is included in U. Let $\rho(\varepsilon) > 0$ be the minimum value of L on the sphere of radius ε, and define the open set $U' = \{x \in D_\varepsilon(0) \mid L(x) < \rho(\varepsilon)\}$. By (i), $U' \neq \varnothing$, $0 \in U'$, and by (ii), no solution starting in U′ can meet the sphere of radius ε (since L is decreasing on integral curves of X). Thus all solutions starting in U′ never leave $D_\varepsilon(0) \subset U$ and therefore by uniformity of time of existence, these solutions can be extended indefinitely in steps of δ time units. This shows 0 is stable.■

Note that if E is finite dimensional, the proof can be done without appeal to **4.1.8**: since the closed ε-ball is compact, solutions starting in U′ exist for all time by **4.1.19.**

4.3.12 Theorem *Let* $X \in \mathfrak{X}^r(M)$, $r \geq 1$, m *be an equilibrium of* X, *and* L *a strict Liapunov function for* X *at* m. *Then* m *is asymptotically stable if any one of the following conditions hold:*

(i) M *is finite dimensional;*

(ii) *solutions starting near* m *stay in a compact set (i.e., trajectories are precompact);*

(iii) *in a chart* $\varphi : V \to E$ *satisfying* (iii) *in definition* **4.3.10** *the following inequality is valid for some constant* $a > 0$

$$X[L](x) \leq -a \, \| X(x) \| .$$

Proof We can assume M = E, and m = 0. By **4.3.11**, 0 is stable, so if t_n is an increasing sequence, $t_n \to \infty$, and x(t) is an integral curve of X starting in U′ (see the proof of **4.3.11**), the sequence $\{x(t_n)\}$ in E has a convergent subsequence in cases (i) and (ii). Let us show that under hypothesis (iii), $\{x(t_n)\}$ is Cauchy, so by completeness of E it is convergent. For $t > s$, the inequality

$$L(x(t)) - L(x(s)) = \int_s^t X[L]\,(x(\lambda))\, d\lambda \leq -a \int_s^t \| X(x(\lambda)) \| < 0 ,$$

implies that

$$L(x(s)) - L(x(t)) \geq a \int_s^t \| X(x(\lambda)) \| \, d\lambda \geq a \left\| \int_s^t X(x(\lambda))\, d\lambda \right\| = a \| x(t) - x(s) \| ,$$

which together with the continuity of $\lambda \mapsto L(x(\lambda))$ shows that $\{x(t_n)\}$ is a Cauchy sequence. Thus, in all three cases, there is a sequence $t_n \to +\infty$ such that $x(t_n) \to x_0 \in D_\varepsilon(0)$, $D_\varepsilon(0)$ being given in the proof of **4.3.11**. We shall prove that $x_0 = 0$. Since L(x(t)) is a *strictly* decreasing function of t by (ii′), $L(x(t)) > L(x_0)$ for all $t > 0$. If $x_0 \neq 0$, let c(t) be the solution of X starting at x_0, so that $L(c(t)) < L(x_0)$, again since $t \mapsto L(x(t))$ is strictly decreasing. Thus, for any solution $\tilde{c}(t)$ starting close to x_0, $L(\tilde{c}(t)) < L(x_0)$ by continuity of L. Now take $\tilde{c}(0) = x(t_n)$ for n large to get the contradiction $L(x(t_n + t)) < L(x_0)$. Therefore $x_0 = 0$ is the only limit point of $\{x(t) \mid t \geq 0\}$ if $x(0) \in U'$, i.e., 0 is asymptotically stable. ■

The method of Theorem **4.3.12** can be used to detect the instability of equilibrium solutions.

4.3.13 Theorem *Let* m *be an equilibrium point of* $X \in \mathfrak{X}^r(M)$, $r \geq 1$. *Assume there is a continuous function* $L : U \to M$ *defined in a neighborhood of* U *of* m, *which is differentiable on* $U \setminus \{m\}$, *and satisfyies* $L(m) = 0$, $X[L] \geq a > 0$ *(respectively* $\leq a < 0$*) on* $U \setminus \{m\}$. *If there exists a sequence* $m_k \to m$ *such that* $L(m_k) > 0$ *(respectively* < 0*), then* m *is unstable.*

Proof. We need to show that there is a neighborhood W of m such that for any neighborhood V of m, $V \subset U$, there is a point m_V whose integral curve leaves W. Since m is an equilibrium, by Corollary **4.1.25**, there is a neighborhood $W_1 \subset U$ of m such that each integral curve starting in W_1 exists for time at least $1/a$. Let $W = \{m \in W_1 \mid L(m) < 1/2\}$. We can assume $M = E$, and $m = 0$. Let $c_n(t)$ denote the integral curve of X with initial condition $m_n \in W$. Then

$$L(c_n(t)) - L(m_n) = \int_0^t X[L]\,(c_n(\lambda))\,d\lambda \geq at$$

so that

$$L(c_n(1/a)) \geq 1 + L(m_n) > 1,$$

i.e. $c_n(1/a) \notin W$. Thus all integral curves starting at the points $m_n \in W$ leave W after time at most $1/a$. Since $m_n \to 0$, the origin is unstable. ■

Note that if M is finite dimensional, the condition $X[L] \geq a > 0$ can be replaced by the condition $X[L] > 0$; this follows, as usual, by local compactness of M.

4.3.14 Examples

A The vector field $X(x, y) = (-y - x^5)(\partial/\partial x) + (x - 2y^3)(\partial/\partial y) \in \mathfrak{X}(\mathbb{R}^2)$ has the origin as an isolated equilibrium. The characteristic exponents of X at $(0, 0)$ are $\pm i$ and so Liapunov's Stability Criterion **4.3.4** does not give any information regarding the stability of the origin. If we suspect that $(0, 0)$ is asymptotically stable, we can try searching for a Liapunov function of the form $L(x, y) = ax^2 + by^2$, so we need to determine the coefficients $a, b, \neq 0$ in such a way that $X[L] < 0$. We have

$$X[L] = 2ax(-y - x^5) + 2by(x - 2y^3) = 2xy(b - a) - 2ax^6 - 4by^4,$$

so that choosing $a = b = 1$, we get $X[L] = -2(x^6 + 2y^4)$ which is strictly negative if $(x, y) \neq (0, 0)$. Thus the origin is asymptotically stable by Theorem **4.3.12**.

B Consider the vector field $X(x, y) = (-y + x^5)(\partial/\partial x) + (x + 2y^3)(\partial/\partial y)$ with the origin as an isolated critical point and characteristic exponents $\pm i$. Again **4.3.4** cannot be applied, so that we search for a function $L(x, y) = ax^2 + by^2$, $a, b \neq 0$ in such a way that $X[L]$ has a definite sign. As above we get

$$X[L] = 2ax(-y + x^5) + 2by(x + 2y^3) = 2xy(b - a) + 2ax^6 + 4by^4,$$

so that choosing $a = b = 1$, it follows that $X[L] = 2(x^6 + y^4) > 0$ if $(x, y) \neq (0, 0)$. Thus by Theorem **4.3.13**, the origin is unstable.

These two examples show that if the spectrum of X lies on the imaginary axis, the stability nature of the equilibrium is determined by the nonlinear terms.

C Consider Newton's equations in $\mathbb{R}^3$, $\ddot{\mathbf{q}} = -(1/m)\nabla V(\mathbf{q})$ written as a first order system $\dot{\mathbf{q}} = \mathbf{v}$, $\dot{\mathbf{v}} = -(1/m)\nabla V(\mathbf{q})$ and so define a vector field X on $\mathbb{R}^3 \times \mathbb{R}^3$. Let $(\mathbf{q}_0, \mathbf{v}_0)$ be an equilibrium of this system, so that $\mathbf{v}_0 = \mathbf{0}$ and $\nabla V(\mathbf{q}_0) = \mathbf{0}$. In Example **4.1.23B** we saw that the total energy $E(\mathbf{q}, \mathbf{v}) = (1/2)m \| \mathbf{v} \|^2 + V(\mathbf{q})$ is conserved, so we try to use E to construct a Liapunov function L. Since $L(\mathbf{q}_0, \mathbf{0}) = 0$, define

$$L(\mathbf{q}, \mathbf{v}) = E(\mathbf{q}, \mathbf{v}) - E(\mathbf{q}_0, \mathbf{0}) = \frac{1}{2} m \| \mathbf{v} \|^2 + V(\mathbf{q}) - V(\mathbf{q}_0),$$

which satisfies $X[L] = 0$ by conservation of energy. If $V(\mathbf{q}) > V(\mathbf{q}_0)$ for $\mathbf{q} \neq \mathbf{q}_0$, then L is a Liapunov function. Thus we have proved **Lagrange's Stability Theorem**: *an equilibrium point* $(\mathbf{q}_0, \mathbf{0})$ *of Newton's equations for a particle of mass* m, *moving under the influence of a potential* V, *which has a local absolute minimum at* $\mathbf{q}_0$, *is stable.*

D *Let* E *be a Banach space and* $L : E \to \mathbb{R}$ *be* C^2 *in a neighborhood of* 0. *If* $\mathbf{D}L(0) = 0$ *and there is a constant* $c > 0$ *such that* $\mathbf{D}^2L(0)(e, e) > c \| e \|^2$ *for all* e, *then* 0 *lies in a potential well for* L *(i.e. (ii) and (iii) of* **4.3.10** *hold).* Indeed, by Taylor's Theorem **2.4.15**,

$$L(h) - L(0) = \frac{1}{2} \mathbf{D}^2L(0)(h, h) + o(h^2) \geq c\| h \|^2/2 + o(h^2).$$

Thus, if $\delta > 0$ is such that for all $\| h \| < \delta$, $| o(h^2) | \leq c\| h \|^2/4$, then $L(h) - L(0) > c\| h \|^2/4$, i.e. $\inf_{\| h \| = \varepsilon}[L(h) - L(0)] \geq c\varepsilon/4$ for $\varepsilon < \delta$. ♦

In many basic infinite dimensional examples, some technical sharpening of the preceding ideas is necessary for them to be applicable. We refer therader to LaSalle [1976], Marsden and Hughes [1983, §6.6], Hale, Magalhães, and Oliva [1984] and Holm, Marsden, Ratiu, and Weinstein [1985] for more information.

Next we turn to cases where the equilibirum need not be stable.

A critical point is called ***hyperbolic*** or ***elementary*** if none of its characteristic exponents has zero real part. A generalization of Liapunov's theorem called the ***Hartman-Grobman theorem*** shows that near a hyperbolic critical point the flow looks like that of its linearization. (See Hartman [1973, Ch.9] and Nelson [1979, Ch. 3], for proofs and discussions.) In the plane, the possible hyperbolic flows near a critical point are summarized in the table below and shown in Fig. 4.3.2. (Remember that for real systems, the characteristic exponents occur in conjugate pairs.)

Eigenvalues	Real Jordan Form	Name	Phase portrait in a coordinate system (y^1, y^2) adapted to the Jordan form (Figure 4.3.2)
$\lambda_1 < 0 < \lambda_2$	$\begin{bmatrix} \lambda_1 & 0 \\ 0 & \lambda_2 \end{bmatrix}$	saddle	(a)
$\lambda_1 < \lambda_2 < 0$		stable focus	(b)
$\lambda_1 = \lambda_2 < 0$		stable node	(c)
$\lambda_1 = \lambda_2 < 0$	$\begin{bmatrix} \lambda_1 & 0 \\ 1 & \lambda_2 \end{bmatrix}$	stable improper node	(d)
$\lambda_1 = a + ib,\ a < 0$ $\lambda_2 = a - ib,\ b \neq 0$	$\begin{bmatrix} a & -b \\ b & a \end{bmatrix}$	stable spiral sink	(e)
$\lambda_1 = ib,\ \lambda_2 = -\,ib,$ $b \neq 0$	$\begin{bmatrix} 0 & -b \\ b & 0 \end{bmatrix}$	center	(f)

Table 4.3.1

In cases 1 to 5, all arrows in the phase portraits are reversed and "stable" is replaced by "unstable," if the signs of λ_1, λ_2, are changed. In the original coordinate system (x^1, x^2) all phase portraits in Figure 4.3.2 are rotated and sheared.

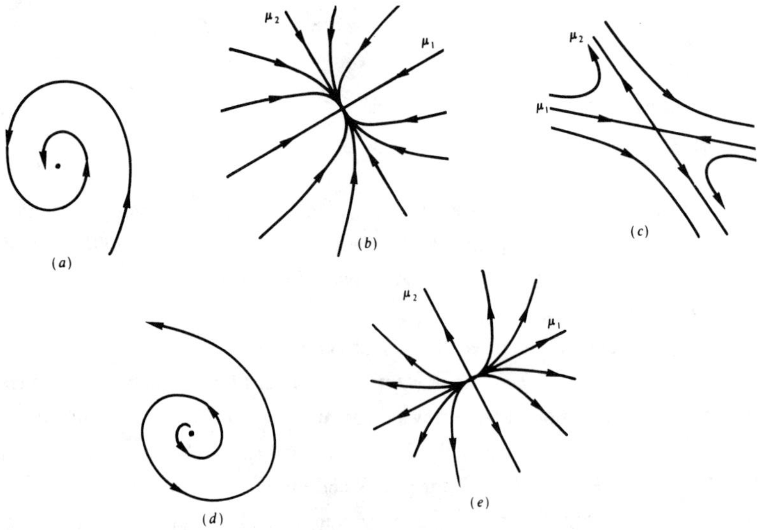

Figure 4.3.2

To understand the higher dimensional case, a little more spectral theory is required.

4.3.15 Lemma *Suppose* $\sigma(T) = \sigma_1 \cup \sigma_2$ *where* $d(\sigma_1, \sigma_2) > 0$. *Then there are unique* T-*invariant subspaces* E_1 and E_2 *such that* $E = E_1 \oplus E_2$, $\sigma(T_i) = \sigma_i$, *where* $T_i = T \mid E_i$; E_i *is called the* ***generalized eigenspace*** *of* σ_i.

The basic idea of the proof is this: let γ_j be a closed curve with σ_j in its interior and σ_k, $k \neq j$, in its exterior; then

$$T_j = \frac{1}{2\pi i} \int_{\gamma_j} \frac{dz}{zI - T} .$$

Note that the eigenspace of an eigenvalue λ is not always the same as the generalized eigenspace of λ. In the finite dimensional case, the generalized eigenspace of T is the subspace corresponding to all the Jordan blocks containing λ in the Jordan canonical form.

4.3.16 Lemma *Let* T, σ_1, *and* σ_2 *be as in Lemma* **4.3.15**; *assume* $d(\exp(\sigma_1), \exp(\sigma_2)) > 0$. *Then for the operator* $\exp(tT)$, *the generalized eigenspace of* $\exp(tT_i)$ is E_i.

Proof Write, according to Lemma **4.3.15**, $E = E_1 \oplus E_2$. Thus

$$e^{tT}(e_1, e_2) = \sum_{n=0}^{\infty} \frac{t^n T^n}{n!} (e_1, e_2) = \sum_{n=0}^{\infty} \left(\frac{t^n T^n}{n!} e_1, \frac{t^n T^n}{n!} e_2 \right)$$

$$= \left(\sum_{n=0}^{\infty} \frac{t^n T^n}{n!} e_1, \sum_{n=0}^{\infty} \frac{t^n T^n}{n!} e_2 \right) = (e^{tT_1} e_1, e^{tT_2} e_2).$$

From this the result follows easily. ■

Now let us discuss the generic nonlinear case; i.e. let m be a hyperbolic equilibrium of the vector field X and let F_t be its flow. Define the ***inset*** of m by

$$\mathrm{In}(m) = \{ m' \in M \mid F_t(m') \to m \text{ as } t \to +\infty \}$$

and similarly, the ***outset*** is defined by

$$\mathrm{Out}(m) = \{ m' \in M \mid F_t(m') \to m \text{ as } t \to +-\infty \}.$$

In the case of a linear system, $\dot{x} = Ax$, where A has no eigenvalue on the imaginary axis (so the origin is a hyperbolic critical point), In(0) is the generalized eigenspace of the eigenvalues with negative real parts, while Out(0) is the generalized eigenspace corresponding to the eigenvalues with positive real parts. Clearly, these are complementray subspaces. The dimension of the linear subspace In(0) is called the ***stability index*** of the critical point. The ***Hartman-Grobman linearization theorem*** states that the phase portrait of X near m is ***topologically conjugate*** to the phase portrait of the linear system $\dot{x} = Ax$, near the origin, where $A = X'(m)$, the linearized vector field at m. This means there is a homeomorphism of the two domains, preserving the oriented trajectories of the respective flows. Thus in this nonlinear hyperbolic case, the inset and outset are C^0 submanifolds. Another important theorem of dynamical systems theory, the ***stable manifold theorem*** (Smale [1967]) says that in addition, these are smooth, injectively immersed submanifolds, intersecting transversally at the critical point m. See Fig. 4.3.3 for an illustration showing part of the inset and outset near the critical point.

It follows from these important results that there are (up to topological conjugacy) only a few essentially different phase portraits, near hyperbolic critical points. These are classified by the dimension of their insets, called the ***stability index***, which is denoted by S(X, m) for m an equilibrium, as in the linear case.

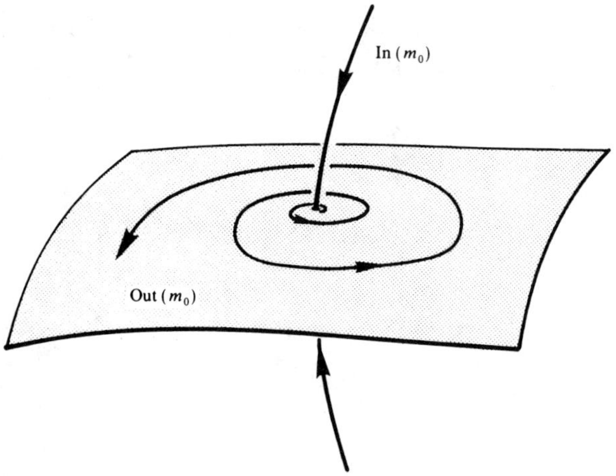

Figure 4.3.3

The word ***index*** comes up in this context with another meaning. If M is finite dimensional and m is a critical point of a vector field X, the ***topological index*** of m is +1 if the number of eigenvalues (counting multiplicities) with negative real part is even and is −1 if it is odd. Let this index be denoted I(X, m), so that $I(X, m) = (-1)^{S(X, m)}$. The ***Poincaré-Hopf index theorem*** states that if M is compact and X only has (isolated) hyperbolic critical points, then

$$\sum_{\substack{\text{m is a critical} \\ \text{point of X}}} I(X, m) = \chi(M)$$

where $\chi(M)$ is the Euler-Poincaré characteristic of M. For isolated nonhyperbolic critical points the index is also defined but requires degree theory for its definition--a kind of generalized winding number; see Section **7.5** or Guillemin and Pollack [1974, p. 133].

We now illustrate these basic concepts about critical points with some classical applications.

4.3.17 Examples

A The ***simple pendulum with linear damping*** is defined by the second-order equation

$$\ddot{x} + c\dot{x} + k \sin x = 0 \qquad (c > 0).$$

This is equivalent to the following dynamical system whose phase portrait is shown in Fig. 4.3.4:

$$\dot{x} = v, \qquad \dot{v} = -cv - k \sin x.$$

The stable focus at the origin represents the motionless, hanging pendulum. The saddle point at $(k\pi, 0)$ corresponds to the motionless bob, balanced at the top of its swing.

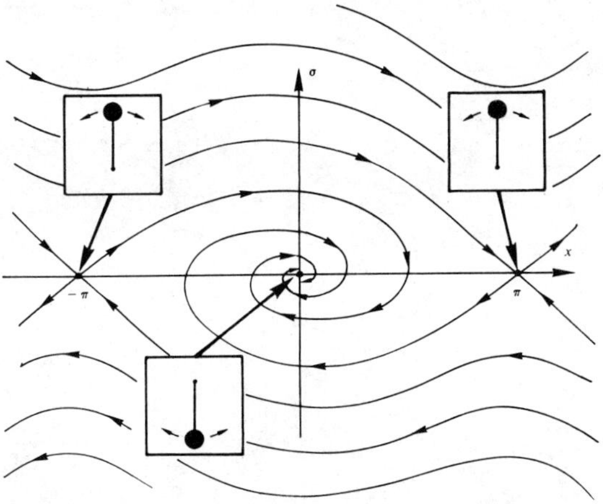

Figure 4.3.4

B Another classical equation models the ***buckling column*** (see Stoker [1950, ch. 3, §10]):

$$m\ddot{x} + c\dot{x} + a_1 x + a_3 x^3 = 0 \quad (a_1 < 0,\ a_3, c > 0),$$

or equivalently, the planar dynamical system

$$\dot{x} = v, \quad \dot{v} = -\frac{cv}{m} - \frac{a_1 x}{m} - \frac{a_3 x^3}{m}$$

with the phase portrait shown in Fig. 4.3.5. This has *two stable foci* on the horizontal axis, denoted m_1 and m_2, corresponding to the column buckling (due to a heavy weight on the top) to either side. The saddle at the origin corresponds to the unstable equilibrium of the straight, unbuckled column.

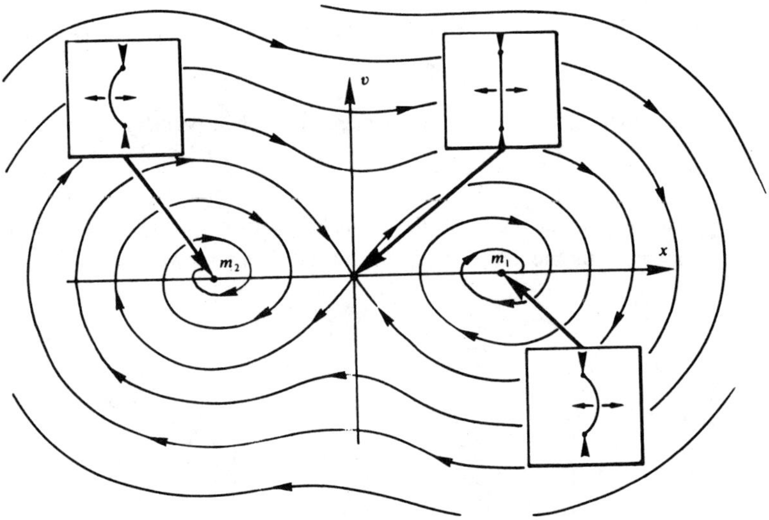

Figure 4.3.5

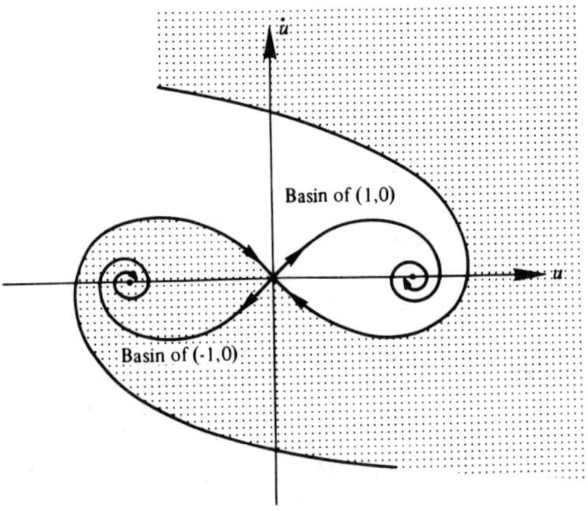

Figure 4.3.6

Note that in this phase portrait, some initial conditions tend toward one stable focus, while some tend toward the other. The two tendencies are divided by the curve, In(0, 0), the inset of the saddle at the origin. This is called the ***separatrix***, as it separates the domain into the two disjoint open sets, $\mathrm{In}(m_0)$ and $\mathrm{In}(m_1)$. The stable foci are called ***attractors***, and their insets are called their ***basins***. See Fig. 4.3.6 for the special case $\ddot{x} + \dot{x} - x + x^3 = 0$. This is a special case of a general theory, which is increasingly important in dynamical systems applications. The attractors are regarded as the principal features of the phase portrait; the size of their basins measures the probability of observing the attractor, and the separatrices help find them. ♦

Another basic ingredient in the qualitative theory is the notion of a ***closed orbit***, also called a ***limit cycle***).

4.3.18 Definition *An orbit* $\gamma(t)$ *for a vector field* X *is called* **closed** *when* $\gamma(t)$ *is not a fixed point and there is a* $\tau > 0$ *such that* $\gamma(t + \tau) = \gamma(t)$ *for all* t. *The* **inset** *of* γ, $\mathrm{In}(\gamma)$, *is the set of points* $m \in M$ *such that* $F_t(m) \to \gamma$ *as* $t \to +\infty$ *(i.e., the distance between* $F_t(m)$ *and the (compact) set* $\{\gamma(t) \mid 0 \le t \le \tau\}$ *tends to zero as* $t \to \infty$.) *Likewise, the* **outset**, $\mathrm{Out}(\gamma)$, *is the set of points tending to* γ *as* $t \to -\infty$.

4.3.19 Example One of the earliest occurrences of an attractive closed orbit in an important application is found in Baron ***Rayleigh's model for the violin string*** (see Rayleigh [1887, vol. 1, §68a]),

$$\ddot{u} + k_1 \dot{u} + k_3 \dot{u}^3 + \omega^2 u = 0, \quad k_1 < 0 < k_3,$$

or equivalently,

$$\begin{cases} \dot{u} = v \\ \dot{v} = -k_1 v - k_3 v^3 - \omega^2 u \end{cases}$$

with the phase portrait shown in Fig. 4.3.7.

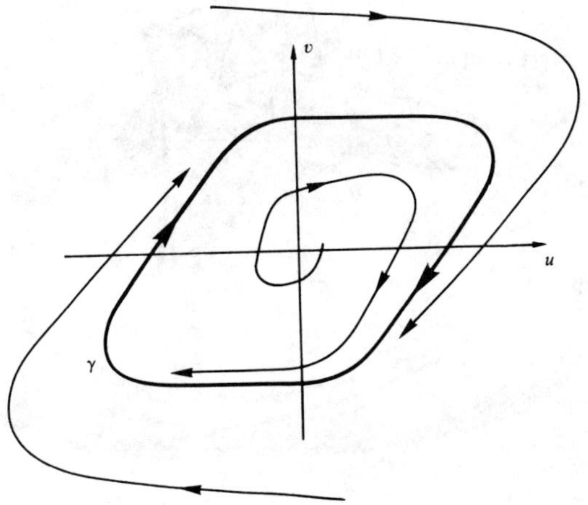

Figure 4.3.7

This has an unstable focus at the origin, with an attractive closed orbit around it. That is, the closed orbit γ is a limiting set for every point in its basin (or inset) $\mathrm{In}(\gamma)$, which is an open set of the domain. In fact the entire plane (excepting the origin) comprises the basin of this closed orbit. Thus every trajectory tends asymptotically to the limit cycle γ and winds around closer and closer to it. Meanwhile this closed orbit is a periodic function of time, in the sense of Definition **4.3.6**. Thus the eventual (asymptotic) behavior of every trajectory (other than the unstable constant trajectory at the origin) is periodic; it is an oscillation.
This picture thus models the *sustained oscillation* of the violin string, under the influence of the moving bow. Related systems occur in electrical engineering under the name van der Pol equation. (See Hirsch and Smale [1974], Chapter 10 for a discussion). ♦

We now proceed toward the analog of Liapunov's theorem for the stability of closed orbits. To do this we need to introduce Poincaré maps and characteristic multipliers.

4.3.20 Definition *Let* X *be a* C^r *vector field on a manifold* M, $r \geq 1$. *A* ***local transversal section*** *of* X *at* $m \in M$ *is a submanifold* $S \subset M$ *of codimension one with* $m \in S$ *and for all* $s \in S$, $X(s)$ *is not contained in* T_sS.

Let X *be a* C^r *vector field on a manifold* M *with* C^r *flow* $F : \mathcal{D}_X \subset M \times \mathbb{R} \to M$, γ *a closed orbit of* X *with period* τ, *and* S *a local transversal section of* X *at* $m \in \gamma$. *A* ***Poincaré map*** *of* γ *is a* C^r *mapping* $\Theta : W_0 \to W_1$ *where:*

PM 1 $W_0, W_1 \subset S$ *are open neighborhoods of* $m \in S$, *and* Θ *is a* C^r *diffeomorphism;*

PM 2 *there is a* C^r *function* $\delta : W_0 \to \mathbb{R}$ *such that for all* $s \in W_0$, $(s, \tau - \delta(s)) \in \mathcal{D}_X$, *and* $\Theta(s) = F(s, \tau - \delta(s))$; *and finally,*

PM 3 *if* $t \in \,]0, \tau - \delta(s)[$, *then* $F(s, t) \notin W_0$ (*see Fig.* 4.3.8).

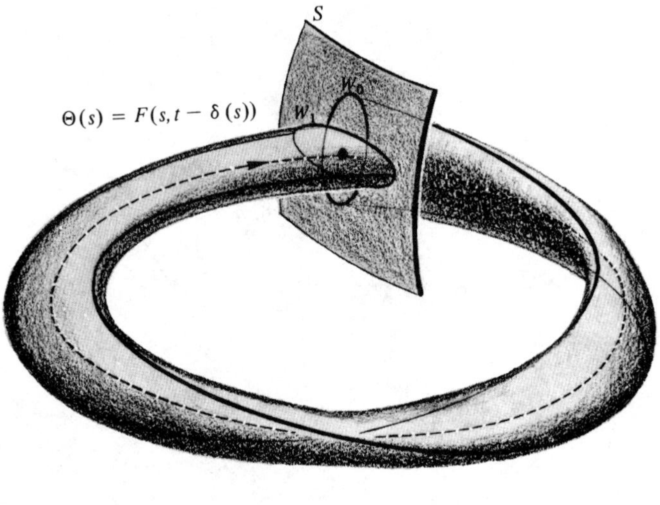

Figure 4.3.8

4.3.21 Existence and Uniqueness of Poincaré Maps

(i) *If* X *is a* C^r *vector field on* M, *and* γ *is a closed orbit of* X, *then there exists a Poincaré map of* γ.

(ii) *If* $\Theta : W_0 \to W_1$ *is a Poincaré map of* γ *(in a local transversal section* S *at* $m \in \gamma$*) and* Θ' *also (in* S′ *at* $m' \in \gamma$*), then* Θ *and* Θ' *are locally conjugate. That is, there are open neighborhoods* W_2 *of* $m \in S$, W_2' *of* $m' \in S'$, *and a* C^r*-diffeomorphism* H $: W_2 \to W_2'$, *such that* $W_2 \subset W_0 \cap W_1$, $W_2' \subset W_0' \cap W_1'$ *and the following diagram commutes:*

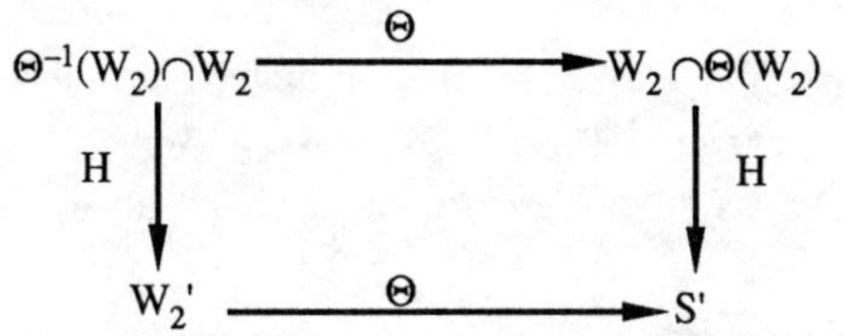

Proof (i) At any point $m \in \gamma$ we have $X(m) \neq 0$, so there exists a flow box chart (U, φ) at m with $\varphi(U) = V \times I \subset \mathbb{R}^{n-1} \times \mathbb{R}$. Then $S = \varphi^{-1}(V \times \{0\})$ is a local transversal section at m. If $F : \mathcal{D}_X \subset M \times \mathbb{R} \to M$ is the integral of X, $\mathcal{D}_X$ is open, so we may suppose $U \times [-\tau, \tau] \subset \mathcal{D}_X$, where τ is the period of γ. As $F_\tau(m) = m \in M$ and F_τ is a homeomorphism, $U_0 = F_\tau^{-1}(U) \cap U$ is an open neighborhood of $m \in M$ with $F_\tau(U_0) \subset U$. Let $W_0 = S \cap U_0$ and $W_2 = F_\tau(W_0)$. Then W_2 is a local transversal section at $m \in M$ and $F_\tau : W_0 \to W_2$ is a diffeomorphism (see Fig. 4.3.9).

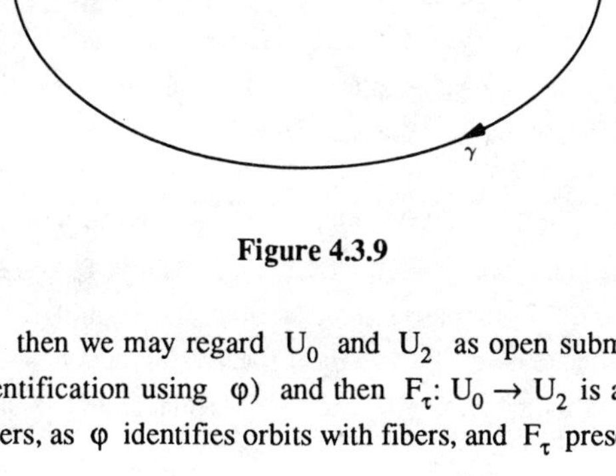

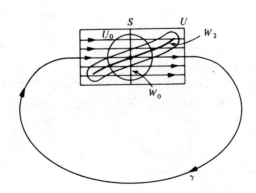

Figure 4.3.9

If $U_2 = F_\tau(U_0)$, then we may regard U_0 and U_2 as open submanifolds of the vector bundle $V \times \mathbb{R}$ (by identification using φ) and then $F_\tau: U_0 \to U_2$ is a C^r diffeomorphism mapping fibers into fibers, as φ identifies orbits with fibers, and F_τ preserves orbits. Thus W_2

is a section of an open subbundle. More precisely, if $\pi : V \times I \to V$ and $\rho : V \times I \to I$ are the projection maps, then the composite mapping is C^r and

$$W_0 \xrightarrow{F_\tau} W_2 \xrightarrow{\varphi} V \times I \xrightarrow{\pi} V \xrightarrow{\varphi^{-1}} S$$

has a tangent that is an isomorphism at each point, and so by the inverse mapping theorem, it is a C^r diffeomorphism onto an open submanifold. Let W_1 be its image, and Θ the composite mapping.

We now show that $\Theta : W_0 \to W_1$ is a Poincaré map. Obviously **PM 1** is satisfied. For **PM 2**, we identify U and $V \times I$ by means of φ to simplify notations. Then $\pi : W_2 \to W_1$ is a diffeomorphism, and its inverse $(\pi \mid W_2)^{-1} : W_1 \to W_2 \subset W_1 \times \mathbb{R}$ is a section corresponding to a smooth function $\sigma : W_1 \to \mathbb{R}$. In fact, σ is defined implicitly by

$$F_\tau(w_0) = (\pi \circ F_\tau(w_0), \rho \circ F_\tau(w_0)) = (\pi \circ F_\tau(w_0), \sigma \circ \pi F_\tau(w_0))$$

or $\rho \circ F_\tau(w_0)) = \sigma \circ \pi F_\tau(w_0)$. Now let $\delta : W_0 \to \mathbb{R}$ be given by $w_0 \mapsto \sigma \circ F_t(w_0)$ which is C^r. Then we have

$$F_{\tau-\delta(w_0)}(w_0) = (F_{-\delta(w_0)} \circ F_\tau)(w_0) = (\pi \circ F_\tau(w_0), \rho \circ F_\tau(w_0) - \delta(w_0)) = (\pi \circ F_\tau(w_0), 0) = \Theta(w_0).$$

Finally, **PM 3** is obvious since (U, φ) is a flow box.

(ii) The proof is burdensome because of the notational complexity in the definition of local conjugacy, so we will be satisfied to prove uniqueness under additional simplifying hypotheses that lead to global conjugacy (identified by italics). The general case will be left to the reader.

We consider first the special case $m = m'$. Choose a flow box chart (U, φ) at m, and *assume* $S \cup S' \subset U$, *and that* S *and* S' *intersect each orbit arc in* U *at most once, and that they intersect exactly the same sets of orbits.* (These three conditions may always be obtained by shrinking S and S'.) Then let $W_2 = S$, $W_2' = S'$, and $H : W_2 \to W_2'$ the bijection given by the orbits in U. As in (i), this is seen to be a C^r diffeomorphism, and $H \circ \Theta = \Theta' \circ H$.

Finally, suppose $m \neq m'$. Then $F_a(m) = m'$ for some $a \in \,]0, \tau[$, and as $\mathcal{D}_X$ is open there is a neighborhood U of m such that $U \times \{a\} \subset \mathcal{D}_X$. Then $F_a(U \cap S) = S''$ is a local transversal section of X at $m' \in \gamma$, and $H = F_a$ effects a conjugacy between Θ and $\Theta'' = F_a \circ \Theta \circ F_a^{-1}$ on S''. By the preceding paragraph, Θ'' and Θ' are locally conjugate, but conjugacy is an equivalence relation. This completes the argument. ■

If γ is a closed orbit of $X \in \mathfrak{X}(M)$ and $m \in \gamma$, the behavior of nearby orbits is given by a Poincaré map Θ on a local transversal section S at m. Clearly $T_m\Theta \in L(T_mS, T_mS)$ is a linear approximation to Θ at m. By uniqueness of Θ up to local conjugacy, $T_{m'}\Theta'$ is similar to $T_m\Theta$, for any other Poincaré map Θ' on a local transversal section at $m' \in \gamma$. Therefore, the spectrum of $T_m\Theta$ is independent of $m \in \gamma$ and the particular section S at m.

4.3.22 Definition *If* γ *is a closed orbit of* $X \in \mathfrak{X}(M)$, *the* ***characteristic multipliers*** *of* X *at* γ *are the points in the spectrum of* $T_m\Theta$, *for any Poincaré map* Θ *at any* $m \in \gamma$.

Another linear approximation to the flow near γ is given by $T_mF_\tau \in L(T_mM, T_mM)$ if $m \in \gamma$ and τ is the period of γ. Note that $F_\tau{}^*(X(m)) = X(m)$, so T_mF_τ always has an eigenvalue 1 corresponding to the eigenvector $X(m)$. The $(n-1)$ remaining eigenvalues (if $\dim(M) = n$) are in fact the characteristic multipliers of X at γ.

4.3.23 Proposition *If* γ *is a closed orbit of* $X \in \mathfrak{X}(M)$ *of period* τ *and* c_γ *is the set of characteristic multipliers of* X *at* γ, *then* $c_\gamma \cup \{1\}$ *is the spectrum of* T_mF_τ, *for any* $m \in \gamma$.

Proof We can work in a chart modeled on E and assume $m = 0$. Let V be the span of $X(m)$ so $E = T_mM \oplus V$. Write the flow $F_t(x, y) = (F_t{}^1(x, y), (F_t{}^2(x, y))$. By definition, we have

$$D_1F_t{}^1(m) = T_m\Theta \quad \text{and} \quad D_2F_\tau{}^2(m) \cdot X(m) = X(m).$$

Thus the matrix of T_mF_τ is of the form

$$\begin{bmatrix} T_m\Theta & 0 \\ A & 1 \end{bmatrix}$$

where $A = D_1F_\tau{}^2(m)$. From this it follows that the spectrum of T_mF_τ is $\{1\} \cup c_\gamma$. ■

If the characteristic exponents of an equilibrium point lie (strictly) in the left half-plane, we know from Liapunov's theorem that the equilibrium is stable. For closed orbits we introduce stability by means of the following definition.

4.3.24 Definition *Let* X *be a vector field on a manifold* M *and* γ *a closed orbit of* X. *An orbit* $F_t(m_0)$ *is said to* ***wind toward*** γ *if* m_0 *is + complete and for any local transversal section* S *to* X *at* $m \in \gamma$ *there is a* $t(0)$ *such that* $F_{t(0)}(m_0) \in S$ *and successive applications of the Poincaré map yield a sequence of points that converges to* m. *If the closed orbit* γ *has a neighborhood* U *such that for any* $m_0 \in U$, *the orbit through* m_0 *winds towards* γ, *then* γ *is called* ***asymptotically stable***.

In other words, orbits starting "close to γ, "converge" to γ; see Fig. 4.3.10.

4.3.25 Proposition *If* γ *is an asymptotically stable periodic orbit of the vector field* X *and* $m_0 \in U$, *the neighborhood given in* **4.3.24**, *then for any neighborhood* V *of* γ, *there exists* $t_0 > 0$ *such that for all* $t \geq t_0$, $F_t(m) \in V$.

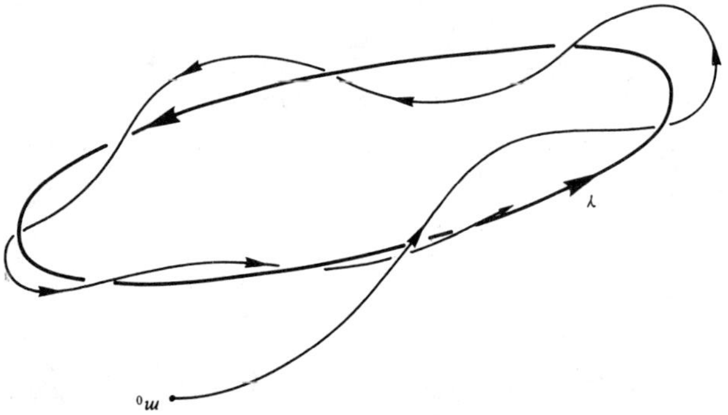

Figure 4.3.10

Proof Define $m_k = \Theta^k(m_0)$, where Θ is a Poincaré map for a local transversal section at m to γ containing m_0. Let $t(n)$ be the "return time" of n, i.e., $t(n)$ is defined by $F_{t(n)}(n) \in S$. If τ denotes the period of γ and $\tau_k = t(m_k)$, then since $m_k \to m$, it follows that $\tau_k \to \tau$ since $t(n)$ is a smooth function of n by **4.3.21**. Let M be an upper bound for the set $\{ |\tau_k| \mid k \in \mathbb{N}\}$. By smoothness of the flow, $F_s(m_k) \to F_s(m)$ as $k \to \infty$, uniformly in $s \in [0, M]$. Now write for any $t > 0$, $F_t(m_0) = F_{T(t)}(m_{k(t)})$, for $T(t) \in [0, M]$ and observe that $k(t) \to \infty$ as $t \to \infty$. Therefore, if W is any neighborhood of $F_{T(t)}(m_0)$ contained in V, since $F_t(m_0) = F_{T(t)}(m_{k(t)})$ converges to $F_{T(t)}(m)$ as $t \to \infty$ it follows that there exists $t_0 > 0$ such that for all $t \geq t_0$, $F_t(m_0) \in W \subset V$. ∎

4.3.26 Liapunov Stability Theorem for Closed Orbits *Let γ be a closed orbit of $X \in \mathfrak{X}(M)$ and let the characteristic multipliers of γ lie strictly inside the unit circle. Then γ is asymptotically stable.*

The proof relies on the following lemma.

4.3.27 Lemma *Let $T : E \to E$ be a bounded linear operator. Let $\sigma(T)$ lie strictly inside the unit circle. Then $\lim_{n\to\infty} T^n e = 0$ for all $e \in E$.*

Proof By lemma **4.3.7** and compactness of $\sigma(T)$, there is a norm $|\ \ |$ on E equivalent to the original norm on E such that $|T| \leq r < 1$. Therefore $|T^n e| \leq r^n |e| \to 0$ as $n \to \infty$. ∎

4.3.28 Lemma *Let $f : S \to S$ be a smooth map on a manifold S with $f(s) = s$ for some s. Let the spectrum of $T_s f$ lie strictly inside the unit circle. Then there is a neighborhood U of s such that if $s' \in U$, $f(s') \in U$ and $f^n(s') \to s$ as $n \to \infty$, where $f^n = f \circ f \circ \ldots \circ f$ (n times).*

Proof We can assume that S is a Banach space E and that $s = 0$. As above, renorm E and find $0 < r < 1$ such that $|T| \leq r$, where $T = \mathbf{D}f(0)$. Let $\varepsilon > 0$ be such that $r + \varepsilon < 1$. Choose a neighborhood V of 0 such that for all $x \in V$

$$|f(x) - Tx| \leq \varepsilon |x|$$

which is possible since f is smooth. Therefore,

$$|f(x)| \leq |Tx| + \varepsilon |x| \leq (r + \varepsilon)|x|.$$

Now, choose $\delta > 0$ such that the ball U of radius δ at 0 lies in V. Then the above inequality implies $|f^n(x)| \leq (r + \varepsilon)^n |x|$ for all $x \in U$ which shows that $f^n(x) \in U$ and $f^n(x) \to 0$. ■

Proof of 4.3.26 The previous lemma applied to $f = \Theta$, a Poincaré map in a transversal slice S to γ at m, implies that there is an open neighborhood V of m in S such that the orbit through every point of V winds toward γ. Thus, the orbit through every point of $U = \{F_t(m) \mid t \geq 0\} \supset \gamma$ winds toward γ. U is a neighborhood of γ since by the Sraightening Out Theorem **4.1.14**, each point of γ has a neighborhood contained in $\{F_t(\Theta(U)) \mid t > -\varepsilon, \varepsilon > 0\}$. ■

4.3.29 Definition *If* $X \in \mathfrak{X}(M)$ *and* γ *is a closed orbit of* X, γ *is called* ***hyperbolic*** *if none of the characteristic multipliers of* X *at* γ *has modulus* 1.

Hyperbolic closed orbits are isolated (see Abraham-Robbin [1967, Ch. 5]). The local qualitative behavior near an hyperbolic closed orbit, γ, may be visualized with the aid of the Poincaré map, $\Theta : W_0 \subset S \to W_1 \subset S$, as shown in Fig. 4.3.8. The qualitiative behavior of this map, under iterations, determines the asymptotic behavior of the trajectories near γ. Let $m \in \gamma$ be the base point of the section, and $s \in S$. Then In(γ), the inset of γ, intersects S in the inset of m under the iterations of Θ. That is, $s \in$ In(γ) if the trajectory $F_t(s)$ winds towards γ, and this is equivalent to saying that $\Theta^k(s)$ tends to m as $k \to +\infty$.

The inset and outset of $m \in S$ are classified by linear algebra, as there is an analogue of the linearization theorem for maps at hyperbolic critical points. The ***linearization theorem for maps*** says that there is a C^0 coordinate chart on S, in which the local representative of Θ is a linear map. Recall that in the hyperbolic case, the spectrum of this linear isomorphism avoids the unit circle. The eigenvalues inside the unit circle determine the generalized eigenspace of contraction i.e., the inset of $m \in S$ under the iterates of Θ. The eigenvalues outside the unit circle similarly determine the outset of $m \in S$. Although this argument provides only local C^0 submanifolds, the ***global stable manifold theorem*** improves this: the inset and outset of a fixed point of a diffeomorphism are smooth, injectively immersed submanifolds meeting transversally at m.

Returning to closed orbits, the inset and outset of $\gamma \subset M$ may be visualized by choosing a section S_m at every point $m \in \gamma$. The inset and outset of γ in M intersect each section S_m in submanifolds of S_m, meeting transversally at $m \in \gamma$. In fact, In(γ) is a cylinder over γ, that is, a bundle of injectively immersed disks. So, likewise, is Out(γ). And these two cylinders intersect transversally in γ, as shown in Fig. 4.3.11. These bundles need not be trivial.

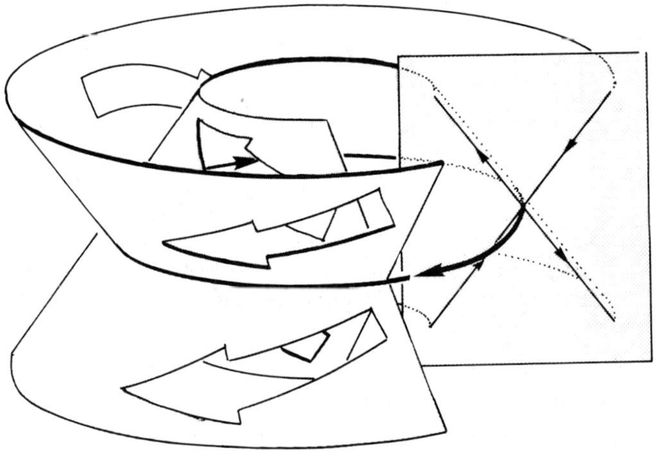

Figure 4.3.11

Another argument is sometimes used to study the inset and outset of a closed orbit, in place of the Poincaré section technique described before, and is originally due to Smale [1967]. The flow F_t leaves the closed orbit invariant. A special coordinate chart may be found in a neighborhood of γ. The neighborhood is a disk bundle over γ, and the flow F_t, is a bundle map. On each fiber, F_t is a linear map of the form $Z_t e^{Rt}$, where Z_t is a constant, and R is a linear map. Thus, if $s = (m, x)$ is a point in the chart, the local representative of F_t is given by the expression

$$F_t(m, x) = (m_t,\ Z_t e^{Rt} \cdot x)$$

called the ***Floquet normal form***. This is the ***linearization theorem for closed orbits***. A related result, the ***Floquet theorem***, eliminates the dependence of Z_t on t, by making a further (time-dependent) change of coordinates (see Hartman [1973, ch. 4, §6], or Abraham and Robbin [1967]). Finally, linear algebra applied to the linear map R in the exponent of the Floquet normal form, establishes the C^0 structure of the inset and outset of γ.

To get an overall picture of a dynamical system in which all critical elements (critical points and closed orbits) are hyperbolic, we try to draw or visualize the insets and outsets of each. Those with open insets are ***attractors***, and their open insets are their ***basins***. The domain is divided into basins by the ***separatrices***, which includes the insets of all the nonattractive (saddle-type) critical elements (and possible other, more complicated limit sets, called ***chaotic attractors***, not described here.)

We conclude with an example of sufficient complexity, which has been at the center of dynamical system theory for over a century.

4.3.30 Example The simple pendulum equation may be "simplified" by approximating $\sin x$ by two terms of its MacLaurin expansion. The resulting system is a model for a ***nonlinear spring with linear damping***,

$$\dot{x} = v, \quad \dot{v} = -cv - kx + \frac{k}{3}x^3.$$

Adding a periodic forcing term, we have

$$\dot{x} = v, \quad \dot{v} = -cv - kx + \frac{k}{3}x^3 + F\cos\omega t.$$

This time-dependent system in the plane is transformed into an autonomous system in a solid ring by adding an angular variable proportional to the time, $\theta = \omega t$. Thus,

$$\dot{x} = v, \quad \dot{v} = -cv - kx + \frac{k}{3}x^3 + F\cos\theta, \quad \dot{\theta} = \omega.$$

Although this was introduced by Baron Rayleigh to study the resonance of tuning forks, piano strings, and so on, in his classic book *Theory of Sound* [1877], this system is generally named the ***Duffing equation*** after Duffing who obtained the first important results (see Duffing [1908] and Stoker [1950]).

Depending on the values of the three parameters (c, k, F) various phase portraits are obtained. One of these is shown in Fig. 4.3.12, adapted from the experiments of Hayashi [1964]. There are three closed orbits: two attracting, one of saddle type. The inset of the saddle is a cylinder topologically, but the whole cylinder revolves around the saddle-type closed orbit. This cylinder is the separatrix between the two basins. For other parameter values the dynamics can be chaotic (see for example, Holmes [1979a,b] and Ueda [1980]).

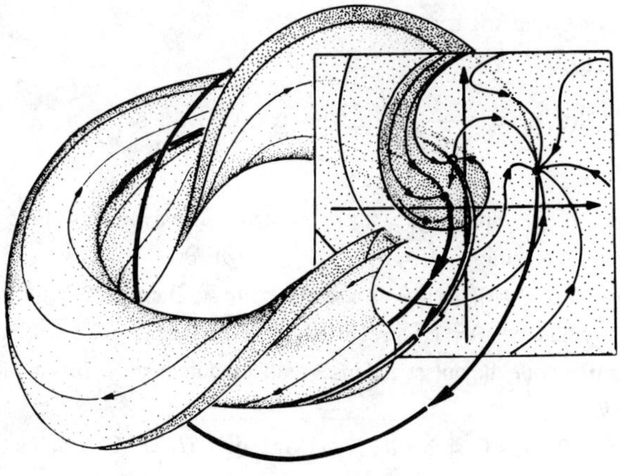

Figure 4.3.12

For further information on dynamical systems, see Abraham and Shaw [1985] and Guckenheimer and Holmes [1983].

Exercises

4.3A (i) Let $E \to M$ be a vector bundle and $m \in M$ an element of the zero section. Show that T_mE is isomorphic to $T_mM \oplus E_m$ in a natural, chart independent way.

(ii) If $\xi : M \to E$ is a section of E, and $\xi(m) = 0$, define $\xi'(m) : T_mM \to E_m$ to be the projection of $T_m\xi$ to E_m. Write out $\xi'(m)$ relative to coordinates.

(iii) Show that if X is a vector field, then $X'(m)$ defined this way coincides with Definition **4.3.1**.

4.3B Prove that the equation $\ddot{\theta} + 2k\,\dot{\theta} - q \sin\theta = 0$ $(q > 0, k > 0)$ has a saddle point at $\theta = 0$, $\dot{\theta} = 0$.

4.3C Consider the differential equations $\dot{r} = ar^3 - br$, $\dot{\theta} = 1$ using polar coordinates in the plane.

(i) Determine those a, b for which this system has an attractive periodic orbit.

(ii) Calculate the eigenvalues of this system at the origin for various a, b.

4.3D Let $X \in \mathfrak{X}(M)$, $\varphi : M \to N$ be a diffeomorphism, and $Y = \varphi_* X$. Show that

(i) $m \in M$ is a critical point of X iff $\varphi(m)$ is a critical point of Y and the characteristic exponents are the same for each;

(ii) $\gamma \subset M$ is a closed orbit of X iff $\varphi(\gamma)$ is a closed orbit of Y and their characteristic multipliers are the same.

4.3E The energy for a symmetric heavy top is

$$H(\theta, p_\theta) = \frac{1}{2I\sin^2\theta}\{p_\psi{}^2(b - \cos\theta)^2 + p_\theta{}^2 \sin^2\theta\} + \frac{p_\psi{}^2}{J} + Mg\ell\cos\theta$$

where $I, J > 0$, b, p_ψ, and $Mg\ell > 0$ are constants. The dynamics of the top is described by the differential equations $\dot{\theta} = \partial H/\partial p_\theta$, $\dot{p}_\theta = -\partial H/\partial\theta$.

(i) Show that $\theta = 0$, $p_\theta = 0$ is a saddle point if $0 < p_\psi < 2(Mg\ell I)^{1/2}$ (a slow top).

(ii) Verify that $\cos\theta = 1 - \gamma\,\mathrm{sech}^2((\beta\gamma)^{1/2}/2)$, where $\gamma = 2 - b^2/\beta$ and $\beta = 2Mg\ell/I$ describe *both* the outset and inset of this saddle point. (This is called a ***homoclinic orbit.***)

(iii) Is $\theta = 0$, $p_\theta = 0$ stable if $p_\psi > (Mg\ell I)^{1/2}$? (*Hint:* Use the fact that H is constant along the trajectories.)

4.3F Let $A \in L(\mathbb{R}^n, \mathbb{R}^n)$ and suppose that $a < \mathrm{Re}(\lambda_i) < b$, for all eigenvalues λ_i, $i = 1, ..., n$ of A. Show that $\mathbb{R}^n$ admits an inner product $\langle\langle\ ,\ \rangle\rangle$ with associated norm $|||\cdot|||$ such that

$$a\,|||\,x\,|||^2 \le \langle\langle Ax, x\rangle\rangle \le b\,|||\,x\,|||^2 .$$

Prove this by following the outline below.

(i) If A is diagonalizable over $\mathbb{C}$, then find a basis in $\mathbb{R}^n$ in which the matrix of A has either the entries on the diagonal, the real eigenvalues of A, or 2×2 blocks of the form

$$\begin{bmatrix} a_j & -b_j \\ b_j & a_j \end{bmatrix}, \text{ for } \lambda_j = a_j + ib_j, \text{ if } b_j \ne 0.$$

Choose the inner product $\langle\langle\ ,\ \rangle\rangle$ on $\mathbb{R}^n$ such that the one and two-dimensional invariant subspaces of A defined by this block-matrix are mutually orthogonal; pick the standard $\mathbb{R}^2$-basis in the two-dimensional spaces.

(ii) If A is not diagonalizable, then pass to the real Jordan form. There are two kinds of Jordan $k_j \times k_j$ blocks:

$$\begin{bmatrix} \lambda_j & 0 & 0 & \cdots & 0 & 0 \\ 1 & \lambda_j & 0 & \cdots & 0 & 0 \\ 0 & 1 & \lambda_j & \cdots & 0 & 0 \\ \cdots & & & & \cdots & \cdots \\ 0 & \cdots & \cdots & & 1 & \lambda_j \end{bmatrix},$$

if $\lambda_j \in \mathbb{R}$, or

$$\begin{bmatrix} \Delta_j & 0 & 0 & \cdots & 0 & 0 \\ I_2 & \Delta_j & 0 & \cdots & 0 & 0 \\ 0 & I_2 & \Delta_j & \cdots & 0 & 0 \\ \cdots & & & & \cdots & \cdots \\ 0 & \cdots & \cdots & & I_2 & \Delta_j \end{bmatrix},$$

where

$$\Delta_j = \begin{bmatrix} a_j & -b_j \\ b_j & a_j \end{bmatrix}, \quad I_2 = \begin{bmatrix} 1 & 0 \\ 0 & 1 \end{bmatrix},$$

if $\lambda_j = a_j + ib_j$, $a_j, b_j \in \mathbb{R}$, $b_j \ne 0$. For the first kind of block, choose the basis $e_1', ..., e_{k(j)}'$ of eigenvectors of the diagonal part D. Then, for $\varepsilon > 0$ small, put $e_r^\varepsilon = e_r'/\varepsilon^{r-1}$, $r = 1, ..., k(j)$ and define $\langle\ ,\ \rangle_\varepsilon$ on the subspace $\mathrm{span}\{e_1^\varepsilon, ..., e_{k(j)}^\varepsilon\} \subset \mathbb{R}^n$ to be the Euclidean inner product given by this basis. Compute the marix of A in this basis and show that

$$\frac{\langle Ax, x\rangle_\varepsilon}{\langle x, x\rangle_\varepsilon} \to \frac{Dx \cdot x}{\|x\|^2}, \quad \text{as } \varepsilon \to 0.$$

Conclude that for ε small the statement holds for the first kind of block. Do the same for the second kind of block.

4.3G Let $A \in L(\mathbb{R}^n, \mathbb{R}^n)$. Show that the following are equivalent.

(i) All eigenvalues of A have strictly negative real part (the origin is called a ***sink*** in this case).

(ii) For any norm $|\cdot|$ on $\mathbb{R}^n$, there exist constants $k > 0$ and $\varepsilon > 0$ such that for all $t \geq 0$, $|e^{tA}| \leq ke^{-t\varepsilon}$.

(iii) There is a norm $|||\cdot|||$ on $\mathbb{R}^n$ and a constant $\delta > 0$ such that for all $t \geq 0$, $|||e^{tA}||| \leq e^{-t\delta}$. (*Hint:* (ii) $\Rightarrow$ (i) by using the real Jordan form: if every solution of $\dot{x} = Ax$ tends to zero as $t \to +\infty$, then every eigenvalue of A has strictly negative real part. For (i) $\Rightarrow$ (iii) use Exercise **4.3E** and observe that if $x(t)$ is a solution of $\dot{x} = Ax$, then we have $(d/dt)\,|||x(t)||| = \langle\langle x(t), Ax(t)\rangle\rangle / |||x(t)|||$, so that we get the following inequality : $at \leq \log |||x(t)||| / \log |||x(0)||| \leq bt$, where $a = \min\{\mathrm{Re}\,(\lambda_i) \mid i = 1, \ldots, n\}$, and $b = \max\{\mathrm{Re}\,(\lambda_i) \mid i = 1, \ldots, n\}$. Then let $-\varepsilon = b$.) Prove a similar theorem if all eigenvalues of A have strictly positive real part; the origin is then called a ***source***.

4.3H Give a proof of Theorem **4.3.4** in the finite dimensional case without using the variation of constants formula (Exercise **4.1E**) and using Exercise **4.3F**. (*Hint:* If $A = X'(0)$ locally show $\lim_{x \to 0} \langle\langle X(x) - Ax, x\rangle\rangle / |||x|||^2 = 0$. Since $\langle\langle Ax, x\rangle\rangle \leq -\varepsilon\, |||x|||^2$, $\varepsilon = \max \mathrm{Re}\,\{\lambda_i | i = 1, \ldots, n$, λ_i eigenvalues of $A\}$, there exists $\delta > 0$ such that if $|||x||| \leq \delta$, then $\langle\langle X(x), x\rangle\rangle \leq -C\,|||x|||^2$, for some $C > 0$. Show that if $x(t)$ is a solution curve in the closed δ–ball, $t \in [0, T]$, then $d|||x(t)|||/dt \leq -C|||x(t)|||$. Conclude $|||x(t)||| \leq \delta$ for all $t \in [0, T]$ and thus by compactness of the δ-ball, $x(t)$ exists for all $t \geq 0$. Finally, show that $|||x(t)||| \leq e^{-t\varepsilon}\,|||x(0)|||$.)

4.3I An equilibrium point m of a vector field $X \in \mathfrak{X}(M)$ is called a ***sink***, if there is a $\delta > 0$ such that all points in the spectrum of $X'(m)$ have real part $< -\delta$.

(i) Show that in a neighborhood of a sink there is no other equilibrium of X.

(ii) If $M = \mathbb{R}^n$ and X is a linear vector field, Exercise **4.3G** shows that provided $\lim_{t \to \infty} m(t) = 0$ for every integral curve $m(t)$ of X, then the eigenvalues of X have all strictly negative real part. Show that this statement is false for general vector fields by finding an example of a non-linear vector field X on $\mathbb{R}^n$ whose integral curves tend to zero as $t \to \infty$, but is such that $X'(0)$ has at least one eigenvalue with zero real part. (*Hint :* See Exercise **4.3C**).

4.3J Hyperbolic Flows An operator $A \in L(\mathbb{R}^n, \mathbb{R}^n)$ is called ***hyperbolic*** if no eigenvalue of A has zero real part. The linear flow $x \mapsto e^{tA}x$ is then called a ***hyperbolic flow.***

(i) Let A be hyperbolic. Show that there is a direct sum decomposition $\mathbb{R}^n = E^s \oplus E^u$, $A(E^s) \subset E^s$, $A(E^u) \subset E^u$, such that the origin is a sink on E^s and a source on E^u; E^s and E^u are called the ***stable*** and ***unstable subspaces*** of $\mathbb{R}^n$. Show that the decomposition is unique. (*Hint:* E^s is the sum of all subspaces defined by the real Jordan form for which the real part of the eigenvalues is negative. For uniqueness, if $\mathbb{R}^n = E'^s \oplus E'^u$ and $v \in E'^s$, then $v = x + y$, $x \in E^s$, $y \in E^u$ with $e^{tA}v \to 0$ as $t \to \infty$, so that $e^{tA}x \to 0$, $e^{tA}y \to 0$ as $t \to \infty$. But since the origin is a source on E'^u, $\| e^{tA}y \| \geq e^{t\varepsilon} \| y \|$ for some $\varepsilon > 0$ by the analogue of Exercise **4.3G**(ii).)

(ii) Show that A is hyperbolic iff for each $x \neq 0$, $\| e^{tA}x \| \to \infty$ as $t \to \pm\infty$.

(iii) Conclude that hyperbolic flows have no periodic orbits.

4.3K Gradient flows (this continues Exercise **4.1H**) Let $f: \mathbb{R}^n \to \mathbb{R}$ be C^1 and let $X = (\partial f/\partial x^1, \ldots, \partial f/\partial x^n)$. Show that at regular points of f, the integral curves of X cross the level surfaces of f orthogonally and that every singular point of f is an equilibrium of X. Show that isolated maxima of f are asymptotically stable. (*Hint:* If x_0 is the isolated maximum of f, then $f(x_0) - f(x)$ is a strict Liapunov function.) Draw the level sets of f and the integral curves of X on the same diagram in $\mathbb{R}^2$, when f is defined by $f(x^1, x^2) = (x^1 - 1)^2 + (x^2 - 2)^2(x^2 - 3)^2$.

4.3L Consider $\ddot{u} + \dot{u} + u^3 = 0$. Show that solutions converge to zero like $C/\sqrt{t}$ as $t \to \infty$ by considering $H(u, \dot{u}) = (u + \dot{u})^2 + u^3 + u^4$. (See Ball and Carr [1976] for more information.)

4.3M Use the method of Liapunov functions to study the stability of the origin for the following vector fields:

(i) $X(x, y) = -(3y + x^3)(\partial/\partial x) + (2x - 5y^3)(\partial/\partial y)$ (asymptotically stable);

(ii) $X(x, y) = -xy^4(\partial/\partial x) + x^6y(\partial/\partial y)$ (stable; look for L of the form $x^4 + ay^6$);

(iii) $X(x, y) = (xy - x^3 + y)(\partial/\partial x) + (x^4 - x^2y + x^3)(\partial/\partial y)$ (stable; look for L of the form $ax^4 + by^2$);

(iv) $X(x, y) = (y + x^7)(\partial/\partial x) + (y^9 - x)(\partial/\partial y)$ (unstable);

(v) $X(x, y, z) = 3y(z + 1)(\partial/\partial x) + x(z + 1)(\partial/\partial y) + yz(\partial/\partial z)$ (stable);

(vi) $X(x, y, z) = (-x^5 + 5x^6 + 2y^3 + xz^2 + xyz)(\partial/\partial x) + (-y - 2z + 3x^6 + 4yz + xz + xy^2)(\partial/\partial y) + (2y - z - 2x^8 - y^2 + xz^2 + xy^3)(\partial/\partial z)$ (asymptotically stable; use $L(x, y, z) = (1/2)x^2 + 5(y^2 + z^2)$).

4.3N Consider the following vector field on $\mathbb{R}^{n+1}$; $X(s, x) = (as^N + f(s) + g(s, x), Ax + F(x) + h(s, x))$ where $s \in \mathbb{R}$, $x \in \mathbb{R}^n$, $f(0) = \ldots = f^{(N)}(0) = 0$, $g(s, x)$, has all derivatives of order ≤ 2 zero at the origin, and $F(x)$, $h(s, x)$ vanish together with their first derivative the origin. Assume the $n \times n$ matrix A has all eigenvalues distinct with strictly negative real part. Prove the following theorem of Liapunov: *if* N *is even or* N *is odd and* $a > 0$, *then the origin is unstable; if* N *is odd and* $a < 0$, *the origin is asymptotically stable.* (*Hint:* for N even, use $L(s, x) = s - a \| x \|^2/2$ and for N odd, $L(s, x) = (s^2 - a \| x \|^2)/2$; show that in both cases the sign of X[L] near the origin is given by the sign of a.)

4.1O Let E be a Banach space and $A : \mathbb{R} \to L(E, E)$ a continuous map. Let $F_{t,s}$ denote the evolution operator of the time–dependent vector field $X(t, x) = A(t)x$ on E.

(i) Show that $F_{t,s} \in \mathcal{GL}(E)$.

(ii) Show that $\| F_{t,s} \| \leq e^{(t-s)\alpha}$, where $\alpha = \sup_{\lambda \in [s, t]} \| A(\lambda) \|$. Conclude that the vector field X(t, x) is complete. (*Hint:* Use Gronwall's inequality and the time dependent version of **4.1.22.**)

Next assume that A is periodic with period T, i.e. $A(t + T) = A(t)$ for all $t \in \mathbb{R}$.

(iii) Show that $F_{t+T, s+T} = F_{t,s}$ for any $t, s \in \mathbb{R}$. (*Hint:* Show that $t \mapsto F_{t+T, s+T}(x)$ satisfies the differential equation $\dot{x} = A(t)x$.)

(iv) Define the ***monodromy operator*** by $M(t) = F_{t+T,t}$. Show that if A is independent of t, then $M(t) = e^{TA}$ is also independent of t. Show that $M(s) = F_{t,s} \circ M(t) \circ F_{s,t}$ for any continuous $A : \mathbb{R} \to L(E, E)$. Conclude that all solutions of $\dot{x} = A(t)x$ are of period T if and only if there is a t_0 such that $M(t_0) = $ identity. Show that the eigenvalues of M(t) are independent of t.

(v) (Floquet) Show that each real eigenvalue λ of $M(t_0)$ determines a solution, denoted $c(t; \lambda, t_0)$ of $\dot{x} = A(t)x$ satisfying $c(t + T; \lambda, t_0) = \lambda c(t; \lambda, t_0)$, and also each complex eigenvalue $\lambda = a + ib$, of $M(t_0)$, where $a, b \in \mathbb{R}$, determines a pair of solutions denoted $c^r(t; \lambda, t_0)$ and $c^i(t; \lambda, t_0)$ satisfying

$$\begin{bmatrix} c^r(t + T; \lambda, t_0) \\ c^i(t + T; \lambda, t_0) \end{bmatrix} = \begin{bmatrix} a & -b \\ b & a \end{bmatrix} \begin{bmatrix} c^r(t; \lambda, t_0) \\ c^i(t; \lambda, t_0) \end{bmatrix}.$$

(*Hint:* Let $c(t; \lambda, t_0)$ denote the solution of $\dot{x} = A(t)x$ with the initial condition $c(t_0; \lambda, t_0) = e$, where e is an eigenvector of $M(t_0)$ corresponding to λ; if λ is complex, work on the complexification of E. Then show that $c(t + T; \lambda, t_0) - \lambda c(t; \lambda, t_0)$ satisfies the same differential equation and its value at t_0 is zero since $c(t_0 + T; \lambda, t_0) = M(t_0)\, c(t_0; \lambda, t_0) = \lambda c(t_0; \lambda, t_0)$.)

(vi) (Floquet) Show that there is a nontrivial periodic solution of period T of $\dot{x} = A(t)x$ if and only if 1 is an eigenvalue of $M(t_0)$ for some $t_0 \in \mathbb{R}$. (*Hint:* If c(t) is such a periodic solution, then $c(t_0) = c(t_0 + T) = M(t_0)c(t_0)$.)

(vii) (Liapunov) Let $P : \mathbb{R} \to \mathcal{GL}(E)$ be a C^1 function which is periodic with period T. Show that the change of variable $y = P(t)x$ transforms the equation $\dot{x} = A(t)x$ into the equation $\dot{y} = B(t)y$, where $B(t) = (P'(t) + P(t)A(t))P(t)^{-1}$. If N(t) is the monodromy operator of $\dot{y} = B(t)y$, show that $N(t) = P(t)M(t)P(t)^{-1}$.

4.1P (i) (Liapunov) Let E be a finite dimensional *complex* vector space and let $A : \mathbb{R} \to$

$L(E, E)$ be a continuous function which is periodic with period T. Let $M(s)$ be the monodromy operator of the equation $\dot{x} = A(t)\,x$ and let $B \in L(E, E)$ be such that $M(s) = e^{TB}$ (see exercise **4.1O**). Define $P(t) = e^{tB}F_{s,t}$ and put $y(t) = P(t)x(t)$. Use (vii) in the previous exercise to show that $y(t)$ satisfies $\dot{y} = By$. Prove that $P(t)$ is a periodic C^1–function with period T. (*Hint:* Use (iii) of the previous exercise and $e^{TB} = M(s) = F_{s+T,\, s}$). Thus, for complex finite dimensional vector spaces, the equation $\dot{x} = A(t)x$, where $A(t + T) = A(t)$, can be transformed via $y(t) = P(t)x(t)$ into the constant coefficient linear equation $\dot{y} = By$.

(ii) Since the general solution of $\dot{y} = By$ is a linear combination of vectors $\exp(t\lambda_i)t^{\ell(i)}u_i$, where $\lambda_1, \ldots, \lambda_m$ are the distinct eigenvalues of B, $u_i \in E$, and $1 \leq \ell(i) \leq$ multiplicity of λ_i, conclude that the general solution of $\dot{x} = A(t)x$ is a linear combination of vectors $\exp(t\lambda_i)t^{\ell(i)}\, P(t)^{-1}u_i$ where $P(t + T) = P(t)$. Show that the eigenvalues of $M(s) = e^{TB}$ are $\exp(T\lambda_i)$. Show that $\mathrm{Re}(\lambda_i) < 0\ (> 0)$ for all $i = 1, \ldots, m$ if and only if all the solutions of $\dot{x} = A(t)x$ converge to the origin as $t \to +\infty\ (-\infty)$, i.e. if and only if the origin is asympotically stable (unstable).

§4.4 *Frobenius' Theorem and Foliations*

The main pillars supporting differential topology and calculus on manifolds are the Implicit Function Theorem, the Existence Theorem for ordinary differential equations, and Frobenius' Theorem, which we discuss briefly here. First some definitions.

4.4.1 Definition *Let* M *be a manifold and let* $E \subset TM$ *be a subbundle of its tangent bundle; i.e.,* E *is a* ***distribution*** *(or a* ***plane field****) on* M.

(i) *We say* E *is* ***involutive*** *if for any two vector fields* X *and* Y *defined on open sets of* M *and which take values in* E, [X, Y] *takes values in* E *as well.*

(ii) *We say* E *is* ***integrable*** *if for any* $m \in M$ *there is a (local) submanifold* $N \subset M$, *called a* ***(local) integral manifold*** *of* E *at* m *containing* m, *whose tangent bundle is exactly* E *restricted to* N.

The situation is shown in Figure 4.4.1.

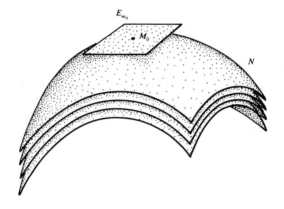

Figure 4.4.1

4.4.2 Examples

A Any subbundle *E* of TM with one dimensional fibers is involutive; *E* is also integrable, which is seen in the following way. Using local bundle charts for TM at $m \in M$ with the subbundle property for *E*, find in an open neighborhood of m, and a vector field that never vanishes and has values in *E*. Its local integral curves through m have as their tangent bundles *E* restricted to these curves. If the vector field can be found globally and has no zeros, then through any point of the manifold there is exactly one maximal integral curve of the vector field, and this integral curve never reduces to a point.

B Let $f : M \to N$ be a submersion and consider the bundle $\ker Tf \subset TM$. This bundle is involutive since for any $X, Y \in \mathfrak{X}(M)$ which take values in $\ker Tf$, we have $Tf([X, Y]) = 0$ by **4.2.25**. The bundle is integrable since for any $m \in M$ the restriction of $\ker Tf$ to the submanifold $f^{-1}(f(m))$ coincides with the tangent bundle of this submanifold (see §**3.5**).

C Let $\mathbb{T}^n$ be the n-dimensional torus, $n \geq 2$. Let $1 \leq k \leq n$ and consider $E = \{(v_1, ..., v_n) \in T\mathbb{T}^n \mid v_{k+1} = \cdots = v_n = 0\}$. This distribution is involutive and integrable; the integral manifold through $(t_1, ..., t_n)$ is $\mathbb{T}^k \times (t_{k+1}, ..., t_n)$.

D $E = TM$ is involutive and integrable; the integral submanifold through any point is M itself.

E An example of a noninvolutive distribution is as follows. Let $M = SO(3)$, the rotation group (see Exercise **3.5S**). The tangent space at $I = \textit{identity}$ consists of the 3×3 skew symmetric matrices. Let

$$E_I = \left\{ A \in T_I SO(3) \;\middle|\; A = \begin{bmatrix} 0 & 0 & -q \\ 0 & 0 & p \\ q & -p & 0 \end{bmatrix} \text{ for some } p, q \in \mathbb{R} \right\},$$

a two dimensional subspace. For $Q \in SO(3)$, let

$$E_Q = \{B \in T_Q SO(3) \mid Q^{-1}B \in E_I\}.$$

Then $E = \bigcup \{E_Q \mid Q \in SO(3)\}$ is a distribution but is not involutive. In fact, one computes that the two vector fields with $p = 1$, $q = 0$ and $p = 0$, $q = 1$ have a bracket that does not lie in E. Further insight into this example is gained after one studies Lie groups (a supplementary chapter).

F Let E be a distribution on M. Suppose that a collection $\mathcal{E}$ of smooth sections of E spans E in the sense that for each section X of E there are vector fields $X_1, ..., X_k$ in $\mathcal{E}$ and smooth functions $a^1, ..., a^k$ on M such that $X = a^i X_i$. Suppose $\mathcal{E}$ is closed under bracketing; i.e., if X and Y have values in $\mathcal{E}$, so does [X, Y]. We claim that *E is involutive.*

To prove this assertion, let X and Y be sections of E and write $X = a^i X_i$ and $Y = b^j Y_j$, where a^i and b^j are smooth functions on M and X_i and Y_j belong to $\mathcal{E}$. We calculate

$$[X, Y] = B^j Y_j - A^i X_i + a^i b^j [X_i, Y_j],$$

where

$$B^j = a^i X_i[b^j] \quad \text{and} \quad A^i = b^j Y_j[a^i].$$

Thus [X, Y] is a section of E, so E is involutive. ♦

Frobenius' Theorem asserts that the two conditions in **4.4.1** are equivalent.

4.4.3 The Local Frobenius Theorem *A subbundle E of TM is involutive if and only if it is integrable.*

Proof Suppose E is integrable. Let X and Y be sections of E and let N be a local integral manifold through $m \in M$. At points of N, X and Y are tangent to N, so define restricted vector fields $X|_N$, $Y|_N$ on N. By **4.2.25** (on φ-relatedness of brackets) applied to the inclusion map, we have

$$[X|_N, Y|_N] = [X, Y]|_N .$$

Since N is a manifold, $[X|_N, Y|_N]$ is a vector field on N, so $[X, Y]$ is tangent to N and hence in E.

Conversely, suppose that E is involutive. By choosing a vector bundle chart, one is reduced to this local situation: E is a model space for the fibers of E, F is a complementary space, and $U \times V \subset E \times F$ is an open neighborhood of $(0, 0)$, so $U \times V$ is a local model for M. We have a map $f : U \times V \to L(E, F)$ such that the fiber of E over (x, y) is

$$E_{(x, y)} = \{(u, f(x, y) \cdot u) \mid u \in E\} \subset E \times F ,$$

and we can assume we are working near $(0, 0)$ and $f(0, 0) = 0$. Let us express involutivity of E in terms of f.

For fixed $u \in E$, let $X_u(x, y) = (u, f(x, y) \cdot u)$. Using the local formula for the Lie bracket (see formula (5) in §**4.2**) one finds that

$$[X_{u_1}, X_{u_2}](x, y) = (0, \mathbf{D}f(x, y) \cdot (u_1, f(x, y) \cdot u_1) \cdot u_2 - \mathbf{D}f(x, y) \cdot (u_2, f(x, y) \cdot u_2) \cdot u_1) . \quad (1)$$

By the involution assumption, this lies in $E_{(x, y)}$. Since the first component vanishes, the local description of $E_{(x, y)}$ above shows that the second must as well; i.e., we get the following identity:

$$\mathbf{D}f(x, y) \cdot (u_1, f(x, y) \cdot u_1) \cdot u_2 = \mathbf{D}f(x, y) \cdot (u_2, f(x, y) \cdot u_2) \cdot u_1 . \quad (1')$$

Consider the time dependent vector fields

$$X_t(x, y) = (0, f(tx, y) \cdot x) \quad \text{and} \quad X_{t, u}(x, y) = (u, f(tx, y) \cdot tu) ,$$

so that by the local formula for the Jacobi-Lie bracket,

$$\begin{aligned}[X_t, X_{t, u}](x, y) &= (0, t\mathbf{D}_2 f(tx, y) \cdot (f(tx, y) \cdot x) \cdot u \\ &\quad - t\mathbf{D}f(tx, y) \cdot (u, f(tx, y) \cdot u) \cdot x - f(tx, y) \cdot u) \\ &= (0, -t\mathbf{D}_1 f(tx, y) \cdot x \cdot u - f(tx, y) \cdot u) , \end{aligned} \quad (2)$$

where the last equality follows from (1′). But $\partial X_{t,u}/\partial t$ equals the negative of the right hand side of (2), i.e.,

$$[X_t, X_{t,u}] + \frac{\partial X_{t,u}}{\partial t} = 0 ,$$

which by **4.2.31** is equivalent to

$$F_t^* X_{t,u} = 0 \tag{3}$$

where $F_t = F_{t,0}$ and $F_{t,s}$ is the evolution operator of the time dependent vector field X_t. Since $X_t(0, 0) = 0$, it follows that F_t is defined for $0 \le t \le 1$ by **4.1.25**.

Since F_0 is the identity, relation (3) implies that

$$F_t^* X_{t,u} = X_{0,u} , \quad \text{i.e.,} \quad TF_1 \circ X_{0,u} = X_{1,u} \circ F_1 . \tag{4}$$

Let $N = F_1(E \times \{0\})$, a submanifold of $E \times F$, the model space of M. If $(x, y) = F_1(e, 0)$, the tangent space at (x, y) to N equals

$$\begin{aligned} T_{(x,y)}N &= \{T_{(e,0)}F_1(u, 0)) \mid u \in E\} = \{T_{(e,0)}F_1(X_{0,u}(e, 0)) \mid u \in E\} \\ &= \{X_{1,u}(F_1(e, 0)) \mid u \in E\} && \text{(by (4))} \\ &= \{(u, f(x, y) \cdot u) \mid u \in E\} = E_{(x,y)} . \quad \blacksquare \end{aligned}$$

The method of using the time-one map of a time-dependent flow to provide the appropriate coordinate change is useful in a number of situations and is called the ***method of Lie transforms***. An abstract version of this method is given in **5.4.7**; we shall use this method again in Chapter 6 to prove the Poincaré Lemma and in Chapter 8 to prove the Darboux Theorem.

Note: The method of Lie transforms is also used in singularity theory and bifurcation theory (see Golubitsky and Schaeffer [1985]). For a proof of the Morse Lemma using this method, see **5.5.8**, which is based on Palais [1969] and Golubitsky and Marsden [1983]. For a proof of the Frobenius Theorem from the Implicit Function Theorem using manifolds of maps in the spirit of Supplement **4.1C**, see Penot [1970]. See Exercise **4.4F** for another proof of the Frobenius Theorem in finite dimensions.

The Frobenius Theorem is intimately connected to the global concept of foliations. Roughly speaking, the integral manifolds N can be glued together to form a "nicely stacked" family of submanifolds filling out M (see Figure 4.4.1 or Example **4.4.2A**).

4.4.4 Definition *Let* M *be a manifold and* $\Phi = \{\mathcal{L}_\alpha\}_{\alpha \in A}$ *a partition of* M *into disjoint connected sets called* ***leaves***. *The partition* Φ *is called a* ***foliation*** *if each point of* M *has a chart* (U, φ), $\varphi : U \to U' \times V' \subset E \oplus F$ *such that for each* $\mathcal{L}_\alpha$ *the connected components* $(U \cap \mathcal{L}_\alpha)^\beta$

of $U \cap \mathcal{L}_\alpha$ *are given by* $\varphi((U \cap \mathcal{L}_\alpha)^\beta) = U' \times \{c_\alpha{}^\beta\}$, *where* $c_\alpha{}^\beta \in F$ *are constants for each* $\alpha \in A$ *and* β. *Such charts are called* ***foliated*** *(or* ***distinguished****) by* Φ. *The* ***dimension*** *(respectively,* ***codimension****)* ***of the foliation*** Φ *is the dimension of* E *(resp.,* F). See Figure 4.4.2.

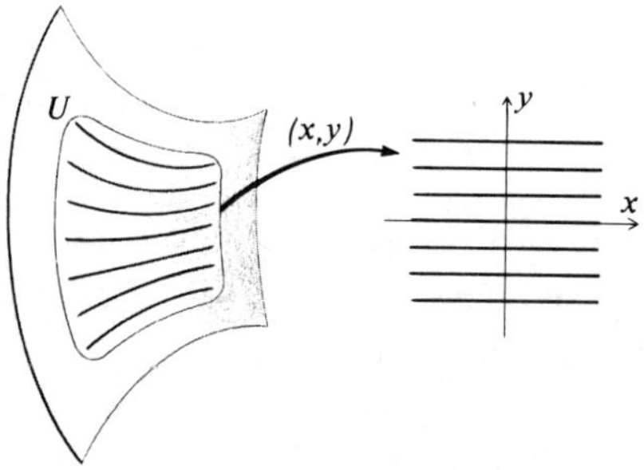

Figure 4.4.2

Note that each leaf $\mathcal{L}_\alpha$ is a connected immersed submanifold. In general, this immersion is *not* an embedding; i.e., the induced topology on $\mathcal{L}_\alpha$ from M does not necessarily coincide with the topology of $\mathcal{L}_\alpha$ (the leaf $\mathcal{L}_\alpha$ may accumulate on itself, for example). A differentiable structure on $\mathcal{L}_\alpha$ is induced by the foliated charts in the following manner. If (U, φ), $\varphi : U \to U' \times V' \subset E \oplus F$ is a foliated chart on M, and $\chi : E \oplus F \to F$ is the canonical projection, then $\chi \circ \varphi$ restricted to $(U \cap \mathcal{L}_\alpha)^\beta$ defines a chart on $\mathcal{L}_\alpha$.

4.4.5 Examples

A The ***trivial foliation*** of a connected manifold M has only one leaf, M itself. It has codimension zero. If M is finite dimensional, both M and the leaf have the same dimension. Conversely, on a finite-dimensional connected manifold M, a foliation of dimension equal to dim(M) is the trivial foliaton.

B The ***discrete foliation*** of a manifold M is the only zero-dimensional foliation; its leaves are all points of M. If M is finite dimensional, the dimension of M is the codimension of this foliation.

C A vector field X that never vanishes on M determines a foliation; its leaves are the maximal integral curves of the vector field X. The fact that this is a foliaton is the Straightening Out Theorem (see §**4.1**).

D Let $f : M \to N$ be a submersion. It defines a foliation on M (of codimension equal to dim(N) if dim(N) is finite) by the collection of all connected components of $f^{-1}(n)$ when n varies throughout N. The fact that this is a foliation is given by Theorem **3.5.4**. In particular, we see that $E \oplus F$ is foliated by the family $\{E \times \{f\}\}_{f \in F}$.

E In the preceding example, let $M = \mathbb{R}^3$, $N = \mathbb{R}$, and $f(x^1, x^2, x^3) = \varphi(r^2)\exp(x^3)$, where $\varphi : \mathbb{R} \to \mathbb{R}$ is a C^∞ function satisfying $\varphi(0) = 0$, $\varphi(1) = 0$, and $\varphi'(s) < 0$ for $s > 0$, and where $r^2 = (x^1)^2 + (x^2)^2$. Since $\mathbf{d}f(x^1, x^2, x^3) = \exp(x^3)[2\varphi'(r^2)x^1\,\mathbf{d}x^1 + 2\varphi'(r^2)x^2\,\mathbf{d}x^2 + \varphi(r^2)\,\mathbf{d}x^3]$ and $\varphi(r^2)$ is a strictly decreasing function of r^2, f is submersion, so its level sets define a codimension one foliation on $\mathbb{R}^3$. Since the only zero of $\varphi(r^2)$ occurs for $r = 1$, $f^{-1}(0)$ equals the cylinder $\{(x^1, x^2, x^3) \mid (x^1)^2 + (x^2)^2 = 1\}$. Since $\varphi(r^2)$ is a positive function for $r \in [0, 1[$, it follows that if $c > 0$, then $f^{-1}(c) = \{(x^1, x^2, \log(c/\varphi(r^2)) \mid 0 \le (x^1)^2 + (x^2)^2 < 1\}$, which is diffeomorphic to the open unit ball in (x^1, x^2)-space via the projection $(x^1, x^2, \log(c/\varphi(r^2)) \mapsto (x^1, x^2)$. Note that the leaves $f^{-1}(c)$, $c > 0$, are asymptotically tangent to the cylinder $f^{-1}(0)$. Finally, since $\varphi(r^2) < 0$ if $r > 1$, for $c < 0$ the leaves given by $f^{-1}(c) = \{(x^1, x^2, \log(c/\varphi(r^2)) \mid (x^1)^2 + (x^2)^2 > 1\}$ are diffeomorphic to the plane minus the closed unit disk, i.e., they are diffeomorphic to cylinders. As before, note that the cylinders $f^{-1}(c)$ are symptotically tangent to $f^{-1}(0)$; see Figure 4.4.3.

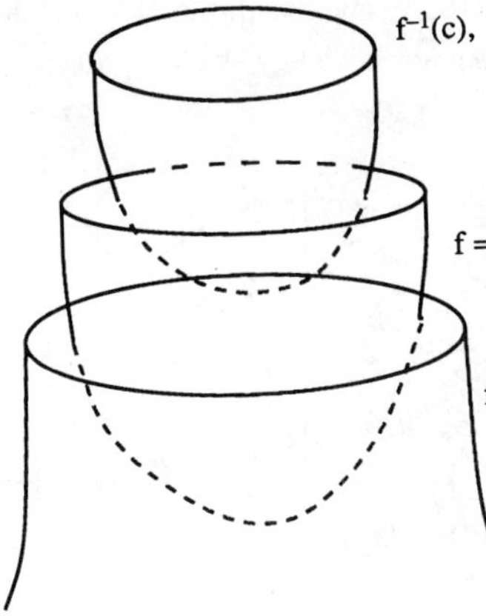

Figure 4.4.3

F The Reeb Foliation on the Solid Torus and the Klein Bottle (Reeb [1952]). We claim that in the previous example, the leaves are in some sense translation invariant. The cylinder $f^{-1}(0)$ is invariant; if $c \neq 0$, invariance is in the sense $f^{-1}(c) + (0, 0, \log t) = f^{-1}(tc)$, for any $t > 0$. Consider the part of the foliation within the solid cylinder $(x^1)^2 + (x^2)^2 \le 1$ and form the solid torus from this cylinder: identify $(a^1, a^2, 0)$ with $(b^1, b^2, 1)$ iff $a^i = b^i$, $i = 1, 2$. The foliation of the solid torus so obtained is called the ***orientable Reeb foliation***. Out of the cylinder one can form the Klein bottle (see Figure 1.4.2) by considering the equivalence relation which identifies $(a^1, a^2, 0)$ with $(b^1, b^2, 1)$ iff $a^1 = b^1$, $a^2 = -b^2$. In this way one obtains the ***nonorientable Reeb foliation***. (This terminology regarding orientability will be explained in §6.5).

G The Reeb Foliation on S^3 (Reeb [1952]). Two orientable Reeb foliations on the solid torus determine a foliation on S^3 in the following way. The sphere S^3 is the union of two solid tori which are identified along their common boundary, the torus $\mathbb{T}^2$, by the diffeomorphism taking meridians of one to parallels of the other and vice-versa. This foliation is called the ***Reeb foliation of*** S^3; it has one leaf diffeomorphic to the torus $\mathbb{T}^2$ and all its other leaves are diffeomorphic to $\mathbb{R}^2$ and accumulate on the torus. Below we describe, pictorially, the decomposition of S^3 in two solid tori. Remove the north pole, (1, 0, 0, 0), of S^3 and stereographically project the rest of S^3 onto $\mathbb{R}^3$. In the plane (x^2, x^4) draw two equal circles centered on the x^2-axis at the points a and b, where $-a = b$. Rotating about the x^4–axis yields the solid torus in $\mathbb{R}^3$. Now draw all of the circles in the (x^2, x^4) plane minus the two discs, centered on the x^4-axis and passing through a and b. Each such circle yields two connected arcs joining the two discs. In addition, consider the two portions of the x^2-axis: the line joining the two discs and the two rays going off from each disk separately. (See Figure 4.4.4.) Now rotate this figure about the x^4-axis. All arcs joining the disks generate smooth surfaces diffeomorphic to $\mathbb{R}^2$ and each such surface meets the solid torus along a parallel. Only the two rays emanating from the disks generate a surface diffeomorphic to the cylinder. Now add the north pole back to S^3 and pull back the whole structure via stereographic projection from $\mathbb{R}^3$ to S^3: the cylinder becomes a torus and all surfaces diffeomorphic to $\mathbb{R}^2$ intersect this torus in meridians. Thus S^3 is the union of two solid tori glued along their common boundary by identifying parallels of one with meridians of the other and vice-versa.

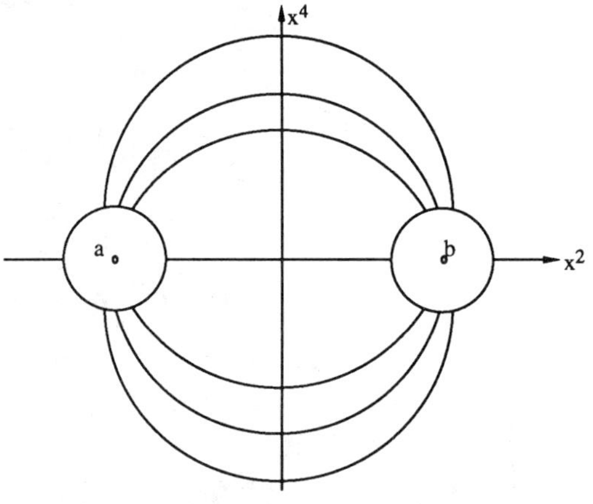

Figure 4.4.4

4.4.6 Proposition *Let* M *be a manifold and* $\Phi = \{\mathcal{L}_\alpha\}_{\alpha \in A}$ *be a foliation on* M. *The set*

$$T(M, \Phi) = \bigcup_{\alpha \in A} \bigcup_{m \in \mathcal{L}_\alpha} T_m\mathcal{L}_\alpha$$

is a subbundle of TM *called the* ***tangent bundle to the foliation****. The quotient bundle, denoted* $\nu(\Phi) = TM/T(M, \Phi)$, *is called the* ***normal bundle*** *to the foliation* Φ. *Elements of* $T(M, \Phi)$ *are called* ***vectors tangent to the foliation*** Φ.

Proof Let (U, φ), $\varphi : U \to U' \times V' \to E \oplus F$ be a foliated chart. Since $T_u\varphi(T_u\mathcal{L}_\alpha) = E \times \{0\}$ for every $u \in U \cap \mathcal{L}_\alpha$, we have $T\varphi(TU \cap T(M,\Phi)) = (U' \times V') \times (E \times \{0\})$. Thus the standard tangent bundle charts induced by foliated charts of M have the subbundle property and naturally induce vector bundle charts by mapping $v_m \in T_m(M, \Phi)$ to $(\varphi(m), T_m\varphi(v_m)) \in U' \times V') \times$

$(E \times \{0\})$. ■

4.4.7 The Global Frobenius Theorem *Let* E *be a subbundle of* TM. *The following are equivalent:*

(i) *There exists a foliation* Φ *on* M *such that* $E = T(M, \Phi)$.

(ii) E *is integrable.*

(iii) E *is involutive.*

Proof The equivalence of (ii) and (iii) was proved in **4.4.3**. Let (i) hold. Working with a foliated chart, E is integrable by **4.4.6**, the integral submanifolds being the leaves of Φ. Thus (ii) holds. Finally, we need to show that (ii) implies (i). Consider on M the family of (local) integral manifolds of E, each equipped with its own submanifold topology. It is straightforward to verify that the family of finite intersections of open subsets of these local integral submanifolds defines a topology on M, finer in general than the original one. Let $\{\mathcal{L}_\alpha\}_{\alpha \in A}$ be its connected components. Then, denoting by $(\mathcal{L}_\alpha \cap U)^\beta$ the connected components of $\mathcal{L}_\alpha \cap U$ in U, we have by definition $E \mid (U \cup \mathcal{L}_\alpha)^\beta = (E$ restricted to $(U \cup \mathcal{L}_\alpha)^\beta)$ equals $T((U \cap \mathcal{L}_\alpha)^\beta)$. Let $(\tau^{-1}(U), \psi)$, $U \subset M$ be a vector bundle chart of TM with the subbundle property for E, and let $\varphi : U \to U_1$ be the induced chart on the base. This means, shrinking U if necessary, that $\varphi : U \to U' \times V' \subset E \oplus F$ and $\psi(\tau^{-1}(U) \cap E) = (U' \times V') \times (E \times \{0\})$. Thus $\varphi((U \cap \mathcal{L}_\alpha)^\beta) = U' \times \{c_\alpha{}^\beta\}$ and so M is foliated by the family $\Phi = \{\mathcal{L}_\alpha\}_{\alpha \in A}$. Because $E \mid (U \cap \mathcal{L}_\alpha)^\beta = T((U \cap \mathcal{L}_\alpha)^\beta$, we also have $T(M, \Phi) = E$. ■

There is an important global topological condition that integrable subbundles must satisfy that was discovered by Bott [1970]. The result, called the ***Bott Vanishing Theorem***, can be found, along with related results, by readers with background in algebraic topology, in Lawson [1977].

The leaves of a foliation are characterized by the following property.

4.4.8 Proposition *Let* Φ *be a foliation on* M. *Then* x *and* y *are in the same leaf if and only if* x *and* y *lie on the same integral curve of a vector field* X *defined on an open set in* M *and which is tangent to the foliation* Φ.

Proof Let X be a vector field on M with values in $T(M, \Phi)$ and assume that x and y lie on the same integral curve of X. Let $\mathcal{L}$ denote the leaf of Φ containing x. Since X is tangent to the foliation, $X \mid \mathcal{L}$ is a vector field on $\mathcal{L}$ and thus any integral curve of X starting in $\mathcal{L}$ stays in $\mathcal{L}$. Since y is on such an integral curve, it follows that $y \in \mathcal{L}$.

Conversely, let $x, y \in \mathcal{L}$ and let c(t) be a smooth non-intersecting curve in $\mathcal{L}$ such that $c(0) = x$, $c(1) = y$, $c'(t) \neq 0$. (This can always be done on a connected manifold by showing that the set of points that can be so joined is open and closed.) Thus $c : [0, 1] \to \mathcal{L}$ is an immersion, and hence by compactness of $[0, 1]$, c is an embedding. Using definition **4.4.4**, there is a neighborhood of the curve c in M which is diffeomorphic to a neighborhood of $[0, 1] \times \{0\} \times \{0\}$ in $\mathbb{R} \times F \times G$ for Banach spaces F and G such that the leaves of the foliation have the local representation $\mathbb{R} \times F \times \{w\}$, for fixed $w \in G$, and the image of the curve c has the local representation $[0, 1] \times \{0\} \times \{0\}$. Thus we can find a vector field X which is defined by $c'(t)$ along c and extends off c to be constant in this local representation. ■

Let R denote the following equivalence relation in a manifold M with a given folation Φ : xRy if x, y belong to the same leaf of Φ. The previous proposition shows that R is an open equivalence relation. It is of interest to know whether M/R is a manifold. Foliations for which R is a regular equivalence relation are called ***regular foliations***. (See §**3.5** for a discussion of regular equivalence relations.) The following is a useful criterion.

4.4.9 Proposition *Let* Φ *be a foliation on a manifold* M *and* R *the equivalence relation in* M *determined by* Φ. R *is regular iff for every* $m \in M$ *there exists a local submanifold* Σ_m *of* M *such that* Σ_m *intersects every leaf in at most one point (or nowhere) and* $T_m\Sigma_m \oplus T_m(M, \Phi) = T_mM$. *(Sometimes* Σ_m *is called a* ***slice*** *or a* ***local cross-section*** *for the foliation.)*

Proof Assume that R is regular and let $\pi : M \to M/R$ be the canonical projection. For Σ_m choose the submanifold using the following construction. Since π is a submersion, in appropriate charts (U, φ), (V, ψ), where $\varphi : U \to U' \times V'$ and $\psi : V \to V'$, the local representative of π, $\pi_{\varphi\psi} : U' \times V' \to V'$, is the projection onto the second factor, and every leaf $\pi^{-1}(v) \subset U$, $v \in V$, is represented in these charts as $U' \times \{v'\}$ where $v' = \psi(v)$. Thus if $\Sigma_m = \psi^{-1}(\{0\} \times V')$, we see that Σ_m satisfies the two required conditions.

Conversely, assume that each point $m \in M$ admits a slice Σ_m. Working with a foliated chart, we are reduced to the following situation: let U, V be open balls centered at the origin in Banach spaces E and F, respectively, let Σ be a submanifold of $U \times V$, $(0, 0) \in \Sigma$, such that $T_{(0,0)}\Sigma = F$, and $\Sigma \cap (U \times \{v\})$ is at most one point for all $v \in V$. If $p_2 : E \oplus F \to F$ is the second projection, since $p_2 \,|\, \Sigma$ has tangent map at $(0, 0)$ equal to the identity, it follows that for V small enough, $p_2 \,|\, \Sigma : \Sigma \to V$ is a diffeomorphism. Shrinking Σ and V if necessary we can assume that $\Sigma \cap (U \times \{v\})$ is exactly one point. Let $q : V \to \Sigma$ be the inverse of p_2 and define the smooth map $s : U \times V \to \Sigma$ by $s(u, v) = q(v)$. Then $\Sigma \cap (U \times \{v\}) = \{q(v)\}$, thus showing that Σ is a slice in the sense of Lemma **3.5.25**. Pulling everything back to M by the foliated chart, the prior argument shows that for each point $m \in M$ there is an open neighborhood U, a submanifold Σ_m of U, and a smooth map $s : U \to \Sigma_m$ such that $\mathcal{L}_u \cap \Sigma_m = \{s(u)\}$, where $\mathcal{L}_u$ is the leaf containing $u \in U$. By the argument following Lemma **3.5.25**, the equivalence relation R is locally regular, i.e., $R_U = R \cap (U \times U)$ is regular. If $U' = \pi^{-1}(\pi(U))$ where $\pi : M \to M/R$ is the projection, the argument at the end of Step 1 in the proof of Theorem **3.5.24** shows that R is regular. Thus, all that remains to be proved is that U can be chosen to equal U′. But this is clear by defining $s' : U' \to \Sigma_m$ by $s'(u') = s(u)$, where $u \in U \cap \mathcal{L}_{u'}$, $\mathcal{L}_{u'}$ being the leaf containing u′; smoothness of s′ follows from smoothness of s by composing it locally with the flow of a vector field given by Proposition **4.4.8**. ■

To get a feeling for the foregoing condition we will study the linear flow on the torus.

4.4.10 Example On the two-torus $\mathbb{T}^2$ consider the global flow $F : \mathbb{R} \times \mathbb{T}^2 \to \mathbb{T}^2$ defined by $F(t, (s_1, s_2)) = (s_1 e^{2\pi i t}, s_2 e^{2\pi i \alpha t})$ for a fixed number $\alpha \in [0, 1[$. By Example **4.4.5C** this defines a foliation on $\mathbb{T}^2$. If $\alpha \in \mathbb{Q}$, notice that every integral curve is closed and that all integral curves have the same period. The condition of the previous theorem is easily verified and we conclude that in this case the equivalence relation R is regular; $\mathbb{T}^2/R = S^1$. If α is irrational, however, the situation is completely different. Let $\varphi(t) = (e^{2\pi i t}, e^{2\pi i \alpha t})$ denote the integral curve through $(1, 1)$. The following argument shows that $\mathrm{cl}(\varphi(\mathbb{R})) = \mathbb{T}^2$; that is, $\varphi(\mathbb{R})$ is *dense* in $\mathbb{T}^2$. Let $p = (e^{2\pi i x}, e^{2\pi i y}) \in \mathbb{T}^2$; then for all $m \in \mathbb{Z}$,

$$\varphi(x + m) - p = (0, e^{2\pi i \alpha x}(e^{2\pi i m \alpha} - e^{2\pi i z}))$$

where $y = \alpha x + z$. It suffices to show that $C = \{e^{2\pi i m \alpha} \in S^1 \mid m \in \mathbb{Z}\}$ is dense in S^1 because then there is a sequence $m_k \in \mathbb{Z}$ such that $\exp(2\pi i m_k \alpha)$ converges to $e^{2\pi i z}$. Hence, $\varphi(x + m_k)$ converges to p. If for each $k \in \mathbb{Z}_+$ we divide S^1 into k arcs of length $2\pi/k$, then, because $\{e^{2\pi i m \alpha} \in S^1 \mid m = 1, 2, \ldots, k + 1\}$ are distinct for some $1 \leq n_k < m_k \leq k + 1$, $\exp(2\pi i m_k \alpha)$ and $\exp(2\pi i n_k \alpha)$ belong to the same arc. Therefore, $|\exp(2\pi i m_k \alpha) - \exp(2\pi i n_k \alpha)| < 2\pi/k$, which implies $|\exp(2\pi i q_k \alpha) - 1| < 2\pi/k$, where $q_k = m_k - n_k$. Because

$$\bigcup_{j \in \mathbb{Z}_+} \{e^{2\pi i \alpha s} \in S^1 \mid s \in [jq_k, (j+1)q_k]\} = S^1 ,$$

every arc of length less than $2\pi/k$ contains some $\exp(2\pi i j q_k)$, which proves $\mathrm{cl}(C) = S^1$. Thus any submanifold Σ_m, $m \in \mathbb{T}^2$ not coinciding with the integral curve through m will have to intersect $\varphi(\mathbb{R})$ infinitely many times; the condition in the previous theorem is violated and so R is *not* regular. ♦

Remark Novikov [1965] has shown that the Reeb foliation is in some sense typical. A foliation Φ on M is said to be ***transversally orientable*** if $TM = T(M, \Phi) \oplus E$, where E is an orientable subbundle of TM (see Exercise **6.5N** for the definition.) A foliation on a three dimensional manifold M is said to have a ***Reeb component*** if it has a compact leaf diffeomorphic to $\mathbb{T}^2$ or $\mathbb{K}$ and if the foliation within this torus or Klein bottle is diffeomorphic to the orientable or non-orientable Reeb foliation in the solid torus or the solid Klein bottle. Novikov has proved the following remarkable result. *Let Φ be a transversally orientable C^2 codimension one foliation of a compact three dimensional manifold M. If $\pi_1(M)$ is finite, then Φ has a Reeb component, which may or may not be orientable.* If $\pi_2(M) \neq 0$ *(with no hypotheses on $\pi_1(M)$) and Φ has no Reeb components, then all the leaves of Φ are compact with finite fundamental group.* We refer the reader to Camacho and Neto [1985] for a proof of this result and to this reference and Lawson [1977] for a study of foliations in general.

Even though foliations encompass "nice" partitions of a manifold into submanifolds, there are important situations when foliations are inappropriate because they are not regular or the leaves jump in dimension from point to point. Consider, for example, $\mathbb{R}^2$ as a union of concentric circles centered at the origin. As this example suggests, one would like to relax the condition that M/R be a manifold, provided that M/R turns out to be a union of manifolds that fit "nicely" together. *Stratifications*, another concept allowing us to "stack" manifolds, turn out to be the natural tool to describe the topology of orbit spaces of compact Lie group actions or non-compact Lie group actions admitting a slice (see, for instance, Bredon [1972], Burghelea, Albu, and Ratiu [1975], Fischer [1970], and Bourguignon [1975]). We shall limit ourselves to the definition in the finite-dimensional case and some simple remarks.

4.4.11 Definition *Let* M *be a locally compact topological space. A* ***stratification*** *of* M *is a partition of* M *into manifolds* $\{M_a\}_{a \in A}$ *called* ***strata****, satisfying the following conditions:*

S1 M_a *are manifolds of constant dimension; they are submanifolds of* M *if* M *is itself a manifold.*

S2 *The family* $\{M_a{}^\alpha\}$ *of connected components of all the* M_a *is a locally finite partition of* M*; i.e., for every* $m \in M$*, there exists an open neighborhood* U *of* m *in* M *intersecting only finitely many* $M_b{}^\beta$.

S3 *If* $M_a{}^\alpha \cap \mathrm{cl}(M_b{}^\beta) \neq \varnothing$ *for* $(a, \alpha) \neq (b, \beta)$, *then* $M_a{}^\alpha \subset M_b{}^\beta$ *and* $\dim(M_a) < \dim(M_b)$.

S4 $\mathrm{cl}(M_a) \backslash M_a$ *is a disjoint union of strata of dimension strictly less than* $\dim(M_a)$.

From the definition it follows that if $M_a{}^\alpha \cap \mathrm{cl}(M_b{}^\beta) \neq \varnothing$ and if $m \in M_a{}^\alpha \subset \mathrm{cl}(M_b{}^\beta)$ has an open neighborhood U in the topology of $M_a{}^\alpha$ such that $U \subset M_a{}^\alpha \cap \mathrm{cl}(M_b{}^\beta)$, then necessarily $M_a{}^\alpha \subset M_b{}^\beta$ and thus $\dim(M_a) < \dim(M_b)$. To see this, it is enough to note that the given hypothesis makes $M_a{}^\alpha \cap \mathrm{cl}(M_b{}^\beta)$ open in $M_a{}^\alpha$. Since it is also closed (by definition of the relative topology) and $M_a{}^\alpha$ is connected, it must equal $M_a{}^\alpha$ itself, whence $M_a{}^\alpha \subset \mathrm{cl}(M_b{}^\beta)$ and by **S3**, $M_a{}^\alpha \subset M_b{}^\beta$ and $\dim(M_a) < \dim(M_b)$.

For nonregular equivalence relations R, M/R is often a stratified space. The intuitive idea is that it is often possible to group together equivalence classes of the same dimension, and this grouping is parametrized by a manifold, which will be a stratum in M/R. A simple example is $\mathbb{R}^2$ partitioned by circles (the equivalence classes for R). The circles of positive radius are parametrized by the interval $]0, \infty[$. Thus M/R is the stratified set $[0, \infty[$ consisting of the two strata $\{0\}$ and $]0, \infty[$.

Exercises

4.4A Let M be a manifold such that $TM = E_1 \oplus \ldots \oplus E_p$, where E_i is involutive. Show that there are local charts (U, φ), $\varphi : U \to E_1 \oplus \ldots \oplus E_p$ such that E_i is described by the equations $e_j = 0$, $j \neq i$.

4.4B In $\mathbb{R}^4$ consider the family of surfaces given by $x^2 + y^2 + z^2 - t^2 = \text{const}$. Show that these surfaces define a stratification. What part of $\mathbb{R}^4$ should be thrown out so as to obtain a regular foliation?

4.4C Let $f : M \to N$ be a C^∞ map and Φ a foliation on N. The map f is said to be ***transversal*** to Φ, denoted $f \pitchfork \Phi$, if for every $m \in M$, $T_m f(T_m M) + T_{f(m)}(N, \Phi) = T_{f(m)}N$ and $(T_m f)^{-1}(T_{f(m)}(N, \Phi))$ splits in $T_m M$. Show that if $\{\mathcal{L}_\alpha\}_{\alpha \in A}$ are the leaves of Φ, the connected components of $f^{-1}(\mathcal{L}_\alpha)$ are leaves of a foliation (denoted by $f^*(\Phi)$) on M, and if Φ has finite codimension in N, then so does the foliation $f^*(\Phi)$ on M and the two codimensions coincide.

4.4D (Bourbaki [1971]). Let M be a manifold and denote by M′ the manifold with underlying set M but with a different differentiable structure. Show that the collection of connected components of M′ defines a foliation of M iff for every $m \in M$, there exists an open set U in M, $m \in U$, a manifold N, and a submersion $\rho : U \to N$ such that the submanifold $\rho^{-1}(n)$ of U is open in M′ for all $n \in N$. (*Hint:* For the "if" part use **3.3.5** and for the "only if" part use Exercise **3.2F** to define a manifold structure on the leaves; the charts of the second structure are $(U \cap \mathcal{L}_\alpha)^\beta \to U'$.)

4.4E On the manifold $\mathbf{SO}(3)$, consider the partition $\mathcal{L}_A = \{QA \mid Q$ is an arbitrary rotation about the z-axis in $\mathbb{R}^3\}$, $A \in \mathbf{SO}(3)$. Show that $\Phi = \{\mathcal{L}_A \mid A \in \mathbf{SO}(3)\}$ is a regular foliation and that the quotient manifold $\mathbf{SO}(3)/R$ is diffeomorphic to S^2.

4.4F (Hirsch and Weinstein). Give another proof of the Frobenius Theorem as follows:

Step 1 Prove it for the abelian case in which all sections of E satisfy $[X, Y] = 0$ by choosing a local basis $X_1, \ldots, X_k$ of sections and successively flowing out by the commuting flows of $X_1, \ldots, X_k$.

Step 2 Given a k-dimensional plane field, locally write it as a "graph" over $\mathbb{R}^k$. Choose k commuting vector fields on $\mathbb{R}^k$ and lift them to the plane field. If E is involutive, the bracket of two of them lies in E and, moreover, since the bracket "pushes down" to $\mathbb{R}^k$ (by "relatedness"), it is zero. (This is actually demonstrated in formula (1)). Now use Step 1.

Chapter 5
Tensors

In the previous chapter we studied vector fields and functions on manifolds. Now these objects are generalized to tensor fields, which are sections of vector bundles built out of the tangent bundle. This study is continued in the next chapter when we discuss differential forms, which are tensors with special symmetry properties. One of the objectives of this chapter is to extend the pull-back and Lie derivative operations from functions and vector fields to tensor fields.

§5.1 *Tensors on Linear Spaces*

Preparatory to putting tensors on manifolds, we first study them on vector spaces. This subject is an extension of linear algebra sometimes called "multilinear algebra." Ultimately our constructions will be done on each fiber of the tangent bundle, producing a new vector bundle.

As in Chapter 2, E, F, ... denote Banach spaces and $L^k(E_1, ..., E_k; F)$ denotes the vector space of continuous k-multilinear maps of $E_1 \times ... \times E_k$ to F. The special case $L(E, \mathbb{R})$ is denoted E^*, the ***dual space*** of E. If $\{e_1, ..., e_n\}$ is an ordered basis of E, there is a unique ordered basis of E^*, the ***dual basis*** $\{e^1, ..., e^n\}$, such that $\langle e^j, e_i\rangle = \delta^j_i$ where $\delta^j_i = 1$ if $j = i$ and 0 otherwise. Furthermore, for each $e \in E$,

$$e = \sum_{i=1}^{n} \langle e^i, e\rangle\, e_i$$

and for each $\alpha \in E^*$,

$$\alpha = \sum_{i=1}^{n} \langle \alpha, e_i\rangle\, e^i ,$$

where $\langle\ ,\ \rangle$ is the pairing between E and E^*. Employing the ***summation convention*** whereby summation is implied when an index is repeated on upper and lower levels, these expressions become $e = \langle e^i, e\rangle\, e_i$ and $\alpha = \langle \alpha, e_i\rangle\, e^i$.

As in Supplement **2.4C**, if E is infinite dimensional, by E^* we will mean another Banach space weakly paired to E; it need not be the full functional analytic dual of E. In particular, E^{**} will *always* be chosen to be E. With these conventions, tensors are defined as follows.

5.1.1 Definition *For a vector space* E *we put* $T^r_s(E) = L^{r+s}(E^*, ..., E^*, E, ..., E; \mathbb{R})$ (r *copies of* E^* *and* s *copies of* E). *Elements of* $T^r_s(E)$ *are called* ***tensors on*** E, ***contravariant of order*** r *and* ***covariant of order*** s; *or simply, of* ***type*** (r, s).

Given $t_1 \in T^{r_1}{}_{s_1}(E)$ *and* $t_2 \in T^{r_2}{}_{s_2}(E)$, *the* ***tensor product*** of t_1 *and* t_2 *is the tensor* $t_1 \otimes t_2 \in T^{r_1+r_2}{}_{s_1+s_2}(E)$ *defined by*

$$\begin{aligned}(t_1 \otimes t_2)(\beta^1, ..., \beta^{r_1}, \gamma^1, ..., \gamma^{r_2}, f_1, ..., f_{s_1}, g_1, ..., g_{s_2})\\ = t_1(\beta^1, ..., \beta^{r_1}, f_1, ..., f_{s_1})\, t_2(\gamma^1, ..., \gamma^{r_2}, g_1, ..., g_{s_2})\end{aligned}$$

where $\beta^j, \gamma^j \in E^*$ *and* $f_j, g_j \in E$.

Replacing $\mathbb{R}$ by a space F gives $T^r_s(E; F)$, the F-valued tensors of type (r, s). The tensor product now requires a bilinear form on the value space for its definition. For $\mathbb{R}$-valued tensors, $\otimes$ is associative, bilinear and continuous; it is *not* commutative. We also have the special cases $T^1{}_0(E) = E$, $T^0{}_1(E) = E^*$, $T^0{}_2(E) = L(E; E^*)$, and $T^1{}_1(E) = L(E; E)$ and make the convention that $T^0{}_0(E; F) = F$.

5.1.2 Proposition *Let* E *be an* n *dimensional vector space. If* $\{e_1, ..., e_n\}$ *is a basis of* E *and* $\{e^1, ..., e^n\}$ *is the dual basis, then*

$$\{e_{i_1} \otimes ... \otimes e_{i_r} \otimes e^{j_1} \otimes ... \otimes e^{j_s} \mid i_1, ..., i_r, j_1, ..., j_s = 1, ..., n\}$$

is a basis of $T^r_s(E)$ *and thus* $\dim(T^r_s(E)) = n^{r+s}$.

Proof We must show that the elements

$$e_{i_1} \otimes ... \otimes e_{i_r} \otimes e^{j_1} \otimes ... \otimes e^{j_s}$$

of $T^r_s(E)$ are linearly independent and span $T^r_s(E)$. Suppose a finite sum

$$t^{i_1...i_r}{}_{j_1...j_s}\, e_{i_1} \otimes ... \otimes e_{i_r} \otimes e^{j_1} \otimes ... \otimes e^{j_s} = 0.$$

Apply this to $(e^{k_1}, ..., e^{k_r}, e_{\ell_1}, ..., e_{\ell_s})$ to get $t^{k_1...k_r}{}_{\ell_1...\ell_s} = 0$. Next, check that for $t \in T^r_s(E)$ we have

$$t = t(e^{i_1}, ..., e^{i_r}, e_{j_1}, ..., e_{j_s})\, e_{i_1} \otimes ... \otimes e_{i_r} \otimes e^{j_1} \otimes ... \otimes e^{j_s}. \blacksquare$$

The coefficients

$$t^{i_1...i_r}{}_{j_1...j_s} = t(e^{i_1}, ..., e^{i_r}, e_{j_1}, ..., e_{j_s})$$

and called the ***components of*** t ***relative to the basis*** $\{e_1, ..., e_n\}$.

5.1.3 Examples

A If t is a $(0, 2)$-tensor on E then t has components $t_{ij} = t(e_i, e_j)$, an $n \times n$ matrix. This is the usual way of associating a bilinear form with a matrix. For instance, in $\mathbb{R}^2$ the bilinear form

$$t(x, y) = Ax_1y_1 + Bx_1y_2 + Cx_2y_1 + Dx_2y_2$$

(where $x = (x_1, x_2)$ and $y = (y_1, y_2)$) is associated to the 2×2 matrix

$$\begin{bmatrix} A & B \\ C & D \end{bmatrix}.$$

B If t is a $(0, 2)$-tensor on E, it makes sense to say that t is ***symmetric;*** i.e., $t(e_i, e_j) = t(e_j, e_i)$. This is equivalent to saying that the matrix $[t_{ij}]$ is symmetric. Symmetric $(0, 2)$-tensors t can be recovered from their quadratic form $Q(e) = t(e, e)$ by $t(e_1, e_2) = [Q(e_1 + e_2) - Q(e_1 - e_2)]/4$, the ***polarization identity***. If $E = \mathbb{R}^2$ and t has the matrix

$$\begin{bmatrix} A & B \\ B & C \end{bmatrix}$$

then $Q(x) = Ax_1{}^2 + 2Bx_1x_2 + Cx_2{}^2$. Symmetric $(0, 2)$-tensors are thus closely related to quadratic forms and arise, for example, in mechanics as moment of inertia tensors and stress tensors.

C In general, a ***symmetric*** $(r, 0)$-***tensor*** is defined by the condition

$$t(\alpha^1, \ldots, \alpha^r) = t(\alpha^{\sigma(1)}, \ldots, \alpha^{\sigma(r)})$$

for all permutations σ of $\{1, \ldots, r\}$, and all elements $\alpha^1, \ldots, \alpha^r \in E^*$. One may associate to t a homogeneous polynomial of degree r, $P(\alpha) = t(\alpha, \ldots, \alpha)$ and as in the case $r = 2$, P and t determine each other. A similar definition holds for $(0, s)$-tensors. It is clear that *a tensor is symmetric iff all its components in an arbitrary basis are symmetric.*

D An inner product $\langle\ ,\ \rangle$ on E is a symmetric $(0, 2)$-tensor. Its matrix has components $g_{ij} = \langle e_i, e_j \rangle$. Thus g_{ij} is symmetric and positive definite. The components of the inverse matrix are written g^{ij}.

E The space $L^k(E_1, \ldots, E_k; F)$ is isometric to $L^k(E_{\sigma(1)}, \ldots, E_{\sigma(k)}; F)$ for any permutation σ of $\{1, \ldots, k\}$, the isometry being given by $A \mapsto A'$, where $A'(e_{\sigma(1)}, \ldots, e_{\sigma(k)}) = A(e_1, \ldots, e_k)$. Thus if $t \in T^r{}_s(E; F)$, the tensor t can be regarded in $C(r + s, s)$ (the number of ways $r + s$ objects chosen s at a time) ways as an $(r + s)$-multilinear F-valued map. For example, if $t \in T^2{}_1(E)$, the standard way is to regard it as a 3-linear map $t : E^* \times E^* \times E \to \mathbb{R}$. There are two more ways to interpret this map, however, namely as $E^* \times E \times E^* \to \mathbb{R}$ and as $E \times E^* \times E^* \to \mathbb{R}$. In finite dimensions, where one writes the tensors in components, this distinction is important and is reflected in the index positions. Thus the three *different* tensors described above are written

$$t^{ij}{}_k e_i \otimes e_j \otimes e^k, \quad t^i{}_k{}^j e_i \otimes e^k \otimes e_j, \quad t_k{}^{ij} e^k \otimes e_i \otimes e_j.$$

F In classical mechanics one encounters the notion of a ***dyadic*** (cf. Goldstein [1980]). A dyadic is the formal sum of a finite number of ***dyads***, a dyad being a pair of vectors $e_1, e_2 \in \mathbb{R}^3$ written in a specific order in the form e_1e_2. The action of a dyad on a pair of vectors, called the ***double dot product*** of two dyads is defined by

$$e_1e_2 : u_1u_2 = (e_1 \cdot u_1)(e_2 \cdot u_2),$$

where $\cdot$ stands for the usual dot product in $\mathbb{R}^3$. In this way dyads and dyadics are nothing but (0, 2)-tensors on $\mathbb{R}^3$; i.e., $e_1e_2 = e_1 \otimes e_2 \in T^0{}_2(\mathbb{R}^3)$, by identifying $(\mathbb{R}^3)^*$ with $\mathbb{R}^3$.

G Higher order tensors arise in elasticity and Riemannian geometry. In elasticity, the stress tensor is a symmetric 2-tensor and the elasticity tensor is a fourth-order tensor (see Marsden and Hughes [1983]). In Riemannian geometry the metric tensor is a symmetric 2-tensor and the curvature tensor is a fourth-order tensor (see Section **5.6**). ♦

The ***interior product*** of a vector $v \in E$ (resp. a form $\beta \in E^*$) with a tensor $t \in T^r{}_s(E; F)$ is the $(r, s-1)$ (resp. $(r-1, s)$) type F-valued tensor defined by

$$(i_v t)(\beta^1, \ldots, \beta^r, v_1, \ldots, v_{s-1}) = t(\beta^1, \ldots, \beta^r, v, v_1, \ldots, v_{s-1})$$
$$(i^\beta t)(\beta^1, \ldots, \beta^{r-1}, v_1, \ldots, v_s) = t(\beta, \beta^1, \ldots, \beta^{r-1}, v_1, \ldots, v_s).$$

Clearly, $i_v : T^r{}_s(E; F) \to T^r{}_{s-1}(E; F)$ and $i^\beta : T^r{}_s(E; F) \to T^{r-1}{}_s(E; F)$ are linear continuous maps, as are $v \mapsto i_v$ and $\beta \mapsto i^\beta$. If $F = \mathbb{R}$ and $\dim(E) = n$, these operations take the following form in components. If e_k (resp. e^k) denotes the k-th basis (resp. dual basis) element of E, we have

$$i_{e_k}(e_{i_1} \otimes \ldots \otimes e_{i_r} \otimes e^{j_1} \otimes \ldots \otimes e^{j_s}) = \delta^{j_1}{}_k\, e_{i_1} \otimes \ldots \otimes e_{i_r} \otimes e^{j_2} \otimes \ldots \otimes e^{j_s}$$

$$i^{e^k}(e_{i_1} \otimes \ldots \otimes e_{i_r} \otimes e^{j_1} \otimes \ldots \otimes e^{j_s}) = \delta^k{}_{i_1}\, e_{i_2} \otimes \ldots \otimes e_{i_r} \otimes e^{j_1} \otimes \ldots \otimes e^{j_s}.$$

By **5.1.2** these formulas and linearity enable us to compute any interior product.

Let $\dim(E) = n$. The contraction of the k-th contravariant with the ℓ-th covariant index, or for short, the (k, ℓ)-***contraction***, is the family of linear maps $C^k{}_\ell : T^r{}_s(E; F) \to T^{r-1}{}_{s-1}(E)$ defined for any pair of natural numbers $r, s \geq 1$ by

$$C^k{}_\ell(t^{i_1\ldots i_s}{}_{j_1\ldots j_r} e_{i_1} \otimes \ldots \otimes e_{i_s} \otimes e^{j_1} \otimes \ldots \otimes e^{j_r})$$
$$= t^{i_1\ldots i_{k-1}p i_{k+1}\ldots i_s}{}_{j_1\ldots j_{\ell-1}p j_{\ell+1}\ldots j_r} e_{i_1} \otimes \ldots \otimes \hat{e}_{i_k} \otimes \ldots \otimes e_{i_s} \otimes e^{j_1} \otimes \ldots \otimes \hat{e}^{j_\ell} \otimes \ldots \otimes e^{j_r},$$

where $\{e_1, \ldots, e_n\}$ is a basis of E, $\{e^1, \ldots, e^n\}$ is the dual basis in E^*, and $\hat{}$ over a vector or covector means that it is omitted. It is straightforward to verify that $C^k{}_\ell$ so defined is independent of the basis. If E is infinite dimensional, contraction is not defined for arbitrary tensors. One introduces the so-called ***contraction class tensors***, analogous to the trace class operators, defines contraction as above in terms of a Banach space basis and its dual, and shows that the contraction class condition implies that the definition is basis independent. We shall not dwell upon these technicalities, refer to Rudin [1973] for a brief discussion of trace class operators, and invite the reader to model the concept of contraction class along these lines. For example, if $E^* = E = \ell^2(\mathbb{R})$, and $e_i = e^i$ equals the sequence with 1 in the i-th place and zero everywhere else, then

$$t = \sum_{n=0}^{\infty} 2^{-n}\, e_n \otimes e^n \in T^1_1(E) \quad \text{and} \quad C^1_1(t) = \sum_{n=0}^{\infty} 2^{-n} = 2.$$

The ***Kronecker delta*** is the tensor $\delta \in T^1{}_1(E)$ defined by $\delta(\alpha, e) = \langle \alpha, e\rangle$. If E is finite dimensional, δ corresponds to the identity $I \in L(E; E)$ under the canonical isomorphism $T^1{}_1(E) \cong L(E; E)$. Relative to any basis, the components of δ are the usual Kronecker symbols $\delta^i{}_j$, i.e., $\delta = \delta^i{}_j e_i \otimes e^j$.

Suppose E is a finite-dimensional real inner product space with a basis $\{e_1, ..., e_n\}$ and corresponding dual basis $\{e^1, ..., e^n\}$ in E^*. Using the inner product, with matrix denoted by $[g_{ij}]$, we get the isomorphism

$$\flat : E \to E^* \text{ given by } x \mapsto \langle x, \cdot\,\rangle \text{ and its inverse } {}^\# : E^* \to E.$$

The matrix of $\flat$ is $[g^{ij}]$; i.e.,

$$(x^\flat)_i = g_{ij}x^j$$

and of $\#$ is $[g^{ij}]$; i.e.,

$$(\alpha^\#)^i = g^{ij}\alpha_j,$$

where x^j and α_j are the components of e and α, respectively. We call $\flat$ the ***index lowering operator*** and $\#$ the ***index raising operator***.

These operators can be applied to tensors to produce new ones. For example if t is a tensor of type (0, 2) we can define an ***associated tensor*** t' of type (1, 1) by

$$t'(e, \alpha) = t(e, \alpha^\#).$$

The components are

$$(t')_i{}^j = g^{jk}t_{ik} \quad \text{(as usual, sum on k).}$$

In the classical literature one writes $t_i{}^j$ for $g^{jk}t_{ik}$, and this is indeed a convenient notation in calculations. However, contrary to the impression one may get from the classical theory of Cartesian tensors, t and t' are *different tensors*.

5.1.4 Examples Let E be a finite-dimensional real vector space with basis $\{e_1, ..., e_n\}$ and dual basis $\{e^1, ..., e^n\}$.

A If $t \in T^2{}_1(E)$ and $x = x^i e_i$, then

$$\begin{aligned} \mathbf{i}_x t &= x^p \mathbf{i}_{e_p}(t^{k\ell}{}_j e_k \otimes e_\ell \otimes e^j) = x^p t^{k\ell}{}_j \, \mathbf{i}_{e_p}(e_k \otimes e_\ell \otimes e^j) \\ &= x^p t^{k\ell}{}_j \delta^j{}_p e_k \otimes e_\ell = x^p t^{k\ell}{}_p \, e_k \otimes e_\ell \,. \end{aligned}$$

Thus, the components of $\mathbf{i}_x t$ are $x^p t^{k\ell}{}_p$. The interior product of the same tensor with $\alpha = \alpha_p e^p$ takes the form

$$\mathbf{i}^\alpha t = \alpha_p t^{k\ell}{}_j \mathbf{i}^{e^p} (e_k \otimes e_\ell \otimes e^j) = \alpha_p t^{k\ell}{}_j \, \delta^p{}_k e_\ell \otimes e^j = \alpha_p t^{p\ell}{}_j e_\ell \otimes e^j \,.$$

B If $t \in T^2{}_3(E)$, the (2,1)-contraction is given by

$$C^2{}_1(t^{ij}{}_{k\ell m} \, e_i \otimes e_j \otimes e^k \otimes e^\ell \otimes e^m) = t^{ij}{}_{j\ell m} \, e_i \otimes e^\ell \otimes e^m \,.$$

C An important particular example of contraction is the ***trace*** of a (1, 1)-tensor. Namely, if $t \in T^1{}_1(E)$, then $\mathrm{trace}(t) = C^1{}_1(t) = t^i{}_i$, where $t = t^i{}_j \, e_i \otimes e^j$.

D The components of the tensor associated to g by raising the second index are $g^{jk} g_{ik} = g^{jk} g_{ki} = \delta^j{}_i$.

E Let

$$t \in T^3{}_2(E), \; t = t^{ijk}{}_{\ell m} e_i \otimes e_j \otimes e_k \otimes e^\ell \otimes e^m \,.$$

Then t has quite a few associated tensors, depending on which index is lowered or raised. For example

$$\begin{aligned} t^{ijk}{}_\ell{}^m &= g^{mp} \, t^{ijk}{}_{\ell p} \,, \\ t^i{}_{jk\ell}{}^m &= g_{ja} \, g_{kb} \, g^{mc} \, t^{iab}{}_{\ell c} \\ t_{ij}{}^{k\ell m} &= g_{ia} \, g_{jb} \, g^{\ell c} \, g^{md} \, t^{abk}{}_{cd} \\ t_i{}^j{}_k{}^\ell{}_m &= g_{ia} \, g_{kb} \, g^{\ell c} \, t^{ajb}{}_{cm} \end{aligned}$$

and so on.

F The positioning of the indices in the components of associated tensors is important. For example, if $t \in T^0{}_2(E)$, we saw earlier that $t_i{}^j = g^{jk} \, t_{ik}$. However, $t^j{}_i = g^{ij} \, t_{ki}$, which is in general *different* from $t_i{}^j$ when t is not symmetric. For example, if $E = \mathbb{R}^3$ with $g_{ij} = \delta_{ij}$ and the nine components of t in the standard basis are $t_{12} = 1, t_{21} = -1$, $t_{ij} = 0$ for all other pairs i, j, then $t_i{}^j = t_{ij}$, $t^j{}_i = t_{ji}$, so that $t_1{}^2 = t_{12} = 1$ while $t^2{}_1 = t_{21} = -1$.

G The ***trace*** of a (2, 0)-tensor is defined to be the trace of the associated (1, 1) tensor; i.e., if $t = t^{ij} e_i \otimes e_j$, then $\mathrm{trace}(t) = t^i{}_i = g_{ik} t^{ik}$. The question naturally arises whether we get the same answer by lowering the first index instead of the second, i.e., if we consider $t_i{}^i$. By symmetry of g_{ij} we have $t^i{}_i = g_{ki} t^{ik} = t_k{}^k$, so that the definition of the trace is independent of which index is lowered. Similarly, if $t \in T^0{}_2(E)$, trace(t) $= t_i{}^i = g^{ik}t_{ik} = t^k{}_k$. In particular trace(g) $= g_i{}^i = g^{ik}{}_{ik} =$ dim(E). ♦

Now we turn to the effect of linear transformations on dual spaces. If $\varphi \in L(E, F)$, the ***transpose*** of φ, denoted $\varphi^* \in L(F^*, E^*)$ is defined by $\langle \varphi^*(\beta), e \rangle = \langle \beta, \varphi(e) \rangle$, where $\beta \in F^*$ and $e \in E$.

Let us analyze the matrices of φ and φ^*. As customary in linear algebra, vectors in a given basis are represented by a column whose entries are the components of the vector. Let $\varphi \in L(E, F)$ and let $\{e_1, ..., e_n\}$ and $\{f_1, ..., f_m\}$ be ordered bases of E and F respectively. Put $\varphi(e_i) = A^a{}_i f_a$. (We use a different dummy index for the F-index to avoid confusion.) This defines the matrix of φ; $\mathbf{A} = [A^a{}_i]$. Thus, for $v = v^i e_i \in E$ the components of $\varphi(v)$ are given by $\varphi(v)^a = A^a{}_i v^i$. Hence, thinking of v and $\varphi(v)$ as column vectors, this formula shows that $\varphi(v)$ *is computed by multiplying* v *on the* ***left*** *by* **A**, *the matrix of* φ, as in elementary linear algebra. Thus the upper index is the row index, while the lower index is the column index. Consequently, $\varphi(e_i)$ represents the i-th column of the matrix of φ. Let us now investigate the matrix of $\varphi^* \in L(F^*, E^*)$. In the dual ordered bases, $\langle \varphi^*(f^a), e_i \rangle = \langle f^a, \varphi(e_i) \rangle = \langle f^a, A^b{}_i f_b \rangle = A^b{}_i \delta^a{}_b = A^a{}_i$, i.e., $\varphi^*(f^a) = A^a{}_i e^i$ and thus $\varphi^*(f^a)$ is the a-th row of **A**. Consequently *the matrix of* φ^* *is the transpose of the matrix of* φ. If $\beta = \beta_a f^a \in F^*$ then $\varphi^*(\beta) = \beta_a \varphi^*(f^a) = \beta_a A^a{}_i e^i$, which says that he i-th component of $\varphi^*(\beta)$ equals $\varphi^*(\beta)_i = \beta_a A^a{}_i$. *Thinking of elements in the dual as rows whose entries are their components in the dual basis, this shows that* $\varphi^*(\beta)$ *is computed by multiplying* β *on the* ***right*** *by* **A**, *the matrix of* φ, again in agreement with linear algebra.

Now we turn to the effect of linear transformations on tensors. We start with an induced map that acts "forward" like φ.

5.1.5 Definition *If* $\varphi \in L(E, F)$ *is an isomorphism, define the* ***push-forward*** *of* φ, $T^r{}_s\varphi = \varphi_* \in L(T^r{}_s(E), T^r{}_s(F))$ by

$$\varphi_* t(\beta^1, ..., \beta^r, f_1, ..., f_s) = t(\varphi^*(\beta^1), ..., \varphi^*(\beta^r), \varphi^{-1}(f_1), ..., \varphi^{-1}(f_s))$$

where $t \in T^r{}_s(E)$, $\beta^1, ..., \beta^r \in F^*$, *and* $f_1, ..., f_s \in F$.

We leave the verification that φ_* is continuous to the reader. Note that $T^0{}_1\varphi = (\varphi^{-1})^*$, which maps "forward" like φ. If E and F are finite dimensional, then $T^1{}_0(E) = E$, $T^1{}_0(F) = F$ and we identify φ with $T^1{}_0\varphi$. The next proposition asserts that the push-forward operation is compatible with compositions and the tensor product.

5.1.6 Proposition *Let* $\varphi : E \to F$ *and* $\psi : F \to G$ *be isomorphisms. Then*

(i) $(\psi \circ \varphi)_* = \psi_* \circ \varphi_*$;

(ii) *if* $i : E \to E$ *is the identity, then so is* $i_* : T^r{}_s(E) \to T^r{}_s(E)$;

(iii) $\varphi_* : T^r{}_s(E) \to T^r{}_s(F)$ *is an isomorphism, and* $(\varphi_*)^{-1} = (\varphi^{-1})_*$.

(iv) *If* $t_1 \in T^{r_1}{}_{s_1}(E)$ *and* $t_2 \in T^{r_2}{}_{s_1}(E)$, *then* $\varphi_*(t_1 \otimes t_2) = \varphi_*(t_1) \otimes \varphi_*(t_2)$.

Proof For (i),

$$\begin{aligned}\psi_*(\varphi_* t)(\gamma^1, \ldots, \gamma^r, g_1, \ldots, g_s) &= \varphi_* t(\psi^*(\gamma^1), \ldots, \psi^*(\gamma^r), \psi^{-1}(g_1), \ldots, \psi^{-1}(g_s)) \\ &= t(\varphi^*\psi^*(\gamma^1), \ldots, \varphi^*\psi^*(\gamma^r), \varphi^{-1}\psi^{-1}(g_1), \ldots, \varphi^{-1}\psi^{-1}(g_s)) \\ &= t((\psi \circ \varphi)^*(\gamma^1), \ldots, (\psi \circ \varphi)^*(\gamma^r), (\psi \circ \varphi)^{-1}(g_1), \ldots, (\psi \circ \varphi)^{-1}(g_s)) \\ &= (\psi \circ \varphi)_* t(\gamma^1, \ldots, \gamma^r, g_1, \ldots, g_s),\end{aligned}$$

where $\gamma^1, \ldots, \gamma^r \in G^*$, $g_1, \ldots, g_s \in G$, and $t \in T^r_s(E)$. We have used the fact that the transposes and inverses satisfy $(\psi \circ \varphi)^* = \varphi^* \circ \psi^*$ and $(\psi \circ \varphi)^{-1} = \varphi^{-1} \circ \psi^{-1}$, which the reader can easily check. Part (ii) is an immediate consequence of the definition and the fact that $i^* = i$ and $i^{-1} = i$. Finally, for (iii) we have $\varphi_* \circ (\varphi^{-1})_* = i_*$, the identity on $T^r_s(F)$, by (i) and (ii). Similarly, $(\varphi^{-1})_* \circ \varphi_* = i_*$ the identity on $T^r_s(E)$. Hence (iii) follows. Finally (iv) is a straightforward consequence of the definitions. ■

Since $(\varphi^{-1})_*$ maps "backward" it is called the ***pull-back*** of φ and is denoted φ^*. The next proposition gives a connection with component notation.

5.1.7 Proposition *Let $\varphi \in L(E, F)$ be an isomorphism of finite dimensional vector spaces. Let $[A^a{}_i]$ denote the matrix of φ in the ordered bases $\{e_1, \ldots, e_n\}$ of E and $\{f_1, \ldots, f_m\}$ of F, i.e., $\varphi(e_i) = A^a{}_i f_a$. Denote by $[B^i{}_a]$ the matrix of φ^{-1}, i.e., $\varphi^{-1}(f_a) = B^i{}_a e_i$. Then $[B^i{}_a]$ is the inverse matrix of $[A^a{}_i]$: $B^i{}_a A^a{}_j = \delta^i{}_j$. Let*

$$t \in T^r_s(E) \textit{ with components } t^{i_1 \ldots i_r}{}_{j_1 \ldots j_s} \textit{ relative to } \{e_1, \ldots, e_n\}$$

and

$$q \in T^r_s(F) \textit{ with components } q^{a_1 \ldots a_r}{}_{b_1 \ldots b_s} \textit{ relative to } \{f_1, \ldots, f_m\}.$$

Then the components of $\varphi_* t$ relative to $\{f_1, \ldots, f_m\}$ and of $\varphi^* q$ relative to $\{e_1, \ldots, e_n\}$ are given respectively by

$$(\varphi_* t)^{a_1 \ldots a_r}{}_{b_1 \ldots b_s} = A^{a_1}{}_{i_1} \ldots A^{a_r}{}_{i_r} t^{i_1 \ldots i_r}{}_{j_1 \ldots j_s} B^{j_1}{}_{b_1} \ldots B^{j_s}{}_{b_s}$$
$$(\varphi_* q)^{i_1 \ldots i_r}{}_{j_1 \ldots j_s} = B^{i_1}{}_{a_1} \ldots B^{i_r}{}_{a_r} q^{a_1 \ldots a_r}{}_{b_1 \ldots b_s} A^{b_1}{}_{j_1} \ldots A^{b_s}{}_{j_s}.$$

Proof We have $e_i = \varphi^{-1}(\varphi(e_i)) = \varphi^{-1}(A^a{}_i f_a) = A^a{}_i \varphi^{-1}(f_a) = A^a{}_i B^j{}_a e_j$, whence $B^j{}_a A^a{}_i = \delta^j{}_i$ for all i, j. Similarly, one shows that $A^b{}_i B^i{}_a = \delta^b{}_a$, so that $[A^a{}_i]^{-1} = [B^j{}_a]$. We have

$$\begin{aligned}(\varphi_* t)^{a_1 \ldots a_r}{}_{b_1 \ldots b_s} &= (\varphi_* t)(f^{a_1}, \ldots, f^{a_r}, f_{b_1}, \ldots, f_{b_s}) \\ &= t(\varphi^*(f^{a_1}), \ldots, \varphi^*(f^{a_r}), \varphi^{-1}(f_{b_1}), \ldots, \varphi^{-1}(f_{b_s}))\end{aligned}$$

$$= t(A^{a_1}{}_{i_1} e^{i_1}, \ldots, A^{a_r}{}_{i_r} e^{i_r}, B^{j_1}{}_{b_1} e_{j_1}, \ldots, B^{j_s}{}_{b_s} e_{i_s})$$

$$= (A^{a_1}{}_{i_1} \ldots A^{a_r}{}_{i_r} t^{i_1 \ldots i_r}{}_{j_1 \ldots j_s} B^{j_1}{}_{b_1} \ldots B^{j_s}{}_{b_s} .$$

To prove the second relation, we need the matrix of $(\varphi^{-1})^* \in L(E^*, F^*)$. We have

$$\langle (\varphi^{-1})^*(e^i), f_a \rangle = \langle e^i, \varphi^{-1}(f_a) \rangle = \langle e^i, B^k{}_a e_k \rangle = B^i{}_a$$

so that $(\varphi^{-1})^*(e^i) = B^i{}_a f^a$. Now proceed with the proof as in the previous case. ■

Note that *the matrix of* $(\varphi^{-1})^* \in L(E^*, F^*)$ *is the transpose of the inverse of the matrix of* φ.

The assumption that φ be an isomorphism for φ_* to exist is quite restrictive but clearly cannot be weakened. However, one might ask if instead of "push-forward," the "pull-back" operation is considered, this restrictive assumption can be dropped. This is possible when working with covariant tensors, even when $\varphi \in L(E, F)$ is arbitrary.

5.1.8 Definition *If* $\varphi \in L(E, F)$ *(not necessarily an isomorphism), define the* ***pull-back*** $\varphi^* \in L(T^0{}_s(F) T^0{}_s(E))$ *by*

$$\varphi^* t(e_1, \ldots, e_s) = t(\varphi(e_1), \ldots, \varphi(e_s)),$$

where $t \in T^0{}_s(E)$ *and* $e_1, \ldots, e_s \in E$.

The next proposition asserts that φ^* is compatible with compositions and the tensor product. Its proof is almost identical to that of **5.1.6** and is left as an exercise for the reader.

5.1.9 Proposition *Let* $\varphi \in L(E; F)$ *and* $\psi \in L(F; G)$.

(i) $(\psi \circ \varphi)^* = \varphi^* \circ \psi^*$.

(ii) *If* $i: E \to E$ *is the identity, then so is* $i^* \in L(T^0{}_s(E), T^0{}_s(E))$.

(iii) *If* φ *is an isomorphism, then so is* φ^* *and* $\varphi^* = (\varphi^{-1})_*$

(iv) *If* $t_1 \in T^0_{s_1}(F)$ *and* $t_2 \in T^0_{s_2}(F)$, *then* $\varphi^*(t_1 \otimes t_2) = (\varphi^* t_1) \otimes (\varphi^* t_2)$.

Finally the components of $\varphi^* t$ are given by the following.

5.1.10 Proposition *Let* E *and* F *be finite-dimensional vector spaces and* $\varphi \in L(E, F)$. *For ordered bases* $\{e_1, \ldots, e_n\}$ *of* E *and* $\{f_1, \ldots, f_m\}$ *of* F, *suppose that* $\varphi(e_i) = A^a{}_i f_a$, *and let* $t \in T^0_s(F)$ *have components* $t_{b_1 \ldots b_s}$. *Then the components of* $\varphi^* t$ *relative to* $\{e_1, \ldots, e_n\}$ *are given by*

$$(\varphi^* t)_{j_1 \ldots j_s} = t_{b_1 \ldots b_s} A^{b_1}{}_{j_1} \ldots A^{b_s}{}_{j_s} .$$

Proof

$$\begin{aligned}(\varphi^* t)_{j_1 \ldots j_s} &= (\varphi^* t)(e_{j_1}, \ldots, e_{j_s}) = t(\varphi(e_{j_1}), \ldots, \varphi(e_{j_s}))\\ &= t(A^{b_1}{}_{j_1} f_{b_1}, \ldots, A^{b_s}{}_{j_s} f_{b_s}) = t(f_{b_1}, \ldots, f_{b_s}) A^{b_1}{}_{j_1} \ldots A^{b_s}{}_{j_s}\\ &= t_{b_1 \ldots b_s} A^{b_1}{}_{j_1} \ldots A^{b_s}{}_{j_s} \quad \blacksquare\end{aligned}$$

5.1.11 Examples

A On $\mathbb{R}^2$ with the standard basis $\{e_1, e_2\}$, let $t \in T^2{}_0(\mathbb{R}^2)$ be given by $t = e_1 \otimes e_1 + 2e_1 \otimes e_2 - e_2 \otimes e_1 + 3e_2 \otimes e_2$ and let $\varphi \in L(\mathbb{R}^2, \mathbb{R}^2)$ have the matrix

$$\mathbf{A} = \begin{bmatrix} 2 & 1 \\ 1 & 1 \end{bmatrix}.$$

Then φ is clearly an isomorphism, since $\mathbf{A}$ has an inverse matrix given by

$$\mathbf{B} = \begin{bmatrix} 1 & -1 \\ -1 & 2 \end{bmatrix}.$$

According to **5.1.7**, the components of $T = \varphi^* t$ relative to the standard basis of $\mathbb{R}^2$ are given by

$$T^{ij} = B^i{}_a B^j{}_b t^{ab}, \text{ with } B^1{}_1 = 1,\ B^2{}_1 = B^1{}_2 = -1 \text{ and } B^2{}_2 = 2$$

so that

$$\begin{aligned}T^{12} &= B^1{}_1 B^2{}_1 t^{11} + B^1{}_1 B^2{}_2 t^{12} + B^1{}_2 B^2{}_1 t^{21} + B^1{}_2 B^2{}_2 t^{22}\\ &= 1 \cdot (-1) \cdot 1 + 1 \cdot 2 \cdot 2 + (-1) \cdot (-1) \cdot (-1) + (-1) \cdot 2 \cdot 3 = -4\end{aligned}$$

$$\begin{aligned}T^{21} &= B^2{}_1 B^1{}_1 t^{11} + B^2{}_1 B^1{}_2 t^{12} + B^2{}_2 B^1{}_1 t^{21} + B^2{}_2 B^1{}_2 t^{22}\\ &= (-1) \cdot 1 \cdot 1 + (-1) \cdot (-1) \cdot 2 + 2 \cdot 1 \cdot (-1) + 2 \cdot (-1) \cdot 3 = -7\end{aligned}$$

$$\begin{aligned}T^{11} &= B^1{}_1 B^1{}_1 t^{11} + B^1{}_1 B^1{}_2 t^{12} + B^1{}_2 B^1{}_1 t^{21} + B^1{}_2 B^1{}_2 t^{22}\\ &= 1 \cdot 1 \cdot 1 + 1 \cdot (-1) \cdot 2 + (-1) \cdot 1 \cdot (-1) + (-1) \cdot (-1) \cdot 3 = 3\end{aligned}$$

$$\begin{aligned}T^{22} &= B^2{}_1 B^2{}_1 t^{11} + B^2{}_1 B^2{}_2 t^{12} + B^2{}_2 B^2{}_1 t^{21} + B^2{}_2 B^2{}_2 t^{22}\\ &= (-1) \cdot (-1) \cdot 1 + (-1) \cdot 2 \cdot 2 + 2 \cdot (-1) \cdot (-1) + 2 \cdot 2 \cdot 3 = 11.\end{aligned}$$

Thus, $\varphi^* t = 3e_1 \otimes e_1 - 4e_1 \otimes e_2 - 7e_2 \otimes e_1 + 11e_2 \otimes e_2$.

B Let $t = e_1 \otimes e^2 - 2e_2 \otimes e^2 \in T^1{}_1(\mathbb{R}^2)$ and consider the same map $\varphi \in L(\mathbb{R}^2, \mathbb{R}^2)$ as in part **A** above. We could compute the components of $\varphi_* t$ relative to the standard basis of $\mathbb{R}^2$ using the formula in **5.1.7** as before. An alternative way to proceed directly using **5.1.6**(iv), that is, the fact that φ_* is compatible with tensor products. Thus

$$\varphi_* t = \varphi_*(e_1 \otimes e^2 - 2e_2 \otimes e^2) = \varphi(e_1) \otimes \varphi_*(e^2) - 2\varphi(e_2) \otimes \varphi_*(e^2).$$

But $\varphi(e_1) = 2e_1 + e_2$, $\varphi(e_2) = e_1 + e_2$, and $\varphi_*(e^2) = -e^1 + 2e^2$, so that

$$\varphi_* t = (2e_1 + e_2) \otimes (-e^1 + 2e^2) - 2(e_1 + e_2) \otimes (-e^1 + 2e^2) = e_2 \otimes e^1 - 2e_2 \otimes e^2 .$$

C Let $t = -2e^1 \otimes e^2 \in T^0{}_2(\mathbb{R}^2)$ and $\varphi \in L(\mathbb{R}^3, \mathbb{R}^2)$ be given by the matrix

$$\mathbf{A} = \begin{bmatrix} 2 & 1 \\ 1 & 1 \end{bmatrix}.$$

Again we shall compute $\varphi^* t \in T^0{}_2(\mathbb{R}^3)$ by using the fact that φ^* is compatible with tensor products and that the matrix of $\varphi^* \in L(\mathbb{R}^2, \mathbb{R}^3)$ is the transpose of $\mathbf{A}$. Recall that $\varphi^*(e^i)$ is the i-th *row*, since matrices act on the right on covectors. Denote by f^1, f^2, f^3 the standard dual basis of $\mathbb{R}^3$. Then $\varphi^*(e^1) = f^1 + 2f^3$ and $\varphi^*(e^2) = -f^2 + f^3$, so that

$$\varphi^*(t) = -2\varphi^*(e^1) \otimes \varphi^*(e^2) = -2(f^1 + 2f^3) \otimes (-f^2 + f^3)$$
$$= 2f^1 \otimes f^2 - 2f^1 \otimes f^3 + 4f^3 \otimes f^2 - 4f^3 \otimes f^3 . \; ♦$$

Exercises

5.1A Compute the interior product of the tensor $t = e_1 \otimes e_1 \otimes e^2 + 3e_2 \otimes e_2 \otimes e^1$ with $e = -e_1 + 2e_2$ and $\alpha = 2e^1 + e^2$. What are the (1, 1) and (2, 1) contractions of t?

5.1B Compute all associated tensors of $t = e_1 \otimes e^2 \otimes e^2 + 2e_2 \otimes e^1 \otimes e^2 - e_2 \otimes e^2 \otimes e^1$ with respect to the standard metric of $\mathbb{R}^2$.

5.1C Let $t = 2e^1 \otimes e^1 - e^2 \otimes e^1 + 3e^1 \otimes e^2$ and $\varphi \in L(\mathbb{R}^2, \mathbb{R}^2)$, $\psi \in L(\mathbb{R}^3, \mathbb{R}^2)$ be given by the matrices

$$\begin{bmatrix} 2 & 1 \\ -1 & 1 \end{bmatrix}, \quad \begin{bmatrix} 0 & 1 & -1 \\ 1 & 0 & 2 \end{bmatrix}.$$

Compute: trace(t), $\varphi^* t$, $\psi^* t$, trace $(\varphi^* t)$, trace $(\psi^* t)$, $\varphi_* t$, and all associated tensors of t, $\varphi^* t$, $\psi^* t$ and $\varphi_* t$ with respect to the corresponding standard inner products in $\mathbb{R}^2$ and $\mathbb{R}^3$.

5.1D Let dim (E) = n and dim (F) = m. Show that $T^r{}_s(E; F)$ is an mn^{r+s}-dimensional real vector space by exhibiting a basis.

§5.2 *Tensor Bundles and Tensor Fields*

We now extend the tensor algebra to local vector bundles, and then to vector bundles. For $U \subset E$ (open) recall that $U \times F$ is a local vector bundle. Then $U \times T^r_s(F)$ is also a local vector bundle in view of **5.1.2**. Suppose $\varphi : U \times F \to U' \times F'$ is a local vector bundle mapping and is an *isomorphism on each fiber*; that is $\varphi_u = \varphi \mid \{u\} \times F \in L(F, F')$ is an isomorphism. Also, let φ_0 denote the restriction of φ to the zero section. Then φ induces a mapping of the local tensor bundles as follows.

5.2.1 Definition *If* $\varphi : U \times F \to U' \times F'$ *is a local vector bundle mapping such that for each* $u \in U$, φ_u *is an isomorphism, let* $\varphi_* : U \times T^r_s(F) \to U' \times T^r_s(F')$ *be defined by*

$$\varphi_*(u, t) = (\varphi_0(u), (\varphi_u)_* t),$$

where $t \in T^r_s(F)$.

Before proceeding, we shall pause and recall some useful facts concerning linear isomorphisms from **2.5.4** and **2.5.5**.

5.2.2 Proposition *Let* $GL(E, F)$ *denote the set of linear isomorphisms from* E *to* F. *Then* $GL(E, F) \subset L(E, F)$ *is open.*

5.2.3 Proposition *Let* $\mathcal{A} : L(E, F) \to L(F^*, E^*)$; $\varphi \mapsto \varphi^*$ *and* $I : GL(E, F) \to GL(F, E)$; $\varphi \mapsto \varphi^{-1}$. *Then* $\mathcal{A}$ *and* I *are of class* C^∞ *and*

$$\mathbf{D}I^{-1}(\varphi)\cdot \psi = -\varphi^{-1} \circ \psi \circ \varphi^{-1}.$$

Smoothness of $\mathcal{A}$ is clear since it is linear.

5.2.4 Proposition *If* $\varphi : U \times F \to U' \times F'$ *is a local vector bundle map and* φ_u *is an isomorphism for all* $u \in U$, *then* $\varphi_* : U \times T^r_s(F) \to U' \times T^r_s(F')$ *is a local vector bundle map and* $(\varphi_u)_* = (\varphi_*)_u$ *is an isomorphism for all* $u \in U$. *Moreover, if* φ *is a local vector bundle isomorphism then so is* φ_*.

Proof That φ_* is an isomorphism on fibers follows from **5.1.6**(iii) and the last assertion follows from the former. By **5.2.1** we need only establish that $(\varphi_u)_* = (\varphi_*)_u$ is of class C^∞. Now φ_u is a smooth function of u, and, by **5.2.3** $\varphi_u{}^*$ and $\varphi_u{}^{-1}$ are smooth functions of u. The map $(\varphi_u)_*$ is a Cartesian product of r factors $\varphi_u{}^*$ and s factors $\varphi_u{}^{-1}$, so is smooth. Hence, from the product rule, $(\varphi_u)_*$ is smooth. ■

This smoothness can be verified also for finite-dimensional bundles by using the standard bases in the tensor spaces as local bundle charts and proving that the components of $\varphi_* t$ are C^∞ functions.

We have the following commutative diagram, which says that φ_* *preserves fibers*:

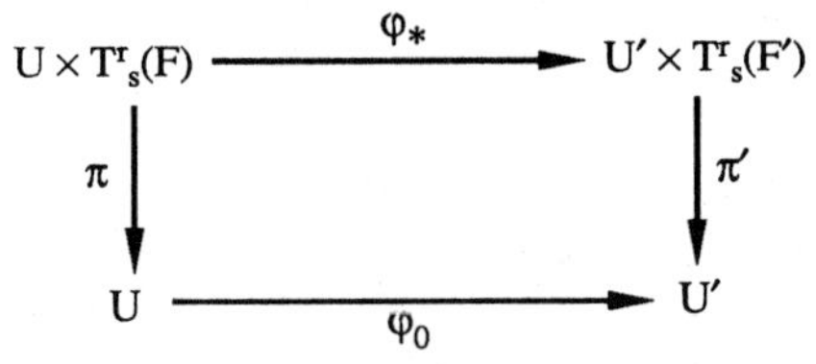

5.2.5 Definition *Let* $\pi : E \to B$ *be a vector bundle with* $E_b = \pi^{-1}(b)$ *denoting the fiber over the point* $b \in B$. *Define* $T^r_s(E) = \bigcup_{b\in B} T^r_s(E_b)$ *and* $\pi^r_s : T^r_s(E) \to B$ *by* $\pi^r_s(e) = b$ *where* $e \in T^r_s(E_b)$. *Furthermore, for a given subset* A *of* B, *we define* $T^r_s(E) \mid A = \bigcup_{b\in A} T^r_s(E_b)$. *If* $\pi' : E' \to B'$ *is another vector bundle and* $(\varphi, \varphi_0) : E \to E'$ *is a vector bundle mapping with* $\varphi_b = \varphi \mid E_b$ *an isomorphism for all* $b \in B$, *let* $\varphi_* : T^r_s(E) \to T^r_s(E')$ *be defined by* $\varphi_* \mid T^r_s(E_b) = (\varphi_b)_*$.

Now suppose that $(E \mid U, \varphi)$ is an admissible local bundle chart of π, where $U \subset B$ is an open set. Then the mapping $\varphi_* \mid [T^r_s(E) \mid U]$ is obviously a bijection onto a local bundle, and thus is a local bundle chart. Further, $(\varphi_*)_b = (\varphi_b)_*$ is a linear isomorphism, so this chart preserves the linear structure of each fiber. We shall call such a chart a ***natural chart*** of $T^r_s(E)$.

5.2.6 Theorem *If* $\pi : E \to B$ *is a vector bundle, then the set of all natural charts of* $\pi^r_s : T^r_s(E) \to B$ *is a vector bundle atlas.*

Proof Condition **VB1** is obvious. For **VB2**, suppose we have two overlapping natural charts, φ_* and ψ_*. For simplicity, let them have the same domain. Then $\alpha = \psi \circ \varphi^{-1}$ is a local vector bundle isomorphism, and by **5.1.6**, $\psi_* \circ (\varphi_*)^{-1} = \alpha_*$, is a local vector bundle isomorphism by **5.2.4**. ■

This atlas of natural charts called the ***natural atlas*** of π^r_s, generates a vector bundle structure, and it is easily seen that the resulting vector bundle is Hausdorff, and all fibers are isomorphic Banachable spaces. *Hereafter,* $T^r_s(E)$ *will denote all of this structure.*

5.2.7 Proposition *If* $f : E \to E'$ *is a vector bundle map that is an isomorphism on each fiber, then* $f_* : T^r_s(E) \to T^r_s(E')$ *is also a vector bundle map that is an isomorphism on each fiber.*

Proof Let (U, φ) be an admissible vector bundle chart of E, and let (V, ψ) be one of E' so that $f(U) \subset V$ and $f_{\varphi\psi} = \psi \circ f \circ \varphi^{-1}$ is a local vector bundle mapping. Then using the natural atlas, we see that

$$(f_*)_{\varphi_*, \psi_*} = (f_{\varphi\psi})_* . \blacksquare$$

5.2.8 Proposition *Let* $f : E \to E'$ *and* $g : E' \to E''$ *be vector bundle maps that are isomorphisms on each fiber. Then so is* $g \circ f$, *and*

(i) $(g \circ f)_* = g_* \circ f_*$;

(ii) *if* $i : E \to E$ *is the identity, then* $i_* : T^r_s(E) \to T^r_s(E)$ *is the identity.*

(iii) *if* $f : E \to E'$ *is a vector bundle isomorphism, then so if* f_* *and* $(f_*)^{-1} = (f^{-1})_*$.

Proof For (i) we examine representatives of $(g \circ f)_*$ and $g_* \circ f_*$. These representatives are the same in view of **5.1.6**. Part (ii) is clear from the definition, and (iii) follows from (i) and (ii) by the same method as in **5.1.6.** $\blacksquare$

We now specialize to the case where $\pi : E \to B$ is the tangent vector bundle of a manifold.

5.2.9 Definition *Let* M *be a manifold and* $\tau_M : TM \to M$ *its tangent bundle. We call* $T^r_s(M) = T^r_s(TM)$ *the* ***vector bundle of tensors contravariant order*** r ***and covariant order*** s, *or simply of* ***type*** (r, s). *We identify* $T^1_0(M)$ *with* TM *and call* $T^0_1(M)$ *the* ***cotangent bundle*** *of* M *also denoted by* $\tau^*_M : T^*M \to M$. *The zero section of* $T^r_s(M)$ *is identified with* M.

Recall that a section of a vector bundle assigns to each base point b a vector in the fiber over b and the addition and scalar multiplication of sections takes place within each fiber. In the case of $T^r_s(M)$ these vectors are called ***tensors***. The C^∞ sections of $\pi : E \to B$ were denoted $\Gamma^\infty(\pi)$, or $\Gamma^\infty(E)$. Recall that $\mathcal{F}(M)$ denotes the set of mappings from M into $\mathbb{R}$ that are of class C^∞ (the standard local manifold structure being used on $\mathbb{R}$) together with its structure as a ring; namely, $f + g$, cf, fg for $f, g \in \mathcal{F}(M)$, $c \in \mathbb{R}$ are defined by $(f + g)(x) = f(x) + g(x)$, $(cf)(x) = c(f(x))$ and $(fg)(x) = f(x)\, g(x)$. Finally, recall that a ***vector field*** on M is an element of $\mathfrak{X}(M) = \Gamma^\infty(TM)$.

5.2.10 Definition *A* ***tensor field of type*** (r, s) *on a manifold* M *is a* C^∞ *section of* $T^r_s(M)$. *We denote by* $\mathcal{T}^r_s(M)$ *the set* $\Gamma^\infty(T^r_s(M))$ *together with its (infinite-dimensional) real vector space structure. A* ***covector field*** *or* ***differential one-form*** *is an element of* $\mathfrak{X}^*(M) = \mathcal{T}^0_1(M)$.

If $f \in \mathcal{F}(M)$ *and* $t \in \mathcal{T}^r_s(M)$, *let* $ft: M \to T^r_s(M)$ *be defined by* $m \mapsto f(m)t(m)$. *If* $X_i \in \mathfrak{X}(M)$, $i = 1, ..., s$, $\alpha^j \in \mathfrak{X}^*(M)$, $j = 1, ..., r$, *and* $t' \in \mathcal{T}^{r'}_{s'}(M)$ *define*

$$t(\alpha^1, ..., \alpha^r, X_1, ..., X_s) : M \to \mathbb{R} \ \ by \ \ m \mapsto t(m)(\alpha^1(m), ..., X_s(m))$$

and

$$t \otimes t' : M \to T^{r+r'}_{s+s'}(M) \ \ by \ \ m \mapsto t(m) \otimes t'(m).$$

5.2.11 Proposition *With* f, t, X_i, α^j, *and* t' *as in* **5.2.10**, $ft \in \mathcal{T}^r_s(M)$, $t(\alpha^1, ..., X_s) \in \mathcal{F}(M)$, *and* $t \otimes t' \in \mathcal{T}^{r+r'}_{s+s'}$.

Proof The differentiability is evident in each case from the product rule in local representation. ∎

For the tangent bundle TM, a natural chart is obtained by taking $T\varphi$, where φ is an admissible chart of M. This in turn induces a chart $(T\varphi)_*$ on T^r_sM. We shall call these the ***natural charts*** of T^r_sM.

Now we turn to the expression of tensor fields in local coordinates. Recall that $\partial/\partial x^i = (T\varphi)^{-1}(e_i)$, for $\varphi : U \to U' \subset \mathbb{R}^n$ a chart on M, is a basis of $\mathfrak{X}(U)$. The vector field $\partial/\partial x^i$ corresponds to the derivation $f \mapsto \partial f/\partial x^i$. Since $dx^i(\partial/\partial x^j) = \partial x^i/\partial x^j = \delta^i_j$, we see that dx^i is the dual basis of $\partial/\partial x^i$ at every point of U, i.e., that $dx^i = \varphi^*(e^i)$, where $\{e^1, ..., e^n\}$ is the dual basis to $\{e_1, ..., e_n\}$. Let

$$t^{i_1 \dots i_r}{}_{j_1 \dots j_s} = t(dx^{i_1}, ..., dx^{i_r}, \partial/\partial x^{j_1}, ..., \partial/\partial x^{j_s}) \in \mathcal{F}(U).$$

Applying **5.1.6**(iv) at every point yields the coordinate expression of an (r, s)-tensor field:

$$t \,|\, U = t^{i_1 \dots i_r}{}_{j_1 \dots j_s}\, \partial/\partial x^{i_1} \otimes \dots \otimes \partial/\partial x^{i_r} \otimes dx^{j_1} \otimes \dots \otimes dx^{j_s}.$$

To discuss the behavior of these components relative to a change of coordinates, assume that $X^i: U \to \mathbb{R}$, i = 1, ..., n is a different coordinate system. We can write $\partial/\partial x^i = a_i^j \partial/\partial X^j$, since both are bases of $\mathfrak{X}(U)$. Applying both sides to X^k yields $a_i^j = \partial X^j/\partial x^i$ i.e., $(\partial/\partial x^i) = (\partial X^j/\partial x^i)(\partial/\partial X^j)$. Thus the dx^i, as dual basis change with the inverse of the Jacobian matrix $[\partial X^j/\partial x^i]$; i.e., $dx^i = (\partial x^i/\partial X^j)dX^j$. Writing t in both coordinate systems and isolating equal terms gives the following change of coordinate formula for the components:

$$T^{k_1 \dots k_r}{}_{\ell_1 \dots \ell_s} = \frac{\partial X^{k_1}}{\partial x^{i_1}} \cdots \frac{\partial X^{k_r}}{\partial x^{i_r}} \frac{\partial x^{j_1}}{\partial X^{\ell_1}} \cdots \frac{\partial x^{j_s}}{\partial X^{\ell_s}}\, t^{i_1 \dots i_r}{}_{j_1 \dots j_s}$$

This formula is known as the ***tensoriality criterion:***

A set of n^{r+s} *functions* $t^{i_1 \dots i_r}{}_{j_1 \dots j_s}$ *defined for each coordinate system on the open set* U *of* M *locally define an* (r, s)*–ensor field iff changes of coordinates have the aforementioned effect on them.* This statement is clear since at every point it assures that the n^{r+s} functions are the components of an (r, s)-tensor in T_uU and conversely.

The algebraic operations on tensors, such as contraction, inner products and traces, all carry over fiberwise to tensor fields. For example, if $\delta_m \in T^1{}_1(T_mM)$ is the Kronecker delta, then $\delta: M \to T^1{}_1(M);\ m \mapsto \delta_m$ is obviously C^∞, and $\delta \in \mathcal{T}^1{}_1(M)$ is called the ***Kronecker delta.*** Similarly, a tensor field of type (0, s) or (r, 0) is called ***symmetric***, if it is symmetric at every point. A basic example of a symmetric covariant tensor field is the following.

5.2.12 Definition *A **weak pseudo-Riemannian metric** on a manifold* M *is defined to be a tensor field* $g \in \mathcal{T}^0{}_2(M)$ *that is symmetric and weakly nondegenerate, i.e., such that at each* $m \in M$,

$g(m)(v_m, w_m) = 0$ *for all* $w_m \in T_mM$ *implies* $v_m = 0$. *A* ***strong pseudo-Riemannian metric*** *is a 2-tensor field that, in addition is* ***strongly nondegenerate*** *for all* $m \in M$; *i.e., the map* $v_m \mapsto g(m)(v_m, \cdot)$ *is an isomorphism of* T_mM *onto* $T_m{}^*M$. *A weak (resp. strong) pseudo-Riemannian metric is called* ***weak (resp. strong) Riemannian*** *if in addition* $g(m)(v_m, v_m) > 0$ *for all* $v_m \in T_mM$, $v_m \neq 0$.

A strong Riemannian manifold is necessarily modeled on a Hilbertizable space; i.e., the model space has an equivalent norm arising from an inner product. For finite-dimensional manifolds weak and strong metrics coincide: indeed T_mM and $T_m{}^*M$ have the same dimension and so a one-to-one map of T_mM to $T_m{}^*M$ is an isomorphism. It is possible to have weak metrics on a Banach or Hilbert manifold that are not strong. For example, the L^2 inner product on $M = C^0([0, 1], \mathbb{R})$ is a weak metric that is not strong. For a similar Hilbert space example, see Exercise **5.2C**.

Any Hilbert space is a Riemannian manifold with a constant metric equal to the inner product. A symmetric bilinear (weakly) nondegenerate two-form on any Banach space provides an example of a (weak) pseudo-Riemannian constant metric. A pseudo-Riemannian manifold used in the theory of special relativity is $\mathbb{R}^4$ with the ***Minkowski pseudo-Riemannian metric***

$$g(x)(v, w) = v^1w^1 + v^2w^2 + v^3w^3 - v^4w^4 ,$$

where $x, v, w \in \mathbb{R}^4$.

As in the algebraic context of §**5.1**, pseudo-Riemannian metrics (and for that matter any strongly nondegenerate bilinear tensor) can be used to define associated tensors. Thus the maps $\sharp$, $\flat$ become vector bundle isomorphisms over the identity $\flat : TM \to T^*M$, $\sharp : T^*M \to TM$; $\sharp$ is the inverse of $\flat$, where $v_m{}^\flat = g(m)(v_m, \cdot\,)$. In particular, they induce isomorphisms of the spaces of sections $\flat : \mathfrak{X}(M) \to \mathfrak{X}^*(M)$, $\sharp : \mathfrak{X}^*(M) \to \mathfrak{X}(M)$. In finite dimensions this is the operation of raising and lowering indices. Thus formulas like the ones in Example **5.1.4E** should be read pointwise in this context. There is, however, a particular index raising operation that requires special attention.

5.2.13 Definition *Let* M *be a pseudo-Riemannian* n*-manifold with metric* g. For $f \in \mathcal{F}(M)$, $\operatorname{grad} f = (df)^\sharp \in \mathfrak{X}(M)$ *is called the* ***gradient*** *of* f.

To find the expression of grad f in local coordinates, we write

$$g_{ij} = g\left(\frac{\partial}{\partial x^i}, \frac{\partial}{\partial x^j}\right), \quad X = X^i \frac{\partial}{\partial x^j}, \quad \text{and} \quad Y = Y^i \frac{\partial}{\partial x^i},$$

so we have

$$\langle X^\flat, Y\rangle = g(X, Y) = X^iY^jg\left(\frac{\partial}{\partial x^i}, \frac{\partial}{\partial x^j}\right) = X^iY^jg_{ij}, \ \text{i.e.,} \ X^\flat = X^ig_{ij}dx^j .$$

If $\alpha \in \mathfrak{X}^*(M)$ has the coordinate expression $\alpha = \alpha_i dx^i$, we have $\alpha^\sharp = \alpha_i g^{ij}\partial/\partial x^j$ where $[g^{ij}]$

is the inverse of the matrix $[g_{ij}]$. Thus for $\alpha = \mathbf{d}f$, the local expression of the gradient is

$$\text{grad } f = g^{ij} \frac{\partial f}{\partial x^j} \frac{\partial}{\partial x^i} \quad \text{or} \quad (\text{grad } f)^i = g^{ij} \frac{\partial f}{\partial x^j}.$$

If $M = \mathbb{R}^n$ with standard Euclidean metric $g_{ij} = \delta_{ij}$, this formula becomes

$$\text{grad } f = \frac{\partial f}{\partial x^i} \frac{\partial}{\partial x^i}; \text{ i.e., } \text{grad } f = \left(\frac{\partial f}{\partial x^1}, \ldots, \frac{\partial f}{\partial x^n} \right),$$

the familiar expression of the gradient from vector calculus.

Now we turn to the effect of mappings and diffeomorphisms on tensor fields.

5.2.14 Definition *If* $\varphi : M \to N$ *is a diffeomorphism and* $t \in T^r_s(M)$, *let* $\varphi_* t = (T\varphi)_* \circ t \circ \varphi^{-1}$, *be the* ***push-forward*** *of* t *by* φ. *If* $t \in T^r_s(N)$, *the* ***pull-back*** *of* t *by* φ *is given by* $\varphi^* t = (\varphi^{-1})_* t$.

5.2.15 Proposition *If* $\varphi : M \to N$ *is a diffeomorphism, and* $t \in \mathcal{T}^r_s(M)$, *then*

(i) $\varphi_* t \in \mathcal{T}^r_s(M)$;

(ii) $\varphi_* : \mathcal{T}^r_s(M) \to \mathcal{T}^r_s(N)$ *is a linear isomorphism*;

(iii) $(\varphi \circ \psi)_* = \varphi_* \circ \psi_*$; *and*

(iv) $\varphi_*(t \otimes t') = \varphi_* t \otimes \varphi_* t'$, *where* $t \in \mathcal{T}^r_s(M)$ *and* $t' \in \mathcal{T}^r_s(M)$.

Proof (i) The differentiability is evident from the composite mapping theorem, together with **5.2.4**. The other three statements are proved fiberwise, where they are consequences of **5.1.6**. ∎

As in the algebraic context, the pull-back of covariant tensors is defined even for maps that are not diffeomorphisms. Globalizing **5.1.8** we get the following.

5.2.16 Definition *If* $\varphi : M \to N$ *and* $t \in \mathcal{T}^0_s(N)$, *then* $\varphi^* t$, *the* ***pull-back*** *of* t *by* φ, *is defined by*

$$(\varphi^* t)(m)(v_1, \ldots, v_s) = t(\varphi(m))(T_m\varphi(v_1), \ldots, T_m\varphi(v_s))$$

for $m \in M$, $v_1, \ldots, v_s \in T_mM$.

The next proposition is similar to **5.2.15** and is proved by globalizing the proof of **5.1.9**.

5.2.17 Proposition *If* $\varphi : M \to N$ *is* C^∞ *and* $t \in \mathcal{T}^0_s(N)$, *then*

(i) $\varphi^* \in \mathcal{T}^0_s(M)$;

(ii) $\varphi^* : \mathcal{T}^0_s(N) \to \mathcal{T}^0_s(M)$ *is a linear map*;

(iii) $(\psi \circ \varphi)^* = \varphi^* \circ \psi^*$ *for* $\psi : N \to P$;

(iv) if φ *is a diffeomorphism then* φ^* *is an isomorphism with inverse* φ^*; *and*

(v) $t_1 \in \mathcal{T}^0_{s_1}(N)$, $t_2 \in \mathcal{T}^0_{s_2}(N)$, then $\varphi^*(t_1 \otimes t_2) = (\varphi^* t_1) \otimes (\varphi^* t_2)$.

For finite-dimensional manifolds the coordinate expressions of the pull-back and push-forward can be read directly from **5.1.7** and **5.1.10**, taking into account that $T\varphi$ is given locally by the Jacobian matrix. This yields the following.

5.2.18 Proposition *Let* M *and* N *be finite-dimensional manifolds,* $\varphi : M \to N$ *a* C^r *map and denote by* $y^j = \varphi^j(x^1, ..., x^m)$ *the local expression of* φ *relative to charts where* $m = \dim(M)$ *and* $j = 1, ..., n = \dim(N)$.

(i) *If* $t \in \mathcal{T}^r_s(M)$ *and* φ *is a diffeomorphism, the coordinates of the push-forward* $\varphi_* t$ *are*

$$(\varphi_* t)^{i_1 \dots i_r}{}_{j_1 \dots j_s} = \frac{\partial y^{i_1}}{\partial x^{k_1}} \cdots \frac{\partial y^{i_r}}{\partial x^{k_r}} \left(\frac{\partial x^{\ell_1}}{\partial y^{j_1}} \circ \varphi^{-1} \right) \cdots \left(\frac{\partial x^{\ell_s}}{\partial y^{j_s}} \circ \varphi^{-1} \right) t^{k_1 \dots k_r}{}_{\ell_1 \dots \ell_s} \circ \varphi^{-1} .$$

If $t \in \mathcal{T}^0_s(N)$ *and* φ *is a diffeomorphism, the coordinates of the pull-back* $\varphi^* t$ *are*

$$(\varphi^* t)^{i_1 \dots i_r}{}_{j_1 \dots j_s} = \left(\frac{\partial x^{i_1}}{\partial y^{\ell_1}} \circ \varphi \right) \cdots \left(\frac{\partial x^{i_r}}{\partial y^{\ell_r}} \circ \varphi \right) \frac{\partial y^{k_1}}{\partial x^{j_1}} \cdots \frac{\partial y^{k_s}}{\partial x^{j_s}} t^{\ell_1 \dots \ell_r}{}_{k_1 \dots k_s} \circ \varphi .$$

(ii) *If* $t \in \mathcal{T}^0_s(N)$ *and* $\varphi : M \to N$ *is arbitrary, the coordinates of the pull-back* $\varphi^* t$ *are*

$$(\varphi^* t)_{j_1 \dots j_s} = \frac{\partial y^{k_1}}{\partial x^{j_1}} \cdots \frac{\partial y^{k_s}}{\partial x^{j_s}} t_{k_1 \dots k_s} \circ \varphi .$$

Notice the similarity between the formulas for coordinate change and pull-back. The situation is similar to the *passive* and *active interpretation* of similarity transformations $\mathbf{PAP}^{-1}$ in linear algebra. Of course it is important not to confuse the two.

5.2.19 Examples A Let $\varphi : \mathbb{R}^2 \to \mathbb{R}^2$ be defined by $\varphi(x, y) = (x + 2y, y)$ and let $t = 3x(\partial/\partial x) \otimes dy + (\partial/\partial y) \otimes dy \in \mathcal{T}^1_1(\mathbb{R}^2)$. The matrix of φ_* on vector fields is

$$\left[\frac{\partial \varphi^i}{\partial x^j} \right] = \begin{bmatrix} 1 & 2 \\ 0 & 1 \end{bmatrix} \text{ and on forms is } \left[\frac{\partial x^i}{\partial \varphi^j} \right] = \begin{bmatrix} 1 & 2 \\ 0 & 1 \end{bmatrix}^{-1} = \begin{bmatrix} 1 & -2 \\ 0 & 1 \end{bmatrix} .$$

In other words,

$$\varphi_* \left(\frac{\partial}{\partial x} \right) = \frac{\partial}{\partial x} , \quad \varphi_* \left(\frac{\partial}{\partial y} \right) = 2 \frac{\partial}{\partial x} + \frac{\partial}{\partial y} .$$

$$\varphi_*(dx) = dx - 2\, dy, \text{ and } \varphi_*(dy) = dy.$$

Noting that $\varphi^{-1}(x, y) = (x - 2y, y)$, we get

$$\varphi_* t = 3(x - 2y)\varphi_*\left(\frac{\partial}{\partial x}\right) \otimes \varphi_*(dy) + \varphi_*\left(\frac{\partial}{\partial y}\right) \otimes \varphi_*(dy)$$

$$= 3(x - 2y)\frac{\partial}{\partial x} \otimes dy + \left(2\frac{\partial}{\partial x} + \frac{\partial}{\partial y}\right) \otimes dy = (3x - 6y + 2)\frac{\partial}{\partial x} \otimes dy + \frac{\partial}{\partial y} \otimes dy.$$

B With the same mapping and tensor, we compute $\varphi^* t$. Since

$$\varphi^*\left(\frac{\partial}{\partial x}\right) = \frac{\partial}{\partial x}, \ \varphi^*\left(\frac{\partial}{\partial y}\right) = -2\frac{\partial}{\partial x} + \frac{\partial}{\partial y}.$$

$$\varphi^*(dx) = dx + 2\,dy, \text{ and } \varphi^*(dy) = dy,$$

we have

$$\varphi^* t = 3(x + 2y)\varphi^*\left(\frac{\partial}{\partial x}\right) \otimes \varphi^*(dy) + \varphi^*\left(\frac{\partial}{\partial y}\right) \otimes \varphi^*(dy)$$

$$= 3(x + 2y)\frac{\partial}{\partial x} \otimes dy + \left(-2\frac{\partial}{\partial x} + \frac{\partial}{\partial y}\right) \otimes dy = (3x + 6y - 2)\frac{\partial}{\partial x} \otimes dy + \frac{\partial}{\partial y} \otimes dy.$$

C Let $\varphi : \mathbb{R}^3 \to \mathbb{R}^2$, $\varphi(x, y, z) = (2x + z, xyz)$ and $t = (u + 2v)du \otimes du + (u)^2\, du \otimes dv \in \mathcal{T}^0_2(\mathbb{R}^2)$. Since $\varphi^*(du) = 2\,dx + dz$ and $\varphi^*(dv) = yz\,dx + xz\,dy + xy\,dz$, we have

$$\varphi^* t = (2x + z + 2xyz)(2\,dx + dz) \otimes (2\,dx + dz)$$
$$+ (2x + z)^2 (2\,dx + dz) \otimes (yz\,dx + xz\,dy + xy\,dz)$$
$$= 2[4x + 2z + 4xyz + (2x + z)^2 yz]\,dx \otimes dx + 2(2x + z)^2 xz\,dx \otimes dy$$
$$+ 2[2x + z + 2xyz + (2x + z)^2 xy]\,dx \otimes dz + [4x + 2z + 4xyz + yz(2x + z)^2]dz \otimes dx$$
$$+ xz(2x + z)^2\,dz \otimes dy + [2x + z + 2xyz + xy(2x + z)^2]dz \otimes dz.$$

D If $\varphi : M \to N$ represents the deformation of an elastic body and g is a Riemannian metric on N, then $C = \varphi^* g$ is called the ***Cauchy-Green tensor***; in coordinates

$$C_{ij} = \frac{\partial \varphi^\alpha}{\partial x^i}\frac{\partial \varphi^\beta}{\partial x^j} g_{\alpha\beta} \circ \varphi.$$

Thus, C measures how φ deforms lengths and angles. ♦

Finally, we describe an alternative approach to tensor fields. Suppose $\mathcal{F}(M)$ and $\mathfrak{X}(M)$ have been defined. With the "scalar multiplication" $(f, X) \mapsto fX$ defined in **5.2.10**, $\mathfrak{X}(M)$ becomes an $\mathcal{F}(M)$-module. That is, $\mathfrak{X}(M)$ is essentially a vector space over $\mathcal{F}(M)$, but the "scalars" $\mathcal{F}(M)$ form only a commutative ring with identity, rather than a field. Define

$$L_{\mathcal{F}(M)}(\mathfrak{X}(M), \mathcal{F}(M)) = \mathcal{X}^*(M)$$

the $\mathcal{F}(M)$-linear mappings on $\mathfrak{X}(M)$, and similarly

$$\mathbf{T}^r{}_s(M) = L^{r+s}{}_{\mathcal{F}(M)}(\mathcal{X}^*(M), ..., \mathfrak{X}(M); \mathcal{F}(M)$$

the $\mathcal{F}(M)$-multilinear mappings. From **5.2.10**, we have a natural mapping $\mathcal{T}^r{}_s(M) \to \mathbf{T}^r{}_s(M)$ which is $\mathcal{F}(M)$-linear.

5.2.20 Proposition *Let* M *be a finite-dimensional manifold or be modeled on a Banach space with norm* C^∞ *away from the origin. Then* $\mathcal{T}^r{}_s(M)$ *is isomorphic to the* $\mathcal{F}(M)$*-multilinear maps from* $\mathcal{X}^*(M) \times ... \times \mathfrak{X}(M)$ *into* $\mathcal{F}(M)$, *regarded as* $\mathcal{F}(M)$*-modules or as real vector spaces. In particular,* $\mathfrak{X}^*(M)$ *is isomorphic to* $\mathcal{X}^*(M)$.

Proof Consider the map $\mathcal{T}^r{}_s(M) \to L^{r+s}{}_{\mathcal{F}(M)}(\mathcal{X}^*(M), ..., \mathfrak{X}(M); \mathcal{F}(M))$ given by $\ell(\alpha^1, ..., \alpha^r, X_1, ..., X_s)(m) = \ell(m)(\alpha^1(m), ..., X_s(m))$. This map is clearly $\mathcal{F}(M)$-linear. To show it is an isomorphism, given such a multilinear map ℓ, define t by $t(m)(\alpha^1(m), ..., X_s(m)) = \ell(\alpha^1, ..., X_s)(m)$. To show this is well-defined we must show that, for each $v_0 \in T_m(M)$, there is an $X \in \mathfrak{X}(M)$ such that $X(m) = v_0$, and similarly for dual vectors. Let (U, φ) be a chart at m and let $T_m\varphi(v_0) = (\varphi(m), v_0')$. Define $Y \in \mathfrak{X}(U')$ by $Y(u) = (u', v_0')$ on a neighborhood V_1 of $\varphi(m)$ where $w = \varphi(n)$. Extend Y to U′ so Y is zero outside V_2, where $cl(V_1) \subset V_2$, $cl(V_2) \subset U'$, by means of a bump function. Define X by $X_\varphi = Y$ on U, and $X = 0$ outside U. Then $X(m) = v_0$. The construction is similar for dual vectors. As in Proposition **4.2.16**, $\mathcal{F}(M)$-linearity of ℓ shows that the definition of t(m) is independent of how the vectors v_0 (and corresponding dual vectors) are extended to fields. The tensor field t(m) so defined is C^∞; indeed, using the chart φ, the local representative of t is C^∞ by Supplement **3.4A**, since ℓ induces a C^∞ map $M \times T^r{}_s(M) \to \mathbb{R}$ (by the composite function theorem), which is (r + s)-linear at every $m \in M$. If M is finite dimensional this last step of the proof can be simplified as follows. In the chart φ with coordinates $(x^1, ..., x^n)$,

$$t = t^{i_1 \dots i_r}{}_{j_1 \dots j_s}\, \partial/\partial x^{i_1} \otimes ... \otimes \partial/\partial x^{i_r} \otimes dx^{j_1} \otimes ... \otimes dx^{j_s}$$

and all components of t are C^∞ by hypothesis. ∎

The preceding proposition can be clearly generalized to the C^k situation. One can also get around the use of a smooth norm on the model space if one assumes that the multilinear maps are ***localizable***, that is, are defined on $\mathfrak{X}^*(U) \times ... \times \mathfrak{X}(U)$ with values in $\mathcal{F}(U)$ for any open set U in a way compatible with restriction to U. We shall take this point of view in the next section.

The direct sum $\mathcal{T}(M)$ of the $\mathcal{T}^r{}_s(M)$, including $\mathcal{T}^0{}_0(M) = \mathcal{F}(M)$, is a real vector space with $\otimes$-product, called the ***tensor algebra*** of M, and if $\varphi : M \to M$ is a diffeomorphism, $\varphi_* : \mathcal{T}(M) \to \mathcal{T}(N)$ is an algebra isomorphism.

The construction of $\mathcal{T}(M)$ and the properties discussed in this section can be generalized to vector bundle valued (r, s)-tensors (resp. tensor fields), i.e., elements (resp. sections) of $L(T^*M \oplus \ldots \oplus T^*M \oplus TM \oplus \ldots \oplus TM, E)$, the vector bundle of vector bundle maps from $T^*M \oplus \ldots \oplus TM$ (with r factors of T^*M and s factors of TM) to the vector bundle E, which cover the identity map of the base M.

Exercises

5.2A Let $\varphi : \mathbb{R}^2 \backslash \{(0, y) \mid y \in \mathbb{R}\} \to \mathbb{R}^2 \backslash \{(x, x) \mid x \in \mathbb{R}\}$ be defined by $\varphi(x, y) = (x^3 + y, y)$ and let

$$t = x\frac{\partial}{\partial x} \otimes dx \otimes dy + y\frac{\partial}{\partial y} \otimes dy \otimes dy.$$

Show that φ is a diffeomorphism and compute φ_*t, φ^*t. Endow $\mathbb{R}^2$ with the standard Riemannian metric. Compute the associated tensors of t, φ_*t, and φ^*t as well as their (1, 1) and (1, 2) contractions. What is the trace of the interior product of t with $\partial/\partial x + x\partial/\partial y$?

5.2B Let $\varphi : \mathbb{R}^2 \to \mathbb{R}^3$, $\varphi(x, y) = (y, x, y + x^2)$ be the deformation of an elastic shell. Compute the Cauchy-Green tensor and its trace.

5.2C Let H be the set of real sequences $\{a_n\}_{n = 1,2\ldots}$ such that $\| a_n \|^2 = \Sigma_{n\geq 1} n^2 a_n^2 < \infty$. Show that H is a Hilbert space. Show that $g(a, b) = \Sigma_{n\geq 1} a_n b_n$ is a weak Riemannian metric on H that is *not* a strong metric.

5.2D Let (M, g) be a Riemannian manifold and let $N \subset M$ be a submanifold. Define $\nu_g(N) = \{v \in T_nM \mid g(n)(v, u) = 0 \text{ for all } u \in T_nN \text{ and all } n \in N\}$. Show that $\nu_g(N)$ is a sub-bundle of $TM \mid N$ isomorphic to both the normal and conormal bundles $\nu(N)$ and $\mu(N)$ defined in exercises **3.4J** and **3.4K**.

§5.3 *The Lie Derivative: Algebraic Approach*

This section extends the Lie derivative $\mathcal{L}_X$ from vector fields and functions to the full tensor algebra. We shall do so in two ways. This section does this algebraically and in the next section, it is done in terms of the flow of X. The two approaches will be shown to be equivalent.

We shall demand certain properties of $\mathcal{L}_X$ such as: if t is a tensor field of type (r, s), so is $\mathcal{L}_X t$, and $\mathcal{L}_X$ should be a derivation for tensor products and contractions. First of all, how should $\mathcal{L}_X$ be defined on covector fields? If Y is a vector field and α is a covector field, then the contraction $\alpha \cdot Y$ is a function, so $\mathcal{L}_X(\alpha \cdot Y)$ and $\mathcal{L}_X Y$ are already defined. (See §**4.2**.) However, if we would like to have the derivation property for contractions, namely

$$\mathcal{L}_X(\alpha \cdot Y) = (\mathcal{L}_X \alpha) \cdot Y + \alpha \cdot (\mathcal{L}_X Y),$$

which forces us to define $\mathcal{L}_X \alpha$ by

$$(\mathcal{L}_X \alpha) \cdot Y = \mathcal{L}_X(\alpha \cdot Y) - \alpha \cdot (\mathcal{L}_X Y) \text{ for all vector fields } Y.$$

Since this defines an $\mathcal{F}(M)$-linear map, $\mathcal{L}_X \alpha$ is a well-defined covector field. The extension to general tensors now proceeds inductively in the same spirit.

5.3.1 Definition *A **differential operator** on the full tensor algebra $\mathcal{T}(M)$ of a manifold M is a collection $\{\mathbf{D}^r{}_s(U)\}$ of maps of $\mathcal{T}^r{}_s(U)$ into itself for each r and $s \geq 0$ and each open set $U \subset M$, any of which we denote merely $\mathbf{D}$ (the r, s and U are to be inferred from the context), such that*

D01 $\mathbf{D}$ *is a **tensor derivation**,* or $\mathbf{D}$ ***commutes with contractions**, i.e.,* $\mathbf{D}$ *is $\mathbb{R}$-linear and if*

$$t \in \mathcal{T}^r{}_s(M),\ \alpha_1, ..., \alpha_r \in \mathfrak{X}^*(M),\ \textit{and}\ X_1, ..., X_s \in \mathfrak{X}(M),$$

then

$$\mathbf{D}(t(\alpha_1, ..., \alpha_r, X_1, ..., X_s)) = (\mathbf{D}t)(\alpha_1, ..., \alpha_r, X_1, ..., X_s)$$

$$+ \sum_{j=1}^{r} t(\alpha_1, ..., \mathbf{D}\alpha_j, ..., \alpha_r, X_1, ..., X_s) + \sum_{k=1}^{s} t(\alpha_1, ..., \alpha_r, X_1, ..., \mathbf{D}X_k, ..., X_s).$$

D02 $\mathbf{D}$ *is **local**, or is **natural with respect to restrictions**. That is, for $U \subset V \subset M$ open sets, and $t \in \mathcal{T}^r{}_s(V)$*

$$(\mathfrak{D}\, t) \mid U = \mathfrak{D}\,(t \mid U) \in \mathcal{T}^r_s(U)$$

or the following diagram commutes

$$\begin{array}{ccc}
\mathcal{T}^r_s(V) & \xrightarrow{\ \mid U\ } & \mathcal{T}^r_s(U) \\
\Big\downarrow \mathcal{D} & & \Big\downarrow \mathcal{D} \\
\mathcal{T}^r_s(V) & \xrightarrow[\ \mid U\ ]{} & \mathcal{T}^r_s(U)
\end{array}$$

Note that we do *not* demand that $\mathfrak{D}$ be natural with respect to push-forward by diffeomorphisms. Indeed, several important differential operators, such as the covariant derivative, are not natural with respect to diffeomorphisms, although the Lie derivative is, as we shall see.

5.3.2 Theorem *Suppose for each open set* $U \subset M$ *we have maps* $\mathfrak{E}_U\colon \mathcal{F}(U) \to \mathcal{F}(U)$ *and* $\mathfrak{F}_U\colon \mathfrak{X}(U) \to \mathfrak{X}(U)$, *which are* ($\mathbb{R}$*-linear) tensor derivations and natural with respect to restrictions. That is*

(i) $\mathfrak{E}_U(f \otimes g) = (\mathfrak{E}_U f) \otimes g + f \otimes \mathfrak{E}_U g$ for $f, g \in \mathcal{F}(U)$;

(ii) *for* $f \in \mathcal{F}(M)$, $\mathfrak{E}_U(f \mid U) = (\mathfrak{E}_M f) \mid U$;

(iii) $\mathfrak{F}_U(f \otimes X) = (\mathfrak{E}_U f) \otimes X + f \otimes \mathfrak{F}_U X$ *for* $f \in \mathcal{F}(U)$, *and* $X \in \mathfrak{X}(U)$;

(iv) *for* $X \in \mathfrak{X}(M)$, $\mathfrak{F}_U f\,(X \mid U) = (\mathfrak{F}_M X) \mid U$.

Then there is a unique differential operator $\mathfrak{D}$ *on* $\mathcal{T}(M)$ *that coincides with* $\mathfrak{E}_U$ *on* $\mathcal{F}(U)$ *and with* $\mathfrak{F}_U$ *on* $\mathfrak{X}(U)$.

Proof Since $\mathfrak{D}$ must be a tensor derivation, define $\mathfrak{D}$ on $\mathfrak{X}^*(U)$ by the formula $(\mathfrak{D}\alpha) \cdot X = \mathfrak{D}(\alpha \cdot X) - \alpha \cdot (\mathfrak{D}X) = \mathfrak{E}_U(\alpha \cdot X) - \alpha \cdot \mathfrak{F}_U X$ for all $X \in \mathfrak{X}(U)$. By properties (i) and (iii), $\mathfrak{D}\alpha$ is $\mathcal{F}(M)$-linear and thus by the remark following **5.2.20**, $\mathfrak{D}$ so defined on $\mathfrak{X}^*(U)$ has values in $\mathfrak{X}^*(U)$. Note also that $\mathfrak{D}(f \otimes \alpha) = (\mathfrak{E}f) \otimes \alpha + f \otimes (\mathfrak{D}\alpha)$ for any $\alpha \in \mathfrak{X}^*(U)$, $f \in \mathcal{F}(U)$. This shows that $\mathfrak{D}$ exists and is unique on $\mathfrak{X}^*(U)$ (by the Hahn-Banach theorem). Define $\mathfrak{D}_U$ on $\mathcal{T}^r_s(U)$ by requiring **D01** to hold:

$$(\mathfrak{D}_U t)(\alpha_1, \ldots, \alpha_r, X_1, \ldots, X_s) = \mathfrak{E}_U(t(\alpha_1, \ldots, \alpha_r, X_1, \ldots, X_s))$$
$$-\sum_{j=1}^{r} t(\alpha_1, \ldots, \mathfrak{D}\alpha_j, \ldots, \alpha_r, X_1, \ldots, X_s) - \sum_{k=1}^{s} t(\alpha_1, \ldots, \alpha_r, X_1, \ldots, \mathfrak{F}_U X_k, \ldots, X_s).$$

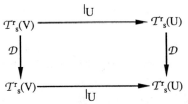

From (i), (iii), and **D01** for $\mathfrak{D}_U$ on $\mathfrak{X}^*(U)$, it follows that $\mathfrak{D}_U t$ is an $\mathcal{F}(M)$-multilinear map, i.e., that $\mathfrak{D}_U t \in \mathcal{T}^r_s(U)$ (see the comment following **5.2.20**). The definition of $\mathfrak{D}_U$ on $\mathcal{T}^r_s(U)$ uniquely determines $\mathfrak{D}_U$ from the property **D01**. Finally, if V is any open subset of U, by (ii) and (iv) it follows that $\mathfrak{D}_V(t \mid V) = (\mathfrak{D}_U t) \mid V$. This enables us to define $\mathfrak{D}$ on $\mathcal{F}(M)$ by $(\mathfrak{D}t)(m) = (\mathfrak{D}_U t)(m)$, where U is any open subset of M containing m. Since $\mathfrak{D}_U$ is unique, so is $\mathfrak{D}$, and so **DO2** is satisfied by the construction of $\mathcal{D}$. ∎

5.3.3 Corollary *We have*
 (i) $\mathcal{D}(t_1 \otimes t_2) = \mathcal{D}t_1 \otimes t_2 + t_1 \otimes \mathcal{D}t_2$, *and*
 (ii) $\mathcal{D}\delta = 0$, *where δ is Kronecker's delta.*

Proof (i) is a direct application of **DO1**. For (ii) let $\alpha \in \mathcal{X}^*(U)$ and $X \in \mathcal{X}(U)$ where U is an arbitrary chart domain. Then

$$(\mathcal{D}\delta)(\alpha, X) = \mathcal{D}(\delta(\alpha, X)) - \delta(\mathcal{D}\alpha, X) - \delta(\alpha, \mathcal{D}X)$$
$$= \mathcal{D}(\alpha \cdot X) - \mathcal{D}\alpha \cdot X - \alpha \cdot \mathcal{D}X = 0.$$

Again the Hahn-Banach theorem assures that $\mathcal{D}\delta = 0$ on U, and thus by **DO2**, $\mathcal{D}\delta = 0$. ∎

Taking $\mathcal{E}_U$ and $\mathcal{F}_U$ to be $L_{X|U}$ we see that the hypotheses of theorem **5.3.2** are satisfied. Hence we can define a differential operator as follows.

5.3.4 Definition *If $X \in \mathcal{X}(M)$, we let L_X be the unique differential operator on $\mathcal{T}(M)$, called the **Lie derivative with respect to** X, such that L_X coincides with L_X as given on $\mathcal{F}(M)$ and $\mathcal{X}(M)$ (see **4.2.6** and **4.2.20**).*

5.3.5 Proposition *Let $\varphi : M \to N$ be a diffeomorphism and X a vector field on M. Then L_X is **natural with respect to push-forward by** φ; that is,*

$$L_{\varphi_* X} \varphi_* t = \varphi_* L_X t \text{ for } t \in \mathcal{T}^r_s(M),$$

or the following diagram commutes:

$$\begin{array}{ccc} \mathcal{T}^r_s(M) & \xrightarrow{\varphi_*} & \mathcal{T}^r_s(N) \\ {\scriptstyle L_X}\downarrow & & \downarrow{\scriptstyle L_{\varphi_* X}} \\ \mathcal{T}^r_s(M) & \xrightarrow{\varphi_*} & \mathcal{T}^r_s(N) \end{array}$$

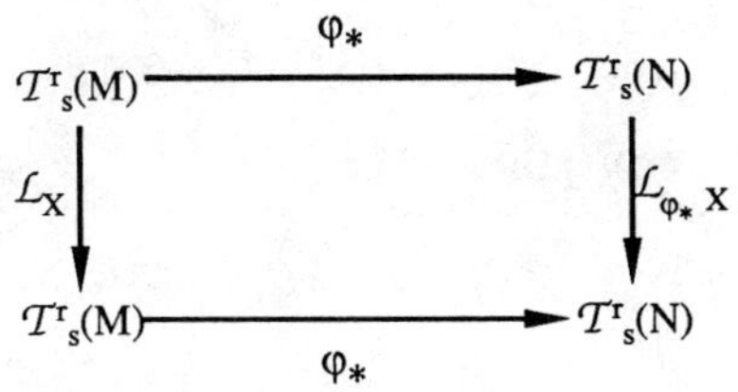

Proof For an open set $U \subset M$ define

$$\mathcal{D} : \mathcal{T}^r_s(U) \to \mathcal{T}^r_s(U) \text{ by } \mathcal{D}t = \varphi^* L_{\varphi_* X | U}(\varphi_* f),$$

where we use the same symbol φ for $\varphi | U$. By naturality on $\mathcal{F}(U)$ and $\mathcal{X}(U)$, $\mathcal{D}$ coincides

with $\mathbf{L}_{X|U}$ on $\mathcal{F}(U)$ and $\mathfrak{X}(U)$. Next, we show that $\mathfrak{D}$ is a differential operator. For **D01**, we use the fact that

$$\varphi_*(t(\alpha_1, ..., \alpha_r, X_1, ..., X_s)) = (\varphi_* t)(\varphi_*\alpha_1, ..., \varphi_*\alpha_r, \varphi_* X_1, ..., \varphi_* X_s),$$

which follows from the definitions. Then for $X, X_1, ..., X_s \in \mathfrak{X}(U)$ and $\alpha_1, ..., \alpha_r \in \mathfrak{X}^*(U)$,

$$\begin{aligned}
\mathfrak{D}(t(\alpha_1, ..., \alpha_r, X_1, ..., X_s)) &= \varphi^* \mathbf{L}_{\varphi_* X}(\varphi_*(t(\alpha_1, ..., \alpha_r, X_1, ..., X_s))) \\
&= \varphi^* \mathbf{L}_{\varphi_* X}((\varphi_* t)(\varphi_*\alpha_1, ..., \varphi_*\alpha_r, \varphi_* X_1, ..., \varphi_* X_s)) \\
&= \varphi^* [(\mathbf{L}_{\varphi_* X}\varphi_* t)(\varphi_*\alpha_1, ..., \varphi_*\alpha_r, \varphi_* X_1, ..., \varphi_* X_s)) \\
&\quad + \sum_{j=1}^{r} (\varphi_* t)(\varphi_*\alpha_1, ..., \mathbf{L}_{\varphi_* X}\, \varphi_*\alpha_j, ..., \varphi_*\alpha_r, \varphi_* X_1, ..., \varphi_* X_s) \\
&\quad + \sum_{k=1}^{s} (\varphi_* t)(\varphi_*\alpha_1, ..., \varphi_*\alpha_r, \varphi_* X_1, ..., \mathbf{L}_{\varphi_* X}\varphi_* X_k, ..., \varphi_* X_s)],
\end{aligned}$$

by **D01** for $\mathbf{L}_X$. Since $\varphi^* = (\varphi^{-1})_*$ by **5.2.14**, this becomes

$$(\mathfrak{D} t)(\alpha_1, ..., \alpha_r, X_1, ..., X_s) + \sum_{j=1}^{r} (\alpha_1, ..., \mathfrak{D}\alpha_j, ..., \alpha_r, X_1, ..., X_s) + \sum_{k=1}^{s} t(\alpha_1, ..., \alpha_r, X_1, ..., \mathfrak{D} X_k, ..., X_s).$$

For **D02**, let $t \in \mathcal{T}^r_s(M)$ and write

$$\begin{aligned}
\mathfrak{D}t \mid U &= [(\varphi_*)^{-1}\, \mathbf{L}_{\varphi_* X}\varphi_* t] \mid U = (\varphi_*)^{-1}[\mathbf{L}_{\varphi_* X}\varphi_* t] \mid U \\
&= (\varphi_*)^{-1}\, \mathbf{L}_{\varphi_* X|U}\, \varphi_* t \mid U \quad (\text{by } \mathbf{D02} \text{ for } \mathbf{L}_X) \\
&= \mathfrak{D}(t \mid U)\, .
\end{aligned}$$

The result now follows by **5.3.2.** ■

Using the same reasoning, a differential operator that is natural with respect to diffeomorphisms on functions and vector fields is natural on all tensors.

Let us now compute the local formula for $\mathbf{L}_X t$ where t is a tensor field of type (r, s). Let $\varphi: U \subset M \to V \subset E$ be a local chart and let X' and t' be the principal parts of the local representatives, $\varphi_* X$ and $\varphi_* t$ respectively. Thus $X': V \to E$ and $t': V \to T^r_s(E)$. Recall from §**4.2** that the local formulas for the Lie derivatives of functions and vector fields are:

$$(\mathbf{L}_X f)'(x) = \mathbf{D}f(x) \cdot X'(x) \tag{1}$$

where f' is the local representative of f and

$$(\mathbf{L}_X Y)'(x) = \mathbf{D}Y'(x) \cdot X'(x) - \mathbf{D}X'(x) \cdot Y'(x). \tag{2}$$

In finite dimensions these become

$$\mathbf{L}_X f = X^i \frac{\partial f}{\partial x^i} \tag{1'}$$

and

$$[X, Y]^i = X^j \frac{\partial Y^i}{\partial x^j} - Y^j \frac{\partial X^i}{\partial x^j}. \tag{2'}$$

Let us first find the local expression for $\mathbf{L}_X \alpha$ where α is a one-form. By **5.3.5**, the local representative of $\mathbf{L}_X \alpha$ is

$$\varphi_*(\mathbf{L}_X \alpha) = \mathbf{L}_{\varphi_* X} \varphi_* \alpha,$$

which we write as $\mathbf{L}_{X'} \alpha'$ where X' and α' are the principal parts of the local representatives, so $X' : V \to E$ and $\alpha' : V \to E^*$. Let $v \in E$ be fixed and regarded as a constant vector field. Then as $\mathbf{L}_X$ is a tensor derivation,

$$\mathbf{L}_{X'}(\alpha' \cdot v) = (\mathbf{L}_{X'} \alpha') \cdot v + \alpha'(\mathbf{L}_{X'} v).$$

By (1) and (2) this becomes

$$\mathbf{D}(\alpha' \cdot v) \cdot X' = (\mathbf{L}_{X'} \alpha') \cdot v - \alpha' \cdot (\mathbf{D}X' \cdot v).$$

Thus

$$(\mathbf{L}_{X'} \alpha') \cdot v = (\mathbf{D}\alpha' \cdot X') \cdot v + \alpha' \cdot (\mathbf{D}X' \cdot v).$$

In the expression $(\mathbf{D}\alpha' \cdot X') \cdot v$, $\mathbf{D}\alpha' \cdot X'$ means the derivative of α' in the direction X'; the resulting element of E^* is then applied to v. Thus we can write

$$\mathbf{L}_{X'} \alpha' = \mathbf{D}\alpha' \cdot X' + \alpha' \cdot \mathbf{D}X'. \tag{3}$$

In finite dimensions, the corresponding coordinate expression is

$$(\mathbf{L}_X \alpha)_i v^i = \frac{\partial \alpha^i}{\partial x^j} X^j v^i + \alpha_j \frac{\partial X^j}{\partial x^i} v^i;$$

i.e.,

$$(\mathbf{L}_X \alpha)_i = X^j \frac{\partial \alpha_i}{\partial x^j} + \alpha_j \frac{\partial X^j}{\partial x^i}. \tag{3'}$$

Now let t be of type (r, s), so $t' : V \to L(E^*, \ldots, E^*, E, \ldots, E; \mathbb{R})$. Let $\alpha^1, \ldots, \alpha^r$ be (constant) elements of E^* and $v_1, \ldots, v_s$ (constant) elements of E. Then again by the derivation property,

$$\mathbf{L}_{X'}[t'(\alpha^1, ..., \alpha^r, v_1, ..., v_s)] = (\mathbf{L}_{X'} t')(\alpha^1, ..., \alpha^r, v_1, ..., v_s)$$

$$+\sum_{i=1}^{r} t'(\alpha^1, ..., \mathbf{L}_{X'}\alpha^i, ..., \alpha^r, v_1, ..., v_s) + \sum_{j=1}^{s} t'(\alpha^1, ..., \alpha^r, v_1, ..., \mathbf{L}_{X'}v_j, ..., v_s) .$$

Now using the local formula (1)-(3) for the Lie derivatives of functions, vector fields, and one-forms, we get

$$(\mathbf{D}t' \cdot X') \cdot (\alpha^1, ..., \alpha^r, v_1, ..., v_s) = (\mathbf{L}_{X'}t')(\alpha^1, ..., \alpha^r, v_1, ..., v_s)$$

$$+\sum_{i=1}^{r} t'(\alpha^1, ..., \alpha^i \cdot \mathbf{D}X', ..., \alpha^r, v_1, ..., v_s) + \sum_{j=1}^{s} t'(\alpha^1, ..., \alpha^r, v_1, ..., -\mathbf{D}X' \cdot v_j, ..., v_s).$$

Therefore,

$$(\mathbf{L}_{X'}t')(\alpha^1, ..., \alpha^r, v_1, ..., v_s) = (\mathbf{D}t' \cdot X')(\alpha^1, ..., \alpha^r, v_1, ..., v_s)$$

$$-\sum_{i=1}^{r} t'(\alpha^1, ..., \alpha^i \cdot \mathbf{D}X', ..., \alpha^r, v_1, ..., v_s) + \sum_{j=1}^{s} t'(\alpha^1, ..., \alpha^r, v_1, ..., \mathbf{D}X' \cdot v_j, ..., v_s).$$

In components, this reads

$$(\mathbf{L}_X t)^{i_1 \dots i_r}{}_{j_1 \dots j_s} = X^k \frac{\partial}{\partial x^k} t^{i_1 \dots i_r}{}_{j_s \dots j_s} - \frac{\partial X^{i_1}}{\partial x^\ell} t^{\ell i_2 \dots i_r}{}_{j_1 \dots j_s} - \text{(all upper indices)}$$

$$+ \frac{\partial X^m}{\partial x^{j_1}} t^{i_1 \dots i_r}{}_{m j_2 \dots j_s} + \text{(all lower indices)} \qquad (4)$$

We deduced the component formulas for $\mathbf{L}_X t$ in the case of a finite-dimensional manifold as corollaries of the general Banach manifold formulas. Because of their importance, we shall deduce them again in a different manner, without appealing to **5.3.5**. Let

$$t = t^{i_1 \dots i_r}{}_{j_1 \dots j_s} \frac{\partial}{\partial x^{i_1}} \otimes \dots \otimes \frac{\partial}{\partial x^{i_r}} \otimes dx^{j_1} \dots \otimes dx^{j_s} \in \mathcal{T}^r{}_s(U),$$

where U is a chart domain on M. If $X = X^k \partial/\partial x^k$, the tensor derivation property can be used to compute $\mathbf{L}_X t$. For this we recall that

$$\mathbf{L}_X(t^{i_1\ldots i_r}{}_{j_1\ldots j_s}) = X^k \frac{\partial t^{i_1\ldots i_r}{}_{j_1\ldots j_s}}{\partial x^k}$$

and that

$$\mathbf{L}_X \frac{\partial}{\partial x^k} = \left[X, \frac{\partial}{\partial x^k}\right] = -\frac{\partial X^i}{\partial x^k}\frac{\partial}{\partial x^i}$$

by the general formula for the bracket components. The formula for $\mathbf{L}_X(dx^k)$ is found in the following way. The relation $\delta^k{}_i = dx^k(\partial/\partial x^i)$ implies by **D01** that

$$0 = \mathbf{L}_X\left(dx^k\left(\frac{\partial}{\partial x^i}\right)\right) = (\mathbf{L}_X(dx^k))\left(\frac{\partial}{\partial x^i}\right) + dx^k\left(\left[X, \frac{\partial}{\partial x^i}\right]\right)$$

$$= (\mathbf{L}_X(dx^k))\left(\frac{\partial}{\partial x^i}\right) + dx^k\left(-\frac{\partial X^\ell}{\partial x^i}\frac{\partial}{\partial x^\ell}\right).$$

Thus

$$(\mathbf{L}_X(dx^k))\left(\frac{\partial}{\partial x^i}\right) = dx^k\left(\frac{\partial X^\ell}{\partial x^i}\frac{\partial}{\partial x^\ell}\right) = \frac{\partial X^k}{\partial x^i},$$

so

$$\mathbf{L}_X(dx^k) = \left(\frac{\partial X^k}{\partial x^i}\right)dx^i.$$

Now one simply applies **D01** and collects terms to get the same local formula for $\mathbf{L}_X t$ found in (4). Note especially that

$$\mathbf{L}_{\frac{\partial}{\partial x^i}}\left(\frac{\partial}{\partial x^j}\right) = 0 \text{ and } \mathbf{L}_{\frac{\partial}{\partial x^i}}(dx^j) = 0, \text{ for all } i, j.$$

5.3.6 Examples

A Compute $\mathbf{L}_X t$, where

$$t = x\frac{\partial}{\partial y} \otimes dx \otimes dy + y\frac{\partial}{\partial y} \otimes dy \otimes dy \text{ and } X = \frac{\partial}{\partial x} + x\frac{\partial}{\partial y}.$$

Solution. *Method* 1 Note that

(i) $\mathbf{L}_X t = \mathbf{L}_{\partial/\partial x + x\partial/\partial y}\, t = \mathbf{L}_{\partial/\partial x} t + \mathbf{L}_{\partial/\partial y} t$

(ii) $$\mathbf{L}_{\partial/\partial x}\, t = \mathbf{L}_{\partial/\partial x}\left\{x\frac{\partial}{\partial y} \otimes dx \otimes dy + y\frac{\partial}{\partial y} \otimes dy \otimes dy\right\}$$

$$= \mathbf{L}_{\partial/\partial x}\left(x\frac{\partial}{\partial y} \otimes dx \otimes dy\right) + \mathbf{L}_{\partial/\partial x}\left(y\frac{\partial}{\partial y} \otimes dy \otimes dy\right) = \frac{\partial}{\partial y} \otimes dx \otimes dy + 0\,.$$

Now note

$$\mathcal{L}_{x\partial/\partial y}\frac{\partial}{\partial y}=0,\ \mathcal{L}_{x\partial/\partial y}\frac{\partial}{\partial x}=-\mathcal{L}_{\partial/\partial x}\left(x\frac{\partial}{\partial y}\right)=-\left\{1\cdot\frac{\partial}{\partial y}+x\cdot 0\right\}=-\frac{\partial}{\partial y},$$

$$\mathcal{L}_{x\partial/\partial y}\,dx=0,\quad\text{and}\quad \mathcal{L}_{x\partial/\partial y}\,dy=dx.$$

thus

$$\text{(iii)}\ \ \mathcal{L}_{x\partial/\partial y}\,t=\mathcal{L}_{x\partial/\partial y}\left\{x\frac{\partial}{\partial y}\otimes dx\otimes dy+y\frac{\partial}{\partial y}\otimes dy\otimes dy\right\}$$
$$=\left(0+0+0+x\frac{\partial}{\partial y}\otimes dx\otimes dx\right)$$
$$+\left(x\frac{\partial}{\partial y}\otimes dy\otimes dy+0+y\frac{\partial}{\partial y}\otimes dx\otimes dy+y\frac{\partial}{\partial y}\otimes dy\otimes dx\right).$$

Thus, substituting (ii) and (iii) into (i), we find

$$\mathcal{L}_X t=\frac{\partial}{\partial y}\otimes dx\otimes dy+x\frac{\partial}{\partial y}\otimes dx\otimes dx+x\frac{\partial}{\partial y}\otimes dy\otimes dy+y\frac{\partial}{\partial y}\otimes dx\otimes dy+y\frac{\partial}{\partial y}\otimes dy\otimes c$$
$$=(y+1)\frac{\partial}{\partial y}\otimes dx\otimes dy+x\frac{\partial}{\partial y}\otimes dx\otimes dx+x\frac{\partial}{\partial y}\otimes dy\otimes dy+y\frac{\partial}{\partial y}\otimes dy\otimes dx.$$

Method 2 Using component notation, t is a tensor of type (1, 2) whose nonzero components are $t^2{}_{12}=x$ and $t^2{}_{22}=y$. The components of X are $X^1=1$ and $X^2=x$. Thus, by the component formula (4),

$$(\mathcal{L}_X t)^i{}_{jk}=X^k\frac{\partial}{\partial x^k}t^i{}_{jk}-t^\ell{}_{jk}\frac{\partial X^i}{\partial x^\ell}+t^i{}_{mk}\frac{\partial X^m}{\partial x^j}+t^i{}_{jp}\frac{\partial X^p}{\partial x^k}.$$

The nonzero components are

$$(\mathcal{L}_X t)^2{}_{12}=1-0+y+0=1+y;\qquad (\mathcal{L}_X t)^2{}_{22}=x-0+0+0=x;$$
$$(\mathcal{L}_X t)^2{}_{11}=0-0+0+x=x;\qquad (\mathcal{L}_X t)^2{}_{21}=0-0+0+y=y,$$

and hence

$$\mathcal{L}_X t=(y+1)\frac{\partial}{\partial y}\otimes dx\otimes dy+x\frac{\partial}{\partial y}\otimes dx\otimes dx+x\frac{\partial}{\partial y}\otimes dy\otimes dy+y\frac{\partial}{\partial y}\otimes dy\otimes dx.$$

The two methods thus give the same answer. It is useful to understand both methods since they both occur in the literature, and depending on the circumstances, one may be easier to apply than the other.

B In Riemannin geometry, vector fields X satisfying $\mathcal{L}_X g=0$ are called ***Killing vector fields***; their geometric significance will become clear in the next section. For now, let us compute the system of equations that the components of a Killing vector field must satisfy. If $X=X^i\partial/\partial x^i$, and $g=g_{ij}dx^i\otimes dx^j$, then

$$\begin{aligned}\mathcal{L}_X g &= (\mathcal{L}_X g_{ij})\, dx^i \otimes dx^j + g_{ij}(\mathcal{L}_X dx^i) \otimes dx^j + g_{ij} dx^i \otimes (\mathcal{L}_X dx^j) \\ &= X^k \frac{\partial g_{ij}}{\partial x^k}\, dx^i \otimes dx^j + g_{ij} \frac{\partial X^i}{\partial x^k}\, dx^k \otimes dx^j + g_{ij} dx^i \otimes \frac{\partial X^j}{\partial x^k}\, dx^k \\ &= \left\{ X^k \frac{\partial g_{ij}}{\partial x^k} + g_{kj} \frac{\partial X^k}{\partial x^i} + g_{ik} \frac{\partial X^k}{\partial x^j} \right\} dx^i \otimes dx^j.\end{aligned}$$

Note that $\mathcal{L}_X g$ is still a symmetric (0, 2)-tensor, as it must be. Hence X is a Killing vector field iff its components satisfy the following system of n partial differential equations, called ***Killing's equations***

$$X^k \frac{\partial g_{ij}}{\partial x^k} + g_{kj} \frac{\partial X^k}{\partial x^i} + g_{ik} \frac{\partial X^k}{\partial x^j} = 0.$$

C In the theory of elasticity, if u represents the ***displacement vector field***, the expression $\mathcal{L}_u g$ is called the ***strain tensor***. As we shall see in the next section, this is related to the Cauchy-Green tensor $C = \varphi^* g$ by linearization of the deformation φ.

D Let us show that $\mathcal{L}_X$ does not necessarily commute with the formation of associated tensors; e.g., that $(\mathcal{L}_X t)_{ij} \neq (\mathcal{L}_X \tau)_{ij}$, where $t = t^i{}_j \partial/\partial x^i \otimes dx^j \in T^1{}_1(M)$ and $\tau = t_{ij} dx^i \otimes dx^j \in T^0{}_2(M)$ is the associated tensor with components $t_{ij} = g_{ij} t^k{}_j$. We have from (4)

$$(\mathcal{L}_X t)^i_j = X^k \frac{\partial t^i{}_j}{\partial k^k} - t^k{}_j \frac{\partial X^i}{\partial x^k} + t^i{}_k \frac{\partial X^k}{\partial x^j},$$

and so

$$(\mathcal{L}_X t)_{ij} = g_{i\ell} \left(X^k \frac{\partial t^\ell{}_j}{\partial x^k} - t^k{}_j \frac{\partial X^\ell}{\partial x^k} + t^\ell{}_k \frac{\partial X^k}{\partial x^j} \right).$$

But also from (4)

$$\begin{aligned}(\mathcal{L}_X \tau)_{ij} &= X^k \frac{\partial t_{ij}}{\partial x^k} + t_{\ell j} \frac{\partial X^\ell}{\partial x^i} + t_{ik} \frac{\partial X^k}{\partial x^j} \\ &= X^k \frac{\partial}{\partial x^k}(g_{i\ell} t^\ell{}_j) + g_{\ell k} t^k{}_j \frac{\partial X^\ell}{\partial x^i} + g_{i\ell} t^\ell{}_k \frac{\partial X^k}{\partial x^j} \\ &= X^k \frac{\partial g_{i\ell}}{\partial x^k} t^\ell{}_j + X^k g_{i\ell} \frac{\partial t^\ell{}_j}{\partial x^k} + g_{\ell k} t^k{}_j \frac{\partial X^\ell}{\partial x^i} + g_{i\ell} t^\ell{}_k \frac{\partial X^k}{\partial x^j}.\end{aligned}$$

Thus, to have equality it is necessary and sufficient that

$$X^k \frac{\partial g_{i\ell}}{\partial x^k} t^\ell{}_j + g_{\ell k} t^k{}_j \frac{\partial X^\ell}{\partial x^i} + g_{i\ell} t^k{}_j \frac{\partial X^\ell}{\partial x^k} = 0$$

for all pairs of indices (i, j), which is a nontrivial system of n^2 linear partial differential equations for g_{ij}. If X is a Killing vector field, then

$$g_{\ell k}\frac{\partial X^\ell}{\partial x^i}+g_{i\ell}\frac{\partial X^\ell}{\partial x^k}=-X^\ell\frac{\partial g_{ik}}{\partial x^\ell},$$

which substituted in the preceding equation, gives zero. The converse statement is proved along the same lines. In other words, *a necessary and sufficient condition that $\mathcal{L}_X$ commute with the formation of associated tensors is that* X *be a Killing vector field for the pseudo-Riemannian metric* g. ♦

As usual, the development of $\mathcal{L}_X$ extends from tensor fields to F-valued tensor fields.

Exercises

5.3A Let $t = xy\partial/\partial x \otimes dx + y\partial/\partial y \otimes dx + \partial/\partial x \otimes dy \in T^1{}_1(\mathbb{R}^2)$. Define the map φ as follows: $\varphi : \{(x, y) \mid y > 0\} \to \{(x, y) \mid x > 0, x^2 < y\}$ and $\varphi(x, y) = (ye^x, y^2e^{2x} + y)$. Show that φ is a diffeomorphism and compute trace(t), φ^*t, φ_*t, $\mathcal{L}_Xt$, $\mathcal{L}_X\varphi^*t$, and $\mathcal{L}_{\varphi_*X}t$, for $X = y\partial/\partial x + x^2\partial/\partial y$.

5.3B Verify explicitly that $\mathcal{L}_X(t^\flat) \neq (\mathcal{L}_Xt)^\flat$ where $^\flat$ denotes the associated tensor with both indices lowered, for X and t in Exercise **5.3A**.

5.3C Compute the coordinate expressions for the Killing equations in $\mathbb{R}^3$ in rectangular, cylindrical, and spherical coordinates. What are the Killing vector fields in $\mathbb{R}^n$?

5.3D Let (M, g) be a finite dimensional pseudo-Riemannian manifold, and $g^\#$ the tensor g with both indices raised. Let $X \in \mathfrak{X}(M)$. Calculate $(\mathcal{L}_Xg^\#)^\flat - \mathcal{L}_Xg$ in coordinates.

5.3E If (M, g) is a finite-dimensional pseudo-Riemannian manifold and $f \in \mathcal{F}(M)$, $X \in \mathfrak{X}(M)$, calculate $\mathcal{L}_X(\nabla f) - \nabla(\mathcal{L}_Xf)$.

5.3F ***Nijenhuis tensor*** (i) Let $t \in \mathcal{T}^1{}_1(M)$. Show that there is a unique tensor field $N_t \in \mathcal{T}^1{}_2(M)$, skew-symmetric in its covariant indices, such that

$$\mathcal{L}_{t\cdot X}t - t\cdot\mathcal{L}_Xt = N_t \cdot X$$

for all $X \in \mathfrak{X}(M)$, where the dots mean contractions, i.e., $(t \cdot X)^i = t^i{}_jX^j$, $(t \cdot s)^i{}_j = t^i{}_ks^k{}_j$, where $t, s \in \mathcal{T}^1{}_1(M)$, and $N_t \cdot X = N^i{}_{jk}X^k$, where $N_t = N^i{}_{jk}\,\partial/\partial x^i \otimes dx^j \otimes dx^k$. N_t is called the ***Nijenhuis tensor***. Generalize to the infinite dimensional case. (*Hint*: Show that

$N^i{}_{jk} = t^\ell{}_k t^i{}_{j,\ell} - t^\ell{}_j t^i{}_{k,\ell} + t^i{}_\ell t^\ell{}_{k,j} - t^i{}_\ell t^\ell{}_{j,k}$.)

(ii) Show that $N_t = 0$ iff

$$[t \cdot X, t \cdot Y] - t \cdot [t \cdot X, Y] = t \cdot [X, t \cdot Y] - t^2 \cdot [X, Y]$$

for all $X, Y \in \mathfrak{X}(M)$, where $t^2 \in \mathcal{T}^1{}_1(M)$ is the tensor field obtained by the composition $t \circ t$, when t is thought of as a map $t : \mathfrak{X}(M) \to \mathfrak{X}(M)$.

§5.4 *The Lie Derivative: Dynamic Approach*

We now turn to the dynamic interpretation of the Lie derivative. In §**4.2** it was shown that $\mathcal{L}_X$ acting on an element of $\mathcal{F}(M)$ or $\mathfrak{X}(M)$, respectively, is the time derivative at zero of that element of $\mathcal{F}(M)$ or $\mathfrak{X}(M)$ Lie dragged along by the flow of X. The same situation holds for general tensor fields. Given $t \in \mathcal{T}^r_s(M)$ and $X \in \mathfrak{X}(M)$, we get a curve through t(m) in the fiber over m by using the flow of X. The derivative of this curve is the Lie derivative.

5.4.1 Lie Derivative Theorem *Let* $X \in \mathfrak{X}^k(M)$, $t \in \mathcal{T}^r_s(M)$ *be of class* C^k, *and* F_λ *be the flow of* X. *Then on the domain of the flow (see Fig.* 5.4.1*) we have*

$$\boxed{\frac{d}{d\lambda} F_\lambda^* t = F_\lambda^* \mathcal{L}_X t}$$

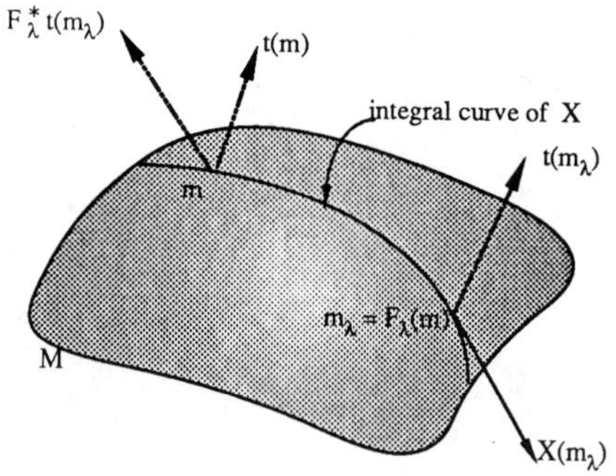

Fig 5.4.1

Proof It suffices to show that

$$\left.\frac{d}{d\lambda}\right|_{\lambda=0} F_\lambda^* t = \mathcal{L}_X t .$$

Indeed, if this is proved then

$$\frac{d}{d\lambda} F_\lambda^* t = \left.\frac{d}{d\mu}\right|_{\mu=0} F_{\mu+\lambda}^* t = \left.\frac{d}{d\mu}\right|_{\mu=0} F_\lambda^* F_\mu^* t = F_\lambda^* \mathcal{L}_X t .$$

Define $\theta_X : \mathcal{T}(M) \to \mathcal{T}(M)$ by

$$\theta_X(t)(m) = \left.\frac{d}{d\lambda}\right|_{\lambda=0} (F_\lambda^* t)(m) .$$

Note that $\theta_X(t)$ is a smooth tensor field of the same type as t, by smoothness of t and F_λ. (We suppress the notational clutter of restricting to the domain of the flow.) Let us apply Theorem **5.3.2.** Clearly θ_X is $\mathbb{R}$-linear and is natural with respect to restrictions. It is a tensor derivation from the product rule for derivatives and the relation

$$(\varphi^*t)(\varphi^*\alpha^1, ..., \varphi^*\alpha^r, \varphi^*X_1, ..., \varphi^*X_s) = \varphi^*(t(\alpha^1, ..., \alpha^r, X_1, ..., X_s))$$

for φ a diffeomorphism. Hence θ_X is a differential operator. It remains to show that θ_X coincides with $\mathcal{L}_X$ on $\mathcal{F}(M)$ and $\mathfrak{X}(M)$. For $f \in \mathcal{F}(M)$, and $X \in \mathfrak{X}(M)$, we have

$$\theta_X f = \frac{d}{d\lambda}\bigg|_{\lambda=0} F_\lambda^* f = \mathcal{L}_X f \quad \text{and} \quad \theta_X Y = \frac{d}{d\lambda}\bigg|_{\lambda=0} F_\lambda^* Y = [X, Y]$$

by **4.2.10** and **4.2.19**, respectively. Thus by **5.3.2** and **5.3.4**, $\theta_X t = \mathcal{L}_X t$ for all $t \in \mathcal{T}(M)$. ■

This theorem can also be verified in finite dimensions by a straightforward coordinate computation. See Exercise **5.4A.**

The identity in this theorem relating flows and Lie derivatives is so basic, some authors like to take it as the *definition* of the Lie derivative (see Exercise **5.4C**).

5.4.2 Corollary *If* $t \in \mathcal{T}(M)$, $\mathcal{L}_X t = 0$ *iff* t *is constant along the flow of* X. *That is,* $t = F_\lambda^* t$.

As an application of **5.4.1**, let us generalize the naturality of $\mathcal{L}_X$ with respect to diffeomorphisms. As remarked in **§5.2**, the pull-back of covariant tensor fields makes sense even when the mapping is not a diffeomorphism. It is thus natural to ask whether there is some analogue of **5.3.5** for pull-backs with no invertibility assumption on the mapping φ. Of course, the best one can hope for, since vector fields can be operated upon only by diffeomorphisms, is to replace the pair X, $\varphi_* X$ be a pair of φ-related vector fields.

5.4.3 Proposition *Let* $\varphi : M \to N$ *be* C^∞, $X \in \mathfrak{X}(M)$, $Y \in \mathfrak{X}(N)$, $X \sim_\varphi Y$ *and* $t \in \mathcal{T}^0_s(N)$. *Then* $\varphi^*(\mathcal{L}_Y t) = \mathcal{L}_X \varphi^* t$.

Proof Recall from **4.2.4** that $X \sim_\varphi Y$ iff $G_\lambda \circ \varphi = \varphi \circ F_\lambda$, where F_λ and G_λ are the flows of X and Y, respectively. Thus by **5.4.1**,

$$\mathcal{L}_X(\varphi^* t) = \frac{d}{d\lambda}\bigg|_{\lambda=0} F_\lambda^* \varphi^* t = \frac{d}{d\lambda}\bigg|_{\lambda=0} (\varphi \circ F_\lambda)^* t = \frac{d}{d\lambda}\bigg|_{\lambda=0} (G_\lambda \circ \varphi)^* t$$

$$= \frac{d}{d\lambda}\bigg|_{\lambda=0} \varphi^* G_\lambda^* t = \varphi^* \frac{d}{d\lambda}\bigg|_{\lambda=0} G_\lambda^* t = \varphi^*(\mathcal{L}_Y t). \blacksquare$$

As with functions and vector fields, the Lie derivative can be generalized to include time-dependent vector fields.

5.4.4 First Time-dependent Lie Derivative Theorem *Let* $X_\lambda \in \mathfrak{X}^k(M)$, $k \geq 1$, *for* $\lambda \in \mathbb{R}$ *and suppose that* $X(\lambda, m)$ *is continuous in* (λ, m). *Then if* $F_{\lambda,\mu}$ *is the evolution operator for* X_λ, *we have*

$$\frac{d}{d\lambda} F_{\lambda,\mu}^* t = F_{\lambda,\mu}^*(\mathcal{L}_{X_\lambda} t)$$

where $t \in \mathcal{T}^r_s(M)$ *is of class* C^k.

Warning. It is *not* generally true for time-dependent vector fields that the right hand-side in **5.4.4** is

$$\mathcal{L}_{X_\lambda} F_{\lambda,\mu}^* t \, .$$

Proof As in **5.4.1**, it is enough to prove the formula at $\lambda = \mu$ where $F_{\lambda,\lambda} = identity$, for then

$$\frac{d}{d\lambda} F_{\lambda,\mu}^* t = \left.\frac{d}{d\rho}\right|_{\rho = \lambda} (F_{\rho,\lambda} \circ F_{\lambda,\mu})^* t = F_{\lambda,\mu}^* \left.\frac{d}{d\rho}\right|_{\rho = \lambda} F_{\rho,\lambda}^* t = F_{\lambda,\mu}^* \mathcal{L}_{X_\lambda} t \, .$$

As in **5.4.1**,

$$\theta_{X_\lambda} t = \left.\frac{d}{d\lambda}\right|_{\lambda = \mu} F_{\lambda,\mu}^* t$$

is a differential operator that coincides with $\mathcal{L}_{X_\lambda}$ on $\mathcal{F}(M)$ and on $\mathfrak{X}(M)$ by **4.2.31**. Thus by **5.3.2**, $\theta_{X_\lambda} = \mathcal{L}_{X_\lambda}$ on all tensors. ∎

Let us generalize the relationship between Lie derivatives and flows one more step. Call a smooth map $t : \mathbb{R} \times M \to \mathcal{T}^r_s(M)$ satisfying $t_\lambda(m) = t(\lambda, m) \in (T_m M)^r_s$ a ***time-dependent tensor field.*** Theorem **5.4.4** generalizes to this context as follows.

5.4.5 Second Time-dependent Lie Derivative Theorem *Let* t_λ *be a* C^k *time-dependent tensor, and* X_λ *be as in* **5.4.4**, $k \geq 1$, *and denote by* $F_{\lambda,\mu}$ *the evolution operator of* X_λ. *Then*

$$\frac{d}{d\lambda} F_{\lambda,\mu}^* t_\lambda = F_{\lambda,\mu}^* \mathcal{L}_{X_\lambda} t_\lambda + F_{\lambda,\mu}^* \frac{\partial t_\lambda}{\partial \lambda}$$

Proof By the product rule for derivatives and **5.4.4** we get

$$\left.\frac{d}{d\lambda}\right|_{\lambda = \sigma} F_{\lambda,\mu}^* t_\lambda = \left.\frac{d}{d\lambda}\right|_{\lambda = \sigma} F_{\lambda,\mu}^* t_\sigma + F_{\sigma,\mu}^* \left.\frac{dt_\lambda}{d\lambda}\right|_{\lambda = \sigma} = F_{\sigma,\mu}^*(\mathcal{L}_{X_\sigma} t_\sigma) + F_{\sigma,\mu}^* \left.\frac{dt_\lambda}{d\lambda}\right|_{\lambda = \sigma} . \quad ∎$$

5.4.6 Examples

A If g is a pseudo-Riemannian metric on M, the Killing equations are $\mathcal{L}_X g = 0$ (see Example **5.3.6B**). By **5.4.2** this says that $F_\lambda{}^* g = g$, where F_λ is the flow of X, i.e., that the flow of X consists of isometries.

B In elasticity the vanishing of the strain tensor means, by Example A, that the body moves as a rigid body. ♦

We close this section with an important technique based on the dynamic approach to the Lie derivative, called the ***Lie transform method***. It has been used already in the proof of the Frobenius theorem (§**4.4**) and we shall see it again in **6.4.14** and **9.1.2**. The method is also used in the theory of *normal forms* (cf. Takens [1974], Guckenheimer and Holmes [1983], and Golubitsky and Schaeffer [1985]).

5.4.7 The Lie Transform Method Let two tensor fields t_0 and t_1 be given on a smooth manifold M. We say they are ***locally equivalent*** at $m_0 \in M$ if there is a diffeomorphism φ of a neighborhood of m_0 to itself, such that $\varphi^* t_1 = t_0$. *One way to show that* t_0 *and* t_1 *are equivalent is to join them with a curve* $t(\lambda)$ *satisfying* $t(0) = t_0$, $t(1) = t_1$ *and to seek a curve of local diffeomorphisms* φ_λ *such that* $\varphi_0 =$ *identity and*

$$\varphi_\lambda^* t(\lambda) = t_0, \quad \lambda \in [0, 1].$$

If this is done, $\varphi = \varphi_1$ is the desired diffeomorphism. A way to find the curve of diffeomorphisms φ_λ satisfying the relation above is to *solve the equation*

$$\mathcal{L}_{X_\lambda} t(\lambda) + \frac{d}{d\lambda} t(\lambda) = 0$$

for X_λ. If this is possible, let $\varphi_\lambda = F_{\lambda,0}$, where $F_{\lambda,\mu}$ is the evolution operator of the time-dependent vector field X_λ. Then by **5.4.5** we have

$$\frac{d}{d\lambda}\varphi_\lambda^* t(\lambda) = \varphi_\lambda^*\left(\mathcal{L}_{X_\lambda} t(\lambda) + \frac{d}{d\lambda} t(\lambda)\right) = 0$$

so that $\varphi_\lambda^* t(\lambda) = \varphi_0^*(0) = t_0$. If we choose X_λ so $X_\lambda(m_0) = 0$, then φ_λ exists for a time ≥ 1 by **4.1.25** and $\varphi_\lambda(m_0) = m_0$.

One often takes $t(\lambda) = (1 - \lambda)t_0 + \lambda t_1$. Also, in applications this method is not always used in exactly this way since the algebraic equation for X_λ might be hard to solve. We shall see this happen in the proof of the Poincaré lemma **6.4.14**. The reader should now also look back at the Frobenius theorem **4.4.5** and recognize the spirit of the Lie transform method in its proof. ♦

We shall next prove a version of the classical Morse lemma in infinite dimensions using the method of Lie transforms. The proof below is due to Golubitsky and Marsden [1983]; see Palais [1969] and Tromba [1976] for the original proofs; Palais′ proof is similar in spirit to the one we give.

5.4.8 The Morse-Palais-Tromba Lemma *Let* E *be a Banach space and* $\langle\, ,\rangle$ *a weakly nondegenerate, continuous, symmetric bilinear form on* E. *Let* $h: U \to \mathbb{R}$ *be* C^k, $k \geq 3$, *where* U *is open in* E, *and let* $u_0 \in U$ *satisfy* $h(u_0) = 0$, $\mathbf{D}h(u_0) = 0$. *Let* $B = \mathbf{D}^2h(u_0) : E \times E \to \mathbb{R}$. *Assume that there is a linear isomorphism* $T : E \to E$ *such that* $B(u, v) = \langle Tu, v\rangle$ *for all* $u, v \in E$ *and that* h *has a* C^{k-1} *gradient* $\langle \nabla h(y), u\rangle = \mathbf{D}h(y) \cdot u$. *Then there is a local* C^{k-2} *diffeomorphism* φ *of* E *with* $\varphi(u_0) = u_0$, $\mathbf{D}\varphi(u_0) = I$, *and*

$$h(\varphi(x)) = \frac{1}{2}B(x - u_0, x - u_0)\,.$$

Proof Symmetry of B implies that T is self-adjoint relative to $\langle\, ,\rangle$. Let $f(y) = (1/2)B(y - u_0, y - u_0)$, $h_1 = h$, and $h_\lambda = f + \lambda p$, where $p(y) = h(y) - (1/2)B(y - u_0, y - u_0)$ is C^k and satisfies $p(u_0) = 0$, $\mathbf{D}p(u_0) = 0$, and $\mathbf{D}^2p(u_0) = 0$. We apply the Lie transform method to h_λ. Thus we have to solve the following equation for a C^{k-2} vector field X_λ

$$\mathbf{L}_{X_\lambda} h_\lambda + \frac{dh_\lambda}{d\lambda} = 0, \quad X_\lambda(u_0) = 0. \tag{1}$$

Then $\varphi_1{}^*h = f$, where $\varphi_\lambda = F_{\lambda,0}$, for $F_{\lambda,\mu}$ the evolution operator of X_λ, and hence $\varphi = \varphi_1$ is a C^{k-2} diffeomorphism of a neighborhood of u_0 satisfying $\varphi_1(u_0) = u_0$. If we can prove that $\mathbf{D}\varphi_1(u_0) = I$, $\varphi_1 = \varphi$ will be the desired diffeomorphism.

To solve (1), differentiate $\mathbf{D}p(x) \cdot e = \langle \nabla p(x), e\rangle$ with respect to x and use the symmetry of the second derivative to conclude that $\mathbf{D}\nabla p(x)$ is symmetric relative to $\langle\, ,\,\rangle$. Therefore,

$$\begin{aligned}\mathbf{D}p(x) \cdot e &= \langle \nabla p(x), e\rangle = \left\langle \int_0^1 \mathbf{D}\nabla p(u_0 + \tau(x - u_0)) \cdot (x - u_0)d\tau,\ e\right\rangle \\ &= \langle T(x - u_0), R(x) \cdot e\rangle \end{aligned} \tag{2}$$

where $R : U \to L(E, E)$ is the C^{k-2} map given by

$$R(x) = T^{-1}\int_0^1 \mathbf{D}p(u_0 + \tau(x - u_0)) \cdot (x - u_0)d\tau$$

which satisfies $R(u_0) = 0$. Thus $p(y)$ has the expression

$$p(y) = \int_0^1 \mathbf{D}p(u_0 + \tau(y - u_0)) \cdot (y - u_0)d\tau = -\langle T(y - u_0), X(y)\rangle$$

where $X : U \to E$ is the C^{k-2} vector field given by

$$X(y) = -\int_0^1 \tau R(u_0 + \tau(y - u_0)) \cdot (y - u_0)d\tau$$

which satisfies $X(u_0) = 0$ and $\mathbf{D}X(u_0) = I$. Therefore

$$(\mathbf{L}_{X_\lambda} h_\lambda)(y) = \mathbf{D}h_\lambda(y) \cdot X_\lambda(y) = B(y - u_0, X_\lambda(y)) + \lambda \mathbf{D}p(y) \cdot X_\lambda(y)$$
$$= \langle T(y - u_0), (I + \lambda R(y)) \cdot X_\lambda(y) \rangle \qquad \text{(by (2))}$$

so that the equation (1) becomes

$$\langle T(y - u_0), (I + \lambda R(y)) \cdot X_\lambda(y) \rangle = \langle T(y - u_0), X(y) \rangle . \qquad \text{(by (3))}$$

Since $R(u_0) = 0$, there exists a neighborhood of u_0, such that the norm of $\lambda R(y)$ is < 1 for all $\lambda \in [0,1]$. Thus for y in this neighborhood, $I + \lambda R(y)$ can be inverted and we can take $X_\lambda(y) = (I + \lambda R(y))^{-1} X(y)$ which is a C^{k-2} vector field defined for all $\lambda \in [0,1]$ and which satisfies $X_\lambda(u_0) = 0$, $\mathbf{D}X_\lambda(u_0) = 0$. Differentiating the relation $(d/d\lambda)\varphi_\lambda(u) = X_\lambda(\varphi_\lambda(u))$ in u at u_0 and using $\varphi_\lambda(u_0) = u_0$ yields $(d/d\lambda)\mathbf{D}\varphi_\lambda(u_0) = \mathbf{D}X_\lambda(\varphi_\lambda(u_0)) \circ \mathbf{D}\varphi_\lambda(u_0) = \mathbf{D}X_\lambda(u_0) \circ \mathbf{D}\varphi_\lambda(u_0) = 0$, i.e., $\mathbf{D}\varphi_\lambda(u_0)$ is constant in $\lambda \in [0,1]$. Since it equals I at $\lambda = 0$, it follows that $\mathbf{D}\varphi_\lambda(u_0) = I$. ■

5.4.9 The Classical Morse Lemma *Let* $h : U \to \mathbb{R}$ *be* C^k, $k \geq 3$, U *open in* $\mathbb{R}^n$, *and let* $u \in U$ *be a nondegenerate critical point of* h, *i.e.,* $h(u) = 0$, $\mathbf{D}h(u) = 0$ *and the symmetric bilinear form* $\mathbf{D}^2h(u)$ *on* $\mathbb{R}^n$ *is nondegenerate. Then there is a local* C^{k-2} *diffeomorphism* ψ *of* $\mathbb{R}^n$ *fixing* u *such that*

$$h(\psi(x)) = (1/2)[(x^1 - u^1)^2 + \ldots + (x - u^{n-i})^2 - (x^{n-i+1} - u^{n-i+1})^2 - \ldots - (x^n - u^n)^2].$$

Proof. In **4.5.8** take $\langle\ ,\ \rangle$ to be the dot-product in $\mathbb{R}^n$ to find a local C^{k-2} diffeomorphism on $\mathbb{R}^n$ fixing u_0 such that $h(\varphi(x)) = (1/2)\mathbf{D}^2h(u)(x - u, x - u)$. Next, apply the Gram-Schmidt procedure to find a basis of $\mathbb{R}^n$ in which the matrix of $\mathbf{D}^2h(u)$ is diagonal with entries ± 1 (see **6.2.9** for a review of the proof of the existence of such a basis). If i is the number of -1's (the ***index***), let φ be the composition of ψ with the linear isomorphism determined by the change of an arbitrary basis of $\mathbb{R}^n$ to the one above. ■

Exercises

5.4A Verify Theorem **5.4.1** by a coordinate computation as follows. Let $F_\lambda(x) = (y^1(\lambda, x), \ldots, y^n(\lambda, x))$ so that $\partial y^i/\partial\lambda = X^i(y)$ and $\partial y^i/\partial x^j$ satisfy the variational equation

$$\frac{\partial}{\partial \lambda} \frac{\partial y^i}{\partial x^j} = \frac{\partial X^i}{\partial x^k} \frac{\partial y^k}{\partial x^j} .$$

Then write

$$(F_\lambda^* t)^{a_1 \ldots a_r}{}_{b_1 \ldots b_s} = \frac{\partial x^{a_1}}{\partial y^{i_1}} \cdots \frac{\partial x^{a_r}}{\partial y^{i_r}} \frac{\partial y^{j_1}}{\partial x^{b_1}} \cdots \frac{\partial y^{j_s}}{\partial x^{b_s}} t^{i_1 \ldots i_r}{}_{j_1 \ldots j_s} .$$

Differentiate this in λ at $\lambda = 0$ and obtain the coordinate expression (4) of Section **5.3** for $\mathcal{L}_X t$.

5.4B Carry out the proof outlined in Exercise **5.4A** for time-dependent vector fields.

5.4C Starting with Theorem **5.4.1** as the definition of $\mathcal{L}_X t$, check that $\mathcal{L}_X$ satisfies **DO1**, **DO2** and the properties (i)-(iv) of **5.3.2.**

5.4D Let $\mathbf{C}$ be a contraction operator mapping $\mathcal{T}^r_s(M)$ to $\mathcal{T}^{r-1}_{s-1}(M)$. Use both **5.4.1** and **DO1** to show that $\mathcal{L}_X(\mathbf{C}t) = \mathbf{C}(\mathcal{L}_X t)$.

5.4E Extend Theorem **5.4.1** to F-valued tensors.

5.4F Let $f(y) = (1/2)y^2 - y^3 + y^5$. Use the Lie transform method to show that there is a local diffeomorphism φ, defined in a neighborhood of $0 \in \mathbb{R}$ such that $(f \circ \varphi)(x) = x^2/2$.

5.4G Let $E = \ell^2(\mathbb{R})$, let

$$\langle x, y \rangle = \sum_{n=1}^{\infty} \frac{1}{n} x_n y_n \text{ and } h(x) = \frac{1}{2} \sum_{n=1}^{\infty} \frac{1}{n} x_n^2 - \frac{1}{3} \sum_{n=1}^{\infty} x_n^3 .$$

Show that h vanishes on (0, 0, ..., 3/2n, 0, ...) which $\to 0$ as $n \to \infty$, so the conclusion of the Morse lemma fails. What hypothesis in **4.5.8** fails?

5.4H (Buchner, Marsden and Schecter [1983]). In the notation of Exercise **2.4O**, show that f has a sequence of critical points approaching 0, so the Morse lemma fails. (The only missing hypothesis is that ∇h is C^1.)

§5.5 *Partitions of Unity*

A partition of unity is a technical device that is often used to piece smooth local tensor fields together to form a smooth global tensor field. Partitions of unity will be useful for studying integration; in this section they are used to study when a manifold admits a Riemannian metric.

5.5.1 Definition *If* t *is a tensor field on a manifold* M, *the* ***carrier*** *of* t *is the set of* $m \in M$ *for which* $t(m) \neq 0$, *and is denoted* carr t. *The* ***support*** *of* t, *denoted* supp t, *is the closure of* carr t. *We say* t ***has compact support*** *if* supp t *is compact in* M. *An open set* $U \subset M$ *is called a* C^r ***carrier*** *if there exists an* $f \in \mathcal{F}^r(M)$, *such that* $f \geq 0$ *and* $U = \text{carr } f$. *A collection of subsets* $\{C_\alpha\}$ *of a manifold* M *(or, more generally, a topological space) is called* ***locally finite*** *if for each* $m \in M$, *there is a neighborhood* U *of such that* $U \cap C_\alpha = \varnothing$ *except for finitely many indices* α.

5.5.2 Definition ***A partition of unity*** *on a manifold* M *is a collection* $\{(U_i, g_i)\}$, *where*

(i) $\{U_i\}$ *is a locally finite open covering of* M;

(ii) $g_i \in \mathcal{F}(M)$, $g_i(m) \geq 0$ *for all* $m \in M$, *and* $\text{supp } g_i \subset U_i$ *for all* i;

(iii) *for each* $m \in M$, $\Sigma_i g_i(m) = 1$. *(By* (i), *this is a finite sum.)*

If $\mathcal{A} = \{(V_\alpha, \varphi_\alpha)\}$ *is an atlas on* M, *a* ***partition of unity subordinate*** *to* $\mathcal{A}$ *is a partition of unity* $\{(U_i, g_i)\}$ *such that each open set* U_i *is a subset of a chart domain* $V_{\alpha(i)}$. *If any atlas* $\mathcal{A}$ *has a subordinate partition of unity, we say* M ***admits partitions of unity***.

Occasionally one works with C^k partitions of unity. They are defined in the same way except g_i are only required to be C^k rather than C^∞.

5.5.3 Patching Construction *Let* M *be a manifold with an atlas* $\mathcal{A} = \{(V_\alpha, \varphi_\alpha)\}$ *where* $\varphi_\alpha : V_\alpha \to V'_\alpha \subset E$ *is a chart. Let* t_α *be a* C^k *tensor field,* $k \geq 1$, *of fixed type* (r,s) *defined on* V'_α *for each* α, *and assume that there exists a partition of unity* $\{(U_i, g_i)\}$ *subordinate to* $\mathcal{A}$. *Let* t *be defined by*

$$t(m) = \sum_i g_i \, \varphi^*_{\alpha(i)} t_{\alpha(i)} \, (m),$$

a finite sum at each $m \in M$. *Then* t *is a* C^k *tensor field of type* (r,s) *on* M.

Proof Since $\{U_i\}$ is locally finite, the sum at every point is a finite sum, and thus t(m) is a type (r, s) tensor for every $m \in M$. Also, t is C^k since the local representative of t in the chart $(V_{\alpha(i)}, \varphi_{\alpha(i)})$ is $\Sigma_j (g_i \circ \varphi_{\alpha(j)}^{-1}) t_{\alpha(j)}$, the summation taken over all indices j such that $V_{\alpha(i)} \cap V_{\alpha(j)} \neq \varnothing$; by local finiteness the number of these j is finite. ■

Clearly this construction is not unique; it depends on the choices of the indices $\alpha(i)$ such that $U_i \subset V_{\alpha(i)}$ and on the functions g_i. As we shall see later, under suitable hypotheses, one can always construct partitions of unity; again the construction is not unique. The same construction (and proof) can be used to patch together local sections of a vector bundle into a global section when the base is a manifold admitting partitions of unity subordinate to any open covering.

To discuss the existence of partitions of unity and consequences thereof, we need some topological preliminaries.

5.5.4 Definition *Let* S *be a topological space. A covering* $\{U_\alpha\}$ *of* S *is called a* ***refinement*** *of a covering* $\{V_i\}$ *if for every* U_α *there is a* V_i *such that* $U_\alpha \subset V_i$. *A topological space is called* ***paracompact*** *if every open covering of* S *has a locally finite refinement of open sets, and* S *is Hausdorff.*

5.5.5 Proposition *Second-countable, locally compact Hausdorff spaces are paracompact.*

Proof By second countability and local compactness of S, there exists a sequence $O_1, ..., O_n, ...$ of open sets with $cl(O_n)$ compact and $\bigcup_{n \in N} O_n = S$. Let $V_n = O_1 \cup ... \cup O_n$, $n = 1,2, ...$ and put $U_1 = V_1$. Since $\{V_n\}$ is an open covering of S and $cl(U_1)$ is compact,

$$cl(U_1) \subset V_{i_1} \cup ... \cup V_{i_r}.$$

Put

$$U_2 = V_{i_1} \cup ... \cup V_{i_r};$$

then $cl(U_2)$ is compact. Proceed inductively to show that S is the countable union of open sets U_n such that $cl(U_n)$ is compact and $cl(U_n) \subset U_{n+1}$. If W_α is a covering of S by open sets, and $K_n = cl(U_n) \setminus U_{n-1}$, then we can cover K_n by a finite number of open sets, each of which is contained in some $W_\alpha \cap U_{n+1}$, and is disjoint from $cl(U_{n-2})$. The union of such collections yields the desired refinement of $\{W_\alpha\}$. ■

Another class of paracompact spaces are the metrizable spaces (see **5.5.15** in Supplement **5.5A**). In particular, Banach spaces are paracompact.

5.5.6 Proposition *Every paracompact space is normal.*

Proof We first show that if A is closed and $u \in S \setminus A$, there are disjoint neighborhoods of u and A (regularity). For each $v \in A$, let U_u, V_v be disjoint neighborhoods of u and v. Let W_α be a locally finite refinement of the covering $\{V_v, S \setminus A \mid v \in A\}$, and $V = \bigcup W_\alpha$, the union over those α with $W_\alpha \cap A \neq \varnothing$. A neighborhood U_0 of u meets a finite number of W_α. Let U denote the intersection of U_0 and the corresponding U_u. Then V and U are the required neighborhoods. The case for two closed sets proceeds somewhat similarly, so we leave the details for the reader. ■

Later we shall give general theorems on the existence of partitions of unity. However, there is a simple case that is commonly used, so we present it first.

5.5.7 Theorem *Let* M *be a second-countable (Hausdorff)* n*-manifold. Then* M *admits partitions of unity.*

Proof The proof of **5.5.5** shows the following. Let M be an n-manifold and $\{W_\alpha\}$ be an open covering. Then there is a locally finite refinement consisting of charts (V_i, φ_i) such that $\varphi_i(V_i)$ is the disk of radius 3, and such that $\varphi_i^{-1}(D_1(0))$ cover M, where $D_1(0)$ is the unit disk, centered at the origin in the model space. Now let $\mathcal{A}$ be an atlas on M and let $\{(V_i, \varphi_i)\}$ be a locally finite refinement with these properties. From **4.2.13**, there is a nonzero function $h_i \in \mathcal{F}(M)$ whose support lies in V_i and $h_j \geq 0$. Let

$$g_i(u) = \frac{h_i(u)}{\Sigma_i h_i(u)}$$

(the sum is finite). These are the required functions. ∎

If $\{V_\alpha\}$ is an open covering of M, we can always find an atlas $\mathcal{A} = \{(U_i, \varphi_i)\}$ such that $\{U_i\}$ is a refinement of $\{V_\alpha\}$ since the atlases generate the topology. Thus, if M admits partitions of unity, we can find partitions of unity subordinate to any open covering.

The case of C^0-partitions of unity differs drastically from the smooth case. Since we are primarily interested in this latter case, we summarize the topological situation, without giving the proofs.

1 *If* S *is a Hausdorff space, the following are equivalent:*

(i) S *is normal;*

(ii) **Urysohn's lemma** *For any two closed nonempty disjoint sets* A, B *there is a continuous function* $f : S \to [0, 1]$ *such that* $f(A) = 0$ *and* $f(B) = 1$

(iii) **Tietze extension theorem** *For any closed set* $A \subset S$ *and continuous function* $g : A \to [a, b]$, *there is a continuous extension* $G : S \to [a, b]$ *of* g

2 *A Hausdorff space is paracompact iff it admits a* C^0 *partition of unity subordinate to any open covering.*

It is clear that if $\{(U_i, g_i)\}$ is a continuous partition of unity subordinate to the given open covering $\{V_\alpha\}$, then by definition $\{U_i\}$ is an open locally finite refinement. The converse--the existence of partitions of unity--is the hard part; the proof of this and of the equivalences of (i), (ii) and (iii) can be found for instance in Kelley [1975] and Choquet [1969; sec. 6]. These results are important for the rich supply of continuous functions they provide. We shall not use these topological theorems in the rest of the book, but we do want their smooth versions on manifolds.

Note that if M is a manifold admitting partitions of unity subordinate to any open covering, then M is paracompact, and thus normal by **5.5.6**. This already enables us to generalize (ii) and

(iii) to the smooth (or C^k) situation.

5.5.8 Proposition *Let* M *be a manifold admitting smooth (or* C^k*) partitions of unity. If* A *and* B *are closed disjoint sets then, there exists a smooth (or* C^k*) function* $f : M \to [0, 1]$ *such that* $f(A) = 0$ *and* $f(B) = 1$.

Proof As we saw, the condition on M implies that M is normal and thus there is an atlas $\{(U_\alpha, \varphi_\alpha)\}$ such that $U_\alpha \cap A \neq \varnothing$ implies $U_\alpha \cap B = \varnothing$. Let $\{(V_i, g_i)\}$ be a subordinate C^k partition on unity and $f = \Sigma\, g_i$, where the sum is over those i for which $V_i \cap B \neq \varnothing$. Then f is C^k, is one on B, and zero on A. ■

5.5.9 Smooth Tietze Extension Theorem *Let* M *be a manifold admitting partitions of unity, and let* $\pi : E \to M$ *be a vector bundle with base* M. *Suppose* $\sigma : A \to E$ *is a* C^k *section defined on the closed set* A *(i.e., every point* $a \in A$ *has a neighborhood* U_a *and a* C^k *section* $\sigma_a : U_a \to E$ *extending* σ*). Then* σ *can be extended to a* C^k *global section* $\Sigma : M \to E$. *In particular, if* $g : A \to F$ *is a* C^k *function defined on the closed set* A, *where* F *is a Banach space, then there is a* C^k *extension* $G : M \to F$; *if* g *is bounded by a constant* R, *i.e.,* $\| g(a) \| \leq R$ *for all* $a \in A$, *then so is* G.

Proof Consider the open covering $\{U_\alpha, M \backslash A \mid a \in A\}$ of M, with U_α given by the definition of smoothness on the closed set A. Let $\{(U_i, g_i)\}$ be a C^k partition of unity subordinate to this open covering and define $\sigma_i : U_i \to E$, by $\sigma_i = \sigma_a \mid U_i$ for all U_i and $\sigma_i \equiv 0$ on all U_i disjoint from U_a, $a \in A$. Then $g_i\sigma_i : U_i \to E$ is a C^k section on U_i and since $\mathrm{supp}(g_i\sigma_i) \subset \mathrm{supp}(g_i) \subset U_i$, it can be extended in a C^k manner to M by putting it equal to zero on $M \backslash U_i$. Thus $g_i\sigma_i : M \to E$ is a C^k-section of $\pi : E \to M$ and hence $\Sigma = \Sigma_i g_i \sigma_i$ is a C^k section; note that the sum is finite in a neighborhood of every point $m \in M$. Finally, if $a \in A$

$$\Sigma(a) = \sum_i g_i(a)\sigma_i(a) = \left(\sum_i g_i(a)\right)\sigma(a) = \sigma(a),$$

i.e., $\Sigma \mid A = \sigma$.

The second part of the theorem is a particular case of the one just proved by considering the trivial bundle $M \times F \to M$ and the section σ defined by $\sigma(m) = (m, g(m))$. The boundedness statement follows from the given construction, since all the g_i have values in [0, 1]. ■

Before discussing general questions on the existence of partitions of unity on Banach manifolds, we discuss the existence of Riemannian metrics. Recall that a Riemannian metric on a Hausdorff manifold M is a tensor field $g \in \mathcal{T}^0{}_2(M)$ such that for all $m \in M$, $g(m)$ is symmetric and positive definite. Our goal is to find topological conditions on an n-manifold that are necessary and sufficient to ensure the existence of Riemannian metrics. The proof of the necessary conditions will be simplified by first showing that any Riemannian manifold is a metric space. For this, define for $m, n \in M$,

$d(m, n) =$

$$\inf\left\{ \ell(\gamma) \mid \gamma\colon [0, 1] \to M \text{ is a continuous piecewise } C^1 \text{ curve with } \gamma(0) = m,\ \gamma(1) = n \right\}.$$

Here $\ell(\gamma)$ is the ***length*** of the curve γ, defined by

$$\ell(\gamma) = \int_0^1 \|\dot\gamma(t)\| dt,$$

where $\dot\gamma(t) = d\gamma/dt$ is the tangent vector at $\gamma(t)$ to the curve γ and $\|\dot\gamma\| = [g_{\gamma(t)}(\dot\gamma(t), \dot\gamma(t))]^{1/2}$ is its length.

5.5.10 Proposition *d is a metric on each connected component of M whose metric topology is the original topology of M. If d is a complete metric, M is called a* ***complete Riemannian manifold.***

Proof Clearly $d(m, m) = 0$, $d(m, n) = d(n, m)$, and $d(m, p) \le d(m, n) + d(n, p)$, by using the definition. Next we will verify that $d(m, n) > 0$ whenever $m \ne n$.

Let $m \in U \subset M$ where (U, φ) is a chart and suppose $\varphi(U) = U' \subset E$. Then for any $u \in U$, $g(u)(v, v)^{1/2}$, defined for $v \in T_uM$, is a norm on T_uM. This is equivalent to the norm on E, under the linear isomorphism $T_m\varphi$. Thus, if g' is the local expression for g, then $g'(u')$ defines an inner product on E, yielding equivalent norms for all $u' \in U'$. Using continuity of g and choosing U' to be an open disk in E, we can conclude that the norms $g'(u')^{1/2}$ and $g'(m)^{1/2}$, where $m' = \varphi(m)$ satisfy: $a\, g'(m')^{1/2} \le g'(u')^{1/2} \le b g'(m')^{1/2}$ for all $u' \in U'$, where a and b are positive constants. Thus, if $\eta\colon [0, 1] \to U'$ is a continuous piecewise C^1 curve, then

$$\ell(\eta) = \int_0^1 g'(\eta(t))(\dot\eta(t), \dot\eta(t))^{1/2} dt \ge a\int_0^1 g'(m')(\dot\eta(t), \dot\eta(t))^{1/2} dt$$

$$\ge a g'(m')\left(\int_0^1 \dot\eta(t)dt, \int_0^1 \dot\eta(t)dt\right)^{1/2} \ge a g'(m')(\eta(1) - \eta(0), \eta(1) - \eta(0))^{1/2}.$$

Here we have used the following property of the Bochner integral:

$$\left\| \int_a^b f(t)dt \right\| \le \int_a^b \|f(t)\| dt,$$

valid for any norm on E (see the remarks following **2.2.7**).

Now let $\gamma\colon [0, 1] \to M$ be a continuous piecewise C^1 curve, $\gamma(0) = m$, $\gamma(1) = n$, $m \in U$, where (U, φ), $\varphi : U \to U' \subset E$ a chart of M, $\varphi(m) = 0$. If γ lies entirely in U, then $\varphi \circ \gamma = \eta$ lies entirely in U' and the previous estimate gives $\ell(\gamma) \ge a g'(m')(n', n')^{1/2} \ge ar$, where r is the radius of the disk U' in E about the origin and, $n' = \varphi(n)$. If γ is not entirely contained in U, then let r be the radius of a disk about the origin and let $c \in \,]0, 1[$ be the smallest number for which $\gamma(c) \cap \varphi^{-1}(\{x \in E \mid \|x\| = r\}) \neq \varnothing$. Then $\ell(\gamma) \geq \ell(\gamma|[0, c]) \geq ag(m')((\varphi \circ \gamma)(c), (\varphi \circ \gamma)(c))^{1/2} \geq ar$. Thus we conclude $d(m, n) \geq ar > 0$.

The equivalence of the original topology of M and of the metric topology defined by d is clear if one notices that they are equivalent in every chart domain U, which in turn is implied by their equivalence in $\varphi(U)$. ■

Notice that the preceding proposition holds in infinite dimensions.

5.5.11 Proposition *A connected Hausdorff* n*-manifold admits a Riemannian metric if and only if it is second countable. Hence for Hausdorff* n*-manifolds (not necessarily connected) paracompactness and metrizability are equivalent.*

Proof If M is second countable, it admits partitions of unity by **5.5.7**. Then the patching construction **5.5.3** gives a Riemannian metric on M by choosing in every chart the standard inner product in $\mathbb{R}^n$.

Conversely, assume M is Riemannian. By **5.5.10** it is a metric space, which is locally compact and first countable, being locally homeomorphic to $\mathbb{R}^n$. By **1.6.14**, it is second countable. ■

The main theorem on the existence of partitions of unity in the general case is as follows.

5.5.12 Theorem *Any second-countable or paracompact manifold modeled on a separable Banach space with a* C^k *norm away from the origin admits* C^k *partitions of unity. In particular paracompact (or second countable) manifolds modeled on separable Hilbert spaces admit* C^∞ *partitions of unity.*

5.5.13 Corollary *Paracompact (or second countable) Hausdorff manifolds modeled on separable real Hilbert spaces admit Riemannian metrics.*

Theorem **5.5.12** will be proved in the following two supplements.

There are Hausdorff nonparacompact n-manifolds. These manifolds are necessarily nonmetrizable and do not admit partitions of unity. The standard example of a one-dimensional nonparacompact Hausdorff manifold is the "long line." In dimensions 2 and 3 such manifolds are constructed from the Prüfer manifolds. Since nonparacompact manifolds occur rarely in applications, we refer the reader to Spivak [1979, vol. 1, Appendix A] for the aforementioned examples.

Partitions of unity are an important technical tool in many proofs. We illustrate this with the sample theorem below which combines differential topological ideas of §**3.5**, the local and global existence and uniqueness theorem for solutions of vector fields, and partitions of unity. More appplications of this sort can be found in the exercises.

5.5.14 Ehresmann Fibration Theorem *A proper submersion* $f : M \to N$ *of finite dimensional manifolds with* M *paracompact is a locally trivial fibration, i.e., for any* $p \in N$ *there exists an open neighborhood* V *of* u *in* N *and a diffeomorphism* $\varphi : V \times f^{-1}(p) \to f^{-1}(V)$ *such that* $f(\varphi(x, u)) = x$ *for all* $x \in V$ *and all* $u \in f^{-1}(p)$.

Proof Since the statement is local we can replace M, N by chart domains and, in particular, we can ssume that $N = \mathbb{R}^n$ and $p = 0 \in \mathbb{R}^n$. We claim that there are smooth vector fields $X_1, ..., X_n$ on M such that X_i is f-related to $\partial/\partial x^i \in \mathfrak{X}(\mathbb{R}^n)$. Indeed, around any point in M such vector fields are easy to obtain using the implicit function theorem (see **3.5.2**). Cover M with such charts, choose a partition of unity subordinate to this covering, and patch these vector fields by means of this partition of unity to obtain $X_1, ..., X_n$, f-related to $\partial/\partial x^1, ..., \partial/\partial x^n$, respectively.

Let $F^k{}_{t(k)}$ denote the flow of X_k with time variable $t(k)$, $k = 1, ..., n$ and let $\mathbf{t} = (t(1), ..., t(n)) \in \mathbb{R}^n$. If $\| \mathbf{t} \| < C$, then then integral curves of each X_k starting in $f^{-1}(\{\mathbf{u} \in \mathbb{R}^n \mid \| \mathbf{u} \| \leq C\})$ stay in $f^{-1}\{\mathbf{v} \in \mathbb{R}^n \mid \| \mathbf{v} \| \leq 2C\}$, since by **4.2.4**

$$(f \circ F^k{}_{t(k)})(y) = (f^1(y), ..., f^k(y) + t(k), ..., f^n(y)). \tag{1}$$

Therefore, since f is proper, **4.1.19** implies that the vector fields $X_1, ..., X_n$ are complete.

Finally, let $\varphi : \mathbb{R}^n \times f^{-1}(0) \to M$ be given by $\varphi(t(1), ..., t(n), u) = (F^1{}_{t(1)} \circ ... \circ F^n{}_{t(n)})(u)$ and note that φ is smooth (see **4.1.17**). The map $\varphi^{-1}: M \to \mathbb{R}^n \times f^{-1}(0)$ given by $\varphi^{-1}(m) = (f(m), (F^n{}_{-t(n)} \circ ... \circ F^1{}_{-t(1)})(m))$ is smooth and is easily checked to be the inverse of φ. Finally, $(f \circ \varphi)(\mathbf{t}, u) = \mathbf{t}$ by (1) since $f(u) = 0$. ∎

☞ SUPPLEMENT 5.5A
Partitions of Unity: Reduction to the Local Case

We begin with some topological preliminaries. Let S be a paracompact space. If $\{U_\beta\}$ is an open covering of S, it can be refined to a locally finite covering $\{W_\beta\}$. The first lemma below will show that we can shrink this covering further to get another one, $\{V_\alpha\}$ such that $cl(V_\alpha) \subset W_\alpha$ with the same indexing set.

A technical device used in the proof is the concept of a well-ordered set. An ordered set A in which any two elements can be compared is called ***well-ordered*** if every subset has a smallest element (see the introduction to Chapter 1).

5.5.15 Shrinking Lemma *Let* S *be a normal space and* $\{W_\alpha\}_{\alpha \in A}$ *a locally finite open covering of* S. *Then there is a locally finite open refinement* $\{V_\alpha\}_{\alpha \in A}$ *(with the same indexing set) such that* $cl(V_\alpha) \subset W_\alpha$.

Proof Well-order the indexing set A and call its smallest element $\alpha(0)$. The set C_0 defined as C_0

$= S \setminus \bigcup_{\alpha>\alpha(0)} W_\alpha$ is closed, so by normality there exists an open set $V_{\alpha(0)}$ such that $C_0 \subset \mathrm{cl}(V_{\alpha(0)}) \subset W_{\alpha(0)}$. If V_γ is defined for all $\gamma < \alpha$, put $C_\alpha = S \setminus \{(\bigcup_{\gamma<\alpha} V_\gamma) \cup (\bigcup_{\gamma>\alpha} W_\gamma)\}$ and by normality find V_α such that $C_\alpha \subset \mathrm{cl}(V_\alpha) \subset W_\alpha$. The collection $\{V_\alpha\}_{\alpha\in A}$ is the desired locally finite refinement of $\{W_\alpha\}_{\alpha\in A}$, provided we can show that it covers S. Given $s \in S$, by local finiteness of the covering $\{W_\alpha\}_{\alpha\in A}$, s belongs to only a finite collection of them, say $W_1, W_2, \ldots, W_n$, corresponding to the elements $\alpha_1, \ldots, \alpha_n$ of the index set. If β denotes the maximum of the elements $\alpha_1, \ldots, \alpha_n$, then $s \notin W_\gamma$ for all $\gamma > \beta$, so that if in addition $s \notin V_\gamma$ for all $\gamma < \beta$, then $s \in C_\beta \subset V_\beta$, i.e., $s \in V_\beta$. ∎

5.5.16 Lemma (A. H. Stone) *Every pseudometric space is paracompact.*

Proof Let $\{U_\alpha\}_{\alpha\in A}$ be an open covering of the pseudometric space S with distance function d. Put $U_{n,\alpha} = \{x \in U_\alpha \mid d(x, S\setminus U_\alpha) \geq 1/2^n\}$. By the triangle inequality we have the inequality $d(U_{n,\alpha}, S\setminus U_{n+1,\alpha}) \geq 1/2^n - 1/2^{n+1} = 1/2^{n+1}$. Well-order the indexing set A and let $V_{n,\alpha} = U_{n,\alpha} \setminus \bigcup_{\beta<\alpha} U_{n+1,\beta}$. If $\gamma, \delta \in A$, we have $V_{n,\gamma} \subset S\setminus U_{n+1,\delta}$, if $\gamma < \delta$, or $V_{n,\delta} \subset S\setminus U_{n+1,\gamma}$ if $\delta < \gamma$. But in both cases we have $d(V_{n,\gamma}, V_{n,\delta}) \geq 1/2^{n+1}$. Define $W_{n,\alpha} = \{s \in S \mid d(s, V_{n,\alpha}) < 1/2^{n+3}\}$, and observe that $d(W_{n,\alpha}, W_{n,\beta}) \geq 1/2^{n+2}$. Thus for a fixed n, every point $s \in S$ has a neighborhood intersecting at most one member of the family $\{W_{n,\alpha} \mid \alpha \in A\}$. Hence $\{W_{n,\alpha} \mid n \in \mathbb{N}, \alpha \in A\}$ is a locally finite open refinement of $\{U_\alpha\}$. ∎

Let us now turn to the question of the existence of partitions of unity subordinate to any open covering.

5.5.17 Proposition (R. Palais) *Let* M *be a paracompact manifold modeled on the Banach space* E. *The following are equivalent:*

(i) M *admits* C^k *partitions of unity;*

(ii) *any open covering of* M *admits a locally finite refinement by* C^k *carriers.*

(iii) *for any open sets* O_1, O_2 *such that* $\mathrm{cl}(O_1) \subset O_2$, *there exists a* C^k *carrier* V *such that* $O_1 \subset V \subset O_2$.

(iv) *every chart domain of* M *admits* C^k *partitions of unity subordinate to any open covering;*

(v) E *admits* C^k *partitions of unity subordinate to any open covering of* E.

Proof ((i) ⇒ (ii)). If $\{(U_i, g_i)\}$ is a C^k-partition of unity subordinate to an open covering, then clearly carr g_i forms a locally finite refinement of the covering by C^k carriers.

((ii) ⇒ (iii)). Let $\{V_\alpha\}_{\alpha\in A}$ be a locally finite refinement of the open covering $\{O_2, S\setminus \mathrm{cl}(O_1)\}$ by C^k carriers and denote by $f_\alpha \in \mathcal{F}^k(M)$, the function for which carr $f_\alpha = V_\alpha$. Let $B = \{\alpha \in A \mid V_\alpha \subset O_2\}$. Put $V = \bigcap_{\beta\in B} V_\beta$, $f = \sum_{\beta\in B} f_\beta$ and remark that $O_1 \subset V \subset O_2$, carr $f = V$.

((iii) ⇒ (iv)). Let U be any chart domain of M. Then U is diffeomorphic to an open set in E which is a metric space, so is paracompact by Stone's theorem **5.5.16**. Let $\{U_\alpha\}$ be an arbitrary open covering of U and $\{V_\beta\}$ be a locally finite refinement. By the shrinking lemma we may assume that $cl(V_\beta) \subset U$. Again by the shrinking lemma, refine further to a locally finite covering $\{W_\beta\}$ such that $cl(W_\beta) \subset V_\beta$. But by (iii) there exists a C^k-carrier O_β such that $W_\beta \subset O_\beta \subset V_\beta$, and so $\{O_\beta\}$ is a locally finite refinement of $\{U_\alpha\}$ by C^k-carriers, whose corresponding functions we denote by f_β. Thus $f = \Sigma_\beta f_\beta$ is a C^k map and $\{(V_\beta, f_\beta/f)\}$ is a C^k partition of unity subordinate to $\{U_\alpha\}$.

((iv) ⇒ (v)). Consider now any open covering $\{U_\alpha\}_{\alpha \in A}$ of E and let (U, φ) be an arbitrary chart of M. Refine first the covering of E by taking the intersections of all its elements with all translates of $\varphi(U)$. Since E is paracompact, refine again to a locally finite open covering $\{V_\beta\}$. The inverse images by translations and φ of these open sets are subsets of U, hence chart domains, and thus by (iv) they admit partitions of unity subordinate to any covering. Thus every V_β admits a C^k partition of unity subordinate to any open covering, for example to $\{V_\beta \cap U_\alpha \mid \alpha \in A\}$; call it $\{g_i^\beta\}$. Then $g = \Sigma_{i,\beta} g_i^\beta$ is a C^k map and the double-indexed set of functions g_i^β/g forms a C^k partition of unity of E.

((v) ⇒ (iv)). If E admits C^k partitions of unity subordinate to any open covering, then so does every open subset by the (already proved) implication (i) ⇒ (ii) applied to $M = E$, which is paracompact by **5.5.16**. Thus if (U, φ) is a chart on M, U admits partitions of unity, since $\varphi(U)$ does.

Finally, we show (iv) implies (i). Choosing a locally finite atlas, this proof repeats the one given in the last part of (iv) ⇒ (v). ■

As an application of this proposition we get the following.

5.5.18 Proposition *Every paracompact* n*-manifold admits* C^∞*-partitions of unity.*

Proof By **5.5.17**(ii) and (v) it suffices to show that every open set in $\mathbb{R}^n$ is a C^∞ carrier. Any open set U is a countable union of open disks D_i. By **4.2.13**, $D_i = \text{carr}\, f_i$, for some C^∞ function $f_i : \mathbb{R}^n \to \mathbb{R}$. Put $M_i = \sup\{\|D^k f_i(x)\| \mid x \in \mathbb{R}^n, k \le i\}$ and let

$$f = \sum_{i=1}^{\infty} f_i / 2^i M_i.$$

By Exercise **2.4J**, f is a C^∞ function for which carr f = U clearly holds. ■

In particular, second-countable n-manifolds admit partitions of unity, recovering **5.5.7**. ✍

☞ SUPPLEMENT 5.5B
Partitions of Unity: The Local Case

Theorem **5.5.17** reduces the problem of the existence of partitions of unity to the local one, namely finding partitions of unity in Banach spaces. This problem has been studied by Bonic and Frampton [1966] for separable Banach spaces.

5.5.19 Proposition (R. Bonic and J. Frampton [1966]). *Let* E *be a separable Banach space. The following are equivalent.*

(i) *Any open set of* E *is a* C^k *carrier*.

(ii) E *admits* C^k *partitions of unity subordinate to any open covering of* E.

(iii) *There exists a bounded nonempty* C^k *carrier in* E.

Proof By **5.5.17**, (i) and (ii) are equivalent since E is paracompact by **5.5.16**. It remains to be shown that (iii) implies (i), since clearly (ii) implies (iii).

This proceeds in several steps. First, we show that any neighborhood contains a C^k carrier. Let U be any open set and let carr f $\subset D_r(0)$ be the bounded carrier given by (iii), $f \in C^k(E)$, $f \geq 0$. Let $e \in U$, fix $e_0 \in$ carr f, and choose $\varepsilon > 0$ such that $D_\varepsilon(e) \subset U$. Define $g \in C^k(E)$, $g \geq 0$ by

$$g(v) = f(K(v - e) + e_0), \quad K > 0,$$

where K remains to be determined from the condition that carr g $\subset D_\varepsilon(e)$. An easy computation shows that if $K > (r + \| e_0 \|)/\varepsilon$, this inclusion is verified. Since $e \in$ carr g, carr g is an open neighborhood of e.

Second, we show that any open set can be covered by a countable locally finite family of C^k carriers. By the first step, the open set U can be covered by a family of C^k carriers. By Lindelöf's theorem, **1.1.6**, $U = \bigcup_n V_n$ where V_n is a C^k carrier, the union being over the positive integers. We need to find a refinement of this covering by C^k carriers. Let $f_n \in C^k(E)$ be such that carr $f_n = V_n$. Define $U_n = \{e \in E \mid f_n(e) > 0, f_i(e) < 1/n$ for all $i < n\}$. Clearly $U_1 = V_1$ and inductively $U_n = V_n \cap [\bigcap_{i<n} f_i^{-1}(]-\infty, 1/n[)]$. By the composite mapping theorem, the inverse image of a C^k carrier is a C^k carrier, so that $f_i^{-1}(]-\infty, 1/n[)$ is a C^k carrier, since $]\infty, 1/n[$ is a C^k carrier in $\mathbb{R}$ (see the proof of **5.5.18**). Finite intersections of C^k carriers is a C^k carrier (just take the product of the functions in question) so that U_n is also a C^k carrier. Clearly $U_n \subset V_n$. We shall prove that $\{U_n\}$ is a locally finite open covering of U. Let $e \in U$. If $e \in V_n$ for all n, then clearly $e \in U_1 = V_1$. If not, then there exists a smallest n, say N, such that $e \in V_N$. Then $f_i(e) = 0$ for $i < N$ and thus $e \in U_N = \{e \in E \mid f_N(e) > 0, f_i(e) < 1/N$ for all $i < N\}$. Thus, the sets U_n cover U. This open covering is also locally finite for if $e \in V_n$ and N is such that $f(e) > 1/N$, then the neighborhood $\{u \in U \mid f_n(e) > 1/N\}$ has empty intersections with all U_m for $m > N$.

Third we show that the open set U is a C^k-carrier. By the second step, $U = \bigcup_n U_n$, with U_n a locally finite open covering of U by C^k carriers. Then $f = \Sigma_n f_n$ is C^k, $f(e) \geq 0$ for all $e \in E$ and carr f = U. ∎

The separability assumption was used only in showing that (iii) implies (i). There is no general theorem known to us for nonseparable Banach spaces. Also, it is not known in general whether Banach spaces admit bounded C^k carriers, for $k \geq 1$. However, we have the following.

5.5.20 Proposition *If the Banach space* E *has a norm* C^k *away from its origin,* $k \geq 1$, *then* E *has bounded* C^k*-carriers.*

Proof By **4.2.13** there exists $\varphi : \mathbb{R} \to \mathbb{R}$, C^∞ with compact support and equal to one in a neighborhood of the origin. If $\|\cdot\| : E \setminus \{0\} \to \mathbb{R}$ is C^k, $k \geq 1$, then $\varphi \circ \|\cdot\| : E \setminus \{0\} \to \mathbb{R}$ is a nonzero map which is C^k, has bounded support $\|\cdot\|^{-1}(\operatorname{supp} \varphi)$, and can be extended in a C^k manner to E. ∎

Theorem **5.5.12** now follows from **5.5.20, 5.5.19** and **5.5.17.**

The situation with regard to Banach subspaces and submanifolds is clarified in the following proposition, whose proof is an immediate consequence of **5.5.19** and **5.5.17.**

5.5.21 Proposition (i) *If* E *is a Banach space admitting* C^k *partitions of unity then so does any closed subspace,*

(ii) *If a manifold admits* C^k *partitions of unity subordinate to any open covering, then so does any submanifold.*

We shall not develop this discussion of partitions of unity on Banach manifolds any further, but we shall end by quoting a few theorems that show how intimately connected partitions of unity are with the topology of the model space. By **5.5.19** and **5.5.20**, for separable Banach spaces one is interested whether the norm is C^k away from the origin. Restrepo [1964] has shown that a separable Banach space has a C^1 norm away from the origin if and only if its dual is separable. Bonic and Reis [1966] and Sundaresan [1967] have shown that if the norms on E and E^* are differentiable on $E \setminus \{0\}$ and $E^* \setminus \{0\}$, respectively, then E is reflexive, for E a real Banach space (not necessarily separable). Moreover, E is a Hilbert space if and only if the norms on E *and* E^* are twice differentiable away from the origin. This result has been strengthened by Leonard and Sundaresan [1973], who show that a real Banach space is isometric to a Hilbert space if and only if the norm is C^2 away from the origin and the second derivative of $e \mapsto \|e\|^2/2$ is bounded by 1 on the unit sphere; see Rao [1972] for a related result. Palais [1956b] has shown that any paracompact Banach manifold admits Lipschitz partitions of unity.

Because of the importance of the differentiability class of the norm in Banach spaces there has been considerable work in the direction of determining the exact differentiability class of concrete function spaces. Thus Bonic and Frampton [1966] have shown that the canonical norms on the spaces $L^p(\mathbb{R})$, $\ell^p(\mathbb{R})$, $p \geq 1$, $p < \infty$ are C^∞ away from the origin if p is even, C^{p-1} with $D(\|\cdot\|^{p-1})$ Lipschitz, if p is odd, and $C^{[p]}$ with $D^{[p]}(\|\cdot\|^p)$ Hölder continuous of order $p - [p]$, if p is not an integer. The space c_0 of sequences of real numbers convergent to zero has an equivalent norm that is C^∞ away from the origin, a result due to Kuiper. Using this result, Frampton and Tromba [1972] show that the Λ-spaces (closures of C^∞ in the Hölder norm) admit a C^∞ norm away from the origin. The standard norm on the Banach space of continuous real valued functions on $[0, 1]$ is nowhere differentiable. Moreover, since $C^0([0, 1], \mathbb{R})$ is separable with nonseparable dual, it is impossible to find an equivalent norm that is differentiable away from the origin. To our knowledge it is still an open problem whether $C^0([0, 1], \mathbb{R})$ admits C^∞ partitions of unity for $k \geq 1$.

Finally, the only results known to us for nonseparable Hilbert spaces are those of Wells [1971], [1973], who has proved that nonseparable Hilbert space admits C^2 partitions of unity. The techniques used in the proof, however, do not seem to indicate a general way to approach this problem. ✍

☞ SUPPLEMENT 5.5C
Simple Connectivity of Fiber Bundles

The goal of this supplement is to discuss the homotopy lifting property for locally trivial continuous fiber bundles over a paracompact base. This theorem is shown to imply an important criterion on the simple connectedness of the total space of fiber bundles with paracompact base.

5.5.21 Homotopy Lifting Theorem *Let* $\pi : E \to B$ *be a locally trivial* C^0 *fiber bundle and let* M *be a paracompact topological space. If* $h : [0, 1] \times M \to B$ *is a continuous homotopy and* $f : M \to E$ *is any continuous map satisfying* $\pi \circ f = h(0, \cdot)$, *there exists a continuous homotopy* $H: [0, 1] \times M \to E$ *satisfying* $\pi \circ H = h$ *and* $H(0, \cdot) = f$. *If in addition,* h *fixes some point* $m \in M$, *i.e.,* $h(t, m)$ *is constant for* t *in a segment* Δ *of* $[0, 1]$, *then* $H(t, m)$ *is also constant for* $t \in \Delta$.

Remark The property in the statement of the Theorem is called the ***homotopy lifting property***. A ***Hurewicz fibration*** is a continuous surjective map $\pi : E \to B$ satisfying the homotopy lifting property relative to any topological space M. Thus the theorem above says that a *locally trivial* C^0 *fiber bundle is a Hurewicz fibration relative to paracompact spaces.*

See Steenrod [1951] and Huebsch [1955] for the proof.

5.5.22 Corollary *Let* $\pi : E \to B$ *be a* C^0 *locally trivial fiber bundle. If the base* B *and the fiber* F *are simply connected, then* E *is simply connected.*

Proof Let $c : [0, 1] \to E$ be aloop, $c(0) = c(1) = e_0$. Then $d = \pi \circ c$ is a loop in B based at $\pi(e_0) = b_0$. Since B is simply connected there is a homotopy $h : [0, 1] \times [0, 1] \to B$ such that $h(0, t) = d(t)$, $h(1, t) = b_0$ for all $t \in [0, 1]$, and $h(s, 0) = h(s, 1) = b_0$ for all $s \in [0, 1]$. By the homotopy lifting theorem there is a homotopy $H: [0, 1] \times [0, 1] \to E$ such that $\pi \circ H = h$, $H(0, \cdot) = c$, and $H(s, 0) = H(0, 0) = c(0) = e_0$, $H(s, 1) = c(1) = e_0$. Since $(\pi \circ H)(1, t) = h(1, t) = b_0$, it follows that $t \mapsto H(1, t)$ is a path in $\pi^{-1}(b_0)$ starting at $H(1, 0) = e_0$ and ending also at $H(1, 1) = e_0$. Since $\pi^{-1}(b_0)$ is simply connected, there is a homotopy $k: [1, 2] \times [0, 1] \to$

$\pi^{-1}(b_0)$ such that $k(1, t) = H(1, t)$, $k(2, t) = e_0$ for all $t \in [0, 1]$ and $k(s, 0) = k(s, 1) = e_0$ for all $s \in [1, 2]$. Define the continuous homotopy $K: [0, 2] \times [0, 1] \to E$ by

$$K(s, t) = \begin{cases} H(s, t), & \text{if } s \in [0, 1] \\ k(s, t), & \text{if } s \in [1, 2] \end{cases}$$

and note that $K(0, t) = H(0, t) = c(t)$, $K(2, t) = k(2, t) = e_0$ for any $t \in [0, 1]$, and $K(s, 1) = e_0$ for any $s \in [0, 2]$. Thus c is contractible to e_0 and E is therefore simply connected. ■

Exercises

5.5A (Whitney) Show that any closed set F in $\mathbb{R}^n$ is the inverse image of 0 by a C^∞ real-valued positive function on $\mathbb{R}^n$. Generalize this to any n-manifold. (*Hint:* Cover $\mathbb{R}^n \backslash F$ with a sequence of open disks D_n and choose for each n a smooth function $\chi_n \ge 0$, satisfying $\chi_n \mid D_n > 0$, with the absolute value of χ_n and all its derivatives $\le 2^n$. Set $\chi = \Sigma_{n\ge 0}\, \chi_n$.)

5.5B In a paracompact topological space, an open subset need not be paracompact. Prove the following.

(i) If every open subset of a paracompact space is paracompact, then any subspace is paracompact.

(ii) Every open submanifold of a paracompact manifold is paracompact. (*Hint:* Use chart domains to conclude metrizability.)

5.5C Let $\pi : E \to M$ be a vector bundle, $E' \subset E$ a subbundle and assume M admits C^k partitions of unity subordinate to any open covering. Show that E' splits in E, i.e., there exists a subbundle E'' such that $E = E'' \oplus E''$. (*Hint:* The result is trivial for local bundles. Construct for every element of a locally finite covering $\{U_i\}$ a vector bundle map f_i whose kernel is the complement of $E' \mid U_i$. For $\{(U_i, g_i)\}$ a C^k partition of unity, put $f = \Sigma_i g_i f_i$ and show that $E = E' \oplus \ker f$.)

5.5D Let $\pi : E \to M$ be a vector bundle over the base M that admits C^k partitions of unity and with the fibers of E modeled on a Hilbert space. Show that E admits a C^k bundle metric, i.e., a C^k map $g : M \to T^0{}_2(E)$ that is symmetric, strongly nondegenerate, and positive definite at every point $m \in M$.

5.5E Let $E \to M$ be a line bundle over the manifold M admitting C^k partitions of unity subordinate to any open covering. Show that $E \times E$ is trivial (*Hint:* $E \times E = L(E^*, E)$ and construct a local base that can be extended.)

5.5F Assume M admits C^k partitions of unity. Show that any submanifold of M diffeomorphic to S^1 is the integral cuve of a C^k vector field on M.

5.5G Let M be a connected paracompact manifold. Show that there exists a C^∞ proper mapping $f: M \to \mathbb{R}^k$. (*Hint*: M is second countable, being Riemmannian. Show the statement for k = 1, where $f = \Sigma_{i\geq 1} i\varphi_i$, and $\{\varphi_i\}$ is a countable partition of unity.)

5.5H Let M be a connected paracompact n-manifold and $X \in \mathfrak{X}(M)$. Show that there exist $h \in \mathcal{F}(M)$, $h > 0$ such that $Y = hX$ is complete. (*Hint:* With f as in Exercise **5.5G** put $h = \exp\{-(X[f])^2\}$ so that $|Y[f]| \leq 1$. Hence $(f \circ c)(]a, b[)$ is bounded for any integral curve c of Y and]a, b[in the domain of c.)

5.5I Let M be a paracompact, non-compact manifold.

(i) Show that there exists a locally finite sequence of open sets $\{U_i \mid i \in \mathbb{Z}\}$ such that $U_i \cap U_{i+1} \neq \varnothing$ unless $j = i-1, i, i+1$, and each U_i is a chart domain diffeomorphic (by the chart map φ_i) with the open unit ball in the model space of M. See Fig. 5.5.1. (*Hint*: Let $\mathcal{V}$ be a locally finite open cover of M with chart domains diffeomorphic by their chart maps to the open unit ball such that no finite subcover of $\mathcal{V}$ covers M, and no two elements of $\mathcal{V}$ include each other. Let U_0, U_1 be distinct elements of $\mathcal{V}$, $U_0 \cap U_1 \neq \varnothing$. Let $U_{-1} \in \mathcal{V}$ be such that $U_{-1} \cap U_0 \neq \varnothing$; $U_{-1} \cap U_1 = \varnothing$; such a U_{-1} exists by local finiteness of $\mathcal{V}$. Now use induction.)

(ii) Use (i) to show that there exists an embedding of $\mathbb{R}$ in M as a closed manifold. (*Hint:* Let c_0 be a smooth curve in U_0 diffeomorphic by φ_0 to]–1, 1[connecting a point in $U_{-1} \cap U_0$ to a point in $U_0 \cap U_1$. Next, extend $c_0 \cap U_1$ smoothly to a curve c_1, diffeomorphic by φ_1 to]0, 2[ending inside $U_1 \cap U_2$; show that $c_1 \cap U_0$ extends the curve $c_0 \cap U_1$ inside $U_0 \cap U_1$. Now use induction.)

(iii) Show that on each non-compact paracompact manifold admitting partitions of unity there exists a non-complete vector field. (*Hint*: Embed $\mathbb{R}$ in M by (ii) and on $\mathbb{R}$ consider the vector field $\dot{x} = x^2$. Extend it to M via partitions of unity.)

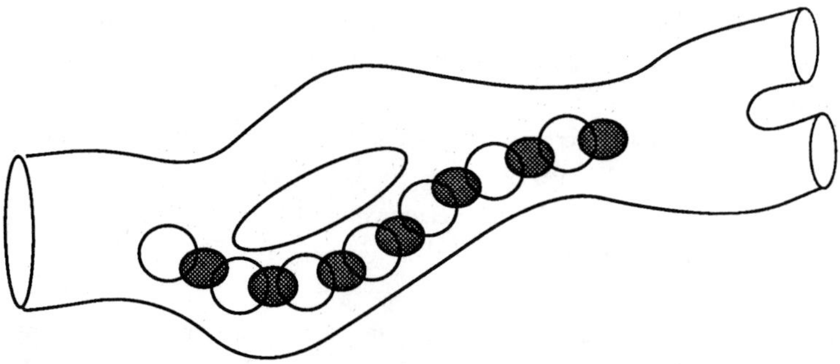

Figure 5.5.1

5.5J Show that every compact n-manifold embeds in some $\mathbb{R}^k$ for k big enough in the following way. If $\{(U_i, \varphi_i)\}_{i=1,\ldots,N}$ is a finite atlas with $\varphi_i(U_i)$ the ball of radius 2 in $\mathbb{R}^n$, let $\chi \in C^\infty(\mathbb{R}^n)$, $\chi = 1$ on the ball of radius 1 and $\chi = 0$ outside the ball of radius 2. Put $f_i = (\chi \circ \varphi_i) \cdot \varphi_i \colon M \to \mathbb{R}^n$, where $f_i = 0$ outside U_i. Show that f_i is C^∞ and that $\psi \colon M \to \mathbb{R}^{Nn} \times \mathbb{R}^N$, $\psi(m) = (f_1(m), \ldots, f_N(m), \chi(\varphi_1(m)), \ldots, \chi(\varphi_N(m)))$ is an embedding.

5.5K Let g be a Riemannian metric on M.

(i) Show that if N is a submanifold, its g-normal bundle $\nu_g(N) = \{v \in T_nM \mid n \in N, v \perp T_nN\}$ is a subbundle of TM.

(ii) Show that $TM \mid N = \nu_g(N) \oplus TN$.

(iii) If h is another Riemannian metric on M, show that $\nu_g(N)$ is a vector bundle iso-- morphic to $\nu_h(N)$.

5.5L Show that if $f \colon M \to N$ is a proper surjective submersion with M paracompact and N connected, then it is a locally trivial fiber bundle. (*Hint*: To show that all fibers are diffeomorphic, connect a fixed point of N with any other point by a smooth path and cover the path with the neighborhoods in N given by **5.5.14**.)

Chapter 6
Differential Forms

Differential k-forms are tensor fields of type (0, k) that are completely antisymmetric. Such tensor fields arise in many applications in physics, engineering, and mathematics. A hint at why this is so is the fact that the classical operations of grad, div, and curl and the theorems of Green, Gauss, and Stokes can all be expressed concisely in terms of differential forms. However, the examples of Hamiltonian mechanics and Maxwell's equations (see Chapter 8) show that their applicability goes well beyond this.

The goal of the chapter is to develop a special calculus of differential forms, due largely to E. Cartan [1945]. The exterior derivative operator **d** plays a central role. The properties of **d** and the expression of the Lie derivative in terms of it will be developed.

§6.1 *Exterior Algebra*

We begin with the exterior algebra of a vector space and extend this fiberwise to a vector bundle. As with tensor fields, the most important case is the tangent bundle of a manifold, which is considered in the next section.

We first recall a few facts about the permutation group on k elements. Proofs of the results that we cite are obtainable from virtually any elementary algebra book. The ***permutation group*** on k elements, denoted S_k, consists of all bijections $\sigma : \{1,..., k\} \to \{1,..., k\}$ usually given in the form of a table

$$\begin{pmatrix} 1 & \cdots & k \\ \sigma(1) & \cdots & \sigma(k) \end{pmatrix},$$

together with the structure of a group under composition of maps. Clearly, S_k has order k!. Letting $\{-1, 1\}$ have its natural multiplicative group structure, there is a homomorphism denoted sign : $S_k \to \{-1, 1\}$; that is, for $\sigma, \tau \in S_k$, $\mathrm{sign}(\sigma \circ \tau) = (\mathrm{sign}\ \sigma)(\mathrm{sign}\ \tau)$. A permutation σ is called ***even*** when sign $\sigma = +1$ and ***odd*** when sign $\sigma = -1$. This homomorphism can be described as follows. A ***transposition*** is a permutation that swaps two elements of $\{1,..., k\}$, leaving the remainder fixed. An even (odd) permutation can be written as the product of an even (odd) number of transpositions. The expression of σ as a product of transpositions is not unique, but the number of transpositions is always even or odd corresponding to σ being even or odd.

If E and F are Banach spaces, an element of $T^0{}_k(E, F) = L^k(E; F)$; i.e., a k-multilinear continuous mapping of $E \times \cdots \times E \to F$ is called ***skew symmetric*** when

$$t(e_1, ..., e_k) = (\text{sign } \sigma) t(e_{\sigma(1)}, ..., e_{\sigma(k)})$$

for all $e_1, ..., e_k \in E$ and $\sigma \in S_k$. This is equivalent to saying that $t(e_1, ..., e_k)$ changes sign when any two of $e_1, ..., e_k$ are swapped. The subspace of skew symmetric elements of $L^k(E; F)$ is denoted $L^k{}_a(E; F)$ (the subscript a stands for "alternating"). Some additional shorthand will be useful. Namely, let $\Lambda^0(E, F) = F$, $\Lambda^1(E, F) = L(E, F)$ and in general, $\Lambda^k(E, F) = L^k{}_a(E; F)$, be the vector space of skew symmetric F-valued multilinear maps or exterior F-valued k-forms on E. If $F = \mathbb{R}$, we write $\Lambda^0(E) = \mathbb{R}$, $\Lambda^1(E) = E^*$ and $\Lambda^k(E) = L^k{}_a(E; \mathbb{R})$; elements of $\Lambda^k(E)$ are called ***exterior*** k-***forms***. Some authors write $\Lambda^k(E^*)$ where we write $\Lambda^k(E)$.

To form elements of $\Lambda^k(E, F)$ from elements of $T^0{}_k(E; F)$, we can skew-symmetrize the latter. For example, if $t \in T^0{}_2(E)$, the two tensor At defined by

$$(At)(e_1, e_2) = (t(e_1, e_2) - t(e_2, e_1))/2$$

is skew symmetric and if t is already skew, At coincides with t. More generally, we make the following definition.

6.1.1 Definition *The* ***alternation mapping*** $A : T^0{}_k(E, F) \to T^0{}_k(E, F)$ *(for notational simplicity we do not index the* A *with* E, F *or* k) *is defined by*

$$At(e_1, \ldots, e_k) = \frac{1}{k!} \sum_{\sigma \in S_k} (\text{sign } \sigma)\, t(e_{\sigma(1)}, \ldots, e_{\sigma(k)}),$$

where the sum is over all k! *elements of* S_k.

6.1.2 Proposition A *is a linear mapping onto* $\Lambda^k(E, F)$, $A \mid \Lambda^k(E, F)$ *is the identity, and* $A \circ A = A$.

Proof Linearity of A is clear from the definition. If $t \in \Lambda^k(E, F)$, then

$$At(e_1, \ldots, e_k) = \frac{1}{k!} \sum_{\sigma \in S_k} (\text{sign } \sigma)\, t\,(e_{\sigma(1)}, \ldots, e_{\sigma(k)}) = \frac{1}{k!} \sum_{\sigma \in S_k} t(e_1, \ldots, e_k) = t(e_1, \ldots, e_k)$$

since S_k has order k!. This proves the first two assertions, and the last follows from them. ■

From $A = A \circ A$, it follows that $\| A \| \le \| A \|^2$, and so, as $A \ne 0$, $\| A \| \ge 1$. From the definition of A, we see $\| A \| \le 1$; thus $\| A \| = 1$. In particular, A is continuous.

6.1.3 Definition *If* $\alpha \in T^0{}_k(E)$ *and* $\beta \in T^0{}_\ell(E)$, *define their* ***wedge product*** $\alpha \wedge \beta \in \Lambda^{k+\ell}(E)$ *by*

$$\alpha \wedge \beta = \frac{(k+\ell)!}{k!\ell!} A(\alpha \otimes \beta).$$

For F-valued forms, we can also define $\wedge$, where $\otimes$ is taken with respect to a given bilinear form $B \in L(F_1, F_2; F_3)$. Since **A** and $\otimes$ are continuous, so is $\wedge$. There are several possible conventions for defining the wedge product $\wedge$. The one here conforms to Spivak [1979], and Bourbaki [1971] but not to Kobayashi and Nomizu [1963] or Guillemin and Pollack [1974]. See Exercise **6.1G** for the possible conventions. Our definition of $\alpha \wedge \beta$ is the one that eliminates the largest number of constants encountered later.

The reader should prove that for exterior forms

$$(\alpha \wedge \beta)(e_1, ..., e_{k+1}) = \sum (\text{sign } \sigma)\alpha(e_{\sigma(1)}, ..., e_{\sigma(k)})\beta(e_{\sigma(k+1)}, ..., e_{\sigma(k+1)}) \quad (1)$$

where the sum is over all (k, ℓ) shuffles σ. The reason for the name "shuffles" is that these are the kind of permutations made when a deck of $k + l$ cards is shuffled, with k cards held in one hand and ℓ in the other); that is, permutations σ of $\{1, 2,..., k + \ell\}$ such that $\sigma(1) < \cdots < \sigma(k)$ and $\sigma(k + 1) < \cdots < \sigma(k + \ell)$. Formula (1) is a convenient way to compute wedge products, as we see in the following examples.

6.1.4 Examples

A If α is a two-form and β is a one-form, then

$$(\alpha \wedge \beta)(e_1, e_2, e_3) = \alpha(e_1, e_2)\beta(e_3) - \alpha(e_1, e_3)\beta(e_2) + \alpha(e_2, e_3)\beta(e_1)$$

Indeed the only (2, 1) shuffles in S_3 are

$$\begin{pmatrix} 1 & 2 & 3 \\ 2 & 2 & 3 \end{pmatrix}, \quad \begin{pmatrix} 1 & 2 & 3 \\ 1 & 3 & 2 \end{pmatrix}, \quad \text{and} \quad \begin{pmatrix} 1 & 2 & 3 \\ 2 & 3 & 1 \end{pmatrix},$$

of which only the second one has sign -1.

B If α and β are one-forms, then

$$(\alpha \wedge \beta)(e_1, e_2) = \alpha(e_1)\beta(e_2) - \alpha(e_2)\beta(e_1)$$

since S_2 consists of two (1, 1) shuffles. ♦

6.1.5 Proposition *For* $\alpha \in T^0{}_k(E)$, $\beta \in T^0{}_\ell(E)$, *and* $\gamma \in T^0{}_m(E)$, *we have*

(i) $\alpha \wedge \beta = A\alpha \wedge \beta = \alpha \wedge A\beta$;

(ii) $\wedge$ *is bilinear;*

(iii) $\alpha \wedge \beta = (-1)^{k\ell}\, \beta \wedge \alpha$;

(iv) $\alpha \wedge (\beta \wedge \gamma) = (\alpha \wedge \beta) \wedge \gamma = \dfrac{(k+\ell+m)!}{k!\,\ell!\,m!} A(\alpha \otimes \beta \otimes \gamma)$.

Proof For (i), first note that if $\sigma \in S_k$ and we define $\sigma t(e_1, ..., e_k) = t(e_{\sigma(1)}, ..., e_{\sigma(k)})$, then $A(\sigma t) = (\text{sign}\, \sigma) At$, because

$$\begin{aligned} A(\sigma t)(e_1, \ldots, e_k) &= \frac{1}{k!} \sum_{\rho \in S_k} (\text{sign}\, \rho) t(e_{\rho\sigma(1)}, \ldots, e_{\rho\sigma(k)}) \\ &= \frac{1}{k!} \sum_{\rho \in S_k} (\text{sign}\, \sigma)(\text{sign}\, \rho\sigma) t(e_{\rho\sigma(1)}, \ldots, e_{\rho\sigma(k)}) \\ &= (\text{sign}\, \sigma) At(e_1, ..., e_k) \end{aligned}$$

since $\rho \mapsto \rho\sigma$ is a bijection. Therefore, since

$$At = \frac{1}{k!} \sum_{\sigma \in S_k} (\text{sign}\, \sigma)\sigma t,$$

we get

$$\begin{aligned} A(A\alpha \otimes \beta) &= A\left(\frac{1}{k!} \sum_{\tau \in S_k} (\text{sign}\, \tau)(\tau\alpha \otimes \beta)\right) \\ &= \frac{1}{k!} \sum_{\tau \in S_k} (\text{sign}\, \tau) A(\tau\alpha \otimes \beta) \qquad \text{(by linearity of } A) \\ &= \frac{1}{k!} \sum_{\tau \in S_k} (\text{sign}\, \tau') A\tau'(\alpha \otimes \beta) \end{aligned}$$

where $\tau' \in S_{k+\ell}$ is defined by

$$\tau'(1, ..., k, ..., k+\ell) = (\tau(1), ..., \tau(k), k+1, ..., k+\ell)$$

so $\text{sign}\, \tau = \text{sign}\, \tau'$ and $\tau\alpha \otimes \beta = \tau'(\alpha \otimes \beta)$. Thus the preceding expression for $A(A\alpha \otimes \beta)$ becomes

$$\frac{1}{k!} \sum_{\tau \in S_k} (\text{sign}\, \tau')(\text{sign}\, \tau') A(\alpha \otimes \beta) = A(\alpha \otimes \beta) \frac{1}{k!} \sum_{\tau \in S_k} 1 = A(\alpha \otimes \beta).$$

Thus $A(A\alpha \otimes \beta) = A(\alpha \otimes \beta)$ which is equivalent to $(A\alpha) \wedge \beta = \alpha \wedge \beta$. The other equality in (i) is similar. (ii) is clear since $\otimes$ is bilinear and A is linear.

For (iii), let $\sigma_0 \in S_{k+\ell}$ be given by $\sigma_0(1, ..., k+\ell) = (k+1, ..., k+\ell, 1, ..., k)$. Then

$(\alpha \otimes \beta)(e_1, \ldots, e_{k+\ell}) = (\beta \otimes \alpha)(e_{\sigma_0(1)}, \ldots, e_{\sigma_0(k+\ell)})$. Hence, by the proof of (i), $A(\alpha \otimes \beta) = (\text{sign } \sigma_0)$ $A(\beta \otimes \alpha)$. But $\text{sign } \sigma_0 = (-1)^{k\ell}$. Finally, for (iv),

$$\begin{aligned}\alpha \wedge (\beta \wedge \gamma) &= \frac{(k+\ell+m)!}{k!(\ell+m)!} A(\alpha \otimes (\beta \wedge \gamma)) \\ &= \frac{(k+\ell+m)!}{k!(\ell+m)!} \frac{(\ell+m)!}{\ell!m!} A(\alpha \otimes A(\beta \otimes \gamma)) \\ &= \frac{(k+\ell+m)!}{k!\ell!m!} A(\alpha \otimes \beta \otimes \gamma)\end{aligned}$$

since $A(\alpha \otimes A\beta) = A(\alpha \otimes \beta)$, which was proved in (i), and by associativity of $\otimes$. We calculate $(\alpha \wedge \beta) \wedge \gamma$ in the same way. ■

Conclusions (i)-(iii) hold (with identical proofs) for F-valued forms when the wedge product is taken with respect to a given bilinear mapping B. Associativity can also be generalized under suitable assumptions on the bilinear mappings, such as requiring F to be an associative algebra under B. Because of associativity, $\alpha \wedge \beta \wedge \gamma$ can be written with no ambiguity.

6.1.6 Examples

A If α^i, $i = 1, \ldots, k$ are one-forms, then

$$(\alpha^1 \wedge \cdots \wedge \alpha^k)(e_1, \ldots, e_k) = \sum_\sigma (\text{sign } \sigma)\alpha^1(e_{\sigma(1)}) \cdots \alpha^k(e_{\sigma(k)}) = \det[\alpha^i(e_j)].$$

Indeed, repeated application of **6.1.5** (iv) gives

$$\alpha_1 \wedge \cdots \wedge \alpha_k = \frac{(d_1 + \cdots + d_k)!}{d_1! \cdots d_k!} A(\alpha_1 \otimes \cdots \otimes \alpha_k) \tag{2}$$

where α_i is a d_i-form on E. In particular, if α_i is a one form, (2) gives

$$\alpha_1 \wedge \cdots \wedge \alpha_k = k!\, A(\alpha_1 \otimes \cdots \otimes \alpha_k) \tag{2'}$$

which yields the stated formula after using the definition of **A**. If $e_1, \ldots, e_n$ and $e^1, \ldots, e^n$ are dual bases, observe that as a special case, $(e^1 \wedge \cdots \wedge e^k)(e_1, \ldots, e_k) = 1$.

B If at least one or α or β is of even degree, then **6.1.5** (iii) says that $\alpha \wedge \beta = \beta \wedge \alpha$. If both are of odd degree, then $\alpha \wedge \beta = -\beta \wedge \alpha$. Thus, if α is a one-form, then $\alpha \wedge \alpha = 0$. But if α is a two-form, then in general $\alpha \wedge \alpha \neq 0$. For example, if $\alpha = e^1 \wedge e^2 + e^3 \wedge e^4 \in \Lambda^2(\mathbb{R}^4)$ where e^1, e^2, e^3, e^4 is the standard dual basis of $\mathbb{R}^4$, then $\alpha \wedge \alpha = 2e^1 \wedge e^2 \wedge e^3 \wedge e^4 \neq 0$.

C The properties listed in **6.1.5** make the computations of wedge products similar to polynomial multiplication, care being taken with commutativity. For example, if $\alpha^1, ..., \alpha^5$ are one forms on $\mathbb{R}^5$,

$$\alpha = 2\alpha^1 \wedge \alpha^3 + \alpha^2 \wedge \alpha^3 - 3\alpha^3 \wedge \alpha^4 \in \Lambda^2(\mathbb{R}^5)$$

and

$$\beta = -\alpha^1 \wedge \alpha^2 \wedge \alpha^5 + 2\alpha^1 \wedge \alpha^3 \wedge \alpha^4 \in \Lambda(\mathbb{R}^5),$$

then the wedge product $\alpha \wedge \beta$ is computed using the bilinearity and commutation properties of $\wedge$:

$$\begin{aligned} \alpha \wedge \beta &= -2(\alpha^1 \wedge \alpha^3) \wedge (\alpha^1 \wedge \alpha^2 \wedge \alpha^5) - (\alpha^2 \wedge \alpha^3) \wedge (\alpha^1 \wedge \alpha^2 \wedge \alpha^5) \\ &\quad + 3(\alpha^3 \wedge \alpha^4) \wedge (\alpha^1 \wedge \alpha^2 \wedge \alpha^5) + 4(\alpha^1 \wedge \alpha^3) \wedge (\alpha^1 \wedge \alpha^3 \wedge \alpha^4) \\ &\quad + 2(\alpha^2 \wedge \alpha^3) \wedge (\alpha^1 \wedge \alpha^3 \wedge \alpha^4) - 6(\alpha^3 \wedge \alpha^4) \wedge (\alpha^1 \wedge \alpha^3 \wedge \alpha^4) \\ &= 3\alpha^3 \wedge \alpha^4 \wedge \alpha^1 \wedge \alpha^2 \wedge \alpha^5 = 3\alpha^1 \wedge \alpha^2 \wedge \alpha^3 \wedge \alpha^4 \wedge \alpha^5 \quad \blacklozenge \end{aligned}$$

To express the wedge product in coordinate notation, suppose E is finite dimensional with basis $e_1, ..., e_n$. The components of $t \in T^0{}_k(E)$ are the real numbers

$$t_{i_1 \cdots i_k} = t(e_{i_1}, \ldots, e_{i_k}), \qquad 1 \le i_1, \ldots, i_k \le n. \tag{3}$$

For $t \in \Lambda^k(E)$, (3) is antisymmetric in its indices $i_1, ..., i_k$. For example, $t \in \Lambda^2(E)$ yields t_{ij}, a skew symmetric $n \times n$ matrix. From definition **6.1.1** of the alternation mapping and (3), we have

$$(\mathbf{A}t)_{i_1 \cdots i_k} = \frac{1}{k!} \sum_{\sigma \in S_k} (\text{sign}\, \sigma) t_{\sigma(i_1) \cdots \sigma(i_k)};$$

i.e., **A**t antisymmetrizes the components of t. For example, if $t \in T^0{}_2(E)$, then

$$(\mathbf{A}t)_{ij} = (t_{ij} - t_{ji})/2.$$

If $\alpha \in \Lambda^k(E)$ and $\beta \in \Lambda^\ell(E)$, then (1) and (3) yield

$$(\alpha \wedge \beta)_{i_1 \cdots i_{k+\ell}} = \sum (\text{sign}\, \sigma) \alpha_{\sigma(i_1) \ldots \sigma(i_k)} \beta_{\sigma(i_{k+1}) \ldots \sigma(i_{k+\ell})}$$

where the sum is over all the (k, ℓ)-shuffles in $S_{k+\ell}$.

6.1.7 Definition *The direct sum of the spaces* $\Lambda^k(E)$ $(i = 0, 1, 2 ...)$ *together with its structure of real vector space and multiplication induced by* $\wedge$, *is called the* ***exterior algebra*** *of* E, *or the* ***Grassmann algebra*** *of* E. *It is denoted by* $\Lambda(E)$.

Thus $\Lambda(E)$ is a ***graded associative algebra***, i.e., an algebra in which every element has a degree (a k-form has degree k), and the degree map is additive on products (by **6.1.2** and **6.1.3**). Elements of $\Lambda(E)$ may be written as finite sums of increasing degree exactly as one writes a polynomial as a sum of monomials. Thus if $a, b, c \in \mathbb{R}$, $\alpha \in \Lambda^1(E)$ and $\beta \in \Lambda^2(E)$ then $a + b\alpha + c\beta$ makes sense in $\Lambda(E)$. The one-form α can be understood as an element of $\Lambda^1(E)$ and also of $\Lambda(E)$, where α is identified with $0 + \alpha + 0 + 0 + \cdots$.

6.1.8 Proposition *Suppose* E *is finite dimensional and* $n = \dim E$. *Then for* $k > n$, $\Lambda^k(E) = \{0\}$, *while for* $0 < k \leq n$, $\Lambda^k(E)$ *has dimension* $n!/(n-k)!k!$. *The exterior algebra over* E *has dimension* 2^n. *If* $\{e_1, \ldots, e_n\}$ *is an (ordered) basis of* E *and* $\{e^1, \ldots, e^n\}$ *its dual basis, a basis of* $\Lambda^k(E)$ *is*

$$\{e^{i_1} \wedge \cdots \wedge e^{i_k} \mid 1 \leq i_1 < i_k < \cdots < i_k \leq n\}. \tag{4}$$

Proof First we show that the indicated wedge products span $\Lambda^k(E)$. If $\alpha \in \Lambda^k(E)$, then from **5.1.2**,

$$\alpha = \alpha(e_{i_1}, \ldots, e_{i_k})e^{i_1} \otimes \cdots \otimes e^{i_k},$$

where the summation convention indicates that this should be summed over all choices of $i_1, \ldots, i_k$ between 1 and n. If the linear operator $\mathbf{A}$ is applied to this sum and (2) is used, we get

$$\alpha = \mathbf{A}\alpha = \alpha(e_{i_1}, \ldots, e_{i_k})\mathbf{A}(e^{i_1} \otimes \cdots \otimes e^{i_k}) = \alpha(e_{i_1}, \ldots, e_{i_k})\,\frac{1}{k!}\,e^{i_1} \wedge \cdots \wedge e^{i_k}$$

The sum still runs over all choices of the $i_1, \ldots, i_k$ and we want only distinct, ordered ones. However, since α is skew symmetric, the coefficient in α is 0 if $i_1, \ldots, i_k$ are not distinct. If they are distinct and $\sigma \in S_k$, then

$$\alpha(e_{i_1}, \ldots, e_{i_k})e^{i_1} \wedge \cdots \wedge e^{i_k} = \alpha(e_{\sigma(i_1)}, \ldots, e_{\sigma(i_k)})e^{\sigma(i_1)} \wedge \cdots \wedge e^{\sigma(i_k)},$$

since both α and the wedge product change by a factor of sign σ. Since there are $k!$ of these rearrangements, we are left with

$$\alpha = \sum_{i_1 < \cdots < i_k} \alpha(e_{i_1}, \ldots, e_{i_k})e^{i_1} \wedge \cdots \wedge e^{i_k}.$$

This shows that (4) spans $\Lambda^k(E)$.

Secondly, we show that the elements in (4) are linearly independent. Suppose that

$$\sum_{i_1 < \cdots < i_k} \alpha_{i_1 \ldots i_k} e^{i_1} \wedge \cdots \wedge e^{i_k} = 0.$$

For fixed $i'_1, \ldots, i'_k$, let $j'_{k+1}, \ldots, j'_n$ denote the complementary set of indices, $j'_{k+1} < \cdots < j'_n$.

Then

$$\sum_{i_1<\cdots<i_k} \alpha_{i_1\ldots i_k} e^{i_1} \wedge \cdots \wedge e^{i_k} \wedge e^{j'_{k+1}} \wedge \cdots \wedge e^{j'_n} = 0.$$

However, this reduces to

$$\alpha_{i'_1\cdots i'_k} e^1 \wedge \cdots \wedge e^n = 0.$$

But $e^1 \wedge \cdots \wedge e^n \neq 0$, as $(e^1 \wedge \cdots \wedge e^n)(e_1, \ldots, e_n) = 1$ by Example **6.1.6A**. Hence the coefficients are zero. ■

6.1.9 Corollary *If* $\dim E = n$, *then* $\dim \Lambda^n(E) = 1$. *If* $\{\alpha^1, \ldots, \alpha^n\}$ *is a basis for* E^*, *then* $\alpha^1 \wedge \cdots \wedge \alpha^n$ *spans* $\Lambda^n(E)$.

Proof This follows from **6.1.8.** ■

6.1.10 Corollary *Let* $\alpha^1, \ldots, \alpha^k \in E^*$. *Then* $\alpha^1, \ldots, \alpha^k$ *are linearly dependent iff* $\alpha^1 \wedge \cdots \wedge \alpha^n$ *spans* $\Lambda^n(E)$.

Proof If $\alpha^1, \ldots, \alpha^k$ are linearly dependent, then

$$\alpha^i = \sum_{j\neq i} c_j \alpha^j$$

for some i. Since $\alpha \wedge \alpha = 0$, for α a one-form, we see that $\alpha^1 \wedge \cdots \wedge \alpha^k = 0$. Conversely, if $\alpha^1 \wedge \cdots \wedge \alpha^k = 0$, then by **6.1.9**, $\alpha^1, \ldots, \alpha^k$ is not a basis for $\mathrm{span}\{\alpha^1, \ldots, \alpha^k\}$. Therefore $k > \dim(\mathrm{span}\{\alpha^1, \ldots, \alpha^k\})$ and so $\alpha^1, \ldots, \alpha^k$ are linearly dependent. ■

6.1.11 Corollary *Let* $\theta \in \Lambda^1(E)$ *and* $\alpha \in \Lambda^k(E)$. *Then* $\theta \wedge \alpha = 0$ *iff there exists* $\beta \in \Lambda^{k-1}(E)$ *such that* $\alpha = \theta \wedge \beta$.

Proof Clearly, if $\alpha = \theta \wedge \beta$, then $\theta \wedge \alpha = 0$. Conversely, assume $\theta \wedge \alpha = 0$, $\theta \neq 0$ and choose a basis $\{e_i\}_{i\in I}$ of E such that for some $k \in I$, $e^k = \theta$. If

$$\sum_{i_1<\cdots<i_k} \alpha_{i_1\ldots i_k} e^{i_1} \wedge \cdots \wedge e^{i_k} = 0.$$

from $\theta \wedge \alpha = 0$ it follows that all summands not involving e^k are zero. Now factor e^k out of the remaining terms and call the resulting (k–1)-form β. ■

6.1.12 Examples

A Let $E = \mathbb{R}^2$, $\{e_1, e_2\}$ be the standard basis of $\mathbb{R}^2$ and $\{e^1, e^2\}$ the dual basis. Any element ω of $\Lambda^1(\mathbb{R}^2)$ can be written uniquely as $\omega = \omega_1 e^1 + \omega_2 e^2$, and any element ω of $\Lambda(\mathbb{R}^2)$ can be written uniquely as $\omega = \omega_{12}\, e^1 \wedge e^2$.

B Let $E = \mathbb{R}^3$, $\{e_1, e_2, e_3\}$ be the standard basis, and $\{e^1, e^2, e^3\}$ the dual basis. Any element $\omega \in \Lambda^1(\mathbb{R}^3)$ can be written uniquely as $\omega = \omega_1 e^1 + \omega_2 e^2 + \omega_3 e^3$. Similarly, any elements $\eta \in \Lambda^2(\mathbb{R}^3)$ and $\xi \in \Lambda^3(\mathbb{R}^3)$ can be uniquely written as $\eta = \eta_{12} e^1 \wedge e^2 + \eta_{13} e^1 \wedge e^3 + \eta_{23} e^2 \wedge e^3$ and $\xi = \xi_{123} e^1 \wedge e^2 \wedge e^3$. ♦

Since $\mathbb{R}^3$, $\Lambda^1(\mathbb{R}^3)$, and $\Lambda^2(\mathbb{R}^3)$ all have the same dimension, they are isomorphic. An isomorphism $\mathbb{R}^3 \cong \Lambda^1(\mathbb{R}^3) = (\mathbb{R}^3)^*$ is the standard one associated to a given basis: $e_i \mapsto e^i$, $i = 1, 2, 3$. An isomorphism of $\Lambda^1(\mathbb{R}^3)$ with $\Lambda^2(\mathbb{R}^3)$ is determined by

$$e^1 \mapsto e^2 \wedge e^3, \quad e^2 \mapsto e^3 \wedge e^1, \quad \text{and} \quad e^3 \mapsto e^1 \wedge e^2.$$

This isomorphism is usually denoted by $*\colon \Lambda^1(\mathbb{R}^3) \mapsto \Lambda^2(\mathbb{R}^3)$; we shall study this map in general in the next section under the name ***Hodge star operator.***

The standard isomorphism of $\mathbb{R}^3$ with $\Lambda^1(\mathbb{R}^3) = (\mathbb{R}^3)^*$ is given by the index lowering action $\flat$ of the standard metric on $\mathbb{R}^3$; i.e., $\flat(e_i) = e^i$. Then $* \circ \flat : \mathbb{R}^3 \to \Lambda^2(\mathbb{R}^3)$ has the following property:

$$(* \circ \flat)(e \times f) = \flat(e) \wedge \flat(f) \tag{5}$$

for all $v, w \in \mathbb{R}^3$, where $\times$ denotes the usual cross-product of vectors; i.e.,

$$v \times w = (v^2w^3 - v^3w^2)e_1 + (v^3w^1 - v^1w^3)e_2 + (v^1w^2 - v^2w^1)e_3.$$

The relation (5) follows from the definitions and the fact that if $\alpha = \alpha_1 e^1 + \alpha_2 e^2 + \alpha_3 e^3$ and $\beta = \beta_1 e^1 + \beta_2 e^2 + \beta_3 e^3$, then

$$\alpha \wedge \beta = (\alpha_2\beta_3 - \alpha_3\beta_2)e^2 \wedge e^3 + (\alpha_3\beta_1 - \alpha_1\beta_3)e^3 \wedge e^1 + (\alpha_1\beta_2 - \alpha_2\beta_1)e^1 \wedge e^2.$$

Exercises

6.1A Compute $\alpha \wedge \alpha$, $\alpha \wedge \beta$, $\beta \wedge \beta$, and $\beta \wedge \alpha \wedge \beta$ for $\alpha = 2e^1 \wedge e^3 - e^2 \wedge e^3 \in \Lambda^2(\mathbb{R}^3)$ and $\beta = -e^1 + e^2 - 2e^3$ where $\{e^1, e^2, e^3\}$ is a basis of $(\mathbb{R}^3)^*$.

6.1B If k! is omitted in the definition of **A (6.1.1)**, show that $\wedge$ fails to be associative.

6.1C Let $v_1, \ldots, v_k$ be linearly dependent vectors. Show that $\alpha(v_1, \ldots, v_k) = 0$ for all $\alpha \in \Lambda^k(E)$.

6.1D Let E be finite dimensional. Show that $\Lambda^k(E^*)$ is isomoprhic to $(\Lambda^k(E))^*$. (*Hint:* Define $\varphi : (\Lambda^k(E))^* \to \Lambda^k(E^*)$ by $\varphi(\sigma)(\alpha^1, ..., \alpha^k) = \sigma(\alpha^1 \wedge \cdots \wedge \alpha^k)$ and construct its inverse using the basis in **6.1.8**(i).)

6.1E Let $\{e_1, ..., e_n\}$ be a basis of E with dual basis $\{e^1, ..., e^n\}$ and let $\{f_1, ..., f_m\}$ be a basis of F. Show the following:

(i) every $\beta \in \Lambda^k(E, F)$ can be uniquely written as $\beta = \Sigma_{1 \le i \le m} \beta_i f_i$ for $\beta_i \in \Lambda^k(E)$, where $(\gamma f)(v_1, ..., v_k) = \gamma(v_1, ..., v_k) f \in F$ for $v_1, ..., v_k \in E$, $f \in F$, and $\gamma \in \Lambda^k(E)$;

(ii) $\{(e^{i_1} \wedge \cdots \wedge e^{i_k}) f_j \mid i_1 < \cdots < i_k\}$ is a basis of $\Lambda^k(E, F)$ and thus $\dim(\Lambda^k(E, F)) = \dfrac{mn!}{(n-k)!k!}$;

(iii) $\dim(\Lambda(E, F)) = m2^n$;

(iv) if $B \in L(\mathbb{R}, F; F)$, where $B(t, f) = tf$ and $\wedge$ is the wedge product defined by B, regarded as a map $\wedge : \Lambda^1(E) \times \Lambda^k(E, F) \to \Lambda^{k+1}(E, F)$ show that

$$\alpha \wedge \beta = \Sigma_{1 \le i \le m} (\alpha \wedge \beta_i) f_i.$$

If $E = \mathbb{R}^3$, $F = \mathbb{R}^2$,

$$\alpha = e^1 \wedge e^2 - 2e^1 \wedge e^3, \quad \text{and} \quad \beta = (e^1 \wedge e^3) f_1 + 2(e^2 \wedge e^3) f_2 - (e^1 \wedge e^2) f_3,$$

compute $\alpha \wedge \beta$.

6.1F Let $\{e_1, ..., e_k\}$ and $\{f_1, ..., f_k\}$ be linearly independent sets of vectors. Show that they span the same k-dimensional subspace iff $f_1 \wedge \cdots \wedge f_k = a e_1 \wedge \cdots \wedge e_k$, where $a \neq 0$. (Give a definition of $f_1 \wedge \cdots \wedge f_k$ as part of your answer.) Show that in fact

$$a = \det \varphi \quad \text{where} \quad \varphi : \text{span}\{e_1, ..., e_k\} \to \text{span}\{f_1, ..., f_k\}$$

is determined by $\varphi(e_i) = f_i$, $i = 1, ..., k$. Use this to relate Λ^k with G_k in **3.1.8 G**.

6.1G (P. Chernoff, J. Robbin). Let $\wedge'$ be another wedge product on forms that is associative and satisfies $\alpha \wedge' \beta = c(k, \ell) \alpha \wedge \beta$, where α is a k-form and β is an l-form, $c(k, \ell)$ is a scalar, and forms of degree zero act as scalars.

(i) Prove the "cocycle identity" $c(k, \ell) c(k + \ell, m) = c(k, \ell + m) c(\ell, m)$.

(ii) Define $\psi(\ell)$ inductively by $\psi(0) = \psi(1) = 1$ and $\psi(\ell + 1) = c(1, \ell)\psi(\ell)$. Show that $c(k, \ell) = \psi(k + \ell)/\psi(k)\psi(\ell)$. Deduce that $c(k, \ell) = c(\ell, k)$; i.e., $\wedge'$ satisfies $\alpha \wedge' \beta = (-1)^{k\ell} \beta \wedge' \alpha$ automatically.

(iii) Show that c given by (iii) yields an associative wedge product. ($\psi(k) = 1/k!$ converts our wedge product convention to that of Kobayashi and Nomizu [1963]).

§6.2 *Determinants, Volumes, and the Hodge Star Operator*

According to linear algebra, the determinant of an $n \times n$ matrix is a skew symmetric function of its rows or columns. Thus, if $x_1, ..., x_n \in \mathbb{R}^n$, then $\omega(x_1, ..., x_n) = \det[x_1, ..., x_n]$ denotes the $n \times n$ matrix whose colums are $x_1, ..., x_n$, is an element of $\Lambda^n(\mathbb{R}^n)$. We also recall from linear algebra that $\det[x_1, ...,x_n]$ is the oriented volume of the parallelipiped spanned by $x_1, ..., x_n$ (Fig. 6.2.1) and that if x_i has components x^j_i, the determinant is given by

$$\det[x_1, ...,x_n] = \sum_{\sigma \in S_n} (\text{sign } \sigma) x^1_{\sigma(1)} \cdots\cdots x^n_{\sigma(s)}.$$

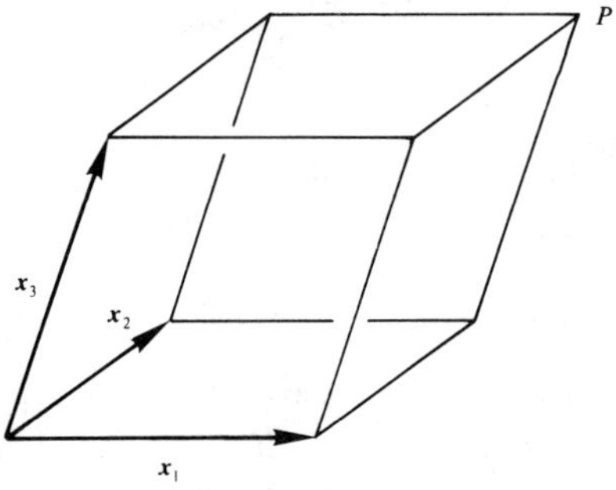

Figure 6.2.1 Volume(P) = $\det[x_1, x_2, x_3]$

In this section determinants and volumes are approached from the point of view of exterior algebra. Throughout this section E is assumed to be a finite-dimensional vector space and we shall denote its dimension by $\dim E = n$.

If $\varphi : \mathbb{R}^3 \to \mathbb{R}^3$ is a linear transformation, it is shown in linear algebra that $\det \varphi$ is the oriented volume of the image of the unit cube under φ (see Fig. 6.2.2). In fact $\det \varphi$ is a measure of how φ changes volumes. In advanced calculus, this fact is the basis for introducing the Jacobian determinant in the change of variables formula for multiple integrals. This background will lead to the development of the Jacobian determinant of a mapping of manifolds.

Recall that the pull-back $\varphi^*\alpha$ of $\alpha \in T^0_k(F)$ by $\varphi \in L(E, F)$ is the element of $T^0_k(E)$ defined by $(\varphi^*\alpha)(e_1, ..., e_k) = \alpha(\varphi(e_1), ..., \varphi(e_k))$. If $\varphi \in GL(E, F)$, then $\varphi_* = (\varphi^{-1})^*$ denotes the push-forward. The following proposition is a consequence of the definitions and **5.1.9.** (The same results hold for Banach space valued forms.)

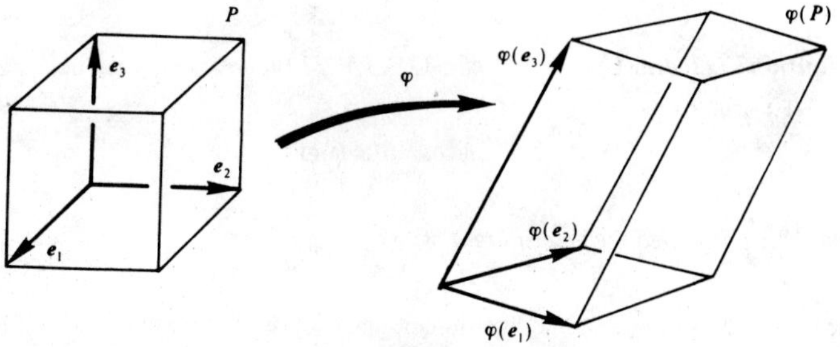

Figure 6.2.2

6.2.1 Proposition *Let* $\varphi \in L(E, F)$, *and* $\psi \in L(F, G)$

(i) $\varphi^* : T^0{}_k(F) \to T^0{}_k(E)$ *is linear, and* $\varphi^*(\Lambda(F)) \subset \Lambda^k(E)$.

(ii) $(\psi \circ \varphi)^* = \varphi^* \circ \psi^*$.

(iii) *If* φ *is the identity, so is* φ^*.

(iv) *If* $\varphi \in GL(E,F)$, *then* $\varphi^* \in GL(T^0{}_k(F), T^0{}_k(F))$, $(\varphi^{-1})^* = \varphi_*$, *and* $(\varphi^*)^{-1} = (\varphi^{-1})_*$; *if* $\psi \in GL(F, G)$, *then* $(\psi \circ \varphi)_* = \psi_* \circ \varphi_*$.

(v) *If* $\alpha \in \Lambda^k(F)$, *and* $\beta \in \Lambda^\ell(F)$, *then* $\varphi^*(\alpha \wedge \beta) = \varphi^*\alpha \wedge \varphi^*\beta$.

For example, if

$$\beta = \beta_{a_1 \cdots a_k} f^{a_1} \wedge \cdots \wedge f^{a_k} \in \Lambda^k(F) \quad \text{(the sum over } a_1 < \cdots < a_k)$$

and $\varphi \in L(E, F)$ is given by the matrix $[A^a{}_i]$, i.e., relative to ordered bases $\{e_1, ..., e_n\}$ of E and $\{f_1, ..., f_m\}$ of F, one has $\varphi(e_i) = A^a{}_i f_a$, then

$$\begin{aligned}(\varphi^*\beta) &= \beta_{a_1 \cdots a_k} \varphi^*(f^{a_1}) \wedge \cdots \wedge \varphi^*(f^{a_k}), \quad \text{(sum over } a_1 < \cdots < a_k) \\ &= \beta_{a_1 \cdots a_k} A^{a_1}{}_{j_1} e^{j_1} \wedge \cdots \wedge A^{a_k}{}_{j_k} e^{j_k} \\ &= \beta_{a_1 \cdots a_k} A^{a_1}{}_{j_1} \cdots A^{a_k}{}_{j_k} e^{j} \wedge \cdots \wedge e^{j_k}, \\ &= k!\beta_{a_1 \cdots a_k} A^{a_1}{}_{j_1} \cdots A^{a_k}{}_{j_k} e^{j_1} \wedge \cdots \wedge e^{j_k}, \; j_1 < \cdots < j_k.\end{aligned}$$

That is

$$\begin{aligned}(\varphi^*\beta) &= \beta_{a_1 \cdots a_k} \varphi^*(f^{a_1}) \wedge \cdots \wedge \varphi^*(f^{a_k}), \quad \text{(sum over } a_1 < \cdots < a_k) \\ &= \beta_{a_1 \cdots a_k} A^{a_1}{}_{j_1} e^{j_1} \wedge \cdots \wedge A^{a_k}{}_{j_k} e^{j_k} \\ &= \beta_{a_1 \cdots a_k} A^{a_1}{}_{j_1} \cdots A^{a_k}{}_{j_k} e^{j} \wedge \cdots \wedge e^{j_k}, \\ &= k!\beta_{a_1 \cdots a_k} A^{a_1}{}_{j_1} \cdots A^{a_k}{}_{j_k} e^{j_1} \wedge \cdots \wedge e^{j_k}, \; j_1 < \cdots < j_k.\end{aligned}$$

Recall that $\varphi^*: \Lambda^n(E) \to \Lambda^n(E)$ is a linear mapping and $\Lambda^n(E)$ is one-dimensional. Thus, if ω_0

is a basis and $\omega = c\omega_0$, then $\varphi^*\omega = c\varphi^*\omega_0 = b\omega$ for some constant b, clearly unique.

6.2.2 Definition *Let* $\dim(E) = n$ *and* $\varphi \in L(E, E)$. *The unique constant* $\det \varphi$, *such that* $\varphi^* : \Lambda^n(E) \to \Lambda^n(E)$ *satisfies*

$$\varphi^*\omega = (\det \varphi)\omega$$

for all $\omega \in \Lambda^n(E)$ *is called the* ***determinant*** *of* φ.

The definition shows that the determinant does not depend on the choice of basis of E, nor does it depend on a norm on E. To compute $\det \varphi$, choose a basis $\{e_1, ..., e_n\}$ of E with dual basis $\{e^1, ..., e^n\}$. Let $\varphi \in L(E, E)$ have the matrix $[A^j_i]$; that is, $\varphi(e_i) = \Sigma_{1 \le j \le n} A^j_i e_j$. By Example **6.1.6A**,

$$\varphi^*(e^1 \wedge \ldots \wedge e^n)(e_1, ..., e_n) = (e^1 \wedge \ldots \wedge e^n)(\varphi(e_1), ..., \varphi(e_n)) = \det[e^j(\varphi(e_i))] = \det[A^j_i].$$

Since $(e^1 \wedge \ldots \wedge e^n)(e_1, ..., e_n) = 1$ we get $\det \varphi = \det[A^j_i]$, the classical expression of the determinant of a matrix with $x_1, ..., x_n$ as columns, where x_i has components A^j_i. Thus the definition of the determinant in **6.2.2** coincides with the classical one.

From properties of pull-back, we deduce corresponding properties of the determinant, all of which are well known from linear algebra.

6.2.3 Proposition *Let* $\varphi, \psi \in L(E, E)$. *Then*

(i) $\det(\varphi \circ \psi) = (\det \varphi)(\det \psi)$;

(ii) *if* φ *is the identity,* $\det \varphi = 1$;

(iii) φ *is an isomorphism iff* $\det \varphi \neq 0$, *and in this case* $\det(\varphi^{-1}) = (\det \varphi)^{-1}$.

Proof For (i), $(\varphi \circ \psi)^*\omega = \det(\varphi \circ \psi)\omega$; but $(\varphi \circ \psi)^*\omega = (\psi^* \circ \varphi^*)\omega$. Hence, $(\varphi \circ \psi)^*\omega = \psi^*(\det \varphi)\omega = (\det \psi)(\det \varphi)\omega$ and (i) follows. (ii) follows at once from the definition. For (iii), suppose φ is an isomorphism with inverse φ^{-1}. Therefore, by (i) and (ii), $1 = \det(\varphi \circ \varphi^{-1}) = (\det \varphi)(\det \varphi^{-1})$, and, in particular, $\det \varphi \neq 0$. Conversely, if φ is not an isomorphism there is an $e_1 \neq 0$ satisfying $\varphi(e_1) = 0$. Extend to a basis $\{e_1, e_2, ..., e_n\}$. Then for all n-forms ω, we have $(\varphi^*\omega)(e_1, ..., e_n) = \omega(0, \varphi(e_2), ..., \varphi(e_n)) = 0$. Hence, $\det \varphi = 0$. ■

In Chapter 2 we saw that if E and F are finite dimensional, one convenient norm giving the topology of L(E, F) is the operator norm:

$$\| \varphi \| = \sup\{\| \varphi(e)\| \mid \| e \| = 1\} = \sup\left\{ \frac{\| \varphi(e) \|}{\| e \|} \,\middle|\, e \neq 0 \right\}$$

where $\| e \|$ is a norm on E. (See §**2.2.**) Hence, for any $e \in E$,

$$\| \varphi(e) \| \le \| \varphi \| \| e \|.$$

6.2.4 Proposition *The map* $\det : L(E, E) \to \mathbb{R}$ *is continuous.*

Proof This is clear from the component formula for det, but let us also give a coordinate free proof. Note that

$$\begin{aligned}\| \omega \| &= \sup\{ \, | \omega(e_1, \ldots, e_n) | \, | \, \| e_1 \| = \cdots = \| e_n \| = 1 \} \\ &= \sup\{ \, | \omega(e_1, \ldots, e_n) | / \| e_1 \| \cdots \| e_n \| \, | \, e_1, \ldots, e_n \neq 0 \}\end{aligned}$$

is a norm on $\Lambda^n(E)$ and $| \omega(e_1, \ldots, e_n) | \le \| \omega \| \| e_1 \| \cdots \| e_n \|$. Then, for $\varphi, \psi \in L(E, E)$,

$$\begin{aligned}&| \det \varphi - \det \psi | \, \| \omega \| \\ &= \| \varphi^*\omega - \psi^*\omega \| \\ &= \sup\{ \, | \omega(\varphi(e_1), \ldots, \varphi(e_n)) - \omega(\psi(e_1), \ldots, \psi(e_n)) | \, | \, \| e_1 \| = \cdots = \| e_n \| = 1 \} \\ &\le \sup\{ \, | \omega(\varphi(e_1) - \psi(e_1), \varphi(e_2), \ldots, \varphi(e_n)) | + \cdots \\ &\qquad + | \omega(\psi(e_1), \psi(e_2), \ldots, \varphi(e_n) - \psi(e_n)) | \, | \, \| e_1 \| = \cdots = \| e_n \| = 1 \} \\ &\le \| \omega \| \| \varphi - \psi \| \{ \| \varphi \|^{n-1} + \| \varphi \|^{n-2} \| \psi \| + \cdots + \| \psi \|^{n-1} \} \\ &\le \| \omega \| \| \varphi - \psi \| (\| \varphi \| + \| \psi \|)^{n-1}.\end{aligned}$$

Consequently, $| \det \varphi - \det \psi | \le \| \varphi - \psi \| (\| \varphi \| + \| \psi \|)^{n-1}$ from which the result follows. ■

In Chapter 2 we saw that the set of isomorphisms of E to F form an open subset of L(E, F). Using the determinant, we can give an alternate proof in the finite-dimensional case.

6.2.5 Proposition *Suppose* E *and* F *are finite-dimensional and let* $GL(E, F)$ *denote those* $\varphi \in L(E, F)$ *that are isomorphisms. Then* $GL(E, F)$ *is an open subset of* L(E, F).

Proof If $GL(E, F) = \varnothing$, the conclusion is true. Otherwise, there is an isomorphism $\psi \in GL(E, F)$. A map φ in L(E, F) is an isomorphism if and only if $\psi^{-1} \circ \varphi$ is also. This happens precisely when $\det(\psi^{-1} \circ \varphi) \neq 0$. Therefore, $GL(E, F)$ is the inverse image of $\mathbb{R} \setminus \{0\}$ under the map taking φ to $\det(\psi^{-1} \circ \varphi)$. Since this is continuous and $\mathbb{R} \setminus \{0\}$ is open, $GL(E, F)$ is also open. ■

The basis elements of $\Lambda^n(E)$ enable us to define orientation or "handedness" of a vector space.

6.2.6 Definition *The nonzero elements of the one-dimensional space* $\Lambda^n(E)$ *are called* ***volume elements.*** *If* ω_1 *and* ω_2 *are volume elements, we say* ω_1 *and* ω_2 *are* ***equivalent*** *iff there is a* c

> 0 *such that* $\omega_1 = c\omega_2$. *An equivalence class* $[\omega]$ *of volume elements on* E *is called an* ***orientation*** *on* E. *An* ***oriented vector space*** $(E, [\omega])$ *is a vector space* E *together with an orientation* $[\omega]$ *on* E; $[-\omega]$ *is called the* ***reverse orientation****. A basis* $\{e_1, ..., e_n\}$ *of the oriented vector space* $(E, [\omega])$ *is called* ***positively*** *(resp.* ***negatively****) oriented, if* $\omega(e_1, ..., e_n) > 0$ *(resp.* $<$ 0).

The last statement is independent of the representative of the orientation $[\omega]$, for if $\omega' \in [\omega]$, then $\omega' = c\omega$ for some $c > 0$, and thus $\omega'(e_1, ..., e_n)$ and $\omega(e_1, ..., e_n)$ have the same sign. Also note that a vector space E has exactly two orientations: one given by selecting an arbitrary dual basis $\{e^1, ..., e^n\}$ and taking $[e^1 \wedge ... \wedge e^n]$; the other is its reverse orientation.

This definition of orientation is related to the concept of orientation from calculus as follows. In $\mathbb{R}^3$, a right-handed coordinate system like the one in Fig. 6.2.1 is by convention positively oriented, as are all other right-handed systems. On the other hand, any left-handed coordinate system, obtained for example from the one in Fig. 6.2.1 by interchanging x_1 and x_2, is by convention negatively oriented. Thus one would call a positive orientation in $\mathbb{R}^3$ the set of all right-handed coordinate systems. The key to the abstraction of this construction for any vector space lies in the observation that the determinant of the change of ordered basis of two right-handed systems in $\mathbb{R}^3$ is always strictly positive. Thus, if E is an n-dimensional vector space, define an equivalence relation on the set of ordered bases in the following way: two bases $\{e_1, ..., e_n\}$ and $\{e_1', ..., e_n'\}$ are equivalent iff $\det \varphi > 0$, where $\varphi \in GL(E)$ is given by $\varphi(e_i) = e_i'$, $i = 1, ..., n$. We can relate n-forms to the bases by associating to a basis $\{e_1, ..., e_n\}$ and its dual basis $\{e^1, ..., e^n\}$ the n-form $\omega = e^1 \wedge ... \wedge e^n$. The following proposition shows that this association gives an identification of the corresponding equivalence classes.

6.2.7 Proposition *An orientation in a vector space is uniquely determined by an equivalence class of ordered bases.*

Proof If $[\omega]$ is an orientation of E there exists a basis $\{e_1, ..., e_n\}$ such that $\omega(e_1, ..., e_n) \neq 0$ since $\omega \neq 0$ in $\Lambda^n(E)$. Changing the sign of e_1 if necessary, we can find a basis that is positively oriented. Let $\{e_1', ..., e_n'\}$ be an equivalent basis and $\varphi \in GL(E)$, defined by $\varphi(e_i) = e_i'$, $i = 1, ..., n$ be the change of basis isomorphism. Then if $\omega' \in [\omega]$, there exists $c > 0$ such that $\omega' = c\omega$, so we get

$$\omega'(e_1', ..., e_n') = c\omega(\varphi(e_1), ..., \varphi(e_n)) = c(\varphi^*\omega)(e_1, ..., e_n) = c(\det \varphi)\omega(e_1, ..., e_n) > 0.$$

That is, $[\omega]$ uniquely determines the equivalence class of $\{e_1, ..., e_n\}$.

Conversely, let $\{e_1, ..., e_n\}$ be a basis of E and let $\omega = e^1 \wedge ... \wedge e^n$, where $\{e^1, ..., e^n\}$ is the dual basis. As before, $\omega'(e_1', ..., e_n') > 0$ for any $\omega' \in [\omega]$ and $\{e_1', ..., e_n'\}$ equivalent to $\{e_1, ..., e_n\}$; thus, the equivalence class of the ordered basis $\{e_1, ..., e_n\}$ uniquely determines the orientation $[\omega]$. ■

Next we shall discuss volume elements in inner product spaces. An important point is that to get a particular volume element on E requires additional structure, although the determinant does not. The idea is based on the fact that in $\mathbb{R}^3$ the volume of the parallelipiped $P = P(x_1, x_2, x_3)$ spanned by three positively oriented vectors x_1, x_2, and x_3 can be expressed independent of any basis as

$$\mathrm{Vol}(P) = (\det[\langle x_i, x_j\rangle])^{1/2},$$

where $[\langle x_i, x_j\rangle]$ denotes the symmetric 3×3 matrix whose entries are $\langle x_i, x_j\rangle$. If x_1, x_2, and x_3 are negatively oriented, $\det[\langle x_i, x_j\rangle] < 0$ and so the formula has to be modified to

$$\mathrm{Vol}(P) = (|\det[\langle x_i, x_j\rangle]|)^{1/2}. \tag{1}$$

This indicates that besides the volumes, there are quantities involving absolute values of volume elements that are also important. This leads to the notion of densities.

6.2.8 Definition *Let* α *be a real number. A continuous mapping* $\rho : E \times \cdots \times E \to \mathbb{R}$ (n *factors of* E *for* E *an* n*-dimensional vector space) is called an* α***-density*** *if* $\rho(\varphi(v_1), \ldots, \varphi(v_n)) = |\det \varphi|^\alpha \rho(v_1, \ldots, v_n)$, *for all* $v_1, \ldots, v_n \in E$ *and all* $\varphi \in L(E, E)$. *Let* $|\Lambda|^\alpha(E)$ *denote the* α*-densities on* E. *With* $\alpha = 1$, 1*-densities on* E *are simply called* ***densities*** *and* $|\Lambda|^1(E)$ *is denoted by* $|\Lambda|(E)$.

The determinant of φ in this definition is taken with respect to any volume element of E. As we saw in **6.2.2**, this is independent of the choice of the volume element. Note that $|\Lambda|^\alpha(E)$ is one-dimensional. Indeed, if ρ_1 and $\rho_2 \in |\Lambda|^\alpha(E)$, $\rho_1 \neq 0$, and $\{e_1, \ldots, e_n\}$ is a basis of E, then $\rho_2(e_1, \ldots, e_n) = a\rho_1(e_1, \ldots, e_n)$, for some constant $a \in \mathbb{R}$. If $v_1, \ldots, v_n \in E$, let $v_i = \varphi(e_i)$, defining $\varphi \in L(E, E)$. Then

$$\rho_2(v_1, \ldots, v_n) = |\det \varphi|^\alpha \rho_2(e_1, \ldots, e_n) = a\,|\det \varphi|^\alpha \rho_1(e_1, \ldots, e_n) = a\rho_1(v_1, \ldots, v_n);$$

i.e., $\rho_2 = a\rho_1$.

Alpha-densities can be constructed from volume elements as follows. If $\omega \in \Lambda^n(E)$, define $|\omega|^\alpha \in |\Lambda|^\alpha(E)$ by $|\omega|^\alpha(e_1, \ldots, e_n) = |\omega(e_1, \ldots, e_n)|^\alpha$ where $e_1, \ldots, e_n \in E$. This association defines an isomorphism of $\Lambda^n(E)$ with $|\Lambda|^\alpha(E)$. Thus one often uses the notation $|\omega|^\alpha$ for α-densities.

We shall construct canonical volume elements (and hence α-densities) for vector spaces carrying a bilinear symmetric nondegenerate covariant two-tensor, and in particular for inner product spaces. First we recall a fact from linear algebra.

6.2.9 Proposition *Let* E *be an* n*-dimensional vector space and* $g = \langle , \rangle \in T^0_2(E)$ *be symmetric and of rank* r; *i.e., the map* $e \in E \mapsto g(e, \cdot) \in E^*$ *has* r*-dimensional range. Then there is an ordered basis* $\{e_1, \ldots, e_n\}$ *of* E *with dual basis* $\{e^1, \ldots, e^n\}$ *such that*

$$g = \sum_{i=1}^{r} c_i e \otimes e^i,$$

where $c_i = \pm 1$ *and* $r \leq n$, *or equivalently, the matrix of* g *is*

$$\begin{bmatrix} c_1 & 0 & 0 & \cdots & 0 & 0 & \cdots & 0 \\ 0 & c_2 & 0 & \cdots & 0 & 0 & \cdots & 0 \\ 0 & 0 & c_3 & \cdots & 0 & 0 & \cdots & 0 \\ & \cdots & \cdots & & \cdots & & & \\ 0 & 0 & 0 & \cdots & c_r & 0 & \cdots & 0 \\ 0 & 0 & 0 & \cdots & 0 & 0 & \cdots & 0 \\ \cdots & \cdots & \cdots & & \cdots & \cdots & & \cdots \\ 0 & 0 & 0 & \cdots & 0 & 0 & \cdots & 0 \end{bmatrix}$$

This basis $\{e_1, ..., e_n\}$ *is called a* **g-orthonormal basis.** *Moreover, the number of basis vectors for which* $g(e_i, e_i) = 1$ *(resp.* $g(e_i, e_i) = -1$*) is unique and equals the maximal dimension of any subspace on which* g *is positive (resp. negative) definite. The number* s *= the number of* + 1's *minus the number of* −1's *is called the* **signature** *of* g. *The number of* −1's *is called the* **index** *of* g *and is denoted* Ind(g).

Proof (Gram-Schmidt argument) Since g is symmetric, the following polarization identity holds:

$$g(e, f) = [(g(e + f, e + f) - g(e - f, e - f))]/4.$$

Thus if $g \neq 0$, there is an $e_1 \in E$ such that $g(e_1, e_1) \neq 0$. Rescaling, we can assume $c_1 = g(e_1, e_1) = \pm 1$. Let E_1 be the span of e_1 and $E_2 = \{e \in E \mid g(e_1, e) = 0\}$. Clearly $E_1 \cap E_2 = \{0\}$. Also, if $z \in E$, then $z - c_1 g(z, e_1)e_1 \in E_2$ so that $E = E_1 + E_2$ and thus $E = E_1 \oplus E_2$. Now if $g \neq 0$ on E_2, there is an $e_2 \in E_2$ such that $g(e_2, e_2) = c_2 = \pm 1$. Continue inductively to complete the proof.

For the second part, in the basis $\{e_1, ..., e_n\}$ just found, let $E_1 = \text{span}\{e_i \mid g(e_i, e_i) = 1\}$, $E_2 = \text{span}\{e_i \mid g(e_i, e_i) = -1\}$ and $\ker g = \{e \mid g(e, e') = 0 \text{ for all } e' \in E\}$. Note that $\ker g = \text{span}\{e_i \mid g(e_i, e_i) = 0\}$ and thus $E = E_1 \oplus E_2 \oplus \ker g$. Let F be any subspace of E on which g is positive definite. Then clearly $F \cap \ker g = \{0\}$. We also have $E_2 \cap F = \{0\}$ since any $v \in E_2 \cap F$, $v \neq 0$, must simultaneously satisfy $g(v, v) > 0$ and $g(v, v) < 0$. Thus $F \cap (E_2 \oplus \ker g) = \{0\}$ and consequently $\dim F \leq \dim E_1$. A similar argument shows that $\dim E_2$ is the maximal dimension of any subspace of E on which g is negative definite. ■

Note that the number of ones in the diagonal representation of g is $(r + s)/2$ and the number of minus-ones is $\mathrm{Ind}(g) = (r - s)/2$. Nondegeneracy of g means that $r = n$. In this case $e \in E$ may be written $e = \Sigma_{i = 1, \ldots, n} [g(e, e_i)/c_i]e_i$, where $c_i = g(e_i, e_i) = \pm 1$ and $\{e_i\}$ is a g-orthonormal basis. For g a positive definite inner product, $r = n$ and $\mathrm{Ind}(g) = 0$; for g a Lorentz inner product $r = n$ and $\mathrm{Ind}(g) = 1$.

6.2.10 Proposition *Let* E *be an* n*-dimensional vector space and* $g \in T^0{}_2(E)$ *be nondegenerate and symmetric.*

(i) *If* $[\omega]$ *is an orientation of* E *there exists a unique volume element* $\mu = \mu(g) \in [\omega]$, *called the* **g-volume,** *such that* $\mu(e_1, \ldots, e_n) = 1$ *for all positively oriented* g*-orthonormal bases* $\{e_1, \ldots, e_n\}$ *of* E. *In fact, if* $\{e^1, \ldots, e^n\}$ *is the dual basis, then* $\mu = e^1 \wedge \ldots \wedge e^n$. *More generally, if* $\{f_1, \ldots, f_n\}$ *is a positively oriented basis with dual basis* $\{f^1, \ldots, f^n\}$, *then*

$$\mu = |\det[g(f_i, f_j)]|^{1/2} f^1 \wedge \ldots \wedge f^n.$$

(ii) *There is a unique* α*-density* $|\mu|^\alpha$, *called the* **g-α-density,** *with the property that* $|\mu|^\alpha(e_1, \ldots, e_n) = 1$ *for all* g*-orthonormal bases* $\{e_1, \ldots, e_n\}$ *of* E. *If* $\{e^1, \ldots, e^n\}$ *is the dual basis, then* $|\mu|^\alpha = |e^1 \wedge \ldots \wedge e^n|^\alpha$. *More generally, if* $v_1, \ldots, v_n \in E$, *then*

$$|\mu|^\alpha(v_1, \ldots, v_n) = |\det[g(v_i, v_j)]|^{\alpha/2}.$$

Proof First we establish a relation between the determinants of the following three matrices: $[g(e_i, e_j)] = \mathrm{diag}(c_1, \ldots, c_n)$ (see **6.2.9**), $[g(f_i, f_j)]$ for an arbitrary basis $\{f_1, \ldots, f_n\}$, and the matrix representation of $\varphi \in GL(E)$ where $\varphi(e_i) = f_i = A^j{}_i e_j$. By **6.2.9** we have

$$g(f_i, f_j) = \left(\sum_{p=1}^{n} c_p e^p \otimes e^p \right) (A^k{}_i e_k, A^\ell{}_j e_\ell) = c_p \delta^p{}_k \delta^p{}_\ell A^k{}_i A^\ell{}_j = c_p A^p{}_i A^p{}_j \,. \quad \text{(sum on p)}$$

Thus

$$\det[g(f_i, f_j)] = (\det \varphi)^2 \det[g(e_i, e_j)]. \tag{2}$$

By **6.2.9**, $|\det[g(e_i, e_j)]| = 1$.

(i) Clearly if $\{e_1, \ldots, e_n\}$ is positively oriented and g-orthonormal, then $\mu(e_1, \ldots, e_n) = 1$ uniquely determines $\mu \in [\omega]$ by multilinearity. Let $\{f_1, \ldots, f_n\}$ be another positively oriented g-orthonormal basis. If $\varphi \in GL(E)$ where $\varphi(e_i) = f_i$, $i = 1, \ldots, n$, then by (2) and **6.2.9** it follows that $|\det \varphi| = 1$. But $0 < \mu(f_1, \ldots, f_n) = (\varphi^*\mu)(e_1, \ldots, e_n) = \det \varphi$, so that $\det \varphi = 1$. The second statement in (i) follows from the third.

For the third statement of (i), note that by (2)

$$\mu(f_1, \ldots, f_n) = \det \varphi = |\det[g(f_i, f_j)]|^{1/2}$$

(ii) follows from (i) and the remarks following **6.2.8.** ∎

A covariant symmetric nondegenerate two-tensor g on E induces one on $\Lambda^k(E)$ for every $k = 1, \ldots, n$ in the following way. Let

$$\alpha = \alpha_{i_1 \cdots i_k} e^{i_1} \wedge \cdots \wedge e^{i_k} \text{ and } \beta = \beta_{i_1 \cdots i_k} e^{i_1} \wedge \cdots \wedge e^{i_k} \in \Lambda^k(E),$$

(sum over $i_1 < \cdots < i_k$) and let

$$\beta^{i_1 \cdots i_k} = g^{i_1 j_1} \cdots g^{i_k j_k} \beta_{j_1 \cdots j_k}$$

(sum over *all* $j_i, \ldots, j_k$) be the components of the associated contravariant k-tensor, where $[g^{kj}]$ denotes the inverse of the matrix with entries $g_{ij} = g(e_i, e_j)$. Then put

$$g^{(k)}(\alpha, \beta) = \sum_{i_1 < \cdots < i_k} \alpha_{i_1 \cdots i_k} \beta^{i_1 \cdots i_k}. \tag{3}$$

If there is no danger of confusion, we will write $\langle \alpha, \beta \rangle = g^{(k)}(\alpha, \beta)$. We now show that this definition does not depend on the basis. If $\{f_1, \ldots, f_n\}$ is another ordered basis of E, let $\alpha = \alpha'_{a_i \cdots a_k} f^{a_1} \wedge \cdots \wedge f^{a_k}$ and $\beta = \beta'_{a_1 \cdots a_k} f^{a_1} \wedge \cdots \wedge f^{a_k}$. Then the identity map on E has matrix representation relative to the bases $\{e_1, \ldots, e_n\}$ and $\{f_1, \ldots, f_n\}$ given by $e_i = A^a{}_i f_a$. If $B = A^{-1}$ we have by **5.1.7**,

$$\begin{aligned} \alpha'_{a_1 \cdots a_k} \beta'^{a_1 \cdots a_k} &= \alpha_{i_1 \cdots i_k} B^{i_1}{}_{a_1} \cdots B^{i_k}{}_{a_k} A^{a_1}{}_{j_1} \cdots A^{a_k}{}_{j_k} \beta^{j_1 \cdots j_k} \\ &= \alpha_{i_1 \cdots i_k} \delta^{i_1}{}_{j_1} \cdots \delta^{i_k}{}_{j_k} \beta^{j_1 \cdots j_k} = \alpha_{i_1 \cdots i_k} \beta^{i_1 \cdots i_k}. \end{aligned}$$

So defined, $g^{(k)}$ is clearly bilinear. It is also symmetric since

$$\begin{aligned} \beta_{i_1 \cdots i_k} \alpha^{i_1 \cdots i_k} &= g_{i_1 j_1} \cdots g_{i_k j_k} \beta^{j_1 \cdots j_k} g^{i_1 \ell_1} \cdots g^{i_k \ell_k} \alpha_{\ell_1 \cdots \ell_k} \\ &= \delta^{\ell_1}_{j_1} \cdots \delta^{\ell_k}_{j_k} \alpha_{\ell_1 \cdots \ell_k} \beta^{j_1 \cdots j_k} = \alpha_{j_1 \cdots j_k} \beta^{j_1 \cdots j_k} \end{aligned}$$

where $[g^{ij}] = [g_{ij}]^{-1}$, and $g_{ij} = g(e_i, e_j)$. Notice that $g^{(k)}$ is also nondegenerate since if $g^{(k)}(\alpha, \beta) = 0$ for all $\beta \in \Lambda^k(E)$, choosing for β all elements of a basis, shows that $\alpha_{i_1 \cdots i_k} = 0$, i.e., that $\alpha = 0$. The following has thus been proved.

6.2.11 Proposition *A nondegenerate symmetric covariant two-tensor* $g = \langle\,,\rangle$ *on the finite-dimensional vector space* E *induces a similar tensor on* $\Lambda^k(E)$ *for all* $k = 1, ..., n$. *Moreover, if* $\{e_1, ..., e_n\}$ *is a* g*-orthonormal basis of* E *in which*

$$g = \sum_{i=1}^{n} c_i e^i \otimes e^i, \quad c_i = \pm 1,$$

then the basis

$$\{e^{i_1} \wedge \cdots \wedge e^{i_k} \mid i_1 < \cdots < i_k\}$$

is orthnormal with respect to $g^{(k)} = \langle\,,\rangle$, *and*

$$\langle e^{i_1} \wedge \cdots \wedge e^{i_k}, e^{i_1} \wedge \cdots \wedge e^{i_k}\rangle = c_{i_1} \cdots c_{i_k} \ (= \pm 1). \tag{4}$$

With the aid of the g-volume μ on E we introduce the ***Hodge star operator.***

6.2.12 Proposition *Let* E *be an oriented* n*-dimensional vector space and* $g = \langle,\rangle \in T^0{}_2(E)$ *a given symmetric and nondegenerate tensor. Let* μ *be the corresponding volume element of* E. *Then there is a unique isomorphism* $* : \Lambda^k(E) \to \Lambda^{n-k}(E)$ *satisfying*

$$\alpha \wedge * \beta = \langle\alpha, \beta\rangle\mu \quad \text{for} \quad \alpha, \beta \in \Lambda^k(E). \tag{5}$$

If $\{e_1, ..., e_n\}$ *is a positively oriented* g*-orthonormal basis of* E *and* $\{e^1, ..., e^n\}$ *is its dual basis, then*

$$*(e^{\sigma(1)} \wedge \ldots \wedge e^{\sigma(k)}) = c_{\sigma(1)} \cdots c_{\sigma(k)} (\text{sign } \sigma)(e^{\sigma(k+1)} \wedge \ldots \wedge e^{\sigma(n)}) \tag{6}$$

where $\sigma(1) < \ldots < \sigma(k)$ *and* $\sigma(k+1) < \ldots < \sigma(n)$.

Proof First uniqueness is proved. Let $*$ satisfy Eq. (5) and let $\beta = e^{\sigma(1)} \wedge \cdots \wedge e^{\sigma(k)}$ and α be one of the g-orthornomal basis vectors

$$e^{i_1} \wedge \cdots \wedge e^{i_k} \text{ of } \Lambda^k(E),\ i_1 < \cdots < i_k.$$

By (5), $\alpha \wedge * \beta = 0$ unless $(i_1, ..., i_k) = (\sigma(1), ..., \sigma(k))$. Thus, $*\beta = a e^{\sigma(k+1)} \wedge \ldots \wedge e^{\sigma(n)}$ for a constant a. But then $\beta \wedge * \beta = a\,\text{sign}(\sigma)\mu$ and by (4), $\langle\beta, \beta\rangle = c_{\sigma(1)} \cdots c_{\sigma(k)}$. Hence $a = c_{\sigma(1)} \cdots c_{\sigma(k)}\text{sign}(\sigma)$ and so $*$ must satisfy (6). Thus $*$ is unique.

Define $*$ by (6), recalling that $e^{\sigma(1)} \wedge \ldots \wedge e^{\sigma(k)}$ for $\sigma(1) < \ldots < \sigma(k)$ forms a $g^{(k)}$-orthonormal basis of $\Lambda^k(E)$. As before, (5) is then verified using this basis. Clearly $*$ defined by (6) is an isomorphism, as it maps the g-orthonormal basis of $\Lambda^k(E)$ to that of $\Lambda^{n-k}(E)$. ■

6.2.13 Proposition *Let* E *be an oriented* n*-dimensional vector space,* $g = \langle\, , \rangle \in T^0{}_2(E)$ *symmetric and nondegenerate of signature* s, *and* μ *the associated* g*-volume of* E. *The Hodge star operator satisfies the following properties for* $\alpha, \beta \in \Lambda^k(E)$:

$$\alpha \wedge * \beta = \beta \wedge * \alpha = \langle \alpha, \beta \rangle \mu, \tag{7}$$

$$* 1 = \mu, \; * \mu = (-1)^{\mathrm{Ind}(g)}, \tag{8}$$

$$* * \alpha = (-1)^{\mathrm{Ind}(g)}(-1)^{k(n-k)}\alpha, \tag{9}$$

$$\langle \alpha, \beta \rangle = (-1)^{\mathrm{Ind}(g)}\langle * \alpha, * \beta \rangle. \tag{10}$$

Proof Equation (7) follows from (5) by symmetry of $\langle \alpha, \beta \rangle$. Equations (8) follow directly from (6), with $k = 0, n$, respectively, and $\sigma =$ *identity* (note that $c_1 \ldots c_n = (-1)^{\mathrm{Ind}(g)}$). To verify (9), it suffices to take $\alpha = e^{\sigma(1)} \wedge \ldots \wedge e^{\sigma(k)}$. By (6),

$$*(e^{\sigma(k+1)} \wedge \ldots \wedge e^{\sigma(n)}) = b e^{\sigma(1)} \wedge \ldots \wedge e^{\sigma(k)}$$

for a constant b. To find b use (5) with $\alpha = \beta = e^{\sigma(k+1)} \wedge \ldots \wedge e^{\sigma(n)}$ to give (see (4))

$$b e^{\sigma(k+1)} \wedge \ldots \wedge e^{\sigma(n)} \wedge e^{\sigma(1)} \wedge \ldots \wedge e^{\sigma(k)} = c_{\sigma(k+1)} \cdots c_{\sigma(n)}\mu.$$

Hence $b = c_{\sigma(k+1)} \cdots c_{\sigma(n)}(-1)^{k(n-k)} \operatorname{sign} \sigma$. Thus (6) implies

$$\begin{aligned} **(e^{\sigma(1)} \wedge \ldots \wedge e^{\sigma(n)}) &= c_{\sigma(1)} \cdots c_{\sigma(k)} (\operatorname{sign} \sigma) *(e^{\sigma(k+1)} \wedge \ldots \wedge e^{\sigma(n)}) \\ &= c_{\sigma(1)} \cdots c_{\sigma(k)} c_{\sigma(k+1)} \cdots c_{\sigma(n)} (\operatorname{sign} \sigma)^2 (-1)^{k(n-k)} e^{\sigma(1)} \wedge \ldots \wedge e^{\sigma(k)} \\ &= (-1)^{\mathrm{Ind}(g)}(-1)^{k(n-k)} e^{\sigma(1)} \wedge \ldots \wedge e^{\sigma(k)}. \end{aligned}$$

Finally for (10), we use (7) and (9) to give

$$\begin{aligned} \langle *\alpha, *\beta \rangle \mu &= * \alpha \wedge **\beta = (-1)^{\mathrm{Ind}(g)}(-1)^{k(n-k)} * \alpha \wedge \beta \\ &= (-1)^{\mathrm{Ind}(g)} \beta \wedge * \alpha = (-1)^{\mathrm{Ind}(g)} \langle \alpha, \beta \rangle \mu. \quad \blacksquare \end{aligned}$$

6.2.14 Examples

A The Hodge operator on $\Lambda^1(\mathbb{R}^3)$ where $\mathbb{R}^3$ has the standard metric and dual basis is given from (6) by $* e^1 = e^2 \wedge e^3$, $* e^2 = -e^1 \wedge e^3$, and $* e^3 = e^1 \wedge e^2$. (This is the isomorphism considered in **6.1.11B**).

B Using (5), we compute $*$ in an arbitrary oriented basis. Write

$$*(e^{i_1} \wedge \cdots \wedge e^{i_k}) = c^{i_1 \cdots i_k}_{j_{k+1} \cdots j_n} e^{j_{k+1}} \wedge \cdots \wedge e^{j_n}$$

(sum over $j_{k+1} < \cdots < j_n$) and apply (5) with

$$\beta = e^{i_1} \wedge \cdots \wedge e^{i_k} \text{ and } \alpha = e^{j_1} \wedge \cdots \wedge e^{j_k}$$

where $\{j_1, ..., j_k\}$ is a complementary set of indices to $\{j_{k+1}, ..., j_n\}$. One gets

$$c^{i_1 \cdots i_k}_{j_{k+1} \cdots j_n} = g^{i_1 j_1} \cdots g^{i_k j_k} \,|\det[g_{ij}]\,|^{1/2} \operatorname{sign}\begin{pmatrix} 1 & \cdots & n \\ j_1 & \cdots & j_n \end{pmatrix}.$$

Hence

$$*(e^{i_1} \wedge \cdots \wedge e^{i_k}) = |\det[g_{ij}]\,|^{1/2} \sum \operatorname{sign}\begin{pmatrix} 1 & \cdots & n \\ j_1 & \cdots & j_n \end{pmatrix} g^{i_1 j_1} \cdots g^{i_k j_k} e^{j_{k+1}} \wedge \cdots \wedge e^{j_n}, \tag{11}$$

where the sum is over all $(k, n-k)$ shuffles $\begin{pmatrix} 1 & \cdots & n \\ j_1 & \cdots & j_n \end{pmatrix}$.

C In particular, if $k = 1$, (11) yields

$$* e^i = |\det([g_{ij}])\,|^{1/2} \sum_{j=1}^{n} (-1)^{j-1} g^{ij} e^1 \wedge \cdots \wedge \hat{e}^j \wedge \cdots \wedge e^n \tag{12}$$

since $\operatorname{sign}(j_1, j_2, ..., j_n) = (-1)^{j-1}$, for $j_2 < \cdots < j_n$, $j_1 = j$, and where $\hat{e}^j$ means that e^j is deleted.

D From **B** we can compute the *components* of $* \alpha$, where $\alpha \in \Lambda^k(E)$, relative to any oriented basis: write $\alpha = \alpha_{i_1 \cdots i_k} e^{i_1} \wedge \cdots \wedge e^{i_k}$ (sum over $i_1 < \cdots < i_k$) and apply (11) to give

$$(*\alpha) = |\det[g_{ij}]\,|^{1/2} \sum \operatorname{sign}\begin{pmatrix} 1 & \cdots & n \\ j_1 & \cdots & j_n \end{pmatrix} \alpha_{i_1 \cdots i_k} g^{i_1 j_1} \cdots g^{i_k j_k} e^{j_{k+1}} \wedge \cdots \wedge e^{j_n}.$$

Hence

$$(*\alpha)_{j_{k+1} \cdots j_n} = |\det[g_{ij}]\,|^{1/2} \sum \alpha_{i_1 \cdots i_k} g^{i_1 j_1} \cdots g^{i_k j_k} \operatorname{sign}\begin{pmatrix} 1 & \cdots & n \\ j_1 & \cdots & j_n \end{pmatrix} \tag{13}$$

for $j_{k+1} < \ldots < j_n$ and where the sum is over all complementary indices $j_1 < \ldots < j_k$.

E Consider $\mathbb{R}^4$ with the Lorentz inner product, which in the standard basis $\{e_1, e_2, e_3, e_4\}$ of $\mathbb{R}^4$ has the matrix

$$\begin{bmatrix} 1 & 0 & 0 & 0 \\ 0 & 1 & 0 & 0 \\ 0 & 0 & 1 & 0 \\ 0 & 0 & 0 & -1 \end{bmatrix}.$$

Let $\{e^1, e^2, e^3, e^4\}$ be the dual basis. The Hodge operator on $\Lambda^1(\mathbb{R}^4)$ is given by

$$* e^1 = e^2 \wedge e^3 \wedge e^4, \quad * e^2 = -e^1 \wedge e^3 \wedge e^4, \quad * e^3 = e^1 \wedge e^2 \wedge e^4, \quad * e^4 = e^1 \wedge e^2 \wedge e^3,$$

and on $\Lambda^2(\mathbb{R}^4)$ by

$$*(e^1 \wedge e^2) = e^3 \wedge e^4, \quad *(e^1 \wedge e^3) = -e^2 \wedge e^4, \quad *(e^2 \wedge e^3) = e^1 \wedge e^4$$

$$*(e^1 \wedge e^4) = -e^2 \wedge e^3, \quad *(e^2 \wedge e^4) = e^1 \wedge e^3, \quad *(e^3 \wedge e^4) = -e^1 \wedge e^2.$$

If $\mathbb{R}^4$ had been endowed with the usual Euclidean inner product, the formulas for $* e^4$, $*(e^1 \wedge e^4)$, $*(e^2 \wedge e^4)$, and $*(e^3 \wedge e^4)$ would have opposite signs. The Hodge $*$ operator on $\Lambda^3(\mathbb{R}^4)$ follows from the formulas on $\Lambda^1(\mathbb{R}^4)$ and the fact that for $\beta \in \Lambda^1(\mathbb{R}^4)$, $** \beta = \beta$ (from formula (9)). Thus we obtain

$$*(e^2 \wedge e^3 \wedge e^4) = e^1, \quad *(e^1 \wedge e^3 \wedge e^4) = -e^2, \quad *(e^1 \wedge e^2 \wedge e^4) = e^3, \quad *(e^1 \wedge e^2 \wedge e^3) = e^4.$$

F If β is a one form and $v_1, v_2, \ldots, v_n$ is a positively oriented orthonormal basis, then

$$(*\beta)(v_2, \ldots, v_n) = \beta(v_1).$$

This follows from (5) taking $\alpha = v^1$, the first element in the dual basis.

Exercises

6.2A Let $\{e^1, e^2, e^3\}$ be the standard dual basis of $\mathbb{R}^3$ and $\alpha = e^1 \wedge e^2 - 2e^2 \wedge e^3 \in \Lambda^2(\mathbb{R}^3)$, $\beta = 3e^1 - e^2 + 2e^3 \in \Lambda^1(\mathbb{R}^3)$, and $\varphi \in L(\mathbb{R}^2, \mathbb{R}^3)$ have the matrix

$$\begin{bmatrix} 1 & 0 \\ 0 & -1 \\ 2 & 1 \end{bmatrix}$$

Compute $\varphi^*\alpha$. With the aid of the standard metrics in $\mathbb{R}^2$ and $\mathbb{R}^3$, compute $* \alpha$, $* \beta$, $*(\varphi^*\alpha)$, and $*(\varphi^*\beta)$. Do you get any equalities? Explain.

6.2B A map $\varphi \in L(E, F)$, where (E, ω), (F, μ) are oriented vector spaces with chosen volume elements, is called ***volume preserving*** if $\varphi^*\mu = \omega$. Show that if E and F have the same (finite) dimension, then φ is an isomorphism.

6.2C A map $\varphi \in L(E, F)$, where $(E, [\mu])$ and $(F, [\omega])$ are oriented vector spaces, is called ***orientation preserving*** if $\varphi^*\mu \in [\omega]$. If $\dim E = \dim F$, and φ is orientation preserving, show that φ is an isomorphism. Given an example for $F = E = \mathbb{R}^3$ of an orientation-preserving but not volume-preserving map.

6.2D Let E and F be n-dimensional real vector spaces with nondegenerate symmetric two-tensors, $g \in T^0{}_2(E)$ and $h \in T^0{}_2(F)$. Then $\varphi \in L(E, F)$ is called an ***isometry*** if $h(\varphi(e), \varphi(e')) = g(e, e')$, for all $e, e' \in E$.

(i) Show that an isometry is an isomorphism.

(ii) Consider on E and F the g- and h-volumes $\mu(g)$ and $\mu(h)$. Show that if φ is an orientation-preserving isometry, then φ^* commutes with the Hodge star operator, i.e., the following diagram commutes:

$$\begin{array}{ccc} \Lambda^k(F) & \xrightarrow{\ \ *\ \ } & \Lambda^{n-k}(F) \\ \Big\downarrow{\scriptstyle \varphi^*} & & \Big\downarrow{\scriptstyle \varphi^*} \\ \Lambda^k(E) & \xrightarrow[\ \ *\ \ ]{} & \Lambda^{n-k}(E) \end{array}$$

If φ is orientation reversing, show that $*(\varphi^*\alpha) = -\varphi^*(*\,\alpha)$ for $\alpha \in \Lambda^k(F)$.

6.2E Let g be an inner product and $\{f_1, f_2, f_3\}$ be a positively oriented basis of $\mathbb{R}^3$. Denote by $\flat$ and $\#$ the index lowering and raising actions defined by g.

(i) Show that for any vectors $u, v \in \mathbb{R}^3$

$$[*(u^\flat \wedge v^\flat)]^\# = \operatorname{sign}\begin{pmatrix} 1 & 2 & 3 \\ i & j & k \end{pmatrix} \,|\det[g(f_a, f_b)]\,|^{1/2}\, u^i v^j g^{k\ell} f_\ell .$$

(ii) Show that if g is the standard dot-product in $\mathbb{R}^3$ the formula in (i) reduces to the cross-product of u and v.

(iii) Generalize (i) to define the cross-product of $n-1$ vectors $u_1, \ldots, u_{n-1}$ in an oriented n-dimensional inner product space (E, g), and find its coordinate expression.

6.2F Let E be an n-dimensional oriented vector space and let $g \in T^0{}_2(E)$ be symmetric and non-degenerate of signature s. Using the g-volume, define the Hodge star operator $* : \Lambda^k(E; F) \to \Lambda^{n-k}(E; F)$, where F is another finite-dimension vector space by

$$*\,\alpha = (*\,\alpha^i) f_i,$$

where $\alpha^i \in \Lambda^k(E)$, $\{f_1, \ldots, f_m\}$ is a basis of F and $\alpha = \alpha^i f_i$. Show the following.

(i) The definition is independent of the basis of F.

(ii) $*\,* = (-1)^{(n-s)/2+k(n-k)}$ on $\Lambda^k(E; F)$.

(iii) If $h \in T^0{}_2(F)$ and if we let $h'(f, \alpha) = (*\,\alpha^i) h(f, f_i)$, then $*\,h'(f, \alpha) = h'(f, *\,\alpha)$.

(iv) If $\wedge$ is the wedge product in $\Lambda(E; F)$ with respect to a given bilinear form on F, then for $\alpha, \beta \in \Lambda^k(E; F)$,

$$(*\,\alpha) \wedge \beta = (*\,\beta) \wedge \alpha \quad \text{and} \quad \alpha \wedge (*\,\beta) = \beta \wedge (*\,\alpha).$$

(v) Show how g and h induce a symmetric nondegenerate covariant two-tensor on $\Lambda^k(E; F)$ and find formulas analogous to (7)-(10).

6.2G Prove the following identities in $\mathbb{R}^3$ using the Hodge star operator:

$$\| u \times v \|^2 = \| u \|^2 \| v \|^2 - (u \cdot v)^2 \quad \text{and} \quad u \times (v \times w) = (u \cdot w)v - (u \cdot v)w.$$

6.2H (i) Prove the following identity for the Hodge star operator:

$$\langle * \alpha, \beta \rangle = \langle \alpha \wedge \beta, \mu \rangle,$$

where $\alpha \in \Lambda^k(E)$ and $\beta \in \Lambda^{n-k}(E)$.

(ii) Prove the basic properties of $*$ using (i) as the definition.

6.2I Let E be an oriented vector space and $S \subset T^0_2(E)$ be the set of nondegenerate symmetric two-tensors of a fixed signature s.

(i) Show that S is open.

(ii) Show that the map $\text{vol} : g \mapsto \mu(g)$ assigning to each $g \in S$ its g-volume element is differentiable and has derivative at g given by $h \mapsto (\text{trace } h)\mu(g)/2$.

§6.3 Differential Forms

The exterior algebra can now be extended from vector spaces to vector bundles and in particular to the tangent bundle. First of all, we need to consider the action of local bundle maps. As in Chapter 3, $U \times F$ denotes a local vector bundle, where U is open in a Banach space E and F is a Banach space. From $U \times F$, we construct the local vector bundle $U \times \Lambda^k(F)$. Now we want to piece these local objects together into a global one.

6.3.1 Definition *Let* $\varphi : U \times F \to U' \times F'$ *be a local vector bundle map that is an isomorphism on each fiber. Then define* $\varphi_* : U \times \Lambda^k(F) \to U' \times \Lambda^k(F')$ *by* $(u, \omega) \mapsto (\varphi_0(u), \varphi_{u*}\omega)$, *where* φ_u *is the second factor of* φ *(an isomorphism for each* u).

6.3.2 Proposition *If* $\varphi : U \times F \to U' \times F'$ *is a local vector bundle map that is an isomorphism on each fiber, then so is* φ_*. *Moreover, if* φ *is a local vector bundle isomorphism, so is* φ_*.

Proof This is a special case of **5.2.4.** ∎

6.3.3 Definition *Suppose* $\pi : E \to B$ *is a vector bundle. Define*

$$\Lambda^k(E) \mid A = \bigcup_{b \in A} \Lambda^k(E_b).$$

where A *is a subset of* B *and* $E_b = \pi^{-1}(b)$ *is the fiber over* $b \in B$. *Let* $\Lambda^k(E) \mid B = \Lambda^k(E)$ *and define* $\Lambda^k(\pi) : \Lambda^k(E) \to B$ *by* $\Lambda^k(\pi)(t) = b$ *if* $t \in \Lambda^k(E_b)$.

6.3.4 Theorem *Assume* $\{E \mid U_i, \varphi_i\}$ *is a vector bundle atlas for the vector bundle* π, *where* $\varphi_i : E \mid U_i \to U'_i \times F'_i$. *Then* $\{\Lambda^k(E) \mid U_i, \varphi_{i*}\}$ *is a vector bundle atlas of* $\Lambda^k(\pi) : \Lambda^k(E) \to B$, *where* $\varphi_{i*} : \Lambda^k(E) \mid U_i \to U'_i \times \Lambda^k(F')$ *is defined by* $\varphi_{i*} \mid E_b = (\varphi_i \mid E_b)_*$.

Proof We must verify **VB1** and **VB2** of **3.4.4**: **VB1** is clear; for **VB2** let φ_i, φ_j be two charts for π, so that $\varphi_i \circ \varphi_j^{-1}$ is a local vector bundle isomorphism on its domain. But then, $\varphi_{i*} \circ \varphi_{j*}^{-1} = (\varphi_i \circ \varphi_j^{-1})_*$, which is a local vector bundle isomorphism by **6.3.2.** ∎

Because of this theorem, the vector bundle structure of $\pi : E \to B$ induces naturally a vector bundle structure on $\Lambda^k(E) \to B$.

We now specialize to the important case when $\pi : E \to B$ is the tangent bundle. If $\tau_M : TM \to M$ is the tangent bundle of a manifold M, let $\Lambda^k(M) = \Lambda^k(TM)$, and $\Lambda^k{}_M = \Lambda^k(\tau_M)$, so $\Lambda^k{}_M : \Lambda^k(M) \to M$ is the vector bundle of exterior k forms on the tangent spaces of M. Also, let

$\Omega^0(M) = \mathcal{F}(M)$, $\Omega^1(M) = \mathcal{T}^0{}_1(M)$, and $\Omega^k(M) = \Gamma^\infty(\Lambda^k{}_M)$, $k = 2, 3, \ldots$.

6.3.5 Proposition *Regarding* $\mathcal{T}^0{}_k(M)$ *as an* $\mathcal{F}(M)$ *module,* $\Omega^k(M)$ *is an* $\mathcal{F}(M)$ *submodule; i.e.,* $\Omega^k(M)$ *is a subspace of* $\mathcal{T}^0{}_k(M)$ *and if* $f \in \mathcal{F}(M)$ *and* $\alpha \in \Omega^k(M)$, *then* $f\alpha \in \Omega^k(M)$.

Proof If $t_1, t_2 \in \Omega^k(M)$ and $f \in \mathcal{F}(M)$, then we must show $ft_1 + t_2 \in \Omega^k(M)$. This follows from the fact that for each $m \in M$, the exterior algebra on T_mM is a vector space. ■

6.3.6 Proposition *If* $\alpha \in \Omega^k(M)$ *and* $\beta \in \Omega^\ell(M)$, $k, \ell = 0, 1, \ldots$, *define* $\alpha \wedge \beta : M \to \Lambda^{k+\ell}(M)$ *by* $(\alpha \wedge \beta)(m) = \alpha(m) \wedge \beta(m)$. *Then* $\alpha \wedge \beta \in \Omega^{k+\ell}(M)$, *and* $\wedge$ *is bilinear and associative.*

Proof First, $\wedge$ is bilinear and associative since it is true pointwise. To show $\alpha \wedge \beta$ is of class C^∞, consider the local representative of $\alpha \wedge \beta$ in natural charts. This is a map of the form $(\alpha \wedge \beta)_\varphi = B \circ (\alpha_\varphi \times \beta_\varphi)$, with α_φ, β_φ, C^∞ and $B = \wedge$, which is bilinear and continuous. Thus $(\alpha \wedge \beta)_\varphi$ is C^∞ by the Leibniz rule. ■

6.3.7 Definition *Let* $\Omega(M)$ *denote the direct sum of* $\Omega^k(M)$, $k = 0, 1, \ldots$, *together with its structure as an (infinite-dimensional) real vector space and with the multiplication* $\wedge$ *extended componentwise to* $\Omega(M)$. *(If* $\dim M = n < \infty$, *the direct sum need only be taken for* $k = 0, 1, \ldots, n$.) *We call* $\Omega(M)$ *the algebra of* ***exterior differential forms*** *on* M. *Elements of* $\Omega^k(M)$ *are called* ***k-forms****. In particular, elements of* $\mathfrak{X}^*(M)$ *are called* ***one-forms****.*

Note that we generally regard $\Omega(M)$ as a real vector space rather than an $\mathcal{F}(M)$ module (as with $\mathcal{T}(M)$). The reason is that $\mathcal{F}(M) = \Omega^0(M)$ is included in the direct sum, and $f \wedge \alpha = f \otimes \alpha = f\alpha$.

6.3.8 Remarks and Examples

A A one-form θ on a manifold M assigns to each $m \in M$ a linear functional on T_mM.

B A two-form ω on a manifold assigns to each $m \in M$ a skew symmetric bilinear map

$$\omega_m : T_mM \times T_mM \to \mathbb{R}.$$

C For an n-manifold M, a tensor field $t \in \mathcal{T}^r{}_s(M)$ has the local expression

$$t(u) = t^{i_1 \cdots i_r}{}_{j_1 \cdots j_s}(u) \frac{\partial}{\partial x^{i_1}} \otimes \cdots \otimes \frac{\partial}{\partial x^{i_r}} \otimes dx^{j_1} \otimes \cdots \otimes dx^{j_s}$$

where $u \in U$, (U, φ) is a local chart on M, and

$$t^{i_1\cdots i_r}{}_{j_1\cdots j_s}(u) = t\left(dx^{i_1},\cdots, dx^{i_r}, \frac{\partial}{\partial x^{j_1}},\cdots, \frac{\partial}{\partial x^{j_s}}\right)(u).$$

The proof of **6.1.8**(ii) gives then the local expression for $\omega \in \Lambda^k(M)$, namely

$$\omega(u) = \omega_{i_1\cdots i_k}(u) dx^{i_1} \wedge \cdots \wedge dx^{i_k}, \quad i_1 < \cdots < i_k,$$

where

$$\omega_{i_1\cdots i_k}(u) = \omega\left(\frac{\partial}{\partial x^{i_1}},\cdots, \frac{\partial}{\partial x^{i_k}}\right)(u).$$

D In $\Omega(M)$, the addition of forms of different degree is "purely formal" as in the case $M = E$. Thus, for example, if M is a two-manifold (a surface) and (x, y) are local coordinates on $U \subset M$, a typical element of $\Omega(M)$ has the local expression $f + g\,dx + h\,dy + k\,dx \wedge dy$, for $f, g, h, k \in \mathcal{F}(U)$.

E As in §**6.1**, we have an isomorphism of vector bundles $* : \Lambda^1(\mathbb{R}^3) \to \Lambda^2(\mathbb{R}^3)$ given by

$$dx^1 \mapsto dx^2 \wedge dx^3, \quad dx^2 \mapsto dx^3 \wedge dx^1, \quad dx^3 \mapsto dx^1 \wedge dx^2.$$

On the other hand, the index lowering action given by the standard Riemannian metric on $\mathbb{R}^3$ defines a vector bundle isomorphism $\flat : T(\mathbb{R}^3) \to T^*(\mathbb{R}^3) = \Lambda^1(\mathbb{R}^3)$. These two isomorphisms applied pointwise define maps

$$* : \mathfrak{X}^*(\mathbb{R}^3) \to \Omega^2, \quad \alpha \mapsto * \alpha$$

and

$$\flat : \mathfrak{X}(\mathbb{R}^3) \to \mathfrak{X}^*(\mathbb{R}^3), \quad X \mapsto X^\flat.$$

Then **6.1.12C** implies

$$*[(X \times Y)^\flat] = X^\flat \wedge Y^\flat$$

for any vector fields $X, Y \in \mathfrak{X}(\mathbb{R}^3)$, where $X \times Y$ denotes the usual cross-product of vector fields on $\mathbb{R}^3$ from calculus. That is,

$$X \times Y = (X^2Y^3 - X^3Y^2)\frac{\partial}{\partial x^1} + (X^3Y^1 - X^1Y^3)\frac{\partial}{\partial x^2} + (X^1Y^2 - X^2Y^1)\frac{\partial}{\partial x^3}$$

for $X = X^i \dfrac{\partial}{\partial x^i}$ and $Y = Y^i \dfrac{\partial}{\partial x^i}$, $i = 1, 2, 3$.

F The wedge product is taken in $\Omega(M)$ in the same way as in the algebraic case. For example, if $M = \mathbb{R}^3$, $\alpha = dx^1 - x^1dx^2 \in \Omega^1(M)$ and $\beta = x^2dx^1 \wedge dx^3 - dx^2 \wedge dx^3$, then

$$\begin{aligned}\alpha \wedge \beta &= (dx^1 - x^1dx^2) \wedge (x^2dx^1 \wedge dx^3 - dx^2 \wedge dx^3)\\ &= 0 - x^1x^2dx^2 \wedge dx^1 \wedge dx^3 - dx^1 \wedge dx^2 \wedge dx^3 + 0\\ &= (x^1x^2 - 1)dx^1 \wedge dx^2 \wedge dx^3. \quad \blacklozenge\end{aligned}$$

6.3.9 Definition *Suppose* $F : M \to N$ *is a* C^∞ *mapping of manifolds. For* $\omega \in \Omega^k(N)$, *define* $F^*\omega : M \to \Lambda^k(M)$ *by* $F^*\omega(m) = (T_mF)^* \circ \omega \circ F(m)$; *i.e.,*

$$(F^*\omega)_m(v_1, \ldots, v_k) = \omega_{F(m)}(T_mF\cdot v_1, \ldots, T_mF\cdot v_k),$$

where $v_1, \ldots, v_k \in T_mM$; *for* $g \in \Omega^0(N)$, $F^*g = g \circ F$. *We say* $F^*\omega$ *is the* ***pull-back*** *of* ω *by* F. (See Fig. 6.3.1.)

6.3.10 Proposition *Let* $F : M \to N$ *and* $G : N \to W$ *be* C^∞ *mappings of manifolds. Then*

(i) $F^* : \Omega^k(N) \to \Omega^k(M)$;

(ii) $(G \circ F)^* = F^* \circ G^*$;

(iii) *if* $H : M \to M$ *is the identity, then* $H^* : \Omega^k(M) \to \Omega^k(M)$ *is the identity;*

(iv) *if* F *is a diffeomorphism, then* F^* *is an isomorphism and* $(F^*)^{-1} = (F^{-1})^*$;

(v) $F^*(\alpha \wedge \beta) = F^*\alpha \wedge F^*\beta$ *for* $\alpha \in \Omega^k(N)$ *and* $\beta \in \Omega^\ell(N)$.

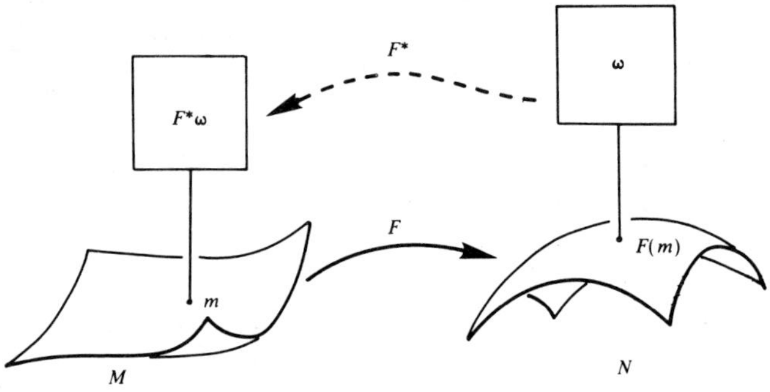

Figure 6.3.1

Proof Choose charts (U, φ), (V, ψ) of M and N so that $F(U) \subset V$. Then the local representative $F_{\varphi\psi} = \psi \circ F \circ \varphi^{-1}$ is of class C^∞, as is $\omega_\psi = (T\psi)^* \circ \omega \circ \psi^{-1}$. The local representative of $F^*\omega$ is

$$(F^*\omega)_\varphi(u) = (T\varphi)^* \circ F^*\omega \circ \varphi^{-1}(u) = (T_uF_{\varphi\psi})^* \circ \omega_\psi \circ F_{\varphi\psi}(u)$$

which is of class C^∞ by the composite mapping theorem.

For (ii), note that it holds for the local representatives; (iii) follows from the definition; (iv) follows in the usual way from (ii) and (iii); and (v) follows from the corresponding pointwise result. ■

We close this section with a few optional remarks about vector-bundle-valued forms. As before, the idea is to globalize vector-valued exterior forms.

6.3.11 Definition *Let* $\pi : E \to B$, $\rho : F \to B$ *be vector bundles over the same base. Define*

$$\Lambda^k(E; F) = L(\Lambda^k(E), F),$$

the vector bundle with base B *of vector bundle homomorphisms over the identity from* $\Lambda^k(E)$ *to* F. *If* $E = TB$, $\Lambda^k(TB; F)$ *is denoted by* $\Lambda^k(B; F)$ *and is called the vector bundle of* F-***valued k-forms on*** M. *If* $F = B \times F$, *we denote it by* $\Lambda^k(B, F)$ *and call its elements* ***vector-valued k-forms on*** M. *The spaces of sections of these bundles are denoted respectively by* $\Omega^k(E; F)$, $\Omega^k(B; F)$ *and* $\Omega^k(B; F)$. *Finally,* $\Omega(E; F)$ *(resp.* $\Omega(B; F)$, $\Omega(B, F)$) *denotes the direct sum of* $\Omega^k(E; F)$, $k = 1, 2, \ldots, n$, *together with its structure of an infinite-dimensional real vector space and* $\mathcal{F}(B)$*-module.*

Thus, $\alpha \in \Omega^k(E; F)$ is a smooth assignment to the points b of B of skew symmetric k-linear maps $\alpha_b : E_b \times \ldots \times E_b \to F_b$. In particular, if all manifolds and bundles are finite dimensional, then $\alpha \in \Omega^k(M, \mathbb{R}^p)$ may be uniquely written in the form $\alpha = \sum_{i=1,\ldots,p} \alpha^i e_j$, where $\alpha^1, \ldots, \alpha^p \in \Omega^k(M)$, and $\{e_1, \ldots, e_p\}$ is the standard basis of $\mathbb{R}^p$. Thus $\alpha \in \Omega^k(E, \mathbb{R}^p)$ is written in local coordinates as

$$(\alpha^1_{i_1 \ldots i_k} dx^{i_1} \wedge \cdots \wedge dx^{i_k}, \cdots, \alpha^p_{i_1 \ldots i_k} dx^{i_1} \wedge \cdots \wedge dx^{i_k})$$

for $i_1 < \ldots < i_k$. Proposition **6.3.10**(i)-(iv) and its proof have straightforward generalizations to vector-bundle-valued forms on M. The wedge product requires additional structure to be defined, namely a smooth assignment $b \mapsto g_b$ of a symmetric bilinear map $g_b : F_b \times F_b \to F_b$ for each $b \in B$. With this structure, **6.3.10**(v) also carries over.

Exercises

6.3A Show that for a vector bundle $\pi : E \to B$, $\Lambda^k(E)$ is a (smooth) subbundle of $T^0{}_k(E)$. Generalize to vector-bundle-valued tensors and forms.

6.3B Let $\varphi : \mathbb{R}^3 \to \mathbb{R}^2$ be given by $\varphi(x, y, z) = (x^2, yz)$. For $\alpha = v^2 du + dv \in \Omega^1(\mathbb{R}^2)$ and $\beta = uv\, du \wedge dv \in \Omega^2(\mathbb{R}^2)$, compute $\alpha \wedge \beta$, $\varphi^*\alpha$, $\varphi^*\beta$ and $\varphi^*(\alpha \wedge \beta)$.

6.3C (E. Cartan's lemma). Let M be an n-manifold and $\alpha^1, \ldots, \alpha^k \in \Omega^1(M)$, $k \leq n$, be pointwise linearly independent. Show that $\beta^1, \ldots, \beta^k \in \Omega^1(M)$ satisfy $\sum_{1 \leq i \leq k} \alpha^i \wedge \beta^i = 0$ iff there exist C^∞ functions $a^j{}_i \in \mathcal{F}(M)$ satisfying $a^j{}_i = a^i{}_j$ such that $\beta^j = a^j{}_i \beta^i$. (*Hint:* work in a local chart and show first that α^i can be chosen to be dx^i; the symmetry of the matrix $[a^j{}_i]$ follows from antisymmetry of $\wedge$ and the given condition.)

6.3D A (strong) ***bundle metric*** g on $\pi : E \to B$ is a smooth section of $L^2{}_s(E; \mathbb{R})$ such that g(b) is an inner product on E_b for every $b \in B$ which is (strongly) nondegenerate, i.e., $e_b \in$

$E_b \mapsto g(b)(e_b, \cdot) \in E_b^*$ is an isomorphism of Banach spaces.

(i) Show that the model of the fiber of E is a Hilbertizable space.

(ii) If $F \subset E$ is a subbundle of E, show that $F^\perp = \bigcup_{b \in B} F_b^\perp$ is a subbundle of E, where we define $F_b^\perp = \{e_b \in E_b \mid g(b)(e_b, f_b) = 0 \text{ for all } f_b \in F_b\}$

(iii) Show that $E = F \oplus F^\perp$.

6.3E Assume the vector bundle $\pi : E \to B$ has a strong bundle metric.

(i) If $\sigma : B \to F$ is a smooth nowhere vanishing section of E, let $F_b = \text{span}\,\{\sigma(b)\}$, $F = \bigcup_{b \in B} F_b$. Show that F is a subbundle of E which is isomorphic to the trivial bundle $E^1{}_B = \mathbb{R} \times B$. Conclude from **6.3D** that $E^1 \oplus (E^1)^\perp = E$.

(ii) Show that a manifold M is parallelizable if and only if TM is isomorphic to a trivial bundle.

(iii) Assume that M is a strong Riemannian manifold, that M admits a nowhere vanishing vector field and that $TM \oplus E^1{}_M$ is isomorphic to a trivial bundle. Let N be another manifold of dimension ≥ 1 such that $TN \oplus E^1{}_N$ is trivial. Show that $M \times N$ is parallelizable. (*Hint:* Use (i) and pull everything back to $M \times N$ by the two projections.)

(iv) Show that if $\dim N = 0$, the conclusion of (ii) is false. (*Hint:* It is know that the only odd dimensional spheres with trivial tangent bundle are S^1, S^3 and S^7. Show that TS^{2n-1} has a nowhere vanishing vector field.)

(v) Show that $S^{a(1)} \times \ldots \times S^{a(n)}$ is parallelizable provided that $a(i) \geq 1$, $i = 1, \ldots, n$ and at least one $a(i)$ is odd. (*Hint*: Use (iii) and Exercise **3.4C**.)

§6.4 The Exterior Derivative, Interior Product, and Lie Derivative

Here we extend the differential of functions to a map $\mathbf{d} : \Omega^k(M) \to \Omega^{k+1}(M)$ for any k. This operator turns out to have marvelous algebraic properties. After studying these we shall show how $\mathbf{d}$ is related to the basic operations of div, grad, and curl on $\mathbb{R}^3$. Then we develop formulas for the Lie derivative. We first develop the exterior derivative $\mathbf{d}$ for finite-dimensional manifolds. Later in the section the infinite-dimensional case is discussed.

6.4.1 Theorem *Let* M *be an n-dimensional manifold. There is a unique family of mappings* $\mathbf{d}^k(U) : \Omega^k(U) \to \Omega^{k+1}(U)$ $(k = 0, 1, 2, \ldots, n,$ *and* U *is open in* $M)$, *which were merely denote by* $\mathbf{d}$, *called the* ***exterior derivative*** *on* M, *such that*

(i) $\mathbf{d}$ *is a* $\wedge$***-antiderivation.*** *That is,* $\mathbf{d}$ *is* $\mathbb{R}$ *linear and for* $\alpha \in \Omega^k(U)$ *and* $\beta \in \Omega^\ell(U)$,

$$\mathbf{d}(\alpha \wedge \beta) = \mathbf{d}\alpha \wedge \beta + (-1)^k \alpha \wedge \mathbf{d}\beta \quad \textit{(product rule)}$$

(ii) *If* $f \in \mathcal{F}(U)$, *then* $\mathbf{d}f$ *is as defined in* **4.2.5**;

(iii) $\mathbf{d}^2 = \mathbf{d} \circ \mathbf{d} = 0$ *(that is,* $\mathbf{d}^{k+1}(U) \circ \mathbf{d}^k(U) = 0)$;

(iv) $\mathbf{d}$ ***is natural with respect to restrictions;*** *that is, if* $U \subset V \subset M$ *are open and* $\alpha \in \Omega^k(V)$, *then* $\mathbf{d}(\alpha \mid U) = (\mathbf{d}\alpha) \mid U$, *or the following diagram commutes:*

$$\begin{array}{ccc} \Omega^k(V) & \xrightarrow{\;|U\;} & \Omega^k(U) \\ \Big\downarrow \mathbf{d} & & \Big\downarrow \mathbf{d} \\ \Omega^{k+1}(V) & \xrightarrow[\;|U\;]{} & \Omega^{k+1}(U) \end{array}$$

As usual, condition (iv) means that $\mathbf{d}$ *is a* ***local operator.***

Proof We first establish uniqueness. Let (U, φ), where $\varphi(u) = (x^1, \ldots, x^n)$, be a chart and

$$\alpha = \alpha_{i_1 \cdots i_k} dx^{i_1} \wedge \cdots \wedge dx^{i_k} \in \Omega^k(U), \; i_1 < \cdots < i_k.$$

If $k = 0$, the local formula $\mathbf{d}\alpha = (\partial\alpha/\partial x^i)dx^i$ applied to the coordinate functions x^i, $i = 1, \ldots, n$ shows that the differential of x^i is the one-form dx^i. From (iii), $\mathbf{d}(dx^i) = 0$, so by (i)

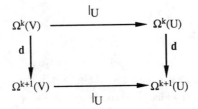

$$d(dx^{i_1} \wedge \cdots \wedge dx^{i_k}) = 0.$$

Thus, again by (i),

$$d\alpha = \frac{\partial \alpha_{i_1 \cdots i_k}}{\partial x^i} dx^i \wedge dx^{i_1} \wedge \cdots \wedge dx^{i_k}, \quad (\text{sum over } i_1 < \cdots < i_k), \tag{1}$$

and so **d** is uniquely determined on U by properties (i)-(iii), and by (iv) on any open subset of M.

For existence, define on every chart (U, φ) the operator **d** by (1). Then (ii) is trivially verified as is $\mathbb{R}$-linearity. If $\beta = \beta_{j_1 \cdots j_\ell} dx^{j_1} \wedge \cdots \wedge dx^{j_\ell} \in \Omega^\ell(U)$, then

$$\begin{aligned}
d(\alpha \wedge \beta) &= d(a_{i_1 \cdots i_k} \beta_{j_1 \cdots j_\ell} dx^{i_1} \wedge \cdots \wedge dx^{i_k} \wedge dx^{j_1} \wedge \cdots dx^{j_\ell}) \\
&= \left(\frac{\partial \alpha_{i_1 \cdots i_k}}{\partial x^i} \beta_{j_1 \cdots j_\ell} + \alpha_{i_1 \cdots i_k} \frac{\partial \beta_{j_1 \cdots j_\ell}}{\partial x^i} \right) dx^i \wedge dx^{i_1} \wedge \cdots \wedge dx^{i_k} \wedge dx^{j_1} \wedge \cdots \wedge dx^{j_\ell} \\
&= \frac{\partial \alpha_{i_1 \cdots i_k}}{\partial x^i} dx^i \wedge dx^{i_1} \wedge \cdots \wedge dx^{i_k} \wedge \beta_{j_1 \cdots j_\ell} dx^{j_1} \wedge \cdots \wedge dx^{j_\ell} \\
&\quad + (-1)^k \alpha_{i_1 \cdots i_k} dx^{i_1} \wedge \cdots \wedge dx^{i_k} \wedge \frac{\partial \beta_{j_1 \cdots j_\ell}}{\partial x^i} dx^i \wedge dx^{j_1} \wedge \cdots \wedge dx^{j_\ell} \\
&= d\alpha \wedge \beta + (-1)^k \alpha \wedge d\beta.
\end{aligned}$$

and (i) is verified. For (iii), symmetry of the second partial derivatives shows that

$$d(d\alpha) = \frac{\partial^2 \alpha_{i_1 \cdots i_k}}{\partial x^i \partial x^j} dx^i \wedge dx^j \wedge dx^{i_1} \wedge \cdots \wedge dx^{i_k} = 0, \; i_1 < \cdots < i_k.$$

Thus, in every chart (U, φ), (1) defines an operator **d** satisfying (i)-(iii). It remains to be shown that these local **d**s define an operator **d** on any open set and (iv) holds. To do this, it is sufficient to show that this definition is chart independent. Let **d′** be the operator given by (1) on a chart (U', φ'), where $U' \cap U \neq \varnothing$. Since **d′** also satisfies (i)-(iii), and local uniqueness has already been proved, $\mathbf{d}'\alpha = \mathbf{d}\alpha$ on $U \cap U'$. The theorem thus follows. ■

6.4.2 Corollary *Let* $\omega \in \Omega^k(U)$, *where* $U \subset E$ *is open. Then*

$$d\omega(u)(v_0, \ldots, v_k) = \sum_{i=0}^{k} (-1)^i D\omega(u) \cdot v_i(v_0, \ldots, \hat{v}_i, \ldots, v_k) \tag{2}$$

where $\hat{v}_i$ *denotes that* v_i *is deleted. Also, we denote elements* (u, v) *of* TU *merely by* v *for brevity. (Note that* $D\omega(u) \cdot v \in L^k_a(E, \mathbb{R})$ *since* $\omega : U \to L^k_a(E, \mathbb{R})$.)

Proof Since we are in the finite dimensional case, we can proceed with a coordinate computation. (An alternative is to check that **d** defined by (2) satisfies properties (i) to (iv). Checking (i) and (iii) is straightforward but lengthy.) Indeed, if the local coordinates of u are $(x^1, ..., x^n)$,

$$\omega(u) = \omega_{i_1, \dots, i_k}(u) dx^{i_1} \wedge \cdots \wedge dx^{i_k}$$

(sum over $i_1 < \cdots < i_k$), then

$$\mathbf{D}\omega(u)\cdot v_i = \frac{\partial \omega_{i_1 \cdots i_k}}{\partial x^j} v_i^j \, dx^{i_1} \wedge \cdots \wedge dx^{i_k}$$

(where the sum is over all j and $i_1 < \cdots <_k$). From (1),

$$\begin{aligned} \mathbf{d}\omega(v_0, \dots, v_k) &= \frac{\partial \omega_{i_1 \cdots i_k}}{\partial x^j} dx^j \wedge dx^{i_1} \wedge \cdots \wedge dx^{i_k}(v_0, \dots, v_k) \\ &= \frac{\partial \omega_{i_1 \cdots i_k}}{\partial x^j} (\text{sign}\, \sigma) v_0^{\sigma(j)} \, v_1^{\sigma(i_1)} \cdots v_k^{\sigma(i_k)} \end{aligned} \tag{2'}$$

(where the sum is over all $i_1 < \cdots < i_k$, j and σ's with $\sigma(j) < \sigma(i_1) < \cdots < \sigma(i_k)$). The right hand side of (2) is

$$(-1)^i \frac{\partial \omega_{i_1 \cdots i_k}}{\partial x^j} v_i{}^j (\text{sign}\, \eta) v_0^{\eta(i_1)} \cdots \hat{v}_i^{\eta(i_j)} \cdots v_k^{\eta(i_k)} \tag{2''}$$

(where the sum is over $i_1 < \cdots < i_k$, j, i, and η's with $\eta(i_1) < \cdots < \eta(i_k)$). Writing σ as a product of a permutation moving j to a designated position and a permutation η, we see that (2') and (2'') coincide. ■

If M is finite dimensional and $\alpha \in \Omega^k(M)$ has the local expression $\alpha = \alpha_{i_1 \dots i_k} dx^{i_k} \wedge \dots \wedge dx^{i_k}$, for $i_1 < \dots < i_k$, it is straightforward to check that the strict components of $\mathbf{d}\alpha$ are given by

$$(\mathbf{d}\alpha)_{j_1 \dots j_{l+1}} = \sum_{p=1}^{k} (-1)^{p-1} \partial \frac{\alpha_{j_1 \dots j_{p-1} j_{p+1} \dots j_{k+1}}}{\partial x^{j_p}} + (-1)^k \partial \frac{\alpha_{j_1 \dots j_k}}{\partial x^{j_{k+1}}}, \quad j_1 < \dots < j_{k+1}.$$

6.4.3 Examples

A On $\mathbb{R}^2$, let $\alpha = f(x, y)dx + g(x, y)dy$. Then $\mathbf{d}\alpha = \mathbf{d}f \wedge dx + f\mathbf{d}(dx) + \mathbf{d}g \wedge dy + g\mathbf{d}(dy)$ by linearity and the product rule. Since $\mathbf{d}^2 = 0$,

$$\mathbf{d}\alpha = \mathbf{d}f \wedge dx + \mathbf{d}g \wedge dy = \left(\frac{\partial f}{\partial x} dx + \frac{\partial f}{\partial y} dy\right) \wedge dx + \left(\frac{\partial g}{\partial x} dx + \frac{\partial g}{\partial y} dy\right) \wedge dy.$$

Since $dx \wedge dx = 0$ and $dy \wedge dy = 0$, this becomes

$$\mathbf{d}\alpha = \frac{\partial f}{\partial y} dy \wedge dx + \frac{\partial g}{\partial x} dx \wedge dy = \left(\frac{\partial g}{\partial x} - \frac{\partial f}{\partial y} \right) dx \wedge dy.$$

B On $\mathbb{R}^3$, let $f(x, y, z)$ be given. Then

$$\mathbf{d}f = \frac{\partial f}{\partial x} dx + \frac{\partial f}{\partial y} dy + \frac{\partial f}{\partial z} dz,$$

so the components of $\mathbf{d}f$ are those of grad f. That is, $(\text{grad } f)^\flat = \mathbf{d}f$, where $\flat$ is the index lowering operator defined by the standard metric of $\mathbb{R}^3$ (see §5.1).

C On $\mathbb{R}^3$, let $F^\flat = F_1(x, y, z)dx + F_2(x, y, z)dy + F_3(x, y, z)dz$. Computing as in Example **A** yields

$$\mathbf{d}F^\flat = \left(\frac{\partial F_2}{\partial x} - \frac{\partial F_1}{\partial y} \right) dx \wedge dy - \left(\frac{\partial F_1}{\partial z} - \frac{\partial F_3}{\partial x} \right) dx \wedge dz + \left(\frac{\partial F_3}{\partial y} - \frac{\partial F_2}{\partial z} \right) dy \wedge dz.$$

Thus associated to each vector field $G = G_1\mathbf{i} + G_2\mathbf{j} + G_3\mathbf{k}$ on $\mathbb{R}^3$ is the one-form $G^\flat$ and to this the two-form $*(G^\flat)$ by

$$*(G^\flat) = G_3 dx \wedge dy - G_2 dx \wedge dz + G_1 dy \wedge dz.$$

where $*$ is the Hodge operator (see §6.2); it is clear the $\mathbf{d}F^\flat = *(\text{curl } F)^\flat$.

D The divergence is obtained from $\mathbf{d}$ by

$$\mathbf{d} * F^\flat = (\text{div } F) dx \wedge dy \wedge dz; \text{ i.e., } * \mathbf{d} * F^\flat = \text{div } F.$$

Thus, by associating to a vector field F on $\mathbb{R}^3$ the one-form $F^\flat$ and the two-form $\mathbf{d} * F^\flat$, gives rise to the operators curl F and div F. From $\mathbf{d}F^\flat = *(\text{curl } F)^\flat$ it is apparent that

$$\mathbf{dd}F^\flat = 0 = \mathbf{d} *(\text{curl } F)^\flat = (\text{div curl } F) dx \wedge dy \wedge dz.$$

That is, $\mathbf{d}^2 = 0$ gives the well-known vector identity div curl F = 0. Likewise, $\mathbf{dd}f = 0$ becomes $\mathbf{d}(\text{grad } f)^\flat = 0$; i.e., $*(\text{curl grad } f)^\flat = 0$. So here $\mathbf{d}^2 = 0$ becomes the identity curl grad f = 0.

We summarize the relationship between the operators in vector calculus and differential forms in the table on the next page.

VECTOR CALCULUS AND DIFFERENTIAL FORMS

1 Sharp and Flat (Using standard coordinates in $\mathbb{R}^3$)

(a) $v^\flat = v^1 dx + v^2 dy + v^3 dz$ = one form corresponding to the vector $v = v^1 e_1 + v^2 e_2 + v^3 e_3$

(b) $\alpha^\# = \alpha_1 e_1 + \alpha_2 e_2 + \alpha_3 e_3$ = vector corresponding to the one form $\alpha = \alpha_1 dx + \alpha_2 dy + \alpha_3 dz$

2 Hodge Star Operator (Equation (6), §6.2)

(a) $*1 = dx \wedge dy \wedge dz$

(b) $*dx = dy \wedge dz, \; *dy = -dx \wedge dz, \; *dz = dx \wedge dy$

(c) $*(dy \wedge dz) = dx, \; *(dx \wedge dz) = -dy, \; *(dx \wedge dy) = dz$

(d) $*(dx \wedge dy \wedge dz) = 1$

3 Cross Product and Dot Product

(a) $\mathbf{v} \times \mathbf{w} = [*(\mathbf{v}^\flat \wedge \mathbf{w}^\flat)]^\#$

(b) $(\mathbf{v} \cdot \mathbf{w}) dx \wedge dy \wedge dz = \mathbf{v}^\flat \wedge *(\mathbf{w}^\flat)$

4 Gradient

$$\nabla f = \operatorname{grad} f = (\mathbf{d}f)^\#$$

5 Divergence

$$\nabla \cdot \mathbf{F} = \operatorname{div} \mathbf{F} = *\mathbf{d}(*\mathbf{F}^\flat)$$

6 Curl

$$\nabla \times \mathbf{F} = \operatorname{curl} \mathbf{F} = [*(\mathbf{d}\mathbf{F}^\flat)]^\#$$

The effect of mappings on $\mathbf{d}$ can now be considered. Recall that $\Omega(M)$ is the direct sum of all the $\Omega^k(M)$. Let $F : M \to N$ be a C^1 map. As $F^* : \Omega^k(N) \to \Omega^k(M)$ is $\mathbb{R}$-linear, it induces a mapping on the direct sums, $F^* : \Omega(N) \to \Omega(M)$.

6.4.4 Theorem *Let* $F : M \to N$ *be of class* C^1. *Then* $F^* : \Omega(N) \to \Omega(M)$ *is a homomorphism of differential algebras; that is,*

(i) $F^*(\psi \wedge \omega) = F^*\psi \wedge F^*\omega$, *and*

(ii) $\mathbf{d}$ *is natural with respect to mappings; that is,* $F^*(\mathbf{d}\omega) = \mathbf{d}(F^*\omega)$, *or the following diagram commutes:*

Proof Part (i) was established in **6.3.10**. For (ii), we shall show that if $m \in M$, then there is a neighborhood U of $m \in M$ such that $\mathbf{d}(F^*\omega \mid U) = (F^*\mathbf{d}\omega) \mid \mid U$, which is sufficient, as F^* and $\mathbf{d}$ are both natural with respect to restriction. Let (V, φ) be a local chart at $F(m)$ and U a neighborhood of $m \in M$ with $F(U) \subset V$. Then for $\omega \in \Omega^k(V)$, we can write

$$\omega = \omega_{i_1\cdots k_k} dx^{i_1} \wedge \cdots \wedge dx^{i_k} \qquad \text{(sum over } i_1 < \cdots i_k)$$

and so $\mathbf{d}\omega = \partial_{i_0}\omega_{i_1\cdots i_k} dx^{i_0} \wedge \cdots \wedge dx^{i_k}$, where $\partial_{i_0} = \dfrac{\partial}{\partial x^{i_0}}$ (sum over i_0 and $i_1 < \cdots < i_k$)

and by (i)

$$F^*\omega \mid U = (F^*\omega_{i_1\cdots i_k})F^*dx^{i_1} \wedge \cdots \wedge F^*dx^{i_k}.$$

If $\psi \in \Omega^0(N)$, then $\mathbf{d}(F^*\psi) = F^*\mathbf{d}\psi$ by the composite mapping theorem, so by (i),

$$\mathbf{d}(F^*\omega \mid U) = F^*(\mathbf{d}\omega_{i_1\cdots i_k}) \wedge F^*dx^{i_1} \wedge \cdots \wedge F^*dx^{i_k} = F^*(\mathbf{d}\omega) \mid U. \quad \blacksquare$$

6.4.5 Corollary *The operator* $\mathbf{d}$ *is natural with respect to push-forward by diffeomorphisms. That is, if* $F : M \to N$ *is a diffeomorphism, then* $F_*\mathbf{d}\omega = \mathbf{d}F_*\omega$, *or the following diagram commutes:*

$$\begin{array}{ccc} \Omega^k(M) & \xrightarrow{F_*} & \Omega^k(N) \\ \downarrow{\scriptstyle \mathbf{d}} & & \downarrow{\scriptstyle \mathbf{d}} \\ \Omega^{k+1}(M) & \xrightarrow[F_*]{} & \Omega^{k+1}(N) \end{array}$$

Proof Since $F_* = (F^{-1})^*$, the result follows from **6.4.4**(ii). $\blacksquare$

6.4.6 Corollary *Let* $X \in \mathfrak{X}(M)$. *Then* $\mathbf{d}$ *is natural with respect to* $\mathbf{L}_X$. *That is, for* $\omega \in \Omega^k(M)$ *we have* $\mathbf{L}_X\omega \in \Omega^k(M)$ *and*

$$\mathbf{d}\mathbf{L}_X\omega = \mathbf{L}_X\mathbf{d}\omega.$$

or the following diagram commutes:

$$\begin{array}{ccc} \Omega^k(M) & \xrightarrow{\mathcal{L}_X} & \Omega^k(M) \\ \downarrow{\scriptstyle \mathbf{d}} & & \downarrow{\scriptstyle \mathbf{d}} \\ \Omega^{k+1}(M) & \xrightarrow[\mathcal{L}_X]{} & \Omega^{k+1}(M) \end{array}$$

Proof Let F_t be the (local) flow of X. Then we know that

$$\mathcal{L}_X\omega(m) = \frac{d}{dt}(F_t^*\omega)(m)\Big|_{t=0}.$$

Since $F_t^*\omega \in \Omega^k(M)$, it follows that $\mathcal{L}_X\omega \in \Omega^k(M)$. Now we have $F_t^*\mathbf{d}\omega = \mathbf{d}(F_t^*\omega)$. Then, since $\mathbf{d}$ is $\mathbb{R}$-linear, it commutes with d/dt and so $\mathbf{d}\mathcal{L}_X\omega = \mathcal{L}_X\mathbf{d}\omega$. ∎

In Chapter 5 contractions of general tensor fields were studied. For differential forms, contractions play a special role.

6.4.7 Definition *Let* M *be a manifold,* $X \in \mathfrak{X}(M)$, *and* $\omega \in \Omega^{k+1}(M)$. *Then define* $\mathbf{i}_X\omega \in \mathcal{T}^0_k(M)$ *by*

$$\mathbf{i}_X\omega(X_1, \ldots, X_k) = \omega(X, X_1, \ldots, X_k).$$

If $\omega \in \Omega^0(M)$, *we put* $\mathbf{i}_X\omega = 0$. *We call* $\mathbf{i}_X\omega$ *the* ***interior product*** *or* ***contraction*** *of* X *and* ω. *(Sometimes* $X \,\lrcorner\, \omega$ *is written for* $\mathbf{i}_X\omega$.)

6.4.8 Theorem *We have* $\mathbf{i}_X\colon \Omega^k(M) \to \Omega^{k-1}(M)$, $k = 1, \ldots, n$, *and if* $\alpha \in \Omega^k(M)$, $\beta \in \Omega^\ell(M)$, *and* $f \in \Omega^0(M)$, *then*

(i) $\mathbf{i}_X$ is a $\wedge$-***antiderivation;*** *that is,* $\mathbf{i}_X$ *is* $\mathbb{R}$-*linear and we have the identity* $\mathbf{i}_X(\alpha \wedge \beta) = (\mathbf{i}_X\alpha) \wedge \beta + (-1)^k\alpha \wedge (\mathbf{i}_X\beta)$;

(ii) $\mathbf{i}_{fX}\alpha = f\mathbf{i}_X\alpha$;

(iii) $\mathbf{i}_X\mathbf{d}f = \mathcal{L}_Xf$;

(iv) $\mathcal{L}_X\alpha = \mathbf{i}_X\mathbf{d}\alpha + \mathbf{d}\mathbf{i}_X\alpha$;

(v) $\mathcal{L}_{fX}\alpha = f\mathcal{L}_X\alpha + \mathbf{d}f \wedge \mathbf{i}_X\alpha$.

Proof That $\mathbf{i}_X\alpha \in \Omega^{k-1}(M)$ follows from the definitions. For (i), $\mathbb{R}$-linearity is clear. For the second part of (i), write

$$\mathbf{i}_X(\alpha \wedge \beta)(X_2, X_3, \ldots, X_{k+1}) = (\alpha \wedge \beta)(X, X_2, \ldots, X_{k+1})$$

and

$$\mathbf{i}_X\alpha \wedge \beta + (-1)^k\alpha \wedge \mathbf{i}_X\beta = \frac{(k+\ell-1)!}{(k-\ell)!\,\ell!}\mathbf{A}(\mathbf{i}_X\alpha \otimes \beta) + (-1)^k\frac{(k+\ell-1)!}{k!(\ell-1)!}\mathbf{A}(\alpha \otimes \mathbf{i}_X\beta).$$

Now write out the definition of **A** in terms of permutations from **6.1.1**. The sum over all permutations in the last term can be replaced by the sum over $\sigma\sigma_0$, where σ_0 is the permutation $(2, 3, \ldots, k+1, 1, k+2, \ldots k+\ell) \mapsto (1, 2, 3, \ldots, k+\ell)$ whose sign is $(-1)^k$. Hence (i) follows. For (ii), we note that α is $\mathcal{F}(M)$-multilinear, and (iii) is just the definition of $\mathcal{L}_Xf$.
For (iv) we proceed by induction on k. First note that for $k = 0$, (iv) reduces to (iii). Now assume that (iv) holds for k. Then a $(k+1)$-form may be written as $\Sigma\mathbf{d}f_i \wedge \omega_i$, where ω_i is a k

form, in some neighborhood of $m \in M$. But $\mathcal{L}_X(df \wedge \omega) = \mathcal{L}_X df \wedge \omega + df \wedge \mathcal{L}_X\omega$ since $\mathcal{L}_X$ is a tensor derivation and commutes with **A** (or from the definition of $\mathcal{L}_X$ in terms of flows), and so

$$\begin{aligned} \mathbf{i}_X\mathbf{d}(df \wedge \omega) + \mathbf{d}\mathbf{i}_X(df \wedge \omega) &= -\mathbf{i}_X(df \wedge \mathbf{d}\omega) + \mathbf{d}(\mathbf{i}_X df \wedge \omega - df \wedge \mathbf{i}_X\omega) \\ &= -\mathbf{i}_X df \wedge \mathbf{d}\omega + df \wedge \mathbf{i}_X\mathbf{d}\omega + \mathbf{d}\mathbf{i}_X df \wedge \omega + \mathbf{i}_X df \wedge \mathbf{d}\omega + df \wedge \mathbf{d}\mathbf{i}_X\omega \\ &= df \wedge \mathcal{L}_X\omega + \mathbf{d}\mathcal{L}_X f \wedge \omega \end{aligned}$$

by the inductive assumption and (iii). Since $\mathbf{d}\mathcal{L}_X f = \mathcal{L}_X \mathbf{d}f$, the result follows.

Finally for (v) we have

$$\mathcal{L}_{fX}\omega = \mathbf{i}_{fX}\mathbf{d}\omega + \mathbf{d}\mathbf{i}_{fX}\omega = f\mathbf{i}_X\mathbf{d}\omega + \mathbf{d}(f\mathbf{i}_X\omega) = f\mathbf{i}_X\mathbf{d}\omega + df \wedge \mathbf{i}_X\omega + f\mathbf{d}\mathbf{i}_X\omega = f\mathcal{L}_X\omega + df \wedge \mathbf{i}_X\omega. \quad \blacksquare$$

Note that the proofs of (i), (ii) and (iii) are valid without change on Banach manifolds. Formula (iv)

$$\mathcal{L}_X\alpha = \mathbf{i}_X\mathbf{d}\alpha + \mathbf{d}\mathbf{i}_X\alpha \tag{3}$$

(a "magic" formual of Cartan) is particularly useful. It can be used in the following way.

6.4.9 Examples

A If α is a k-form such that $\mathbf{d}\alpha = 0$ and X is a vector field such that $\mathbf{d}\mathbf{i}_X\alpha = 0$, then $F_t^*\alpha = \alpha$, where F_t is the flow of X. Indeed,

$$\frac{d}{dt}F_t^*\alpha = F_t^*\mathcal{L}_X\alpha = F_t^*(\mathbf{i}_X\mathbf{d}\alpha + \mathbf{d}(\mathbf{i}_X\alpha)) = 0.$$

so $F_t^*\alpha$ is constant in t. Since $F_0 = \textit{identity}$, $F_t^*\alpha = \alpha$ for all t.

B Let $M = \mathbb{R}^3$, suppose $\operatorname{div} X = 0$, and let $\alpha = dx \wedge dy \wedge dz$. Thus $\mathbf{d}\alpha = 0$. Also

$$\mathbf{i}_X\alpha = \mathbf{i}_X(dx \wedge dy \wedge dz) = X^1 dy \wedge dz - X^2 dx \wedge dz + X^3 dx \wedge dy = *X^\flat.$$

so $\mathbf{d}\mathbf{i}_X\alpha = \mathbf{d} * X^\flat = (\operatorname{div} X)\alpha = 0$. Thus by Example **A**, $F_t^*(dx \wedge dy \wedge dz) = dx \wedge dy \wedge dz$. As we shall see in the next section in a more general context, this means that the flow of X is volume preserving. Of course this can be proved directly as well (see for example Chorin and Marsden [1979]). For related applications to fluid mechanics, see §**8.2**. ♦

The behavior of contractions under mappings is given by the following proposition. (The statement and proof also hold for Banach manifolds.)

6.4.10 Proposition *Let* M *and* N *be manifolds and* $f: M \to N$ *a* C^1 *mapping. If* $\omega \in \Omega^k(N)$, $X \in \mathfrak{X}(N)$, $Y \in \mathfrak{X}(M)$, *and* Y *is* f-*related to* X, *then*

$$\mathbf{i}_X f^*\Omega = f^*\mathbf{i}_X\omega.$$

In particular, if f *is a diffeomorphism, then*

$$\mathbf{i}_{f^*X} f^*\omega = f^*\mathbf{i}_X\omega.$$

That is, inerior products are natural with respect to diffeomorphisms and the following diagram commutes:

$$\begin{CD}
\Omega^k(N) @>{f^*}>> \Omega^k(M)\\
@V{\mathbf{i}_X}VV @VV{\mathbf{i}_{f^*X}}V\\
\Omega^{k-1}(N) @>>{f^*}> \Omega^{k-1}(M)
\end{CD}$$

Similarly for $Y \in \mathfrak{X}(M)$ *we have the following commutative diagram:*

$$\begin{CD}
\Omega^k(M) @>{f_*}>> \Omega^k(N)\\
@V{\mathbf{i}_Y}VV @VV{\mathbf{i}_{f_*Y}}V\\
\Omega^{k-1}(M) @>>{f_*}> \Omega^{k-1}(N)
\end{CD}$$

Proof Let $v_1, \ldots, v_{k-1} \in T_m(M)$ and $n = f(m)$. Then

$$\begin{aligned}
\mathbf{i}_Y f^*\omega(m)\cdot(v_1, \ldots, v_{k-1}) &= f^*\omega(m)\cdot(Y(m), v_1, \ldots, v_{k-1})\\
&= \omega(n)\cdot((Tf \circ Y)(m), Tf(v_1), \ldots, Tf(v_{k-1}))\\
&= \omega(n)((X \circ f)(m), Tf(v_1), \ldots, Tf(v_{k-1}))\\
&= \mathbf{i}_X\omega(n)\cdot(Tf(v_1), \ldots, Tf(v_{k-1}))\\
&= f^*\mathbf{i}_X\omega(m)\cdot(v_1, \ldots, v_{k-1}). \quad \blacksquare
\end{aligned}$$

The next proposition expresses **d** in terms of the Lie derivatives (see Palais [1954]).

6.4.11 Proposition *Let* $X_i \in \mathfrak{X}(M)$, $i = 0, \ldots, k$, *and* $\omega \in \Omega^k(M)$. *Then we have*

$$\text{(i)}\ (\mathbf{L}_{X_0}\omega)(X_1, \ldots, X_k) = \mathbf{L}_{X_0}(\omega(X_1, \ldots, X_k)) - \sum_{i=1}^{k} \omega(X_1, \ldots, \mathbf{L}_{X_0}X_\ell, \ldots, X_k)$$

and

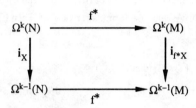

$$\text{(ii)}\quad \mathbf{d}\,\omega(X_0, X_1, \ldots, X_k) = \sum_{\ell=0}^{k} (-1)^{\ell} \mathcal{L}_{X_\ell}(\omega(X_0, \ldots, \hat{X}_\ell, \ldots, X_k))$$

$$+ \sum_{0 \le i < j \le k} (-1)^{i+j} \omega(\mathcal{L}_{X_i}(X_j), X_0, \ldots, \hat{X}_1, \ldots, \hat{X}_j, \ldots, X_k)$$

where $\hat{X}_\ell$ *denotes that* X_ℓ *is deleted.*

Proof Part (i) is condition **DO1** in **5.3.1**. For (ii) we proceed by induction. For $k = 0$, it is merely $\mathbf{d}\,\omega(X_0) = \mathcal{L}_{X_0}\omega$. Assume the formula for $k - 1$. Then if $\omega \in \Omega^k(M)$, we have, by Cartan's formula (3)

$$\begin{aligned}
\mathbf{d}\,\omega(X_0, X_1, \ldots, X_k) &= (\mathbf{i}_{X_0}\mathbf{d}\,\omega)(X_1, \ldots, X_k) \\
&= (\mathcal{L}_{X_0}\omega)(X_1, \ldots, X_k) - (\mathbf{d}(\mathbf{i}_{X_0}\omega))(X_1, \ldots, X_k) \\
&= \mathcal{L}_{X_0}(\omega(X_1, \ldots, X_k)) - \sum_{\ell=1}^{k} \omega(X_1, \ldots, \mathcal{L}_{X_0}X_\ell, \ldots, X_k) \\
&\quad - (\mathbf{d}\mathbf{i}_{X_0}\omega)(X_1, \ldots, X_k) \qquad \text{(by (i))}.
\end{aligned}$$

But $\mathbf{i}_{X_0}\omega \in \Omega^{k-1}(M)$ and we may apply the induction assumption. This gives, after a permutation

$$\begin{aligned}
(\mathbf{d}(\mathbf{i}_{X_0}\omega))(X_1, \ldots, X_k) &= \sum_{\ell=1}^{k} (-1)^{\ell-1} \mathcal{L}_{X_\ell}(\omega(X_0, X_1, \ldots, \hat{X}_\ell, \ldots, X_j)) \\
&\quad - \sum_{1 \le i < j \le k} (-1)^{i+j} \omega(\mathcal{L}_{X_i}X_j, X_0, X_1, \ldots, \hat{X}_i, \ldots, \hat{X}_j, \ldots, X_k).
\end{aligned}$$

Substituting this into the foregoing yields the result. ■

Note that the proof of (i) and of the first formula in the next corollary holds as well for infinite-dimensional manifolds.

6.4.12 Corollary *Let* $X, Y \in \mathfrak{X}(M)$. *Then*

$$[\mathcal{L}_X, \mathbf{i}_Y] = \mathbf{i}_{[X,Y]} \quad \textit{and} \quad [\mathcal{L}_X, \mathcal{L}_Y] = \mathcal{L}_{[X,Y]}$$

In particular, $\mathbf{i}_X \circ \mathcal{L}_X = \mathcal{L}_X \circ \mathbf{i}_X$.

Proof It is sufficient to check the first formula on any k-form $\omega \in \Omega^k(U)$ and any $X_1, \ldots, X_{k-1} \in \mathfrak{X}(U)$ for any open set U of M. We have by **6.4.11**(i)

$$(\mathbf{i}_Y \mathcal{L}_X \omega)(X_1, \ldots, X_{k-1}) = (\mathcal{L}_X \omega)(Y, X_1, \ldots, X_{k-1})$$

$$= \mathcal{L}_X(\omega(Y, X_1, \ldots, X_{k-1})) - \sum_{\ell=1}^{k-1} \omega(Y, X_1, \ldots, [X, X_\ell], \ldots, X_{k-1}) - \omega([X, Y], X_1, \ldots, X_{k-1})$$

$$= \mathcal{L}_X((\mathbf{i}_Y \omega)(X_1, \ldots, X_{k-1})) - \sum_{\ell=1}^{k-1} (\mathbf{i}_Y \omega)(X_1, \ldots, [X, X_\ell], \ldots, X_{k-1}) - (\mathbf{i}_{[X,Y]} \omega)(X_1, \ldots, X_{k-1})$$

$$= (\mathcal{L}_X \mathbf{i}_Y \omega)(X_1, \ldots, X_{k-1})) - (\mathbf{i}_{[X,Y]} \omega)(X_1, \ldots, X_{k-1}),$$

One proves $[\mathcal{L}_X, \mathcal{L}_Y] = \mathcal{L}_{[X,Y]}$ using the first relation and Cartan's formula (3). ■

☞ SUPPLEMENT 6.4A
The Exterior Derivative on Infinite-dimensional Manifolds

Now we discuss the exterior derivative on infinite-dimensional manifolds. Theorem **6.4.1** is rather awkward, primarily because we cannot, without a lot of technicalities, pass from, for example, one-forms to two-forms by linear combinations of decomposable two-forms, i.e., two-forms of the type $\alpha \wedge \beta$. However, there is a simpler alternative available.

1. Adopt formula **6.4.11**(ii) as the definition of **d** on any open subset of M. Note that at first it is defined as a multilinear function on vector fields and note that $\mathcal{L}_X$ is already defined.
2. In charts, **6.4.11**(ii) reduces to the local formula (2). This or a direct computation shows that $\mathbf{d} : \Omega^k(M) \to \Omega^{k+1}(M)$ is well defined, depending only on the point values of the vector fields.
3. One checks the basic properties of **d**. This can be done in two ways: directly, using the local formula, or using the definition and the following lemma, easily deducible from the Hahn-Banach theorem: *if a* k*-form* ω *is zero on any set* $X_1, \ldots, X_k \in \mathfrak{X}(U)$ *for all open sets* U *in* M, *then* $\omega \equiv 0$. This second method is slightly faster if one first proves formula (3), which in turn implies **6.4.12**.

Proof of Formula (3) Let α be a k-form and $X_1, \ldots, X_k$ a set of k vector fields defined on some open subset of M. Writing $X_0 = X$,

$$(\mathbf{i}_X \mathbf{d}\alpha + \mathbf{d}\mathbf{i}_X \alpha)(X_1, \ldots, X_k) = \mathbf{d}\alpha(X, X_1, \ldots, X_k) + \mathbf{d}(\mathbf{i}_X \alpha)(X_1, \ldots, X_k)$$

$$= \sum_{\ell=1}^{k} (-1)^\ell \mathcal{L}_{X_i}(\alpha(X_0, \ldots, \hat{X}_\ell, \ldots, X_k))$$

$$+ \sum_{0 \le i < j \le k} (-1)^{i+j} \alpha(\mathcal{L}_{X_i} X_j, X_0, \ldots, \hat{X}_i, \ldots, \hat{X}_j, \ldots, X_k)$$

$$+\sum_{\ell=1}^{k}(-1)^{\ell-1}\mathcal{L}_{X_i}(\alpha(X_0, X_1, \ldots, \hat{X}_\ell, \ldots, X_k))$$

$$-\sum_{1\le i<j\le k}(-1)^{i+j}\alpha(\mathcal{L}_{X_0}X_j, X_1, \ldots, \hat{X}_i, \ldots, \hat{X}_j, \ldots, X_k)$$

$$= \mathcal{L}_{X_0}(\alpha(X_1, \ldots, X_k)) + \sum_{j=1}^{k}(-1)^j\alpha(\mathcal{L}_{X_0}X_j, X_1, \ldots, \hat{X}_j, \ldots, X_k)$$

$$= (\mathcal{L}_X\alpha)(X_1, \ldots, X_k) \text{ by } \mathbf{6.4.11}\text{(ii)}. \blacksquare$$

This and **6.4.12** will allow us to give a proof of the infinite-dimensional version of **6.4.6**:

$$\mathcal{L}_X \circ \mathbf{d} = \mathbf{d} \circ \mathcal{L}_X.$$

For functions f this formula is proved as follows. By **6.4.11**(i),

$$\begin{aligned}(\mathcal{L}_X\mathbf{d}f)(Y) &= \mathcal{L}_X(\mathbf{d}f(Y)) - \mathbf{d}f([X, Y]) = X[Y[f]] - [X, Y][f] \\ &= Y[X[f]] = \mathbf{d}(X[f])(Y) = (\mathbf{d}\mathcal{L}_Xf)(Y).\end{aligned}$$

Inductively, assume the formula holds for $(k-1)$-forms. Then for any k-form α and any vector field Y defined on an open subset of M, $\mathbf{d}\mathcal{L}_X\mathbf{i}_Y\alpha = \mathcal{L}_X\mathbf{d}\mathbf{i}_Y\alpha$. Thus by **6.4.12**,

$$\begin{aligned}\mathbf{i}_Y\mathbf{d}\mathcal{L}_X\alpha &= \mathcal{L}_Y\mathcal{L}_X\alpha - \mathbf{d}\mathbf{i}_Y\mathcal{L}_X\alpha = \mathcal{L}_X\mathcal{L}_Y\alpha - \mathcal{L}_{[X,Y]}\alpha + \mathbf{d}\mathbf{i}_{[X,Y]}\alpha - \mathbf{d}\mathcal{L}_X\mathbf{i}_Y\alpha \\ &= \mathcal{L}_X\mathcal{L}_Y\alpha - \mathcal{L}_X\mathbf{d}\mathbf{i}_Y\alpha - \mathbf{i}_{[X,Y]}\mathbf{d}\alpha = \mathcal{L}_X\mathbf{i}_Y\mathbf{d}\alpha - \mathbf{i}_{[X,Y]}\mathbf{d}\alpha = \mathbf{i}_Y\mathcal{L}_X\mathbf{d}\alpha.\end{aligned}$$

Hence $\mathbf{d} \circ \mathcal{L}_X = \mathcal{L}_X \circ \mathbf{d}$.

Next, the remaining properties of $\mathbf{d}$ are checked in the following way. $\mathbb{R}$-linearity and **6.4.1**(iv) are immediate consequences of the definition. For **6.4.1**(ii), note that $\mathbf{d}f(X) = \mathbf{i}_X\mathbf{d}f = \mathcal{L}_Xf - \mathbf{d}\mathbf{i}_Xf = \mathcal{L}_Xf = X[f]$. To show that $\mathbf{d}^2 = 0$, first observe that

$$\mathbf{i}_X \circ \mathbf{d} \circ \mathbf{d} = \mathcal{L}_X \circ \mathbf{d} - \mathbf{d} \circ \mathbf{i}_X \circ \mathbf{d} = \mathbf{d} \circ \mathcal{L}_X - \mathbf{d} \circ \mathcal{L}_X + \mathbf{d} \circ \mathbf{d} \circ \mathbf{i}_X = \mathbf{d} \circ \mathbf{d} \circ \mathbf{i}_X,$$

so that for any k-form α and any vector fields $X_1, \ldots, X_{k+2}$, we have

$$(\mathbf{d}\mathbf{d}\alpha)(X_1, \ldots, X_{k+2}) = \mathbf{i}_{X_{k+2}} \cdots \mathbf{i}_{X_1}\mathbf{d}\mathbf{d}\alpha = \mathbf{i}_{X_{k+2}}\mathbf{d}\mathbf{d}\mathbf{i}_{X_{k+1}} \cdots \mathbf{i}_{X_1}\alpha = \mathbf{i}_{X_{k+2}}\mathbf{d}\mathbf{d}\mathbf{i}_{X_{k+1}}(\alpha(X_1, \ldots, X_k)) = 0.$$

The antiderivation property of $\mathbf{d}$ is proved by induction using (3) and the antiderivation property for the interior products. Finally, the formula $f^* \circ \mathbf{d} = \mathbf{d} \circ f^*$ for a map f follows by definition and the properties $f^*(\mathcal{L}_X\omega) = \mathcal{L}_X(f^*\omega)$, $(f^*\omega)(X_1, \ldots, X_k) = f^*(\omega(X'_1, \ldots, X'_k))$, $f^*[X, Y] = [X', Y']$ if $X_i \sim_f X'$, $i = 1, \ldots, k$, $X \sim_f X'$ and $Y \sim_f Y'$.

Thus, with the preceding procedure, $\mathbf{d}$ is defined on Banach manifolds and satisfies all the key properties that it does in the finite-dimensional case. These key properties are summarized in Table 6.4.1 below.

For vector-valued k-forms, we adopt, as in the preceding box, Palais' formula, **6.4.1**(ii), as the definition on an open subset of M. Note again that this definition uses the fact that $\mathcal{L}_X$ is defined for vector-valued tensors, and again one has to prove that the local formula **6.4.2** holds. Then all properties in Table 6.4.1 are verified in the same manner as previously.
For vector-valued forms we have however an additional formula on $\Omega^k(M; F)$

$$\mathbf{d} \circ A = A \circ \mathbf{d}$$

for any $A \in L(F, F')$. If F is finite dimensional, the definition and properties of $\mathbf{d}$ become quite obvious; one notices that if

$$\omega = \sum_{j=1}^{n} \omega_j f_j \in \Omega^k(M; F), \text{ where } \omega_j \in \Omega^k(M),$$

then $\mathbf{d}\omega$ is given by

$$\mathbf{d}\omega = \sum_{j=1}^{n} \mathbf{d}\omega_j f_j$$

and this formula can be taken as the definition of $\mathbf{d}$ in this case. This method does not work for vector-***bundle***-valued forms. Additional structure on the bundle is required to be able to lift $\mathcal{L}_X$.

Next we discuss the important concepts of closed and exact forms and the Poincaré lemma. This lemma is a generalization and unification of two well-known facts in vector calculus:

1. if curl $\mathbf{F} = \mathbf{0}$, then locally $\mathbf{F} = \nabla f$;
2. if div $\mathbf{F} = 0$, then locally $\mathbf{F} = \text{curl}\, \mathbf{G}$.

6.4.13 Definition *We call* $\omega \in \Omega^k(M)$ ***closed*** *if* $\mathbf{d}\omega = 0$, ***and exact*** *if there is an* $\alpha \in \Omega^{k-1}(M)$ *such that* $\omega = \mathbf{d}\alpha$.

6.4.14 Theorem (i) *Every exact form is closed.*

(ii) **Poincaré Lemma** *If* ω *is closed, then for each* $m \in M$, *there is a neighborhood* U *of* m *for which* $\omega \mid U \in \Omega^k(U)$ *is exact.*

Proof Part (i) is clear since $\mathbf{d} \circ \mathbf{d} = 0$. Using a local chart it is sufficient to consider the case $\omega \in \Omega^k(U)$, where $U \subset E$ is a disk about $0 \in E$, to prove (ii). On U we construct an $\mathbb{R}$-linear mapping $\mathbf{H} : \Omega^k(U) \to \Omega^{k-1}(U)$ such that $\mathbf{d} \circ \mathbf{H} + \mathbf{H} \circ \mathbf{d}$ is the identity on $\Omega^k(U)$. This will give the result, for $\mathbf{d}\omega = 0$ implies $\mathbf{d}(\mathbf{H}\omega) = \omega$. For $e_1, \ldots, e_k \in E$, define

$$\mathbf{H}\omega(u)(e_1, \ldots, e_{k-1}) = \int_0^1 t^{k-1}\omega(tu)(u, e_1, \ldots, e_{k-1})dt.$$

Then by **6.4.2**

$$\begin{aligned}\mathbf{dH}\omega(u)\cdot(e_1, \ldots, e_k) &= \sum_{i=1}^{k}(-1)^{i+1}\mathbf{DH}\omega(u)\cdot e_1(e_1, \ldots, \hat{e}_i, \ldots, e_k)\\ &= \sum_{i=1}^{k}(-1)^{i+1}\int_0^1 t^{k-1}\omega(tu)(e_i, e_1, \ldots, \hat{e}_i, \ldots, e_k)dt\\ &+ \sum_{i=1}^{k}(-1)^{i+1}\int_0^1 t^k\mathbf{D}\omega(tu)\cdot e_i(u, e_1, \ldots, \hat{e}_i, \ldots, e_k)dt.\end{aligned}$$

(The interchange of $\mathbf{D}$ and f is permissible, as ω is smooth and bounded as a function of $t \in [0, 1]$.) However, we also have, by **6.4.2**

$$\begin{aligned}\mathbf{Hd}\omega(u)\cdot(e_1, \ldots, e_k) &= \int_0^1 t^k\mathbf{d}\omega(tu)(u, e_1, \ldots, e_k)dt\\ &= \int_0^1 t^k\mathbf{D}\,\omega(tu)\cdot u(e_1, \ldots, e_k)dt + \sum_{i=1}^{k}(-1)^i\int_0^1 t^k\mathbf{D}\,\omega(tu)\cdot e_i(u, e_1, \ldots, \hat{e}_i, \ldots, e_k)dt.\end{aligned}$$

Hence

$$\begin{aligned}[\mathbf{dH}\omega(u) + \mathbf{Hd}\omega(u)](e_1, \ldots, e_k) &= \int_0^1 kt^{k-1}\omega(tu)\cdot(e_1, \ldots, e_k)dt + \int_0^1 t^k\mathbf{D}\,\omega(tu)\cdot u(e_1, \ldots, e_k)dt\\ &= \int_0^1 \frac{d}{dt}[t^k\omega(tu)\cdot(e_1, \ldots, e_k)]dt = \omega(u)\cdot(e_1, \ldots, e_k). \quad \blacksquare\end{aligned}$$

There is another proof of the Poincaré lemma based on the Lie transform method **5.4.7**. It will help the reader master the proof of Darboux' theorem in §**8.1** and is similar in spirit to the proof of Frobenius' theorem, **4.4.3**.

Alternative Proof of the Poincaré Lemma Again let U be a ball about 0 in E. Let, for $t > 0$, $F_t(u) = tu$. Thus F_t is a diffeomorphism and, starting at $t = 1$, is generated by the time-dependent vector field

$$X_t(u) = u/t;$$

that is, $F_1(u) = u$ and $\mathbf{d}F_t(u)/dt = X_t(F_t(u))$. Therefore, since ω is closed,

$$\frac{d}{dt}F_t^*\omega = F_t^*\mathbf{L}_{X_t}\omega = \mathbf{d}(F_t^*\mathbf{i}_{X_t}\omega).$$

For $0 < t_0 \leq 1$, we get

$$\omega - F_{t_0}^*\omega = \mathbf{d}\int_{t_0}^1 F_t^*\mathbf{i}_{X_t}\omega\, dt.$$

Letting $t_0 \to 0$, we get $\omega = \mathbf{d}\beta$, where

$$\beta = \int_0^1 F_t^* \mathbf{i}_{X_t}\omega \, dt.$$

Explicitly,

$$\beta_u(e_1, \ldots, e_{k-1}) = \int_0^1 t^{k-1}\omega_{tu}(u, e_1, \ldots, e_{k-1})dt.$$

(Note that this formula for β agrees with that in the previous proof.) ■

It is not true that closed forms are always exact (for example, on $\mathbb{R}^2 \setminus \{(0, 0)\}$ or on a sphere). In fact, the quotient groups of closed forms by exact forms (called the ***deRham cohomology groups*** of M) are important algebraic objects attached to a manifold; they are discussed in §7.6. Below we shall prove that on smoothly contractible manifolds, closed forms are always exact.

6.4.15 Definition *Let* $r \geq 1$. *Two* C^r *maps* $f, g : M \to N$ *are said to be* ***(properly) C^r-homotopic****, if there exists* $\varepsilon > 0$ *and a* C^r *(proper) mapping* $F :]-\varepsilon, 1 + \varepsilon[\times M \to N$ *such that* $F(0, m) = f(m)$, *and* $F(1, m) = g(m)$ *for all* $m \in M$. *The manifold* M *is called* ***C^r-contractible*** *if there exists a point* $m_0 \in M$ *and* C^r*-homotopy* F *of the constant map* $m \mapsto m_0$ *with the identity map of* M; F *is called a* ***C^r-contraction*** *of* M *to* m_0.

The following theorem represents a verification of the ***homotopy axiom for the deRham cohomology.***

6.4.16 Theorem *Let* $f, g : M \to N$ *be two (properly)* C^r*-homotopic maps and* $\alpha \in \Omega^k(N)$ *a closed* k*-form (with compact support) on* N. *Then* $g^*\alpha - f^*\alpha \in \Omega^k(M)$ *is an exact* k*-form on* M *(with compact support).*

The proof is based on the following.

6.4.17 Deformation Lemma *For a* C^r*-manifold* M *let the* C^r *mapping* $i_t : \to]-\varepsilon, 1 + \varepsilon[\times M$ *be given by* $i_t(m) = (t, m)$. *Define* $\mathbf{H} : \Omega^{k+1}(]-\varepsilon, 1 + \varepsilon[\times M) \to \Omega^k(M)$ *by*

$$\mathbf{H}\alpha = \int_0^1 i_s^*(\mathbf{i}_{\partial/\partial t}\alpha)ds.$$

Then $\mathbf{d} \circ \mathbf{H} \circ \mathbf{d} = i_1^* - i_0^*$.

Proof Since the flow of the vector field $\partial/\partial t \in \mathfrak{X}^r(]-\varepsilon, 1 + \varepsilon[\times M)$ is given by $F_\lambda(s, m) = (s + \lambda, m)$, i.e., $i_{s+\lambda} = F_\lambda \circ i_s$, for any form $\beta \in \Omega^\ell(]-\varepsilon, 1 + \varepsilon[\times M)$ we get

$$i_s^* \mathbf{L}_{\partial/\partial t}\beta = i_s^* \frac{d}{d\lambda}\bigg|_{\lambda=0} F_\lambda^*\beta = \frac{d}{d\lambda}\bigg|_{\lambda=0} i_s^* F_\lambda^*\beta = \frac{d}{d\lambda}\bigg|_{\lambda=0} i_{s+\lambda}^*\beta = \frac{d}{ds} i_s^*\beta.$$

Therefore, since the integrand in the formual for **H** is smooth, **d** and the integral sign commute, so that by Cartan's formula (3) and the above formula we get

$$\mathbf{d}(\mathbf{H}\alpha) + \mathbf{H}(\mathbf{d}\alpha) = \int_0^1 i_s^*(\mathbf{d}\mathbf{i}_{\partial/\partial t} + \mathbf{i}_{\partial/\partial t}\mathbf{d})\alpha \, ds = \int_0^1 i_s^* \mathbf{L}_{\partial/\partial t}\alpha \, ds = \int_0^1 \frac{d}{ds} i_s^* \alpha \, ds = i_1^*\alpha - i_0^*\alpha. \ \blacklozenge$$

Proof of Theorem 6.4.16 Define $G = \mathbf{H} \circ F^*$, where **H** is given in the Deformation Lemma **6.4.17** and F is the (proper) homotopy between f and g. Since F^* commutes with **d** we get $\mathbf{d} \circ G + G \circ \mathbf{d} = g^* - f^*$, so that if the term $\alpha \in \Omega^k(N)$ is closed (and has compact support), $(g^* - f^*)(\alpha) = \mathbf{d}(G\alpha)$ (and $G\alpha$ has compact support). ■

6.4.18 Poincaré Lemma for Contractible Manifolds *Any closed form on a smoothly contractible manifold is exact.*

Proof Apply the previous theorem with g = *identity* on M, and $f(m) = m_0$. ■

The naturality of the exterior derivative has been investigated by Palais [1959]. He proves the following result. Let M be a connected paracompact n-manifold and assume that $A : \Omega^p(M) \to \Omega^q(M)$ is a linear operator commuting with pull-back, i.e., $A \circ \varphi^* = \varphi^* \circ A$ for any diffeomorphism $\varphi : M \to M$. Then

$$A = \begin{cases} 0, \text{ if } 0 \le p \le n, \ 0 < q \le n, \ q \ne p, \ p+1, \\ a\,(\text{Identity}), \text{ if } 0 < q = p \le n, \\ b\mathbf{d}, \text{ if } 0 \le p \le n-1, \ q = p+1, \end{cases}$$

for some real constants a, b. If M is compact, then in addition we have

$$A = \begin{cases} 0, \text{ if } q = 0, \ 0 < p < n \\ c\,(\text{Identity}), \text{ if } p = q = 0 \\ 0, \text{ if } q = 0, p = n, \ M \text{ is non-orientable or orientable and reversible}, \\ d\int_M, \text{ if } q = 0, \ p = n, \ M \text{ is orientable and non-reversible}, \end{cases}$$

for some real constants c, d. (Orientability and reversibility will be defined in the next section whereas integration will be the subject of Chapter 7.)

☞ SUPPLEMENT 6.4B
Differential Ideals and Pfaffian Systems

This box discusses a reformulation of the Frobenius theorem in terms of differential ideals in the spirit of E. Cartan. Recall that a subbundle $E \subset TM$ is called ***involutive*** if for all pairs (X, Y) of local sections of E defined on some open subset of M, the bracket $[X, Y]$ is also a local section of E. The subbundle E is called ***integrable*** if at every point $m \in M$ there is a local submanifold N of M such that $T_mN = E_m$. Frobenius' theorem states that E is integrable iff it is involutive (see §**4.4**).

Before starting the general theory let us show by a simple example how forms and involutive subbundles are interconnected. Let $\omega \in \Omega^2(M)$ and assume that $E_\omega = \{v \in TM \mid \mathbf{i}_v\omega = 0\}$ is a subbundle of TM. If X and Y are two sections of E_ω, then

$$\mathbf{i}_{[X,Y]}\omega = \mathbf{L}_X\mathbf{i}_Y\omega - \mathbf{i}_Y\mathbf{L}_X\omega = -\mathbf{i}_Y\mathbf{d}\mathbf{i}_X\omega - \mathbf{i}_Y\mathbf{i}_X\mathbf{d}\omega = \mathbf{i}_X\mathbf{i}_Y\mathbf{d}\omega.$$

Thus, E_ω is involutive iff $\mathbf{i}_X\mathbf{i}_Y\mathbf{d}\omega = 0$. In particular, if ω is closed, then E_ω is involutive. In this supplement we shall express conditions such as $\mathbf{i}_X\mathbf{i}_Y\mathbf{d}\omega = 0$ in terms of one forms for subbundles E of TM that are not necessarily defined by exterior forms.

For any subbundle E, the k-***annihilator*** of E defined by

$$E^0(k) = \{\alpha \in \Lambda^k{}_m(M) \mid \alpha(m)(v_1, \ldots, v_k) = 0 \text{ for all } v_1, \ldots, v_k \in E_m,\ m \in M\}$$

is a subbundle of the bundle $\Lambda^k(M)$ of k-forms. Denote by $\Gamma(U, E)$ the C^∞ sections of E over the open set U of M and notice that

$$I(E) = \bigoplus_{0 \le k < \infty} \Gamma(M, E^0(k)$$

is an ***ideal*** of $\omega(M)$; i.e., if $\omega_1, \omega_2 \in I(E)$ and $\rho \in \Omega(M)$, then $\omega_1 + \omega_2 \in I(E)$ and $\rho \wedge \omega_1 \in I(E)$.

6.4.15 Proposition *The subbundle* E *of* TM *is involutive if for all open subsets* U *of* M *and all* $\omega \in \Gamma(U, E^0(1))$, *we have* $\mathbf{d}\omega \in \Gamma(U, E^0(2))$. *If* E *is involutive, then* $\omega \in \Gamma(U, E^0(k))$ *implies* $\mathbf{d}\omega \in \Gamma(U, E^0(k + 1)$.

Proof For any $\alpha \in \Gamma(U, E^0(1))$ and $X, Y \in \Gamma(U, E)$, **6.4.11**(ii) yields

$$\mathbf{d}\alpha(X, Y) = X[\alpha(Y)] - Y[\alpha(X)] - \alpha([X, Y]) = -\alpha([X, Y]).$$

Thus E is involutive iff $\mathbf{d}\alpha(X, Y) = 0$; i.e., $\mathbf{d}\alpha \in \Gamma(U, E^0(2))$. ■

The Frobenius theorem in terms of differential forms takes the following form.

6.4.20 Corollary *The subbundle* $E \subset TM$ *is integrable if for all open subsets* U of M, $\omega \in \Gamma(U, E^0(1))$ *implies* $\mathbf{d}\omega \in \Gamma(U, E^0(2))$.

The following considerations are strictly finite dimensional. They can be generalized to infinite-dimensional manifolds under suitable splitting assumptions. We restrict ourselves to the finite-dimensional situation due to their importance in applications and for simplicity of presentation.

6.4.21 Definition *Let* M *be an* n*-manifold and* $I \subset \Omega(M)$ *be an ideal. We say that* I *is* ***locally generated by*** $n-k$ ***independent one-forms,*** *if every point of* M *has a neighborhood* U *and* $n-k$ *pointwise linearly independent one-forms* $\omega_1, \ldots, \omega_{n-k} \in \Omega^1(U)$ *such that:*

(i) *if* $\omega \in I$, *then* $\omega \,|\, U = \sum_{i=1}^{n-k} \theta_i \wedge \omega_i$ *for some* $\theta_i \in \Omega(M)$;

(ii) *if* $\omega \in \Omega(M)$ *and* M *is covered by open sets* U *such that for each* U *in this cover,*

$$\omega \,|\, U = \sum_{i=1}^{n-k} \theta_i \wedge \omega_i \text{ for some } \theta_i \in \Omega(M),$$

then $\omega \in I$.

The ideal $I \subset \Omega(M)$ *is called a* ***differential ideal*** *if* $\mathbf{d}I \subset I$.

Finitely generated ideals of $\Omega(M)$ are characterized by being of the form $I(E)$. More precisely, we have the following.

6.4.22 Proposition *Let* I *be an ideal of* $\Omega(M)$ *and let* $n = \dim(M)$. *The ideal* I *is locally generated by* $n-k$ *linearly independent one-forms iff there exists a subbundle* $E \subset TM$ *with* k*-dimensional fiber such that* $I = I(E)$. *Moreover, the bundle* E *is uniquely determined by* I.

Proof If E has k-dimensional fiber, let $X_{n-k+1}, \ldots, X_n$ be a local basis of $\Gamma(U, E)$. Complete this to a basis of $\mathfrak{X}(U)$ and let $\omega_i \in \Omega^1(U)$ be the dual basis. Then clearly $\omega_1, \ldots, \omega_{n-k}$ are linearly independent and locally generate $I(E)$.

Conversely, let $\omega_1, \ldots, \omega_{n-k}$ generate I over U and define $E_m = \{v \in T_mM \mid \omega_i(m)(v) = 0, i = 1, \ldots, n-k\}$. E_m is clearly independent of the generators of I over U so that $E = \cap_{m \in M} E_m$ is a subbundle of TM. It is straightforward to check that $I = I(E)$. Finally, E is unique since $E \neq E'$ implies $I(E) \neq I(E')$ by construction. ■

Differential ideals are characterized among finitely generated ones by the following.

6.4.23 Proposition *Let I be an ideal of $\Omega(M)$ locally generated by $n-k$ linearly independent forms $\omega_1, \ldots, \omega_{n-k} \in \Omega^1(U)$, $n = \dim(M)$, and let $\omega_1 \wedge \cdots \wedge \omega_{n-k} = \omega \in \Omega^{n-k}(U)$. Then the following are equivalent:*

(i) *I is a differential ideal;*

(ii) $\mathbf{d}\omega = \Sigma_{j=1}^{n-k} \omega_{ij} \wedge \omega_j$ *for some* $\omega_{ij} \in \Omega^1(U)$ *and for every* U *as in the hypothesis;*

(iii) $\mathbf{d}\omega_i \wedge \omega = 0$ *for all open sets* U, *as in the hypothesis;*

(iv) *there exists* $\theta \in \Omega^1(U)$ *such that* $\mathbf{d}\omega = \theta \wedge \omega$ *for all open sets* U, *as in the hypothesis.*

Proof That conditions (i) and (ii) are equivalent and (ii) implies (iv) follows from the definitions. Condition (iv) means that

$$\sum_{i=1}^{n-k} (-1)^i \mathbf{d}\, \omega_i \wedge \omega_1 \wedge \cdots \wedge \hat{\omega}_i \wedge \cdots \wedge \omega_{n-k} = \theta \wedge \omega_1 \wedge \cdots \wedge \omega_{n-k},$$

so that multiplying by ω_i we get (iii). Finally, we show that (iii) implies (ii). Let $\omega_1, \ldots, \omega_n \in \Omega^1(U)$ be a basis such that $\omega_1, \ldots, \omega_{n-k}$ generates I over U. Then

$$\mathbf{d}\omega_i = \sum_{j<\ell} \alpha_{ij\ell}\omega_j \wedge \omega_\ell, \text{ where } \alpha_{ij\ell} \in \mathcal{F}(U).$$

But

$$0 = \mathbf{d}\, \omega_i \wedge \omega = \sum_{n-k<j<\ell} \alpha_{ij\ell}\omega_j \wedge \omega_\ell \wedge \omega_1 \wedge \cdots \wedge \omega_{n-k}$$

and thus $\alpha_{ij\ell} = 0$ for $n-k<j<\ell$. Hence

$$\mathbf{d}\omega_i = \sum_{j=1}^{n-k} \left(-\sum_{\ell=j+1}^{n} \alpha_{ij\ell}\omega_\ell \right) \wedge \omega_j. \quad \blacksquare$$

Assembling the preceding results, we get the following version of the Frobenius theorem.

6.4.24 Theorem *Let* M *be an* n*-manifold and* $E \subset TM$ *be a subbundle with* k*-dimensional fiber, and* $I(E)$ *the associated ideal. The following are equivalent:*

(i) *E is integrable;*

(ii) *E is involutive;*

(iii) *$I(E)$ is a differential ideal locally generated by* $n-k$ *linearly independent one-forms* $\omega_1, \ldots, \omega_{n-k} \in \Omega^1(U)$;

(iv) *for every point of* M *there exists an open set* U *and* $\omega_1, \ldots, \omega_{n-k} \in \Omega^1(U)$ *generating* $I(E)$ *such that*

$$d\omega_i = \sum_{j=1}^{n-k} \omega_{ij} \wedge \omega_j \textit{ for some } \omega_{ij} \in \Omega^1(U);$$

(v) *same as* (iv) *but with the condition on* ω_i *being:* $d\omega_i \wedge \omega_1 \wedge \cdots \wedge \omega_{n-k} = 0$;

(vi) *same as* (iv) *but with the condition on* ω_i *being: there exists* $\theta \in \Omega^1(U)$ *such that* $d\omega = \theta \wedge \omega$, *where* $\omega = \omega_1 \wedge \cdots \wedge \omega_{n-k}$.

6.4.25 Examples

A In classical texts (such as Cartan [1945] and Flanders [1963]), a system of equations

$$\omega_1 = 0, \ldots, \omega_{n-k} = 0, \text{ where } \omega_i \in \Omega^1(U) \text{ and } U \subset \mathbb{R}^n$$

is called a ***Pfaffian system.*** A solution of this system is a k-dimensional submanifold N of U given by $x^i = x^i(u^1, \ldots, u^k)$ such that if one plugs in these values of x^i in the system, the result is identically zero. Geometrically, this means that $\omega_1, \ldots, \omega_{n-k}$ annihilate TN. Thus, finding solutions of the Pfaffian system reverts to finding integral manifolds of the subbundle $E = \{v \in TU \mid \omega_i(v) = 0, i = 1, \ldots, n-k\}$ for which Frobenius' theorem is applicable; thus we must have $d\omega_i \wedge \omega_1 \wedge \cdots \wedge \omega_{n-k} = 0$. This condition is equivalent to the existence of smooth functions a_{ij}, b_j on U such that

$$\omega_i = \sum_{j=1}^{n-k} a_{ij} db_j.$$

To see this, recall that by the Frobenius theorem there are local coordinates $b_1, \ldots, b_n$ on U such that the integral manifolds of E are given by $b_1 = \text{constant}, \ldots, b_{n-k} = \text{constant}$, so that db_i, $i = 1, \ldots, n-k$ annihilate the tangent spaces to these submanifolds. Thus the ideal I generated by $db_1, \ldots, db_{n-k}$; i.e., $\omega_i = \Sigma_{j=1,\ldots,n-k} a_{ij} db_j$ for some smooth functions a_{ij} on U.

B Let us analyze the case of one Pfaffian equation in $\mathbb{R}^2$. Let $\omega = P(x, y)dx + Q(x, y)dy \in \Omega^1(\mathbb{R}^2)$ using standard (x, y) coordinates. We seek a solution to $\omega = 0$. This is equivalent to $dy/dx = -P(x, y)/Q(x, y)$, so existence and uniqueness of solutions for ordinary differential equations assures the local existence of a function f(x, y) such that f(x, y) = constant give the integral curves y(x). In other words, f(x, y) = constant is an integral manifold of $\omega = 0$. The same result could have been obtained by means of the Frobenius theorem. Since $d\omega \wedge \omega \in \Omega^3(\mathbb{R}^2)$, we get $d\omega \wedge \omega = 0$, so integral manifolds exist and are unique. In texts on differential equations, this problem is often solved with the aid of integrating factors. More precisely, if ω is not (locally) exact, can a function f and a function g, called an ***integrating factor***, be found, such that $g\omega = df$? The answer is "yes" by **6.4.24**(iii) for choosing f as above, $E = \ker df$ locally. Thus g is found by solving the partial differential equation,

$$\frac{\partial(gP)}{\partial y} = \frac{\partial(gQ)}{\partial x}.$$

This always has a solution and the connection between g and f is given by

$$g = \frac{1}{P}\frac{\partial f}{\partial x} = \frac{1}{Q}\frac{\partial f}{\partial y},$$

$f(x, y)$ = constant being the solution of $\omega = 0$.

C Let us analyze a Pfaffian equation $\omega = 0$ in $\mathbb{R}^n$. As in **B**, we would like to be able to write $g\omega = \mathbf{d}f$ with $\mathbf{d}f \neq 0$ on $U \subset \mathbb{R}^n$, for then $f(x^1, ..., x^n)$ = constant gives the (n – 1)-dimensional integral manifolds; i.e., the bundle defined by ω is integrable. Conversely, if the bundle defined by ω is integrable, then by **B**, $g\omega = \mathbf{d}f$. Integrability is (by the Frobenius theorem) equivalent to $\mathbf{d}\omega \wedge \omega = 0$, which, as we have seen in **B**, is always verified for $n = 2$. For $n \geq 3$, however, this is a genuine condition. If $n = 3$, let $\omega = P(x, y, z)dx + Q(x, y, z)dy + R(x, y, z)dz$. Then

$$\mathbf{d}\omega \wedge \omega = \left[P\left(\frac{\partial R}{\partial y} - \frac{\partial Q}{\partial z}\right) + Q\left(\frac{\partial P}{\partial z} - \frac{\partial R}{\partial x}\right) + R\left(\frac{\partial Q}{\partial x} - \frac{\partial P}{\partial y}\right)\right] dx \wedge dy \wedge dz;$$

so $\omega = 0$ is integrable iff the term in the square brackets vanishes.

D The Frobenius theorem is often used in overdetermined systems of partial differential equations to answer the question of existence and uniqueness of solutions. Consider for instance the following system of A. Mayer [1872] in $\mathbb{R}^{p+q} = \{(x^1, ..., x^p, y^1, ..., y^q)\}$:

$$\frac{dy^\alpha}{dx^i} = A^\alpha{}_i(x^1, \ldots, x^p, y^1, \ldots, y^q), \qquad i = 1, \ldots, p, \quad \alpha = 1, \ldots, q.$$

We ask whether there is a solution $y = f(x, c)$ for any choice of initial conditions c such that $f(0, c) = c$. The system is equivalent to the following Pfaffian system:

$$\omega^\alpha = dy^\alpha - A^\alpha{}_i dx^i = 0.$$

Since the existence of a solution is equivalent to the existence of p-dimensional integral manifolds, Frobenius' theorem asserts that the existence and uniqueness is equvalent to $\mathbf{d}\omega^\alpha = \Sigma_{\beta=1,\ldots,q}\omega^{\alpha\beta} \wedge \omega^\beta$ for some one-forms $\omega^{\alpha\beta}$. A straightforward computation shows that

$$\mathbf{d}\omega^\alpha = C^\alpha{}_{jk} dx^j \wedge dx^k + \frac{\partial A^\alpha{}_i}{\partial y^\beta} dx^i \wedge \omega^\beta,$$

where

$$C^\alpha{}_{jk} = \frac{\partial A^\alpha{}_j}{\partial x^k} - \frac{\partial A^\alpha{}_k}{\partial x^j} + \frac{\partial A^\alpha{}_j}{\partial y^\beta} A^\beta{}_k - \frac{\partial A^\alpha{}_k}{\partial y^\beta} A^\beta{}_j.$$

Since $dx^1, \ldots, dx^p, \omega^1, \ldots, \omega^q$ are a basis of $\Omega^1(\mathbb{R}^{p+q})$, we see that

$$\mathbf{d}\omega^\alpha = \Sigma_{\beta=1,\ldots,q}\omega^{\alpha\beta} \wedge \omega^b \textit{ for some one-forms } \omega^{\alpha\beta} \textit{ iff } C^\alpha{}_{jk} = 0.$$

Thus *the Mayer system is integrable iff* $C^\alpha{}_{jk} = 0$. ♦

In **§8.4** and **§8.5** we shall give some applications of Frobenius' theorem to problems in constraints and control theory. Many of these applications may alternatively be understood in terms of Pfaffian systems; see for example, Hermann [1977, Ch. 18].

IDENTITIES FOR VECTOR FIELDS AND FORMS

1. Vector fields on M with the bracket [X, Y] form a Lie algebra; that is, [X, Y] is real bilinear, skew symmetric, and Jacobi's identity holds:

$$[[X, Y], Z] + [[Z, X], Y] + [[Y, Z], X] = 0.$$

   Locally,

$$[X, Y] = \mathbf{D}Y \cdot X - \mathbf{D}X \cdot Y$$

   and on functions, $[X, Y][f] = X[Y[f]] - Y[X[f]]$.
2. For diffeomorphisms φ, ψ, $\varphi_*[X, Y] = [\varphi_*X, \varphi_*Y]$ and $(\varphi \circ \psi)_*X = \varphi_*\psi_*X$.
3. The forms on a manifold are a real associative algebra with $\wedge$ as multiplication. Furthermore, $\alpha \wedge \beta = (-1)^{k\ell}\beta \wedge \alpha$ for k and ℓ-forms α and β, respectively.
4. For maps φ, ψ, $\varphi^*(\alpha \wedge \beta) = \varphi^*\alpha \wedge \varphi^*\beta$, $(\varphi \circ \psi)^*\alpha = \psi^*\varphi^*\alpha$.
5. **d** is a real linear map on forms and

$$\mathbf{dd}\alpha = 0, \quad \mathbf{d}(\alpha \wedge \beta) = \mathbf{d}\alpha \wedge \beta + (-1)^k\, \alpha \wedge \mathbf{d}\beta \text{ for } \alpha \text{ a k-form.}$$

6. For α a k-form and $X_0, \ldots, X_k$ vector fields:

$$\mathbf{d}\alpha(X_0, \ldots, X_k) = \sum_{i=0}^{k}(-1)^i X_i[\alpha(X_0, \ldots, \hat{X}_i, \ldots, X_k)] + \sum_{0\le i<j\le k}(-1)^{i+j}\alpha([X_i, X_j], X_0, \ldots, \hat{X}_i, \ldots, \hat{X}_j, \ldots, X_k)$$

   If M is finite dimensional and $\alpha = \alpha_{i_1 \ldots i_k} dx^{i_1} \wedge \cdots \wedge dx^{i_k}$, $i_1 < \cdots < i_k$, then

$$(\mathbf{d}\alpha)_{j_1 \ldots j_{k+1}} = \sum_{p=1}^{k} (-1)^{p-1} \frac{\partial}{\partial x^{j_k}} \alpha_{j_1 \cdots j_{p-1} j_{p+1} \cdots j_{k+1}} + (-1)^k \frac{\partial}{\partial x^{j_{k+1}}} \alpha_{j_1 \cdots j_k}, \text{ for } j_1 < \cdots < j_{k+1}.$$

Locally,

$$\mathbf{d}\omega(x)(v_0, \ldots, v_k) = \sum_{i=0}^{k} (-1)^i \mathbf{D}\omega(x) \cdot v_i(v_0, \ldots, \hat{v}_i, \ldots, v_k).$$

7. For a map φ, $\varphi^* \mathbf{d}\alpha = \mathbf{d}\varphi^* \alpha$.
8. (Poincaré lemma) If $\mathbf{d}\alpha = 0$, then α is locally exact; that is, there is a neighborhood U about each point on which $\alpha = \mathbf{d}\beta$ for some form β defined on U. The same result holds globally on a contractible manifold.
9. $\mathbf{i}_X \alpha$ is real bilinear in X, α and for $h : M \to \mathbb{R}$, $\mathbf{i}_{hX}\alpha = h\mathbf{i}_X\alpha = \mathbf{i}_X h\alpha$. Also $\mathbf{i}_X \mathbf{i}_X \alpha = 0$, and

$$\mathbf{i}_X(\alpha \wedge \beta) = \mathbf{i}_X\alpha \wedge \beta + (-1)^k \alpha \wedge \mathbf{i}_X \beta$$

for α a k-form.

10. For a diffeomorphism φ, $\varphi^* \mathbf{i}_X \alpha = \mathbf{i}_{\varphi^* X} \varphi^* \alpha$; if $f : M \to N$ is a mapping and Y is f-related to X, then $\mathbf{i}_Y f^* \alpha = f^* \mathbf{i}_X \alpha$.
11. $\mathbf{L}_X \alpha$ is real bilinear in X, α and $\mathbf{L}_X(\alpha \wedge \beta) = \mathbf{L}_X \alpha \wedge \beta + \alpha \wedge \mathbf{L}_X \beta$.
12. $\mathbf{L}_X \alpha = \mathbf{d}\mathbf{i}_X \alpha + \mathbf{i}_X \mathbf{d}\alpha$.
13. For a diffeomorphism φ, $\varphi^* \mathbf{L}_X \alpha = \mathbf{L}_{\varphi^* X} \varphi^* \alpha$; if $f : M \to N$ is a mapping and Y is f-related to X, then $\mathbf{L}_Y f^* \alpha = f^* \mathbf{L}_X \alpha$.

14. $$(\mathbf{L}_X \alpha)(X_1, \ldots, X_k) = X[\alpha(X_1, \ldots, X_k)] - \Sigma^k_{i=1} \alpha(X_1, \ldots, [X, X_i], \ldots, X_k).$$

Locally,

$$(\mathbf{L}_X \alpha)_x \cdot (v_1, \ldots, v_k) = \mathbf{D}\alpha_x \cdot X(x) \cdot (v_1, \ldots, v_k) + \sum_{i=1}^{k} \alpha_x(v_1, \ldots, \mathbf{D}X_x \cdot v_i, \ldots, v_k).$$

15. The following identities hold:
    (a) $\mathbf{L}_{fX}\alpha = f\mathbf{L}_X \alpha + \mathbf{d}f \wedge \mathbf{i}_X \alpha$
    (b) $\mathbf{L}_{[X,Y]}\alpha = \mathbf{L}_X \mathbf{L}_Y \alpha - \mathbf{L}_Y \mathbf{L}_X \alpha$
    (c) $\mathbf{i}_{[X,Y]}\alpha = \mathbf{L}_X \mathbf{i}_Y \alpha - \mathbf{i}_Y \mathbf{L}_X \alpha$
    (d) $\mathbf{L}_X \mathbf{d}\alpha = \mathbf{d}\mathbf{L}_X \alpha$
    (e) $\mathbf{L}_X \mathbf{i}_X \alpha = \mathbf{i}_X \mathbf{L}_X \alpha$.

16. If M is a finite dimensional manifold, $X = X^\ell \partial/\partial x^\ell$ and $\alpha = \alpha_{i_1 \cdots i_k} dx^{i_1} \wedge \cdots \wedge dx^{i_k}$, $i_1 < \cdots < i_k$, the following local formulas hold:

$$
\begin{aligned}
\mathbf{d}\alpha &= \frac{\partial \alpha_{i_1 \dots i_k}}{\partial x^\ell} dx^\ell \wedge dx^{i_1} \wedge \cdots \wedge dx^{i_k}, \\
\mathbf{i}_X\alpha &= X^\ell \alpha_{\ell i_2 \cdots i_k} dx^{i_2} \wedge \cdots \wedge dx^{i_k}, \\
\boldsymbol{L}_X\alpha &= X^\ell \frac{\partial \alpha_{i_1 \dots i_k}}{\partial x^\ell} dx^{i_1} \wedge \cdots \wedge dx^{i_k} \\
&\quad + \alpha_{i_1 i_2 \cdots i_k} \frac{\partial X^{i_1}}{\partial x^\ell} dx^\ell \wedge dx^{i_2} \wedge \cdots \wedge dx^{i_k} \\
&\quad + \alpha_{i_1 i_2 \cdots i_k} \frac{\partial X^{i}_{2}}{\partial x^\ell} dx^{i_1} \wedge dx^\ell \wedge dx^{i_3} \wedge \cdots \wedge dx^{i_k} + \cdots
\end{aligned}
$$

Exercises

6.4A Compute the exterior derivative of the following differential forms on $\mathbb{R}^3$:

$$\alpha = x^3 dx + y^3 dx \wedge dy + xyz dx \wedge dz;$$

$$\beta = 3e^x dx \wedge dy + 8\cos(xy) dx \wedge dy \wedge dz.$$

6.4B Using **6.4.3** and the properties of $\mathbf{d}$ and $*$, prove the following formulas in $\mathbb{R}^3$ for $f, g : \mathbb{R}^3 \to \mathbb{R}$ and $\mathbf{F}, \mathbf{G} \in \mathfrak{X}(\mathbb{R}^3)$:

(i) $\mathrm{grad}(fg) = (\mathrm{grad}\, f)g + f(\mathrm{grad}\, g)$

(ii) $\mathrm{curl}(f\mathbf{F}) = (\mathrm{grad}\, f) \times \mathbf{F} + f(\mathrm{curl}\, \mathbf{F})$

(iii) $\mathrm{div}(f\mathbf{F}) = \mathrm{grad}(f) \cdot \mathbf{F} + f\, \mathrm{div}\, \mathbf{F}$

(iv) $\mathrm{div}(\mathbf{F} \times \mathbf{G}) = \mathbf{G} \cdot \mathrm{curl}\, \mathbf{F} - \mathbf{F} \cdot \mathrm{curl}\, \mathbf{G}$

(v) $\boldsymbol{L}_{\mathbf{F}}\mathbf{G} = (\mathbf{F} \cdot \nabla)\mathbf{G} - (\mathbf{G} \cdot \nabla)\mathbf{F} = \mathbf{F}\, \mathrm{div}\, \mathbf{G} - \mathbf{G}\, \mathrm{div}\, \mathbf{F} - \mathrm{curl}(\mathbf{F} \times \mathbf{G})$

(vi) $\mathrm{curl}(\mathbf{F} \times \mathbf{G}) = (\mathrm{div}\, \mathbf{G})\mathbf{F} - (\mathrm{div}\, \mathbf{F})\mathbf{G} + (\mathbf{G} \cdot \nabla)\mathbf{F} - (\mathbf{F} \cdot \nabla)\mathbf{G}$

(vii) $\mathrm{curl}(\mathrm{curl}\, \mathbf{F}) = \mathrm{grad}(\mathrm{div}\, \mathbf{F}) - \Delta\mathbf{F}$, where $(\Delta\mathbf{F})^i = \partial^2 F^i/\partial x^2 + \partial^2 F^i/\partial y^2 + \partial^2 F^i/\partial z^2$ is the usual Laplacian

(viii) $\nabla(\mathbf{F} \cdot \mathbf{F}) = 2(\mathbf{F} \cdot \nabla)\mathbf{F} + 2\mathbf{F} \times \mathrm{curl}\, \mathbf{F}$

6.4C Let $\varphi : S^1 \times \mathbb{R}_+ \to \mathbb{R}^2$ be defined by $\varphi(\theta, r) = (r\cos\theta, r\sin\theta)$. Compute $\varphi^*(dx \wedge dy)$ from the definitions and verify that it equals $\mathbf{d}(\varphi^* x) \wedge \mathbf{d}(\varphi^* y)$.

6.4D On S^1 find a closed one-form α that is not exact. (*Hint:* On $\mathbb{R}^2 \setminus \{0\}$ consider $\alpha = (y\, dx - x\, dy)/(x^2 + y^2)$.)

6.4E Show that the following properties uniquely characterize $\mathbf{i}_X$ on finite-dimensional manifolds

(i) $\mathbf{i}_X : \Omega^k(M) \to \Omega^{k-1}(M)$ is a $\wedge$ antiderivation.

(ii) $\mathbf{i}_X f = 0$ for $f \in \mathcal{F}(M)$;

(iii) $\mathbf{i}_X \omega = \omega(X)$ for $\omega \in \Omega^1(M)$;

(iv) $\mathbf{i}_X$ is natural with respect to restrictions.

Use these properties to show $\mathbf{i}_{[X,Y]} = \mathbf{\pounds}_X \mathbf{i}_Y - \mathbf{i}_Y \mathbf{\pounds}_X$. Finally, show $\mathbf{i}_X \circ \mathbf{i}_X = 0$.

6.4F Show that a derivation mapping $\Omega^k(M)$ to $\Omega^{k+1}(M)$ for all $k \geq 0$ is zero (note that $\mathbf{d}$ and $\mathbf{i}_X$ are *anti*derivations).

6.4G Let $s : T^2M \to T^2M$ be the canonical involution of the second tangent bundle (see exercise **3.4D**).

(i) If X is a vector field on M, show that $s \circ TX$ is a vector field on TM.

(ii) If F_t is the flow of X, prove that TF_t is a flow on TM generated by $s \circ TX$.

(iii) If μ is a one-form on M, $\mu' : TM \to \mathbb{R}$ is the corresponding function, and $w \in T^2M$, then show that

$$\mathbf{d}\mu'(sw) = \mathbf{d}\mu(\tau_{TM}(w), T\tau_M(w)) + \mathbf{d}\mu'(w).$$

6.4H Prove the following ***relative Poincaré lemma***. Let ω be a closed k-form on a manifold M and let $N \subset M$ be a closed submanifold. Assume that the pull-back of ω to N is zero. Then there is a $(k-1)$-form α on a neighborhood of N such that $\mathbf{d}\alpha = \omega$ and α vanishes on N. If ω vanishes on N, then α can be chosen so that all its first partial derivaties vanish on N. (*Hint:* Let φ_t be a homotopy of a neighborhood of N to N and construct an $\mathbf{H}$ operator as in the Poincaré lemma using φ_t.)

6.4I Angular variables Let S^1 denote the circle identified as $S^1 \approx \mathbb{R}/(2\pi) \approx \{z \in \mathbb{C} \mid |z| = 1\}$. Let $\gamma : \mathbb{R} \to S^1$; $x \mapsto e^{ix}$ be the exponential map. Show that γ induces a isomorphism $TS^1 \approx S^1 \times \mathbb{R}$. Let M be a manifold and let ω be an "angular variable," that is a smooth map $\omega : M \to S^1$. Define $\mathbf{d}\omega$, a one-form on M by taking the $\mathbb{R}$-projection of $T\omega$. Show that (i) if $\omega : M \to S^1$, then $\mathbf{dd}\omega = 0$; and (ii) if $f : M \to N$ is smooth, then $f^*(\mathbf{d}\omega) = \mathbf{d}(f^*\omega)$, where $f^*\omega = \omega \circ f$.

6.4J Prove the identity

$$\mathbf{\pounds}_X \mathbf{i}_Y - \mathbf{\pounds}_Y \mathbf{i}_X - \mathbf{i}_{[X,Y]} = [\mathbf{d}, \mathbf{i}_X \circ \mathbf{i}_Y].$$

6.4K (i) Let $X = (X^1, X^2, 0)$ be a vector field defined on the plane $S = \{(x, y, 0) \mid x, y \in \mathbb{R}\}$ in $\mathbb{R}^3$. Show that there exists $Y \in \mathfrak{X}(\mathbb{R}^3)$ such that $X = \operatorname{curl} Y$ on S. (*Hint:* Let $Y(x, y, z) = (zX^2(x, y), -zX^1(x, y), 0)$.)

(ii) Let S be a closed surface in $\mathbb{R}^3$ and $X \in \mathfrak{X}(S)$. Show that there exists $Y \in \mathfrak{X}(\mathbb{R}^3)$ such that $X = \operatorname{curl} Y$ on S. (*Hint:* By **5.5.9** extend X to $\tilde{X} \in \mathfrak{X}(\mathbb{R}^3)$ and put $\omega = * \tilde{X}^\flat$. Locally find α such that $\mathbf{d}\alpha = \omega$ by (i). Use a partition of unity $\{\varphi_i\}$ to write $\omega = \Sigma\varphi_i\omega$ and let $\mathbf{d}\alpha_i = \varphi_i\omega$, $\alpha = \Sigma\alpha_i$.)

(iii) Generalize this to forms on a closed submanifold of a manifold admiting C^k-partitions of unity.

6.4L Let M be a manifold and $\alpha \in \Omega^k(M)$. If $\tau_M : TM \to M$ denotes the tangent bundle projection, let $\alpha' = \tau^*\alpha \in \Omega^k(TM)$. A k-form Γ on TM for which there is an $\alpha \in \Omega^k(M)$ such that $\alpha' = \Gamma$ is called ***basic***. A vector field $x \in \mathfrak{X}(TM)$ is said to be ***vertical*** if $T\tau_M \circ X = 0$. Show that $\Gamma \in \Omega^k(TM)$ is basic if and only if $\mathbf{i}_X\Gamma = 0$, $\mathbf{L}_X \Gamma = 0$ for any vertical vector field X on TM. Conclude that if Γ is closed it is basic if and only if $\mathbf{i}_X\Gamma = 0$ for every vertical vector field X on TM. (*Hint*: Since X and the zero vector field on M are τ_M-related, if Γ is vertical, the two identities follow. Conversely, if F_t is the flow of X, then $F_t^*\Gamma = \Gamma$. Define $\alpha \in \Omega^k(TM)$ by $\alpha(m)(v_1, ..., v_k) = \Gamma(u)(V_1, ..., V_k)$, where $\tau(u) = m$, $T_m\tau_M(V_i) = v_i$, $i = 1, ..., k$. Show that this definition is independent of the choices of u, $V_1, ..., V_k$ in the following way. Let $\tau(u') = m$, $T_m\tau_M(V_i') = v_i$, $i = 1, ..., k$, $w = u - u'$. Consider the local flow F_t in a vector bundle chart of TM containing T_mM which occurs only in the fibers and which on T_mM itself is translation by tw. The vector field it generates is vertical so that $T_t^*\Gamma = \Gamma$ and $F_1(u) = u'$. Let $T_{u'}F_1(V_i') = V''_i \in T_u(TM)$ and show $T_u\tau(V''_i) = v_i$ since $\tau \circ \varphi_t = \tau$; i.e., $V''_i - V_i$ is a vertical vector. Now use the fact that $V''_i - V_i$ contracts with Γ to give zero to prove inductively that $\Gamma(u)(V_1, ..., V_k) = \Gamma(u')(V'_i, ..., V'_k)$.)

6.4M Show that on $\mathbb{R}^4$, the ideal generated by $\omega_1 = x^2dx^1 + x^3dx^4$, $\omega_2 = x^3dx^2 + x^2dx^3$ is a differential ideal. Find its integral manifolds.

§6.5 *Orientation, Volume Elements, and the Codifferential*

This section globalizes the setting of §**6.2** from linear spaces to manifolds. All manifolds in this section are finite dimensional. (For infinite-dimensional analogues of orientability, see Elworthy and Tromba [1970b].)

6.5.1 Definition *A* ***volume form*** *on an* n*-manifold* M *is an* n*-form* $\mu \in \Omega^n(M)$ *such that* $\mu(m) \neq 0$ *for all* $m \in M$; M *is called* ***orientable*** *if there exists some volume form on* M. *The pair* (M, μ) *is called a* ***volume manifold.***

Thus, μ assigns an orientation, as defined in **6.2.5**, to each fiber of TM. For example, $\mathbb{R}^3$ has the standard volume form $\mu = dx \wedge dy \wedge dz$.

6.5.2 Proposition *Let* M *be a connected* n*-manifold.*

(i) M *is orientable iff there is an element* $\mu \in \Omega^n(M)$ *such that every other* $\nu \in \Omega^n(M)$ *may be written* $\nu = f\mu$ *for some* $f \in \mathcal{F}(M)$.

(ii) *If* M *is orientable then it has an atlas* $\{(U_i, \varphi_i)\}$, *where* $\varphi_i : U_i \to U_i' \subset \mathbb{R}^n$, *such that the Jacobian determinant of the overlap maps is positive (the Jacobian determinant being the determinant of the derivative, a linear map from* $\mathbb{R}^n$ *into* $\mathbb{R}^n$*). The converse is true if* M *is paracompact.*

Proof For (i) assume first that M is orientable, with a volume form μ. Let ν be any other element of $\Omega^n(M)$. Now each fiber of $\Lambda^n(M)$ is one-dimensional, so we may define a map $f : M \to \mathbb{R}$ by

$$\nu'(m) = f(m)\mu'(m) \quad \text{where} \quad \mu(m) = \mu'(m)dx^1 \wedge \cdots \wedge dx^n$$

and similarly for ν'. Since $\mu'(m) \neq 0$ for all $m \in M$, $f(m) = \nu'(m)/\mu'(m)$ is of class C^∞. Conversely, if $\Omega^n(M)$ is generated by μ, then $\mu'(m) \neq 0$ for all $m \in M$ since each fiber is one-dimensional.

To prove (ii), let $\{(U_i, \varphi_i)\}$ be an atlas with $\varphi_i(U_i) = U_i' \subset \mathbb{R}^n$. We may assume that all U_i' are connected by taking restrictions if necessary. Now $\varphi_{i*}\mu = f_i dx^1 \wedge \cdots \wedge dx^n = f_i\mu_0$, where μ_0 is the standard volume form on $\mathbb{R}^n$. By means of a reflection if necessary, we may assume that $f_i(u') > 0$ ($f_i \neq 0$ since μ is a volume form). However, a continuous real-valued function on a connected space that is not zero is always > 0 or always < 0. Hence, for overlap maps we have

$$(\varphi_i \circ \varphi_j^{-1})_* dx^1 \wedge \cdots \wedge dx^n = \varphi_{i*} \circ \varphi_{j*}{}^{-1} dx^1 \wedge \cdots \wedge dx^n = \frac{f_i}{f_j \circ \varphi_j \circ \varphi_i^{-1}} dx^1 \wedge \cdots \wedge dx^n.$$

But

$$\psi^*(u)(\alpha^1 \wedge \cdots \wedge \alpha^n) = \mathbf{D}\psi(u)^* \cdot \alpha^1 \wedge \mathbf{D}\psi(u)^* \cdot \alpha^2 \wedge \cdots \wedge \mathbf{D}\psi(u)^* \cdot \alpha^n,$$

where $\mathbf{D}\psi(u)^* \cdot \alpha^1(e) = \alpha^1(\mathbf{D}\psi(u) \cdot e)$. Hence, by definition of determinant,

$$\det(\mathbf{D}(\varphi_j \circ \varphi_i^{-1})) = \frac{f_i(u)}{f_j[(\varphi_j \circ \varphi_i^{-1})(u)]} > 0.$$

(We leave as an exercise the fact that the canonical isomorphism $L(E, E) \approx L(E^*, E^*)$, used before does not affect determinants.)

For the converse of (ii), suppose $\{(V_\alpha, \psi_\alpha)\}$ is an atlas with the given property, and let $\{(U_i, \varphi_i, g_i)\}$ a subordinate parititon of unity. Let

$$\mu_i = \varphi_i^*(dx^1 \wedge \cdots \wedge dx^n) \in \Omega^n(U_i)$$

and let $\tilde{\mu}_i(m) = g_i(m)\mu_i(m)$ if $m \in U_i$ and $\tilde{\mu}_i = 0$ if $m \notin U_i$. Since $\operatorname{supp} g_i \subset U_i$, we have $\tilde{\mu}_i \in \Omega^n(M)$. Let $\mu = \Sigma_i \tilde{\mu}_i$. Since this sum is finite in some neighborhood of each point, it is clear from local representatives that $\mu \in \Omega^n(M)$. To show that μ is a volume form on M, notice that on $U_i \cap U_j \neq \varnothing$ we have

$$\begin{aligned} \mu_j &= \varphi_j^*(dx^1 \wedge \cdots \wedge dx^n) = \varphi_i^*(\varphi_j \circ \varphi_i^{-1})^*(dx^1 \wedge \cdots \wedge dx^n) \\ &= [\det \mathbf{D}(\varphi_j \circ \varphi_i^{-1}) \circ \varphi_i]\varphi_i^*(dx^1 \wedge \cdots \wedge dx^n) = [\det \mathbf{D}(\varphi_j \circ \varphi_i^{-1}) \circ \varphi_i]\mu_i = a_{ji}\mu_i \end{aligned}$$

where $a_{ji} \in \mathcal{F}(U_i \cap U_j)$, $a_{ji} > 0$. By local finiteness of the covering $\{U_i\}$, a given point $m \in M$ belongs only to a finite number of open sets $U_{i_0}, U_{i_1}, \ldots, U_{i_N}$. Thus

$$\mu(M) = \sum_{k=0}^{N} \mu_{i_k}(m) = \left[\sum_{k=1}^{N} (1 + a_{i_k i_0}(m))\right] \mu_{i_0}(m) \neq 0$$

since $\mu_{i_0}(m) \neq 0$ and each $a_{i_k i_0}(m) > 0$. It follows that $\mu(m) \neq 0$ for each $m \in M$. ■

Thus, if (M, μ) is a volume manifold we get a map from $\Omega^n(M)$ to $\mathcal{F}(M)$; namely, for each $\nu \in \Omega^n(M)$, there is a unique $f \in \mathcal{F}(M)$ such that $\nu = f\mu$.

6.5.3 Definition *Let M be an orientable manifold. Two volume forms μ_1 and μ_2 on M are called* ***equivalent*** *if there is an $f \in \mathcal{F}(M)$ with $f(m) > 0$ for all $m \in M$ such that $\mu_1 = f\mu_2$. (This is clearly an equivalence relation.) An* ***orientation*** *of M is an equivalence class $[\mu]$ of volume* *forms on* M. *An* ***oriented manifold*** (M, [μ]), *is an orientable manifold* M *together with an orientation* [μ] *on* M.

If [μ] *is an orientation of* M, *then* [−μ], *(which is clearly another orientation) is called the* ***reverse orientation.***

The next proposition tells us when [μ] and [−μ] are the only two orientations.

6.5.4 Proposition *Let* M *be an orientable manifold. Then* M *is connected iff* M *has exactly two orientations.*

Proof Suppose M is connected, and μ, ν are two volume forms with $\nu = f\mu$. Since M is connected, and $f(m) \neq 0$ for all $m \in M$, $f(m) > 0$ for all m or else $f(m) < 0$ for all m. Thus ν is equivalent to μ or −μ. Conversely, if M is not connected, let $U \neq \varnothing$ or M, be a subset that is both open and closed. If μ is a volume form on M, define ν by letting $\nu(m) = \mu(m)$ if $m \in U$ and $\nu(m) = -\mu(m)$ if $m \notin U$. Obviously, ν is a volume form on M, and $\nu \notin [\mu] \cup [-\mu]$. ■

A simple example of a nonorientable manifold is the Möbius band (see Figure 6.5.1 and Exercise **6.5L**). For other examples, see Exercises **6.5K, M.**

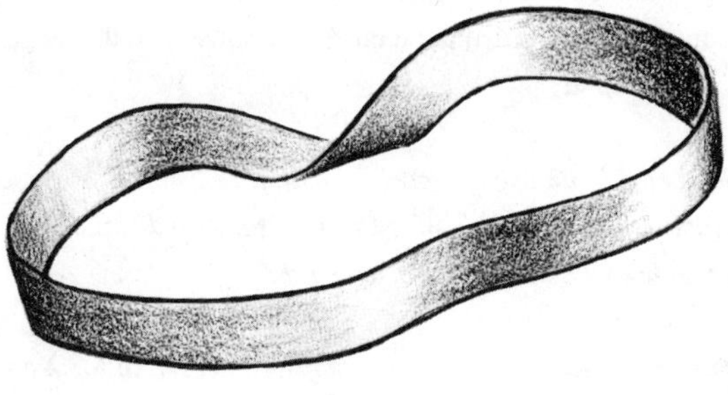

Figure 6.5.1

6.5.5 Proposition *The equivalence relation in* **6.5.3** *is natural with respect to mappings and diffeomorphisms. That is, if* $f : M \to N$ *is of class* C^∞, μ_N *and* ν_N *are equivalent volume forms on* N, *and* $f^*(\mu_N)$ *is a volume form on* M, *then* $f^*(\nu_N)$ *is an equivalent volume form. If* f *is a diffeomorphism and* μ_M *and* ν_M *are equivalent volume forms on* M, *then* $f_*(\mu_M)$ *and* $f_*(\nu_M)$ *are equivalent volume forms on* N.

Proof This follows from the fact that $f^*(g\omega) = (g \circ f)f^*\omega$, which implies $f^*(g\omega) = (g \circ f^{-1})f^*\omega$ when f is a diffeomorphism. ■

6.5.6 Definition *Let* M *be an orientable manifold with orientation* [μ]. *A chart* (U, φ) *with* $\varphi(U) = U' \subset \mathbb{R}^n$ *is called* ***positively oriented*** *if* $\varphi^*(\mu \mid U)$ *is equivalent to the standard volume* *form* $dx^1 \wedge \cdots \wedge dx^n \in \Omega^n(U')$.

From **6.5.5** we see that this definition does not depend on the choice of the representative from $[\mu]$.

If M is orientable, we can find an atlas in which every chart has positive orientation by choosing an atlas of connected charts and, if a chart has negative orientation, by composing it with a reflection. Thus, in **6.5.2**(ii), the atlas consists of positively oriented charts.

If M is not orientable, there is an orientable manifold $\tilde{M}$ and a 2-to-1 C^∞ surjective local diffeomorphism $\pi : \tilde{M} \to M$. The manifold $\tilde{M}$ is called the ***orientable double covering*** and is useful for reducing certain facts to the orientable case. The double covering is constructed as follows. Let $\tilde{M} = \{(m, [\mu_m]) \mid m \in M, [\mu_m]$ an orientation of $T_mM\}$. Define a chart at $(m, [\mu_m])$ in the following way. Fix an orientation $[\omega]$ of $\mathbb{R}^n$ and an orientation reversing isomorphism A of $\mathbb{R}^n$, e.g., the isomorphism given by $A(e_1) = -e_1$, $A(e_i) = e_i$, $i = 2, ..., n$, where $\{e_1, ..., e_n\}$ is the standard basis of $\mathbb{R}^n$. If $\varphi : U \to U' \subset \mathbb{R}^n$ is a chart of M at m, setting $U^\pm = \{(u, [\mu_u]) \mid u \in U, [\varphi_*(\mu_u)] = [\pm\omega]\}$, and defining $\varphi^\pm : U^\pm \to U'$ by $\varphi^+(u, [\mu_u]) = \varphi(u)$, $\varphi^-(u, [\mu_u]) = (A \circ \varphi)(u)$, we get charts $(U^\pm, \varphi^\pm)$ of $\tilde{M}$. It is straightforward to check that the family $\{(U^\pm, \varphi^\pm)\}$ constructed in this way forms an atlas, thus making $\tilde{M}$ into a differentiable n-manifold. Define $\pi : \tilde{M} \to M$ by $\pi(m, [\mu_m]) = m$. In local charts, π is the identity mapping, so that π is a surjective local diffeomorphism. Moreover $\pi^{-1}(m) = \{(m, [\mu_m]), (m, [-\mu_m])\}$, so that π is a twofold covering of M. Finally, $\tilde{M}$ is orientable, since the atlas formed by the charts $(U^\pm, \varphi^\pm)$ is orientable. A natural orientation on M is induced on the tangent space to $\tilde{M}$ at the point $(m, [\mu_m])$ by $[(T_m\pi)^* \mu_m]$.

6.5.7 Proposition *Let* M *be a connected* n*-manifold. Then* $\tilde{M}$ *is connected iff* M *is nonorientable. In fact,* M *is orientable iff* $\tilde{M}$ *consists of two disjoint copies of* M, *one with the given orientation, the other with the reverse orientation.*

Proof The *if* part of the second statement is a reformulation of **6.5.4** and it also proves that if $\tilde{M}$ is connected, then M is nonorientable. Conversely if M is a connected manifold and if $\tilde{M}$ is disconnected, let C be a connected component of $\tilde{M}$. Then since π is a local diffeomorphism, $\pi(C)$ is open in M. We shall prove that $\pi(C)$ is closed. Indeed, if $m \in \mathrm{cl}(\pi(C))$, let $\tilde{m}_1, \tilde{m}_2 \in \tilde{M}$ be such that $\pi(\tilde{m}_1) = \pi(\tilde{m}_2) = m$. If there exist neighborhoods $\tilde{U}_1$, $\tilde{U}_2$, of $\tilde{m}_1, \tilde{m}_2$ such that $\tilde{U}_1 \cap C = \varnothing$ and $\tilde{U}_2 \cap C = \varnothing$, then shrinking $\tilde{U}_1$ and $\tilde{U}_2$ if necessary, the open neighborhoods $\pi(\tilde{U}_1)$ and $\pi(\tilde{U}_2)$ of m have empty intersection with $\pi(C)$, contradicting the fact that $m \in \mathrm{cl}(\pi(C))$. Thus at least one of $\tilde{m}_1, \tilde{m}_2$ is in $\mathrm{cl}(C) = C$; i.e., $m \in \pi(C)$ and hence $\pi(C)$ is closed. Since M is connected, $\pi(C) = M$. But π is a double covering of M so that $\tilde{M}$ can have at most two components, each of them being diffeomorphic to M. Hence M is orientable, the orientation being induced from one of the connected components via π. ∎

Another criterion for orientability is the following.

6.5.8 Proposition *Suppose* M *is an orientable* n*-manifold and* V *is a submanifold of* *codimension* k *with trivial normal bundle. That is, there are* C^∞ *maps* $N_i : V \to TM$, $i = 1, \ldots, k$ *such that* $N_i(v) \in T_v(M)$, *and* $N_i(v)$ *span a subspace* W_v *such that* $T_vM = T_vV \oplus W_v$ *for all* $v \in V$. *Then* V *is orientable.*

Proof Let μ be a volume form on M. Consider the restriction $\mu \mid V : V \to \Lambda^n(M)$. Let us first note that $\mu \mid V$ is a smooth mapping of manifolds. This follows by using charts with the submanifold property, where the local representation is a restriction to a subspace. Now define $\mu_0 : V \to \Lambda^{n-k}(V)$ as follows. For $X_1, \ldots, X_{n-k} \in \mathfrak{X}(V)$, put

$$\mu_0(v)(X_1(v), \ldots, X_{n-k}(v)) = \mu(v)(N_1(v), \ldots, N_k(v), X_1(v), \ldots, X_{n-k}(v)).$$

It is clear that $\mu_0(v) \neq 0$ for all v. It remains only to show that μ_0 is smooth, but this follows from the fact that $\mu \mid V$ is smooth. ■

If g is a Riemannian metric, then $g^\flat : TM \to T^*M$ denotes the index-lowering operator and we write $g^\# = (g^\flat)^{-1}$. For $f \in \mathcal{F}(M)$, grad $f = g^\#(df)$ is called the ***gradient*** of f. Thus, grad $f \in \mathfrak{X}(M)$. In local coordinates, if $[g_{ij}] = [g(\partial/\partial x^i, \partial/\partial x^j)]$ and $[g^{ij}]$ is the inverse matrix, then

$$(\text{grad } f)^i = g^{ij}\frac{\partial f}{\partial x^j}. \tag{1}$$

6.5.9 Corollary *Suppose* M *is an orientable paracompact manifold,* $H \in \mathcal{F}(M)$ *and* $c \in \mathbb{R}$ *is a regular value of* H. *Then* $V = H^{-1}(c)$ *is an orientable submanifold of* M *of codimension one, if it is nonempty.*

Proof Suppose c is a regular value of H and $H^{-1}(c) = V \neq \varnothing$. Then V is a submanifold of codimension one. Let g be a Riemannian metric of M and $N = \text{grad}(H) \mid V$. Then $N(v) \notin T_vV$ for $v \in V$, because T_vV is the kernel of $\mathbf{d}H(v)$, and $\mathbf{d}H(v)[N(v)] = g(N, N)(v) > 0$ as $\mathbf{d}H(v) \neq 0$ by hypothesis. Then **6.5.8** applies, and so V is orientable. ■

Thus if we interpret V as the "energy surface," we see that it is an oriented submanifold for "almost all" energy values by Sard's theorem.

6.5.10 Definition *Let* M *and* N *be two orientable* n*-manifolds with volume forms* μ_M *and* μ_N, *respectively. Then we call a* C^∞ *map* $f : M \to N$ ***volume preserving*** *(with respect to* μ_M *and* μ_N*) if* $f^*\mu_N = \mu_M$, ***orientation preserving*** *if* $f^*(\mu_N) \in [\mu_M]$, *and* ***orientation reversing*** *if* $f^*(\mu_N) \in [-\mu_M]$. *An orientable manifold admitting (not admitting) an orientation reversing diffeomorphism is called* ***reversible*** *(respectively* ***non-reversible****).*

From **6.5.5**, $[f^*\mu_N]$ depends only on $[\mu_N]$. Thus the first part of the definition explicitly depends on μ_M and μ_N, while the last four parts depend only on the orientations $[\mu_M]$ and $[\mu_N]$.

Furthermore, we see from **6.5.5** that if f is volume preserving with respect to μ_M and μ_N, then f is volume preserving with respect to $h\mu_M$ and $g\mu_N$ iff $h = g \circ f$. It is also clear that if f is volume preserving with respect to μ_M and μ_N, then f is orientation preserving with respect to $[\mu_M]$ and $[\mu_N]$.

6.5.11 Proposition *Let* M *and* N *be* n-*manifolds with volume forms* μ_M *and* μ_N, *respectively. Suppose* $f : M \to N$ *is of class* C^∞. *Then* $f^*(\mu_N)$ *is a volume form iff* f *is a local diffeomorphism; that is, for each* $m \in M$, *there is a neighborhood* V *of* m *such that* $f \mid V : V \to f(V)$ *is a diffeomorphism. If* M *is connected, then* f *is a local diffeomorphism iff* f *is orientation preserving or orientation reversing.*

Proof If f is a local diffeomorphism, then clearly $f^*(\mu_N)(m) \neq 0$, by **6.2.3**(ii). Conversely, if $f^*(\mu_N)$ is a volume form, then the determinant of the derivative of the local representative is not zero, and hence the derivative is an isomorphism. The result then follows by the inverse function theorem. The second statement follows at once from the first and **6.5.4**. ∎

Next we consider the global analog of the determinant.

6.5.12 Definition *Suppose* M *and* N *are* n-*manifolds with volume forms* μ_M *and* μ_N, *respectively. If* $f : M \to N$ *is of class* C^∞, *the unique* C^∞ *function* $J(\mu_M, \mu_N)f \in \mathcal{F}(M)$ *such that* $f^*\mu_N = (J(\mu_M, \mu_N)f)\mu_M$ *is called the* ***Jacobian determinant*** *of* f *(with respect to* μ_M *and* μ_N*). If* $f : M \to M$ *we write* $J_\mu f = J(\mu, \mu)f$.

Note that $J(\mu_M, \mu_N)f(m) = \det(T_m f)$, the determinant being taken with respect to the volume forms $\mu_M(m)$ on T_mM and $\mu_N(f(m))$ on $T_{f(m)}N$. The basic properties of determinants that were developed in **§6.2** also hold in the global case, as follows. First, we have the following consequences of **6.5.11**.

6.5.13 Proposition f *is a local diffeomorphism iff* $J(\mu_M, \mu_N)f(m) \neq 0$ *for all* $m \in M$.

Second, we have consequences of the definition and properties of pull-back.

6.5.14 Proposition *Let* (M, μ) *be a volume manifold.*

(i) *If* $f : M \to M$ *and* $g : M \to M$ *are of class* C^∞, *then*

$$J_\mu(f \circ g) = [(J_\mu f) \circ g][J_\mu g].$$

(ii) *If* $h : M \to M$ *is the identity, then* $J_\mu h = 1$.

(iii) *If* $f : M \to M$ *is a diffeomorphism, then*

$$J_\mu(f^{-1}) = \frac{1}{[(J_\mu f) \circ f^{-1}]}.$$

Proof For (i),

$$J_\mu(f \circ g)\mu = (f \circ g)^*\mu = g^*f^*\mu = g^*(J_\mu f)\mu = ((J_\mu f) \circ g)g^*\mu = ((J_\mu f) \circ g)(J_\mu g)\mu.$$

Part (ii) follows since h^* is the identity. For (iii) we have

$$J_\mu(f \circ f^{-1}) = 1 = ((J_\mu f) \circ f^{-1})(J_\mu f^{-1}). \quad \blacksquare$$

6.5.15 Proposition *Let* $(M, [\mu_M])$ *and* $(N, [\mu_N])$ *be oriented manifolds and* $f : M \to N$ *be of class* C^∞. *Then* f *is orientation preserving iff* $J(\mu_M, \mu_N)f(m) > 0$ *for all* $m \in M$, *and orientation reversing if* $J(\mu_M, \mu_N)f(m) < 0$ *for all* $m \in M$. *Also,* f *is volume preserving with respect to* μ_M *and* μ_N *iff* $J(\mu_M, \mu_N)f = 1$.

This proposition follows from the definitions. Note that the first two assertions depend only on the orientations $[\mu_M]$ and $[\mu_N]$ since

$$J(h\mu_M, g\mu_N)f = \left(\frac{g \circ f}{h}\right)J(\mu_M, \mu_N)f,$$

which the reader can easily check. Here $g \in \mathcal{F}(N)$, $h \in \mathcal{F}(M)$, $g(n) \neq 0$, and $h(m) \neq 0$ for all $n \in N$, $m \in M$.

We have seen that in $\mathbb{R}^3$ the divergence of a vector field is expressible in terms of the standard volume element $\mu = dx \wedge dy \wedge dz$ by the use of the metric in $\mathbb{R}^3$ (see Example **6.4.3D**). There is, however, a second characterization of the divergence that does not require a metric but only a volume form μ, namely

$$\mathbf{L}_F\mu = (\text{div } F)\mu,$$

as a simple computation shows. This can now be generalized.

6.5.16 Definition *Let* (M, μ) *be a volume manifold; i.e.,* M *is an orientable manifold with a volume form* μ. *Let* X *be a vector field on* M. *The unique function* $\text{div}_\mu X \in \mathfrak{X}(M)$, *such that* $\mathbf{L}_X\mu = (\text{div}_\mu X)\mu$ *is called the* **divergence** *of* X. *We say* X *is* **incompressible** *(with respect to* μ*) if* $\text{div}_\mu X = 0$.

6.5.17 Proposition *Let* (M, μ) *be a volume manifold and* X *a vector field on* M.

(i) *If* $f \in \mathcal{F}(M)$ *and* $f(m) \neq 0$ *for all* $m \in M$, *then*

$$\mathrm{div}_{f\mu} X = \mathrm{div}_\mu X + \frac{\mathbf{L}_X f}{f}$$

(ii) *For* $g \in \mathcal{F}(M)$, $\mathrm{div}_\mu gX - g\,\mathrm{div}_\mu X + \mathbf{L}_X g$.

Proof Since $\mathbf{L}_X$ is a derivation,

$$\mathbf{L}_X(f\mu) = (\mathbf{L}_X f)\mu + f\mathbf{L}_X\mu.$$

As $f\mu$ is a volume form, $(\mathrm{div}_{f\mu}X)(f\mu) = (\mathbf{L}_X f)\mu + f(\mathrm{div}_\mu X)\mu$. Then (i) follows. For (ii), $\mathbf{L}_{gX}\mu = g\mathbf{L}_X\mu + dg \wedge \mathbf{i}_X\mu$, and from the antiderivation property of $\mathbf{i}_X$, $dg \wedge \mathbf{i}_X\mu = -\,\mathbf{i}_X(dg \wedge \mu) + \mathbf{i}_X dg \wedge \mu$. But $dg \wedge \mu \in \Omega^{n+1}(M)$, and hence $dg \wedge \mu = 0$. Also, $\mathbf{i}_X dg = \mathbf{L}_X g$ and so $\mathbf{L}_{gX}\mu = g\mathbf{L}_X\mu + (\mathbf{L}_X g)\mu$. The result follows from this. ■

6.5.18 Proposition *Let* (M, μ) *be a volume manifold and* X *a vector field on* M. *Then* X *is incompressible (with respect to* μ) *if the flow of* X *is volume preserving: that is, the local diffeomorphism* $F_t : U \to V$ *is volume preserving with respect to* $\mu \mid U$ *and* $\mu \mid V$.

Proof Since X is incompressible, $\mathbf{L}_X\mu = 0$, and so μ is constant along the flow of X; $\mu(m) = (F_t^*\mu)(m)$. Thus F_t is volume preserving. Conversely, if $(F_t^*\mu)(m) = \mu(m)$, then $\mathbf{L}_X\mu = 0$. ■

6.5.19 Corollary *Let* (M, μ) *be a volume manifold and* X *a vector field with flow* F_t *on* M. *Then* X *is incompressible iff* $J_\mu F_t = 1$ *for all* $t \in \mathbb{R}$.

The considerations regarding the Jacobian and the divergence can also be carried out for one-densities. If $|\mu_M|, |\mu_N|$ are one-densities and $f : M \to N$ is C^∞, we shall write $f^*|\mu_N| = J(|\mu_M|, |\mu_N|, f)\,|\mu_M|$, where the pull back is defined as for forms. Then **6.5.13** and **6.5.14** go through for one-densities. The Lie derivative of a one-density is defined by

$$\mathbf{L}_X|\mu| = \left.\frac{d}{dt}\right|_{t=0} F_t^*|\mu|$$

and one defines the divergence of X with respect to $|\mu|$ as in **6.5.16**. Then it is easy to check that **6.5.17-6.15.19** have analogues for one-densities.

We shall now globalize the concepts pertaining to Riemannian volume forms and densities, as well as the Hodge star operator discussed in §**6.2**.

6.5.20 Proposition *Let* (M, g) *be a pseudo-Riemannian manifold of signature* s; i.e., $g(m)$ *has signature* s *for all* $m \in M$.

(i) *If* M *is orientable, then there exists a unique volume element* $\mu = \mu(g)$ *on* M, *called the* **g-volume** *(or* **pseudo-Riemannian volume of** g), *such that* μ *equals one on all positively* *oriented orthonormal bases on the tangent spaces to* M. *If* $X_1, ..., X_n$ *is such a basis in an open set* U *of* M *and if* $\xi^1, ..., \xi^n$ *is the dual basis, then* $\mu = \xi^1 \wedge \cdots \wedge \xi^n$. *More generally, if* $v_1, ..., v_n \in T_xM$ *are positively oriented, then* $\mu(x)(v_1, ..., v_n) = |\det[g(x)(v_i, v_j)]|^{1/2}$.

(ii) *For every* $\alpha \in \mathbb{R}$ *there exists a unique* α*-density* $|\mu|^\alpha$, *called the* **g-α-density** *(or the* ***pseudo-Riemannian α-density of*** **g**), *such that* $|\mu|^\alpha$ *equals* 1 *on all orthonormal bases of the tangent spaces to* M. *If* $X_1,..., X_n$ *is such a basis in an open set* U of M *with dual basis* $\xi^1, ..., \xi^n$, *then* $|\mu|^\alpha = |\xi^1 \wedge \cdots \wedge \xi^n|^\alpha$. *More generally, if* $v_1, ..., v_n \in T_xM$, *then*

$$|\mu|^\alpha(x)(v_1, ..., v_n) = |\det[g(x)(v_i, v_j)]|^{\alpha/2}.$$

The proof is straightforward from **6.2.10** and the fact that μ and $|\mu|^\alpha$ are smooth. Also note that in an oriented chart $(x^1, ..., x^n)$ on M, we have $\mu = |\det[g_{ij}]|^{1/2} dx^1 \wedge \cdots \wedge dx^n$.
As in the vector space situation, g induces a pseudo-Riemannian metric on $\Lambda^k(M)$ by

$$\langle \alpha, \beta \rangle_x = \alpha_{i_1 \cdots i_k} \beta^{i_1 \cdots i_k}, \text{ the sum over } i_1 < \cdots < i_k \tag{2}$$

where $\alpha, \beta \in \Lambda^k(M)_x$ and $\beta^{i_1 \cdots i_k}$ are the components of the associated contravariant tensor to β g(x). As in **6.2.11**, if $X_1, ..., X_n$ is an orthonormal basis in $U \subset M$ with dual basis $\xi^1,..., \xi^n$, then the elements $\xi^{i_1} \wedge \cdots \wedge \xi^{i_k}$, where $i_1 < \cdots < i_k$ form an orthonormal basis of $\Lambda^k(U)$.

The Hodge operator is defined pointwise on an orientable pseudo-Riemannian manifold with pseudo-Riemannian volume form μ by $*: \Omega^k(M) \to \Omega^{n-k}(M)$, $(*\alpha)(x) = *\alpha(x)$, i.e., $\alpha \wedge *\beta = \langle \alpha, \beta \rangle \mu$ for $\alpha, \beta \in \Omega^k(M)$. The properties in Propositions **6.2.12** and **6.2.13** carry over since they hold pointwise.

The exterior derivative and the Hodge star operator enable us to introduce the following linear operator. (The reason for the strange-looking factor (–1) in the definition is so a later integration by parts formula, proved in **7.2.13**, will come out simple.)

6.5.21 Definition *The* ***codifferential*** $\delta : \Omega^{k-1}(M) \to \Omega^k(M)$ *is defined by* $\delta(\Omega^0(M)) = 0$ *and on* k+1 *forms* β *by*

$$\delta\beta = (-1)^{nk+1+\text{Ind}(g)} *\mathbf{d}*\beta$$

Notice that since $\mathbf{d}^2 = 0$ and $**$ is a multiple of the identity, $\delta^2 = 0$.

For example, in $\mathbb{R}^3$, let $\alpha = a dy \wedge dz - b dz \wedge dz + c dx \wedge dy$. Then

$$*\alpha = a dx + b dy + c dz,$$

so

$$\mathbf{d}*\alpha = (b_x - a_y) dx \wedge dy + (c_x - a_z) dx \wedge dz + (c_y - b_z) dy \wedge dz$$

and

$$*\mathbf{d}*\alpha = (b_x - a_y) dz - (c_x - a_z) dy + (c_y - b_z) dx.$$

Thus, as $nk + 1 + \mathrm{Ind}(g) = 4$ is even,

$$\delta\alpha = (c_y - b_z)\mathbf{d}x + (a_z - c_x)\mathbf{d}y + (b_x - a_y)\mathbf{d}z.$$

The formula for $\delta\beta$ in coordinates is given by

$$(\delta\beta)_{i_1 \dots i_k} =$$

$$\frac{1}{k+1} |\det[g_{ij}]|^{1/2} g_{i_1 r_1} \cdots g_{i_k r_k} \frac{\partial}{\partial x^{\ell}} \left(\sum_{p=1}^{k+1} (-1)^p g^{r_1 j_1} \dots g^{r_{p-1} j_{p-1}} g^{\ell j_r} g^{r_p j_{p\neq 1}} \dots g^{r_k j_{k\neq 1}} \beta_{j_1 \dots j_{k\neq 1}} |\det[g_{ij}]|^{1/2} \right)$$

or as a contravariant tensor

$$(\delta\beta)^{r_1 \dots r_k} = \frac{1}{k+1} |\det[g_{ij}]|^{-1/2} \frac{\partial}{\partial x^{\ell}} \left(\sum_{p=1}^{k\neq 1} (-1)^p \beta^{r_1 \dots r_{p-1} \ell r_p \dots r_k} |\det[g_{ij}]|^{1/2} \right);$$

where

$$\beta = \beta_{r_1 \cdots r_k r_{k+1}} dx^{r_1} \wedge \cdots \wedge dx^{r_{k+1}},$$

(sum over $r_1 < \cdots < r_{k+1}$) is the usual coordinate expression for β. Formula (3) is messy to prove directly. However it follows fairly readily from integration by parts in local coordinates and the fact that δ is the adjoint of $\mathbf{d}$, a fact that will be proved in Chapter 7 (see **7.2.13** and Exercise **7.5G**).

We now shall express the divergence of a vector field $X \in \mathfrak{X}(M)$ in terms of δ. We define the divergence $\mathrm{div}_g(X)$ of X with respect to a pseudo-Riemannian metric g to be the divergence of X with respect to the pseudo-Riemannian volume $\mu = \mu(g)$ of g; i.e., $\mathbf{L}_X\mu = \mathrm{div}_g(X)\mu$. To compute the divergence, we prepare a lemma.

6.5.22 Lemma $\mathbf{i}_X\mu = * X^\flat$.

Proof At points where X vanishes this relation is trivial. So let $X(x) \neq 0$ and choose $v_2, \dots, v_n \in T_xM$ such that $\{X(x)/g(X, X)(x), v_2, \dots, v_n\}$ is a positively oriented orthonormal basis of T_xM. Then by **6.2.14**,

$$(\mathbf{i}_X\mu)(x)(v_2, \dots, v_n) = \mu(X(x), v_2, \dots, v_n) = g(X, X)(x) = X^\flat(X)(x) = (*X^\flat)(x)(v_2, \dots, v_n) \quad \blacktriangledown$$

This may also be seen in coordinates using formula (12) of §**6.2**.

6.5.23 Proposition *Let* g *be a pseudo-Riemannian metric on the orientatable* n*-manifold* M. *Then*

$$\mathrm{div}_g(X) = -\delta X^\flat. \tag{4}$$

In (positively oriented) local coordinates

$$\operatorname{div}_g(X) = |\det[g_{ij}]|^{-1/2} \frac{\partial}{\partial x^k}(|\det[g_{ij}]|^{1/2} X^k) \tag{5}$$

Proof Let $\mathbf{L}_X\mu = \mathbf{d}\mathbf{i}_X\mu = \mathbf{d} * X^\flat$ by the lemma. But then $(\operatorname{div}_g X)\mu = -*\,\delta X^\flat = -(\delta X^\flat) * 1$ by the definition of δ and the formula for $**$. Since $*1 = \mu$, we get (4). To prove formula (5), write $\mu = |\det[g_{ij}]|^{1/2} dx^1 \wedge \cdots \wedge dx^n$ and compute $\mathbf{L}_X\mu = \mathbf{d}\mathbf{i}_X\mu$ in these coordinates. We have

$$\mathbf{i}_X\mu = |\det[g_{ij}]^{1/2} X^k (-1)^k dx^1 \wedge \cdots \wedge \widehat{dx^k} \wedge \cdots \wedge dx^n \text{ and so}$$

$$\begin{aligned}\mathbf{d}\mathbf{i}_X\mu &= \left(\frac{\partial}{\partial x^k}|\det[g_{ij}]|^{1/2} X^k\right) dx^1 \wedge \cdots \wedge dx^n \\ &= \frac{1}{|\det[g_{ij}]|^{1/2}} \frac{\partial}{\partial x^k}(|\det[g_{ij}]|^{1/2} X^k)\mu. \quad \blacksquare\end{aligned}$$

6.5.24 Definition *The **Laplace-Beltrami operator** on functions on a orientable pseudo-Riemannian manifold is defined by* $\nabla^2 = \operatorname{div} \circ \operatorname{grad}$.

Thus in a positively oriented chart, equation (5) gives

$$\nabla^2 f = |\det[g_{ij}]|^{-1/2} \frac{\partial}{\partial x^k}\left(g^{\ell k} |\det[g_{ij}]|^{1/2} \frac{\partial f}{\partial x^\ell}\right). \tag{6}$$

Exercises

6.5A Let $f : \mathbb{R}^n \to \mathbb{R}^n$ be a diffeomorphism with positive Jacobian and $f(0) = 0$. Prove that there is a continuous curve f_t of diffeomorphisms joning f to the identity. (*Hint:* First join f to $\mathbf{D}f(0)$ by $g_t(x) = f(tx)/t$.)

6.5B If t is a tensor density of M, that is, $t = t' \otimes \mu$, where μ is a volume form, show that

$$\mathbf{L}_X t = (\mathbf{L}_X t') \otimes \mu + (\operatorname{div}_\mu X) t' \otimes \mu.$$

6.5C A map $A : E \to E$ is said to be *derived from a variational principle* if there is a function $L : E \to \mathbb{R}$ such that

$$\mathbf{d}L(x) \cdot v = \langle A(x), v \rangle,$$

where $\langle\, ,\rangle$ is an inner product on E. Prove *Vainberg's theorem: A comes from a variational principle if and only if* $\mathbf{D}A(x)$ *is a symmetric linear operator.* Do this by applying the Poincaré lemma to the one-form $\alpha(x) \cdot v = \langle A(x)\ , v\rangle$ (see Hughes and

Marsden [1977]).

6.5D Show in three different ways that the sphere S^n is orientable by using **6.5.2** and the two charts given in **3.1.2**, by constructing an explicit n-form, and by using **6.5.9**.

6.5E Use formula (6) to show that in polar coordinates (r, θ) in $\mathbb{R}^2$,

$$\nabla^2 f = \frac{\partial^2 f}{\partial r^2} + \frac{1}{r^2} + \frac{\partial^2 f}{\partial \theta^2} + \frac{1}{r}\frac{\partial f}{\partial r}$$

and that in spherical coordinates (ρ, θ, φ) in $\mathbb{R}^3$,

$$\nabla^2 f = \frac{\partial}{\partial \mu}\left((1-\mu^2)\frac{\partial f}{\partial \mu}\right) + \frac{1}{1-\mu^2}\frac{\partial^2 f}{\partial \theta^2} + \rho\frac{\partial^2 (f)}{\partial \rho^2}$$

where $\mu = \cos\varphi$.

6.5F Let (M, μ) be a volume manifold. Prove the identity

$$\mathrm{div}_\mu[X, Y] = X[\mathrm{div}_\mu Y] - Y[\mathrm{div}_\mu X].$$

6.5G Let $f : M \to N$ be a diffeomorphism of connected oriented manifolds with boundary. Assuming that $T_m f : T_m M \to T_{f(m)}N$ is orientation preserving for some $m \in \mathrm{Int}(M)$, show that $J(f) > 0$ on M; i.e., f is orientation preserving.

6.5H Let g be a pseudo-Riemannian metric on M and define $g_\lambda = \lambda g$ for $\lambda > 0$. Let $*_\lambda$ be the Hodge-star operator defined by g_λ and set $*_1 = *$. Show that if $\alpha \in \Omega^k(M)$, $*_\lambda \alpha = \lambda^{(n/2)-k} * \alpha$.

6.5I In $\mathbb{R}^3$ equipped with the standard Euclidean metric show that for any vector field F and any function f we have: $\mathrm{div}\, F = -\delta F^\flat$, $\mathrm{curl}\, F = (\delta * F^\flat)^\#$, and $\mathrm{grad}\, f = -(*\delta * f)^\#$.

6.5J Show that if M and N are orientable, then so is $M \times N$.

6.5K (i) Let $\sigma : M \to M$ be an involution of M, i.e., $\sigma \circ \sigma =$ identity, and assume that the equivalence relation defined by σ is regular, i.e., there exists a surjective submersion $\pi : M \to N$ such that $\pi^{-1}(n) = \{m, \sigma(m)\}$, where $\pi(m) = n$. Let $\Omega_\pm M) = \{\alpha \in \Omega(M) \mid \sigma^* \alpha = \pm\alpha\}$ be the ± 1 eigenspaces of $\sigma*$. Show that $\pi^* : \Omega(N) \to \Omega_+(M)$ is an isomorphism. (*Hint:* To show that range $\pi^* = \Omega_+(M)$, note that $\pi \circ \sigma = \pi$ implies range $\pi^* \subset \Omega_+(M)$. For the converse inclusion, note that $T_m\pi$ is an isomorphism, so a form at $\pi(m)$ uniquely determines a form at m. Show that this resulting form is smooth by working in a chart on M diffeomorphic to a chart on N.)

(ii) Show that $\mathbb{RP}^n$ is orientable if n is odd and is not orientable if n is even. (*Hint:* In (i) take $M = S^n \subset \mathbb{R}^{n+1}$, $\sigma(x) = x$, and $N = \mathbb{RP}^n$. Let ω be a volume element on S^n induced by a volume element of $\mathbb{R}^{n+1}$. Show $\sigma^*\omega = (-1)^{n+1}\omega$. Now apply (i) to orient $\mathbb{RP}^n$ for n odd. If n is even, let ν be an n-form on $\mathbb{RP}^n$; then $\pi^*\nu = f\omega$. Show that $f(x) = -f(-x)$ so f must vanish at a point of S^n.)

6.5L In Example **3.4.8C**, the ***Möbius band*** $\mathbb{M}$ was defined as the quotient of $\mathbb{R}^2$ by the equivalence relation $(x, y) \sim (x+k, (-1)^k y)$ for any $k \in \mathbb{Z}$.

(i) Show that this equivalence relation is regular. Show that $\mathbb{M}$ is a non-compact, connected, two-manifold.

(ii) Define $\sigma : \mathbb{R}^2 \to \mathbb{R}^2$ by $\sigma(x, y) = (x + 1, -y)$. Show $\pi \circ \sigma = \sigma$, where $\sigma : \mathbb{R}^2 \to \mathbb{M}$, is the canonical projection. If $\nu \in \Omega^2(\mathbb{M})$ define $f \in \mathcal{F}(\mathbb{R}^2)$ by $\pi^*\nu = f\omega$, where ω is an area form on $\mathbb{R}^2$. Show that $f(x + 1, -y) = -f(x, y)$.

(iii) Conclude that f must vanish at a point of $\mathbb{R}^2$, and that this implies $\mathbb{M}$ is not orientable.

6.5M The ***Klein bottle*** $\mathbb{K}$ is defined as the quotient of $\mathbb{R}^2$ by the equivalence relation defined by $(x, y) \sim (x + n, (-1)^n y + m)$ for any $n, m \in \mathbb{Z}$.

(i) Show that this equivalence realtion is regular. Show that $\mathbb{K}$ is a compact, connected, smooth, two-manifold.

(ii) Use **6.5L**(ii), (iii) to show that $\mathbb{K}$ is non-orientable.

6.5N Orientation in vector bundles. Let $\pi : E \to B$ be a vector bundle with finite dimensional fiber modeled on a vector space E, and assume B is connected. The vector bundle is said to be ***orientable*** if the line bundle $L = E^* \wedge \cdots \wedge E^*$ (dim (E) times) has a global nowhere vanishing section. An ***orientation*** of E is an equivalence class of global nowhere vanishing sections of L under the equivalence relation: $\sigma_1 \sim \sigma_2$ iff there exists $f \in \mathcal{F}(B)$, $f > 0$ such that $\sigma_2 = f\sigma_1$.

(i) Prove that E is orientable iff L is a trivial line bundle. Show that E admits exactly two orientations. Showthat an orientation $[\sigma]$ of E induces an orientation in each fiber of E.

(ii) Show that a manifold M is orientable iff its tangent bundle is an oriented vector bundle.

(iii) Let E, F be vector bundles over the same base. Show that if two of E, F and $E \oplus F$ are orientable, so is the third.

(iv) Let E, F be vector bundles (over possibly different bases) Show that $E \times F$ is orientable if and only if E and F are both orientable. Conclude that if M, N are finite dimensional manifolds, then $M \times N$ is orientable if and only if M and N are both orientable.

(v) Show that $E \oplus E^*$ is an orientable vector bundle if E is any vector bundle. (*Hint:*

Consider the section $\Omega(x)((e_1, \alpha_1), (e_2, \alpha_2)) = \langle \alpha_2, e_1\rangle - \langle \alpha_1, e_2\rangle$ of $(E \oplus E^*) \wedge (E \oplus E^*)$.)

(vi) Choose an orientation of the vector space E and assume B admits partitions of unity. Show that the vector bundle atlas all of whose change of coordinate maps have positive determinant relative to the orientation of E, when restricted to the fiber. (*Hint:* If E is oriented and $\psi : \pi^{-1}(U) \to U' \times E$ is a vector bundle chart with U' open in the model space of B, and U is connected in B, define $\varphi : \pi^{-1}(U) \to U' \times F$ by $\varphi(e) = \psi(e)$ if the linear map $\psi_b : \pi^{-1}(b) \to F$ is orientation preserving and $\varphi(e) = (\alpha \circ \psi)(e)$, where $\alpha : F \to F$ is an orientation reversing isomorphism of F, if ψ_b is orientation reversing. For the converse, choose a volume form ω on F and define on a vector bundle chart (V, φ) of E, with U connected in B, $\pi^{-1}(U) = V$, $\omega(U) : U \to L \mid U$ by $\omega(U)(b)(e_1, ..., e_r) = \omega(\varphi_0(b))\,(\varphi_b(e_1), ..., \varphi_b(e_r))$, where $r = \dim F$, $b \in B$, $e_i \in \pi^{-1}(b)$, $i = 1, ..., r$, and $\varphi_0 : U \to U'$ is the induced chart on B. Show that if $b \in U_1 \cap U_2$, then $\omega(U_1)(b) = \det_\omega(\varphi_b^{\,1} \circ (\varphi_b^{\,2})^{-1})\omega(U_2)$ where $(\pi^{-1}(U_i), \varphi_i)$ are vector bundle charts, U_i connected, $i = 1, 2$. Next, glue the $\omega(U)$'s together using a partition of unity.)

(vii) Use (iv) to show that if E and F are oriented, then there exists a vector bundle atlas on E such that all $\varphi_b : \pi^{-1}(b) \to F$ are orientation preserving isomorphisms. Such an atlas is called ***positively oriented.***

(viii) Let E, F, B be finite dimensional, E oriented by σ and B oriented by ω. Show that $\pi^*\omega \wedge \sigma$ is a volume form on E. Conclude that an orientation of B and an orientation of F uniquely determine an orientation of E as a manifold. This orientation is called the ***local product orientation.***

(ix) Show that any vector bundle $\pi : E \to B$ with finite dimensional fiber has an oriented double cover $\tilde{\pi} : \tilde{E} \to \tilde{B}$, where $\tilde{B} = \{(b, [\mu_b]) \mid b \in B, [\mu_b]$ is an orientation of $\pi^{-1}(b)\}$, $p : \tilde{B} \to B$ is the map $p(b, [\mu_b]) = b$, and $\tilde{E} = p^*E$. Find the vector bundle charts of $\tilde{E}$ and show that the fiber at $(b, [\mu_b])$ is oriented by $[\tilde{p}^*(\mu_b)]$, where $\tilde{p} : \tilde{E} \to E$ is the mapping induced by p on the pull-back bundle $\tilde{E}$. If $E = TB$, what is $\tilde{E}$?

6.50 Let M be a compact manifold and $\mathcal{M}$ the space of Riemannian metrics on M. Let $\mathcal{T}$ be a space of tensor fields on M of a fixed type. A mapping $\Phi : \mathcal{M} \to \mathcal{T}$ is called ***covariant*** if for every diffeomorphism $\varphi : M \to M$, we have $\Phi(\varphi^*g) = \varphi^*\Phi(g)$.

(i) Show that covariant maps satisfy the identity

$$D\Phi(g) \cdot \mathcal{L}_X g = \mathcal{L}_X \Phi(g)$$

for every vector field X. (Assume Φ is differentiable and $\mathcal{M}$, $\mathcal{T}$ are given suitable Banach space topologies.)

(ii) Show that if M is oriented, then the map $g \mapsto \mu(g)$, the volume element of g, is covariant. Is the identity in (i) anything interesting?

6.5P Let X be a vector field density on the oriented n-manifold M; i.e., $X = F \otimes \mu$, where F is a vector field and μ is a density. Use Exercise **6.5B** to define div X and to show it makes intrinsic sense.

6.5Q Show that an orientable line bundle over a base admitting partitions of unity is trivial. (*Hint:* Since the bundle is orientable there exist local charts which when restricted to each fiber give positive functions. Regard these as local sections and then glue.)

6.5R Let $\pi : E \to S^1$ be a vector bundle with n-dimensional fiber. If E is orientable show that it is isomorphic to a trivial bundle over S^1. Show that if $\rho : F \to S^1$ is a non-orientable vector bundle with n-dimensional fiber and if E is non-orientable, then E and F are isomorphic. Conclude that there are exactly two isomorphism classes of vector bundles with n-dimensional fiber over S^1. Construct a representative for the class corresponding to the non-orientable case. (*Hint:* Construct a non-orientable vector bundle like the Möbius band: the equivalence relation has a factor $(-1)^{2n-1}$ if the dimension of the fiber is 2n–1 or 2n. To prove non-orientability, proceed as in **6.5L**.)

6.5S Let $\{e_1, ..., e_{n+1}\}$ be the standard basis of $\mathbb{R}^{n+1}$ and $\Omega_{n+1} = e_1 \wedge \cdots \wedge e_{n+1}$ be the induced volume form. On S^n define $\omega_n \in \Omega^n(S^n)$ by

$$\omega_n(s)(v_1, ..., v_n) = \Omega_{n+1}(s, v_1, ..., v_n)$$

for $s \in S^n$, $v_1, ..., v_n \in T_sS^n$.

(i) Use Proposition **6.5.8** to show that ω_n is a volume form on S^n; ω_n is called the ***standard volume form*** on S^n.

(ii) Let $f : \mathbb{R}_+ \times S^n \to \mathbb{R}^{n+1} \setminus \{0\}$ be given by $f(t, s) = ts$, where $\mathbb{R}_+$ is defined to be the set $\{t \in \mathbb{R} \mid t > 0\}$. Show that if $\mathbb{R}_+$ is oriented by dt, S^n by ω_n, and $\mathbb{R}^{n+1}$ by Ω_{n+1}, then $(Jf)(t, s) = t^n$. Conclude that f is orientation preserving.

Chapter 7
Integration on Manifolds

The integral of an n-form on an n-manifold is defined by piecing together integrals over sets in $\mathbb{R}^n$ using a partition of unity subordinate to an atlas. The change-of-variables theorem guarantees that the integral is well defined, independent of the choice of atlas and partition of unity. Two basic theorems of integral calculus, the change-of-variables theorem and Stokes' theorem, are discussed in detail along with some applications.

§7.1 *The Definition of the Integral*

The aim of this section is to define the integral of an n-form on an oriented n-manifold M and prove a few of its basic properties. We begin with a summary of the basic results in $\mathbb{R}^n$. Suppose $f : \mathbb{R}^n \to \mathbb{R}$ is continuous and has compact support. Then $\int f\, dx^1 \cdots dx^n$ is defined to be the Riemann integral over any rectangle containing the support of f.

7.1.1 Definition *Let* $U \subset \mathbb{R}^n$ *be open and* $\omega \in \Omega^n(U)$ *have compact support. If, relative to the standard basis of* $\mathbb{R}^n$,

$$\omega(x) = \frac{1}{n!}\, \omega_{i_1 \cdots i_n}(x)\, dx^{i_1} \wedge \cdots \wedge dx^{i_n} = \omega_{1 \cdots n}(x)\, dx^1 \wedge \cdots \wedge dx^n,$$

where the components of ω *are given by*

$$\omega_{i_1 \cdots i_n}(x) = \omega(x)(e_{i_1}, \ldots, e_{i_n}),$$

then we define

$$\int_U \omega = \int_{\mathbb{R}^n} \omega_{1 \cdots n}(x)\, dx^1 \cdots dx^n.$$

Recall that if ζ is any integrable function and $f : \mathbb{R}^n \to \mathbb{R}^n$ is any diffeomorphism, the ***change of variables*** theorem states that $\zeta \circ f$ is integrable and

$$\int_{\mathbb{R}^n} \zeta(x^1, \ldots, x^n)\, dx^1 \ldots dx^n = \int_{\mathbb{R}^n} |\, J_\Omega f(x^1, \ldots, x^n)\,|\, (\zeta \circ f)(x^1, \ldots, x^n)\, dx^1 \ldots dx^n, \tag{1}$$

where $\Omega = dx^1 \wedge \ldots \wedge dx^n$ is the standard volume form on $\mathbb{R}^n$, and $J_\Omega f$ is the Jacobian determinant of f relative to Ω. This charge of variables theorem can be rephrased in terms of pull backs in the following form.

7.1.2 Change of Variables in $\mathbb{R}^n$ *Let* U *and* V *be open subsets of* $\mathbb{R}^n$ *and suppose* $f : U \to V$ *is an orientation-preserving diffeomorphism. If* $\omega \in \Omega^n(V)$ *has compact support, then* $f^*\omega \in \Omega^n(U)$ *has compact support as well and*

$$\int_U f^*\omega = \int_V \omega \tag{2}$$

Proof If $\omega = \omega_{1 \ldots n} dx^1 \wedge \ldots \wedge dx^n$, then $f^*\omega = (\omega_{1 \ldots n} \circ f)(J_\Omega f)\Omega$, where the n-form $\Omega = dx^1 \wedge \ldots \wedge dx^n$ is the standard volume form on $\mathbb{R}^n$. As discussed in **§6.5**, $J_\Omega f > 0$. Since f is a diffeomorphism, the support of $f^*\omega$ is $f^{-1}(\text{supp } \omega)$, which is compact. Then from (1),

$$\int_U f^*\omega = \int_{\mathbb{R}^n} (\omega_{1 \ldots n} \circ f)(J_\Omega f) dx^1 \cdots dx^n = \int_{\mathbb{R}^n} \omega_{1 \ldots n} dx^1 \cdots dx^n = \int_V \omega. \blacksquare$$

Suppose that (U, φ) is a chart on a manifold M and $\omega \in \Omega^n(M)$. If $\text{supp } \omega \subset U$, we may form $\omega \mid U$, which has the same support. Then $\varphi_*(\omega \mid U)$ has compact support, so we may state the following.

7.3.13 Definition *Let* M *be an orientable* n*-manifold with orientation* $[\Omega]$*. Suppose* $\omega \in \Omega^n(M)$ *has compact support* $C \subset U$*, where* (U, φ) *is a positively oriented chart. Then we define*

$$\int_{(\varphi)} \omega = \varphi_*(\omega|U).$$

7.1.4 Proposition *Suppose* $\omega \in \Omega^n(M)$ *has compact support* $C \subset U \cap V$*, where* (U, φ), (V, ψ) *are two positively oriented charts on the oriented manifold* M. *Then*

$$\int_{(\varphi)} \omega = \int_{(\psi)} \omega$$

Proof By **7.1.2** $\int \varphi_*(\omega \mid U) = \int (\psi \circ \varphi^{-1})_* \varphi_*(\omega \mid U)$. Hence $\int \varphi_*(\omega \mid U) = \int \psi_*(\omega \mid U)$. [Recall that for diffeomorphisms $f_* = (f^{-1})^*$ and $(f \circ g)_* = f_* \circ g_*$.] $\blacksquare$

Thus we may define $\int_U \omega = \int_{(\varphi)} \omega$, where (U, φ) is any positively oriented chart containing the compact support of ω (if one exists). More generally, we can define $\int_M \omega$ where ω has compact support as follows.

7.1.5 Definition *Let* M *be an oriented manifold and* $\mathcal{A}$ *an atlas of positively oriented charts. Let* $P = \{(U_\alpha, \varphi_\alpha, g_\alpha)\}$ *be a partition of unity subordinate to* $\mathcal{A}$. *Define* $\omega_\alpha = g_\alpha \omega$ *(so* ω_α *has compact support in some* U_i*) and let*

$$\int_P \omega = \sum_\alpha \int \omega_\alpha. \tag{3}$$

7.1.6 Proposition (i) *The sum* (3) *contains only a finite number of nonzero terms.*

(ii) *For any other atlas of positively oriented charts and subordinate partition of unity* Q *we have* $\int_P \omega = \int_Q \omega$.

The common value is denoted $\int_M \omega$, *and is called the* ***integral of*** $\omega \in \Omega^n(M)$.

Proof For any $m \in M$, there is a neighborhood U such that only a finite number of g_α are nonzero on U. By compactness of supp ω, a finite number of such neighborhoods cover the support of ω. Hence only a finite number of g_α are nonzero on the union of these U. For (ii), let $P = \{(U_\alpha, \varphi_\alpha, g_\alpha)\}$ and $Q = \{(V_\beta, \psi_\beta, h_\beta)\}$ be two partitions of unity with positively oriented charts. Then the functions $\{g_\alpha h_\beta\}$ satisfy $g_\alpha h_\beta(m) = 0$ except for a finite number of indices (α, β), and $\Sigma_\alpha \Sigma_\beta g_\alpha h_\beta(m) = 1$, for all $m \in M$. Since $\Sigma_\beta h_\beta = 1$, we get

$$\int_P \omega = \sum_\alpha \int g_\alpha \omega = \sum_\beta \sum_\alpha \int h_\beta g_\alpha \omega = \sum_\alpha \sum_\beta \int g_\alpha h_\beta \omega = \int_Q \omega. \quad \blacksquare$$

7.1.7 Change of Variables Theorem *Suppose* M *and* N *are oriented* n*-manifolds and* $f : M \to N$ *is an orientation-preserving diffeomorphism. If* $\omega \in \Omega^n(N)$ *has compact support, then* $f^*\omega$ *has compact support and*

$$\int_N \omega = \int_M f^*\omega. \tag{4}$$

Proof First, supp $f^*\omega = f^{-1}(\text{supp}\ \omega)$, which is compact. To prove (2), let $\{(U_i, \varphi_i)\}$ be an atlas of positively oriented charts of M and let $P = \{g_i\}$ be a subordinate partition of unity. Then $\{(f(U_i), \varphi_i \circ f^{-1})\}$ is an atlas of positively oriented charts of N and $Q = \{g_i \circ f^{-1}\}$ is a partition of unity subordinate to the covering $\{f(U_i)\}$. By **7.1.6**,

$$\int_M f^*\omega = \sum_i \int_M g_i f^*\omega = \sum_i \int_{\mathbb{R}^n} \varphi_{i*}(g_i f^*\omega) = \sum_i \int_{\mathbb{R}^n} \varphi_{i*}(f^{-1})_*(g_i \circ f^{-1})\omega$$

$$= \sum_i \int_{\mathbb{R}^n} (\varphi_i \circ f^{-1})_*(g_i \circ f^{-1})\omega = \int_N \omega. \quad \blacksquare$$

This result is summarized by the following commutative diagram:

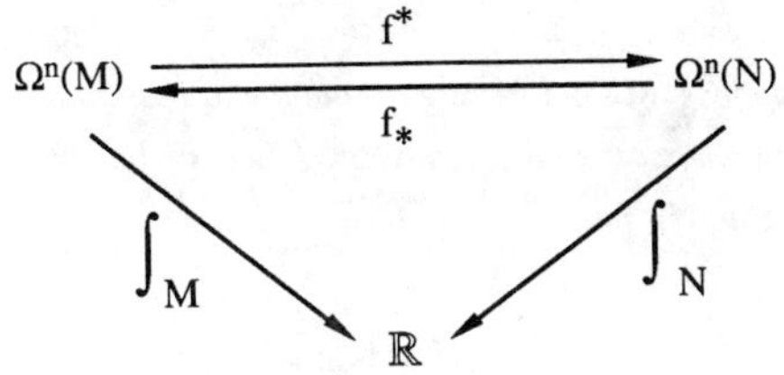

7.1.8 Definition *Let* (M, μ) *be a volume manifold. Suppose* $f \in \mathcal{F}(M)$ *and* f *has compact support. Then we call* $\int_M f\mu$ ***the integral of*** f ***with respect to*** μ.

The reader can check that since the Riemann integral is $\mathbb{R}$-linear, so is the integral just defined.

The next theorem will show that the integral defined by (4) can be obtained in a unique way from a measure on M. (The reader unfamiliar with measure theory can find the necessary background in Royden [1968]; this result will not be essential for future sections.) The integral we have described can clearly be extended to all continuous functions with compact support. Then we have the following.

7.1.9 Riesz Representation Theorem *Let* (M, μ) *be a volume manifold. Let* $\mathcal{B}$ *denote the Borel sets of* M, *the* σ*-agebra generated by the open (or closed, or compact) subsets of* M. *Then there is a unique measure* m_μ *on* $\mathcal{B}$ *such that for every continuous function of compact support*

$$\int_M f\,dm_\mu = \int_M f\mu. \tag{5}$$

Proof Existence of such a measure m_μ is proved in Royden [1968]. For uniqueness, it is enough to consider bounded open sets (by the Hahn extension theorem). Thus, let U be open in M, and let C_U be its characteristic function. We construct a sequence of C^∞ functions of compact support φ_n such that $\varphi_n \downarrow C_U$, pointwise. Hence from the monotone convergence theorem, $\int \varphi_n \mu = \int \varphi_n\, dm_\mu \to \int C_U\, dm_\mu = m_\mu(U)$. Thus, m_μ is unique. ■

The space $L^p(M, \mu)$, $p \in \mathbb{R}$, consists of all measurable functions f such that $|f|^p$ is integrable. For $p \geq 1$, the norm $\|f\|_p = (\int |f|^p dm_\mu)^{1/p}$ makes $L^p(M, \mu)$ into a Banach space (functions that differ only on a set of measure zero are identified). The use of these spaces in studying objects on M itself is discussed in **§7.4.** The next propositions give an indication of some of the ideas. If $F : M \to N$ is a measurable mapping and m_M is a measure on M, then $F_* m_M$ is the measure on N defined by $F_* m_M(A) = m_M(F^{-1}(A))$. If F is bijective, we set $F^*(m_N) = (F^{-1})_* m_N$. If $f : M \to \mathbb{R}$ is an integrable function, then fm_M is the measure on M defined by

$$(fm_M)(A) = \int_A f\, dm_M$$

for every measurable set A in M.

7.1.10 Proposition *Suppose* M *and* N *are orientable* n-*manifolds with volume forms* μ_M *and* μ_N *and corresponding measures* m_M *and* m_N. *Let* F *be an orientation preserving* C^1 *diffeomorphism of* M *onto* N. *Then*

$$F^* m_N = (J_{(\mu_M, \mu_N)} F) m_M. \tag{6}$$

Proof Let f be any C^∞ function with compact support on M. By **7.1.7**,

$$\int_N f\, dm_N = \int_N f\, \mu_N = \int_M F^*(f\, \mu_N) = \int_M (f \circ F)(J_{(\mu_M, \mu_N)} F) \mu_M$$

$$= \int_M (f \circ F)(J_{(\mu_M, \mu_N)} F) dm_M.$$

As in the proof of **7.1.9**, this relation holds for f the characteristic function of F(A). That is,

$$m_N(F(A)) = \int_A (J_{(\mu_M, \mu_N)} F) dm_M. \quad \blacksquare$$

In preparation for the next result, we notice that on a volume manifold (M, μ), the flow F_t of any vector field X is orientation preserving for each $t \in \mathbb{R}$ (regard this as a statement on the domain of the flow, if the vector field is not complete). Indeed, since F_t is a diffeomorphism, $J_\mu(F_t)$ is nowhere zero; since it is continuous in t and equals one at $t = 0$, it is positive for all t.

7.1.11 Proposition *Let* M *be an orientable manifold with volume form* μ *and corresponding measure* m_μ. *Let* X *be a (possibly time-dependent)* C^1 *vector field on* M *with flow* F_t. *The following are equivalent (if the flow of* X *is not complete, the statements involving it are understood to hold on its domain):*

(i) $\operatorname{div}_\mu X = 0$;
(ii) $J_\mu F_t = 1$ *for all* $t \in \mathbb{R}$;
(iii) $F_{t*} m_\mu = m_\mu$ *for all* $t \in \mathbb{R}$;
(iv) $F_t^* \mu = \mu$ *for all* $t \in \mathbb{R}$;
(v) $\int_M f dm_\mu = \int_M (f \circ F_t) dm_\mu$ *for all* $f \in L^1(M, \mu)$ *and all* $t \in \mathbb{R}$.

Proof Statement (i) is equivalent to (ii) by **6.5.19**. Statement (ii) is equivalent to (iii) by (6) and to (iv) by definition. We shall prove that (ii) is equivalent to (v). If $J_\mu F_t = 1$ for all $t \in \mathbb{R}$ and f is continuous with compact support, then

$$\int_M (f \circ F_t)\mu = \int_M (f \circ F_t)(F_t^* \mu) = \int_M F_t^*(f\mu) = \int_M f\mu.$$

Hence, by uniqueness in **7.1.9**, we have $\int_M f\, dm_\mu = \int_M (f \circ F_t) dm_\mu$ for all integrable f, and so (ii) implies (v). Conversely, if

$$\int_M (f \circ F_t) dm_\mu = \int_M f\, dm_\mu$$

then taking f to be continuous with compact support, we see that

$$\int_M (f \circ F_t)\mu = \int_M f\mu = \int_M F_t^*(f\mu) = \int_M (f \circ F_t) F_t^* \mu = \int_M (f \circ F_t)(J_\mu F_t)\mu.$$

Thus, for every integrable f, $\int_M (f \circ F_t) dm_\mu = \int_M (f \circ F_t)(J_\mu F_t) dm_\mu$. Hence $J_\mu F_t = 1$, and so (v) implies (ii). ■

The following result is central to continuum mechanics (see below and **§8.2** for applications).

7.1.12 Transport Theorem *Let* (M, μ) *be a volume manifold and* X *a vector field on* M *with flow* F_t. *For* $f \in \mathcal{F}(M \times \mathbb{R})$ *and letting* $f_t(m) = f(m, t)$, *we have*

$$\frac{d}{dt} \int_{F_t(U)} f_t\, \mu = \int_{F_t(U)} \left(\frac{\partial f}{\partial t} + \operatorname{div}_\mu(f_t X) \right) \mu \tag{7}$$

for any open set U *in* M.

Proof By the flow characterization of Lie derivatives and **6.5.17**(ii), we have

$$\begin{aligned} \frac{d}{dt} F_t^*(f_t\, \mu) &= F_t^*\left(\frac{\partial f}{\partial t} \mu \right) + F_t^* \mathbf{L}_X(f_t\, \mu) \\ &= F_t^*\left(\frac{\partial f}{\partial t} \mu \right) + F_t^*[(\mathbf{L}_X f_t)\mu + f_t(\operatorname{div}_\mu X)\mu] \\ &= F_t^*\left[\left(\frac{\partial f}{\partial t} + \operatorname{div}_\mu(f_t\, X) \right) \mu \right]. \end{aligned}$$

Thus by the change-of-variables formula,

$$\frac{d}{dt} \int_{F_t(U)} f_t\, \mu = \frac{d}{dt} \int_U F_t^*(f_t\, \mu) = \int_U F_t^*\left[\left(\frac{\partial f}{\partial t} + \operatorname{div}_\mu(f_t X) \right) \mu \right] = \int_{F_t(U)} \left(\frac{\partial f}{\partial t} + \operatorname{div}_\mu(f_t X) \right) \mu. \ ■$$

7.1.13 Example Let ρ(x, t) be the density of an ideal fluid moving in a compact region of $\mathbb{R}^3$ with smooth boundary. One of the basic assumptions of fluid dynamics is ***conservation of mass***: the mass of the fluid in the open set U remains unchanged during the motion described by a flow F_t. This means that

$$\frac{d}{dt} \int_{F_t(U)} \rho(x, t) dx = 0 \tag{8}$$

for all open sets U. By the transport theorem, (8) is equivalent to the ***equation of continuity***

$$\frac{\partial \rho}{\partial t} + \operatorname{div}(\rho \mathbf{u}) = 0, \tag{9}$$

where **u** represents the velocity of the fluid particles. We shall return to this example in §**8.2**. As another application of **7.1.11**, we prove the following.

7.1.14 Poincaré Recurrence Theorem *Let* (M, μ) *be a volume manifold,* m_μ *the corresponding measure, and* X *a time-independent divergence-free vector field with flow* F_t. *Suppose* A *is a measurable set in* M *such that* $m_\mu(A) < \infty$, $F_t(x)$ *exists for all* $t \in \mathbb{R}$ *if* $x \in A$, *and* $F_t(A) \subset A$. *Then for each measurable subset* B *of* A *and* $T \geq 0$, *there exists* $S \geq T$ *such that* $B \cap F_S(B) \neq \varnothing$. *Therefore, a trajectory starting in* B *returns infinitely often to* B.

Proof By **7.1.11**, the sets B, $F_T(B)$, $F_{2T}(B)$, ... all have the same finite measure. Since $m_\mu(A) < \infty$, they cannot all be disjoint, so there exist integers $k > \ell > 0$ satisfying $F_{kT}(B) \cap F_{\ell T}(B) \neq \varnothing$. Since $F_{kT} = (F_T)^k$ (as X is time-independent), we get $F_{(k-\ell)T}(B) \cap B \neq \varnothing$. ■

The Poincaré recurrence theorem is one of the forerunners of ergodic thoery, a topic that will be discussed briefly in §**7.4**. A related result is the following.

7.1.15 Schwarzschild Capture Theorem *Let* (M, μ) *be a volume manifold,* X *a time-independent divergence-free vector field with flow* F_t, *and* A *a measurable subset of* M *with finite measure. Assume that for every* $x \in A$, *the trajectory* $t \mapsto F_t(x)$ *exists for all* $t \in \mathbb{R}$. *Then for almost all* $x \in A$ *(relative to* m_μ*) the following are equivalent:*

(i) $F_t(x) \in A$ *for all* $t \geq 0$;

(ii) $F_t(x) \in A$ *for all* $t \leq 0$.

Proof Let $A_1 = \bigcap_{t\geq 0} F_t(A)$ be the set of points in A which have their future trajectory completely in A. Similarly, consider $A_2 = \bigcap_{t\leq 0} F_t(A)$. By **7.1.11**, for any $\tau \geq 0$,

$$\mu(A_1) = \mu(F_{-\tau}(A_1)) = \mu\left(\bigcap\nolimits_{t\geq -\tau} F_\tau(A)\right)$$

which shows, by letting $\tau \to \infty$, that

$$\mu(A_1) = \mu\left(\bigcap\nolimits_{t\in\mathbb{R}} F_t(A)\right) = \mu(A_1 \cap A_2).$$

Reasoning similarly for A_2, we get $\mu(A_1) = \mu(A_1 \cap A_2) = \mu(A_2)$, so that

$$\mu(A_1 \backslash (A_1 \cap A_2)) = \mu(A_2 \backslash (A_1 \cap A_2)) = 0.$$

Let $S = (A_1 \backslash (A_1 \cap A_2)) \cup (A_2 \backslash (A_1 \cap A_2))$; then $m_\mu(S) = 0$ and $S \subset A$. Moreover, we have $A_1 \backslash S = A_1 \cap A_2 = A_2 \backslash S$ which proves the desired equivalence. ■

So far only integration on orientable manifolds has been discussed. A similar procedure can be carried out to define the integral of a one-density (see **§6.5**) on any manifold, orientable or not. The only changes needed in the foregoing definitions and propositions are to replace the Jacobians with their absolute values and to use the definition of divergence with respect to a given density as discussed in **§6.5**. All definitions and propositions go through with these modifications.

F-valued one-forms and one-densities can also be integrated in the following way. If $\omega = \Sigma^{\ell}_{i=1}\omega^i f_i$, where $f_1, \ldots, f_n$ is an ordered basis of F, then we set $\int_M \omega = \Sigma^{\ell}_{i=1}(\int_M \omega^i) f_i \in F$. It is easy to see that this definition is independent of the chosen basis of F and that all the basic properties of the integral remain unchanged. On the other hand, *the integral of vector-bundle-valued* n*-forms on* M *is not defined* unless additional special structures (such as triviality of the bundle) are used. In particular, integration of vector or general tensor fields is not defined.

Exercises

7.1A Let M be an n-manifold and μ a volume form on M. If X is a vector field on M with flow F_t show that

$$\frac{d}{dt}(J_\mu(F_t)) = J_\mu(F_t)(\operatorname{div}_\mu X \circ F_t).$$

(*Hint:* Compute $(d/dt)F_t^*\mu$ using the Lie derivative formula.)

7.1B Prove the following generalization of the transport theorem

$$\frac{d}{dt}\int_{F_t(V)} \omega_t = \int_{F_t(V)} \left(\frac{\partial \omega_t}{\partial t} + \mathbf{L}_X \omega_t\right),$$

where ω_t is a time-dependent k-form on M and V is a k-dimensional submanifold of M.

7.1C (i) Let $\varphi : S^1 \to S^1$ be the map defined by $\varphi(e^{i\theta}) = e^{2i\theta}$, where $\theta \in [0, 2\pi]$. Let, by abuse of notion, $d\theta$ denote the standard volume of S^1. Show that the following identity holds: $\int_{S^1} \varphi^*(d\theta) = 2\int_{S^1} d\theta$.

(ii) Let $\varphi : M \to N$ be a smooth surjective map. Then φ called a k-***fold covering map*** if every $n \in N$ has an open neighborhood V such that $\varphi^{-1}(V) = U_1 \cup \cdots \cup U_k$, are disjoint open sets each of which is diffeomorphic by φ to V. Generalize (i) in the following way. If $\omega \in \Omega^n(N)$ is a volume form, show that $\int_M \varphi^*\omega = k\int_N \omega$.

7.1D Define the integration of Banach space valued n-forms on an n-manifold M. Show that if the Banach space is $\mathbb{R}^l$, you recover the coordinate definition given at the end of this section. If E, F are Banach spaces and $A \in L(E, F)$, define $A_* \in L(\Omega(M, E), \Omega(M, F))$ by $(A_*\alpha)(m) = A(\alpha(m))$. Show that $\left(\int_M\right) \circ A_* = A \circ \left(\int_M\right)$ on $\Omega^n(M, E)$.

7.1E Let M and N be oriented manifolds and endow $M \times N$ with the product orientation. Let $p_M : M \times N \to M$ and $p_N : M \times N \to N$ be the projections. If $\alpha \in \Omega^{\dim M}(M)$ and $\beta \in \Omega^{\dim N}(N)$ have compact support show that

$$\alpha \times \beta := (p_M{}^*\alpha) \wedge (p_N{}^*\beta)$$

has compact support and is a $(\dim M + \dim N)$-form on $M \times N$. Prove ***Fubini's Theorem***

$$\int_{M\times N} \alpha \times \beta = \left(\int_M \alpha\right)\left(\int_N \beta\right).$$

7.1F Fiber Integral Let $\varphi : M \to N$ be a surjective submersion, $\dim M = m$ and $\dim N = n$. The map φ is said to be ***orientable*** if there exists $\eta \in \Omega^p(M)$, where $p = m - n$, such that for each $y \in N$, $j_y{}^*\eta$ is a volume form on $\varphi^{-1}(y)$, where $j_y : \varphi^{-1}(y) \to M$ is the inclusion. An ***orientation*** of φ is an equivalence class of p-forms under the relation: $\eta_1 \sim \eta_2$ iff there exists $f \in \mathcal{F}(M)$, $f > 0$ such that $\eta_2 = f\eta_1$.

(i) If $\varphi : M \to N$ is a vector bundle, show that orientability of φ is equivalent to orientability of the vector bundle as defined in Exercise **6.5M**.

(ii) If φ is oriented by η and N by ω, show that $\varphi^*\omega \wedge \eta$ is a volume form on M. The orientation on M defined by this volume is called the ***local product orientation*** of M (compare with Exercise **6.5M**(vi)).

(iii) Let

$$\Omega_\varphi{}^k(M) := \{\alpha \in \Omega^k(M) \mid \varphi^{-1}(K) \cap \operatorname{supp} \alpha \text{ is compact, for any compact set } K \subset N\},$$

the ***fiber-compactly supported*** k-forms on M. Show that $\Omega_\varphi{}^k(M)$ is an $\mathcal{F}(M)$-submodule of $\Omega^k(M)$, and is invariant under the interior product, exterior differential, and Lie derivative.

(iv) If $\alpha \in \Omega_\varphi{}^{k+p}(M)$, $k \geq 0$ and $y \in N$, define a p-form α_y on $\varphi^{-1}(y)$, with values in $T_y{}^*N \wedge \cdots \wedge T_y{}^*N$ (k times) by

$$[\alpha_y(x)(u_1, ..., u_p)](T_x\varphi(v_1), ..., T_x\varphi(v_k)) = \alpha(x)(v_1, ..., v_k, u_1, ..., u_p),$$

where $\varphi(x) = y$, $x \in M$, $v_1, ..., v_k \in T_xM$, and $u_1, ..., u_p \in \ker(T_x\varphi) = T_x(\varphi^{-1}(y))$. Assume φ is oriented. Define the ***fiber integral***

$$\int_{\text{fib}} : \Omega_\varphi^{k+p}(M) \to \Omega^k(N) \quad \text{by} \quad \left(\int_{\text{fib}} \alpha\right)(x) = \int_{\varphi^{-1}(y)} \alpha_y, \text{ if } \varphi(x) = y;$$

the right-hand side is understood as the integral of a vector-valued p-form on the oriented p-manifold $\varphi^{-1}(y)$. Prove that $\int_{\mathrm{fib}}\alpha$ is a smooth k-form on N. (*Hint:* Use charts in which φ is a projection and apply the theorem of smoothness of the integral with respect to parameters.) Show that if $\varphi : M \to N$ is a locally trivial fiber bundle with N paracompact, $\int_{\mathrm{fib}}$ is surjective.

(v) Let $\beta \in \Omega^{\ell}(N)$ have compact support and let $\alpha \in \Omega_{\varphi}^{k+p}(M)$. Show that $\varphi^*\beta \wedge \alpha \in \Omega_{\varphi}^{k+\ell+p}(M)$ and that

$$\int_{\mathrm{fib}}(\varphi^*\beta \wedge \alpha) = \beta \wedge \int_{\mathrm{fib}} \alpha.$$

(*Hint:* Let $E = T_y{}^*N \wedge \cdots \wedge T_y{}^*N$ (k times) and let F be the wedge product $\ell + k$ times. Define $A \in L(E, F)$ by $A(\gamma) = \beta(y) \wedge \gamma$ and show that $(\varphi^*\beta \wedge \alpha)_y = A_*(\alpha_y)$ using the notation of Exercise **7.1D**. Then apply $\int_{\mathrm{fib}}$ to this identity and use **7.1D**).

(vi) Assume N is paracompact and oriented, φ is oriented, and endow M with the local product orientation. Prove the following ***iterated integration (Fubini-type) formula***

$$\int_M = \int_N \circ \int_{\mathrm{fib}}$$

by following the three steps below.

Step 1: Using a partition of unity, reduce to the case $M = N \times P$ where $\varphi : M \to N$ is the projection and M, N, P are Euclidean spaces.

Step 2: Use (v) and Exercise **7.1E** to show that for $\beta \in \Omega^n(N)$ and $\gamma \in \Omega^p(P)$ with compact support,

$$\int_N \int_{\mathrm{fib}} (\beta \times \gamma) = \int_M (\beta \times \gamma)$$

Step 3: Since M, N, and P are ranges of coordinate patches, show that any $\omega \in \Omega^m(M)$ with compact support is of the form $\beta \times \gamma$.

(vii) Let $\varphi : M \to N$ and $\varphi' : M' \to N'$ be oriented surjective submersions and let $f : M \to M'$, and $f_0 : N \to N'$ be smooth maps satisfying $f_0 \circ \varphi = \varphi' \circ f$. Show that $\int_{\mathrm{fib}} \circ f^* = f_0{}^* \circ \int'_{\mathrm{fib}}$, where $\int'_{\mathrm{fib}}$ denotes the fiber integral of φ'.

(viii) Let $\varphi : M \to N$ be an oriented surjective submersion and assume $X \in \mathfrak{X}(M)$ and $Y \in \mathfrak{X}(N)$ are φ-related. Prove that

$$\int_{\mathrm{fib}} \circ\, \mathbf{i}_X = \mathbf{i}_Y \circ \int_{\mathrm{fib}}, \quad \int_{\mathrm{fib}} \circ\, \mathbf{d} = \mathbf{d} \circ \int_{\mathrm{fib}}, \quad \int_{\mathrm{fib}} \circ\, \boldsymbol{\mathcal{L}}_X = \boldsymbol{\mathcal{L}}_Y \circ \int_{\mathrm{fib}}$$

(For more information on the fiber integral see Bourbaki [1971] and Greub, Halperin, and Vanstone [1973], Vol. I.)

7.1G Let $\varphi : M \to N$ be a smooth orientation preserving map, where M and N are volume manifolds of dimension m and n respectively. For $\alpha \in \Omega^k(M)$ with compact support, define the linear functional $\varphi_*\alpha : \Omega^{m-k} \to \mathbb{R}$ by

$$(\varphi_*\alpha)(\beta) = \int_M \varphi^*\beta \wedge \alpha$$

for all $\beta \in \Omega^{m-k}(N)$; i.e., $\varphi_*\alpha$ is a ***distributional*** **k-*form*** on N. If $m < k$, set $\varphi_*\alpha = 0$. If there is a $\gamma \in \Omega^{n-m+k}(N)$ satisfying $(\varphi_*\alpha)(\beta) = \int_M \beta \wedge \gamma$, identify $\varphi_*\alpha$ with γ and say $\varphi_*\alpha$ is of *form-type*. Prove the following statements.

(i) If φ is a diffeomorphism, then $\varphi_*\alpha$ is the usual push-forward.

(ii) If α is a volume form, this definition corresponds to that for the push-forward of measures.

(iii) If φ is an oriented surjective submersion, show that $\varphi_*\alpha = \int_{\text{fib}} \alpha$, as defined in Exercise **7.1F**(iv). (*Hint:* Prove the identity $\int_M \varphi^*\beta \wedge \alpha = \int_N (\beta \wedge \int_{\text{fib}} \alpha)$ using Exercise **7.1F**(v) and (vi).)

7.1H Let (M, μ) be a paracompact n-dimensional volume manifold.

(i) If (N, ν) is another paracompact n-dimensional volume manifold and $f : M \to N$ is an orientation reversing diffeomorphism, show that $\int_N \omega = -\int_M f^*\omega$ for any $\omega \in \Omega^n(N)$ with compact support. *(Hint:* Use the proof of **7.1.12**.)

(ii) If $\eta \in \Omega^n(M)$ has compact support and $-M$ denotes the manifold M endowed with the orientation $[-\mu]$, show that $\int_{-M} \eta = -\int_M \eta$. (*Hint:* If $\mathcal{A} = \{(U_i, \varphi_i)\}$ is an oriented atlas for $(M, [\mu])$, then $-\mathcal{A} = \{(U_i, \varphi_i \circ \psi_i)\}$, $\psi_i(x^1, \ldots, x^n) = (-x^1, x^2, \ldots, x^n)$ is an oriented atlas for $(M, [-\mu])$.)

7.1I Let ω_n be the standard volume form on S^n. Show that

$$\int_{S^n} \omega_n = 2^{m+1}\pi^m/(2m-1)!!, \text{ if } n = 2m,\ m \geq 1$$

and

$$\int_{S^n} \omega_n = 2\pi^{m+1}/m! \quad , \text{ if } n = 2m+1,\ m \geq 0$$

by the following steps below.

(i) Let $M \subset \mathbb{R}^{n+1}$ be the annulus $\{x \in \mathbb{R}^{n+1} \mid 0 < a < \|x\| < b < \infty\}$ and let $f :]a, b[\times S^n \to A$ be the diffeomorphism $f(t, s) = ts$. Use Exercise **6.5S**(ii) to show that for $x \in \mathbb{R}^{n+1}$,

$$f^*(e^{-\|x\|^2}\Omega_{n+1}) = t^n e^{-t^2}(dt \times \omega_n)$$

where $\Omega_{n+1} = e_1 \wedge \cdots \wedge e_{n+1}$ for $\{e_1, \ldots, e_{n+1}\}$ the standard basis of $\mathbb{R}^{n+1}$, and where $dt \times \omega_n$ denotes the product volume form on $]a, b[\times S^n$.

(ii) Deduce the equality

$$\int_{\mathbf{R}^{n+1}} e^{-\|x\|^2}\Omega_{n+1} = \int_a^b t^n e^{-t^2} dt \int_{S^n} \omega_n.$$

(iii) Let $a \downarrow 0$ and $b \uparrow \infty$ to deduce the equality

$$\int_0^\infty t^n e^{-t^2} dt \int_{S^n} \omega_n = \left(\int_{-\infty}^{+\infty} e^{-u^2} du\right)^{n+1}.$$

Prove that

$$\int_{-\infty}^{\infty} e^{-u^2} du = \sqrt{\pi}, \quad \int_0^\infty t^{2m} e^{-t^2} dt = (2m-1)!!\,\sqrt{\pi}/2^{m+1}, \quad \text{and} \quad \int_0^\infty t^{2m+1} e^{-t^2} dt = m!/2,$$

to deduce the required formula for $\int_{S^n} \omega_n$.

§7.2 *Stokes' Theorem*

Stokes' theorem states that if α is an (n–1)-form on an orientable n-manifold M, then the integral of $\mathbf{d}\alpha$ over M equals the integral of α over ∂M, the boundary of M. As we shall see in the next section, the classical theorems of Gauss, Green, and Stokes are special cases of this result. Before stating Stokes' theorem formally, we need to discuss manifolds with boundary and their orientations.

7.2.1 Definition *Let* E *be a Banach space and* $\lambda \in E^*$. *Let* $E_\lambda = \{x \in E \mid \lambda(x) \geq 0\}$, *called a* ***half-space*** *of* E, *and let* $U \subset E_\lambda$ *be an open set (in the topology induced on* E_λ *from* E). *Call* $\text{Int}\, U = U \cap \{x \in E \mid \lambda(x) > 0\}$ *the* ***interior*** *of* U *and* $\partial U = U \cap \ker \lambda$ *the* ***boundary*** *of* U. *If* $E = \mathbb{R}^n$ *and* λ *is the projection on the* j*-th factor, then* E_λ *is denoted by* $\mathbb{R}^n{}_j$ *and is called* ***positive*** j*-th* ***half-space***. $\mathbb{R}^n{}_n$ *is also denoted by* $\mathbb{R}^n{}_+$.

We have $U = \text{Int}\, U \cup \partial U$, Int U is open in U, ∂U is closed in U (*not* in E), and $\partial U \cap \text{Int}\, U = \varnothing$. The situation is shown in Fig. 7.2.1. Note that ∂U is *not* the topological boundary of U in E, but it *is* the topological boundary of U intersected with that of E_λ. This inconsistent use of the notation ∂U is temporary.

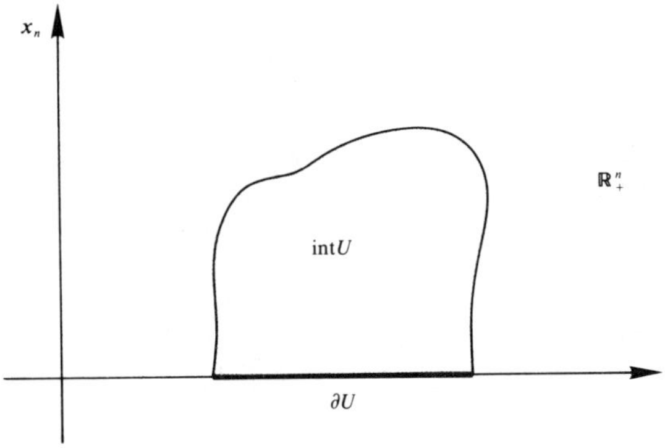

Figure 7.2.1

A manifold with boundary will be obtained by piecing together sets of the type shown in the figure. To carry this out, we need a notion of local smoothness to be used for overlap maps of charts.

7.2.2 Definition *Let* E *and* F *be Banach spaces,* $\lambda \in E^*$, $\mu \in F^*$, U *be an open set in* E_λ, *and* V *be an open set in* F_μ. *A map* $f : U \to V$ *is called* ***smooth*** *if for each point* $x \in U$ *there are* *open neighborhoods* U_1 *of* x *in* E *and* V_1 *of* f(x) *in* F *and a smooth map* $f_1 : U_1 \to V_1$ *such that* $f \mid U \cap U_1 = f_1 \mid U \cap U_1$. *We define* $\mathbf{D}f(x) = \mathbf{D}f_1(x)$. *The map* f *is a* ***diffeomorphism*** *if there is a smooth map* $g : V \to U$ *which is an inverse of* f. (*In this case* $\mathbf{D}f(x)$ *is an isomorphism of* E *with* F.)

We must prove that this definition of **D**f is independent of the choice of f_1, that is, we have to show that if $\varphi : W \to E$ is a smooth map with W open in E such that $\varphi \mid W \cap E_\lambda = 0$, then $\mathbf{D}\varphi(x) = 0$ for all $x \in W \cap E_\lambda$. If $x \in \mathrm{Int}(W \cap E_\lambda)$, this fact is obvious. If $x \in \partial(W \cap E_\lambda)$, choose a sequence $x_n \in \mathrm{Int}(W \cap E_\lambda)$ such that $x_n \to x$; but then $0 = \mathbf{D}\varphi(x_n) \to \mathbf{D}\varphi(x)$ and hence $\mathbf{D}\varphi(x) = 0$, which proves our claim.

7.2.3 Lemma *Let* $U \subset E_\lambda$ *be open,* $\varphi : U \to F_\mu$ *be a smooth map, and assume that for some* $x_0 \in \mathrm{Int}\, U$, $\varphi(x_0) \in \partial F_\mu$. *Then* $\mathbf{D}\varphi(x_0)(E) \subset \partial F_\mu = \ker \mu$.

Proof The quotient $F/\ker \mu$ is isomorphic to $\mathbb{R}$, so that fixing f with $\mu(f) > 0$, the element [f] $\in F/\ker \mu$ forms a basis. Therefore [f] determines the isomrophism $T_f : F/\ker \mu \to \mathbb{R}$ given by $T_f([y]) = t$, where $t \in \mathbb{R}$ is the unique number for which $t[f] = [y]$. This isomorphism in turn defines the isomorphism $S_f : \ker \mu \oplus \mathbb{R} \to F$ given by $S_f(y, t) = y + tf$ which induces diffeomorphisms (in the sense of **7.2.2**) of $\ker \mu \times [0, \infty[$ with F_μ and of $\ker \mu \times]-\infty, 0]$ with $\{y \in F \mid \mu(y) \le 0\}$. Denote by $p : F \to \mathbb{R}$ the linear map given by S_f^{-1} followed by the projection $\ker \mu \oplus \mathbb{R} \to \mathbb{R}$, so that $y \in F_\mu$ (respectively $\ker \mu$, $\{y \in F \mid \mu(y) \le 0\}$) if and only if $p(y) \ge 0$ (respectively $= 0$, ≤ 0).

Notice that the relation

$$\varphi(x_0 + tx) = \varphi(x_0) + \mathbf{D}\varphi(x_0) \cdot tx + o(tx),$$

where $\lim_{t\to 0} o(tx)/t = 0$, together with the hypothesis $(p \circ \varphi)(x) \ge 0$ for all $x \in U$, implies that $0 \le (p \circ \varphi)(x_0 + tx) = 0 + (p \circ \mathbf{D}\varphi)(x_0) \cdot tx + p(o(tx))$, whence for $t > 0$

$$0 \le (p \circ \mathbf{D}\,\varphi)(x_0) \cdot x + p\left(\frac{o(tx)}{t}\right).$$

Letting $t \to 0$, we get $(p \circ \mathbf{D}\varphi)(x_0) \cdot x \ge 0$ for all $x \in E$. Similarly, for $t < 0$ and letting $t \to 0$, we get $(p \circ \mathbf{D}\varphi)(x_0) \cdot x \le 0$ for all $x \in E$. The conclusion is

$$(\mathbf{D}\varphi)(x_0)(E) \subset \ker \mu. \quad \blacksquare$$

Intuitively, this says that if φ preserves the condition $\lambda(x) \ge 0$ and maps an interior point to the boundary, then the derivative must be zero in the normal direction. The reader may also wish to prove **7.2.3** from the implicit mapping theorem. Now we carry this idea one step further.

7.2.3 Lemma *Let* U *be open in* E_λ, V *be open in* F_μ, *and* $f : U \to V$ *be a diffeomorphism. Then* f *restricts to diffeomorphisms* $\mathrm{Int}\, f : \mathrm{Int}\, U \to \mathrm{Int}\, V$ *and* $\partial f : \partial U \to \partial V$.

Proof Assume first that $\partial U = \varnothing$, that is, that $U \cap \ker \lambda = \varnothing$. We shall show that $\partial V = \varnothing$ and hence we take $\text{Int } f = f$. If $\partial V \neq \varnothing$, there exists $x \in U$ such that $f(x) \in \partial V$ and hence by definition of smoothness there are open neighborhoods $U_1 \subset U$ and $V_1 \subset F$, such that $x \in U_1$ and $f(x) \in V_1$, and smooth maps $f_1 : U_1 \to V_1$, $g_1 : V_1 \to U_1$ such that $f \mid U_1 = f_1$, $g_1 \mid V \cap V_1 = f^{-1} \mid V \cap V_1$. Let $x_n \in U_1$, $x_n \to x$, $y_n \in V_1 \backslash \partial V$, and $y_n = f(x_n)$. We have

$$\mathbf{D}f(x) \circ \mathbf{D}g_1(f(x)) = \lim_{y_n \to f(x)} (\mathbf{D}f(g_1(y_n)) \circ \mathbf{D}g_1(y_n)) = \lim_{y_n \to f(x)} \mathbf{D}(f \circ g_1)(y_n) = \text{Id}_F$$

and similarly

$$\mathbf{D}g_1(f(x)) \circ \mathbf{D}f(x) = \text{Id}_E$$

so that $\mathbf{D}f(x)^{-1}$ exists and equals $\mathbf{D}g_1(f(x))$. But by **7.2.3**, $\mathbf{D}f(x)(E) \subset \ker \mu$, which is impossible, $\mathbf{D}f(x)$ being an isomorphism.

Assume that $\partial U \neq \varnothing$. If we assume $\partial V = \varnothing$, then, working with f^{-1} instead of f, the above argument leads to a contradiction. Hence $\partial V \neq \varnothing$. Let $x \in \text{Int } U$ so that x has a neighborhood $U_1 \subset U$, $U_1 \cap \partial U = \varnothing$, and hence $\partial U_1 = \varnothing$. Thus, by the preceding argument, $\partial f(U_1) = \varnothing$, and $f(U_1)$ is open in $V \backslash \partial V$. This shows that $f(\text{Int } U) \subset \text{Int } V$. Similarly, working with f^{-1}, we conclude that $f(\text{Int } U) \supset \text{Int } V$ and hence $f : \text{Int } U \to \text{Int } V$ is a diffeomorphism. But then $f(\partial U) = \partial V$ and $f \mid \partial U : \partial U \to \partial V$ is a diffeomorphism as well. ■

Now we are ready to define a manifold with boundary.

7.2.5 Definition *A **manifold with boundary** is a set* M *together with an **atlas** of **charts with boundary** on* M; ***charts with boundary** are pairs* (U, φ) *where* $U \subset M$ *and* $\varphi(U) \subset E_\lambda$ *for some* $\lambda \in E^*$ *and an **atlas** on* M *is a family of charts with boundary satisfying* **MA1** *and* **MA2** *of* **3.1.1**, *with smoothness of overlap maps* φ_{ji} *understood in the sense of* **7.2.2**. *See Fig.* 7.2.2. *If* $E = \mathbb{R}^n$, M *is called an* n-***manifold with boundary.***

Define $\text{Int } M = \bigcup_U \varphi^{-1}(\text{Int}(\varphi(U)))$ *and* $\partial M = \bigcup_U \varphi^{-1}(\partial(\varphi(U)))$ *called, respectively, the **interior** and **boundary** of* M.

The definition of Int M and ∂M makes sense in view of Lemma **7.2.4**. Note that

1 Int M is a manifold (with atlas obtained from (U, φ) by replacing $\varphi(U) \subset E_\lambda$ by the set $\text{Int } \varphi(U) \subset E$);

2 ∂M is a manifold (with atlas obtained from (U, φ) by replacing $\varphi(U) \subset E_\lambda$ by $\partial\varphi(U) \subset \partial E = \ker \lambda$);

3 ∂M is the topological boundary of Int M in M (although Int M is *not* the topological interior of M).

Summarizing, we have proved the following.

7.2.6 Proposition *If* M *is a manifold with boundary, then its interior* Int M *and its boundary*

∂M *are smooth manifolds without boundary. Moreover, if* $f : M \to N$ *is a diffeomorphism,* N *being another manifold with boundary, then* f *induces, by restriction, two diffeomorphisms* $\text{Int } f : \text{Int } M \to \text{Int } N$ *and* $\partial f : \partial M \to \partial N$. *If* $n = \dim M$, *then* $\dim(\text{Int } M) = n$ *and* $\dim(\partial M) = n - 1$.

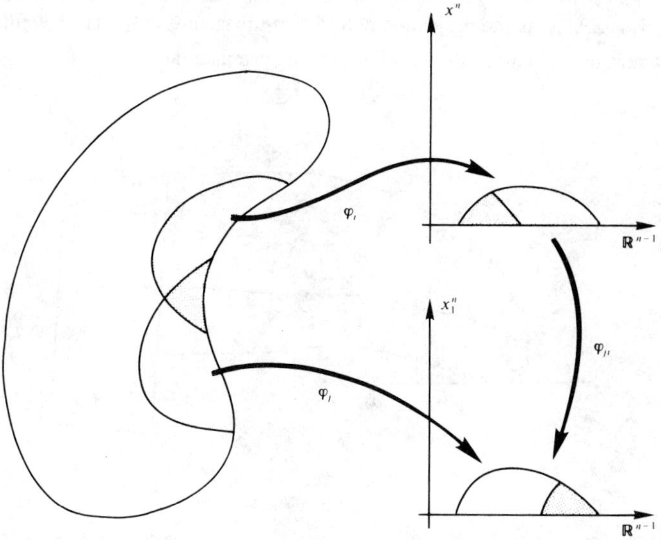

Figure 7.2.2

To integrate a differential n-form over an n-manifold M, M must be oriented. If Int M is oriented, we want to choose an orientation on ∂M compatible with it. In the classical Stokes theorem for surfaces, it is crucial that the boundary curve be oriented , as in Fig. 7.2.3.

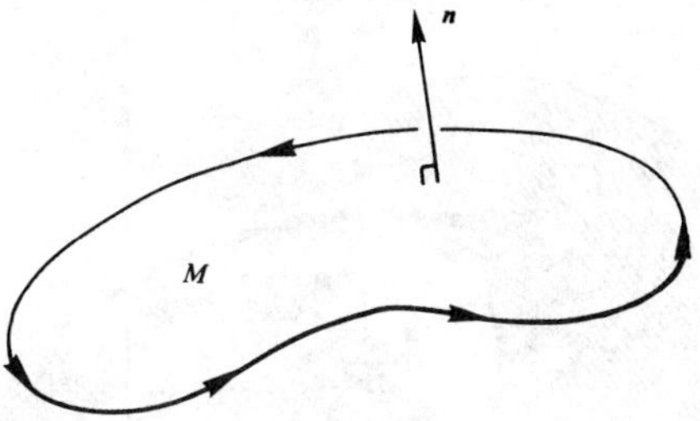

Figure 7.2.3

The tangent bundle to a manifold with boundary is defined in the same way as for manifolds without boundary. Recall that any tangent vector in T_xM has the form $[dc(t)/dt]\,|_{t=\tau}$, where $c : [a, b] \to M$ is a C^1 curve, $a < b$, and $\tau \in [a, b]$. If $x \in \partial M$, we consider curves $c : [a, b] \to M$ such that $c(b) = x$. If $\varphi : U \to U' \subset E_\lambda$ is a chart at m, then $[d(\varphi \circ c)(t)/dt]\,|_{t=b}$ in general points *out* of U′, as in Fig. 7.2.4. Therefore, T_xM is isomorphic to the model space E of M even if $x \in \partial M$ (see Fig. 7.2.5). It is because of this result that tangent vectors are derivatives of C^1-curves defined on *closed* intervals. Had we defined tangent vectors as derivatives of C^1-curves defined on *open* intervals, T_yM for $y \in \partial M$ would be isomorphic to $\ker \lambda$ and not to E. In Fig. 7.2.4, $E = \mathbb{R}^n$, λ is the projection onto the n-th factor, and c(t) is defined on a closed interval whereas the C^1-curve d(t) is defined on an open interval.

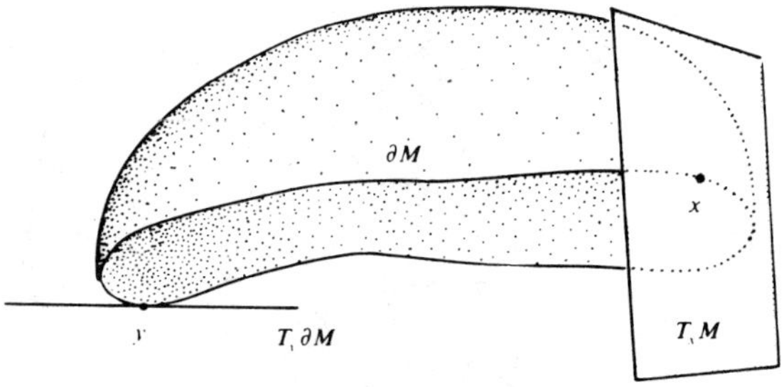

Figure 7.2.4

Having defined the tangent bundle, all of our previous constructions including tensor fields and exterior forms as well as operations on them such as the Lie derivative, interior product, and exterior derivative carry over directly to manifolds with boundary. One word of caution though: the fundamental relation between Lie derivatives and flows still holds if one is careful to take into account that a vector field on M has integral curves which could run into the boundary in finite time and with finite velocity. (If the vector field is tangent to ∂M, this will not happen.)

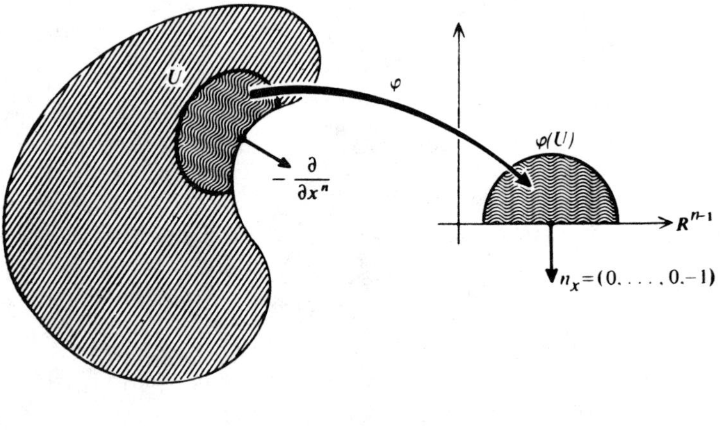

Figure 7.2.5

Next, we turn to the problem of orientation. As for manifolds without boundary a ***volume form*** on an n-manifold with boundary M is a nowhere vanishing n-form on M. Fix an orientation on $\mathbb{R}^n$. Then a chart (U, φ) is called ***positively oriented*** if $T_u\varphi : T_uM \to \mathbb{R}^n$ is orientation preserving for all $u \in U$. If M is paracompact this latter condition is equivalent to orientability of M (the proof is as in **6.5.2.**) Therefore, for paracompact manifolds, an orientation on M is just a smooth choice of orientations of all the tangent spaces, "smooth" meaning that for all the charts of a certain atlas, called the ***oriented charts***, the maps $D(\varphi_j \circ \varphi_i^{-1})(x) : \mathbb{R}^n \to \mathbb{R}^n$ are orientation preserving.

The reader may wonder why for finite dimensional manifolds we did not choose a "standard' half-space, like $x^n \geq 0$ to define the charts at the boundary. Had we done that, the very definition of an oriented chart would be in jeopardy. For example, consider $M = [0, 1]$ and agree that all charts must have range in $\mathbb{R}_+ = \{x \in \mathbb{R} \mid x \geq 0\}$. Then an example of an orientation reversing chart at $x = 1$ is $\varphi(x) = 1 - x$; in fact, every chart at $x = 1$ would be orientation reversing. However, if we admit *any* half-space of $\mathbb{R}$, so charts can be also in $\mathbb{R}_- = \{x \in \mathbb{R} \mid x \leq 0\}$, then a positively oriented chart at 1 is $\varphi(x) = x - 1$. See Figure 7.2.6.

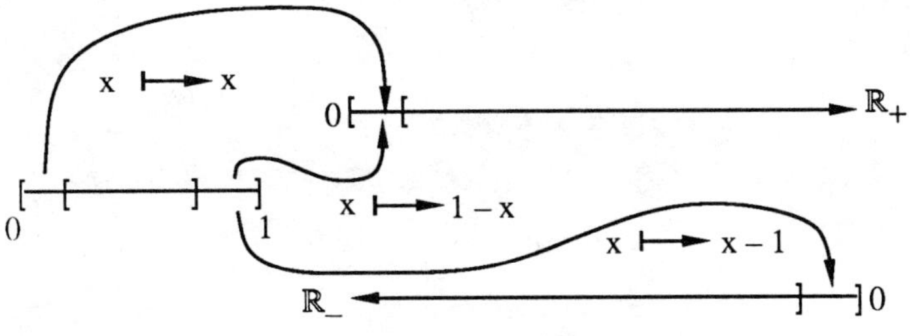

Figure 7.2.6

Once oriented charts and atlases are defined, the theory of integration for oriented paracompact manifolds with boundary proceeds as in §**7.1**.

Finally we define the ***boundary orientation*** of ∂M. At every $x \in \partial M$, the linear space $T_x(\partial M)$ has codimension one in T_xM so that there are (in a chart on M intersecting ∂M) exactly two kinds of vectors not in $\ker \lambda$: those for which their representatives v satisfy $\lambda(v) > 0$ or $\lambda(v) < 0$, i.e., the ***inward*** and ***outward*** pointing vectors. By Lemma **7.2.4**, a change of chart does not affect the property of a vector being outward or inward (see Figure 7.2.4). If $\dim M = n$, these considerations enable us to define the induced orientation of ∂M in the following way.

7.2.7 Definition *Let* M *be an oriented* n*-manifold with boundary,* $x \in \partial M$ *and* $\varphi : U \to \mathbb{R}^n_\lambda$ *a positively oriented chart, where* $\lambda \in \mathbb{R}^{n*}$. *A basis* $\{v_1, ..., v_{n-1}\}$ *of* $T_x(\partial M)$ *is called* ***positively oriented*** *if* $\{(T_x\varphi)^{-1}(n), v_1, ..., v_{n-1}\}$ *is positively oriented in the orientation of* M, *where* n *is any outward pointing vector to* $\mathbb{R}^n_\lambda$ *at* $\varphi(x)$.

For example, we could choose for n the outward pointing vector to $\mathbb{R}^n_\lambda$ and perpendicular to ker λ. If $\lambda : \mathbb{R}^n \to \mathbb{R}$ is the projection on the n-th factor, then $(T_x\varphi)^{-1}(n) = -\partial/\partial x^n$ and the situation is illustrated in Figure 7.2.5.

7.2.8 Stokes' Theorem *Let* M *be an oriented smooth paracompact* n-*manifold with boundary and* $\alpha \in \Omega^{n-1}(M)$ *have compact support. Let* $i : \partial M \to M$ *be the inclusion map so that* $i^*\alpha \in \Omega^{n-1}(\partial M)$. *Then*

$$\int_{\partial M} i^*\alpha = \int_M d\alpha \tag{1}$$

or for short,

$$\int_{\partial M} \alpha = \int_M d\alpha \tag{1'}$$

If $\partial M = \varnothing$, *the left hand side of* (1) *or* (1') *is set equal to zero.*

Proof Since integration was constructed with partitions of unity subordinate to an atlas and both sides of the equation to be proved are linear in α, we may assume without loss of generality that α is a form on $U \subset \mathbb{R}^n_+$ with compact support. Write

$$\alpha = \sum_{i=1}^{n} (-1)^{i-1}\alpha^i dx^1 \wedge \cdots \wedge (dx^i)^\wedge \wedge \cdots \wedge dx^n, \tag{2}$$

where $\hat{}$ above a term means that it is deleted. Then

$$d\alpha = \sum_{i=1}^{n} \frac{\partial\alpha^i}{\partial x^i} dx^1 \wedge \cdots \wedge dx^n \tag{3}$$

and thus

$$\int_U d\alpha = \sum_{i=1}^{n} \int_{\mathbb{R}^n} \frac{\partial\alpha^i}{\partial x^i} dx^1 \cdots dx^n. \tag{4}$$

There are two cases: $\partial U = \varnothing$ and $\partial U \neq \varnothing$. If $\partial U = \varnothing$, we have $\int_{\partial U}\alpha = 0$. The integration of the i-th term in the sum occurring in (4) is

$$\int_{\mathbb{R}^{n-1}} \left(\int_{\mathbb{R}} \frac{\partial\alpha^i}{\partial x^i} dx^i \right) dx^1 \cdots (dx^i)^\wedge \cdots dx^n \ \text{(no sum)} \tag{5}$$

and $\int_{-\infty}^{+\infty}(\partial\alpha^i/\partial x^i)dx = 0$ since α^i has a compact support. Thus (4) is zero as desired.

If $\partial U \neq \varnothing$, then we can do the same trick for each term except the last, which is, by the fundamental theorem of calculus,

$$\int_{\mathbb{R}^{n-1}}\left(\int_0^\infty \frac{\partial \alpha^n}{\partial x^n}\, dx^n\right) dx^1 \cdots dx^{n-1} = -\int_{\mathbb{R}^{n-1}} \alpha^n(x^1, \dots, x^{n-1}, 0)dx^1 \cdots dx^{n-1}. \qquad (6)$$

since α^n has compact support. Thus

$$\int_U d\alpha = -\int_{\mathbb{R}^{n-1}} \alpha^n(x^1, \dots, x^{n-1}, 0)dx^1 \cdots dx^{n-1} \qquad (7)$$

On the other hand,

$$\int_{\partial U} \alpha = \int_{\partial \mathbb{R}^n_+} \alpha = \int_{\partial \mathbb{R}^n_+} (-1)^{n-1}\alpha^n(x^1, \dots, x^{n-1}, 0)dx^1 \wedge \cdots \wedge dx^{n-1}. \qquad (8)$$

But $\mathbb{R}^{n-1} = \partial\mathbb{R}^n_+$ and the usual orientation on $\mathbb{R}^{n-1}$ is *not* the boundary orientation. The outward unit normal is $-e_n = (0, \dots, 0, -1)$ and hence the boundary orientation has the sign of the ordered basis $\{-e_n, e_1, \dots, e_{n-1}\}$, which is $(-1)^n$. Thus (8) becomes

$$\begin{aligned}\int_{\partial U} \alpha &= \int_{\partial \mathbb{R}^n_+} (-1)^{n-1}\alpha^n(x^1, \dots, x^{n-1}, 0)dx^1 \wedge \cdots \wedge dx^{n-1} \\ &= (-1)^{2n-1} \int_{\mathbb{R}^{n-1}} \alpha^n(x^1, \dots, x^{n-1}, 0)dx^1 \cdots dx^{n-1}. \qquad (9)\end{aligned}$$

Since $(-1)^{2n-1} = -1$, combining (7) and (9), we get the desired result. ∎

This basic theorem reduces to the usual theorems of Green, Stokes, and Gauss in $\mathbb{R}^2$ and $\mathbb{R}^3$, as we shall see in the next section. For forms with less smoothness or without compact support, the best results are somewhat subtle. See Gaffney [1954], Morrey [1966], Yau [1976], Karp [1981] and the remarks at the end of Supplement **7.2B**.

Next we draw some important consequences from Stokes' theorem.

7.2.9 Gauss' Theorem *Let* M *be an oriented paracompact* n*-manifold with boundary and* X *a vector field on* M *with compact support. Let* μ *be a volume form on* M. *Then*

$$\int_M (\operatorname{div} X)\mu = \int_{\partial M} \mathbf{i}_X\mu. \qquad (10)$$

Proof Recall that

$$(\operatorname{div} X)\mu = \mathbf{L}_X\mu = \mathbf{d}\mathbf{i}_X\mu + \mathbf{i}_X\mathbf{d}\mu = \mathbf{d}\mathbf{i}_X\mu.$$

The result is thus a consequence of Stokes' theorem. ∎

If M carries a Riemannian metric, there is a unique outward-pointing unit normal $n_{\partial M}$ along

∂M, and M and ∂M carry corresponding uniquely determined volume forms μ_M and $\mu_{\partial M}$ and associated measures m_M and $m_{\partial M}$. Then Gauss' theorem reads as follows.

7.2.10 Corollary

$$\int_M (\operatorname{div} X) dm_M = \int_{\partial M} \langle X, n_{\partial M} \rangle dm_{\partial M},$$

where $\langle X, n_{\partial M} \rangle$ *is the inner product of* X *and* $n_{\partial M}$.

Proof Let $\mu_{\partial M}$ denote the volume element on ∂M induced by the Riemannian volume element $\mu_M \in \Omega^n(M)$; i.e., for any positively oriented basis $v_1, \ldots, v_{n-1} \in T_x(\partial M)$, and charts chosen so that $n_{\partial M} = -\partial/\partial x^n$ at the point x,

$$\mu_{\partial M}(x)(v_1, \ldots, v_{n-1}) = \mu_M(x)\left(-\frac{\partial}{\partial x^n}, v_1, \ldots, v_{n-1}\right).$$

Since

$$\begin{aligned}(i_X\mu_M)(x)(v_1, \ldots, v_{n-1}) &= \mu_M(x)\left(X^i(x)v_i + X^n(x)\frac{\partial}{\partial x^n}, v_1, \ldots, v_{n-1}\right)\\ &= -X^n(x)\mu_{\partial M}(x)(v_1, \ldots, v_{n-1})\end{aligned}$$

and $X^n = -\langle X, n_{\partial M} \rangle$, the corollary follows by Gauss' theorem. ■

7.2.11 Corollary *If* X *is divergence-free on a compact boundaryless manifold with a volume element* μ, *then* X *as an operator is skew-symmetric; that is, for* f *and* $g \in \mathcal{F}(M)$,

$$\int_M X[f] g\mu = -\int_M f X[g]\mu.$$

Proof Since X is divergence free, $\mathbf{L}_X(h\mu) = (\mathbf{L}_X h)\mu$ for any $h \in \mathcal{F}(M)$. Thus

$$X[f]g\mu + fX[g]\mu = \mathbf{L}_X(fg)\mu = \mathbf{L}_X(fg\mu).$$

Integration and the use of Stokes' theorem gives the result. ■

7.2.12 Corollary *If* M *is compact without boundary* $X \in \mathfrak{X}(M)$, $\alpha \in \Omega^k(M)$ *and* $\beta \in \Omega^{n-k}(M)$, *then*

$$\int_M \mathbf{L}_X\alpha \wedge \beta = -\int_M \alpha \wedge \mathbf{L}_X\beta.$$

Proof Since $\alpha \wedge \beta \in \Omega^n(M)$, the formula follows by integrating both sides of the relation $\mathbf{d}i_X(\alpha \wedge \beta) = \mathbf{L}_X(\alpha \wedge \beta) = \mathbf{L}_X\alpha \wedge \beta + \alpha \wedge \mathbf{L}_X\beta$ and using Stokes' theorem. ■

7.2.13 Corollary *If* M *is a compact orientable, boundaryless* n*-dimensional pseudo-Riemannian manifold with a matric* g *of index* Ind(g), *then* **d** *and* δ *are adjoints, i.e.,*

$$\langle \mathbf{d}\alpha, \beta\rangle = \int \mathbf{d}\alpha \wedge *\beta = \int \alpha \wedge *\delta\beta = \langle \alpha, \delta\beta\rangle$$

for $\alpha \in \Omega^k(M)$ *and* $\beta \in \Omega^{k+1}(M)$.

Proof Recall from **6.5.21** that $\delta\beta = (-1)^{nk+1+\mathrm{Ind}(g)} *\mathbf{d}*\beta$, so that

$$\begin{aligned}\mathbf{d}\alpha \wedge *\beta - \alpha \wedge *\delta\beta &= \mathbf{d}\alpha \wedge *\beta + (-1)^{nk+\mathrm{Ind}(g)}\alpha \wedge **\mathbf{d}*\beta \\ &= \mathbf{d}\alpha \wedge *\beta + (-1)^{nk+\mathrm{Ind}(g)+k(n-k)+\mathrm{Ind}(g)}\alpha \wedge \mathbf{d}*\beta \\ &= \mathbf{d}\alpha \wedge *\beta + (-1)^{k}\alpha \wedge \mathbf{d}*\beta \\ &= \mathbf{d}(\alpha \wedge *\beta)\end{aligned}$$

since $k^2 + k$ is an even number for any integer k. Integrating both sides of the equation and using Stokes' theorem gives the result. ■

The same identity $\langle \mathbf{d}\alpha, \beta\rangle = \langle \alpha, \delta\beta\rangle$ holds for noncompact manifolds, possibly with boundary, provided either α or β has compact support in Int M.

☞ SUPPLEMENT 7.2A
Stokes' Theorem for Nonorientable Manifolds

Let M be a nonorientable paracompact n-manifold with a smooth boundary ∂M and inclusion map $i : \partial M \to M$. We would like to give meaning to the formula

$$\int_M \mathbf{d}\rho = \int_{\partial M} i^*\rho.$$

in Stokes' theorem. Clearly, both sides makes sense if $\mathbf{d}\rho$ and $i^*(\rho)$ are defined in such a way that they are densities on M and ∂M, respectively. Here **d** should be some operator analogous to the exterior differential, and ρ should be a section of some bundle over M analogous to $\Lambda^{n-1}(M)$. Denote the as yet unknown bundle analogous to $\Lambda^k(M)$ by $\Lambda^k_t(M)$ and its space of sections $\Omega^k_\tau(M)$. Then we desire an operator $\mathbf{d} : \Omega^k_\tau(M) \to \Omega^{k+1}_\tau(M)$, $k = 0, \ldots, n$, and desire $\Lambda^n_\tau(M)$ to be isomorphic to $|\Lambda|(M)$.

To guess what $\Lambda^k_\tau(M)$ might be, let us first discuss $\Lambda^n_\tau(M)$. The key difference between an n-form ω and a density ρ is their transformation property under a linear map $A : T_mM \to T_mM$ as follows:

$$\omega(m)(A(v_1), \ldots, A(v_n)) = (\det A)\omega(m)(v_1, \ldots, v_n)$$

$$\rho(m)(A(v_1), \ldots, A(v_n)) = |\det A|\, \rho(m)(v_1, \ldots, v_n)$$

for $m \in M$ and $v_1, \ldots, v_n \in T_mM$. If $v_1, \ldots, v_n$ is a basis, then $\det(A) > 0$ if A preserves the orientation given by $v_1, \ldots, v_n$ and $\det(A) < 0$ if A reverses this orientation. Thus ρ can be thought of as an object behaving like an n-form at every $m \in M$ once an orientation of T_mM is given; i.e., ρ should be thought of as an n-form with values in some line bundle (a bundle with one-dimensional fibers) associated with the concept of orientation. This definition would then generalize to any k; $\Lambda^k_\tau(M)$ will be line-bundle-valued k-forms on M. We shall now construct this line bundle.

At every point of M there are two orientations. Using them, we construct the oriented double covering $\tilde{M} \to M$ (see **6.5.7**). Since $\tilde{M}$ is not a line bundle, some other construction is in order. At every $m \in M$, a line is desired such that the positive half-line should correspond to one orientation of T_mM and the negative half-line to the other. The fact that must be taken into account is that multiplication by a negative number switches these two half-lines. To incorporate this idea, identify $(m, [\mu], a)$ with $(m, [-\mu], -a)$ where $m \in M$, $a \in \mathbb{R}$, and $[\mu]$ is an orientation of T_mM. Thus, define the ***orientation line bundle*** $\sigma(M) = \{(m, [\mu], a) \mid m \in M, a \in \mathbb{R}, \text{ and } [\mu] \text{ is an orientation of } T_mM)\}/\sim$ where $\sim$ is the equivalence relation $(m, [\mu], a) \sim (m, [-\mu], -a)$. Denote by $\langle m, [\mu], a\rangle$ the elements of $\sigma(M)$. It can be checked that the map $\pi : \sigma(M) \to M$ defined by $\pi(\langle m, [\mu], a\rangle) = m$ is a line bundle with bundle charts given by $\psi : \pi^{-1}(U) \to \varphi(U) \times \mathbb{R}$, $\psi(\langle m, [\mu], a\rangle) = (\varphi(m), \varepsilon a)$, where $\varphi : U \to \mathbb{R}^n$ is a chart for M at m, and $\varepsilon = +1$ if $T_m\varphi : (T_mM, [\mu]) \to (\mathbb{R}^n, [\omega])$ is orientation preserving and -1 if it is orientation reversing, $[\omega]$ being a fixed orientation of $\mathbb{R}^n$. The change of chart map of the line bundle $\sigma(M)$ is given by $(x, a) \in U' \times \mathbb{R} \subset \mathbb{R}^n \times \mathbb{R} \mapsto ((\varphi_j \circ \varphi_i)^{-1}(x), \operatorname{sign}(\det \mathbf{D}(\varphi_j \circ \varphi_i^{-1})(x))) \in U' \times \mathbb{R}$. If M is paracompact, then $\sigma(M)$ is an orientable vector bundle (see exercise **6.5N**). If in addition M is also connected, then M is orientable if and only if $\sigma(M)$ is trivial line bundle; the proof is similar to that of Proposition **6.5.7**.

7.2.14 Definition *A* ***twisted*** k-***form*** *on* M *is a* $\sigma(M)$-*valued* k-*form on* M. *The bundle of twisted* k-*forms is denoted by* $\Lambda^k_\tau(M)$ *and sections of this bundle are denoted* $\Omega^k_\tau(M)$ *or* $\Gamma^\infty(\Lambda^k_\tau(M))$.

Locally, a section $\rho \in \Omega^k_\tau(M)$ can be written as $\rho = \alpha\xi$ where $\alpha \in \Omega^k(U)$ and ξ is an orientation of U regarded as a locally constant section of $\sigma(M)$ over U. The operators

$$\mathbf{d} : \Omega^k_\tau(M) \to \Omega^{k+1}_\tau(M) \text{ and}$$
$$\mathbf{i}_X : \Omega^k_\tau(M) \to \Omega^{k-1}_\tau(M), \text{ where } X \in \mathfrak{X}(M),$$

are defined to be the unique operators such that if $\rho = \alpha\xi$ in the neighborhood U, then $\mathbf{d}\rho = (\mathbf{d}\alpha)\xi$ and $\mathbf{i}_X\rho = (\mathbf{i}_X\alpha)\xi$. One has $\mathcal{L}_X = \mathbf{i}_X \circ \mathbf{d} + \mathbf{d} \circ \mathbf{i}_X$. Note that if M is orientable, $\Lambda^k_\tau(M)$ coincides with $\Lambda^k(M)$.

Next we show that the line bundles $|\Lambda(M)|$ and $\Lambda^n_\tau(M)$ are isomorphic. If $\lambda \in |\Lambda(M)|_m$ and $v_1, \ldots, v_n \in T_mM$, define $\varphi(\lambda) : T_mM \times \cdots \times T_mM \to \sigma(M)_m$ by setting $\Phi(\lambda)(v_1, \ldots, v_n) = \langle m, [\sigma(v_1, \ldots, v_n)], \lambda(v_1, \ldots, v_n)\rangle$, if $\{v_1, \ldots, v_n\}$ is a basis of T_mM, and setting it equal to

0, if $\{v_1, ..., v_n\}$ are linearly dependent, where $[\sigma(v_1, ..., v_n)]$ denotes the orientation of T_mM given by the ordered basis $\{v_1, ..., v_n\}$. $\Phi(\lambda)$ is skew symmetric and homogeneous with respect to scalar multiplication since if $\{v_1, ..., v_n\}$ is a basis and $a \in \mathbb{R}$, we have

$$\begin{aligned}\Phi(\lambda)(v_2, v_1, v_3, ..., v_n) &= \langle m, [\sigma(v_2, v_1, v_3, ..., v_n)], \lambda(v_2, v_1, v_3, ..., v_n)\rangle \\ &= \langle m, [-\sigma(v_1, v_2, ..., v_n)], \lambda(v_1,..., v_n)\rangle \\ &= \langle m, [\sigma(v_1, ..., v_n)], -\lambda(v_1, ..., v_n)\rangle \\ &= -\Phi(\lambda)(v_1, ..., v_n)\end{aligned}$$

and

$$\begin{aligned}\Phi(\lambda)(av_1, v_2, ..., v_n) &= \langle m, [\sigma(av_1, v_2, ..., v_n)], \lambda(av_1, ..., v_n)\rangle \\ &= \langle m, [(\text{sign } a)\sigma(v_1, ..., v_n)], |a| \lambda(v_1, ..., v_n)\rangle \\ &= \langle m, [\sigma(v_1, ..., v_n)], a\lambda(v_1, ..., v_n)\rangle \\ &= a\Phi(\lambda)(v_1, ..., v_n).\end{aligned}$$

The proof of additivity is more complicated. Let $v_1, v_1', v_2, ..., v_n \in T_mM$. If both $\{v_1, ..., v_n\}$ and $\{v_1', v_2, ..., v_n\}$ are linearly dependent, then so are $\{v_1 + v_1', v_2, ..., v_n\}$ and the additivity property of $\Phi(\lambda)$ is trivially verified. So assume that $\{v_1, ..., v_n\}$ is a basis of T_mM and write $v_1' = a^1v_1 + \cdots + a^nv_n$. Therefore

$$\begin{aligned}\lambda(v_1', v_2, ..., v_n) &= |a^1| \lambda(v_1, ..., v_n), \text{ and} \\ \lambda(v_1 + v_1', v_2, .., v_n) &= |1 + a^1| \lambda(v_1, ..., v_n).\end{aligned}$$

Moreover, if

(i) $a^1 > 0$, then $[\sigma(v_1, ..., v_n)] = [\sigma(v_1', v_2, ..., v_n)] = [\sigma(v_1 + v_1', v_2, ..., v_n)]$;
(ii) $a^1 = 0$, then $[\sigma(v_1, ..., v_n)] = [\sigma(v_1 + v_1', v_2, ..., v_n)]$ and $\Phi(\lambda)(v_1', v_2, ..., v_n) = 0$;
(iii) $-1 < a^1 < 0$, then $[\sigma(v_1, ..., v_n)] = [-\sigma(v_1', v_2, ..., v_n)] = [\sigma(v_1 + v_1', v_2, ..., v_n)]$;
(iv) $a^1 = -1$, then $[\sigma(v_1, ..., v_n)] = [-\sigma(v_1', v_2, ..., v_n)]$ and $\Phi(\lambda)(v_1 + v_1', v_2, ..., v_n) = 0$;
(v) $a^1 < -1$, then $[\sigma(v_1, ..., v_n)] = [-\sigma(v_1', v_2, ..., v_n)] = [-\sigma(v_1 + v_1', v_2, ..., v_n)]$.

Additivity is now checked in all five cases separately. For example, in case (iii) we have

$$\begin{aligned}\Phi(\lambda)(v_1 + v_1', v_2, ..., v_n) &= \langle m, [\sigma(v_1 + v_1', v_2, ..., v_n], \lambda(v_1 + v_1', ..., v_n)\rangle \\ &= \langle m, [\sigma(v_1, ..., v_n)], (1 + a^1)\lambda(v_1, ..., v_n)\rangle \\ &= \langle m, [\sigma(v_1, ..., v_n)], \lambda(v_1, ..., v_n)\rangle \\ &\qquad + \langle m, [-\sigma(v_1', v_2, ..., v_n)], -|a^1| \lambda(v_1, ..., v_n)\rangle \\ &= \Phi(\lambda)(v_1, ..., v_n) + \Phi(\lambda)(v_1', v_2, ..., v_n).\end{aligned}$$

Thus Φ has values in $\Lambda^n_\tau(M)$. The map Φ is clearly linear and injective and thus is an isomorphism of $|\Lambda(M)|$ with $\Lambda^n_\tau(M)$. Denote also by Φ the induced isomorphism of $|\Omega(M)|$

with $\Omega^n_\tau(M)$.

The integral of $\rho \in \Omega^n_\tau(M)$ is defined to be the integral of the density $\Phi^{-1}(\rho)$ over M. In local coordinates the expression for Φ is

$$\Phi(a \mid dx^1 \wedge \cdots \wedge dx^n \mid) = (a\, dx^1 \wedge \cdots \wedge dx^n)\xi^n{}_0,$$

where $\xi^n{}_0$ is the basis element of the space sections of $\sigma(U)$ given by $\xi^n{}_0(u)(v_1, \ldots, v_n) = \langle u, [\sigma(v_1, \ldots, v_n)], \operatorname{sign}(\det(v_j{}^i))\rangle$, where $(v_j{}^i)$ are the components of the vector v_j in the coordinates $(x^1, \ldots, x^n)$ of U. Therefore

$$\Phi^{-1}((a\, dx^1 \wedge \cdots \wedge dx^n)\xi) = \frac{a\xi}{\xi_0^n} \mid dx^1 \wedge \cdots \wedge dx^n$$

and

$$\int_U (a\, dx^1 \wedge \cdots \wedge dx^n) b\xi^n_0 = \int_U ab\, dx^1 \wedge \cdots \wedge dx^n|$$

for any smooth functions $a, b : U \to \mathbb{R}$.

Finally, for the formulation of Stokes' Theorem, if $i : \partial M \to M$ is the inclusion and $\rho \in \Omega^{n-1}{}_\tau(M)$, the induced twisted $(n-1)$-form $i^*\rho$ on ∂M is defined by setting $(i^*\rho)(m)(v_1, \ldots, v_{n-1}) = \langle m, [\operatorname{sign}[\mu_n]\, \sigma(-\partial/\partial x^n, v_1, \ldots, v_{n-1})], \rho'(m)(v_1, \ldots, v_{n-1})\rangle$, if $v_1, \ldots, v_{n-1}$ are linearly independent and setting it equal to zero, if $v_1, \ldots, v_{n-1}$ are linearly dependent, where $(x^1, \ldots, x^n)$ is a coordinate system at m with ∂M described by $x^n = 0$ and $\rho(m)(v_1, \ldots, v_{n-1}) = \langle m, \operatorname{sign}[\mu_m], \rho'(m)(v_1, \ldots, v_{n-1})\rangle$ with $\rho'(m)$ skew symmetric; moreover $\operatorname{sign}[\mu_m] = +1$ (respectively -1) if $[\mu_m]$ and $[\sigma(-\partial/\partial x^n, v_1, \ldots, v_{n-1})]$ define the same (respectively opposite) orientation of T_mM. If $M = U$, where U is open in $\mathbb{R}^n{}_+$ and $\rho = \alpha\, a\xi_0 \in \Omega^{n-1}{}_\tau(U)$, then $i^*\rho = (-1)^n i^*(a\alpha)\xi^{n-1}{}_0$. In particular, if

$$\zeta = \sum_{i=1}^{n} \zeta_i dx^1 \wedge \ldots \wedge (dx^i)^\wedge \wedge \ldots \wedge dx^n,$$

we have

$$\begin{aligned}(i^*\rho)(x^1, \ldots, x^{n-1}) \\ &= (-1)^n a(x^1, \ldots, x^{n-1}, 0)(-1)^{n-1}\alpha^n(x^1, \ldots, x^{n-1}, 0)\, dx^1 \wedge \cdots \wedge (dx^i)^\wedge \wedge \cdots \wedge dx^n \xi^{n-1}{}_0 \\ &= -a(x^1, \ldots, x^{n-1}, 0)\alpha^n(x^1, \ldots, x^{-1}, 0)\, dx^1 \wedge \cdots \wedge (dx^i)^\wedge \wedge \cdots \wedge dx^n \xi^{n-1}{}_0.\end{aligned}$$

With this observation, the proof of Stokes' Theorem **7.2.8** gives the following.

7.2.15 Nonorientable Stokes' Theorem *Let* M *be a paracompact nonorientable* n*-manifold with smooth boundary* ∂M *and* $\rho \in \Omega^{n-1}{}_\tau(M)$, *a twisted* $(n-1)$*-form with compact support. Then*

$$\int_M d\rho = \int_{\partial M} i^*\rho.$$

The same statement holds for vector-valued twisted (n–1)-forms and all corollaries go through replacing everywhere (n–1)-forms with twisted (n–1)-forms. For example, we have the following.

7.2.16 Non-Orientable Gauss Theorem *Let* M *be a nonorientable Riemannian* n*-manifold with associated density* μ_M. *Then for* $X \in \mathfrak{X}(M)$ *with compact support*

$$\int_M \operatorname{div}(X)\mu_M = \int_{\partial M} (X \cdot n)\mu_{\partial M}$$

where n *is the outward unit normal of* ∂M, $\mu_{\partial M}$ *is the induced Riemannian density of* ∂M *and* $\mathcal{L}_X\mu_M = (\operatorname{div} X)\mu_M$.

For a concrete situation in $\mathbb{R}^3$ involving these ideas, see Exercise **7.3I**

SUPPLEMENT 7.2B
Stokes' Theorem on Manifolds with Piecewise Smooth Boundary

The statement of Stokes' theorem we have given does not apply when M is, say a cube or a cone, since these sets do not have a smooth boundary. If the singular portion of the boundary (the four vertices and 12 edges in case of the cube, the vertex and the base circle in case of the cone), is of Lebesgue measure zero (within the boundary) it should not contribue to the boundary integral and we can hope that Stokes' theorem still holds. This supplement discusses such a version of Stokes' theorem inspired by Holmann and Rummler [1972]. (See Lang [1972] for an alternative approach.)

First we shall give the definition of a manifold with piecewise smooth boundary. A glance at the definition of a manifold with boundary makes it clear that one could define a manifold with *corners*, by choosing charts that make regions near the boundary diffeomorphic to open subsets of a finite intersection of positive closed half-spaces. Unfortunately, singular points on the boundary--such as the vertex of a cone--need not be of this type. Thus, instead of trying to classify the singular points up to diffeomorphism and then make a formal intrinsic definition, it is simpler to consider manifolds already embedded in a bigger manifold. Then we can impose a condition on the boundary to insure the validity of Stokes' theorem.

7.2.17 Definition *Let* $U \subset \mathbb{R}^{n-1}$ *be open and* $f : U \to \mathbb{R}$ *be continuous. A point* p *on the graph of* f, $\Gamma_f = \{(x, f(x)) \mid x \in U\}$, *is called* ***regular*** *if there is an open neighborhood* V *of* p *such that* $V \cap \Gamma_f$ *is an* $(n-1)$ *dimensional smooth submanifold of* V. *Let* ρ_f *denote the set of regular points. Any point in* $\sigma_f = \Gamma_f \setminus \rho_f$ *is called* ***singular.*** *The mapping* f *is called* ***piecewise smooth*** *if* ρ_f *is Lebesgue measurable,* $\pi(\sigma_f)$ *has measure zero in* U *(where* $\pi : U \times \mathbb{R} \to U$ *is the projection) and* $f \mid \pi(\sigma_f)$ *is locally Hölder; i.e., for each compact set* $K \subset \pi(\sigma_f)$ *there are* *constants* $c(K) > 0$, $0 < \alpha(K) \le 1$ *such that* $|f(x) - f(y)| \le c(K) \, \| x - y \|^{\alpha(K)}$ *for all* $x, y \in K$.

Note that ρ_f is open in Γ_f and that $\mathrm{Int}(\Gamma_f^-) \cup \rho_f$, where $\Gamma_f^- = \{(x, y) \in U \times \mathbb{R} \mid y \le f(x)\}$, is a manifold with boundary ρ_f. Thus ρ_f has positive orientation induced from the standard orientation of $\mathbb{R}^n$. This will be called the ***positive orientation*** of Γ_f. We are now ready to define manifolds with piecewise smooth boundary.

7.2.18 Definition *Let* M *be an* n*-manifold. A closed subset* N *of* M *is said to be a* ***manifold with piecewise smooth boundary*** *if for every* $p \in N$ *there exists a chart* (U, φ) *of* M *at* p, $\varphi(U) = U' \times U'' \subset \mathbb{R}^{n-1} \times \mathbb{R}$, *and a piecewise smooth mapping* $f : U' \to \mathbb{R}$ *such that*

$$\varphi(\mathrm{bd}(N) \cap U) = \Gamma_f \cap \varphi(U)$$

and $\varphi(N \cap U) = \Gamma_f^- \cap \varphi(U)$. *See Fig.* 7.2.7.

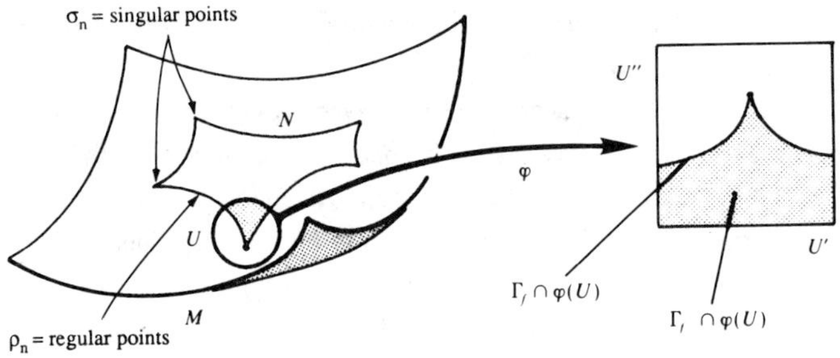

Figure 7.2.7

It is readily verified that the condition on N is chart independent, using the fact that the composition of a piecewise smooth map with a diffeomorphism is still piecewise smooth. Thus, regular and singular points of bd(N) make intrinsic sense and are defined in terms of an arbitrary chart satisfying the conditions of the preceding definition. Let ρ_N and σ_N denote the regular and singular part of the boundary bd(N) of N in M.

To formulate Stokes' theorem, we define $\int_N \eta$, for η an n-form (respectively, density) on M with compact support. This is done as usual via a partition of unity; ρ_N and σ_N play no role since they have Lebesgue measure zero in every chart: ρ_N because it is an $(n-1)$-manifold and σ_N by definition.

It is not so simple to define $\int_{\mathrm{bd}(N)} \zeta$ for $\zeta \in \Omega^{n-1}(M)$ (respectively, a density). First a lemma is needed.

7.2.19 Lemma *Let* $\zeta \in \Omega^{n-1}(U \times \mathbb{R})$, *where* U *is open in* $\mathbb{R}^{n-1}$, $\operatorname{supp}(\zeta)$ *is compact and* $f : U \to \mathbb{R}$ *is a piecewise smooth mapping. Then there is a smooth bounded function* $a : \rho_f \to \mathbb{R}$, *such that* $i^*\zeta = a\lambda$ *where* $i : \rho_f \to U \times \mathbb{R}$ *is the inclusion and* $\lambda \in \Omega^{n-1}(\rho_f)$ *is the boundary volume form induced by the canonical volume form of* $U \times \mathbb{R} \subset \mathbb{R}^n$ *on* $\operatorname{Int}(\Gamma_f^-) \cup \rho_f$.

Proof The existence of the function a on ρ_f follows since $\Omega^{n-1}(\rho_f)$ is one-dimensional with a basis element λ. We prove that a is bounded. Let $p \in \rho_f$ and $v_1, \ldots, v_{n-1} \in T_p(\rho_f)$ be an orthonormal basis with respect to the Riemannian metric on ρ_f induced from the standard metric of $\mathbb{R}^n$, and denote by n the outward unit normal. Then

$$a(p) = a(p)(dx^1 \wedge \cdots \wedge dx^n)(p)(n, v_1, \ldots, v_{n-1}) = a(p)\lambda(p)(v_1, \ldots, v_{n-1}) = \zeta(p)(v_1, \ldots, v_{n-1}).$$

Let $v_i = v_i^j \left.\dfrac{\partial}{\partial x^j}\right|_p$. Since $\left.\dfrac{\partial}{\partial x^1}\right|_p, \ldots, \left.\dfrac{\partial}{\partial x^n}\right|_p$ and $n, v_1, \ldots, v_{n-1}$ are orthonormal bases of $T_p(U \times \mathbb{R})$, we must have $|v_i^j| \le 1$ for all i, j. Hence if

$$\zeta = \sum_{i=1}^{n} \zeta_i dx^1 \wedge \ldots \wedge (dx^i)^\wedge \wedge \ldots \wedge dx^n,$$

then

$$|a(p)| = |\zeta(p)(v_1, \ldots, v_{n-1})| = \left| \sum_{i=1}^{n} \zeta_i(p) \sum_{\sigma \in S_{n-1}} (-1)^i (\operatorname{sign} \sigma) v_1^{\sigma(1)} \cdots v_{n-1}^{\sigma(n-1)} \right|$$

$$= \sum_{i=1}^{n} |\zeta_i(p)| (n-1)!$$

which is bounded, since ζ has compact support. ▼

In view of this lemma and the fact that σ_f has measure zero, we can define

$$\int_{\Gamma_f} \zeta = \int_{\rho_f} i^*\zeta = \int_{\rho_f} a\lambda.$$

Now we can define, via a partition of unity, the integral of $\eta \in \Omega^{n-1}(M)$ (or a twisted (n – 1)-form) by

$$\int_{\mathrm{bd}(N)} \eta = \int_{\rho_N} \eta.$$

7.2.20 Piecewise Smooth Stokes Theorem *Let* M *be a paracompact* n*-manifold and* N *a closed submanifold of* M *with piecewise smooth boundary. If*

(i) M *is orientable and* $\omega \in \Omega^{n-1}(M)$ *has compact support, or*

(ii) M *is nonorientable and* $\omega \in \Omega^{n-1}_{\tau}(M)$ *is a twisted* $(n-1)$*-form (see the preceding* *supplement) which has compact support, then*

$$\int_N d\omega = \int_{bd(N)} \omega.$$

The proof of this theorem reduces via a partition of unity to the local case. Thus it suffices to prove that if U is open in $\mathbb{R}^{n-1}$, $\omega \in \Omega^{n-1}(U \times \mathbb{R})$ has compact support, and $f : U \to \mathbb{R}$ is a piecewise smooth mapping, then

$$\int_{\Gamma_f^-} d\omega = \int_{\Gamma_f} \omega. \tag{1}$$

The left-hand side of (1) is to be understood as the integral over the compact measurable set $\Gamma_f^- \cap \text{supp}(\omega)$. For the proof of (1) we use three lemmas.

7.2.21 Lemma *Equation* (1) *holds if* ω *vanishes in a neighborhood of* σ_f *in* $U \times \mathbb{R}$.

Proof Let V be an open neighborhood of σ_f in $U \times \mathbb{R}$ on which ω vanishes and let W be another open neighborhood of σ_f (which is closed in V) such that $\text{cl}(W) \cap (U \times \mathbb{R}) \subset V$. The set $O = (U \times \mathbb{R}) \setminus \text{cl}(W)$ is open and since it is disjoint from σ_f, $\Gamma_f^- \subset O$ is an n-dimensional submanifold of O with $\text{bd}(\Gamma_f^- \cap O) = \Gamma_f \cap O$. Since $\text{supp}(d\omega) \cap \Gamma_f^- \subset \Gamma_f^- \cap O$ and $\text{supp}(\omega) \cap \Gamma_f \subset \Gamma_f \cap O$, by the usual Stokes theorem, we have

$$\int_{\Gamma_f} d\omega = \int_{\Gamma_f^- \cap O} d\omega = \int_{\Gamma_f \cap O} \omega = \int_{\Gamma_f} \omega. \quad \blacktriangledown$$

The purpose of the next two lemmas is to construct approximations to $d\omega$ and ω if ω does not vanish near σ_f. For this we need translates of bump functions with control on their derivatives.

7.2.22 Lemma *Let* C *be a box (rectangular parallelipiped) in* $\mathbb{R}^n$ *of edge lengths* $2\ell_i$ *and let* D *be the box with the same center as* C *but of edge lengths* $4\ell_i/3$. *There exists a* C^∞ *function* $\varphi : \mathbb{R}^n \to [0, 1]$ *which is* 1 *on* $\mathbb{R}^n \setminus C$, 0 *on* D *and* $|\partial\varphi/\partial x^i| \le A/\ell_i$, *for a constant* A *independent of* ℓ_i.

Proof Assume we have found such a function $\varphi : \mathbb{R} \to [0, 1]$ for $n = 1$. Then $\psi(x^1,\ldots, x^n) = \varphi(x^1) \cdots \varphi(x^n)$ is the desired function.

The function φ is found in the following way. Let $a = 2\ell/3$, $\varepsilon = \ell/3$ and choose an integer N such that $2/N < \varepsilon$. Let $h : \mathbb{R} \to [0, 1]$ be a bump function that is equal to 1 for $|t| < 1/2$ and that vanishes for $|t| > 1$. Then $f : \mathbb{R} \to [0, 1]$, defined by $f(t) = 1 - h(t)$ is a C^∞ function vanishing for $|t| < 1/2$ and equal to 1 for $|t| > 1$. Let $f_n(t) = f(nt)$ for all positive integers n and note that

$$|f_n'(t)| = n\,|f'(nt)| \le Cn.$$

Define the C^∞ function

$$\varphi(t) = \prod f_N\left(t - \frac{z}{2N}\right),$$

where the product is taken over integers z such that $|z| < 2Na + 1$. Note that if $|t| < a + 1/4N$ and $z \in \mathbb{Z}$ is chosen such that $|t - z/2N| < 1/4N$, then $f_N(t - z/2N) = 0$ and $|z| \le 2N|t| + 1/2 < 2Na + 1$, so that $\varphi(t) = 0$. Similarly if $|t| > a + 2/N$ and $|z| < 2Na + 1$, then $|t - z/2N| \ge |t| - |z|/2N > 1/N$ so that $\varphi(t) = 1$.

Finally, let $|t_0 - a| < 2/N$ and let $z_0 \in \mathbb{Z}$ be such that $|t_0 - z_0/2N| < 1/N$. All factors $f_N(t_0 - z/2N)$ are one in a neighborhood of t_0, unless $|t_0 - z/2N| \le 1/N$. In that case we have the inequality $|z - z_0| \le |z - 2Nt_0| + |2Nt_0 - z_0| \le 3$. Hence at most seven factors in the product are not identically 1 in a neighborhood of t_0. Hence

$$|\varphi'(t_0)| \le 7CN = A/\varepsilon. \ \blacktriangledown$$

7.2.23 Lemma *Let* K *be a compact subset of* σ_f, *the singular set of* f. *For every* $\varepsilon > 0$ *there is a neighborhood* U_ε *of* K *in* $U \times \mathbb{R}$ *and a* C^∞ *function* $\varphi_\varepsilon : U \times \mathbb{R} \to [0, 1]$, *which vanishes on a neighborhood of* K *in* U_ε, *is one on the complement of* U_ε, *and is such that*

$$\text{(i)}\ \ \operatorname{vol}(U_\varepsilon)\left[\sup_{x \in \mathbb{R}^n}\left|\frac{\partial \varphi_\varepsilon(x)}{\partial x^i}\right|\right] \le \varepsilon, \quad i = 1, \ldots, n, \ \ \textit{and}$$

(ii) $\operatorname{vol}(U_\varepsilon) \le \varepsilon$ *and* $q(U_\varepsilon \cap \rho_f) \le \varepsilon$, *where* q *is the measure on* ρ_f *associated with the volume form* $\lambda \in \Omega^{n-1}(\rho_f)$, *and* $\operatorname{vol}(U_\varepsilon)$ *is the Lebesgue measure of* U_ε *in* $\mathbb{R}^n$.

Proof Partition $\mathbb{R}^{n-1}$ by closed cubes D of edge length $4\ell/3$, $\ell \le 1$. At most 2^n such cubes can meet at a vertex. The set $\pi(K)$, where $\pi : U \times \mathbb{R} \to U$ is the projection, can be covered by finitely many open cubes C of edge length 2ℓ, each one of these cubes containing a cube D and having the same center as C. Since $\pi(K)$ and K have measure zero, choose ℓ so small that for given $\delta > 0$,

(i) the $(n-1)$-dimensional volume of $\bigcup_{i=1,\ldots,L} C_i$ is smaller than or equal to δ; and

(ii) $q(\pi^{-1}(\bigcup_{i=1,\ldots,L} C_i) \cap \rho_f) \le \delta$.

Since f is locally Hölder and $\pi(K)$ is compact, there exist constants $0 < \alpha \le 1$ and $k > 0$ such that $|f(x) - f(y)| \le k\|x - y\|^\alpha$ for $x, y \in \pi(K)$. We can assume $k \ge 1$ without loss of generality. In each of the sets $\pi^{-1}(C_i) = C_i \times \mathbb{R}$, choose a box P_i with base C_i and height $(2k\ell)^{1/\alpha}$ such that $\pi(K)$ is covered by parallelipeds P_i' with the same center as P_i and edge lengths equal to two-thirds of the edge lengths of P_i.

Let $V = \bigcup_{i=1,\ldots,L} P_i$. Then $\pi(V) = \bigcup_{i=1,\ldots,L} C_i$ and since at most 2^n of the P_i intersect

$$\operatorname{vol}(V) = 2k\ell 2^n \operatorname{vol}(\pi(V)) \le 2^{n+1}k\ell\delta \le 2^{n+1}k\delta$$

and

$$q(V \cap \rho_f) \le \delta.$$

By the previous lemma, for each P_i there is a C^∞ function $\varphi_i : U \times \mathbb{R} \to [0, 1]$ that vanishes on P_i', is equal to 1 on the complement of P_i, and $\sup_{x \in \mathbb{R}^n} \|\partial\varphi_i/\partial x^j\| \le A/\ell$. Let $\varphi = \Pi_{i=1,\ldots,L}\varphi_i$. Clearly $\varphi : U \times \mathbb{R} \to [0, 1]$ is C^∞, vanishes in a neighborhood of K and equals one in the complement of V. But at most 2^n of the P_i can intersect, so that

$$\left|\frac{\partial\varphi}{\partial x^j}\right| = \left|\sum_{i=1}^{L} \frac{\partial\varphi_i}{\partial x^j} \prod_{k \ne i} \varphi_k\right| \le 2^n A/\ell, \; j = 1, \ldots, n.$$

Hence

$$\operatorname{vol}(V)\left[\sup_{x \in \mathbb{R}^n} \left|\frac{\partial\varphi}{\partial x^j}\right|\right] \le 2^{n+1} k \ell \delta 2^n A/\ell = 2^{2n+1} k \delta A.$$

Now let $\delta = \min\{\varepsilon, \varepsilon/2^{2n+1}kA\}$, $\varphi_\varepsilon = \varphi$, and $U_\varepsilon = V$. ▼

Proof of Equation (1) Let

$$\omega = \sum_{i=1}^{n} \omega^i \, dx^1 \wedge \cdots \wedge (dx^1)^\wedge \wedge \cdots \wedge dx^n, \quad \mathbf{d}\omega = b\mathbf{d}x^1 \wedge \cdots \wedge dx^n,$$

and $i^*\omega = a\lambda$. Then ω^i, b, and a are continuous and bounded on $U \times \mathbb{R}$ and ρ_f respectively; i.e., $|\omega^i(x)| \le M$, $|b(x)| \le N$ for $x \in U \times \mathbb{R}$ and $|a(y)| \le N$ for $y \in \rho_f$, where $M, N > 0$ are constants. Let U_ε and φ_ε be given by the previous lemma applied to $\operatorname{supp}(\omega) \cap \sigma_f$. But $\varphi_\varepsilon\omega$ vanishes in a neighborhood of σ_f and lemma **7.2.21** is applicable; that is

$$\int_{\Gamma_f^-} \mathbf{d}(\varphi_\varepsilon\omega) = \int_{\Gamma_f} \varphi_\varepsilon\omega. \tag{2}$$

We have

$$\left|\int_{\Gamma_f} \omega - \int_{\Gamma_f} \varphi_\varepsilon\omega\right| \le \left|\int_{\rho_f} a(1 - \varphi_\varepsilon)\lambda\right| \le N q(U_\varepsilon \cap \rho_t) \le N\varepsilon$$

and

$$\left|\int_{\Gamma_f^-} \mathbf{d}\omega - \int_{\Gamma_f^-} \mathbf{d}(\varphi_\varepsilon\omega)\right| \le \left|\int_{\Gamma_f^-} (\mathbf{d}\omega - \varphi_\varepsilon \mathbf{d}\omega)\right| + \left|\int_{\Gamma_f^-} \mathbf{d}\varphi_\varepsilon \wedge \omega\right|$$

$$\le \left|\int_{\Gamma_f^-} b(1 - \varphi_\varepsilon) dx^1 \wedge \cdots \wedge dx^n\right| + \sum_{i=1}^{n} \int_{\Gamma_f^-} |\omega^i| \left|\frac{\partial\varphi_\varepsilon}{\partial x^i}\right| dx^1 \wedge \cdots \wedge dx^n$$

$$\le N \operatorname{vol}(U_\varepsilon) + M\left[\sum_{i=1}^{n} \sup_{x \in \mathbb{R}} \left|\frac{\partial\varphi_\varepsilon(x)}{\partial x^i}\right|\right] \operatorname{vol}(U_\varepsilon) \le N\varepsilon + Mn\varepsilon. \tag{4}$$

From (2)-(4) we get

$$\left| \int_{\Gamma_f^-} d\omega - \int_{\Gamma_f} \omega \right| \le (2N + nM)\varepsilon$$

for all $\varepsilon > 0$, which proves the equality. ■

In analysis it can be useful to have hypotheses on the smoothness of ω as well as on the boundary that are as weak as possible. Our proofs show that ω need only be C^1. An effective strategy for sharper results is to approximate ω by smooth forms ω_k so that both sides of Stokes' theorem converge as $k \to \infty$. A useful class of forms for which this works are those in Sobolev spaces, function spaces encountered in the study of partial differential equations. The Hölder nature of the boundary of N in Stokes' theorem is exactly what is needed to make this approximation process work. The key ingredients are approximation properties in M (which are obtained from those in $\mathbb{R}^n$) and the ***Calderón extension theorem*** to reduce approximations in N to those in $\mathbb{R}^n$. (Proofs of these facts may be found in Stein [1970], Marsden [1973], and Adams [1975].) ✍

☞ SUPPLEMENT 7.2C
Stokes' Theorem on Chains

In algebraic topology it is of interest to integrate forms over images of simplexes. This box adapts Stokes' theorem to this case. The result could be obtained as a corollary of the piecewise smooth Stokes Theorem, but we shall give a self-contained and independent proof.

7.2.24 Definition *The **standard** p-**simplex** is the closed set*

$$\Delta_p = \left\{ x \in \mathbb{R}^p \;\middle|\; 0 \le x^i \le 1, \sum_{i=1}^{p} x^i \le 1 \right\}.$$

*The **vertices** of* Δ_p *are the* p+1 *points* $v_0 = (0, ..., 0)$ $v_1 = (1, 0, ..., 0), ..., v_p = (0, ..., 0, 1)$. *Opposite to each* v_i *there is the* i-*th **face*** $\Phi_{p-1,i} : \Delta_{p-1} \to \Delta_p$ *given by (see Fig. 7.2.8):*

$$\Phi_{p-1,0}(y^1, ..., y^{p-1}) = \left(1 - \sum_{i=1}^{p-1} y^i, y^1, ..., y^{p-1} \right), \quad \textit{if} \quad i = 0$$

and

$$\Phi_{p-1,i}(y^1, ..., y^{p-1}) = (y^1, ..., y^{i-1}, 0, y^i, ..., y^{p-1}), \quad \textit{if} \ i \ne 0.$$

A C^k-***singular*** p-***simplex*** *on a* C^r-*manifold* M, $1 \le k \le r$, *is a* C^k-*map* $s : U \to M$, *where* U *is an open neighborhood of* Δ_p *in* $\mathbb{R}^p$. *The points* $s(v_0), ..., s(v_p)$ *are the **vertices** of the singular* p-*simplex* s *and the map* $s \circ \Phi_{p-1,i} : V \to M$, *for* V *an open neighborhood of* Δ_{p-1} *in*

$\mathbb{R}^{p-1}$ *and* $\Phi_{p-1,i}$ *extended by the same formula from* Δ_{p-1} *to* V, *is called the* i*-th* ***face*** *of the singular* p*-simplex* s. *A* C^k***-singular*** p***-chain*** *on* M *is a finite formal linear combination with real coefficients of* C^k*-singular* p*-simplexes. The* ***boundary*** *of a singular* p*-simplex* s *is the singular* (p–1)*-chain* ∂s *defined by*

$$\partial s = \sum_{i=0}^{p} (-1)^i\, s \circ \Phi_{p-1,i}$$

and that of a singular p*-chain is obtained by extending* ∂ *from the simplexes by linearity to chains. It is straightforward to verify that* $\partial \circ \partial = 0$ *using the relation* $\Phi_{p-1,j} \circ \Phi_{p-2,j} = \Phi_{p-2,i-1} \circ \Phi_{p-2,i-1}$ *for* $j < i$.

If $s : U \to M$, $\Delta_p \subset U$, *is a singular p-simplex,* $\omega \in \Omega^p(M)$, *and* $s^*\omega = a\, dx^1 \wedge \cdots \wedge dx^p \in \Omega^p(U)$, *the* ***integral*** *of* ω ***over*** s *is defined by*

$$\int_s \omega = \int_{\Delta_p} a\, dx^1 \cdots dx^p,$$

where the integral on the right is the usual integral in $\mathbb{R}^p$. *The* ***integral of*** ω ***over a*** p***-chain*** *is obtained by linear extension.*

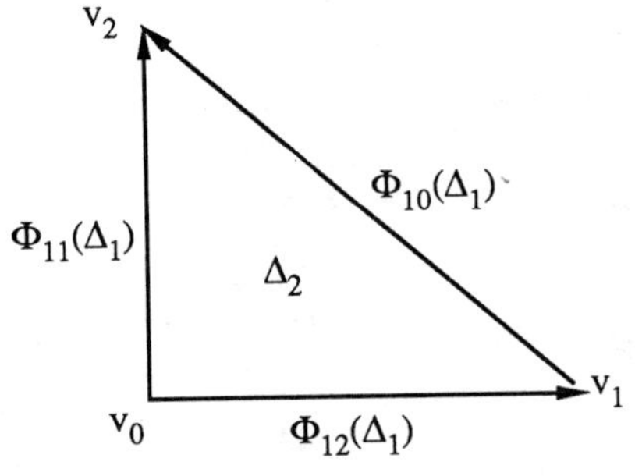

Figure 7.2.8

7.2.25 Stokes' Theorem on Chains *If* c *is any singular* p*-chain and* $\omega \in \Omega^{p-1}(M)$, *then*

$$\int_c d\omega = \int_{\partial c} \omega.$$

Proof By linearity it suffices to prove the formula if c = s, a singular p-simplex. If

$$s^*\omega = \sum_{j=1}^{p} (-1)^{j-1} \omega^j\, dx^1 \wedge \cdots \wedge \widehat{(dx^j)} \wedge \cdots \wedge dx^p,$$

then

$$d(s^*\omega) = \sum_{j=1}^{p-1} \frac{\partial \omega^j}{\partial x^j} dx^1 \wedge \cdots \wedge dx^p$$

and denoting the coordinates in a neighborhood V of Δ_{p-1} by $(y^1, \ldots, y^{p-1}) = y$, we get

$$\Phi^*_{p-1,0} s^* \omega(y) = \sum_{j=1}^{p} \omega^j \left(1 - \sum_{i=1}^{p-1} y^i, y^1, \ldots, y^{p-1} \right) dy^1 \wedge \cdots \wedge dy^{p-1}, \text{ if } i = 0$$

and

$$\Phi^*_{p-1,i} s^* \omega(y) = (-1)^{i-1} \omega^i (y^1, \ldots, y^{i-1}, 0, y^i, \ldots, y^{p-1}) dy^1 \wedge \cdots \wedge dy^{p-1}, \text{ if } i \neq 0.$$

Thus the formula in the statement becomes

$$\sum_{j=1}^{p} \int_{\Delta_p} \frac{\partial \omega^j(x)}{\partial x^j} dx^1 \ldots dx^p = \sum_{j=1}^{p} \int_{\Delta_{p-1}} \left[\omega^j \left(1 - \sum_{i=1}^{p-1} y^i, y^1, \ldots, y^{p-1} \right) - \omega^j (y^1, \ldots, y^{j-1}, 0, y^j, \ldots, y^{p-1}) \right] dy^1 \ldots dy^{p-1}. \quad (1)$$

By Fubini's theorem, each summand on the left hand side of (1) equals

$$\int_{\Delta_p} \frac{\partial \omega^j(x)}{\partial x^j} dx^1 \ldots dx^p = \int_{\Delta_{p-1}} \left(\int_0^{1-\sum_{k\neq j} x^k} \frac{\partial \omega^j}{\partial x^j} dx^j \right) dx^1 \ldots (dx^j)^\wedge \ldots dx^p$$

$$= \int_{\Delta_{p-1}} \left[\omega^j \left(x^1, \ldots, x^{j-1}, 1 - \sum_{k \neq j} x^k, x^{j+1}, \ldots, x^p \right) - \omega^j (x^1, \ldots, x^{j-1}, 0, x^{j+1}, \ldots, x^p) \right] dx^1 \ldots (dx^j)^\wedge \ldots dx^p.$$

Break up this integral as a difference of two terms. In the first term perform the change of variables $(y^1, \ldots, y^{p-1}) \mapsto (x^2, \ldots, x^{j-1}, 1 - \Sigma_{k \neq j} x^k, x^{j+1}, \ldots, x^p)$ which has Jacobian equal to $(-1)^j$, use the change of variables formula from calculus in the multiple integral involving the absolute value of the Jacobian, and note that $x^1 = 1 - \Sigma_{i=1, \ldots, p-1} y^i$. In the second term perform the change of variables $(y^1, \ldots, y^{p-1}) \mapsto (x^1, \ldots, x^{j-1}, x^{j+1}, \ldots, x^p)$ which has Jacobian equal to one. Then we get

$$\int_{\Delta_p} \frac{\partial \omega^j(x)}{\partial x^j} dx^1 \ldots dx^p = \int_{\Delta_{p-1}} \left[\omega^j \left(1 - \sum_{i=1}^{p-1} y^i, y^1, \ldots, y^{p-1} \right) - \omega^j (y^1, \ldots, y^{j-1}, 0, y^j, \ldots, y^{p-1}) \right] dy^1 \ldots dy^{p-1}$$

and formula (1) is thus proved for each corresponding summand. ■

Instead of singular p-chains one can consider ***infinite singular*** p-***chains*** defined as infinite formal sums with real coefficients $\sum_{i \in I} a_i S_i$ such that for each $i \in I$ the family of sets $\{S_i(\Delta_p) \mid a_i \neq 0\}$ is locally finite, i.e., each $m \in M$ has a neighborhood intersecting only finitely many (or no) sets of this family. On compact manifolds only finitely many coefficients in an infinite singular p-chain are non-zero and thus infinite singular p-chains are singular p-chains. The statement and the proof of Stokes' theorem on chains remain unchanged if c is an infinite p-chain and $\omega \in \Omega^{\pi-1}(M)$ has compact support.

Exercises

7.2A Let M and N be oriented n-manifolds with boundary and $f : M \to N$ an orientation-preserving diffeomorphism. Show that the change-of-variables formula and Stokes' theorem imply that $f^* \circ \mathbf{d} = \mathbf{d} \circ f^*$.

7.2B Let M be a compact orientable boundaryless n-manifold and $\alpha \in \Omega^{n-1}(M)$. Show that $\mathbf{d}\alpha$ vanishes at some point.

7.2C Let M be a compact $(n + 1)$-dimensional manifold with boundary, $f : \partial M \to N$ a smooth map and $\omega \in \Omega^n(N)$ where $\mathbf{d}\omega = 0$. Show that if f extends to M, then $\int_{\partial M} f^*\omega = 0$.

7.2D Let (M, μ) be a volume manifold with $\partial M = \varnothing$.

(i) Show that the divergence of a vector field X is uniquely determined by the condition

$$\int_M f(\operatorname{div} X)\mu = -\int_M (\mathcal{L}_X f)\mu \qquad (*)$$

for any f with compact support.

(ii) What does (*) become if M is compact with boundary?

(iii) $X(x, y, z) = (y, -x, 0)$ defines a vector field on S^2. Calculate div X.

7.2E Let M be a paracompact manifold with boundary. Show that there is a positive smooth function $f : M \to [0, \infty[$ with 0 a regular value, such that $\partial M = f^{-1}(0)$. (*Hint:* First do it locally and then patch the local functions together with a partition of unity.)

7.2F Let M be a boundaryless manifold and $f : M \to \mathbb{R}$ a C^∞ mapping having a regular value a. Show that $f^{-1}([a, \infty[)$ is a manifold with boundary $f^{-1}(a)$.

7.2G Let $f : M \to N$ be a C^∞ mapping, $\partial M \neq \varnothing$, $\partial N \neq \varnothing$, and let $P \subset N$ be a submanifold of N. Assume that $f \pitchfork P$, $(f \mid \partial M) \pitchfork P$ and that in addition one of the following conditions hold.

(i) P is boundaryless and $P \subset \text{Int}\, N$; or

(ii) $\partial P \neq \varnothing$ and $\partial P \subset \partial N$; or

(iii) $\partial P \neq \varnothing$, $f \pitchfork \partial P$, and $(f \mid \partial M) \pitchfork \partial P$.

Show that $f^{-1}(P)$ is a submanifold of M whose boundary equals $\partial f^{-1}(P) = f^{-1}(P) \cap \partial M$, in case (i), and $\partial f^{-1}(P) = f^{-1}(\partial P)$ in cases (ii) and (iii). If all manifolds are finite dimensional, show that $\dim M - \dim f^{-1}(P) = \dim N - \dim P$. Formulate and prove the statement replacing this equality between dimensions for infinite dimensional manifolds. (*Hint:* At the boundary, work with a boundary chart using the technique in the proof of **3.5.11**.)

7.2H Without some kind of transversality conditions on $f \mid \partial M$ for $f : M \to N$ a smooth map, even if $f \pitchfork P$, where P is a submanifold of N (like the ones in the previous exercise), $f^{-1}(P)$ is in general *not* a submanifold. For example, let $M = \mathbb{R}^2_+$, $N = \mathbb{R}$, $P = \{0\}$ and $f(x, y) = y + \chi(x)$ for a smooth function $\chi : \mathbb{R} \to \mathbb{R}$. Show that f is a smooth surjective submersion. Find the conditions under which $f \mid \partial M$ has 0 as a regular value. Construct a smooth function χ for which these conditions are violated and $f^{-1}(0)$ is not a manifold. (*Hint:* Take for χ a smooth function which has infinitely many zeros converging to zero.)

7.2I (i) Show that if M is a boundaryless manifold, there is a connected manifold N with $M = \partial N$. (*Hint:* Think of semi-infinite cylinders.)

(ii) Construct an example for (i) in which M is compact but N cannot be chosen to be compact. (*Hint:* Assume $\dim M = 0$.)

7.2J Let M be a manifold, X a smooth vector field on M with flow F_t and $\alpha \in \Omega^k(M)$. We call α an ***invariant* k-*form*** of X when $\mathbf{L}_X\alpha = 0$. Prove the following.

Poincaré-Cartan Theorem α *is an invariant* k-*form of* X *iff for all oriented compact* k-*manifolds with boundary* $(V, \partial V)$ *and* C^∞ *mappings* $\varphi : V \to M$, *such that the domain of* F_t *contains* $\varphi(V)$, $0 \le t \le T$, *we have* $\int_V (F_t \circ \varphi^*\alpha = \int_V \varphi^*\alpha$. (*Hint:* For the converse show that the equality between integrals implies $(F_t \circ \varphi^*\alpha = \varphi^*\alpha$; then differentiate relative to t.)

7.2K Let X be a vector field on a manifold M and α, β invariant forms of X. (See Exercise **7.2J**.) Prove the following.

(i) $\mathbf{i}_X\alpha$ is an invariant form of X.

(ii) $\mathbf{d}\alpha$ is an invariant form of X.

(iii) $\mathbf{L}_X\gamma$ is closed iff $\mathbf{d}\gamma$ is an invariant form, for any $\gamma \in \Omega^k(M)$.

(iv) $\alpha \wedge \beta$ is an invariant form of X.

(v) Let $\mathcal{A}_X$ denote the invariant forms of X. Then $\mathcal{A}_X$ is a $\wedge$ subalgebra of $\Omega(M)$, which is closed under $\mathbf{d}$ and $\mathbf{i}_X$.

7.2L Let X be a vector field on a manifold M with flow F_t and $\alpha \in \Omega^k(M)$. Then α is called a ***relatively invariant*** **k-*form of*** X if $L_X\alpha$ is closed. Prove the following
Poincaré-Cartan Theorem *α is a relatively invariant (k–1)-form of X iff for all oriented compact k-manifolds with boundary $(V, \partial V)$ and C^∞ maps $\varphi : V \to M$ such that the domain of F_t contains $\varphi(V)$ for $0 \le t \le T$, we have*

$$\int_{\partial V} (F_t \circ \varphi \circ i)^* \alpha = \int_{\partial V} (\varphi \circ i)^* \alpha.$$

where $i : \partial V \to V$ is the inclusion map.

7.2M If $X \in \mathfrak{X}(M)$, let $\mathcal{A}_X$ be the set of all invariant forms of X, $\mathcal{R}_X$ the set of all relatively invariant forms of X, $\mathcal{C}$ the set of all closed forms in $\Omega(M)$, and $\mathcal{E}$ the set of all exact forms in $\Omega(M)$. Show that

(i) $\mathcal{A}_X \subset \mathcal{R}_X$, $\mathcal{E} \subset \mathcal{C} \subset \mathcal{R}_X$, $\mathcal{A}_X$ is a differential subalgebra of $\Omega(M)$, but $\mathcal{R}_X$ is only a real vector subspace.

(ii) $0 \to \mathcal{A}_X \xrightarrow{i} \Omega(M) \xrightarrow{L_X} \Omega(M) \xrightarrow{\pi} \Omega(M)/\mathrm{Im}(L_X) \to 0$ is exact.

(iii) $0 \to \mathcal{C} \xrightarrow{i} \mathcal{R}_X \xrightarrow{\mathbf{d}} \mathcal{A}_X \xrightarrow{\pi} \mathcal{A}_X/\mathcal{E} \cap \mathcal{A}_X \to 0$ is exact.

(iv) $\mathbf{d}(\mathcal{A}_X) \subset \mathcal{A}_X$ and $\mathbf{i}_X(\mathcal{A}_X) \subset \mathcal{A}_X$.

7.2N Smale-Sard Theorem for manifolds with boundary: *Let* M *and* N *be* C^k *manifolds, where* M *is Lindelöf, having a boundary* ∂M, *and* N *is boundaryless. Let* $f : M \to N$ *be a* C^k *Fredholm map and let* $\partial f = f \mid \partial M$. *If* $k > \max(0, \mathrm{index}(T_x f))$ *for every* $x \in M$, *show that* $\mathcal{R}_f \cap \mathcal{R}_{\partial f}$ *is residual in* N.

7.2O The Boundaryless Double Let M be a manifold with boundary. Show that the topological space obtained by identifying the points of ∂M in the disjoint union of M with itself is a boundaryless manifold in which M embeds, called the ***boundaryless double*** of M. (*Hint:* Glue together the two boundary charts.)

7.2P Let M be a manifold with $\partial M \neq \varnothing$. Assume M admits partitions of unity. Show that ∂M is orientable and hence by Exercise **6.5Q**, the algebraic normal bundle $\nu(\partial M) = (TM \mid \partial M)/T(\partial M)$ is trivial. (*Hint:* Use Proposition **6.5.8**. Locally $n(m) = \partial/\partial x^n$ for $m \in \partial M$; glue these together.)

7.2Q Collars Let M be a manifold with boundary. A ***collar*** for M is a diffeomorphism of $\partial M \times [0, 1[$ onto an open neighborhood of ∂M in M that is the identity on ∂M.

(i) Show that a manifold with boundary and admitting partitions of unity has a collar. (*Hint:* Via a partition of unity, construct a vector field on M that points inward when restricted to ∂M. Then look at the integral curves starting on ∂M to define the collar.)

(ii) Let $\varphi_1 : \partial M \times [0, 1[\to M$, $i = 0, 1$ be two collars. Show that φ_0 and φ_1 are ***isotopic***, i.e., there is a smooth map $H :]-\varepsilon, 1+\varepsilon[\times \partial M \times [0, 1[\to M$ such that $H(0, m, t) = \varphi_0(m, t)$, $H(1, m, t) = \varphi_1(m, t)$ for all $(m, t) \in \partial M \times [0, 1[$ and that $H(s, \cdot, \cdot)$ is an embedding for all $s \in]-\varepsilon, 1+\varepsilon[$. (*Hint:* Let U_i be the image of φ_i, an open set in M containing ∂M. Let $X_i = \varphi_i{}^*(0, \partial/\partial t)$ and look at the flow of $(1-s)X_0 + sX_1$ on $U_0 \cap U_1$.) [Remark. It can be shown that φ_0 and φ_1 are diffeotopic using Thom's theorem of embedding of isotopies into diffeotopies. This then provides the basis of gluing manifolds together along their boundaries; see Hirsch [1976], Chapter 8, for proofs and the preamble to the next exercise for a discussion.]

(iii) Let N be a submanifold of M such that $\partial N = N \cap \partial M$ and T_nN is not a subset of $T_n(\partial M)$ for all $n \in \partial N$. Show that ∂M has a collar which restricts to a collar of ∂N in N.

7.2R Let M and N be manifolds with boundary and let $\varphi : \partial M \to \partial N$ be a diffeomorphism. Form the topological space $M \cup_\varphi N$ which is the quotient of the disjoint union of M with N by the equivalence relation which identifies m with $\varphi(m)$. Let V be the image of ∂M and ∂N in $M \cup_\varphi N$.

(i) Use collars to construct a homeomorphism of a neighborhood U of V with the space $]-1, 1[\times V$ which maps V pointwise to $V \times \{0\}$ and which maps $V \cap M$ and $V \cap N$ diffeomorphically onto $V \times]0, 1[$ and $V \times]-1, 0[$ respectively. Construct a differentiable structure out of those on M, N, and U.

The "uniqueness theorem of glueing" states that the differentiable structures on the space $M \cup_\varphi N$ obtained in (i) by making various choices are all diffeomorphic. The rest of this exercise uses this fact.

(ii) Two compact boundaryless manifolds M_1, M_2 are called ***cobordant*** if there is a compact manifold with boundary N, called the ***cobordism***, such that ∂N equals the disjoint union of M_1 with M_2. Show that "cobordism" is an equivalence relation. (*Hint:* For transitivity, glue the manifolds along one common component of their boundaries.)

(iii) Show that the operation of disjoint union induces the structure of an abelian group on the set of cobordism classes in which each element has order two. (*Hint:* The zero element is the class of any compact manifold which is the boundary of another compact manifold.)

(iv) Show that the operation of taking products of manifolds induces a multiplicative law on the set of cobordism classes, thus making this set $\mathfrak{N}$ a ring.

(v) Repeat parts (ii) and (iii) for oriented manifolds obtaining a graded ring (that is, $[M, \mu]\cdot[N, \nu] = (-1)^{\dim M + \dim N}[N, \nu]\cdot[M, \mu]$), the graded ring of oriented cobordism classes $\mathfrak{O}$. Are the elements of $\mathfrak{O}$ still of order two relative to addition?

(vi) Denote by $\mathfrak{N}^n$, $\mathfrak{O}^n$, the cobordism classes of a given dimension. Show that $\mathfrak{N}^0 =$

$\mathbb{Z}/2\mathbb{Z}$, $\mathfrak{O}^0 = \mathbb{Z}$, $\mathfrak{N}^1 = \mathfrak{O}^1 = 0$.

(vii) Assume M and N are boundaryless manifolds, M compact, and P a boundaryless submanifold of N. Assume $f, g : M \to N$ are smoothly homotopic maps such that $f \pitchfork P$ and $g \pitchfork P$. Show that $f^{-1}(P)$ and $g^{-1}(P)$ are cobordant. (*Hint:* Choose a smooth homotopy H transverse to P. What is $\partial H^{-1}(P)$?)

7.2S (i) Let χ be a vector field density on a finite dimensional manifold M, i.e., $\chi = X \otimes \rho$ for $X \in \mathfrak{X}(M)$ and $\rho \in |\Omega(M)|$. Recall from exercise **6.5P** that the density $\operatorname{div}\chi$, defined to be $(\operatorname{div}_\rho X)\rho$, is independent of the representation of χ as $X \otimes \rho$. Show that for any $f \in \mathcal{F}(M)$, $Y \in \mathfrak{X}(M)$, and any diffeomorphism $\varphi : M \to M$ we have

$$\varphi^*(\operatorname{div}\chi) = \operatorname{div}(^*\chi),\ \ \mathbf{L}_Y(\operatorname{div}\chi) = \operatorname{div}(\mathbf{L}_Y\chi),$$

and

$$\operatorname{div}(f\chi) = \mathbf{d}f \cdot \chi + f \operatorname{div}\chi.$$

(ii) If M is paracompact and Riemannian, phrase Gauss' theorem for vector field densities.

(iii) Let $\alpha \in \Omega^{k-1}(M)$ and τ be a tensor density of type (k, 0) which is completely antisymmetric. Let $\alpha \cdot \tau$ denote the contraction of α with the first k–1 indices of the tensor part of τ producing a vector field density. Define the ***contravariat exterior derivative*** $\partial\tau$ by requiring the following relation for all $\alpha \in \Omega^{k-1}(M)$:

$$\operatorname{div}(\alpha \cdot \tau) = \mathbf{d}\alpha \cdot \tau + \alpha \cdot \partial\tau,$$

where $\mathbf{d}\alpha \cdot \tau$ and $\alpha \cdot \partial\tau$ means contraction on all indices. Show that if $\tau = t \otimes \rho$, where locally

$$t = t^{i_1 \cdots i_k} \frac{\partial}{\partial x^{i_1}} \wedge \cdots \wedge \frac{\partial}{\partial x^{i_k}}, \quad \text{and} \quad \rho = |dx^1 \wedge \cdots \wedge dx^n|,$$

then the local expression of $\partial\tau$ is

$$\partial\tau = \frac{\partial}{\partial x^j}(t^{i_1 \cdots i_{k-1} j}) \frac{\partial}{\partial x^{i_1}} \wedge \cdots \wedge \frac{\partial}{\partial x^{i_{k-1}}} \otimes |dx^1 \wedge \cdots \wedge dx^n|$$

Show that $\partial^2 = 0$.

(iv) Prove the following properties of ∂:

$$\partial(\tau \wedge \sigma) = \partial\tau \wedge \sigma + (-1)^k \tau \wedge \partial\sigma, \ \text{and}\ \ \mathbf{L}_X \partial\tau = \partial \mathbf{L}_X \tau,\ \ \varphi^*\partial\tau = \partial\varphi^*\tau$$

where τ is a (k , 0)-tensor density, σ is a $(\ell, 0)$-tensor density, $X \in \mathfrak{X}(M)$, and $\varphi : M \to M$ is a diffeomorphism. Show that if χ is a vector field density, $\partial\chi =$

div χ.

(v) Let $\mathbf{j}_X\tau = X \wedge \tau$ for $X \in \mathfrak{X}(M)$. Show that $\mathbf{i}_X\alpha \cdot \tau = \alpha \cdot \mathbf{j}_X\tau$ for any $\alpha \in \Omega^{k+1}(M)$, $X \in \mathfrak{X}(M)$ and τ is a completely antisymmetric (k, 0) tensor density. Prove the analog of **Cartan's formula:** $\mathbf{L}_X = \mathbf{j}_X \circ \partial + \partial \circ \mathbf{j}_X$. (*Hint:* Integrate the defining relation in (iii) for α any form with support in Int M and extend the formula by continuity to ∂M.)

(vi) Formulate and prove a global formula for $\partial\tau(\alpha_1, ..., \alpha_{k-1})$, τ a (k, 0)-tensor density, analogous to Palais' formula **6.4.11**(ii).

7.2T **Prüfer Manifold** Let $\mathbb{R}_d$ denote the set $\mathbb{R}$ with the discrete topology; it is thus a zero dimensional manifold. Let $P = (\mathbb{R}^2_+ \times \mathbb{R}_d)/R$, where $\mathbb{R}^2_+ = \{(x, y) \mid y \geq 0\}$ and R is the equivalence relation: (x, y, a) R (x′, y′, a′) iff $[(y = y' > 0$ and $a + xy = a' + x'y')$ or $(y = y' = 0$ and $a = a'$, $x = x')]$. Show that P is a Hausdorff two dimensional manifold, $\partial P \neq \varnothing$, and ∂P is a disjoint union of uncountably many copies of $\mathbb{R}$. Show that P is not paracompact.

7.2U In the notation of Supplement **7.2C**, verify that $\partial \circ \partial = 0$.

§7.3 *The Classical Theorems of Green, Gauss, and Stokes*

This section obtains these three classical theorems as a consequence of Stokes' theorem for differential forms. We begin with Green's theorem, which relates a line integral along a closed piecewise smooth curve C in the plane $\mathbb{R}^2$ to a double integral over the region D enclosed by C. (Piecewise smooth means that the curve C has only finitely many corners.) Recall from advanced calculus that the ***line integral*** of a one-form $\omega = P\,dx + Q\,dy$ along a curve C parametrized by γ: $[a, b] \to \mathbb{R}^2$ is defined by

$$\int_C \omega = \int_b^a \{P(\gamma_1(t), \gamma_2(t))\gamma_1'(t) + Q(\gamma_1(t), \gamma_2(t))\gamma_2'(t)\}\,dt; \quad \text{i.e.} \quad \int_C \omega = \int_a^b \gamma^*\omega$$

7.3.1 Green's Theorem *Let* D *be a closed bounded region in* $\mathbb{R}^2$ *bounded by a closed positively oriented piecewise smooth curve* C. *(Positively oriented means the region* D *is on your left as you traverse the curve in the positive direction.) Suppose* $P : D \to \mathbb{R}$ *and* $Q : D \to \mathbb{R}$ *are* C^1. *Then*

$$\int_C P\,dx + Q\,dy = \iint_D \left(\frac{\partial Q}{\partial x} - \frac{\partial P}{\partial y}\right) dx\,dy.$$

Proof We assume the boundary $C = \partial D$ is smooth. (The piecewise smooth case follows from the generalization of Stokes' theorem outlined in Supplement **7.2B**).

Let $\omega = P(x, y)dx + Q(x, y)dy \in \Omega^1(D)$. Since $\mathbf{d}\omega = (\partial Q/\partial x - \partial P/\partial y)\,dx \wedge dy$ and the measure associated with the volume $dx \wedge dy$ on $\mathbb{R}^2$ is the usual Lebesgue measure dx dy, the formula of the theorem is a restatement of Stokes' theorem for this case. ■

This theorem may be phrased in terms of the divergence and the outward unit normal. If C is given parametrically by $t \mapsto (x(t), y(t))$, then the outward unit normal is

$$\mathbf{n} = (y'(t), -x'(t))/\sqrt{x'(t)^2 + y'(t)^2} \qquad (1)$$

and the infinitesimal arc–length (the volume element of C) is $ds = \sqrt{x'(t)^2 + y'(t)^2}\,dt$. (See Fig. 7.3.1.) If $X = P\partial/\partial x + Q\partial/\partial y \in \mathfrak{X}(D)$, recall that $\operatorname{div} X = *\mathbf{d}{*}X^\flat = \partial P/\partial x + \partial Q/\partial y$.

7.3.2 Corollary *Let* D *be a region in* $\mathbb{R}^2$ *bounded by a closed piecewise smooth curve* C. *If* $X \in \mathfrak{X}(D)$, *then*

$$\int_C (\mathbf{X} \cdot \mathbf{n})\,ds = \iint_D (\operatorname{div} X)\,dx\,dy.$$

where $\int_C f\,ds$ *denotes the line integral of the function* f *over the positively oriented curve* C *and* $\mathbf{X} \cdot \mathbf{n}$ *is the dot product.*

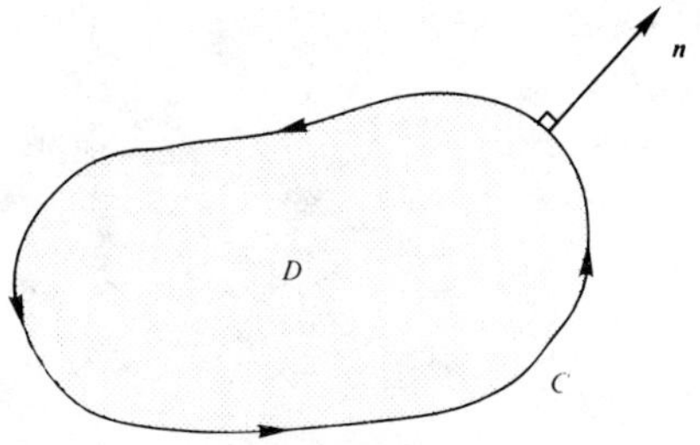

Figure 7.3.1

Proof Using formula (1) for $\mathbf{n}$ we have

$$\int_C (\mathbf{X} \cdot \mathbf{n})\,ds = \int_a^b [P(x(t)y(t))y'(t) - Q(x(t), y(t))x'(t)]dt = \int_C P\,dy - Q\,dx \qquad (2)$$

by the definition of the line integral. By **7.3.1**, (2) equals

$$\iint_D \left(\frac{\partial P}{\partial x} + \frac{\partial Q}{\partial y}\right) dx\,dy = \iint_D (\text{div } X)\,dx\,dy. \quad \blacksquare$$

Taking $P(x, y) = x$ and $Q(x, y) = y$ in Green's theorem, we get the following.

7.3.3 Corollary *Let* D *be a region in* $\mathbb{R}^2$ *bounded by a closed piecewise smooth curve* C. *The area of* D *is given by*

$$\text{area}(D) = \frac{1}{2}\int_C x\,dy - y\,dx.$$

The classical Stokes theorem for surfaces relates the line integral of a vector field around a simple closed curve C in $\mathbb{R}^3$ to an integral over a surface S for which $C = \partial S$. Recall from advanced calculus that the ***line integral*** of a vector field $\mathbf{X}$ in $\mathbb{R}^3$ over the curve $\sigma : [a, b] \to \mathbb{R}^3$ is defined by

$$\int_\sigma \mathbf{X} \cdot d\mathbf{s} = \int_a^b X(\sigma(t)) \cdot \sigma'(t)dt. \qquad (3)$$

The surface integral of a compactly supported two-form ω in $\mathbb{R}^3$ is defined to be the integral of the pull-back of ω to the oriented surface. If S is an oriented surface, $\mathbf{n}$ is called the ***outward unit normal*** at $x \in S$ if $\mathbf{n}$ is perpendicular to T_xS and $\{\mathbf{n}, \mathbf{e}_1, \mathbf{e}_2\}$ is a positively oriented basis of $\mathbb{R}^3$ whenever $\{\mathbf{e}_1, \mathbf{e}_2\}$ is a positively oriented basis of T_xS. Thus S is orientable iff the normal bundle to S, which has one-dimensional fiber, is trivial. In particular, the area element ν of S is given by **6.5.8**. That is

$$\nu(x)(\mathbf{v}_1, \mathbf{v}_2) = \mu(x)(\mathbf{n}, \mathbf{v}_1, \mathbf{v}_2), \text{ where } \mathbf{v}_1, \mathbf{v}_2 \in T_xS, \tag{4}$$

and $\mu = dx \wedge dy \wedge dz$. We want to express $\int_S \omega$ in a form familiar from vector calculus. Let $\omega = P\,dy \wedge dz + Q\,dy \wedge dz + R\,dx \wedge dy$ so that $\omega = *\,\mathbf{X}^\flat$, where

$$\mathbf{X} = P\frac{\partial}{\partial x} + Q\frac{\partial}{\partial y} + R\frac{\partial}{\partial z}.$$

Recall that $\alpha \wedge *\beta = \langle \alpha, \beta \rangle\, \mu$ so that letting $\alpha = \mathbf{n}^\flat$, and $\beta = \mathbf{X}^\flat$, we get $\mathbf{n}^\flat \wedge *\mathbf{X}^\flat = (\mathbf{X} \cdot \mathbf{n})\mu$. Applying both sides to $(\mathbf{n}, \mathbf{v}_1, \mathbf{v}_2)$ and using (2) gives

$$(\mathbf{n}^\flat \wedge *\, \mathbf{X}^\flat)(\mathbf{n}, \mathbf{v}_1, \mathbf{v}_2) = (\mathbf{X} \cdot \mathbf{n})\, \nu\, (\mathbf{v}_1, \mathbf{v}_2) \tag{5}$$

(the base point x is suppressed). The left side of (4) is $*\mathbf{X}^\flat(\mathbf{v}_1, \mathbf{v}_2)$ since $\mathbf{n}^\flat$ is one on $\mathbf{n}$ and zero on $\mathbf{v}_1$ and $\mathbf{v}_2$. Thus (4) becomes

$$*\,\mathbf{X}^\flat = (\mathbf{X} \cdot \mathbf{n})\nu\,. \tag{6}$$

Therefore
$$\int_S \omega = \int_S (\mathbf{X} \cdot \mathbf{n})\, dS = \int_S \mathbf{X} \cdot \mathbf{dS},$$

where dS, the measure on S defined by ν, is identified with a surface integral familiar from vector calculus.

A physical interpretation of $\int_S (\mathbf{X} \cdot \mathbf{n})dS$ may be useful. Think of $\mathbf{X}$ as the velocity field of a fluid, so $\mathbf{X}$ is pointing in the direction in which the fluid is moving across the surface S and $\mathbf{X} \cdot \mathbf{n}$ measures the volume of fluid passing through a unit square of the tangent plane to S in unit time. Hence the integral $\int_S (\mathbf{X} \cdot \mathbf{n})dS$ *is the net quantity of fluid flowing across the surface per unit time, i.e., the rate of fluid flow.* Accordingly, this integral is also called the ***flux*** of $\mathbf{X}$ across the surface.

7.3.4 Classical Stokes Theorem *Let* S *be an oriented compact surface in* $\mathbb{R}^3$ *and* $\mathbf{X}$ *a* C^1 *vector field defined on* S *and its boundary. Then*

$$\int_S (\operatorname{curl} \mathbf{X}) \cdot \mathbf{n}\, dS = \int_{\partial S} \mathbf{X} \cdot \mathbf{ds}.$$

where $\mathbf{n}$ *is the outward unit normal to* S (Fig. 7.3.2).

Proof First extend $\mathbf{X}$ via a bump function to all of $\mathbb{R}^3$ so that the extended $\mathbf{X}$ still has compact support. By definition, $\int_{\partial S} \mathbf{X} \cdot \mathbf{ds} = \int_{\partial S} \mathbf{X}^\flat$ where $^\flat$ denotes the index lowering action defined by the standard metric in $\mathbb{R}^3$. But $\mathbf{dX}^\flat = *(\text{curl } \mathbf{X})^\flat$ (see **6.4.3**(C)) so that by (3), (6), and Stokes' theorem,

$$\int_{\partial S} \mathbf{X} \cdot \mathbf{ds} = \int_{\partial S} \mathbf{dX}^\flat = \int_S *(\text{curl } \mathbf{X})^\flat = \int_S (\text{curl } \mathbf{X} \cdot \mathbf{n}) dS. \blacksquare$$

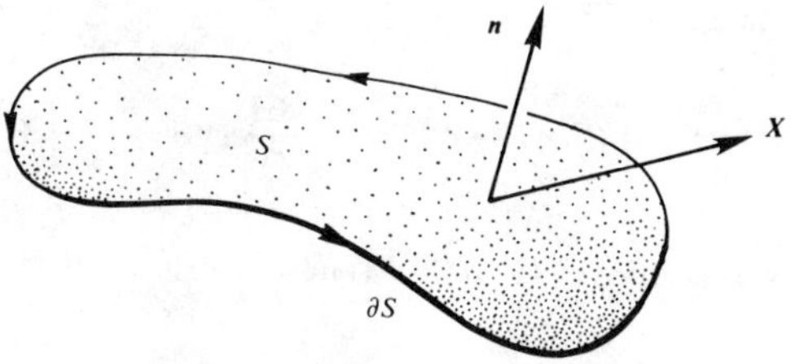

Figure 7.3.2

7.3.5 Examples

A The historical origins of Stokes' formula (6) are connected with Faraday's law, which is discussed in Chapter 8 and examble **B** below. In fluid dynamics, (5) is related to Kelvin's circulation theorem, to be discussed in §**8.2**. Here we concentrate on a physical interpretation of the curl operator. Suppose $\mathbf{X}$ represents the velocity vector field of a fluid. Let us apply Stokes' theorem to a disk D_r of radius r at a point $P \in \mathbb{R}^3$ (Fig. 7.3.3). We get

$$\int_{\partial D_r} \mathbf{X} \cdot \mathbf{ds} = \int_{\partial D_r} (\text{curl } \mathbf{X}) \cdot \mathbf{n} \, ds = (\text{curl } \mathbf{X} \cdot \mathbf{n})(Q)\pi r^2,$$

the last equality coming from the mean value theorem for integrals; here $Q \in D_r$ is some point given by the mean value theorem and πr^2 is the area of D_r. Thus

$$((\text{curl } \mathbf{X}) \cdot \mathbf{n})(P) = \lim_{r \to 0} \frac{1}{\pi r^2} \int_{\partial D_r} \mathbf{X} \cdot \mathbf{ds}. \tag{7}$$

The number $\int_C \mathbf{X} \cdot \mathbf{ds}$ is called the ***circulation*** of $\mathbf{X}$ around the closed curve C. It represents the net amount of turning of the fluid in a counterclockwise direction around C.

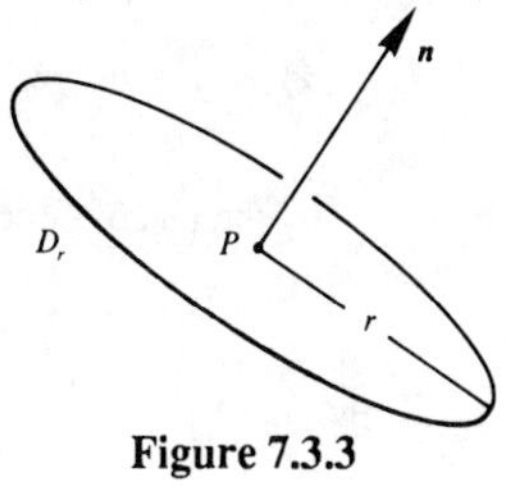

Figure 7.3.3

Formula (7) gives the following physical interpretation for curl $\mathbf{X}$, namely: (curl $\mathbf{X}$) • $\mathbf{n}$ *is the circulation of* $\mathbf{X}$ *per unit area on a surface perpendicular to* $\mathbf{n}$. The magnitude of (curl $\mathbf{X}$) • $\mathbf{n}$ is clearly maximized when $\mathbf{n} = (\text{curl } \mathbf{X})/\|\text{curl } \mathbf{X}\|$. Curl $\mathbf{X}$ is called the ***vorticity vector***.

B One of Maxwell's equations of electromagnetic theory states that if $\mathbf{E}(x, y, z, t)$ and $\mathbf{H}(x, y, z)$ represent the electric and magntic fields at time t, then $\nabla \times \mathbf{E} = -\partial\mathbf{H}/\partial t$, where $\nabla \times \mathbf{E}$ is computed by holding t fixed, and $\partial\mathbf{H}/\partial t$ is computed by holding x, y, and z constant. Let us use Stokes' theorem to determine what this means physically. Assume S is a surface to which Stokes' theorem applies. Then

$$\int_{\partial S} \mathbf{E} \cdot \mathbf{ds} = \int_S (\nabla \times \mathbf{E}) \cdot \mathbf{dS} = -\int_S \frac{d\mathbf{H}}{\partial t} \cdot \mathbf{dS} = -\frac{\partial}{\partial t}\int_S \mathbf{H} \cdot \mathbf{dS}.$$

(The last equality may be justified if $\mathbf{H}$ is C^1.) Thus we obtain

$$\int_{\partial S} \mathbf{E} \cdot \mathbf{ds} = -\frac{\partial}{\partial t}\int_S \mathbf{H} \cdot \mathbf{dS}. \tag{8}$$

Equality (8) is known as ***Faraday's law***. The quantity $\int_{\partial S} \mathbf{E} \cdot \mathbf{ds}$ represents the "voltage" around ∂S, and if ∂S were a wire, a current would flow in proportion to this voltage. Also $\int_S \mathbf{H} \cdot \mathbf{dS}$ is called the ***flux*** of $\mathbf{H}$, or the magnetic flux. Thus, Faraday's law says that *the voltage around a loop equals the negative of the rate of change of magnetic flux through the loop.*

C Let $\mathbf{X} \in \mathfrak{X}(\mathbb{R}^3)$. Since $\mathbb{R}^3$ is contractible, the Poincaré lemma shows that curl $\mathbf{X} = 0$ iff $\mathbf{X} = \text{grad } f$ for some function $f \in \mathcal{F}(\mathbb{R}^3)$. This in turn is equivalent (by Stokes' theorem) to either of the following: (i) for any oriented simple closed curve C, $\int_C \mathbf{X} \cdot \mathbf{ds} = 0$, or (ii) for any oriented simple curves C_1, C_2 with the same end points $\int_{C_1} \mathbf{X} \cdot \mathbf{ds} = \int_{C_2} \mathbf{X} \cdot \mathbf{ds}$.

The function f can be found in the following way:

$$f(x, y, z) = \int_0^x X^1(t, 0, 0)dt + \int_0^y X^2(x, t, 0)dt + \int_0^z X^3(x, y, t)dt. \tag{9}$$

Thus, for example, if

$$\mathbf{X} = y\frac{\partial}{\partial x} + (z\cos(yz) + x)\frac{\partial}{\partial y} + y\cos(yz)\frac{\partial}{\partial z},$$

then curl $\mathbf{X} = 0$ and so $\mathbf{X} = \text{grad } f$, for some f. Using the formula (9), one finds

$$f(x, y, z) = xy + \sin yz.$$

D The same arguments apply in $\mathbb{R}^2$ by the use of Green's theorem. Namely if

$$\mathbf{X} = X^1\frac{\partial}{\partial x} + X^2\frac{\partial}{\partial y} \in \mathfrak{X}(\mathbb{R}^2) \text{ and } \frac{\partial X^2}{\partial x} = \frac{\partial X^1}{\partial y},$$

then $\mathbf{X} = \text{grad } f$, for some $f \in \mathcal{F}(\mathbb{R}^2)$ and conversely.

E The following statement is again a reformulation of the Poincaré lemma: let $\mathbf{X} \in \mathfrak{X}(\mathbb{R}^3)$, then $\text{div } \mathbf{X} = 0$ iff $\mathbf{X} = \text{curl } \mathbf{Y}$ for some $\mathbf{Y} \in \mathfrak{X}(\mathbb{R}^3)$. ♦

7.3.6 Classical Gauss Theorem *Let Ω be a compact set with nonempty interior in $\mathbb{R}^3$ bounded by a surface S that is piecewise smooth. Then for $\mathbf{X}$ a C^1 vector field on $\Omega \cup S$,*

$$\int_\Omega (\text{div } \mathbf{X})\, dV = \int_S (\mathbf{X} \cdot \mathbf{n})\, dS, \tag{10}$$

where dV denotes the standard volume element (Lebesgue measure) in $\mathbb{R}^3$ (Fig. 7.3.4).

Proof Either use **7.2.10** or argue as in **7.3.4**. By (6),

$$\int_S (\mathbf{X} \cdot \mathbf{n})\, dS = \int_S * \mathbf{X}^\flat.$$

By Stokes' theorem, this equals

$$\int_\Omega \mathbf{d} * \mathbf{X}^\flat = \int_\Omega (\text{div } \mathbf{X})\, dV.$$

since $\mathbf{d} * \mathbf{X}^\flat = (\text{div } \mathbf{X})\, dx \wedge dy \wedge dz$. ■

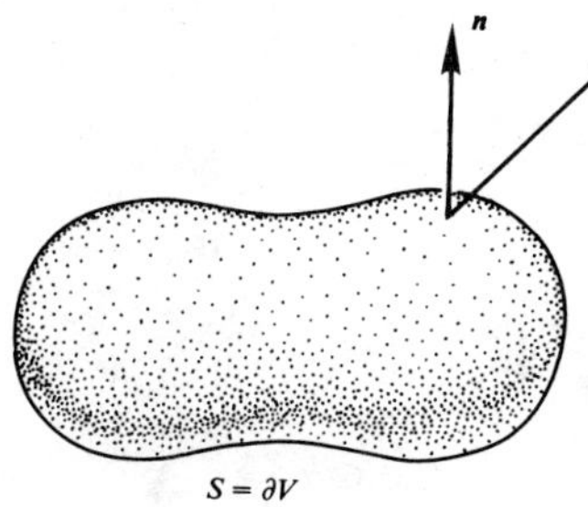

Figure 7.3.4

7.3.7 Example We shall use the preceding theorem to prove ***Gauss' law***

$$\int_{\partial\Omega} \frac{\mathbf{r} \cdot \mathbf{n}}{r^3}\, dS = \begin{cases} 4\pi, & \text{if } 0 \in \Omega \\ 0, & \text{if } 0 \notin \Omega \end{cases} \tag{11}$$

where Ω is a compact set in $\mathbb{R}^3$ with nonempty interior, $\partial\Omega$ is the surface bounding Ω, which is assumed to be piecewise smooth, $\mathbf{n}$ is the outward unit normal, $\mathbf{0} \notin \partial\Omega$, and where

$$r = (x^2 + y^2 + z^2)^{1/2}, \quad \mathbf{r} = (x, y, z).$$

If $\mathbf{0} \notin \Omega$, apply **7.3.6** and the fact that $\operatorname{div}(\mathbf{r}/r^3) = 0$ to get the result. If $\mathbf{0} \in \Omega$, surround $\mathbf{0}$ inside Ω by a ball D_ε of radius ε (Fig. 7.3.5). Since the orientation of ∂D_ε induced from $\Omega \setminus D_\varepsilon$ is the opposite of that induced from D_ε (namely it is given by the *inward* unit normal), Gauss' theorem gives

$$\int_{\partial\Omega} \frac{\mathbf{r} \cdot \mathbf{n}}{r^3} dS + \int_{\partial D_\varepsilon} \frac{\mathbf{r} \cdot \mathbf{n}}{r^3} dS = \int_{\partial(\Omega \setminus D_\varepsilon)} \frac{\mathbf{r} \cdot \mathbf{n}}{r^3} dS = 0 \tag{12}$$

since $\mathbf{0} \notin \Omega \setminus D_\varepsilon$ and thus on $\Omega \setminus D_\varepsilon$, $\operatorname{div}(\mathbf{r}/r^3) = 0$. But on ∂D_ε, $r = \varepsilon$ and $\mathbf{n} = -\mathbf{r}/\varepsilon$, so that

$$\int_{\partial D_\varepsilon} \frac{\mathbf{r} \cdot \mathbf{n}}{r^3} dS = -\int_{\partial D_\varepsilon} \frac{\varepsilon^2}{\varepsilon^4} dS = -\frac{1}{\varepsilon^2} 4\pi\varepsilon^2 = -4\pi$$

since

$$\int_{\partial D_\varepsilon} dS = 4\pi\varepsilon^2, \text{ the area of the sphere of radius } \varepsilon. \quad \blacklozenge$$

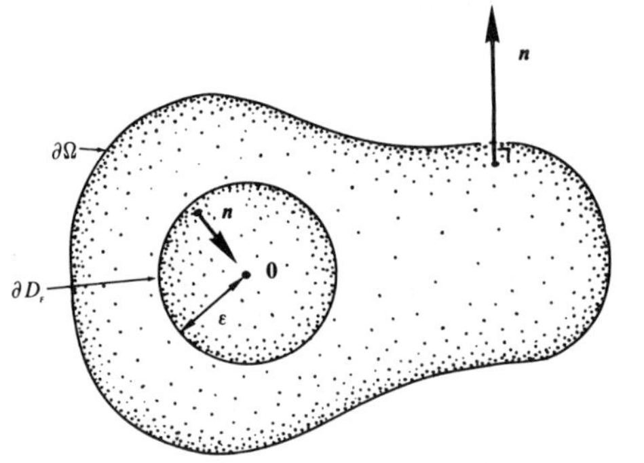

Figure 7.3.5

In electrostatics Gauss' law (11) is used in the following way. The potential due to a point charge q at $\mathbf{0} \in \mathbb{R}^3$ is given by $q/(4\pi r)$, $r = (x^2 + y^2 + z^2)^{1/2}$. The corresponding electric field is defined to be minus the gradient of this potential; i.e., $\mathbf{E} = q\mathbf{r}/(4\pi r^3)$. Thus Gauss' law states that *the total electric flux* $\int_{\partial\Omega} \mathbf{E} \cdot \mathbf{n}\, dS$ *equals* q if $\mathbf{0} \in \Omega$ *and equals zero, if* $\mathbf{0} \notin \Omega$.

A continuous charge distribution in Ω described by a ***charge density*** ρ is related to $\mathbf{E}$ by $\rho = \operatorname{div} \mathbf{E}$. By Gauss' theorem the electric flux $\int_{\partial\Omega} \mathbf{E} \cdot \mathbf{n}\, dS = \int_\Omega \rho\, dV$, which represents the total charge inside Ω. Thus, the relationship $\operatorname{div} \mathbf{E} = \rho$ may be phrased as follows. *The flux out of a surface of an electric field equals the total charge inside the surface.*

Exercises

7.3A Use Green's theorem to show that

(i) The area of the ellipse $x^2/a^2 + y^2/b^2 = 1$ is πab.

(ii) The area of the hypocycloid $x = a\cos^3\theta$, $y = b\sin^3\theta$ is $3\pi a^2/8$.

(iii) The area of one loop of the four-leaved rose $r = 3\sin 2\theta$ is $9\pi/8$.

7.3B Why does Green's theorem fail in the unit disk for $-y\,dx/(x^2+y^2) + x\,dy/(x^2+y^2)$?

7.3C For an oriented surface S and a fixed vector **a**, show that

$$2\int_S \mathbf{a}\cdot\mathbf{n}\,dS = \int_{\partial S}(\mathbf{a}\times\mathbf{r})\cdot d\mathbf{S}.$$

7.3D Let the components of the vector field $\mathbf{X} \in \mathfrak{X}(\mathbb{R}^3)$ be homogeneous of degree one; i.e., **X** satisfies $X^i(tx, ty, tz) = tX^i(x, y, z)$, $i = 1, 2, 3$. Show that if $\operatorname{curl}\mathbf{X} = \mathbf{0}$, then $\mathbf{X} = \operatorname{grad} f$, where $f = (xX^1 + yX^2 + zX^3)/2$.

7.3E Let S be the surface of a region Ω in $\mathbb{R}^3$. Show that

$$\operatorname{volume}(\Omega) = (1/3)\int_S \mathbf{r}\cdot\mathbf{n}\,dS.$$

Give an intuitive argument why this should be so. (*Hint:* Think of cones.)

7.3F Let S be a closed (i.e., compact boundaryless) oriented surface in $\mathbb{R}^3$.

(i) Show in two ways that $\int_S(\operatorname{curl}\mathbf{X})\cdot\mathbf{n}\,dS = 0$.

(ii) Let $\mathbf{X} = \mathfrak{X}(S)$ and $f \in \mathcal{F}(S)$. Make sense of and show that

$$\int_S(\operatorname{grad} f)\mathbf{X}\cdot d\mathbf{S} = -\int_S(f\operatorname{curl}\mathbf{X})dS$$

where grad, curl, and div are taken in $\mathbb{R}^3$.

7.3G Let **X** and **Y** be smooth vector fields on an open set $D \subset \mathbb{R}^3$, with ∂D smooth, and cl(D) compact. Show that

$$\int_D \mathbf{Y}\cdot\operatorname{curl}\mathbf{X}\,dV = \int_D \mathbf{X}\cdot\operatorname{curl}\mathbf{Y}\,dV + \int_{\partial D}(\mathbf{X}\times\mathbf{Y})\cdot\mathbf{n}\,dS.$$

where **n** is the outward unit normal to ∂D and dS the induced surface measure on ∂D. (*Hint:* Show that $\mathbf{Y}\cdot\operatorname{curl}\mathbf{X} - \mathbf{X}\cdot\operatorname{curl}\mathbf{Y} = \operatorname{div}(\mathbf{X}\times\mathbf{Y})$.)

7.3H If C is a closed curve bounding a surface S show that

$$\int_C f(\text{grad } g) \cdot \mathbf{ds} = \int_S (\text{grad } f \times \text{grad } g) \cdot \mathbf{n} \ dS = -\int_C g(\text{grad } f) \cdot \mathbf{ds}$$

where f and g are C^2 functions.

7.3I (A. Lenard) Faraday's law relates the line integral of the electric field around a loop C to the surface integral of the rate of change of the magnetic field over a surface S with boundary C. Regarding the equation $\nabla \times \mathbf{E} = -\partial\mathbf{H}/\partial t$ as the basic equation, Faraday's law is a consequence of Stokes' theorem, as we have seen in Example **7.3.5B**. Suppose we are given electric and magnetic fields in space that satisfy the equation $\nabla \times \mathbf{E} = -\partial\mathbf{H}/\partial t$. Suppose C is the boundary of the Möbius band shown in Fig. 6.5.1. Since the Möbius band cannot be oriented, Stokes' theorem does not apply. What becomes of Faraday's law? Resolve the issue in two ways: (i) by using the results of Supplement **7.2A** or a direct reformulation of Stokes' theorem for nonorientable surfaces, and (ii) realizing C as the boundary of an orientable surface. If $\partial\mathbf{H}/\partial t$ is arbitrary, in general does a current flow around C or not?

§7.4 *Induced Flows on Function Spaces and Ergodicity*

This section requires some results from functional analysis. Specifically *we shall require a knowledge of Stone's theorem and self-adjoint operators.* The required results may be found in Supplement **7.4A** and **7.4B** at the end of this section.

Flows on manifolds induce flows on tangent bundles, tensor bundles, and spaces of tensor fields by means of push-forward. In this section we shall be concerned mainly with the induced flow on the space of functions. This induced flow is sometimes called the ***Liouville flow***.

Let M be a manifold and μ a volume element on M; i.e., (M, μ) is a volume manifold. If F_t is a (volume-preserving) flow on M, then F_t induces a *linear* one-parameter group (of isometries) on the Hilbert space $H = L^2(M\ \mu)$ by

$$U_t(f) = f \circ F_{-t}.$$

The association of U_t with F_t replaces a *nonlinear finite-dimensional* problem with a *linear infinite-dimensional* one.

There have been several theorems that relate properties of F_t and U_t. The best known of these is the result of Koopman [1931], which shows that U_t has one as a simple eigenvalue for all t if and only if F_t is ergodic. (If there are no other eigenvalues, then F_t is called *weakly mixing*.) A few basic results on ergodic theory are given below. We refer the reader to the excellent texts of Halmos [1956], Arnold and Avez [1967], and Bowen [1975] for more information.

We shall first present a result of Povzner [1966], which relates the completeness of the flow of a divergence-free vector field X to the skew-adjointness of X as an operator. (The hypothesis of divergence free is removed in Exercises **7.4A-C**.) We begin with a lemma due to Ed Nelson.

7.4.1 Lemma *Let* A *be an (unbounded) self-adjoint operator on a complex Hilbert space* $\mathcal{H}$. *Let* $D_0 \subset D(A)$ *(the domain of* A*) be a dense linear subspace of* $\mathcal{H}$ *and suppose* $U_t = e^{itA}$ *(the unitary one-parameter group generated by* A*) leaves* D_0 *invariant. Then* $A_0 :=$ (A *restricted to* D_0) *is essentially self-adjoint; that is, the closure of* A_0 *is* A.

Proof Let $\mathbf{A}$ denote the closure of A_0. Since A is closed and extends A_0, A extends $\mathbf{A}$. We need to prove that $\mathbf{A}$ extends A.

For $\lambda > 0$, $\lambda - iA$ is surjective with a bounded inverse. First of all, we prove that $\lambda - iA_0$ has dense range. If not, there is a $v \in \mathcal{H}$ such that

$$\langle v, \lambda x - iA_0 x\rangle = 0 \text{ for all } x \in D_0.$$

In particular, since D_0 is U_t-invariant,

$$\frac{d}{dt}\langle v, U_t x\rangle = \langle v, iAU_tx\rangle = \lambda\langle v, U_tx\rangle$$

so

$$\langle v, U_tx\rangle = e^{\lambda t}\langle v, x\rangle.$$

Since D_0 is dense, this holds for all $x \in \mathcal{H}$. Thus $\| U_t \| = 1$ and $\lambda > 0$ imply $v = 0$. Therefore $(\lambda - iA_0)^{-1}$ makes sense and $(\lambda - iA)^{-1}$ is its closure. It follows from Supplement **7.4A** that A is the closure of A_0 (see **7.4.15** and the remarks following it). ■

7.4.2 Proposition *Let* X *be a* C^∞ *divergence-free vector field on* (M, μ) *with a complete flow* F_t. *Then* iX *is an essentially self-adjoint operator on* C^∞_c = C^∞ *functions with compact support in the complex Hilbert space* $L^2(M, \mu)$.

Proof Let $U_tf = f \circ F_{-t}$ be the unitary one-parameter group induced from F_t. A straightforward convergence argument shows that U_tf is continuous in t in $L^2(M, \mu)$. In Lemma **7.4.1**, choose $D_0 = C^\infty$ functions with compact support. This is clearly invariant under U_t. If $f \in D_0$, then

$$\left.\frac{d}{dt}U_tf\right|_{t=0} = \left.\frac{d}{dt}f \circ F_{-t}\right|_{t=0} = -\,df \cdot X,$$

so the generator of U_t is an extension of $-X$ (as a differential operator) on D_0. The corresponding essentially self-adjoint operator is therefore iX. ■

Now we prove the converse of **7.4.2**. That is, if iX is essentially self-adjoint, then X has an almost everywhere complete flow. This then gives a functional-analytic characterization of completeness.

7.4.3 Theorem *Let* M *be a manifold with a volume element* μ *and* X *be a* C^∞ *divergence-free vector field on* M. *Suppose that, as an operator on* $L^2(M, \mu)$, iX *is essentially self-adjoint on the* C^∞ *functions with compact support. Then, except possibly for a set of points* x *of measure zero, the flow* $F_t(x)$ *of* X *is defined for all* $t \in \mathbb{R}$.

We shall actually prove that, if the defect index of iX is zero in the upper half-plane (i.e., if $(iX + i)(C^\infty_c)$ is dense in L^2), then the flow is defined, except for a set of measure zero, for all $t > 0$. Similarly, if the defect index of iX is zero in the lower half-plane, the flow is essentially complete for $t < 0$. The converses of these more general results can be established along the lines of the proof of **7.4.1**.

Proof (E. Nelson–private communication) Suppose that there is a set E of finite positive measure such that if $x \in E$, $F_t(x)$ fails to be defined for t sufficiently large. Let E_T be the set of $x \in E$

for which $F_t(x)$ is undefined for $t \geq T$. Since $E = \bigcup_{T\geq 1} E_T$, some E_T has positive measure. Replacing E by E_T, we may assume that all points of E "move to infinity" in a time $\leq T$.

If f is any function on M, we adopt the convention that $f(F_t(x)) = 0$ if $F_t(x)$ is undefined. For any $x \in M$, and $t < -T$, $F_t(x)$ must be either in the complement of E or undefined; otherwise it would be a point of E that did not move to infinity in time T. Hence $\chi_E(F_t(x)) = 0$ for $t < -T$, where χ_E is the characteristic function of E. We now define a function on M by

$$g(x) = \int_{-\infty}^{\infty} e^{-\tau}\chi_E(F_\tau(x))d\tau.$$

Note that the integral converges because the integrand vanishes for $t < -T$. In fact, we have

$$0 \leq g(x) \leq \int_{-T}^{\infty} e^{-\tau}d\tau = e^T.$$

Moreover, g is in L^2. Indeed, because F_t is measure-preserving, where defined, denoting by $\| \, \|_2$ the L^2 norm, we have $\| \chi_E \circ F_\tau \|_2 \leq \| \chi_E \|_2$, so that

$$\| g \|_2 \leq \int_{-T}^{\infty} e^{-\tau} \| \chi_E \circ F_\tau \|_2 d\tau \leq \| \chi_E \|_2 e^T.$$

The function g is nonzero because E has positive measure.

Fix a point $x \in M$. Then $F_t(x)$ is defined for t sufficiently small. It is easy to see that in this case $F_\tau(F_t(x))$ and $F_{\tau+t}(x)$ are defined or undefined together, and in the former case they are equal. Hence we have $\chi_E(F_\tau(F_t(x)) = \chi_E(F_{\tau+t}(x))$ for t sufficiently small. Therefore, for t sufficiently small

$$g(F_t(x)) = \int_{-\infty}^{\infty} e^{-\tau}\chi_E(F_{\tau+t}(x))d\tau = \int_{-\infty}^{\infty} e^{t-\tau}\chi_E(F_\tau(x))d\tau = e^t g(x).$$

Now if φ is C^∞ with compact support, we have

$$\int g(x)X[\varphi](x)d\mu = \lim_{t\to 0} \int g(x)\frac{\varphi(F_t(x)) - \varphi(x)}{t}d\mu = \lim_{t\to 0} \int \frac{g(F_{-t}(x)) - g(x)}{t}\varphi(x)d\mu$$

$$= \lim_{t\to 0} \int \frac{e^{-t} - 1}{t} g(x)\varphi(x)d\mu = -\int g(x)\varphi(x)d\mu.$$

These equalities are justified because on the support of φ the flow F_t exists for sufficiently small t and is measure-preserving. Thus g is orthogonal to the range of $X + 1$, and therefore the defect index of iX in the upper half-plane is nonzero.

The case of completeness for $t < 0$ is similar. ■

Methods of functional analysis applied to $L^2(M, \mu)$ can, as we have seen, be used to obtain theorems relevant to flows on M. Related to this is a measure-theoretic analogue of the fact that any automorphism of the algebra $\mathcal{F}(M)$ is induced by a diffeomorphism of M (see Supplement **4.2C**). This result, due to Mackey [1962], states that if U_t is a linear isometry on $L^2(M, \mu)$, which is multiplicative (i.e., $U_t(fg) = (U_tf)(U_tg)$, where defined), then U_t is induced by some measure preserving flow F_t on M. This may be used to give another proof of theorem **7.4.3**.

A central notion in statistical mechanics is that of ergodicity; this is intended to capture the idea that a flow may be random or chaotic. In dealing with the motion of molecules, the founders of statistical mechanics, particularly Boltzmann and Gibbs, made such hypotheses at the outset. One of the earliest precise defintions of randomness of a dynamical system was ***minimality***: the orbit of almost every point is dense. In order to prove useful theorems, von Neumann and Birkhoff in the early 1930s required the strong assumption of ergodicity, defined as follows.

7.4.4 Definition *Let* S *be a measure space and* F_t *a (measurable) flow on* S. *We call* F_t ***ergodic*** *if the only invariant measurable sets are* $\varnothing$ *and all of* S.

Here, invariant means $F_t(A) = A$ for all $t \in \mathbb{R}$ and we agree to write $A = B$ if A and B differ by a set of measure zero. (It is not difficult to see that ergodicity implies minimality if we are on a second countable Borel space.)

A function $f : S \to \mathbb{R}$ will be called a ***constant of the motion*** if $f \circ F_t = f$ a.e. (almost everywhere) for each $t \in \mathbb{R}$.

7.4.5 Proposition *A flow* F_t *on* S *is ergodic iff the only constants of the motion are constant a.e.*

Proof If F_t is ergodic and f is a constant of the motion, the two sets $\{x \in S \mid f(x) \geq a\}$ and $\{x \in S \mid f(x) \leq a\}$ are invariant, so f must be constant a.e. The converse follows by taking f to be a characteristic function. ■

The first major step in ergodic theory was taken by J. von Neumann [1932], who proved the mean ergodic theorem which remains as one of the most important basic theorems. The setting is in Hilbert space, but we shall see how it applies to flows of vector fields in **7.4.7**.

7.4.6 Mean Ergodic Theorem *Let* H *be a real or complex Hilbert space and* $U_t : H \to H$ *a strongly continuous one-parameter unitary group (i.e.,* U_t *is unitary for each* t, *is a flow on* H *and for each* $x \in H$, $t \mapsto U_tx$ *is continuous).*

Let the closed subspace H_0 *be defined by*

$$H_0 = \{x \in H \mid U_tx = x \text{ for all } t \in \mathbb{R}\}$$

and let $\mathbb{P}$ *be the orthogonal projection onto* H_0. *Then for any* $x \in H$,

$$\lim_{t\to\infty} \left\| \frac{1}{t}\int_0^t U_s x \, ds - \mathbb{P}x \right\| = 0.$$

The point av(x) *so defined is called the* ***time average*** *of* x.

Proof of 7.4.6 (F. Riesz [1944]) We must show that

$$\lim_{t\to\infty} \left\| \frac{1}{t}\int_0^t U_s x \, ds - \mathbb{P}x \right\| = 0.$$

If $\mathbb{P}x = x$, this means $x \in H_0$, so $U_s(x) = x$; the result is clearly true in this case. We can therefore suppose that $\mathbb{P}x = 0$ by decomposing $x = \mathbb{P}x + (x - \mathbb{P}x)$. Note that

$$\{U_t y - y \mid y \in H, t \in \mathbb{R}\}^{\perp} = H_0$$

where $\perp$ denotes the orthogonal complement. This is an easy verification using unitarity of U_t and $U_t^{-1} = U_{-t}$. It follows that $\ker \mathbb{P}$ is the closure of the space spanned by elements of the form $U_s y - y$. Indeed $\ker \mathbb{P} = H_0^{\perp}$, and if A is any set in H, and $B = A^{\perp}$, then $B^{\perp}$ is the closure of the span of A. Therefore, for any $\varepsilon > 0$, there exists $t_1, ..., t_n$ and $x_1, ..., x_n$ such that

$$\left\| x - \sum_{j=1}^{n} (U_{t_j} x_j - x_j) \right\| < \varepsilon.$$

It follows from this, again using unitarity of U_t, that it is enough to prove our assertion for x of the form $U_\tau y - y$. Thus we must establish

$$\lim_{t\to\infty} \frac{1}{t}\int_0^t U_s(U_\tau y - y)\, ds = 0.$$

For $t > \tau$ we may estimate this integral as follows:

$$\left\| \frac{1}{t}\int_0^t (U_s U_\tau y - U_s y) ds \right\| = \left\| -\frac{1}{t}\int_0^\tau U_s(y) ds + \frac{1}{t}\int_t^{t+\tau} U_s(y) ds \right\|$$

$$\le \frac{1}{t}\int_0^\tau \|y\|\, ds + \frac{1}{t}\int_t^{t+\tau} \|y\|\, ds = \frac{2\tau \|y\|}{t} \to 0 \quad \text{as } t \to \infty. \quad \blacksquare$$

To apply **7.4.6** to a measure-preserving flow F_t on S, we consider the unitary one-parameter group $U_t(f) = f \circ F_t$ on $L^2(S, \mu)$. We only require a minimal amount of continuity on F_t here, namely, we assume that if $s \to t$, $F_s(x) \to F_t(x)$ for a.e. $x \in S$. We shall also assume $\mu(S) < \infty$ for convenience. Under these hypotheses, U_t is a strongly continuous unitary one-parameter group. The verification is an exercise in the use of the dominated convergence theorem.

7.4.7 Corollary *In the hypotheses above* F_t *is ergodic if and only if for each* $f \in L^2(S)$ *its time average*

$$\mathrm{av}(f)(x) = \lim_{t \to \pm\infty} \frac{1}{t} \int_0^t (f \circ F_s)(x)\, ds$$

(the limit being in the L^2*-mean) is constant a.e. In this case the time average* $\mathrm{av}(f)$ *necessarily equals the* ***space average*** $\int_S f d\mu/\mu(S)$ *a.e.*

Proof Ergodicity of F_t is equivalent by Proposition **7.4.5** to $\dim H_0 = 1$, where H_0 is the closed subspace of $L^2(S)$ given in theorem **7.4.6**. If $\dim H_0 = 1$, $\mathbb{P}(f) = \int_S f d\mu/\mu(S)$ so the equality of $\mathrm{av}(f)$ with $\mathbb{P}(f)$ a.e. is a consequence of Theorem **7.4.6**. Conversely if any $f \in L^2(S)$ has a.e. constant time average $\mathrm{av}(f)$ then taking f to be a constant of motion, it follows that $f = \mathrm{av}(f)$ is constant a.e. Therefore $\dim H_0 = 1$. ■

Thus, if F_t is ergodic, the time average of a function is constant a.e. and equals its space average. A refinement of this is the *individual ergodic theorem* of G. D. Birkoff [1931], in which one obtains convergence almost everywhere. Also, if $\mu(S) = \infty$ but $f \in L^1(S) \cap L^2(S)$, one still concludes a.e. convergence of the time average. (If f is only L^2, mean convergence to zero is still assured by **7.4.5**.)

Modern work in dynamical systems, following the ideas in §**4.3**, has shown that for many interesting flows arising in the physical sciences, the motion can be "chaotic" on large regions of phase space without being ergodic. Much current research is focused on trying to prove analogues of the ergodic theorems for such cases. (See, for instance, Guckenheimer and Holmes [1985] and Eckmann and Ruelle [1985] and references therein.)

A particularly important example of an ergodic flow is the irrational flow on the torus.

7.4.8 Definition *The flow* $F_t : \mathbb{T}^n \to \mathbb{T}^n$ *given by* $F_t([\varphi]) = [\varphi + \nu t]$, *for* $\nu \in \mathbb{R}^n$ *is called the* ***quasiperiodic*** *or* ***linear flow*** *on* $\mathbb{T}^n$ *determined by* ν. *The quasiperiodic flow is called* ***irrational*** *if the components* $(\nu^1, \ldots, \nu^n)$ *of* ν *are linearly independent over* $\mathbb{Z}$ *(or, equivalently, over* $\mathbb{Q}$*), i.e.,* $\mathbf{k} \cdot \nu = 0$ *for* $\mathbf{k} \in \mathbb{Z}^n$ *implies* $\mathbf{k} = 0$.

7.4.9 Proposition *The linear flow* F_t *on* $\mathbb{T}^n$ *determined by* $\nu \in \mathbb{R}^n$ *is ergodic if and only if it is irrational.*

Proof Assume the flow is irrational and let $f \in L^2(\mathbb{T}^n)$ be a constant of the motion. Expand $f([\varphi])$ and $(f \circ F_t)([\varphi])$ in Fourier series:

$$f([\varphi]) = \sum_{\mathbf{k}\in\mathbb{Z}^n} a_{\mathbf{k}} e^{i\mathbf{k}\cdot\varphi} \quad \text{and} \quad (f\circ F_t)([\varphi]) = \sum_{\mathbf{k}\in\mathbb{Z}^n} b_{\mathbf{k}}(t)e^{i\mathbf{k}\cdot\varphi}$$

where the convergence is in L^2 and the Fourier coefficients are given by

$$a_{\mathbf{k}} = \int_{\mathbb{T}^n} e^{i\mathbf{k}\cdot\varphi} f([\varphi])\, d\varphi$$

$$b_{\mathbf{k}}(t) = \int_{\mathbb{T}^n} e^{i\mathbf{k}\cdot\varphi} f([\varphi + \nu t])\, d\varphi = \int_{\mathbb{T}^n} e^{-i\mathbf{k}\cdot(\varphi-\nu t)} f([\varphi])\, d\varphi = e^{i\mathbf{k}\cdot\nu t} a_{\mathbf{k}}.$$

(The measure $d\varphi$ is chosen such that the total volume of $\mathbb{T}^n$ equals one.) Since f is a constant of the motion, $a_{\mathbf{k}} = b_{\mathbf{k}}(t)$ for all $\mathbf{k} \in \mathbb{Z}^n$ and all $t \in \mathbb{R}$ which implies that $e^{\mathbf{k}\cdot\nu t} = 1$ for all $\mathbf{k} \in \mathbb{Z}^n$, $t \in \mathbb{R}$. Thus $\mathbf{k}\cdot\nu = 0$ which by hypothesis forces $\mathbf{k} = 0$. Consequently all $a_{\mathbf{k}} = 0$ with the exception of a_0 and thus $f = a_0$ a.e.

Conversely, assume F_t is ergodic and that $\mathbf{k}\cdot\nu = 0$ for some $\mathbf{k} \in \mathbb{Z}^n\setminus\{\mathbf{0}\}$. Then the set $A = \{[\psi] \in \mathbb{T}^n \mid \mathbf{k}\cdot\psi = 0\}$ is closed and hence measurable and invariant under F_t. But clearly $A \neq \varnothing$ and $A \neq \mathbb{T}^n$ which shows that F_t is not ergodic. ■

7.4.10 Corollary *Let* F_t *be an irrational flow on* $\mathbb{T}^n$ *determined by* ν. *Then every trajectory of* F_t ***is uniformly distributed*** *on* $\mathbb{T}^n$, *i.e., for any measurable set* A *in* $\mathbb{T}^n$,

$$\lim_{t\to\pm\infty} (\text{measure } A(t)/t) = \text{measure } A$$

where $A(t) = \{s \in [0, t] \mid F_s([\psi]) \in A\}$ *and the measure of* $\mathbb{T}^n$ *is assumed to be equal to one.*

Proof Let χ_A be the characteristic function of A. Then

$$\text{av}(\chi_A)([\psi]) = \lim_{t\to\pm\infty} \frac{1}{t}\int_0^t \chi_A(F_s([\psi]))\, ds = \lim_{t\to\pm\infty} \frac{1}{t}(\text{measure } A(t)) = \int_{\mathbb{T}^n} (\chi_A([\psi])\, d\varphi = \text{measure } A$$

by **7.4.7** and **7.4.9.** ■

7.4.11 Corollary *Every trajectory of a quasiperiodic flow* F_t *on* $\mathbb{T}^n$ *is dense if and only if the flow is irrational.*

Proof By translation of the initial condition it is easily seen that every trajectory is dense on $\mathbb{T}^n$ if and only if the trajectory through $[0]$ is dense. Assume first that the flow is irrational. If $\{F_t([\mathbf{0}]) \mid t \in \mathbb{R}\}$ is not dense in $\mathbb{T}^n$ then there is an open set U in $\mathbb{T}^n$ not containing any point of this trajectory. Thus, in the notation of corollary **7.4.10,** $U(t) = \varnothing$. This contradicts **7.4.10** since the measure of U is strictly positive.

Conversely, assume that the trajectory through $[\mathbf{0}]$ is dense and let f be a continuous constant of the motion for F_t. This implies that f is a constant. Since continuous functions are dense in L^2, this in turn implies that any L^2-constant of the motion is constant a.e. By Proposition **7.4.5**, F_t is ergodic and by Proposition **7.4.9**, F_t is irrational. ■

☞ SUPPLEMENT 7.4A
Unbounded and Self Adjoint Operators.
(Written in collaboration with P. Chernoff.)

In many applications involving differential equations, the operators one meets are not defined on the whole Banach space E and are not continuous. Thus we are led to consider a linear transformation $A : D_A \subset E \to E$ where D_A is a linear subspace of E (the domain of A). If D_A is dense in E, we say A is ***densely defined***. We speak of A as an ***operator*** and this shall mean ***linear*** operator unless otherwise specified.

Even though A is not usually continuous, it might have the important property of being closed. We say A is ***closed*** if its graph Γ_A

$$\Gamma_A = \{(x, Ax) \in E \times E \mid x \in D_A\}$$

is a closed subset of $E \times E$. This is equivalent to

$$(x_n \in D_A, x_n \to x \in E \text{ and } Ax_n \to y \in E) \text{ implies } (x \in D_A \text{ and } Ax = y).$$

An operator A (with domain D_A) is called ***closable*** if $\mathrm{cl}(\Gamma_A)$, the closure of the graph of A, is the graph of an operator, say, $\mathbf{A}$. We call $\mathbf{A}$ the ***closure*** of A. It is easy to see that A is closable iff $\{(x_n \in D_A, x_n \to 0 \text{ and } Ax_n \to y) \text{ implies } y = 0\}$. Clearly $\mathbf{A}$ is a closed operator that is an ***extension*** of A; i.e., $D_{\mathbf{A}} \supset D_A$ and $\mathbf{A} = A$ on D_A. One writes this as $\mathbf{A} \supset A$.

The closed graph theorem asserts that an everywhere defined closed operator is bounded. (See §**2.2**.) However, if an operator is only densely defined, "closed" is weaker than "bounded." If A is a closed operator, the map $x \mapsto (x, Ax)$ is an isomorphism between D_A and the closed subspace Γ_A. Hence if we set

$$|||\, x \,|||^2 = \|\, x \,\|^2 + \|\, Ax \,\|^2,$$

D_A becomes a Banach space. We call the norm $|||\ |||$ on D_A the ***graph norm***.

Let A be an operator on a real or complex Hilbert space H with dense domain D_A. The ***adjoint*** of A is the operator A^* with domain D_{A^*}, defined as follows:

$$D_{A^*} = \{y \in H \mid \text{there is a } z \in H \text{ such that } \langle Ax, y\rangle = \langle x, z\rangle \text{ for all } x \in D_A\}$$

and

$$A^* : D_{A^*} \to H, \quad y \mapsto z.$$

From the fact that D_A is dense, we see that A^* is indeed well defined (there is at most one such z for any $y \in H$). It is easy to see that if $A \supset B$ then $B^* \supset A^*$.

If A is everywhere defined and bounded, it follows from the Riesz representation theorem (Supplement **2.2A**) that A^* is everywhere defined; moreover it is not hard to see that, in this case, $\| A^* \| = \| A \|$.

An operator A is ***symmetric*** (***Hermitian*** in the complex case) if $A^* \supset A$; i.e., $\langle Ax, y \rangle = \langle x, Ay \rangle$ for all $x, y \in D_A$. If $A^* = A$ (this includes the condition $D_{A^*} = D_A$), then A is called ***self-adjoint.*** An everywhere defined symmetric operator is bounded (from the closed graph theorem) and so is self-adjoint. It is also easy to see that a self-adjoint operator is closed.

One must be aware that, for technical reasons, it is the notion of self-adjoint rather than symmetric, which is important in applications. Correspondingly, verifying self-adjointness is often difficult while verifying symmetry is usually trivial.

Sometimes it is useful to have another concept at hand, that of essential self-adjointness. First, it is easy to check that any symmetric operator A is closable. The closure $\mathbf{A}$ is easily seen to be symmetric. One says that A is ***essentially self-adjoint*** when its closure $\mathbf{A}$ is self-adjoint.

Let A be a self-adjoint operator. A dense subspace $C \subset H$ is said to be a ***core*** of A if $C \subset D_A$ and the closure of A restricted to C is again A. Thus if C is a core of A one can recover A just be knowing A on C.

We shall now give a number of propositions concerning the foregoing concepts, which are useful in applications. Most of this is classical work of von Neumann. We begin with the following.

7.4.12 Proposition *Let* A *be a closed symmetric operator of a complex Hilbert space* H. *If* A *is self-adjoint then* $A + \lambda I$ *is surjective for every complex number* λ *with* $\operatorname{Im} \lambda \neq 0$ (I *is the identity operator*).

Conversely, if A *is symmetric and* $A - iI$ *and* $A + iI$ *are both surjective then* A *is self-adjoint.*

Proof Let A be self-adjoint and $\lambda = \alpha + i\beta$, $\beta \neq 0$. For $x \in D_A$ we have

$$\begin{aligned} \|(A + \lambda)x\|^2 &= \|(A + \alpha)x\|^2 + i\beta\langle x, Ax\rangle - i\beta\langle Ax, x\rangle + \beta^2 \|x\|^2 \\ &= \|(A + \alpha)x\|^2 + \beta^2\|x\|^2 \geq \beta^2 \|x\|^2, \end{aligned}$$

where $A + \lambda$ means $A + \lambda I$. Thus we have the inequality

$$\|(A + \lambda)x\| \geq |\operatorname{Im} \lambda| \, \|x\|. \tag{1}$$

Since A is closed, it follows from (1) that the range of $A + \lambda$ is a closed set for $\operatorname{Im} \lambda \neq 0$. Indeed, let $y_n = (A + \lambda)x_n \to y$. By the inequality (1), $\|x_n - x_m\| \leq \|y_n - y_m\| / |\operatorname{Im} \lambda|$ so x_n converges to, say x. Also Ax_n converges to $y - \lambda x$; thus $x \in D_A$ and $y - \lambda x = Ax$ as A is closed.

Now suppose y is orthogonal to the range of $A + \lambda I$. Thus

$$\langle Ax + \lambda x, y\rangle = 0 \text{ for all } x \in D_A, \text{ or } \langle Ax, y\rangle = -\langle x, \lambda y\rangle.$$

By definition, $y \in D_{A^*}$ and $A^*y = -\overline{\lambda}y$; since $A = A^*$, $y \in D_A$ and $Ay = -\overline{\lambda}y$, we obtain $(A + \lambda I)y = 0$. Thus the range of $A + \lambda I$ is all of H.

Conversely, suppose $A + i$ and $A - i$ are onto. Let $y \in D_{A^*}$. Thus for all $x \in D_A$,

$$\langle (A + i)x, y\rangle = \langle x, (A^* - i)y\rangle = \langle x, (A - i)z\rangle$$

for some $z \in D_A$ since $A - i$ is onto. Thus

$$\langle (A + i)x, y\rangle = \langle (A + i)x, z\rangle$$

and it follows that $y = z$. This proves that $D_{A^*} \subset D_A$ and so $D_A = D_{A^*}$. The result follows. ■

If A is self-adjoint then for $\operatorname{Im}\lambda \neq 0$, $\lambda I - A$ is onto and from (1) is one-to-one. Thus $(\lambda I - A)^{-1} : H \to H$ exists, is bounded, and we have

$$\|(\lambda I - A)^{-1}\| \leq \frac{1}{|\operatorname{Im}\lambda|}. \tag{2}$$

This operator $(\lambda I - A)^{-1}$ is called the ***resolvent*** of A. Notice that even though A is an unbounded operator, the resolvent is bounded. The same argument used to prove **7.4.12** shows the following.

7.4.13 Proposition *A symmetric operator* A *is essentially self-adjoint iff the ranges of* $A + iI$ *and* $A - iI$ *are dense.*

If A is a (closed) symmetric operator then the ranges of $A + iI$ and $A - iI$ are (closed) subspaces. The dimensions of their orthogonal complements are called the ***deficiency indices*** of A. Thus **7.4.12** and **7.4.13** can be restated as: *a closed symmetric operator (resp., a symmetric operator) is self-adjoint (resp., essentially self-adjoint) iff it has deficiency indices* (0, 0).

If A is a closed symmetric operator then from (1), $A + iI$ is one-to-one and we can consider the inverse $(A + iI)^{-1}$, defined on the range of $A + iI$. One calls $(A - iI)(A + iI)^{-1}$ the ***Cayley transform*** of A. It is always isometric, as is easy to check. Thus A *is self-adjoint iff its Cayley transform is unitary.*

Let us return to the graph of an operator A for a moment. The adjoint can be described entirely in terms of its graph and this is often convenient. Define an isometry $J : H \oplus H \to H \oplus H$ by $J(x, y) = (-y, x)$; note that $J^2 = -I$.

7.4.14 Proposition *Let* A *be densely defined. Then* $(\Gamma_A)^\perp = J(\Gamma_{A^*})$ *and* $-\Gamma_{A^*} = J(\Gamma_A)^\perp$. *In particular,* A^* *is closed, and if* A *is closed, then*

$$H \oplus H = \Gamma_A \oplus J(\Gamma_{A^*}),$$

where $H \oplus H$ *carries the usual inner product:* $\langle (x_1, x_2), (y_1, y_2)\rangle = \langle x_1, y_1\rangle + \langle x_2, y_2\rangle$.

Proof Let $(z, y) \in J(\Gamma_{A^*})$, so $y \in D_{A^*}$ and $z = -A^*y$. Let $x \in D_A$. We have

$$\langle (x, Ax), (-A^*y, y)\rangle = \langle x, -A^*y\rangle + \langle Ax, y\rangle = 0,$$

and so $J(\Gamma_{A^*}) \subset \Gamma_A^\perp$.

Conversely if $(z, y) \in (\Gamma_A)^\perp$, then $\langle x, z\rangle + \langle Ax, y\rangle = 0$ for all $x \in D_A$. Thus by definition, $y \in D_{A^*}$ and $z = -A^*y$. This proves the opposite inclusion. ■

Thus if A is a closed operator, the statement $H \oplus H = \Gamma_A \oplus J(\Gamma_{A^*})$ means that given $e, f \in H$, the equations

$$x - A^*y = e \quad \text{and} \quad Ax + y = f$$

have exactly one solution (x, y). If A is densely defined and symmetric, then its closure satisfies $\mathbf{A} \subset A^*$ since A^* is closed. There are other important consequences of **7.4.14** as well.

7.4.15 Corollary *For* A *densely defined and closeable, we have*

(i) $\mathbf{A} = A^{**}$, *and* (ii) $\mathbf{A}^* = A^*$.

Proof (i) Note that $\Gamma_{A^{**}} = -J\{(\Gamma_{A^*})^\perp\} = -(J(\Gamma_{A^*}))^\perp$ since J is an isometry. But

$$-(J(\Gamma_{A^*}))^\perp = -(J^2\Gamma_A^\perp)^\perp = \Gamma_A^{\perp\perp} = \mathrm{cl}(\Gamma_A) = \Gamma_{\mathbf{A}}.$$

(ii) follows since $\Gamma_{\mathbf{A}}^\perp = \mathrm{cl}(\Gamma_A^\perp)$. ■

Suppose $A : D_A \subset H \to H$ is one-to-one. Then we get an operator A^{-1} defined on the range of A. In terms of graphs:

$$\Gamma_{A^{-1}} = K(\Gamma_A),$$

where $K(x, y) = (y, x)$; note that $K^2 = I$, K is an isometry and $KJ = -JK$. It follows for example that *if* A *is self-adjoint, so is* A^{-1}, since

$$\Gamma_{(A^{-1})^*} = -J(\Gamma_{A^{-1}}^\perp) = -J(K\Gamma_A^\perp) = KJ\Gamma_A^\perp = K\Gamma_{(A^*)} = \Gamma_{A^{*-1}}.$$

Next we consider possible self-adjoint extensions of a symmetric operator.

7.4.16 Proposition *Let* A *be a symmetric densely defined operator on* H *and* $\mathbf{A}$ *its closure. The following are equivalent:*

(i) A *is essentially self-adjoint.*

(ii) A^* *is self-adjoint.*

(iii) $A^{**} \supset A^*$.

(iv) A *has exactly one self-adjoint extension.*

(v) $\mathbf{A} = A^*$.

Proof By definition, (i) means $\mathbf{A}^* = \mathbf{A}$. But we know $\mathbf{A}^* = A^*$ and $\mathbf{A} = A^{**}$ by **7.4.15**. Thus (i), (ii), (v) are equivalent. These imply (iii). Also (iii) implies (ii) since $A \subset \mathbf{A} \subset A^* \subset A^{**} = \mathbf{A}$ and so $A^* = A^{**}$. To prove (iv) is implied let Y be any self-adjoint extention of A. Since Y is closed, $Y \supset \mathbf{A}$. But $\mathbf{A} = A^*$ so Y extends the self-adjoint operator A^*; i.e., $Y \supset A$. Taking adjoints, $A^* = A \supset Y^* = Y$ so $Y = A$.

To prove that (iv) implies the others is a bit more complicated. We shall in fact give a more general result in **7.4.18** below. First we need some notation. Let

$$D_+ = \text{range}(A + iI)^\perp \subset H \quad \text{and} \quad D_- = \text{range}(A - iI)^\perp \subset H$$

called the ***positive*** and ***negative defect spaces***. Using the argument in **7.4.12** it is easy to check that

$$D_+ = \{x \in D_{A^*} \mid A^*x = ix\} \quad \text{and} \quad D_- = \{x \in D_{A^*} \mid A^*x = -ix\}.$$

7.4.17 Lemma *Using the graph norm on* D_{A^*}, *we have the orthogonal direct sum*

$$D_{A^*} = D_{\mathbf{A}} \oplus D_+ \oplus D_-.$$

Proof Since D_+, D_- are closed in H they are closed in D_{A^*}. Also $D_{\mathbf{A}} \subset D_{A^*}$ is closed since A^* is an extention of A and hence of $\mathbf{A}$. It is easy to see that the indicated spaces are orthogonal. For example let $x \in D_{\mathbf{A}}$ and $y \in D_-$. Then using the inner product

$$\langle\langle x, y\rangle\rangle = \langle x, y\rangle + \langle A^*x, A^*y\rangle$$

gives

$$\langle\langle x, y\rangle\rangle = \langle x, y\rangle + \langle A^*x, -iy\rangle = \langle x, y\rangle - i\langle A^*x, y\rangle.$$

Since $x \in D_{\mathbf{A}} = D_{A^{**}}$, by **7.4.16**(v), we get

$$\langle\langle x, y\rangle\rangle = \langle x, y\rangle - i\langle x, A^*y\rangle = \langle x, y\rangle - \langle x, y\rangle = 0.$$

To see that $D_{A^*} = D_{\mathbf{A}} \oplus D_+ \oplus D_-$ it suffices to show that the orthogonal complement of $D_{\mathbf{A}} \oplus D_+ \oplus D_-$ is zero. Let $u \in (D_{\mathbf{A}} \oplus D_+ \oplus D_-)^{\perp}$, so

$$\langle\langle u, x\rangle\rangle = \langle\langle u, y\rangle\rangle = \langle\langle u, z\rangle\rangle = 0$$

for all $x \in D_{\mathbf{A}}$, $y \in D_+$, $z \in D_-$. From $\langle\langle u, x\rangle\rangle = 0$ we get $\langle u, x\rangle + \langle A^*u, A^*x\rangle = 0$ or $A^*u \in D_{A^*}$ and $A^*A^*u = -u$. It follows that $(I - iA^*)u \in D_+$. But from $\langle\langle u, y\rangle\rangle = 0$ we have $\langle (I - iA^*)u, y\rangle = 0$ and so $(I - iA^*)u = 0$. Hence $u \in D_-$. Taking $z = u$ gives $u = 0$. ▼

7.4.18 Proposition *The self-adjoint extensions of a symmetric densely defined operator* A *(if any) are obtained as follows. Let* $T : D_+ \to D_-$ *be an isometry mapping* D_+ *onto* D_- *and let* $\Gamma_T \subset D_+ \oplus D_-$ *be its graph. Then the restriction of* A^* *to* $D_{\mathbf{A}} \oplus \Gamma_T$ *is a self-adjoint extension of* A.

Thus A has self-adjoint extensions iff its defect indices ($\dim D_+$, $\dim D_-$) are equal and these extensions are in one-to-one correspondence with all isometries of D_+ onto D_-. Assuming this result for a moment, we give the following.

Completion of Proof of 7.4.16 If there is only one self-adjoint extension it follows from **7.4.18** that $D_+ = D_- = \{0\}$ so by **7.4.13** A is essentially self-adjoint. ■

Proof of 7.4.18 Let B be a self-adjoint extension of **A**. Then $\mathbf{A}^* = A^* \supset B$ so B is the restriction of A^* to some subspace containing $D_{\mathbf{A}}$. We want to show that these subspaces are of the form $D_{\mathbf{A}} \oplus \Gamma_T$ as stated.

Suppose first that $T : D_+ \to D_-$ is an isometry onto and let $\mathcal{A}$ be the restriction of A^* to $D_{\mathbf{A}} \oplus \Gamma_T$. First of all, one proves that $\mathcal{A}$ is symmetric: i.e., for $u, x \in D_{\mathbf{A}}$ and $v, y \in D_+$ that

$$\langle Ax + A^*y + A^*Ty, u + v + Tv\rangle = \langle x + y + Ty, Au + A^*v + A^*Tv\rangle.$$

This is a straightforward computation using the definitions.

To show that $\mathcal{A}$ is self-adjoint, we show that $D_{\mathcal{A}^*} \subset D_{\mathcal{A}}$. If this does not hold there exists a nonzero $z \in D_{\mathcal{A}^*}$ such that either $\mathcal{A}^*z = iz$ or $\mathcal{A}^*z = -iz$. This follows from Lemma **7.4.17** applied to the operator $\mathcal{A}$. (Observe that $\mathcal{A}$ is a closed operator--this easily follows.) Now $\mathcal{A} \supset A$ so $A^* \supset \mathcal{A}^*$. Thus $z \in D_+$ or $z \in D_-$. Suppose $z \in D_+$. Then $z + Tz \in D_{\mathcal{A}}$ so as $\langle\langle D_{\mathcal{A}}, z\rangle\rangle = 0$ ($\langle\langle\, ,\rangle\rangle$ denotes the inner product relative to $\mathcal{A}$),

$$0 = \langle\langle z + Tz, z\rangle\rangle = \langle\langle z, z\rangle\rangle + \langle\langle Tz, z\rangle\rangle = 2\langle z, z\rangle,$$

since $Tz \in D_-$. Hence $z = 0$. In a similar way one sees that if $z \in D_-$ then $z = 0$. Hence $\mathcal{A}$ is self-adjoint.

We will leave the details of the converse to the reader (they are similar to the foregoing). The idea is this: if $\mathcal{A}$ is restriction of A^* to a subspace $D_{\mathbf{A}} \oplus V$ for $V \subset D_+ \oplus D_-$ and $\mathcal{A}$ is symmetric, then V is the graph of a map $T : W \subset D_+ \to D_-$ and $\langle Tu, Tv\rangle = \langle u, v\rangle$, for a subspace $W \subset D_+$. Then self-adjointness of $\mathcal{A}$ implies that in fact $W = D_+$ and T is onto. ■

A convenient test for establishing the equality of the deficiency indices is to show that T commutes with a conjugation U; i.e., an antilinear isometry $U : H \to H$ satisfying $U^2 = I$; antilinear means

$$U(\alpha x) = \bar{\alpha}Ux \text{ for complex scalars } \alpha \text{ and } U(x + y) = Ux + Uy \text{ for } x, y \in H.$$

It is easy to see that U is the isometry required from D_+ to D_- (use $D_+ = \text{range }(A + iI)^\perp$).

As a corollary, we obtain an important classical result of von Neumann: *Let* H *be* L^2 *of a measure space and let* A *be a (closed) symmetric operator that is real in the sense that it commutes with complex conjugation. Then* A *admits self-adjoint extensions.* (Another sufficient condition of a different nature, due to Friedrichs, is given below.) This result applies to many quantum mechanical operators. However, one is also interested in essential self-adjointness, so that the self-adjoint extension will be unique. Methods for proving this for specific operators in quantum mechanics are given in Kato [1966] and Reed and Simon [1974]. For corresponding questions in elasticity, see Marsden and Hughes [1983].

We now give some additional results that illustrate methods for handling self-adjoint operators.

7.4.19 Proposition *Let* A *be a self-adjoint and* B *a bounded self-adjoint operator. Then* A + B *(with domain* D_A*) is self-adjoint. If* A *is essentially self-adjoint on* D_A *then so is* A + B.

Proof A + B is certainly symmetric on D_A. Let $y \in D_{(A+B)^*}$ so that for all $x \in D_A$,

$$\langle (A + B)x, y\rangle = \langle x, (A + B)^*y\rangle.$$

The left side is

$$\langle Ax, y\rangle + \langle Bx, y\rangle = \langle Ax, y\rangle + \langle x, By\rangle$$

since B is everywhere defined. Thus

$$\langle Ax, y\rangle = \langle x, (A + B)^*y - By\rangle.$$

Hence $y \in D_{A^*} = D_A$ and $Ay = A^*y = (A + B)^*y - By$. Hence $y \in D_{A+B} = D_A$.

Let $\mathbf{A}$ be the closure of A. For the second part, it suffices to show that the closure of A + B equals $\mathbf{A}$ + B. But if $x \in D_{\mathbf{A}}$ there is a sequence $x_n \in D_A$ such that $x_n \to x$, and $Ax_n \to \mathbf{A}x$. Then $Bx_n \to Bx$ as B is bounded so x belongs to the domain of the closure of A + B. ■

In general, the sum of two self-adjoint operators need not be self-adjoint. (See Nelson [1959] and Chernoff [1974] for this and related examples.)

7.4.20 Proposition *Let* A *be a symmetric operator. If the range of* A *is all of* H *then* A *is self-adjoint.*

Proof We first observe that A is one-to-one. Indeed let $Ax = 0$. Then for any $y \in D_A$, $0 = \langle Ax, y\rangle = \langle x, Ay\rangle$. But A is onto and so $x = 0$. Thus A admits an everywhere defined inverse A^{-1}, which is therefore self-adjoint. Hence A is self-adjoint (we proved earlier than the inverse of a self-adjoint operator is self-adjoint). ■

We shall use these results to prove a theorem that typifies the kind of techniques one uses.

7.4.21 Proposition *Let* A *be a symmetric operator on* H *and suppose* $A \le 0$; *that is* $\langle Ax, x\rangle \le 0$ *for* $x \in D_A$. *Suppose* $I - A$ *has dense range. Then* A *is essentially self-adjoint.*

Proof Note that

$$\langle (I - A)u, u\rangle = \langle u, u\rangle - \langle Au, u\rangle \ \ge \| u \|^2$$

and so by the Schwarz inequality we have $\|(I - A)u \| \ge \| u \|$. It follows that the closure of $I - A$ which equals $I - \mathbf{A}$ has closed range, which by hypothesis must be all of H. By **7.4.20** $I - A$ is self-adjoint and so by **7.4.21** $\mathbf{A}$ is self-adjoint. ■

7.4.22 Corollary *If* A *is self-adjoint and* $A \le 0$, *then for any* $\lambda > 0$, $\lambda - A$ *is onto,* $(\lambda - A)^{-1}$ *exists and*

$$\| (\lambda - A)^{-1} \| \le 1/\lambda. \tag{3}$$

Proof As before we have

$$\langle (\lambda - A)u, u\rangle \ge \lambda \| u \|^2$$

which yields

$$\|(\lambda - A)u \| \ge \lambda \| u \|.$$

As A is closed, this implies that the range of $\lambda - A$ is closed. If we can show it is dense, the result will follow. Suppose y is orthogonal to the range:

$$\langle (\lambda - A)u, y\rangle = 0 \ \text{ for all } \ u \in D_A.$$

This means that $(\lambda - A)^* y = 0$, or since A is self-adjoint, $y \in D_A$. Making the choice $u = y$ gives $0 = \langle (\lambda - A)y, y\rangle \ge \lambda \| y \|^2$ so $y = 0$. ■

Note that *an operator* A *has dense range iff* A^* *is one-to-one*; i.e., $A^* w = 0$ implies $w = 0$.

For a given symmetric operator A, we considered the general problem of self-adjoint extensions of A and classified these in terms of the defect spaces. Now, under different hypotheses, we construct a special self-adjoint extension (even though A need not be essentially self-adjoint). This result is useful in many applications, including quantum mechanics.

A symmetric operator A on H is called ***lower semi-bounded*** if there is a constant $c \in \mathbb{R}$ such that $\langle Ax, x\rangle \geq c\| x \|^2$ for all $x \in D_A$. ***Upper semi-bounded*** is defined similarly. If A is either upper or lower semi-bounded then A is called ***semi-bounded***. Observe that if A is positive or negative then A is semi-bounded.

As an example, let $A = -\nabla^2 + V$ where ∇^2 is the Laplacian and let V be a real valued continuous function and bounded below, say $V(x) \geq \alpha$. Let $H = L^2(\mathbb{R}^n, \mathbb{C})$ and D_A the C^∞ functions with compact support. Then $-\nabla^2$ is positive so

$$\langle Af, f\rangle = \langle -\nabla^2 f, f\rangle + \langle Vf, f\rangle \geq \alpha\langle f, f\rangle,$$

and thus A is semi-bounded.

We already know that this operator is real so has self-adjoint extensions by von Neumann's theorem. However, the self-adjoint extension constructed below (called the ***Friedrichs extension***) is "natural." Thus the actual construction is as important as the statement.

7.4.23 Theorem *A semi-bounded symmetric (densely defined) operator admits a self-adjoint extension.*

Proof After multiplying by -1 if necessary and replacing A by $A + (1 - \alpha)I$ we can suppose $\langle Ax, x\rangle \geq \| x \|^2$. Consider the inner product on D_A given by $\langle\langle x, y\rangle\rangle = \langle Ax, y\rangle$. (Using symmetry of A and the preceding inequality one easily checks that this is an inner product.)

Let H^1 be the completion of D_A in this inner product. Since the H^1-norm is stronger than the H-norm, we have $H^1 \subset H$ (i.e., the injection $D_A \subset H$ extends uniquely to the completion).

Now let H^{-1} be the dual of H^1. We have an injection of H into H^{-1} defined as follows: if y is fixed and $x \mapsto \langle x, y\rangle$ is a linear functional on H, it is also continuous on H^1 since

$$|\langle x, y\rangle| \leq \| x \| \| y \| \leq ||| x ||| \| y \|,$$

where $||| \cdot |||$ is the norm of H^1. Thus $H^1 \subset H \subset H^{-1}$.

Now the inner product on H^1 defines an isomorphism $B : H^1 \to H^{-1}$. Let C be the operator with domain $D_C = \{x \in H^1 \mid B(x) \in H\}$, and $Cx = Bx$ for $x \in D_C$. Thus C is an extension of A. This will be the extension we sought. We shall prove that C is self-adjoint. By definition, C is surjective; in fact $C : D_C \to H$ is a linear isomorphism. Thus by **7.4.20** it suffices to show that C is symmetric. Indeed for $x, y \in D_C$ we have, by definition,

$$\overline{\langle Cx, y\rangle} = \overline{\langle\langle x, y\rangle\rangle} = \langle\langle y, x\rangle\rangle = \langle Cy, x\rangle = \langle x, Cy\rangle. \quad \blacksquare$$

The self-adjoint extension C can be alternatively described as follows. Let H^1 be as before and let C be the restriction of A^* to $D_{A^*} \cap H^1$. We leave the verification as an exercise. ✍

☞ SUPPLEMENT 7.4B
Stone's Theorem
(written in collaboration with P. Chernoff)

Here we give a self-contained proof of Stone's theorem for unbounded self-adjoint operators A on a complex Hilbert space H. This guarantees that the one-parameter group e^{itA} of unitary operators exists. In fact, there is a one-to-one correspondence between self-adjoint operators and continuous one-parameter unitary groups. A ***continuous one-parameter unitary group*** is a homomorphism $t \mapsto U_t$ from $\mathbb{R}$ to the group of unitary operators on H, such that for each $x \in H$ the map $t \mapsto U_t x$ is continuous. The ***infinitesimal generator*** A of U_t is defined by

$$iAx = \left.\frac{d}{dt} U_t x\right|_{t=0} = \lim_{h \to 0} \frac{U_h(x) - x}{h},$$

its domain D consisting of those x for which the indicated limit exists. We insert the factor i for convenience; iA is often called the ***generator.***

7.4.24 Stone's Theorem. *Let U_t be a continuous one-parameter unitary group. Then the generator A of U_t is self-adjoint. (In particular, by Supplement* **7.4A,** *it is closed and densely defined.) Conversely, let A be a given self-adjoint operator. Then there exists a unique one-parameter unitary group U_t whose generator is A.*

Before we begin the proof, let us note that if A is a *bounded* self-adjoint operator then one can form the series

$$U_t = e^{itA} = I + itA + \frac{1}{2!}(itA)^2 + \frac{1}{3!}(itA)^3 + \cdots$$

which converges in the operator norm. It is straightforward to verify that U_t is a continuous one-parameter unitary group and that A is its generator. Because of this, one often writes e^{itA} for the unitary group whose generator is A even if A is unbounded. (In the context of the so-called "operational calculus" for self-adjoint operators, one can show that e^{itA} really *is* the result of applying the function $e^{it(\cdot)}$ to A; however, we shall not go into these matters here.)

Proof of Stone's Theorem (first half) Let U_t be a given continuous unitary group. In a series of lemmas, we shall show that the generator A of U_t is self-adjoint.

7.4.25 Lemma *The domain* D *of* A *is invariant under each* U_t, *and moreover* $AU_t x = U_t Ax$ *for each* $x \in D$.

Proof Suppose $x \in D$. Then

$$\frac{1}{h}(U_hU_tx - U_tx) = U_t\left(\frac{1}{h}(U_hx - h)\right).$$

which converges to $U_t(iAx) = iU_tAx$ as $h \to 0$. The lemma follows by the definition of A. ∎

7.4.26 Corollary A *is closed.*

Proof If $x \in D$ then, by **7.4.25**

$$\frac{d}{dt}U_tx = iAU_tx = iU_tAx.$$

Hence
$$U_tx = x + i\int_0^t U_\tau Ax\, d\tau. \tag{1}$$

Now suppose that $x_n \in D$, $x_n \to x$, and $Ax_n \to y$. Then we have, by (1),

$$U_tx = \lim_{n\to\infty} U_tx_n = \lim_{n\to\infty}\left\{x_n + i\int_0^t U_\tau Ax_n d\tau\right\}.$$

Thus
$$U_tx = x + i\int_0^t U_\tau y\, d\tau. \tag{2}$$

(Here we have taken the limit under the integral sign because the convergence is uniform; indeed $\|U_\tau Ax_n - U_\tau y\| = \|Ax_n - y\| \to 0$ independent of $\tau \in [0, t]$.) Then, by (2),

$$\left.\frac{d}{dt}U_tx\right|_{t=0} = iy.$$

Hence $x \in D$ and $y = Ax$. Thus A is closed. ∎

7.4.27 Lemma A *is densely defined.*

Proof Let $x \in H$, and let φ be a C^∞ function with compact support on $\mathbb{R}$. Define

$$x_\varphi = \int_{-\infty}^{\infty} \varphi(t)U_tx\, dt.$$

We shall show that x_φ is in D, and that $x = \lim_{n\to\infty} x_{\varphi_n}$ for suitable sequence $\{\varphi_n\}$. To take the latter point first, let $\varphi_n(t)$ be nonnegative, zero outside the interval $[0, 1/n]$, and such that $\varphi_n(t)$ has integral 1. By continuity, if $\varepsilon > 0$ is given, one can find N so large that $\|U_tx - x\| < \varepsilon$ if $|t| < 1/N$. Suppose that $n > N$. Then

$$\| x_{\varphi_n} - x \| = \left\| \int_{-\infty}^{\infty} \varphi_n(t)(U_t x - x)\, dt \right\| = \left\| \int_0^{1/n} \varphi_n(t)(U_t x - x)\, dt \right\|$$

$$\leq \int_0^{1/n} \varphi_n(t) \, \| U_t x - x \| \, dt \leq \varepsilon \int_0^{1/n} \varphi_n(t)\, dt = \varepsilon.$$

Finally, we show that $x_\varphi \in D$; moreover, we shall show that $iAx_\varphi = -x_{\varphi'}$. Indeed,

$$- \int_0^t U_\tau x_{\varphi'}\, d\tau = - \int_0^t U_\tau\, d\tau \int_{-\infty}^{\infty} \varphi'(\sigma) U_\sigma\, d\sigma$$

$$= - \int_{-\infty}^{\infty} d\sigma \cdot \varphi'(\sigma) \cdot \int_0^t U_{\tau+\sigma} x\, d\tau = - \int_{-\infty}^{\infty} d\sigma \cdot \varphi'(\sigma) \cdot \int_\sigma^{\sigma+t} U_\tau x\, d\tau.$$

Integrating by parts and using the fact that φ has compact support, we get

$$- \int_0^t U_\tau x_{\varphi'}\, d\tau = \int_{-\infty}^{\infty} (U_{\sigma+t} x - U_\sigma x)\varphi(\sigma)\, d\sigma = (U_t - I) \int_{-\infty}^{\infty} U_\sigma x_\varphi(\sigma)\, d\sigma.$$

That is,

$$- \int_0^t U_\tau x_{\varphi'}\, d\tau = U_t x_\varphi - x_\varphi ,$$

from which the assertion follows. ■

Thus far we have made no significant use of the unitarity of U_t. We shall now do so.

7.4.28 Lemma A *is symmetric.*

Proof Take $x, y \in D$. Then we have

$$\langle Ax, y \rangle = \frac{1}{i} \frac{d}{dt} \langle U_t x, y \rangle \Big|_{t=0} = \frac{1}{i} \frac{d}{dt} \langle x, U_t^* y \rangle \Big|_{t=0}$$

$$= \frac{1}{i} \frac{d}{dt} \langle x, U_{-t} y \rangle \Big|_{t=0} = - \frac{1}{i} \frac{d}{dt} \langle x, U_t y \rangle \Big|_{t=0}$$

$$= - \frac{1}{i} \langle x, iAy \rangle = \langle x, Ay \rangle. \ ■$$

To complete the proof that A is self-adjoint, let $y \in D^*$ and $x \in D$. By **7.4.25, 28,**

$$\langle U_t y, x \rangle = \langle y, U_{-t} x \rangle = \langle y, x \rangle + \left\langle y, i \int_0^{-t} U_\tau Ax\, d\tau \right\rangle$$

$$= \langle y, x \rangle - i \int_0^{-t} \langle y, U_\tau Ax \rangle d\tau = \langle y, x \rangle - i \int_0^{-t} \langle y, AU_\tau x \rangle\, d\tau$$

$$= \langle y, x\rangle - i\int_0^{-t} \langle U_{-\tau} A^* y, x\rangle\, d\tau = \langle y, x\rangle + i\int_0^{t} \langle U_\tau A^* y, x\rangle\, d\tau$$

$$= \left\langle y + i\int_0^t U_\tau A^* y\, d\tau, x\right\rangle.$$

Because D is dense, it follows that

$$U_t y = y + i\int_0^t U_\tau A^* y\, d\tau.$$

Hence, differentiating, we see that $y \in D$ and $A^* y = Ay$. Thus $A = A^*$.

Proof of Stone's theorem (second half) We are now given a self-adjoint operator A. We shall construct a continuous unitary group U_t whose generator is A.

7.4.29 Lemma *If* $\lambda > 0$, *then* $I + \lambda A^2 : D_{A^2} \to H$ *is bijective,* $(I + \lambda A^2)^{-1} : H \to D_{A^2}$ *is bounded by* 1, *and* D_{A^2}, *the domain of* A^2, *is dense.*

Proof If A is self-adjoint, so is $\sqrt{\lambda}\, A$. It is therefore enough to establish the lemma for $\lambda = 1$. First we establish surjectivity.

By **7.4.14** and **7.4.26**, if $z \in H$ is given there exists a unique solution (x, y) to the equations

$$x - Ay = 0, \quad Ax + y = z.$$

From the first equation, $x = Ay$. The second equation then yields $A^2 y + y = z$, so $I + A^2$ is surjective.

For $x \in D_{A^2}$, note that $\langle (I + A^2)x, x\rangle \geq \| x \|^2$, so $\| (I + A^2)x \| \geq \| x \|$. Thus $I + A^2$ is one-to-one and $\| (I + A^2)^{-1} \| \leq 1$. Now suppose that u is orthogonal to D_{A^2}. We can find a v such that $u = v + A^2 v$. Then

$$0 = \langle u, v\rangle = \langle v + A^2 v, v\rangle = \| v \|^2 + \| Av \|^2,$$

whence $v = 0$ and therefore $u = 0$. Consequently D_{A^2} is dense in H. ■

For $\lambda > 0$, define an operator A_λ by $A_\lambda = A(I + \lambda A^2)^{-1}$. Note that A_λ is defined on all of H because if $x \in H$ then $(I + \lambda A^2)^{-1} x \in D_{A^2} \subset D$, so $A(I + \lambda A^2)^{-1} x$ makes sense.

7.4.30 Lemma A_λ *is a bounded self-adjoint operator. Also* A_λ *and* A_μ *commute for all* $\lambda, \mu > 0$.

Proof Pick $x \in H$. Then by **7.4.29**

$$\begin{aligned}\lambda\| A_\lambda x \|^2 &= \langle \lambda A(I+\lambda A^2)^{-1}x, A(I+\lambda A^2)^{-1}x\rangle \\ &= \langle \lambda A^2(I+\lambda A^2)^{-1}x, (I+\lambda A^2)^{-1}x\rangle \\ &\le \langle (I+\lambda A^2)(I+\lambda A^2)^{-1}x, (I+\lambda A^2)^{-1}x\rangle \\ &\le \|(I+\lambda A^2)^{-1}x\|^2 \le \|x\|^2,\end{aligned}$$

so $\| A_\lambda \| \le 1/\sqrt{\lambda}$, and thus A_λ is bounded.

We now show that A_λ is self-adjoint. First we shall show that if $x \in D$, then

$$A_\lambda x = (I+\lambda A^2)^{-1}Ax.$$

Indeed, if $x \in D$ we have $A_\lambda x \in D_{A^2}$ by **7.4.29** and so

$$(I+\lambda A^2)A_\lambda x = (I+\lambda A^2)A(I+\lambda A^2)^{-1}x = A(I+\lambda A^2)(I+\lambda A^2)^{-1}x = Ax.$$

Now suppose $x \in D$ and y is arbitrary. Then

$$\begin{aligned}\langle A_\lambda x, y\rangle &= \langle (I+\lambda A^2)^{-1}Ax, y\rangle = \langle (I+\lambda A^2)^{-1}Ax, (I+\lambda A^2)(I+\lambda A^2)^{-1}y\rangle \\ &= \langle Ax, (I+\lambda A^2)^{-1}y\rangle = \langle x, A_\lambda y\rangle.\end{aligned}$$

Because D is dense and A_λ bounded, this relation must hold for all $x \in H$. Hence A_λ is self-adjoint. The proof that $A_\lambda A_\mu = A_\mu A_\lambda$ is a calculation that we leave to the reader. ■

Since A_λ is bounded, we can form the continuous one-parameter unitary groups $U_t^\lambda = e^{itA_\lambda}$, $\lambda > 0$ using power series or the results of §**4.1**. Since A_λ and A_μ commute, it follows that U_s^λ and U_t^μ commute for every s and t.

7.4.31 Lemma *If* $x \in D$ *then* $\lim_{\lambda\to 0} A_\lambda x = Ax$.

Proof If $x \in D$ we have $A_\lambda x - Ax = (I+\lambda A^2)^{-1}Ax - Ax = -\lambda A^2(I+\lambda A^2)^{-1}Ax$. It is therefore enough to show that for every $y \in H$, $\lambda A^2(I+\lambda A^2)^{-1}y \to 0$. From the inequality $\|(I+\lambda A^2)y\|^2 \ge \|\lambda A^2 y\|^2$, valid for $\lambda \ge 0$, we see that $\|\lambda A^2(I+\lambda A^2)^{-1}\| \le 1$. Thus it is even enough to show the preceding equality for all y in some dense subspace of H. Suppose $y \in D_{A^2}$, which is dense by **7.4.29**. Then

$$\|\lambda A^2(I+\lambda A^2)^{-1}y\| = \lambda\|(I+\lambda A^2)^{-1}A^2 y\| \le \lambda\|A^2 y\|,$$

which indeed goes to zero with λ. ■

7.4.32 Lemma *For each* $x \in H$, $\lim_{\lambda\to 0} U_t^\lambda x$ *exists. If we call the limit* $U_t x$, *then* $\{U_t\}$ *is a continuous one-parameter unitary group.*

Proof We have

$$U_t^\lambda x - U_t^\mu x = \int_0^t \frac{d}{d\tau}(U_\tau^\lambda U_{t-\tau}^\mu)x\, d\tau = \int_0^t [iA_\lambda U_\tau^\lambda U_{t-\tau}^\mu x - U_\tau^\lambda i A_u U_{t-\tau}^\mu x]\, d\tau$$

$$= i\int_0^t U_\tau^\lambda U_{t-\tau}^\mu (A_\lambda x - A_\mu x)\, d\tau,$$

whence

$$\| U_t^\lambda x - U_t^\mu x \| \le |t|\, \| A_\lambda x - A_\mu x \|. \tag{3}$$

Now suppose that $x \in D$. Then by **7.4.31**, $A_\lambda x \to Ax$, so that $\| A_\lambda x - A_\mu x \| \to 0$ as λ, $\mu \to 0$. Because of (3) it follows that $\{U_t^\lambda \xi\}_{\lambda > 0}$ is uniformly Cauchy as $\lambda \to 0$ on every compact t-interval. It follows that $\lim_{\lambda \to 0} U_t^\lambda x = U_t x$ exists and is a continuous function of t. Moreover, since D is dense and all of the U_t^λ have norm 1, an easy approximation argument shows that the preceding conclusion holds even if $x \notin D$.

It is obvious that each U_t is a linear opeator. Furthermore,

$$\langle U_t x, U_t y\rangle = \lim_{\lambda \to 0} \langle U_t^\lambda x, U_t^\lambda y\rangle = \lim_{\lambda \to 0} \langle x, y\rangle = \langle x, y\rangle$$

so U_t is isometric. Trivially, $U_0 = I$. Finally,

$$\begin{aligned}\langle U_s U_t x, y\rangle &= \lim_{\lambda\to 0} \langle U_s^\lambda U_t x, y\rangle = \lim_{\lambda\to 0} \langle U_t x, U_{-s}^\lambda y\rangle \\ &= \lim_{\lambda\to 0} \langle U_t^\lambda x, U_{-s}^\lambda y\rangle = \lim_{\lambda\to 0} \langle U_{s+t}^\lambda x, y\rangle = \langle U_{s+t} x, y\rangle,\end{aligned}$$

so $U_s U_t = U_{s+t}$.

Thus, U_s has an inverse, namely U_{-s}, and so U_s is unitary. ■

7.4.33 Lemma *If* $x \in D$, *then*

$$\lim_{t\to 0} \frac{U_t x - x}{t} = iAx.$$

Proof We have

$$U_t^\lambda x - x = i\int_0^t U_\tau^\lambda A_\lambda x\, d\tau. \tag{4}$$

Now

$$U_\tau^\lambda A_\lambda x - U_\tau A x = U_\tau^\lambda (A_\lambda x - Ax) + U_\tau^\lambda Ax - U_\tau Ax \to 0$$

uniformly for $\tau \in [0, t]$ as $\lambda \to 0$. Thus letting $\lambda \to 0$ in (4), we get

$$U_t x - x = i\int_0^t U_\tau Ax\, d\tau \tag{5}$$

for all $x \in D$. The lemma follows directly from (5). ■

7.4.34 Lemma *If*

$$\lim_{t\to 0} \frac{U_t x - x}{t} = iw$$

exists, then $x \in D$.

Proof It suffices to show that $x \in D^*$, the domain of A^*, since $D = D^*$. Let $y \in D^*$. Then by 7.4.33,

$$\langle x, iAy\rangle = \lim_{t\to 0} \left\langle x, \frac{U_{-t}y - y}{-t}\right\rangle = -\lim_{t\to 0} \left\langle \frac{U_t x - x}{t}, y\right\rangle = -\langle iw, y\rangle.$$

So $\langle x, Ay\rangle = \langle w, y\rangle$. Thus $x \in D^*$ and so as A is self-adjoint, $x \in D$. ■

Let us finally prove uniqueness. Let $c(t)$ be a differentiable curve in H such that $c(t) \in D$ and $c'(t) = iA(c(t))$. We claim that $c(t) = U_t c(0)$. Indeed consider, $h(t) = U_{-t}c(t)$. Then

$$\| h(t+\tau) - h(t) \| = \| U_{-t-\tau}c(t+\tau) - U_{-t-\tau}U_\tau c(t) \| = \| c(t+\tau) - U_\tau c(t) \|$$

$$= \|(c(t+\tau) - c(t)) - (U_\tau c(t) - c(t)) \|.$$

Hence

$$\frac{h(t+\tau) - h(t)}{\tau} \to 0$$

as $\tau \to 0$, so h is constant. But $h(t) = h(0)$ means $c(t) = U_t c(0)$. ■

From the proof of Stone's theorem, one can deduce the following Laplace transform expression for the resolvent, which we give for the sake of completeness.

7.4.35 Corollary *Let* $\operatorname{Re}\lambda > 0$. *Then for all* $x \in H$,

$$(\lambda - iA)^{-1}x = \int_0^\infty e^{-\lambda t}U_t x \, dt.$$

Proof The foregoing is formally an identity if one thinks of U_t as e^{itA}. Indeed, if A is *bounded* then it follows just by manipulation of the power series: One has $e^{-\lambda t}e^{itA} = e^{-t(\lambda - iA)}$, as one can see by expanding both sides; next

$$\int_0^R e^{-t(\lambda - iA)}x \, dt = (\lambda - iA)^{-1}\,[x - e^{-R(\lambda - iA)}x]$$

(integrate the series term by term). Letting $R \to \infty$, one has the result.

Now for arbitrary A we know that $U_t x = \lim_{\mu \to 0} U_t^\mu x$, uniformly on bounded intervals. It follows that

$$\int_0^\infty e^{-\lambda t} U_t x \, dt = \lim_{\mu \to 0} \int_0^\infty e^{-\lambda t} U_t^\mu x \, dt = \lim_{\mu \to 0} (\lambda - iA_\mu)^{-1} x.$$

It remains to show that this limit is $(\lambda - iA)^{-1}x$. Now

$$(\lambda - iA)^{-1}x - (\lambda - iA_\mu)^{-1}x = (\lambda - iA_\mu)^{-1}[(\lambda - iA_\mu)(\lambda - iA)^{-1}x - x].$$

But $(\lambda - iA)^{-1}x \in D$ (see Proposition **7.4.12**) and so by **7.4.31**,

$$(\lambda - iA_\mu)(\lambda - iA)^{-1}x \to (\lambda - iA)(\lambda - iA)^{-1}x = x \text{ as } \mu \to 0.$$

Because $\|(\lambda - iA_\mu)^{-1}\| \le |\operatorname{Re}\lambda|^{-1}$ it follows that

$$\|(\lambda - iA)^{-1}x - (\lambda - iA_\mu)^{-1}x\| \to 0. \quad \blacksquare$$

In closing, we mention that many of the results proved have generalizations to continuous one-parameter groups or semi-groups of linear operators in Banach spaces (or on locally convex spaces). The central result, due to Hille and Yosida, characterizes generators of semi-groups. Our proof of Stone's theorem is based on methods that can be used in the more general context. Expositions of this more general context are found in, for example, Kato [1966] and Marsden and Hughes [1983, ch 6].

Exercises

(Exercises **7.4A-C** form a unit.)

7.4A Given a manifold M, show that the space of half-densities on M carries a natural inner product. Let its completion be denoted $\mathfrak{H}(M)$, which is called the ***intrinsic Hilbert space*** of M. If μ is a density on M, define a bijection of $L^2(M, \mu)$ with $\mathfrak{H}(M)$ by $f \mapsto f\sqrt{\mu}$. Show that it is an isometry.

7.4B If F_t is the (local) flow of a smooth vector field X, show that F_t induces a flow of isometries on $\mathfrak{H}(M)$. (Make no assumption that X is divergence-free.) Show that the generator $iX' = \mathcal{L}_X$ of the induced flow on $\mathfrak{H}(M)$ is

$$iX'(f\sqrt{\mu}) = \left(X[f] + \frac{1}{2}(\operatorname{div}_\mu X)f\right)\sqrt{\mu}$$

and check directly that X' is a symmetric operator on the space of half-densities with compact support.

7.4C Prove that F_t is complete a.e. if and only if X' is essentially self-adjoint.

7.4D Consider the flow in $\mathbb{R}^2$ associated with a reflecting particle: for $t > 0$, set

$$F_t(q, p) = q + tp \text{ if } q > 0,\ q + tp > 0$$

and

$$F_t(q, p) = -q - tp \text{ if } q > 0,\ q + tp < 0$$

and set

$$F_t(-q, p) = -F_t(q, p) \text{ and } F_{-t} = F_t^{-1}.$$

What is the generator of the induced unitary flow? Is it essentially self-adjoint on the C^∞ functions with compact support away from the line $q = 0$?

7.4E Let M be an oriented Riemannian manifold and $L^2(\Lambda^k(M))$ the space of L^2 k-forms with inner product $\langle \alpha, \beta \rangle = \int \alpha \wedge *\beta$. If X is a Killing field on M with a complete flow F_t, show that $i\pounds_X$ is a self-adjoint operator on $L^2(\Lambda^k(M))$.

§7.5 *Introduction to Hodge-DeRham Theory and Topological Applications of Differential Forms*

Recall that a k-form α is called ***closed*** if $\mathbf{d}\alpha = 0$ and ***exact*** if $\alpha = \mathbf{d}\beta$ for some $k-1$ form β. Since $\mathbf{d}^2 = 0$, every exact form is closed, but the converse need not hold. Let

$$H^k(M) = \frac{\ker \mathbf{d}^k}{\text{range } \mathbf{d}^{k-1}},$$

(where $\mathbf{d}^k$ denotes the exterior derivative on k-forms), and call it the k-th ***deRham cohomology group of*** M. (The group structure here is a real vector space.) The celebrated ***deRham theorem*** states that for a finite-dimensional compact manifold, these groups are isomorphic to the singular cohomology groups (with real coefficients) defined in algebraic topology; the isomorphism is given by integration. For "modern" proofs, see Singer and Thorpe [1967] or Warner [1971]. The original books of Hodge [1952] and deRham [1955] remain excellent sources of information as well. A special but important case of the deRham theorem is proved in Supplement **7.5B**.

The scope of this section is to informally discuss the Hodge decomposition theory based on differential operators and to explain how it is related to the deRham cohomology groups. In addition, some topological applications of the theory are given, such as the Brouwer fixed-point theorem, and the degree of a map is defined. In the sequel, M will denote a compact oriented Riemannian manifold, and δ the codifferential operator. At first we assume M has no boundary. Later we will discuss the case in which M has a boundary.

7.5.1 Definition *The **Laplace-deRham operator** $\Delta : \Omega^k(M) \to \Omega^k(M)$ is defined by $\Delta = \mathbf{d}\delta + \delta\mathbf{d}$. A form for which $\Delta\alpha = 0$ is called **harmonic**. Let $\mathcal{H}^k = \{\alpha \in \Omega^k(M) \mid \Delta\alpha = 0\}$ denote the vector space of harmonic k-forms.*

If $f \in \mathcal{H}^0(M)$, then $\Delta f = \mathbf{d}\delta f + \delta\mathbf{d}f = \delta\mathbf{d}f = -\operatorname{div}\operatorname{grad} f$. So $\Delta f = -\nabla^2 f$, where ∇^2 is the Laplace-Beltrami operator. This minus sign can be a source of confusion and one has to be careful.

Recall that the L^2-inner product in $\Omega^k(M)$ is defined by

$$\langle \alpha, \beta \rangle = \int_M \alpha \wedge *\beta$$

and that $\mathbf{d}$ and δ are adjoints with respect to this inner product. That is, $\langle \mathbf{d}\alpha, \beta \rangle = \langle \alpha, \delta\beta \rangle$ for all $\alpha \in \Omega^{k-1}(M)$, $\beta \in \Omega^k(M)$. Thus is follows that for $\alpha, \beta \in \Omega^k(M)$, we have

$$\langle\Delta\alpha, \beta\rangle = \langle \mathbf{d}\delta\alpha, \beta\rangle + \langle \delta\mathbf{d}\alpha , \beta\rangle = \langle \delta\alpha, \delta\beta\rangle + \langle \mathbf{d}\alpha, \mathbf{d}\beta\rangle$$
$$= \langle\alpha, \mathbf{d}\delta\beta\rangle + \langle\alpha, \delta\mathbf{d}\beta\rangle = \langle\alpha, \Delta\beta\rangle,$$

and thus Δ is symmetric. This computation also shows that $\langle\Delta\alpha, \alpha\rangle \geq 0$ for all $\alpha \in \Omega^k(M)$.

7.5.2 Proposition *Let* M *be a compact boundaryless oriented Riemannian manifold and* $\alpha \in \Omega^k(M)$. *Then* $\Delta\alpha = 0$ *iff* $\delta\alpha = 0$ *and* $\mathbf{d}\alpha = 0$.

Proof It is obvious from the expression $\Delta\alpha = \delta\mathbf{d}\alpha + \delta\mathbf{d}\alpha$ that if $\mathbf{d}\alpha = 0$ and $\delta\alpha = 0$, then $\Delta\alpha = 0$. Conversely, the previous computation shows that $0 = \langle\Delta\alpha, \alpha\rangle = \langle\mathbf{d}\alpha, \mathbf{d}\alpha\rangle + \langle\delta\alpha, \delta\alpha\rangle$, so the result follows. ∎

7.5.3 The Hodge Decomposition Theorem *Let* M *be a compact, boundaryless, oriented, Riemannian manifold and let* $\omega \in \Omega^k(M)$. *Then there is an* $\alpha \in \Omega^{k-1}(M)$, $\beta \in \Omega^{k+1}(M)$ *and* $\gamma \in \Omega^k(M)$ *such that* $\omega = \mathbf{d}\alpha + \delta\beta + \gamma$ *and* $\Delta(\gamma) = 0$. *Furthermore,* $\mathbf{d}\alpha$, $\delta\beta$, *and* γ *are mutually* L^2 *orthogonal and so are uniquely determined. That is,*

$$\Omega^k(M) = \mathbf{d}\Omega^{k-1}(M) \oplus \delta\Omega^{k+1}(M) \oplus \mathcal{H}^k. \tag{1}$$

We can easily check that the spaces in the Hodge decomposition are orthogonal. For example, $\mathbf{d}\Omega^{k-1}(M)$ and $\delta\Omega^{k+1}(M)$ are orthogonal since

$$\langle\mathbf{d}\alpha, \delta\beta\rangle = \langle\mathbf{dd}\alpha, \beta\rangle = 0,$$

δ being the adjoint of $\mathbf{d}$ and $\mathbf{d}^2 = 0$.

The basic idea behind the proof of the Hodge theorem can be abstracted as follows. We consider a linear operator T on a Hilbert space E and assume that $T^2 = 0$. In our case $T = \mathbf{d}$ and E is the space of L^2 forms. (We ignore the fact that T is only densely defined.) Let T^* be the adjoint of T. Let $\mathcal{H} = \{x \in E \mid Tx = 0 \text{ and } T^*x = 0\}$. We assert that

$$E = \text{cl(range T)} \oplus \text{cl(range } T^*) \oplus \mathcal{H} \tag{2}$$

which, apart from technical points on understanding the closures, is the essential content of the Hodge decomposition. To prove (2), note that the ranges of T and T^* are orthogonal because

$$\langle Tx, T^*y\rangle = \langle Tx^2, y\rangle = 0.$$

If $\mathcal{C}$ denotes the orthogonal complement of cl(range T) $\oplus$ cl(range T^*), then $\mathcal{H} \subset \mathcal{C}$. If $x \in \mathcal{C}$ then $\langle Ty, x\rangle = 0$ for all y implies $T^*x = 0$. Similarly, $Tx = 0$, so $\mathcal{C} \subset \mathcal{H}$ and hence $\mathcal{C} = \mathcal{H}$.

The complete proof of the Hodge theorem requires elliptic estimates and may be found in Morrey [1966]. For more elementary expositions, consult Fladers [1963] and Warner [1971].

7.5.4 Corollary *Let* $\mathcal{H}^k$ *denote the space of harmonic* k*-forms. Then the vector spaces* $\mathcal{H}^k$ *and* H^k (= ker $\mathbf{d}^k$/range $\mathbf{d}^{k-1}$) *are isomorphic.*

Proof Map $\mathcal{H}^k \to \ker \mathbf{d}^k$ by inclusion and then to H^k by projection. We need to show that this map is an isomorphism. Suppose $\gamma \in \mathcal{H}^k$ and $[\gamma] = 0$ where $[\gamma] \in H^k$ is the class of γ. But $[\gamma] = 0$ means that γ is exact; $\gamma = \mathbf{d}\beta$. But since $\delta\gamma = 0$, γ is orthogonal to $\mathbf{d}\beta$; i.e., γ is orthogonal to itself, so $\gamma = 0$. Thus the map $\gamma \mapsto [\gamma]$ is one-to-one. Next let $[\omega] \in H^k$. We can, by the Hodge theorem, decompose $\omega = \mathbf{d}\alpha + \delta\beta + \gamma$, where $\gamma \in \mathcal{H}^k$. Since $\mathbf{d}\omega = 0$, $\mathbf{d}\delta\beta = 0$ so $0 = \langle \beta, \mathbf{d}\delta\beta \rangle = \langle \delta\beta, \delta\beta \rangle$, so $\delta\beta = 0$. Thus $\omega = \mathbf{d}\alpha + \gamma$ and hence $[\omega] = [\gamma]$, so the map $\gamma \mapsto [\gamma]$ is onto. ■

The space $\mathcal{H}^k \cong H^k$ is finite dimensional. Again the proof relies on elliptic theory (the kernel of an elliptic operator on a compact manifold is finite dimensional).

The Hodge theorem plays a fundamental role in incompressible hydrodynamics, as we shall see in §**8.2**. It allows the introduction of the pressure for a given fluid state. It has applications to many other areas of mathematical physics and engineering as well; see for example, Fischer and Marsden [1979] and Wyatt et al. [1978].

Below we shall state a generalization of the Hodge theorem for some decomposition theorems for general elliptic operators (rather than the special case of the Laplacian). However, we first pause to discuss what happens if a boundary is present. This theory was worked out by Kodaira [1949], Duff and Spencer [1952], and Morrey [1966, Ch. 7]. Differentiability across the boundary is very delicate, but important. Some of the best results in this regard are due to Morrey.

Note that $\mathbf{d}$ and δ may not be adjoints in this case, because boundary terms arise when we intregrate by parts (see Exercise **7.5E**). Hence we must impose certain boundary conditions. Let $\alpha \in \Omega^k(M)$. Then α is called *parallel* or *tangent* to ∂M if the normal part, defined by $n\alpha = i^*(*\alpha)$ is zero where $i : \partial M \to M$ is the inclusion map. Analogously, α is *perpendicular* to ∂M if its tangent part, defined by $t\alpha = i^*(\alpha)$ is zero.

Let X be a vector field on M. Using the metric, we know when X is tangent or perpendicular to ∂M. Now X corresponds to the one-form $X^\flat$ and also to the (n–1)-form $\mathbf{i}_X\mu = *X^\flat$ (μ denotes the Riemannian volume form). One checks that X *is tangent to* ∂M *if and only if* $X^\flat$ *is tangent to* ∂M *iff* $\mathbf{i}_X\mu$ *is normal to* ∂M. *Similarly* X *is normal to* ∂M *iff* $\mathbf{i}_X\mu$ *is tangent to* ∂M. Set

$$\Omega^k_t(M) = \{\alpha \in \Omega^k(M) \mid \alpha \text{ is tangent to } \partial M\}$$
$$\Omega^k_n(M) = \{\alpha \in \Omega^k(M) \mid \alpha \text{ is perpdendicular to } \partial M\}, \text{ and}$$
$$\mathcal{H}^k(M) = \{\alpha \in \Omega^k(M) \mid \mathbf{d}\alpha = 0, \delta\alpha = 0\}.$$

The condition that $\mathbf{d}\alpha = 0$ and $\delta\alpha = 0$ is now stronger than $\Delta\alpha = 0$. One calls elements of $\mathcal{H}^k$ ***harmonic fields***, after Kodaira [1949].

7.5.5 The Hodge Theorem for Manifolds with Boundary *Let* M *be a compact oriented Riemannian manifold with boundary. The following decomposition holds:*

$$\Omega^k(M) = \mathbf{d}\Omega^{k-1}{}_t(M) \oplus \delta\Omega^{k+1}{}_n(M) \oplus \mathcal{H}^k.$$

One can easily check from the following formula (obtained from Stokes' theorem):

$$\langle \mathbf{d}\alpha, \beta\rangle = \langle \alpha, \delta\beta\rangle + \int_{\partial M} \alpha \wedge *\beta$$

(see Exercise **7.5E**), that the summands in this decomposition are orthogonal.

Two other closely related decompositions of interest are

(i) $$\Omega^k(M) = \mathbf{d}\Omega^{k-1}(M) \oplus D^k{}_t$$

where

$$D^k{}_t = \{\alpha \in \Omega^k{}_t(M) \mid \delta\alpha = 0\}$$

are the ***co-closed*** k-***forms tangent*** to ∂M and, dually

(ii) $$\Omega^k(M) = \delta(\Omega^{k+1}(M)) \oplus C^k{}_n$$

where $C^k{}_n$ are the ***closed*** k-***forms normal*** to ∂M.

To put the Hodge theorem in a general context, we give a brief discussion of differential operators and their symbols. (See Palais [1965a], Wells [1980], and Marsden and Hughes [1983] for more information and additional details on proofs.) Let E and F be vector bundles of M and let $C^\infty(E)$ denote the C^∞ sections of E. Assume M is Riemannian and that the fibers of E and F have inner products. A k-***th order differential operator*** is a linear map $D : C^\infty(E) \to C^\infty(F)$ such that, if $f \in C^\infty(E)$ and f vanishes to k-th order at $x \in M$, then $D(f)(x) = 0$. It is not difficult to see that vanishing to k-th order makes intrinsic sense independent of charts and that D is a k-th order differential operator iff in local charts D has the form

$$D(f) = \sum_{0 \le |j| \le k} \alpha_j \frac{\partial^{|j|} f}{\partial x^{j_1} \cdots \partial x^{j_s}}.$$

where $j = (j_1, \ldots, j_s)$ is a multi-index and α_j is a C^∞ matrix-valued function of x (the matrix corresponding to linear maps of E to F).

The operator D has an ***adjoint operator*** D^* given in charts (with the standard Euclidean inner product on fibers) by

$$\mathsf{D}^*(h) = \sum_{0 \le |j| \le k} (-1)^{|j|} \frac{1}{\rho} \frac{\partial^{|j|}}{\partial x^{j_1} \cdots \partial x^{j_s}} (\rho \alpha_j^t h),$$

where $\rho\, dx^1 \wedge \cdots \wedge dx^n$ is the volume element on M and $\alpha_j^{\,t}$ is the transpose of α_j. The crucial property of D^* is

$$\langle g, \mathsf{D}^* h\rangle = \langle \mathsf{D} g, h\rangle,$$

where $\langle\, ,\rangle$ denotes the L^2 inner product, $g \in C^\infty_c(E)$, and $h \in C^\infty_c(F)$. That is, g and h are C^∞ sections with compact support. For example, we have the operators

$$\begin{array}{lll} \mathbf{d}\colon C^\infty(\Lambda^k) \to C^\infty(\Lambda^{k+1}) & \text{(first order)} \\ \delta\colon C^\infty(\Lambda^k) \to C^\infty(\Lambda^{k-1}) & \text{(first order)} \\ \Delta\colon C^\infty(\Lambda^k) \to C^\infty(\Lambda^k) & \text{(second order)} \end{array}$$

where $\mathbf{d}^* = \delta$, $\delta^* = \mathbf{d}$ and $\Delta^* = \Delta$. The ***symbol*** of D assigns to each $\xi \in T^*_xM$, a linear map $\sigma(\xi) : E_x \to F_x$ defined by

$$\sigma(\xi)(e) = \mathbf{D}\left(\frac{1}{k!}(g - g(x))^k f\right)(x),$$

where $g \in C^\infty(M, \mathbb{R})$, $\mathbf{d}g(x) = \xi$ and $f \in C^\infty(E)$, $f(x) = e$. By writing this out in coordinates one sees that $\sigma(\xi)$ so defined is independent of g and f and is a homogeneous polynomial expression in ξ of degree k obtained by substituting each ξ_j in place of $\partial/\partial x^j$ in the highest order terms. For example, if

$$\mathsf{D}(f) = \sum \alpha^{ij} \frac{\partial^2 f}{\partial x^i \partial x^j} + \text{(lower order terms)} \quad \text{then} \quad \sigma(\xi) = \sum \alpha^{ij} \xi_i \xi_j$$

(α^{ij} is for each i, j a map of E_x to F_x). For real-valued functions, the classical definition of an elliptic operator is that the foregoing quadratic form be *definite*. This can be generalized as follows: D is called ***elliptic*** if $\sigma(\xi)$ is an isomorphism for each $\xi = 0$. To see that $\Delta : C^\infty(\Lambda^k) \to C^\infty(\Lambda^k)$ is elliptic one uses the following facts:

1 The symbol of $\mathbf{d}$ is $\sigma(\xi) = \xi \wedge$.
2 The symbol of δ is $\sigma(\xi) = \mathbf{i}_{\xi^\#}$.
3 The symbol is multiplicative: $\sigma(\xi)(\mathsf{D}_1 \circ \mathsf{D}_2) = \sigma(\xi)(\mathsf{D}_1) \circ \sigma(\xi)(\mathsf{D}_2)$.

From these, it follows by a straightforward calculation that the symbol of Δ is given by $\sigma(\xi)\alpha = \|\xi\|^2\alpha$, so Δ is elliptic. (Compute $\xi \wedge (\mathbf{i}_{\xi^\#}\alpha) + \mathbf{i}_{\xi^\#}(\xi \wedge \alpha)$ applied to $(v_1, \ldots, v_k)$, noting that all but one term cancel.)

7.5.6 Elliptic Splitting Theorem or **Fredholm Alternative** *Let* D *be an elliptic operator as above. Then*

$$C^\infty(F) = D(C^\infty(E)) \oplus \ker D^*$$

Indeed this holds if it is merely assumed that either D *or* D^* *has injective symbol.*

The proof of this leans on elliptic estimates that are not discussed here. As in the Hodge theorem, the idea is that the L^2 orthogonal complement of range D is ker D^*. This yields an L^2 splitting and we get a C^∞ splitting via elliptic estimates. The splitting in case D (resp. D^*) has injective symbol relies on the fact that then D^*D (resp. DD^*) is elliptic.

For example, the equation $Du = f$ is soluble iff f is orthogonal to ker D^*. More specifically, $\Delta u = f$ is soluble if f is orthogonal to the constants; i.e., $\int f d\mu = 0$.

The Hodge Theorem is derived from the elliptic splitting theorem as follows. Since Δ is elliptic and symmetric

$$C^\infty(\Lambda^k(M)) = \text{range}\,\Delta \oplus \ker \Delta = \text{range}\,\Delta \oplus \mathcal{H}$$

Now write a k-form ω as $\omega = \Delta\rho + \gamma = \mathbf{d}\delta\rho + \delta\mathbf{d}\rho + \gamma$, so to get **7.5.3**, we can choose $\alpha = \delta\rho$ and $\beta = \mathbf{d}\rho$.

☞ SUPPLEMENT 7.5A
Introduction to Degree Theory

One of the purposes of degree theory is to provide algebraic measures of the number of solutions of nonlinear equations. Its development rests on Stokes' theorem. It links beautifully calculus on manifolds with ideas on differential and algebraic topology.

All manifolds in this section are assumed to be finite dimensional, paracompact and Lindelöf. We begin with an extandability result.

7.5.7 Proposition *Let* V *and* N *be orientable manifolds,* $\dim(V) = n + 1$ *and* $\dim(N) = n$. *If* $f : \partial V \to N$ *is a smooth proper map that extends to a smooth map of* V *to* N, *then for every* $\omega \in \Omega^n(N)$ *with compact support,* $\int_{\partial V} f^*\omega = 0$.

Proof Let $F : V \to N$ be a smooth extension of f. Then by Stokes' theorem

$$\int_{\partial V} f^*\omega = \int_{\partial V} F^*\omega = \int_V \mathbf{d}F^*\omega = \int_V F^*\mathbf{d}\omega = 0,$$

since $\mathbf{d}\omega = 0$. ■

This proposition will be applied to the case $V = [0, 1] \times M$. For this purpose let us recall the product orientation (see Exercise **6.5N**). If N and M are orientable manifolds (at most one of which has a boundary), then $N \times M$ is a manifold (with boundary), which is orientable in the following way. Let $\pi_1 : N \times M \to N$ and $\pi_2 : N \times M \to M$ be the canonical projections and $[\omega]$, $[\eta]$ orientations on N and M respectively. Then the orientation of $N \times M$ is defined to be $[\pi_1{}^*\omega \wedge \pi_1{}^*\eta]$. Alternatively, if $v_1, ..., v_n \in T_xN$ and $w_1, ..., w_m \in T_yM$ are positively oriented bases in the respective tangent spaces, then $(v_1, 0), ..., (v_n, 0), (0, w_1), ..., (0, w_m) \in T_{(x,y)}(N \times M)$ is defined to be a positively oriented basis in their product. Thus, for $[0, 1] \times M$, a natural orientation will be given at every point $(t, x) \in [0, 1] \times M$ by $(1, 0), (0, v_1), ..., (0, v_m)$, where $v_1, ..., v_m \in T_xM$ is a positively oriented basis.

The boundary orientation of $[0, 1] \times M$ is determined according to definition **7.2.7**. Since $\partial([0, 1] \times M) = (\{0\} \times M) \cup (\{1\} \times M)$, every element of this union is oriented by the orientation of M. On the other hand, this union is oriented by the boundary orientation of $[0, 1] \times M$. Since the outward normal at $(1, x)$ is $(1, 0)$, we see that a positively oriented basis of $T_{(1,x)}(\{1\} \times M)$ is given by $(0, v_1), ..., (0, v_m)$ for $v_1, ..., v_m \in T_xM$ a positvely oriented basis. However, since the outward normal at $(0, x)$ is $(-1, 0)$, a positively oriented basis of $T_{(0,x)}(\{0\} \times M)$ must consist of elements $(0, w_1), ..., (0, w_m)$ such that $(-1, 0), (0, w_1), ..., (0, w_m)$ is positively oriented in $[0, 1] \times M$, i.e., defines the same orientation as $(1, 0), (0, v_1), ..., (0, v_m)$, for $v_1, ..., v_m \in T_xM$ a positively oriented basis. This means that $w_1, ..., w_m \in T_xM$ is *negatively* oriented (see Fig. 7.5.1.). Thus *the oriented manifold* $\partial([0, 1] \times M)$ *is the disjoint union of* $\{0\} \times M$, *where* M *is negatively oriented, with* $\{1\} \times M$, *where* M *is positively oriented.*

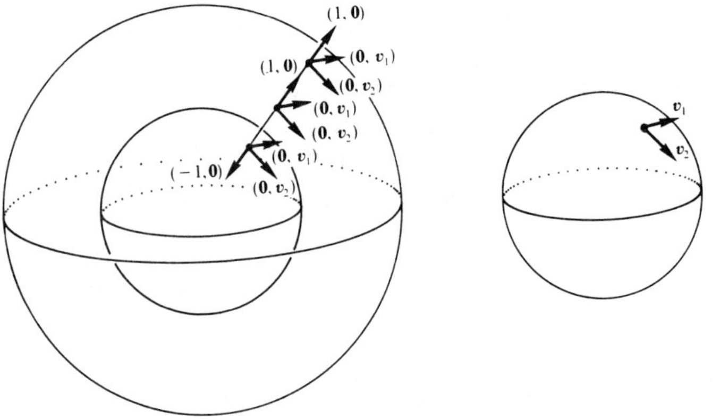

Figure 7.5.1

7.5.8 Definition *Two smooth mappings* $f, g : M \to N$, *are called* C^r-***homotopic*** *if there is a* C^r-*map* $F : [0,1] \times M \to N$ *such that* $F(0, m) = f(m)$ *and* $F(1, m) = g(m)$, *for all* $m \in M$. *The homotopy* F *is called* ***proper*** *if* F *is a proper map; in this case* f *and* g *are said to be* ***properly*** C^r-***homotopic*** *maps.*

Note that if f and g are properly homotopic then necessarily f and g are proper as restrictions of a proper map to the closed sets $\{0\} \times M$ and $\{1\} \times M$, respectively.

7.5.9 Proposition *Let* M *and* N *be orientable* n-*manifolds, with* M *boundaryless,* $\omega \in \Omega^n(N)$ *with compact support, and* $f, g : M \to N$ *be properly smooth homotopic maps. Then*

$$\int_M f^* \omega = \int_M g^* \omega.$$

Proof There are two ways to do this.

Method 1. Let $F : [0, 1] \times M \to N$ be the proper homotopy between f and g. By the remarks preceding definition **7.5.8**, we have

$$\int_{\{0\}\times M} f^* \omega = -\int_M f^* \omega \quad \text{and} \quad \int_{\{1\}\times M} g^* \omega = \int_M g^* \omega,$$

so that

$$\int_M g^* \omega - \int_M f^* \omega = \int_{\partial([0,1]\times M)} (F \mid \partial([0, 1] \times M))^* \omega = 0$$

by proposition **7.5.7**.

Method 2. By **6.4.16**, $f^*\omega - g^*\omega = d\eta$ for some $\eta \in \Omega^{n-1}(M)$ which has compact support since the homotopy between f and g is a proper map. Then by Stokes' theorem

$$\int_M f^* \omega - \int_M g^* \omega = \int_M d\eta = 0. \quad \blacksquare$$

Remark Properness of f and g does not suffice in the hypothesis of Proposition **7.5.9.** For example, if $M = N = \mathbb{R}$, $\omega = a dx$ with $a \geq 0$ a C^∞ function satisfying $\operatorname{supp}(a) \subset\,]-\infty, 0[$, then $f(x) = x$ and $g(x) = x^2$ are smoothly but not properly homotopic via $F(t, x) = (1 - t)x + tx^2$ and $\int_{-\infty}^{+\infty} f^* \omega > 0$, while $\int_{-\infty}^{+\infty} g^* \omega = 0$ since $g^* \omega = 0$.

7.5.10 Degree Theorem *Let* M *and* N *be oriented manifolds,* N *connected,* M *boundaryless, and* $f : M \to N$ *a smooth proper map. Then there is an integer* deg(f) *constant on the proper homotopy class of* f, *called the* ***degree of*** f *such that for any* $\eta \in \Omega^n(N)$ *with compact support,*

$$\int_M f^* \eta = \deg(f) \int_N \eta. \tag{1}$$

If $x \in M$ *is a regular point of* f, *let* $\operatorname{sign}(T_x f)$ *be* 1 *or* -1 *depending on whether the isomorphism* $T_x f : T_x M \to T_{f(x)} N$ *preserves or reverses orientation. The integer* deg(f) *is given by*

$$\deg(f) = \sum_{x \in f^{-1}(y)} \text{sign}(T_x f), \tag{2}$$

where y *is an arbitrary regular value of* f; *if* $y \notin f(M)$ *the right hand side is by convention equal to zero.*

Proof By proposition **7.5.9**, $\int_M f^*\eta$ depends only on the proper homotropy class of f (and on η). By Sard's Theorem, there is a regular value y of f. There are two possibilities: either $\mathcal{R}_f = N \setminus f(M)$ or not. If $\mathcal{R}_f = N \setminus f(M)$, then $T_x f$ is never onto for all $x \in M$. For any $v_1, \ldots, v_n \in T_x M$,

$$(f^*\eta)(x)(v_1, \ldots, v_n) = \eta(f(x))(T_x f(v_1), \ldots, T_x f(v_n)) = 0,$$

since $T_x f(v_1), \ldots, T_x f(v_n)$ are linearly dependent. Thus deg(f) exists and equals zero.

Assume $\mathcal{R}_f \cap f(M) \neq \varnothing$ and let $y \in \mathcal{R}_f \cap f(M)$. Since M and N have the same dimension, $f^{-1}(y)$ is a zero - dimensional submanifold of M, hence discrete. Properness of f implies that $f^{-1}(y)$ is also compact, i.e., $f^{-1}(y) = \{x_1, \ldots, x_{k+\ell}\}$, where $T_{x_i} f$ is orientation preserving for $i = 1, \ldots, k$ and orientation reversing for $i = k + 1, \ldots, k + \ell$. The inverse function theorem implies that there are open neighborhoods U_i of x_i and V of y such that $f^{-1}(V) = U_1 \cup \cdots \cup U_{k+\ell}$, $U_i \cap U_j = \varnothing$ and if $f \mid U_i : U_i \to V$ is a diffeomorphism. If $\text{supp}(\eta) \subset V$, then by the change of variables formula

$$\int_M f^*\eta = \sum_{i=1}^{k+\ell} \int_{U_i} f^*\eta = (k - \ell)\int_V \eta = (k - \ell) \int_N \eta \tag{3}$$

and so the theorem is proved for $\text{supp}(\eta) \subset V$.

To deal with a general η proceed in the following way. For the open neighborhood V of η, consider the collection of open subsets of N, $\mathcal{S} = \{\varphi(V) \mid \varphi$ is a diffeomorphism properly homotopic to the identity$\}$. We shall prove that $\mathcal{S}$ covers N. Let $n \in N$; we will show that there is a diffeomorphism φ properly homotopic to the identity such that $\varphi(n) = y$. Let $c : [0, 1] \to N$ be a smooth curve with $c(0) = n$ and $c(1) = y$. As in Proposition **5.5.9**, use a partition of unity to extend $c'(t)$ to a smooth vector field $X \in \mathfrak{X}(N)$ such that X vanishes outside a compact neighborhood of $c([0, 1])$. The flow F_t of X is complete by Proposition **4.1.20** and is the identity outside the above compact neighborhood of $c([0, 1])$. Thus the restriction $F : [0, 1] \times N \to N$ is proper. Then $\varphi = F_1$ is a proper diffeomorphism properly homotopic to the identity on N and $\varphi(n) = F_1(n) = c(1) = y$.

Since $\mathcal{S}$ covers N, choose a partition of unity $\{(V_\alpha, h_\alpha)\}$ subordinate to $\mathcal{S}$ and let $\eta_\alpha = h_\alpha \eta$; thus $\text{supp}(\eta) \subset V_\alpha \subset \varphi_\alpha(V)$ for some φ_α. Since all φ_α are orientation preserving, the change of variables formula and (3) give

$$(k-\ell)\int_N \eta = (k-\ell)\sum_\alpha \int_{V_\alpha} \eta_\alpha = (k-\ell)\sum_\alpha \int_V \varphi_\alpha^* \eta_\alpha = \sum_\alpha \int_M f^* \varphi_\alpha^* \eta_\alpha .$$

Since φ_α is properly homotopic to the identity and f is proper, it follows that $\varphi_\alpha \circ f$ is properly homotopic to f. Thus by Proposition **7.5.9**, $\int_M (\varphi_\alpha \circ f)^* \eta_\alpha = \int_M f^* \eta_\alpha$ and therefore,

$$(k-\ell)\int_N \eta = \sum_\alpha \int_M f^* \eta_\alpha = \int_M f^* \eta. \quad \blacksquare$$

Notice that by construction, if $\deg(f) \neq 0$, then f is onto, so $f(x) = y$ is solvable for x given y.

7.5.11 Corollary *Let* V *and* N *be orientable manifolds with* $\dim(V) = n + 1$, *and* $\dim(N) = n$. *If* $f : \partial V \to N$ *extends to* V, *then* $\deg(f) = 0$.

This is a reformulation of **7.5.7**. Similarly, **7.5.9** is equivalent to the following.

7.5.12 Corollary *Let* M, N *be orientable* n*-manifolds,* N *connected,* M *boundaryless, and let* $f, g : M \to N$ *be smooth properly homotopic maps. Then* $\deg(f) = \deg(g)$.

This corollary is useful in three important applications. The first concerns vector fields on spheres.

7.5.13 Hairy Ball Theorem *Every vector field on an even dimensional sphere has a critical point.*

Proof Let S^{2n} be embedded as the unit sphere in $\mathbb{R}^{2n+1}$ and $X \in \mathfrak{X}(S^{2n})$. Then X defines a map $f : S^{2n} \to \mathbb{R}^{2n+1}$ with components $f(x) = (f^1(x), \ldots, f^{2n+1}(x))$ satisfying $f^1(x)x^1 + \ldots + f^{2n+1}(x)x^{2n+1} = 0$; $f^i(x)$ are the components of X in $\mathbb{R}^{2n+1}$.

Assume that X has no critical point. Replacing f by $f/\|f\|$, we can assume that $f : S^{2n} \to S^{2n}$. The map

$$F : [0, 1] \times S^{2n} \to S^{2n}, \quad F(t, x) = (\cos \pi t)x + (\sin \pi t)f(x)$$

is a smooth proper homotopy between $F(0, x) = x$ and $F(1, x) = -x$. That is, the identity *Id* is homotopic to the antipodal map $A : S^{2n} \to S^{2n}$, $A(x) = -x$. Thus by corollary **7.5.12**, $\deg A = 1$. However, since the Jacobain of A is -1 (this is the place where we use evenness of the dimension of the sphere), A is orientation reversing and thus by the Degree Theorem **7.5.10**, $\deg A = -1$, which is a contradiction. ■

The second application is to prove the existence of fixed points for maps of the unit ball to itself.

7.5.14 Brouwer's Fixed-Point Theorem *Any smooth mapping of the closed unit ball of* $\mathbb{R}^n$ *into itself has a fixed point.*

Proof Let B denote the closed unit ball in $\mathbb{R}^n$ and let $S^{n-1} = \partial B$ be its boundary, the unit sphere. If $f : B \to B$ has no fixed point, define $g(x) \in S^{n-1}$ to be the intersection of the line starting at f(x) and going through x with S^{n-1}. The map $g : B \to S^{n-1}$ so defined is smooth and for $x \in S^{n-1}$, $g(x) = x$. If $n = 1$ this already gives a contradiction, since g must map $B = [-1, 1]$ onto $\{-1, 1\} = S^0$, which is disconnected. For $n \geq 2$, define a smooth proper homotopy $F : [0, 1] \times S^{n-1} \to S^{n-1}$ by $F(t, x) = g(tx)$. Thus F is a homotopy between the constant map $c : S^{n-1} \to S^{n-1}$, $c(x) = g(0)$ and the identity of S^{n-1}. But $c^*\omega = 0$ for any $\omega \in \Omega^{n-1}(S^{n-1})$, so that by **7.5.10**, deg c = 0. On the other hand, by **7.5.12**, deg c = 1, which is false. ■

The Brouwer fixed point theorem is valid for continuous mappings and is proved in the following way. If f has no fixed points, then by compactness there exists a positive constant $K > 0$ such that $\| f(x) - x \| > K$ for all $x \in B$. Let $\varepsilon < \min(K, 2)$ and choose $\delta > 0$ such that $2\delta/(1 + \delta) < \varepsilon$; i.e., $\delta < \varepsilon/(2 - \varepsilon)$. By the Weierstrass approximation theorem (see, for example, Marsden [1974a]) there exists a polynomial mapping $q : \mathbb{R}^n \to \mathbb{R}^n$ such that $\| f(x) - q(x) \| < \delta$ for all $x \in B$. The image q(B) lies inside the closed ball centered at 0 of radius $1 + \delta$, so that $p \equiv q/(1 + \delta) : B \to B$ and

$$\| f(x) - p(x) \| \leq \left\| f(x) - \frac{f(x)}{1+\delta} \right\| + \left\| \frac{f(x)}{1+\delta} - \frac{q(x)}{1+\delta} \right\| \leq \frac{2\delta}{1+\delta} < \varepsilon$$

for all $x \in B$. Since p is smooth by **7.5.14**, it has a fixed point, say $x_0 \in B$. Then

$$0 < K < \| f(x_0) - x_0 \| \leq \| f(x_0) - p(x_0) \| + \| p(x_0) - x_0 \| \leq \varepsilon,$$

which contradicts the choice $\varepsilon < K$.

Brouwer's fixed point theorem is *false* in an open ball, for the open ball is diffeomorphic to $\mathbb{R}^n$ and translation provides a counterexample.

The proof we have given is not "constructive." For example, it is not clear how to base a numerical search on this proof, nor is it obvious that the fixed point we have found varies continuously with f. For these aspects, see Chow, Mallet-Paret and Yorke [1978].

A third application of **7.5.12** is a topological proof of the fundamental theorem of algebra.

7.5.15 The Fundamental Theorem of Algebra *Any polynomial* $p : \mathbb{C} \to \mathbb{C}$ *of degree* $n > 0$ *has a root.*

Proof Assume without loss of generality that $p(z) = z^n + a_{n-1}z^{n-1} + \cdots + a_0$, where $a_i \in \mathbb{C}$, and regard p as a smooth map from $\mathbb{R}^2$ to $\mathbb{R}^2$. If p has no root, then we can define the smooth map

$f(z) = p(z)/|p(z)|$ whose restriction to S^1 we denote by $g : S^1 \to S^1$.

Let $R > 0$ and define for $t \in [0, 1]$ and $z \in S^1$,

$$p_t(z) = (Rz)^n + t[a_{n-1}(Rz)^{n-1} + \cdots + a_0].$$

Since $p_t(z)/(Rz)^n = 1 + t[a_{n-1}(Rz) + \cdots + a_0/(Rz)^n]$ and the coefficient of t converges to zero as $R \to \infty$, we conclude that for sufficiently large R, none of the p_t has zeros on S^1. Thus

$$F : [0, 1] \times S^1 \to S^1 \text{ defined by } F(t, z) = \frac{p_t(z)}{|p_t(z)|}$$

is a smooth proper homotopy of $d_n(z) = z^n$ with $g(Rz)$ which in turn is properly homotopic to $g(z)$.

On the other hand, $G : [0, 1] \times S^1 \to S^1$ defined by $G(t, z) = f(tz)$ is a proper homotopy of the constant mapping $c : S^1 \to S^1$, $c(z) = f(0)$ with g. Thus d_n is properly homotopic to a constant map and hence $\deg d_n = 0$ by corollary **7.5.12**. However, if S^1 is parameterized by arc length θ, $0 \le \theta \le 2\pi$, then d_n maps the segment $0 \le \theta \le 2\pi/n$ onto the segment $0 \le \theta \le 2\pi$ since d_n has the effect $e^{i\theta} \mapsto e^{in\theta}$. If ω denotes the corresponding volume form on S^1, the change of variables formula thus gives

$$\int_{S^1} d_n^* \omega = n\int_{S^1} \omega = 2\pi n, \text{ i.e., } \deg d_n = n,$$

which for $n \neq 0$ is a contradiction. ■

The fundamental theorem of algebra shows that any polynomial $p : \mathbb{C} \to \mathbb{C}$ of degree n can be written as

$$p(z) = c(z - z_1)^{k_1} \cdots (z - z_m)^{k_m},$$

where $z_1, \ldots, z_m$ are the distinct roots of p, $k_1, \ldots, k_m$ are their multiplicities, $k_1 + \cdots + k_m = n$, and $c \in \mathbb{C}$ is the coefficient of z^n in $p(z)$. The fundamental theorem of algebra can be refined to take into account multiplicities of roots in the following way.

7.5.16 Proposition *Let* D *be a compact subset of* $\mathbb{C}$ *with open interior and smooth boundary* ∂D. *Assume that the polynomial* $p : \mathbb{C} \to \mathbb{C}$ *has no zeros on* ∂D. *Then the total number of zeros of* p *which lie in the interior of* D, *counting multiplicities, equals the degree of the map* $p/|p| : \partial D \to S^1$.

Proof Let $z_1, \ldots, z_m$ be the roots of p in the interior of D with multiplicities $k(1), \ldots, k(m)$. Around each z_i construct an open disk D_i centered at z_i, $D_i \subset D$, such that $\partial D \cap \partial D_i = \varnothing$, and $\partial D_i \cap \partial D_j = \varnothing$, for all $i \neq j$. Then $V = D \setminus (D_1 \cup \cdots \cup D_m)$ is a smooth compact two-dimensional manifold whose boundary is $\partial D \cup \partial D_1 \cup \cdots \partial D_m$. The boundary orientation of ∂D_i induced by V is *opposite* to the usual boundary orientation of ∂D_i as the boundary of the disk D_i; see Figure 7.5.2. Since $p/|p| $ is defined on all of V, corollary **7.5.11** implies that the degree of $p/|p| : \partial V \to S^1$ is zero. But the degree of a map defined on a disjoint union of manifolds is the sum of the individual degrees and thus the degree of $p/|p|$ on ∂D equals the sum of the degrees of $p/|p|$ on all ∂D_j. The proposition is therefore proved if we show that the degree of $p/|p|$ on ∂D_i is the multiplicity $k(i)$ of the root z_i.

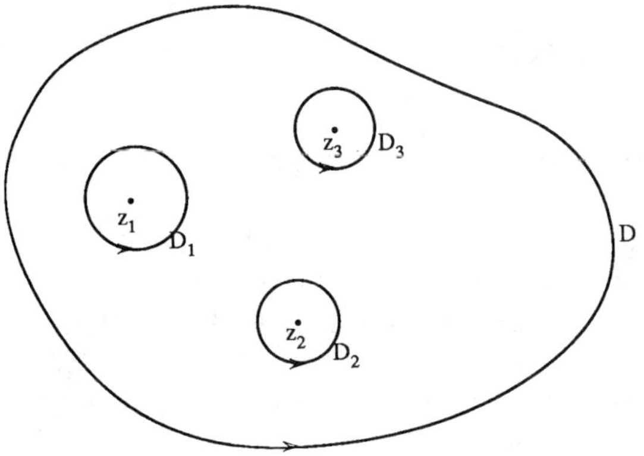

Figure 7.5.2

Let

$$r(z) = c \prod_{j=1,\, j\neq i}^{m} (z - z_j)^{k(j)} \quad \text{so} \quad p(z) = (z - z_i)^{k(i)} r(z)$$

and the only zero of $p(z)$ in the disk D_i is z_i. Then $\varphi : z \in S^1 \to z_i + R_i z \in \partial D_i$, where R_i is the radius of D_i, is a diffeomorphism and therefore the degree of $p/|p| : \partial D_i \to S^1$ equals the degree of $(p \circ \varphi)/|p \circ \varphi| : S^1 \to S^1$. The homotopy $H : [0, 1] \times S^1 \to S^1$ of $z^{k(i)} \arg r(z_i)$ with $(p \circ \varphi)/|p \circ \varphi|$ given by

$$H(t, z) = \frac{z^{k(i)} r(z_i + tR_i z)}{|r(z_i + tR_i z)|}$$

is proper and smooth, since $z_i + tR_i z \in D_i$ for all $z \in S^1$, $t \in [0, 1]$. Thus in ∂D_i we have $\deg(p/|p|) = \deg((p \circ \varphi)/|p \circ \varphi|) = \deg z^{k(i)} = k(i)$. ■

A variant of the fundamental theorem of algebra is the following.

7.5.17 Proposition *Let* U *be an open subset of* $\mathbb{R}^n$ *and* $f : U \to \mathbb{R}^n$ *be a* C^1 *proper map. Assume there is a closed subset* $K \subset U$ *such that for all* $x \in U \setminus K$, *the Jacobian* $J(f)(x)$ *does* *not change sign and is not identically equal to zero. Then* f *is surjective.*

Proof The map f cannot be constant since the Jacobian J(f)(x) is not identically zero for all $x \in U \setminus K$. For the same reason, f has a regular value $y \in f(U \setminus K)$, for if all values in $f(U \setminus K)$ are singular, J(f) will vanish on $U \setminus K$. If $y \in f(U \setminus K)$ is a regular value of f then $\text{sign}(T_x f)$ does not change for all $x \in f^{-1}(y)$ so by the degree theorem **7.5.10**, $\deg f \neq 0$, which then implies that f is onto. ■

The orientation preserving character of proper diffeomorphisms is characterized in terms of the degree as follows.

7.5.18 Proposition *Let* M *and* N *be oriented boundaryless connected manifolds and* $f : M \to N$ *a proper local diffeomorphism. Then* $\deg f = 1$, *if and only if* f *is an orientation preserving diffeomorphism.*

Proof If f is an orientation preserving diffeomorphism then $\deg f = 1$ by theorem **7.5.10.** Conversely, let f be a proper local diffeomorphism with $\deg f = 1$. Define $U_\pm = \{m \in M \mid \text{sign}\, T_m f = \pm 1\}$. Since f is a local diffeomorphism, $U_\pm$ are open in M. Connectedness of M and $M = U_+ \cup U_-$, $U_+ \cap U_- = \varnothing$ imply that $M = U_+$ or $M = U_-$. Let us show that $U_- = \varnothing$. Since $\deg f = 1$, f is onto and hence if $n \in N$, $f^{-1}(n) \neq \varnothing$ is a discrete submanifold of M. Properness of f implies that $f^{-1}(n) = \{m(1), \ldots, m(k)\}$. Since f is a local diffeomorphism of a neighborhood U_i of m(i) onto a neighborhood V of n, $\text{sign}\, T_{m(i)} f$ is the same for all $i = 1, \ldots, k$ (for otherwise J(f) must vanish somewhere). Thus $\deg f = \pm k$ according to whether $T_{m(i)} f$ preserves or reverses orientation. Since $\deg f = 1$, this implies $U_- = \varnothing$ and $k = 1$, i.e., f is injective. Thus f is a bijective local diffeomorphism, i.e., a diffeomorphism. ■

SUPPLEMENT 7.5B
Zero and n-Dimensional Cohomology

Here we compute $H^0(M)$ and $H^n(M)$ for a connected n-manifold M. Recall that the ***cohomology groups*** are defined by $H^k(M) = \ker \mathbf{d}^k / \text{range}\, \mathbf{d}^{k-1}$, where $\mathbf{d}^k : \Omega^k(M) \to \Omega^{k+1}(M)$ is the exterior differential. If $\Omega^k_c(M)$ denotes the k-forms with compact support, then $\mathbf{d}^k : \Omega^k_c(M) \to \Omega^{k+1}_c(M)$ and one forms in the same manner $H^k_c(M)$, the ***compactly supported cohomology groups*** of M.

Thus $H^0(M) = \{f \in \mathcal{F}(M) \mid \mathbf{d}f = 0\} \cong \mathbb{R}$ since any locally constant function on a connected space is constant. If M were not connected, then $H^0(M) = \mathbb{R}^c$, where c is the number of connected components of M. By the Poincaré lemma, if M is contractible, then $H^q(M) = 0$ for $q \neq 0$.

The rest of this supplement is devoted to the proof and applications of the following *special case of deRham's theorem.* (We will provide a supplementary chapter with further development of these ideas on request.)

7.5.19 Theorem *Let* M *be a boundaryless connected* n*-manifold.*

(i) *If* M *is orientable, then* $H^n_c(M) \cong \mathbb{R}$, *the isomorphism being given by integration:* $[\omega] \mapsto \int_M \omega$. *In particular* $\omega \in \Omega^n_c(M)$ *is exact iff* $\int_M \omega = 0$.

(ii) *If* M *is nonorientable, then* $H^n_c(M) = 0$.

(iii) *If* M *is non-compact, orientable or not, then* $H^n(M) = 0$.

Before starting the actual proof, let us discuss (i). The integration mapping $\int_M : \Omega^n(M) \to \mathbb{R}$ is linear and onto. To see that it is onto, let ω be an n-form with support in a chart in which the local expression is $\omega = f\, dx^1 \wedge \ldots \wedge dx^n$ with f a bump function. Then $\int_M \omega = \int_{\mathbb{R}^n} f(x)\, dx > 0$. Since we can multiply ω by any scalar, the integration map is onto. Any ω with nonzero integral cannot be exact by Stokes' theorem. This last remark also shows that integration induces a mapping, which we shall still call integration, $\int_M : H^n_c(M) \to \mathbb{R}$, which is linear and onto. Thus, in order to show that it is an isomorphism as (i) states, it is necessary and sufficient to prove it is injective, i.e., to show that if $\int_M \omega = 0$ for $\omega \in \Omega^n(M)$, then ω is exact. The proof of this will be done in the following lemmas.

7.5.20 Lemma *Theorem* **7.5.19** *holds for* $M = S^1$.

Proof Let $p : \mathbb{R} \to S^1$ be given by $p(t) = e^{it}$ and $\omega \in \Omega^1(S^1)$. Then $p^*\omega = f\, dt$ for $f \in \mathcal{F}(\mathbb{R})$ a 2π-periodic function. Let F be an antiderivative of f. Since

$$0 = \int_{S^1} \omega = \int_t^{t+2\pi} f(s)\, ds = F(t + 2\pi) - F(t)$$

for all $t \in \mathbb{R}$, we conclude that F is also 2π-periodic, so it induces a unique map $G \in \mathcal{F}(S^1)$, determined by $p^*G = F$. Hence $p^*\omega = \mathbf{d}F = p^*\mathbf{d}G$ implies $\omega = \mathbf{d}G$ since p is a surjective submersion. ▼

7.5.21 Lemma *Theorem* **7.5.19** *holds for* $M = S^n$, $s > 1$.

Proof This will be done by induction on n, the case $n = 1$ being the previous lemma. Write $S^n = N \cup S$, where $N = \{x \in S^n \mid x^{n+1} \geq 0\}$ is the closed northern hemisphere and $S = \{x \in S^n \mid x^{n+1} \leq 0\}$ the closed southern hemisphere. Then $N \cap S = S^{n-1}$ is oriented in two *different* ways as the boundary of N and S, respectively. Let $O_N = \{x \in S^n \mid x^{n+1} > -\varepsilon\}$, $O_S = \{x \in S^n \mid x^{n+1} < \varepsilon\}$ be open contractible neighborhoods of N and S, respectively. Thus by the Poincaré lemma, there exist $\alpha_N \in \Omega^{n-1}(O_N)$, $\alpha_S \in \Omega^{n-1}(O_S)$ such that $\mathbf{d}\alpha_N = \omega$ on O_N, $\mathbf{d}\alpha_S = \omega$ on O_S. Hence by hypothesis and Stokes' theorem,

$$0 = \int_{S^n} \omega = \int_N \omega + \int_S \omega = \int_N \mathbf{d}\,\alpha_N + \int_S \mathbf{d}\alpha_S = \int_{\partial N} i^*\alpha_N + \int_{\partial S} i^*\alpha_S = \int_{S^{n-1}} i^*\alpha_N - \int_{S^{n-1}} i^*\alpha_S$$
$$= \int_{S^{n-1}} i^*(\alpha_N - \alpha_S)$$

where $i : S^{n-1} \to S^n$ is the inclusion of S^{n-1} as the equator of S^n; the minus sign appears on the second integral because the orientations of S^{n-1} and ∂S are opposite. By induction, $i^*(\alpha_N - \alpha_S) \in \Omega^{n-1}(S^{n-1})$ is exact.

Let $O = O_N \cap O_S$ and note that the map $r : O \to S^{n-1}$, sending each $x \in S$ to $r(x) \in S^{n-1}$, the intersection of the meridian through x with the equator S^{n-1}, is smooth. Then $r \circ i$ is the identity on S^{n-1}. Also, $i \circ r$ is homotopic to the identity of O, the homotopy being given by sliding $x \in O$ along the meridian to $r(x)$. Since $\mathbf{d}(\alpha_N - \alpha_S) = \omega - \omega = 0$ on O, by Theorem **6.4.16** we conclude that $(\alpha_N - \alpha_S) - r^*i^*(\alpha_N - \alpha_S)$ is exact on O. But we just showed that $i^*(\alpha_N - \alpha_S) \in \Omega^{n-1}(S^{n-1})$ is exact, and hence $r^*i^*(\alpha_N - a_S) \in \Omega^{n-1}(O)$ is also exact. Hence $\alpha_N - \alpha_S \in \Omega^{n-1}(O)$ is exact. Thus, there exists $\beta \in \Omega^{n-2}(O)$ such that $\alpha_N - \alpha_S = \mathbf{d}\beta$ on O. Now use a bump function to extend β to a form $\gamma \in \Omega^{n-2}(S^n)$ so that on O, $\beta = \gamma$ and $\gamma = 0$ on $S^n \setminus V$, where V is an open set such that $\mathrm{cl}(U) \subset V$. Then

$$\lambda(x) = \begin{cases} \alpha_N(x), & \text{if } x \in N \\ \alpha_S(x) + \mathbf{d}\gamma, & \text{if } x \in S \end{cases}$$

is by construction C^∞ and $\mathbf{d}\lambda = \omega$. ▼

7.5.22 Lemma *A compactly supported* n*-form* $\omega \in \Omega^n(\mathbb{R}^n)$ *is the exterior derivative of a compactly supported* $(n-1)$*-form on* $\mathbb{R}^n$ *iff* $\int_{\mathbb{R}^n} \omega = 0$.

Proof Let $\sigma : S^n \to \mathbb{R}^n$ be the stereographic projection from the north pole $(0, ..., 1) \in S^n$ onto $\mathbb{R}^n$ and assume without loss of generality that $(0, ..., 1) \notin \sigma^{-1}(\mathrm{supp}\,\omega)$. By the previous lemma, $\sigma^*\omega = \mathbf{d}\alpha$, for some $\alpha \in \Omega^{n-1}(S^n)$ since $0 = \int_{\mathbb{R}^n} \omega = \int_{S^n} \sigma^*\omega$ by the change-of-variables formula. But $\sigma^*\omega = \mathbf{d}\alpha$ is zero in a contractible neighborhood U of the north pole, so that by the Poincaré lemma, $\alpha = \mathbf{d}\beta$ on U, where $\beta \in \Omega^{n-2}(U)$. Now extend β to an $(n-2)$-form $\gamma \in \Omega^{n-2}(S^n)$ such that $\beta = \gamma$ on U and $\gamma = 0$ outside a neighborhood of cl(U). But then $\sigma_*(\alpha - \mathbf{d}\gamma)$ is compactly supported in $\mathbb{R}^n$ and $\mathbf{d}\sigma_*(\alpha - \mathbf{d}\gamma) = \sigma_*\mathbf{d}\alpha = \omega$. ▼

7.2.23 Lemma *Let* M *be a boundaryless connected* n*-manifold. Then* $H^n_c(M)$ *is at most one-dimensional.*

Proof Let (U_0, φ_0) be a chart on M such that $\varphi_0(U_0)$ is the open unit ball B in $\mathbb{R}^n$. Let $\omega \in \Omega^n_c(M)$, satisfying $\mathrm{supp}\,\omega \subset U_1$, be the pull-back of a form $f\,dx^1 \wedge \cdots \wedge dx^n \in \Omega^n(B)$ where $f \geq 0$ and $\int_{\mathbb{R}^n} f(x)\,dx = 1$. To prove the lemma, it is sufficient to show that for every $\eta \in \Omega^n_c(M)$ there exists a number $c \in \mathbb{R}$ such that $\eta - c\omega = \mathbf{d}\zeta$ for some $\zeta \in \Omega^{n-1}_c(M)$.

First assume $\eta \in \Omega^n{}_c(M)$ has $\mathrm{supp}(\eta)$ entirely contained in a chart (U, φ) and let U_0, $U_1, \ldots, U_k$ be a finite covering of a curve starting in U_0 and ending on $U_k = U$ such that $U_i \cap U_{i+1} = \varnothing$. Let $\alpha_i \in \Omega^n{}_c(U_i)$, $i = 1, \ldots, k-1$ be non-negative n-forms such that $\mathrm{supp}(\alpha_i) \subset U_i$, $\mathrm{supp}(\alpha_i) \cap U_{i+1} \neq \varnothing$, and $\int_{\mathbb{R}^n} \varphi_{i*}(\alpha_i) = 1$. Let $\alpha_0 = \omega$, and $\alpha_k = \eta$. But then

$$\int_{\mathbb{R}^n} \varphi_{i*}(\alpha_{i-1}) \neq 0, \quad i = 1, \ldots, k$$

by the change-of-variables formula, so that with $c_i = -1/\int_{\mathbb{R}^n}\varphi_{i*}(\alpha_{i-1})$ we have $\int_{\mathbb{R}^n}\varphi_{i*}(\alpha_i) - c_i\alpha_{i-1}) = 0$. Thus by the previous lemma $\varphi_{i*}(\alpha_i - c_i\alpha_{i-1})$ is the differential of an $(n-1)$-form supported in B. That is, there exists $\beta_i \in \Omega^{n-1}{}_c(M)$, β_i vanishing outside U_i such that

$$\alpha_i - c_i\alpha_{i-1} = \mathbf{d}\beta_i, \quad i = 1, \ldots, k.$$

Put $c = c_k \cdots c_1$ and

$$\beta = \beta_k + (c_k\beta_{k-1}) + (c_kc_{k-1}\beta_{k-2}) + \cdots + (c_{k-1} \cdots c_2\beta_1) \in \Omega^{n-1}(M).$$

Then

$$\begin{aligned}\eta - c\omega &= \alpha_k - c\alpha_0 = \alpha_k - c_k\alpha_{k-1} + c_k(\alpha_{k-1} - c_{k-1}\alpha_{k-2}) + \cdots + (c_k \cdots c_2)(\alpha_1 - c_1\alpha_0)\\ &= \mathbf{d}\beta_k + c_k\mathbf{d}\beta_{k-1} + \cdots + (c_k \cdots c_2)\mathbf{d}\beta_1 = \mathbf{d}\beta.\end{aligned}$$

Let $\eta \in \Omega^n{}_c(M)$ be arbitrary and $\{\chi_i \mid i = 1, \ldots, k\}$ a partition of unity subordinate to the given atlas $\{(U_i, \varphi_i) \mid i = 1, \ldots, k\}$. Then $\chi_i\eta$ is compactly supported in U_i and hence there exist constants c_i and forms $\alpha_i \in \Omega^{n-1}{}_c(M)$ such that $\chi_i\eta - c_i\omega = \mathbf{d}\alpha_i$. If

$$c = \sum_{i=1}^{k} c_i \quad \text{and} \quad \alpha = \sum_{i=1}^{k} \alpha_i \in \Omega^{n-1}{}_c(M),$$

then

$$\eta - c\omega = \sum_{i=1}^{k} (\chi_i\eta - c_i\omega) = \sum_{i=1}^{k} \mathbf{d}\alpha_i = \mathbf{d}\alpha. \ \blacktriangledown$$

Proof of theorem 7.5.19 (i) By the preceding lemma, $H^n{}_c(M)$ is zero- or one-dimensional. We have seen that $\int_M : H^n{}_c(M) \to \mathbb{R}$ is linear and onto so that necessarily $H^n{}_c(M)$ is one-dimensional; i.e., $\int_M\omega = 0$ iff ω is exact.

(ii) Let $\tilde{M}$ be the oriented double covering of M and $\pi : \tilde{M} \to M$ the canonical projection. Define $\pi^\# : H^n(M) \to H^n(\tilde{M})$ by $\pi^\#[\alpha] = [\pi^*\alpha]$. We shall first prove that $\pi^\#$ is the zero map. Let $\{U_i\}$ be an open covering of M by chart domains and $\{\chi_i\}$ a subordinate partition of unity. Let $\pi^{-1}(U_i) = U_i^1 \cup U_i^2$. Then $\{U_i^j \mid j = 1, 2\}$ is an open covering of $\tilde{M}$ by chart domains and the maps $\psi_i^j = \chi_i \circ \pi/2 : U_i^j \to \mathbb{R}$, $j = 1, 2$, form a subordinate partition of unity on M. Let $\alpha \in \Omega^n{}_c(M)$. Then

$$\int_{\tilde{M}} \pi^*\alpha = \sum_{i,j} \int_{U_i^j} \psi_i^j \pi^*\alpha = \sum_{i=1}^{k} \left(\int_{U_i^j} \psi_i^1 \pi^*\alpha + \int_{U_i^2} \psi_i^2 \pi^*\alpha \right) = 0,$$

each term vanishing since their push-forwards by the coordinate maps coincide on $\mathbb{R}^n$ and U_i^1 and U_i^2 have opposite orientations. By (i), we conclude that $\pi^*\alpha = \mathbf{d}\beta$ for some $\beta \in \Omega^{n-1}(\tilde{M})$; i.e., $\pi^\#[\alpha] = [\pi^*\alpha] = [0]$ for all $[\alpha] \in H^n_c(M)$.

We shall now prove that $\pi^\#$ is injective, which will show that $H^n_c(M) = 0$. Let $\alpha \in \Omega^n_c(M)$ be such that $\pi^*\alpha = \mathbf{d}\beta$ for some $\beta \in \Omega^{n-1}_c(\tilde{M})$ and let $r : \tilde{M} \to \tilde{M}$ be the diffeomorphism associating to $(m, [\omega]) \in \tilde{M}$ the point $(m, [-\omega]) \in \tilde{M}$. Then clearly $\pi \circ r = \pi$ so that $\mathbf{d}(r^*\beta) = r^*(\mathbf{d}\beta) = r^*\pi^*\alpha = (\pi \circ r)^*\alpha = \pi^*\alpha = \mathbf{d}\beta$. Define $\tilde{\gamma} \in \Omega^{n-1}{}_c(M)$ by setting $\tilde{\gamma} = (1/2)(\beta + r^*\beta)$ and note that $r^*\tilde{\gamma} = \tilde{\gamma}$ and $\mathbf{d}\tilde{\gamma} = (\mathbf{d}\beta + \mathbf{d}r^*\beta)/2 = \mathbf{d}\beta = \pi^*\alpha$. But $\tilde{\gamma}$ projects to a well-defined form $\gamma \in \Omega^{n-1}{}_c(M)$ such that $\pi^*\gamma = \tilde{\gamma}$, since $r^*\tilde{\gamma} = \tilde{\gamma}$. Thus $\pi^*\alpha = \mathbf{d}\tilde{\gamma} = \mathbf{d}\pi^*\gamma = \pi^*\mathbf{d}\gamma$, which implies that $\alpha = \mathbf{d}\gamma$, since π is a surjective submersion.

(iii) Assume first that $\omega \in \Omega^n_c(M)$ has its support contained in a relatively compact chart domain U_1 of M. Then out of a finite open relatively compact covering of $\mathrm{cl}(U_1)$ by chart domains, pick a relatively compact chart domain U_2 which does not intersect $\mathrm{supp}(\omega)$. Working with $\mathrm{cl}(U_2) \setminus U_1$, find a relatively compact chart domain U_3 such that $U_1 \cap U_3 = \varnothing$, $U_2 \cap U_3 \neq \varnothing$, $U_3 \cap (M \setminus (U_1 \cup U_2)) \neq \varnothing$. Proceed inductively to find a sequence $\{U_n\}$ of relatively compact chart domains such that $U_n \cap U_{n+1} \neq \varnothing$, $U_n \cap U_{n-1} \neq \varnothing$, $U_n \cap U_m = \varnothing$ for all $m \neq n-1, n, n+1$, and such that $\mathrm{supp}(\omega) \subset U_1$, $U_2 \cap \mathrm{supp}(\omega) = \varnothing$. Since M is not compact, this sequence can be chosen to be infinite; see Figure 7.5.3.

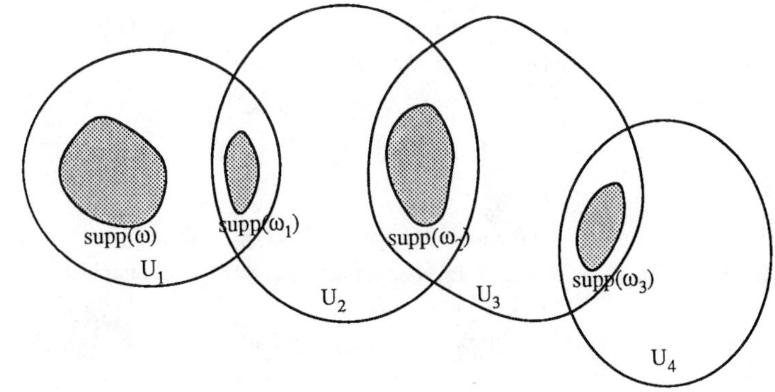

Figure 7.5.3

Now choose in each $U_n \cap U_{n+1}$ an n-form ω_n with compact support such that

$$\int_{U_1} \omega = \int_{U_1} \omega_1 = \int_{U_2} \omega_2 = \cdots = \int_{U_n} \omega_n = \cdots .$$

Since $H^n_c(U_n) = \mathbb{R}$ by (i), U_n being orientable, ω_{n-1} and ω_n define the same cohomology class, i.e., there is $\eta_n \in \Omega^{n-1}{}_c(U_n)$ such that $\omega_{n-1} = \omega_n + \mathbf{d}\eta_n$. If we let $\omega_0 = \omega$, we get recursively

$$\omega = \mathbf{d}\eta_1 + \omega_1 = \mathbf{d}(\eta_1 + \eta_2) + \omega_2 = \cdots = \mathbf{d}\left(\sum_{i=1}^{n} \eta_i\right) + \omega_n = \cdots .$$

We claim that $\omega = \mathbf{d}(\Sigma_{n\geq 1}\eta_n)$, where the sum is finite since any point of the manifold belongs to at most two U_n's. Thus, if $p \in \bigcup_{n\geq 1} U_n$, let $p \in U_n$ so that $\mathbf{d}(\Sigma_{n\geq 1}\eta_n)(p) = \mathbf{d}\eta_{n-1}(p) + \mathbf{d}\eta_n(p) + \mathbf{d}\eta_{n+1}(p) = \mathbf{d}\eta_{n-1}(p) + \omega_{n-1}(p) - \omega_{n+1}(p) = \mathbf{d}\eta_{n-1}(p) + \omega_{n-1}(p) - \omega_{n+2}(p) - \mathbf{d}\eta_{n+2}(p)$ with the convention $\eta_0 = 0$. Since $U_n \cap U_{n+2} = \varnothing$ and $\operatorname{supp} \omega_{n+2}$, $\operatorname{supp} \eta_{n+2} \subset U_{n+2}$, it follows that the last two terms vanish. Thus $\mathbf{d}(\Sigma_{n\geq 1}\eta_n)(p) = \mathbf{d}\eta_{n-1}(p) + \omega_{n-1}(p)$. If $n = 1$, this proves the desired equality. If $n \geq 2$, then $\mathbf{d}(\Sigma_{n\geq 1}\eta_n)(p) = \mathbf{d}\eta_{n-1}(p) + \omega_{n-1}(p) = \omega_{n-2}(p)$ and $U_n \cap U_{n-2} = \varnothing$ implies that $\mathbf{d}(\Sigma_{n\geq 1}\eta_n) = 0$. Since also $\omega(p) = 0$ in this case, the desired equality holds again. Finally, if $p \notin \bigcup_{n\geq 1} U_n$, then both sides of the equality are zero and we showed that ω is exact, $\omega = \mathbf{d}(\Sigma_{n\geq 1}\eta_n)$, with $\operatorname{supp}(\Sigma_{n\geq 1}\eta_n) \subset \bigcup_{n\geq 1} U_n$.

Now if $\omega \in \Omega^n(M)$, let $\{(U_i, g_i)\}$ be a partition of unity subordinate to a locally finite atlas of M whose chart domains are relatively compact. Thus $\operatorname{supp}(g_i\omega) \subset U_i$ and by what we just proved, $g_i\omega = \mathbf{d}\eta_i$, with $\operatorname{supp}(\eta_i)$ contained in the union of the chain of open sets $\{U_n{}^i\}$, $U_1{}^i = U_i$, as described above. Refine each such chain, such that all its elements are one of the U_j's. Since at most two of the $U_n{}^i$ intersect for each fixed i, it follows that the sum $\Sigma_i\eta_i$ is locally finite and therefore $\eta = \Sigma_i\eta_i \in \Omega^{n-1}(M)$. Finally, $\omega = \Sigma_i g_i\omega = \Sigma_i \mathbf{d}\eta_i = \mathbf{d}\eta$, thus showing that ω is exact and hence $H^n(M) = 0$. ■

One can use this result as an alternative method to introduce the ***degree*** of a proper map $f : M \to N$ between oriented manifolds; i.e., that integer $\deg(f)$ such that

$$\int_M f^*\eta = \deg(f) \int_N \eta$$

for any $\eta \in \Omega^n{}_c(N)$. Indeed, since the isomorphism $H^n{}_c(N) \cong \mathbb{R}$ is given by $[\eta] \mapsto \int_N \eta$, the linear map $[\eta] \mapsto \int_M f^*\eta$ of $H^n{}_c(N)$ to $\mathbb{R}$ must be some multiple of this isomorphism:

$$\int_M f^*\eta = \deg(f) \int_N \eta \quad \text{for all } \eta \in \Omega^n{}_c(N).$$

To prove that $\deg(f)$ is an integer in this context and that the formula (ii) for $\deg(f)$ is independent on the regular value y, note that if y is any regular value of f and $x \in f^{-1}(y)$, then there exist compact neighborhoods V of y and U of x such that $f \mid U : U \to V$ is a diffeomorphism. Since $f^{-1}(y)$ is compact and discrete, it must be finite, say $f^{-1}(y) = \{x_1, \ldots, x_k\}$. This shows that $f^{-1}(V) = U_1 \cup \cdots \cup U_k$ with all U_i disjoint and the sum in the degree formula is finite. Shrink V if necessary to lie in a chart domain. Now choose $\eta \in \Omega^n{}_c(N)$ satisfying $\operatorname{supp}(\eta) \subset V$. Then

$$\int_M f^*\eta = \sum_{x_i \in f^{-1}(y)} \int_{U_i} f^*\eta = \left\{ \sum_{x_i \in f^{-1}(y)} (\mathrm{sign}(T_{x_i} f)) \right\} \int_N \eta$$

by the change-of-variables formula in $\mathbb{R}^n$, so the claim follows.

Degree theory can be extended to infinite dimensions as well and has important applications to partial differential equations and bifurcations. This theory is similar in spirit to the above and was developed by Leray and Schauder in the 1930s. See Chow and Hale [1982], Choquet-Bruhat et al. [1977], Nirenberg [1974], and Elworthy and Tromba [1970b] for modern accounts.

Exercises

7.5A Poincaré duality Show that $*$ induces an isomorphism $* : H^k \to H^{n-k}$ and $H^k{}_c \to H^{n-k}{}_c$.

7.5B (For students knowing some algebraic topology) Develop some basic properties of deRham cohomology groups such as homotopy invariance, exact sequences, Mayer-Vietoris sequences and excision. Use this to compute the cohomology of some standard simple spaces (tori, spheres, projective spaces).

7.5C (i) Show that any smooth vector field X on a compact Riemannian manifold (M, g) can be written uniquely as

$$X = Y + \mathrm{grad}\, p$$

where Y has zero divergence (and is parallel to ∂M if M has boundary).

(ii) Show directly that the equation

$$\Delta p = -\,\mathrm{div}\, X, \quad (\mathrm{grad}\, p)\cdot n = X\cdot n$$

is formally soluble using the ideas of the Fredholm alternative.

7.5D Show that any symmetric two-tensor h on a compact Riemannian manifold (M, g) can be uniquely decomposed in the form

$$h = \mathbf{L}_X g + k.$$

where $\delta k = 0$, δ being the divergence of g, defined by $\delta k = (\mathbf{L}_{(\cdot)} g)^* k$, where $(\mathbf{L}_{(\cdot)} g)^*$ is the adjoint of the operator $X \mapsto \mathbf{L}_X g$. (See Berger and Ebin [1969] and Cantor [1981] for more information.)

7.5E Let $\alpha \in \Omega^{k-1}(M)$, $\beta \in \Omega^k(M)$, where M is a compact oriented Riemannian manifold with boundary. Show that

(i)
$$\langle \mathbf{d}\alpha, \beta \rangle - \langle \alpha, \mathbf{d}\beta \rangle = \int_{\partial M} \alpha \wedge *\beta.$$

(*Hint*: Show that $*\,\delta\beta = (-1)^k \mathbf{d}*\beta$ and use Stokes theorem or **7.2.13**.)

(ii)
$$\langle \mathbf{d}\delta\alpha, \beta \rangle - \langle \mathbf{d}\alpha, \mathbf{d}\beta \rangle = \int_{\partial M} \delta\alpha \wedge *\beta$$

$$\langle \mathbf{d}\alpha, \mathbf{d}\beta \rangle - \langle \alpha, \delta\mathbf{d}\beta \rangle = \int_{\partial M} \alpha \wedge *\mathbf{d}\beta$$

(iii) (*Green's formula*)

$$\langle \Delta\alpha, \beta \rangle - \langle \alpha, \Delta\beta \rangle = \int_{\partial M} (\delta\alpha \wedge *\beta - \mathbf{d}\beta \wedge *\alpha + \alpha \wedge *\mathbf{d}\beta - \beta \wedge *\mathbf{d}\alpha).$$

(*Hint:* Show first that

$$\langle \Delta\alpha, \beta \rangle - \langle \mathbf{d}\alpha, \mathbf{d}\beta \rangle - \langle \delta\alpha, \delta\beta \rangle = \int_{\partial M} (\delta\alpha \wedge *\beta - \beta \wedge *\mathbf{d}\alpha).)$$

7.5F (For students knowing algebraic topology) Define relative cohomology groups and relate them to the Hodge decomposition for manifolds with boundary.

7.5G Prove the local formulas

$$(\delta\alpha)_{i_1 \dots i_k} = \frac{1}{k+1} |\det[g_{rs}]|^{-1/2} g_{i_1 r_1} \cdots {}_{i_k r_k} \frac{\partial}{\partial x^\ell} \left(\sum_{p=1}^{k+1} (-1)^p g^{r_1 j_1} \cdots g^{r_{p-1} j_{p-1}} \right.$$
$$\left. g^{\ell j_p} g^{r_p j_{p+1}} \cdots g^{r_k j_{k+1}} \beta_{j_1 \cdots j_{k+1}} |\det[g_{rs}]|^{1/2} \right)$$

$$(\delta\alpha)^{r_1 \cdots r_k} = \frac{1}{k+1} |\det[g_{ij}]|^{-1/2} \frac{\partial}{\partial x^\ell} \left(\sum_{p=1}^{k+1} (-1)^p \alpha^{r_1 \cdots r_{p-1} \ell r_p \dots r_k} |\det[g_{ij}]|^{1/2} \right)$$

where $i_1 < \cdots < i_k$ and $\alpha \in \Omega^{k+1}(M)$ according to the following guidelines. Prove first the second formula. Work in a chart (U, φ) with $\varphi(U) = B_3(0)$ = open ball of radius 3, and prove the formula on $\varphi^{-1}(B_1(0))$. For this, choose a function χ on $\mathbb{R}^n$ with $\operatorname{supp}(\chi) \subset B_3(0)$ and $\chi \mid B_1(0) \equiv 1$. Then extend $\chi\varphi_*\alpha$ to $\mathbb{R}^n$, denote it by α' and consider the set $B_4(0)$.

(i) Show from **7.5E**(i) that $\langle \mathbf{d}\beta, \alpha' \rangle = \langle \beta, \delta\alpha' \rangle$ for any $\beta \in \Omega^{k+1}(B_4(0))$.

(ii) In the explicit expression for $\langle \mathbf{d}\beta, \alpha' \rangle$, perform an integration by parts and justify it.

(iii) Find the expression for $\delta\alpha'$ by comparing $\langle\beta, \delta\alpha'\rangle$ with the expression found in (iii) and argue that it must hold on $\varphi^{-1}(B_1(0))$.

7.5H Let $\varphi : M \to M$ be a diffeomorphism of an oriented Riemannian manifold (M, g) and let δ_g denote the codifferential corresponding to the metric g and $\langle\, ,\rangle_g$ the inner product on $\Omega^k(M)$ corresponding to the metric g. Show that (i) $\langle\alpha, \beta\rangle_g = \langle\varphi^*\alpha, \varphi^*\beta\rangle_{\varphi^*g}$ for $\alpha, \beta \in \Omega^k(M)$ and (ii) $\delta_{\varphi^*g}(\varphi^*\alpha) = \varphi^*(\delta_g\alpha)$ for $\alpha \in \Omega^k(M)$. (*Hint*: Use the fact that $\mathbf{d}$ and δ are adjoints.)

7.5I (i) Let c_1 and c_2 be two differentiably homotopic curves and $\omega \in \Omega^1(M)$ a closed one-form. Show that

$$\int_{c_1} \omega = \int_{c_2} \omega$$

(ii) Let M be simply connected. Show that $H^1(M) = 0$. (*Hint:* For $m_0 \in M$, let c be a curve from m_0 to $m \in M$. Then $f(m) = \int_c \omega$ is well defined by (i) and $\mathbf{d}f = \omega$.)

(iii) Show that $H^1(S^1) \neq 0$ by exhibiting a closed one-form that is not exact.

7.5J The ***Hopf degree theorem*** states that f and $g : M^n \to S^n$ are homotopic iff they have the same degree. By consulting references if necessary, prove this theorem in the context of Supplements **7.5A**, **B**. (*Hint*: Consult Guillemin and Pollack [1974] and Hirsch [1976].)

7.5K What does the degree of a map have to do with Exercise **7.2D** on integration over the fiber? Give some examples and a discussion.

7.5L Show that the equations

$$z^{13} + \sin(|z|^2)z^7 + 3z^4 + 2 = 0$$

$$z^8 + \cos(|z|^2)z^5 + 5\log(|z|^2)z^4 + 53 = 0$$

have a root.

7.5M Let $f : M \to N$ where M and N are compact orientable boundaryless manifolds and N is contractible. Show that $\deg(f) = 0$. Conclude that the only contractible compact manifold (orientable or not) is the one-point space. (*Hint:* Show that the oriented double covering of a contractible non-orientable manifold is contractible.)

7.5N Show that every smooth map $f : S^n \to \mathbb{T}^n$, $n > 1$ has degree zero. (*Hint:* Show that f is homotopic to a constant map.) Conclude that S^n and $\mathbb{T}^n$ are not diffeomorphic if $n > 1$.

Chapter 8
Applications

This chapter presents some applications of manifold theory and tensor analysis to physics and engineering. Our selection is a of limited scope and depth, with the intention of providing an introduction to the techniques. There are many other applications of the ideas of this book as well. We list below a few *selected* references for further reading in the same spirt.

1. Arnold [1978], Abraham and Marsden [1978], Chernoff and Marsden [1974], Weinstein [1977], and Marsden [1981] for Hamiltonian mechanics.
2. Marsden and Hughes [1983] for elasticity theory.
3. Flanders [1963], von Westenholtz [1981], and Schutz [1980] for applications to control theory.
4. Hermann [1980], Knowles [1981], and Schutz [1980] for diverse applications.
5. Bleecker [1981] for Yang-Mills theory.
6. Misner, Thorne, and Wheeler [1973] and Hawking and Ellis [1973] for general relativity.

§8.1 *Hamiltonian Mechanics*

Our starting point is ***Newton's second law*** in $\mathbb{R}^3$, which states that a particle which has mass $m > 0$, and is moving in a given potential field $V(\mathbf{x})$ where $\mathbf{x} \in \mathbb{R}^3$, moves along a curve $\mathbf{x}(t)$ satisfying the equation of motion $m\ddot{\mathbf{x}} = -\operatorname{grad} V(\mathbf{x})$. If we introduce the momentum $\mathbf{p} = m\dot{\mathbf{x}}$ and the energy $H(\mathbf{x}, \mathbf{p}) = (1/2m)\|\mathbf{p}\|^2 + V(\mathbf{x})$ then Newton's law becomes ***Hamilton's equations***:

$$\dot{x}^i = \frac{\partial H}{\partial p_i}, \quad \dot{p}_i = -\frac{\partial H}{\partial x^i} \qquad i = 1, 2, 3.$$

To study this system of first-order equations for given H, we introduce the matrix

$$\mathbb{J} = \begin{bmatrix} 0 & I \\ -I & 0 \end{bmatrix},$$

where I is the 3×3 identity; note that the equations become $\dot{\xi} = \mathbb{J} \operatorname{grad} H(\xi)$ where $\xi = (\mathbf{x}, \mathbf{p})$.

In complex notation, setting $z = x + ip$, they may be written as

$$\dot{z} = 2i \frac{\partial H}{\partial \bar{z}}$$

Suppose we make a change of coordinates, $w = f(\xi)$, where $f : \mathbb{R}^6 \to \mathbb{R}^6$ is smooth. If $\xi(t)$ satisfies Hamilton's equations, the equations satisfied by $w(t)$ are $\dot{w} = A\dot{\xi} = AJ\, \mathrm{grad}_\xi H(\xi) = AJA^* \mathrm{grad}_w H(\xi(w))$, where $A^i{}_j = (\partial w^i/\partial \xi^j)$ is the Jacobian matrix of f, A^* is the transpose of A and $\xi(w)$ denotes the inverse function of f. The equations for w will be Hamiltonian with energy $K(w) = H(\xi(w))$ if $AJA^* = J$. A transformation satisfying this condition is called ***canonical*** or ***symplectic***. One of the things we do in this chapter is to give a coordinate free treatment of this and related concepts.

The space $\mathbb{R}^3 \times \mathbb{R}^3$ of the ξ's is called the ***phase space***. For a system of N particles we would use $\mathbb{R}^{3N} \times \mathbb{R}^{3N}$. However, many fundamental physical systems have a phase space that is a manifold rather than Euclidean space, so doing mechanics soley in the context of Euclidean space is to constraining. For example, the phase space for the motion of a rigid body about a fixed point is the tangent bundle of the group $\mathbf{SO}(3)$ of 3×3 orthogoanl matrices with determinant +1. This manifold is diffeomorphic to $\mathbb{RP}^3$ and is topologically nontrivial. To generalize the notion of a Hamiltonian system to the context of manifolds, we first need to geometrize the symplectic matrix J. In infinite dimensions a few technical points need attention before proceeding.

Let E be a Banach space and $B : E \times E \to \mathbb{R}$ a continuous bilinear mapping. Then B induces a continuous map $B^\flat : E \to E^*$, $e \mapsto B^\flat(e)$ defined by $B^\flat(e) \cdot f = B(e, f)$. We call B ***weakly nondengerate*** if $B^\flat$ is injective, i.e., $B(e, f) = 0$ for all $f \in E$ implies $e = 0$. We call B ***nondengerate*** or ***strongly nondegenerate*** if $B^\flat$ is an isomorphism. By the open mapping theorm it follows that B *is nondegenrate iff* B *is weakly nondegenerate and* $B^\flat$ *is onto.*

If E is finite dimensional there is no difference between strong and weak nondegeneracy. However, if infinite dimensions the distinction is important to bear in mind, and the issue does come up in basic examples, as we shall see in Supplement **8.1A**.

Let M be a Banach manifold. By a ***weak Riemannian structure*** we mean a smooth assignment $g : x \mapsto \langle\, ,\rangle_x = g(x)$ of a weakly nondegenerate inner product (not necessarily complete) to each tangent space T_xM. Here smooth means that in a local chart $U \subset E$, the mapping $g : x \mapsto \langle\, ,\rangle_x \in L^2(E, E; \mathbb{R})$ is smooth, where $L^2(E, E; \mathbb{R})$ denotes the Banach space of bilinear maps of $E \times E$ to $\mathbb{R}$. Equivalently, smooth means g is smooth as a section of the vector bundle $L^2(TM, TM; \mathbb{R})$ whose fiber at $x \in M$ is $L^2(T_xM, T_xM; \mathbb{R})$. By a ***Riemannian manifold*** we mean a weak Riemannian manifold in which $\langle\, ,\rangle_x$ is nondegenerate. Equivalently, the topology of $\langle\, ,\rangle_x$ is complete on T_xM, so that the model space E must be isomorphic to a Hilbert space.

For example the L^2 inner product

$$\langle f, g\rangle = \int_0^1 f(x)g(x)dx \text{ on } E = C^0([0, 1], \mathbb{R})$$

is a weak Riemannian metric on E but is not a Riemannian metric.

8.1.1 Definition *Let* P *be a manifold modeled on a Banach space* E. *By a* ***symplectic form*** *we mean a two-form* ω *on* P *such that*

(i) ω *is closed, i.e.,* $\mathbf{d}\omega = 0$;

(ii) *for each* $z \in P$, $\omega_x : T_zP \times T_zP \to \mathbb{R}$ *is weakly nondegenerate.*

If ω_z *in* (ii) *is nondegenerate, we speak of a* ***strong symplectic form***. *If* (ii) *is dropped we refer to* ω *as a* ***presymplectic*** *form. (For the moment the reader may wish to assume* P *is finite dimensional, in which case the weak – strong distinction vanishes.)*

The first result is referred to as ***Darboux's theorem***. Our proof follows Moser [1965] and Weinstein [1969].

8.1.2 The Darboux Theorem *Let* ω *be a strong symplectic form on the Banach manifold* P. *Then for each* $x \in P$ *there is a local coordinate chart about* x *in which* ω *is constant.*

Proof The proof proceeds by the Lie transform method **5.4.7**. We can assume $P = E$ and $x = 0 \in E$. Let ω_1 be the constant form equaling $\omega_0 = \omega(0)$. Let $\Omega = \omega_1 - \omega$ and $\omega_t = \omega + t\Omega$, for $0 \le t \le 1$. For each t, $\omega_t(0) = \omega(0)$ is nondegenerate. Hence by openness of the set of linear isomorphisms of E to E^*, there is a neighborhood of 0 on which ω_t is nondegenerate for all $0 \le t \le 1$. We can assume that this neighborhood is a ball. Thus by the Poincaré lemma, $\Omega = \mathbf{d}\alpha$ for some one-form α. We can suppose $\alpha(0) = 0$. Define a smooth vector field X_t by

$$\mathbf{i}_{X_t}\omega_t = -\alpha,$$

which is possible since ω_t is strongly non-degenerate. Since $X_t(0) = 0$, by corollary **4.1.25**, there is a sufficiently small ball on which the integral curves of X_t will be defined for time at least one. Let F_t be the flow of X_t starting at $F_0 = \textit{identity}$. By the Lie derivative formula for time-dependent vector fields (Theorem **5.4.4**) we have

$$\frac{d}{dt}(F_t^*\omega_t) = F_t^*(\mathbf{L}_{X_t}\omega_t) + F_t^*\frac{d}{dt}\omega_t = F_t^*\mathbf{d}\mathbf{i}_{X_t}\omega_t + F_t^*\Omega = F_t^*(\mathbf{d}(-\alpha) + \Omega) = 0.$$

Therefore, $F_1^*\omega_1 = F_0^*\omega_0 = \omega$, so F_1 provides the chart transforming ω to the constant form ω_1. ■

We note without proof that such a result is not true for Riemannian structures unless they are flat. Also, the analogue of Darboux's theorem is known to be not valid for weak symplectic forms. (For the example, see Abraham and Marsden [1978], Exercise 3.2H and for conditions under which it is valid, see Marsden [1981].)

8.1.3 Corollary *If* P *is finite dimensional and* ω *is a symplectic form, then*

(i) P *is even dimensional, say* $\dim P = 2n$;

(ii) *locally about each point there are coordinates* $x^1, \ldots, x^n, y^1, \ldots, y^n$ *such that*

$$\omega = \sum_{i=1}^{n} dx^i \wedge dy^i.$$

Such coordinates are called ***canonical***..

Proof By elementary linear algebra, any skew symmetric bilinear form that is nondegenerate has the canonical form

$$\begin{bmatrix} 0 & I \\ -I & 0 \end{bmatrix}$$

where I is the $n \times n$ identity. (This is proved by the same method as **6.2.9.**) This is the matrix version of (ii) pointwise on P. The result now follows from Darboux's theorem. ■

As a bilinear form, ω is given in canonical coordinates by $\omega((x_1, y_1), (x_2, y_2)) = \langle y_2, x_1 \rangle - \langle y_1, x_2 \rangle$. In complex notation with $z = x + iy$ it reads $\omega(z_1, z_2) = -\operatorname{Im}\langle z_1, z_2 \rangle$. This form for canonical coordinates extends to infinite dimensions (see Cook [1966], Chernoff and Marsden [1974] and Abraham and Marsden [1978], section 3.1 for details).

Now we are ready to consider *canonical symplectic forms.*

8.1.4 Definition *Let* Q *be a manifold modeled on a Banach space* E. Let T^*Q *be its contangent bundle, and* $\pi : T^*Q \to Q$ *the projection. Define the* ***canonical one-form*** θ *on* T^*Q *by*

$$\theta(\alpha)w = \alpha \cdot T\pi(w),$$

where $\alpha \in T_q{}^*Q$ *and* $w \in T_\alpha(T^*Q)$. *The* ***canonical two-form*** *is defined by* $\omega = -\,\mathbf{d}\theta$.

In a chart $U \subset E$, the formula for θ becomes

$$\theta(x, \alpha) \cdot (e, \beta) = \alpha(e),$$

where $(x, \alpha) \in U \times E^*$ and $(e, \beta) \in E \times E^*$. If Q is finite dimensional, this formula may be written

$$\theta = p_i\, dq^i,$$

where $q^1, \ldots, q^n, p_1, \ldots, p_n$ are coordinates for T^*Q and the summation convention is enforced. Using the local formula for $\mathbf{d}$ from formula (6) in Table **6.4.1**,

$$\omega(x, \alpha)((e_1, \alpha_1), (e_2, \alpha_2)) = \alpha_2(e_1) - \alpha_1(e_2),$$

or, in the finite-dimensional case,

$$\omega = dq^i \wedge dp_i.$$

In the infinite-dimensional case one can check that ω is weakly nondegenerate and is strongly nondegenerate iff E is reflexive (Marsden [1968b]).

If $\langle \, , \, \rangle_x$ is a weak Riemannian (or pseudo-Riemannian) metric on Q, the smooth vector bundle map $\varphi = g^\flat : TQ \to T^*Q$ defined by $\varphi(v_x) \cdot w_x = \langle v_x, w_x \rangle_x$, $x \in Q$, is injective on fibers. If $\langle \, , \, \rangle$ is a strong Riemannian metric, then φ is a vector bundle isomorphism of TQ onto T^*Q. In any case, set $\Omega = \varphi^*\omega$ where ω is the canonical two-form on T^*Q. Clearly Ω is exact since $\Omega = -\mathbf{d}\Theta$ where $\Theta = \varphi^*\theta$.

In the finite-dimensional case, the formulas for Θ and Ω become

$$\Theta = g_{ij} \, \dot{q}^j dq^i,$$

and

$$\Omega = g_{ij} dq^i \wedge d\dot{q}^j + \frac{\partial g_{ij}}{\partial q^k} \dot{q}^j dq^i \wedge dq^k,$$

where $q^1, \ldots, q^n$, $\dot{q}^1, \ldots, \dot{q}^n$ are coordinates for TQ. This follows by substituting $p_i = g_{ij} \, \dot{q}^j$ into $\omega = dq^i \wedge dp_i$.

In the infinite-dimensional case, if $\langle \, , \, \rangle_x$ is a weak metric, then ω is a weak symplectic form locally given by

$$\Theta(w, e)(e_1, e_2) = -\langle e, e_1 \rangle_x,$$

and

$$\Omega(x, e)((e_1, e_2), (e_3, e_4)) = \mathbf{D}_x\langle e, e_1\rangle_x e_3 - \mathbf{D}_x\langle e, e_3\rangle_x e_1 + \langle e_4, e_1\rangle_x - \langle e_2, e_3\rangle_x,$$

where $\mathbf{D}_x$ denotes the derivative with respect to x. One can also check that if $\langle \, , \, \rangle_x$ is a strong metric and Q is modeled on a reflexive space, then Ω is a strong symplectic form.

8.1.5 Definition *Let* (P, ω) *be a symplectic manifold. A (smooth) map* $f : P \to P$ *is called* ***canonical*** *or* ***symplectic*** *when* $f^*\omega = \omega$.

If follows that $f^*(\omega \wedge \cdots \wedge \omega) = \omega \wedge \cdots \wedge \omega$ (k times). If P is 2n-dimensional, then $\mu = \omega \wedge \cdots \wedge \omega$ (n times) is nowhere vanishing, so is a volume form; for instance by a computation one finds μ to be a multiple of the standard Euclidean volume in canonical coordinates. In paticular, note symplectic manifolds are orientable. We call μ the ***phase volume*** or the ***Liouville form***. Thus a symplectic map preserves the phase volume, and so is necessarily a local diffeomorphism. A map $f : P_1 \to P_2$ between symplectic manifolds (P_1, ω_1) and (P_2, ω_2) is called ***symplectic*** if $f^*\omega_2 = \omega_1$. As above, if P_1 and P_2 have the same dimension, then f is a local diffeomoorphism and preserves the phase volume.

We now discuss symplectic maps induced by maps on the base space of a cotangent bundle.

8.1.6 Proposition *Let* $f : Q_1 \to Q_2$ *be a diffeomorphism; define the* **cotangent lift** *of* f *by*

$$T^*f : T^*Q_2 \to T^*Q_1\ ;\ T^*f(\alpha_q) \cdot v = \alpha_q \cdot Tf(v),$$

where $q \in Q_2$, $\alpha_q \in T_q{}^*Q_2$ *and* $v \in T_{f^{-1}(q)}\, Q_1$; i.e., T^*f *is the* **pointwise adjoint** *of* Tf. *Then* T^*f *is symplectic and in fact* $(T^*f)^*\theta_1 = \theta_2$ *where* θ_i *is the canonical one-form on* Q_i, $i = 1, 2$.

Proof Let $\pi_i : T^*Q_i \to Q_i$ be the cotangent bundle projection, i = 1, 2. For w in the tangent space to T^*Q_2 at α_q, we have

$$\begin{aligned}(T^*f)^*\theta_1(\alpha_q)(w) &= \theta_1(T^*f(\alpha_q))(TT^*f \cdot w) = T^*f(\alpha_q) \cdot (T\pi_1 \cdot TT^*f \cdot w)\\ &= T^*f\,(\alpha_q) \cdot (T(\,\pi_1 \circ T^*f) \cdot w) = \ \alpha_q \cdot (T(f \circ \pi_1\ \circ T^*f) \cdot w\\ &= \alpha_q \cdot (T\pi_2\ \cdot w) = \theta_2(\alpha_q) \cdot w\end{aligned}$$

since, by construction, $f \circ \pi_1 \circ T^*f = \pi_2$. ■

In coordinates, if we write $f(q^1, ..., q^n) = (Q^1, ..., Q^n)$, then T^*f has the effect

$$(q^1, ..., q^n, p_1, ..., p_n) \mapsto (Q^1, ..., Q^n, P_1, ..., P_n),$$

where

$$p_j = \frac{\partial Q^i}{\partial q^j}\, P_i$$

(evaluated at the corresponding points). That this transformation is always canonical and in fact preserves the canonical one-form may be verified directly:

$$P_i dQ^i = P_i\, \frac{\partial Q^i}{\partial q^k} dq^k = p_k\, dq^k.$$

Sometimes one refers to canonical transformations of this type as "point transformations" since they arise from general diffeomorphisms of Q_1 to Q_2. Notice that lifts of diffeomorphisms satisfy

$$f \circ \pi_2 = \pi_1\ \circ T^*f;$$

that is, the following diagram commutes:

$$\begin{array}{ccc} T^*Q_2 & \xrightarrow{T^*f} & T^*Q_1 \\ \Big\downarrow{\scriptstyle \pi_2} & & \Big\downarrow{\scriptstyle \pi_1} \\ Q_2 & \xrightarrow[f]{} & Q_1 \end{array}$$

Notice also that

$$T^*(f \circ g) = T^*g \circ T^*f$$

and compare with

$$T(f \circ g) = Tf \circ Tg.$$

8.1.7 Corollary *If* Q_1 *and* Q_2 *are Riemannian (or pseudo-Riemannian) manifolds and* $f : Q_1 \to Q_2$ *is an isometry, then* $Tf : TQ_1 \to TQ_2$ *is symplectic, and in fact* $(Tf)^*\Theta_2 = \Theta_1$.

Proof This follows from the identity $Tf = g_2{}^\# \circ (T^*f)^{-1} \circ g_1{}^\flat$. All maps in this composition are symplectic and thus Tf is as well. ■

So far no mention has been made of Hamilton's equations. Now we are ready to consider them.

8.1.8 Definition *Let* (P, ω) *be a symplectic manifold. A vector field* $X : P \to TP$ *is called* ***Hamiltonian*** *if there is a* C^1 *function* $H : P \to \mathbb{R}$ *such that*

$$\mathbf{i}_X\omega = \mathbf{d}H.$$

We say X *is* ***locally Hamiltonian*** *if* $\mathbf{i}_X\omega$ *is closed.*

We write $X = X_H$ because usually in examples one is given H and then one constructs the Hamiltonian vector field X_H. If ω is only weakly nondegenerate, then given a smooth function $H : P \to \mathbb{R}$, X_H need not exist on all of P. Rather than being a pathology, this is quite essential in infinite dimensions, for the vector fields then correspond to partial differential equations and are only densely defined. The condition

$$\mathbf{i}_{X_H}\omega = \mathbf{d}H$$

is equivalent to

$$\omega_z(X_H(z), v) = \mathbf{d}H(z) \cdot v,$$

for $z \in P$ and $v \in T_zP$. Let us express this condition in canonical coordinates $(q^1, ..., q^n, p_1, ..., p_n)$ on a $2n$-dimensional symplectic manifold P, i.e., when $\omega = dq^i \wedge dp_i$. If $X = A^i\partial/\partial q^i + B^i\partial/\partial p_i$, then

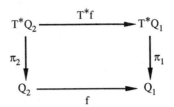

$$\mathbf{i}_{X_H}\omega = \mathbf{i}_{X_H}(dq^i \wedge dp_i) = (\mathbf{i}_{X_H}dq^i)dp_i - (\mathbf{i}_{X_H}dp_i)dq^i = (A^i dp_i - B^i dq^i).$$

This equals

$$\mathbf{d}H = \frac{\partial H}{\partial q^i}dq^i + \frac{\partial H}{\partial p_i}dp_i$$

iff

$$A^i = \frac{\partial H}{\partial p_i}, \text{ and } B^i = -\frac{\partial H}{\partial q^i}; \text{ i.e.,}$$

$$X_H = \sum_{i=1}^{n}\left(\frac{\partial H}{\partial p_i}\frac{\partial}{\partial q^i} - \frac{\partial H}{\partial q^i}\frac{\partial}{\partial p_i}\right).$$

If

$$\mathbb{J} = \begin{bmatrix} 0 & I \\ -I & 0 \end{bmatrix},$$

where I is the $n \times n$ identity matrix, the formula for X_H can be expressed as

$$X_H = \left(\frac{\partial H}{\partial p_i}, -\frac{\partial H}{\partial q^i}\right) = \mathbb{J} \text{ grad } H.$$

More intrinsically, one can write $X_H = \omega^{\#}\, \mathbf{d}H$, so one sometimes says that X_H is the ***symplectic gradient*** of H. Note that the formula $X_H = \mathbb{J}$ grad H is a little misleading in this respect, since no metric structure is actually needed and it is really the differential and not the gradient that is essential.

From the local expression for X_H we see that $(q^i(t), p_i(t))$ is an integral curve of X_H iff ***Hamilton's equations*** hold;

$$\dot{q}^i = \frac{\partial H}{\partial p_i}, \quad \dot{p}_i = -\frac{\partial H}{\partial q^i}.$$

We now give a couple of simple properties of Hamiltonian systems. The proofs are a bit more technical for densely defined vector fields, so for purposes of these theorems, we work with C^∞ vector fields.

8.1.9 Theorem *Let X_H be a Hamiltonian vector field on the (weak) symplectic manifold* (P, ω) *and let F_t be the flow of X_H. Then*

(i) F_t *is symplectic, i.e.* $F_t{}^*\omega = \omega$, *and*

(ii) *energy is conserved, i.e.* $H \circ F_t = F_t$.

Proof Since $F_0 = $ *identity*, it suffices to show that $(d/dt)F_t{}^*\omega = 0$. But by the basic connection between Lie derivatives and flows (§**5.4** and §**6.4**):

$$\frac{d}{dt} F_t^* \omega(x) = F_t^* (\mathcal{L}_{X_H} \omega)(x) = F_t^* (\mathbf{d} i_{X_H} \omega)(x) + F_t^* (i_{X_H} \mathbf{d}\omega)(x).$$

The first term is zero because it is $\mathbf{d}^2 H = 0$ and the second is zero because ω is closed.

(ii) By the chain rule,

$$\frac{d}{dt}(H \circ F_t)(x) = \mathbf{d}H(F_t(x)) \cdot X_H(F_t(x)) = \omega(F_t(x))(X_H(F_t(x)), X_H(F_t(x))).$$

But this is zero in view of the skew symmetry of ω. ∎

A corollary of (i) in finite dimensions is ***Liouville's theorem***: F_t *preserves the phase volume*. This is seen directly in canonical coordinates by observing that X_H is divergence-free.

We shall now generalize (ii). Define for any functions $f, g : U \to \mathbb{R}$, U open in P, their ***Poisson bracket*** by

$$\{f, g\} = \omega(X_f, X_g).$$

Since

$$\mathcal{L}_{X_f} g = i_{X_f} \mathbf{d} g = i_{X_f} i_{X_g} \omega = \omega(X_g, X_f) = -\omega(X_f, X_g) = -\mathcal{L}_{X_g} f,$$

we see that

$$\{f, g\} = \mathcal{L}_{X_g} f = -\mathcal{L}_{X_f} g.$$

If $\varphi : P_1 \to P_2$, is a diffeomorphism where (P_1, ω_1) and (P_2, ω_2) are symplectic manifolds, then by the property $\varphi^* (\mathcal{L}_X \alpha) = \mathcal{L}_{\varphi^* X} \varphi^* \alpha$ of pull-back, we have

$$\varphi^* \{f, g\} = \varphi^* (\mathcal{L}_{X_f} g) = \mathcal{L}_{\varphi^* X_f} \varphi^* g,$$

and

$$\{\varphi^* f, \varphi^* g\} = \mathcal{L}_{X_{\varphi^* f}} \varphi^* g.$$

Thus φ *preserves the Poisson bracket of any two functions defined on some open set of* P_2 *iff* $\varphi^* X_f = X_{\varphi^* f}$ *for all* C^∞ *functions* $f : U \to \mathbb{R}$ *where* U *is open in* P_2. This says that φ preserves the Poisson bracket iff it preserves Hamilton's equations. We have

$$i_{X_{\varphi^* f}} \omega = \mathbf{d}(\varphi^* f) = \varphi^* (\mathbf{d} f) = \varphi^* i_{X_f} \omega = i_{\varphi^* X_f} \varphi^* \omega,$$

so that by the (weak) nondegeneracy of ω and the fact that any $v \in T_z P$ equals some $X_h(z)$ for a C^∞ function h defined in a neighborhood of z, we conclude that φ is symplectic iff $\varphi^* X_f = X_{\varphi^* f}$ for all C^∞ functions $f : U \to \mathbb{R}$, where U is open in P_2. We have thus proved the following.

8.1.10 Proposition *Let* (P_1, ω_1) *and* (P_2, ω_2) *be symplectic manifolds and* $\varphi : P_1 \to P_2$ *a diffeomorphism. The following are equivalent:*

(i) φ *is symplectic*

(ii) φ *preserves the Poisson bracket of any two locally defined functions.*

(iii) $\varphi^* X_f = X_{\varphi^* f}$ *for any local* $f : U \to \mathbb{R}$, *where* U *is open in* P_2 *(i.e.,* φ *locally preserves Hamilton's equations).*

Conservation of energy is generalized in the following way.

8.1.11 Corollary (i) *Let* X_H *be a Hamiltonian vector field on the (weak) symplectic manifold* (P, ω) *with (local) flow* F_t. *Then for any* C^∞ *function* $f : U \to \mathbb{R}$, U *open in* P, *we have*

$$\frac{d}{dt}(f \circ F_t) = \{f, H\} \circ F_t = \{f \circ F_t, H\}.$$

(ii) *The curve* $c(t)$ *satisfies Hamilton's equations defined by* H *if and only if*

$$\frac{d}{dt} f(c(t)) = \{f, H\}\,(c(t))$$

for any C^∞ *function* $f : U \to \mathbb{R}$, *where* U *is open in* P.

Proof (i)

$$\frac{d}{dt}(f \circ F_t) = \frac{d}{dt} F_t^* f = F_t^* \mathbf{L}_{X_H} f = F_t^* \{f, H\} = \{F_t^* f, H\}$$

by the formula for Lie derivatives and the previous proposition.

(ii) Since $df(c(t))/dt = \mathbf{d}f(c(t)) \cdot (dc/dt)$ and

$$\{f, H\}(c(t)) = (\mathbf{L}_{X_H} f)(c(t)) = \mathbf{d}f(c(t)) \cdot X_H(c(t)),$$

the equation in the statement of the proposition holds iff $c'(t) = X_H(c(t))$ by the Hahn-Banach theorem and **4.2.14.** ■

One writes $\dot{f} = \{f, H\}$ to stand for the equation in (ii). This equation is called the ***equation of motion in Poisson bracket formulation.***

Two functions $f, g : P \to \mathbb{R}$ are said to be ***in involution*** or to ***Poisson commute*** if $\{f, g\} = 0$. Any function Poisson commuting with the Hamiltonian of a mechanical system is, by **8.1.11**, necessarily constant along on the flow of the Hamiltonian vector field. This is why such functions are called ***constants of the motion***. A classical theorem of Liouville states that *in a mechanical system with a 2n-dimensional phase space admitting* k *constants of the motion in involution and independent almost everywhere (that is, the differentials are independent on an open dense set) one* *can reduce the dimension of the phase space to* $2(n-k)$. In particular, if $k = n$, the equations of motion can be "explicitly" integrated. In fact, under certain additional hypotheses, the trajectories of the mechanical system are straight lines on high-dimensional cylinders or tori. If the motion takes place on tori, the explicit integration of the equations of motion goes under the name of finding ***action-angle variables***. See Arnold [1978], and Abraham and Marsden [1978, pp. 392-400] for details and Exercise **8.1D** for an example. In infinite-dimensional systems the situation is considerably more complicated. A famous example is the Korteweg-deVries (KdV) equation; for this example we also refer to Abraham and Marsden [1978, pp. 462-72] and references therein. The following supplement gives some elementary but still interesting examples of infinite-dimensional Hamiltonian systems.

☞ Supplement 8.1A
Two Infinite-Dimensional Examples

8.1.12 The Wave Equation as a Hamiltonian System The wave equation for a function $u(x, t)$, where $x \in \mathbb{R}^n$ and $t \in \mathbb{R}$ is given by

$$\frac{d^2u}{dt^2} = \nabla^2 u + m^2 u, \quad \text{(where } m \geq 0 \text{ is a constant)},$$

with u and $\dot{u} = \partial u/\partial t$ given at $t = 0$. The energy is

$$H(u, \dot{u}) = \frac{1}{2}\left(\int |\dot{u}|^2\, dx + \int \|\nabla u\|^2 dx\right).$$

We define H on pairs $(u, \dot{u})$ of finite energy by setting

$$P = H^1(\mathbb{R}^n) \times L^2(\mathbb{R}^n),$$

where H^1 consists of functions in L^2 whose first (distributional) derivatives are also in L^2. (The Sobolev spaces H^s defined this way are Hilbert spaces that arise in many problems involving partial differential equations. We only treat them informally here.) Let $D = H^2 \times H^1$ and define $X_H : D \to P$ by

$$X_H(u, \dot{u}) = (\dot{u}, \nabla^2 u + m^2 u).$$

Let the symplectic form be associated with the L^2 metric as in the discussion following **8.1.4**, namely

$$\omega((u, \dot{u}), (v, \dot{v})) = \int v\dot{u}\, dx - \int u\dot{v}\, dx.$$

It is now an easy verification using integration by parts, to show that X_H, ω and H are in the proper relation, so in this sense the wave equation is Hamiltonian. That the wave equation has a flow on P follows from (the real form of) Stone's theorem (see Supplement **7.4A** and Yosida [1980]).

8.1.13 The Schrödinger Equation Let $P = \mathcal{H}$ a complex Hilbert space with $\omega = -2\mathrm{Im}\langle\,,\,\rangle$. Let H_{op} be a self-adjoint operator with domain D and let

$$X_H(\varphi) = iH_{op} \cdot \varphi$$

and

$$H(\varphi) = \langle H_{op}\varphi, \varphi\rangle, \quad \varphi \in D.$$

Again it is easy to check that ω, X_H and H are in the correct relation. Thus, X_H is Hamiltonian. Note that $\psi(t)$ is an integral curve of X_H if

$$\frac{1}{i}\frac{d\psi}{dt} = H_{op}\psi,$$

which is the *abstract Schrödinger equation* of quantum mechanics. That X_H has a flow is a special case of Stone's theorem. We know from general principles that the flow $e^{itH_{op}}$ will be symplectic. The additional structure needed for unitarity is exactly complex linearity. ✍

Turning our attention to geodesics and to Lagrangian systems, let M be a (weak) Riemannian manifold with metric $\langle\,,\,\rangle_x$ on the tangent space T_xM. The ***spray*** $S: TM \to T^2M$ of the metric $\langle\,,\,\rangle_x$ is the vector field on TM defined locally by $S(x, v) = ((x, v), (v, \gamma(x, v)))$, for $(x, v) \in T_xM$, where γ is defined by

$$\langle\gamma(x, v), w\rangle_x \equiv \frac{1}{2}\mathbf{D}_x\langle v, v\rangle_x \cdot \omega - \mathbf{D}_x\langle v, w\rangle_x \cdot v \tag{1}$$

and $\mathbf{D}_x\langle v, v\rangle_x \cdot w$ means the derivative of $\langle v, v\rangle_x$ with respect to x in the direction of w. If M is finite dimensional, the ***Christoffel symbols*** are defined by putting $\gamma^i(x, v) = -\Gamma^i{}_{jk}(x)v^j v^k$. Equation (1) is equivalent to

$$-\Gamma^i{}_{jk}v^jv^kw_i = \frac{1}{2}\frac{\partial g_{ij}}{\partial x^k}v^iv^jw^k - \frac{\partial g_{ij}}{\partial x^k}v^iw^jv^k;\ \text{i.e.,}$$

$$\Gamma^i{}_{jk} = \frac{1}{2}g^{hi}\left(\frac{\partial g_{hk}}{\partial x^j} + \frac{\partial g_{jh}}{\partial x^k} - \frac{\partial g_{jk}}{\partial x^h}\right).$$

The verification that S is well-defined independent of the charts is not too difficult. Notice that γ is quadratic in v. We will show below that S is the Hamiltonian vector field on TM associated with the kinetic energy $\langle v, v\rangle/2$. The projection of the integral curves of S to M are called ***geodesics***. Their local equations are thus

$$\ddot{x} = \gamma(x, \dot{x}),$$

which in the finite-dimensional case becomes

$$\ddot{x}^i + \Gamma^i_{jk} \dot{x}^j \dot{x}^k = 0, \; i = 1, \ldots, n.$$

The definition of γ in (1) makes sense in the infinite as well as the finite-dimensional case, whereas the coordinate definition of Γ^i_{jk} makes sense only in finite dimensions. This then provides a way to deal with geodesics in infinite-dimensional spaces.

Let $t \mapsto (x(t), v(t))$ be an integral curve of S. That is,

$$\dot{x}(t) = v(t) \quad \text{and} \quad \dot{v}(t) = \gamma(x(t), v(t)). \tag{2}$$

As we remarked, these will shortly be shown to be Hamilton's equations of motion in the absence of a potential. To include a potential, let $V : M \to \mathbb{R}$ be given. At each x, we have the differential of V, $\mathbf{d}V(x) \in T_x{}^*M$, and we define $\text{grad}\, V(x)$ by

$$\langle \text{grad}\, V(x), w \rangle_x = \mathbf{d}V(x) \cdot w. \tag{3}$$

(In infinite dimensions, it is an extra assumption that grad V exists, since the map $T_xM \to T_x{}^*M$ induced by the metric is not necessarily bijective.)

The equations of motion in the potential field V are given by

$$\dot{x}(t) = v(t); \qquad \dot{v}(t) = \gamma(x(t), v(t)) - \text{grad}\, V(x(t))). \tag{4}$$

The total energy, kinetic plus potential, is given by $H(v_x) = (1/2)\, \|v_x\|^2 + V(x)$. The vector field X_H determined by H relative to the symplectic structure on TM induced by the metric, is given by (4). This will be part of a more general derivation of Lagrange's equations given below.

☞ Supplement 8.1B
Geodesics

Readers familiar with Riemannian geometry can reconcile the present approach to geodesics based on Hamiltonian mechanics to the standard one in the following way. Define the ***covariant derivative*** $\nabla : \mathfrak{X}(M) \times \mathfrak{X}(M) \to \mathfrak{X}(M)$ locally by

$$(\nabla_X Y)(x) = \gamma_x(X(x), Y(x)) + \mathbf{D}Y(x) \cdot X(x),$$

where X(x) and Y(x) are the local representatives of X and Y in the model space E of M and $\gamma_x : E \times E \to E$ denotes the symmetric bilinear continuous mapping defined by polarization of the quadratic form $\gamma(x, v)$. In finite dimensions, if $E = \mathbb{R}^n$, then $\gamma(x, v)$ is an $\mathbb{R}^n$-valued quadratic form on $\mathbb{R}^n$ determined by the Christoffel symbols Γ^i_{jk}. The defining relation for $\nabla_X Y$ becomes

$$\nabla_X Y = X^j Y^k \Gamma^i_{jk} \frac{\partial}{\partial x^i} + X^j \frac{\partial Y^k}{\partial x^j} \frac{\partial}{\partial x^k},$$

where locally

$$X = X^i \frac{\partial}{\partial x^i} \text{ and } Y = Y^k \frac{\partial}{\partial x^k}.$$

It is a straightforward exercise to show that the foregoing definition of $\nabla_X Y$ is chart independent and that ∇ satisfies the following conditions defining an ***affine connection:***

(i) ∇ is $\mathbb{R}$-bilinear,
(ii) for $f : M \to \mathbb{R}$ smooth, $\nabla_{fX} Y = f \nabla_X Y$ and $\nabla_X fY = f \nabla_X Y + X[f]Y$,
(iii) $(\nabla_X Y - \nabla_Y X)(x) = \mathbf{D}Y(x) \cdot X(x) - \mathbf{D}X(x) \cdot Y(x) = [X, Y](x)$,

by the local formula for the Jacobi-Lie bracket of two vector fields. (The equivalence of sprays and affine connections was introduced by Palais and Singer [1960].)

If c(t) is a curve in M and $X \in \mathfrak{X}(M)$, the ***covariant derivative of*** X ***along*** c is defined by

$$\frac{DX}{dt} = \nabla_{\dot{c}} X,$$

where $\dot{c}$ is a vector field coinciding with $\dot{c}(t)$ at the points c(t). Locally, using the chain rule, this becomes

$$\frac{DX}{dt}(c(t)) = -\gamma_{c(t)}(X(c(t)), X(c(t))) + \frac{d}{dt} X(c(t)),$$

which also shows that the definition of DX/dt depends only on $\dot{c}(t)$ and not on how $\dot{c}$ is extended to a vector field. In finite dimensions, the coordinate form of the preceding equation is

$$\left(\frac{DX}{dt}\right)^i = \Gamma^i_{jk}(c(t))\, X^j(c(t))\, \dot{c}^k(t) + \frac{d}{dt} X^i(c(t)),$$

where $\dot{c}(t)$ denotes the tangent vector to the curve at c(t).

The vector field X is called ***autoparallel*** or is ***parallel-transported along*** c if $DX/dt = 0$. Thus $\dot{c}$ is autoparallel along c iff in any coordinate system we have

$$\ddot{c}(t) - \gamma_{c(t)}(\dot{c}(t), \dot{c}(t)) = 0$$

or, in finite dimensions

$$\ddot{c}^i(t) + \Gamma^i_{jk}(c(t))\, \dot{c}^j(t)\, \dot{c}^k(t) = 0.$$

That is, $\dot{c}$ *is autoparallel along* c *iff* c *is a geodesic.*

There is feedback between Hamiltonian systems and Riemannian geometry. For example, conservation of energy for geodesics is a direct consequence of their Hamiltonian character but can also be checked directly. Moreover, the fact that the flow of the geodesic spray on TM consists of canonical transformations is also useful in geometry, for example, in the study of closed geodesics (cf. Klingenberg [1978]). On the other hand, Riemannian geometry raises questions (such as parallel transport and curvature) that are useful in studying Hamiltonian systems. ✍

We now generalize the idea of motion in a potential to that of a Lagrangian system; these are, however, still special types of Hamiltonian systems. We begin with a manifold M and a given function $L : TM \to \mathbb{R}$ called the ***Lagrangian***. In case of motion in a potential, take

$$L(v_x) = \frac{1}{2}\langle v_x, v_x\rangle - V(x),$$

which differs from the energy in that $-V$ is used rather than $+V$.

The Lagranian L defines a map called the ***fiber derivative***, $\mathbb{F}L : TM \to T^*M$ as follows: let $v, w \in T_xM$, and set

$$\mathbb{F}L(v) \cdot w \equiv \frac{d}{dt} L(v + tw)\Big|_{t=0} .$$

That is, $\mathbb{F}L(v) \cdot w$ is the derivative of L along the fiber in direction w. In the case of $L(v_x) = (1/2)\langle v_x, v_x\rangle_x - V(x)$, we see that $\mathbb{F}L(v_x) \cdot w_x = \langle v_x, w_x\rangle_x$, so we recover the usual map $g^\flat$: $TM \to T^*M$ associated with the bilinear form $\langle\, , \,\rangle_x$.

Since T^*M carries a canonical symplectic form ω, we can use $\mathbb{F}L$ to obtain a closed two-form ω_L on TM:

$$\omega_L = (\mathbb{F}L)^*\omega .$$

A local coordinate computation yields the following local formula for ω_L : if M is modeled on a linear space E, so locally TM looks like $U \times E$ where $U \subset E$ is open, then $\omega_L(u, e)$ for $(u, e) \in U \times E$ is the skew symmetric bilinear form on $E \times E$ given by

$$\begin{aligned}\omega_L(u, e) \cdot ((e_1, e_2), (f_1, f_2)) &= \mathbf{D}_1(\mathbf{D}_2L(u, e) \cdot e_1) \cdot f_1 - \mathbf{D}_1(\mathbf{D}_2L(u, e) \cdot f_1) \cdot e_1 \\ &\quad + \mathbf{D}_2(\mathbf{D}_2L(u, e) \cdot e_1) \cdot f_2 - \mathbf{D}_2(\mathbf{D}_2L(u, e) \cdot f_1) \cdot e_2 , \qquad (5)\end{aligned}$$

where $\mathbf{D}_1$ and $\mathbf{D}_2$ denote the indicated partial derivatives of L. In finite dimensions this reads

$$\omega_L = \frac{\partial^2 L}{\partial \dot{q}^i \partial q^j} dq^i \wedge dq^j + \frac{\partial^2 L}{\partial \dot{q}^i \partial \dot{q}^j} dq^i \wedge d\dot{q}^j,$$

where $(q^i, \dot{q}^i)$ are the standard local coordinates on TQ.

It is easy to see that ω_L is (weakly) nondegenerate if $\mathbf{D}_2\mathbf{D}_2L(u, e)$ is (weakly) nondegenerate;

in this case L is called ***(weakly) nondegenerate***. In the case of motion in a potential, nondegeneracy of ω_L amounts to nondegeneracy of the metric $\langle\ ,\ \rangle_x$. The ***action*** of L is defined by $A : TM \to \mathbb{R}$, $A(v) = \mathbb{F}L(v) \cdot v$, and the ***energy*** of L is $E = A - L$. In charts,

$$E(u, e) = D_2L(u, e) \cdot e - L(u, e),$$

and in finite dimensions, E is given by the expression

$$E(q, \dot{q}) = \frac{\partial L}{\partial \dot{q}^i} \dot{q}^i - L(q, \dot{q}).$$

Given L, we say that a vector field Z on TM is a ***Lagrangian vector field*** or a ***Lagrangian system*** for L if the ***Lagrangian condition*** holds:

$$\omega_L(v)(Z(v), w) = dE(v) \cdot w \tag{6}$$

for all $v \in T_qM$, and $w \in T_v(TM)$. Here dE denotes the differential of E. We shall see that for motion in a potential, this leads to the same equations of motion as we found before.

If ω_L were a weak symplectic form there would exist at most one such Z, which would be the Hamiltonian vector field for the Hamiltonian E. The dynamics is obtained by find the integral curves of Z; that is, the curves $t \mapsto v(t) \in TM$ satisfying $(dv/dt)(t) = Z(v(t))$. From the Lagranian condition it is easy to check that energy is conserved (even though L may be degenerate).

8.1.14 Proposition *Let* Z *be a Lagrangian vector field for* L *and let* $v(t) \in TM$ *be an integral curve of* Z. *Then* $E(v(t))$ *is constant in* t.

Proof By the chain rule,

$$\frac{d}{dt}E(v(t)) = dE(v(t)) \cdot v'(t) = dE(v(t)) \cdot Z(v(t)) = \omega_L(v(t))(Z(v(t)), Z(v(t)) = 0$$

by skew symmetry of ω_L. ■

We now generalize our previous local expression for the spray of a metric, and the equations of motion in the presence of a potential. In the general case the equations are called Lagrange's equations.

8.1.15 Proposition. *Let* Z *be a Lagrangian system for* L *and suppose* Z *is a second-order equation (that is, in a chart* $U \times E$ *for* TM, $Z(u, e) = (u, e, e, Z_2(u, e))$ *for some map* $Z_2 \colon U \times E \to E$). *Then in the chart* $U \times E$, *an integral curve* $(u(t), v(t)) \in U \times E$ *of* Z *satisfies **Lagrange's equations**: that is,*

$$\frac{du}{dt}(t) = v(t), \qquad \frac{d}{dt}(\mathbf{D}_2 L(u(t), v(t)) \cdot w = \mathbf{D}_1 L(u(t), v(t)) \cdot w \tag{7}$$

for all $w \in E$. If $\mathbf{D}_2\mathbf{D}_2 L$, *or equivalently* ω_L, *is weakly nondegenerate, then* Z *is automatically second order.*

In case of motion in a potential, (7) reduces to the equations (4).

Proof From the definition of the energy E we have locally

$$\mathbf{D}E(u, e) \cdot (f_1, f_2) = \mathbf{D}_1(\mathbf{D}_2 L(u, e) \cdot e) \cdot f_1 + \mathbf{D}_2(\mathbf{D}_2 L(u, e) \cdot e) \cdot f_2 - \mathbf{D}_1 L(u, e) \cdot f_1$$

(a term $\mathbf{D}_2 L(u, e) \cdot f_2$ has cancelled). Locally we may write

$$Z(u, e) = (u, e, Y_1(u, e), Y_2(u, e)).$$

Using formula (5) for ω_L, the condition on Z may be written

$$\begin{aligned} &\mathbf{D}_1(\mathbf{D}_2 L(u, e) \cdot Y_1(u, e)) \cdot f_1 - \mathbf{D}_1(\mathbf{D}_2 L(u, e) \cdot f_1) \cdot Y_1(u, e) \\ &\quad + \mathbf{D}_2(\mathbf{D}_2 L(u, e) \cdot Y_1(u, e)) \cdot f_2 - \mathbf{D}_2(\mathbf{D}_2 L(u, e) \cdot f_1) \cdot Y_2(u, e) \\ &\quad = \mathbf{D}_1(\mathbf{D}_2 L(u, e) \cdot e) \cdot f_1 - \mathbf{D}_1 L(u, e) \cdot f_1 + \mathbf{D}_2(\mathbf{D}_2 L(u, e) \cdot e) \cdot f_2. \end{aligned} \tag{9}$$

Thus if ω_L is a weak symplectic form, then $\mathbf{D}_2\mathbf{D}_2 L(u, e)$ is weakly nondenerage, so setting $f_1 = 0$ we get $Y_1(u, e) = e$, i.e., Z is a second-order equation. In any case, if we assume that Z is second order, then condition (9) becomes

$$\mathbf{D}_1 L(u, e) \cdot f_1 = \mathbf{D}_1(\mathbf{D}_2 L(u, e) \cdot f_1) \cdot e + \mathbf{D}_2(\mathbf{D}_2 L(u, e) \cdot f_1) \cdot Y_2(u, e)$$

for all $f_1 \in E$. If $(u(t), v(t))$ is an integral curve of Z and using dots to denote time differentiation, then $\dot{u} = v$ and $\ddot{u} = Y_2(u, \dot{u})$, so

$$\mathbf{D}_1 L(u, \dot{u}) \cdot f_1 = \mathbf{D}_1(\mathbf{D}_2 L(u, \dot{u}) \cdot f_1) \cdot \dot{u} + \mathbf{D}_2(\mathbf{D}_2 L(u, \dot{u}) \cdot f_1) \cdot \ddot{u} = \frac{d}{dt}\mathbf{D}_2 L(u, \dot{u}) \cdot f_1$$

by the chain rule. ■

The condition of being second order is intrinsic; Z is second order iff $T\tau_M \circ Z = \text{identity}$, where $\tau_M : TM \to M$ is the projection. See Exercise **8.1D**.

In finite dimensions Lagrange's equations (7) take the form

$$\frac{dq^i}{dt} = \dot{q}^i, \quad \text{and} \quad \frac{d}{dt}\left(\frac{\partial L}{\partial \dot{q}^i}\right) = \frac{\partial L}{\partial q^i}, \quad i = 1, \dots, n.$$

8.1.16 Proposition *Assume* $\varphi : Q \to Q$ *is a diffeomorphism which leaves a weakly nondegenerate Lagrangian* L *invariant, i.e.,* $L \circ T\varphi = L$. *Then* c(t) *is an integral curve of the*

Lagrangian vector field Z *if and only if* $T\varphi \circ c$ *is also an integral curve.*

Proof Invariance of L under φ implies $\mathbb{F}L \circ T\varphi = T^*\varphi^{-1} \circ \mathbb{F}L$ so that

$$(T\varphi)^*\omega_L = (\mathbb{F}L \circ T\varphi)^*\omega = (T^*\varphi^{-1} \circ \mathbb{F}L)^*\omega = (\mathbb{F}L)^* (T^*\varphi^{-1})^* \omega = (\mathbb{F}L)^*\omega = \omega_L$$

by Proposition **8.1.6**. We also have for any $v \in TQ$,

$$A(T\varphi(v)) = \mathbb{F}L(T\varphi(v)) \cdot T\varphi(v) = \mathbb{F}L(v)$$

and thus relation (6) implies

$$dE = (T\varphi)^*dE = (T\varphi)^*\mathbf{i}_Z\omega_L = \mathbf{i}_{(T\varphi)^*Z}(T\varphi)^*\omega_L = \mathbf{i}_{(T\varphi)^*Z}\,\omega_L\,.$$

Weak nondegeneracy of L yields then $(T\varphi)^*Z = Z$ which by Proposition **4.2.4** is equivalent to the statement in the proposition. ∎

8.1.17 Geodesics on the Poincaré Upper Half Plane Let $Q = \{(x, y) \in \mathbb{R}^2 \mid y > 0\}$ so that $TQ = Q \times \mathbb{R}^2$. Define the ***Poincaré metric*** g on Q by

$$g(x, y)((u^1, u^2), (v^1, v^2)) = (u^1v^1 + u^2v^2)/y^2$$

and consider the Lagrangian $L(x, y, v^1, v^2) = [(v^1)^2 + (v^2)^2]/y^2$ defined by g. L is nondegenerate and thus by Proposition **8.1.15**, the Lagrangian vector field Z defined by L is a second order equation. By local existence and uniqueness of integral curves, for every point $(x_0, y_0) \in Q$ and every vector $(v^1{}_0, v^2{}_0) \in T_{(x,y)}Q$, there is a unique geodesic $\gamma(t)$ satisfying $\gamma(0) = (x_0, y_0)$, $\gamma'(0) = (v^1{}_0, v^2{}_0)$. We shall determine the geodesics of g by taking advantage of invariance properties of L.

Note that the reflection $r : (x, y) \in Q \mapsto (-x, y) \in Q$ leaves L invariant. Furthermore, consider the homographies $h(z) = (az + b)/(cz + d)$ for $a, b, c, d \in \mathbb{R}$ satisfying $ad - bc = 1$, where $z = x + iy$. Since $\mathrm{Im}[h(z)] = \mathrm{Im}[z/(cz + d)^2]$, it follows that $h(Q) = Q$ and since $T_zh(v) = v/(cz + d)^2$, where $v = v^1 + iv^2$, it follows that h leaves L invariant. Therefore, by Proposition **8.1.16**, γ is a geodesic if and only if $r \circ \gamma$ and $h \circ \gamma$ are. In particular, if $\gamma(0) = (0, y_0)$, $\gamma'(0) = (0, u_0)$, then $(r \circ \gamma)(0) = (0, y_0)$ and $(r \circ \gamma)'(0) = (0, u_0)$, i.e., $\gamma = r \circ \gamma$ and thus γ is the

semiaxis $y > 0$, $x = 0$. Since for any $q_0 \in Q$ and tangent vector v_0 to Q at q_0 there exists a homography h such that $h(iy_0) = q_0$, and the tangent of h at iy_0 in the direction iu_0 is v_0, it follows that the geodesics of g are images by h of the semiaxis $\{(0, y) \mid y > 0\}$. If $c = 0$ or $d = 0$, this image equals the ray $\{(b/d, y) \mid y > 0\}$ or the ray $\{(a/c, y) \mid y > 0\}$. If both $c \neq 0$, $d \neq 0$, then the image equals the arc of the circle centered at $((ad + bc)/2cd, 0)$ of radius $1/(2cd)$. Thus *the geodesics of the Poincaré upper half plane are either rays parallel to the* y*-axis or arcs of circles centered on the* x*-axis.* (See Fig. 8.1.1). The Poincaré upper half-plane is a model of the Lobatchevski geometry. Two geodesics in Q are called ***parallel*** if they do not intersect in Q. Given either a ray parallel to Oy or a semicircle centered on Ox and a point not on this geodesic, there are infinitely many semicircles passing through this point and not intersecting the geodesics, i.e., *through a point not on a geodesic there are infinitely many geodesics parallel to it.* ♦

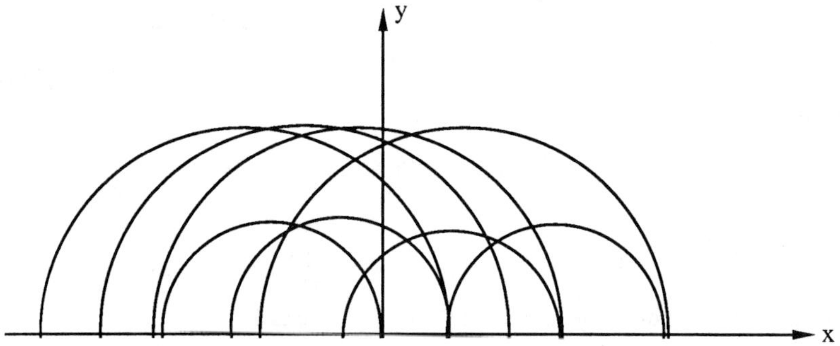

Figure 8.1.1 Geodesics in the Poincaré Upper Half Plane

We close with a completeness result for Lagrangian systems generalizing Example **4.1.23B**.

8.1.18 Definition. *A* C^2 *function* $V_0 : [0, \infty] \to \mathbb{R}$ *is called* ***positively complete*** *if it is decreasing and for any*

$$e > \sup_{i \geq 0}\{V_0(t)\}, \textit{ it satisfies } \int_x^{\infty} [e - V_0(t)]^{-1/2}\, dt\, dt = +\infty, \textit{ where } x \geq 0.$$

The last condition is independent of e. Examples of positively complete functions are $-t^{\alpha}$, $-t[\log(1 + t)]^{\alpha}$, $-t\log(1 + t)[\log(\log(1 + t) + 1)]^{\alpha}$, etc.

8.1.19 Theorem (Weinstein and Marsden [1970]) *Let* Q *be a complete weak Riemannian manifold,* $V: Q \to \mathbb{R}$ *be a* C^2 *function and let* Z *be the Lagrangian vector field for* $L(v) = \| v \|^2/2 - V(\tau(v))$, *where* $\tau : TQ \to Q$ *is the tangent bundle projection. Suppose there is a positively complete function* v_0 *and a point* $q' \in Q$ *such that* $V(q) \geq V_0(d(q, q'))$ *for all* $q \in Q$. *Then* Z *is complete.*

Proof Let c(t) be an integral curve of Z and let $q(t) = (\tau \circ c)(t)$ be its projection in Q. Let $q_0 = q(0)$ and consider the differential equation on $\mathbb{R}$

$$f''(t) = -\frac{dV_0}{df}(f(t)) \tag{10}$$

with initial conditions $f(0) = d(q', q_0)$, $f'(0) = \sqrt{2(\beta - V_0(f(0))}$, where $\beta = E(c(t)) = E(c(0)) \geq V(q_0)$. We can assume $\beta > V(q_0)$, for if $\beta = V(q_0)$, then $\dot{q}(0) = 0$; now if $\dot{q}(t) = 0$, the conclusion is trivially satisfied, so we need to work under the assumption that there exists a t_0 for which $\dot{q}(t_0) \neq 0$; by time translation we can assume $t_0 = 0$.

We show that the solution f(t) of (10) is defined for all $t \geq 0$. Multiplying both sides of (10) by $f'(t)$ and integrating yields

$$\frac{1}{2}f'(t)^2 = \beta - V_0(f(t)), \text{ i.e., } t(s) = \int_{d(q', q_0)}^{s} [2(\beta - V_0(u))]^{-1/2}\, du .$$

By hypothesis, the integral on the right diverges and hence $t(s) \to +\infty$ as $s \to +\infty$. This shows that f(t) exists for all $t \geq 0$.

For $t \geq 0$, conservation of energy and the estimate on the potential V imply

$$\begin{aligned} d(q(t), q') &\leq d(q(t), q_0) + d(q_0, q') \leq d(q', q_0) + \int_0^t \|\dot{q}(s)\|\, ds \\ &= d(q', q_0) + \int_0^t [2(\beta - V_0(q(s))]^{1/2}\, ds \leq d(q', q_0) + \int_0^t [2(\beta - V_0(d(q(s), q'))]^{1/2}\, ds. \end{aligned}$$

Since

$$f(t) = d(q', q_0) + \int_0^1 [2(\beta - V_0(f(s))]^{1/2}\, ds$$

it follows that $d(q(t), q') \leq f(t)$; see Exercise **4.1I**(v) or the reasoning in Example **4.1.23B** plus an approximation of $d(q(t), q')$ by C^1 functions. Hence if Q is finite dimensional, q(t) remains in a compact set for finite t-intervals, $t \geq 0$. Therefore c(t) does as well, V(q(t)) being bounded below on such a finite t-interval. Proposition **4.1.9** implies that c(t) exists for all $t \geq 0$. The proof in infinite dimensions is done in Supplement **8.1C**. If F_t is the local flow of Z, from $\tau(F_{-t}v)) = \tau(F_t(-v))$ (reversibility), it follows that c(t) exists also for all $t \leq 0$ and so the theorem is proved. ■

☞ Supplement 8.1.C
Completeness of Lagrangian Vector Fields on Hilbert Manifolds

This supplement provides the proof of theorem **8.1.19** for infinite dimensional Riemannian manifolds. We start with a few facts of general interest.

Let (Q, g) be a Riemannian manifold and $\tau : TQ \to Q$ the tangent bundle projection. For $v \in T_qQ$, the subspace $V_v = \ker T_v\tau = T_v(T_qQ) \subset T_v(TQ)$ is called the ***vertical subspace*** of $T_v(TQ)$. The local expression of the covariant derivative ∇ defined by g in Supplement **8.1B** shows that $\nabla_Y X$ depends only on the point values of Y and thus it defines a linear map $(\nabla X)(q)$: $v \in T_qQ \mapsto (\nabla_Y X)(q) \in T_qQ$ where $Y \in \mathfrak{X}(Q)$ is any vector field satisfying $Y(q) = v$. Let $j_v : T_qQ \to T_v(T_qQ) = V_v$ denote the isomorphism identifying the tangent space to a linear space with the linear space itself and consider the map $j_v \circ (\nabla X)(q) : T_qQ \to V_v$. Define the horizontal map $h_v : T_qQ \to T_v(TQ)$ by

$$h_v = T_qX - j_v \circ (\nabla X)(q)$$

where $v \in T_qQ$ and $X \in \mathfrak{X}(Q)$ satisfies $X(q) = v$. Locally, if E is the model of Q, h_v has the expression

$$h_v : (x, u) \in U \times E \mapsto (x, v, u - \gamma_x(u, v)) \in U \times E \times E \times E.$$

This shows that h_v is a linear continuous injective map with split image. The image of h_v is called the ***horizontal subspace*** of $T_v(TQ)$ and is denoted by H_v. It is straightforward to check that $T_v(TQ) = V_v \oplus H_v$ and that $T_v\tau \mid H_v : H_v \to T_qQ$, $j_v : T_qQ \to V_v$ are Banach space isomorphisms. Declaring them to be isometries and H_v perpendicular to V_v gives a metric g^T on TQ. We have proved that if (Q, q) *is a (weak) Riemannian manifold, then* g *induces a metric* g^T *on* TQ. The following result is taken from Ebin [1970].

8.1.20 Proposition *If* (Q, g) *is a complete (weak) Riemannian manifold then so is* (TQ, g^T).

Proof. Let $\{v_n\}$ be a Cauchy sequence in TQ and let $q_n = \tau(v_n)$. Since τ is distance decreasing it follows that $\{q_n\}$ is a Cauchy sequence in Q and therefore convergent to $q \in Q$ by completeness of Q. If E is the model of Q, E is a Hilbert space, again by completeness of Q. Let (U, φ) be a chart at q and assume that U is a closed ball in the metric defined by g of radius 3ε. Also, assume that $T_vT\varphi : T_v(TM) \to E \times E$ is an isometry for all $v \in T_qQ$ which implies that for ε small enough there is a $C > 0$ such that $\|TT\varphi(w)\| \leq C\,\|w\|$ for all $w \in TTU$. This means that all curves in TU are stretched by $T\varphi$ by a factor of C. Let $V \subset U$ be the closed ball of radius ε centered at q and let n, m be large enough so the distance between v_n and v_m is smaller than ε and $v_n, v_m \in TV$. If γ is a path from v_n to v_m of length $< 2\varepsilon$, then $\tau \circ \gamma$ is a path from q_n to q_m of length $< 2\varepsilon$ and therefore $\tau \circ \gamma \subset U$, which in turn implies that $\gamma \subset TU$. Moreover, $T\varphi \circ \gamma$ has length $< 2C\varepsilon$ and therefore the distance between $T\varphi(v_n)$ and $T\varphi(v_m)$ in E is at most $2C\varepsilon$. This shows that $\{T\varphi(v_n)\}$ is a Cauchy sequence in E and hence convergent. Since $T\varphi$ is a diffeomorphism, $\{v_n\}$ is convergent. ■

In general, completeness of a vector field on M implies completeness of the first variation equation on TM.

Proof of 8.1.19. Let $c:]a,b[\to TQ$ be a maximal integral curve of Z. We shall prove that $\lim_{t\uparrow b} c(t)$ exists in TQ which implies, by local existence and uniqueness, that c can be continued beyond b, i.e., that $b = +\infty$. One argues similarly for a. We have shown that $q(t) = (\tau \circ c)(t)$ is bounded on finite t-intervals. Since $V(q(t))$ is bounded on such a finite t-interval, it follows that $\dot{q}(t) = c(t)$ is bounded in the metric defined by g^T on TQ. By the mean value inequality it follows that if $t_n \uparrow b$, then $\{q(t_n)\}$ is a Cauchy sequence and therefore convergent since Q is complete.

Next we show by the same argument that if $t_n \uparrow b$, then $\{c(t_n)\}$ is Cauchy, i.e., we will show that $\dot{c}(t)$ is bounded on bounded t-intervals. Write $\dot{c}(t) = Z(c(t)) = S(c(t)) + V(c(t))$, where S is the spray of g and represents the horizontal part of Z and V is the vertical part of

Z. Since $V(c(t))$ depends only on $q(t)$ and since $q(t)$ extends continuously to $q(b)$, it follows that $\|V(c(t))\|$ is bounded as $t \uparrow b$. Since $\|S(c(t))\| = \|c(t)\|$ and $\|\dot{c}(t)\|^2 = \|S(c(t))\|^2 + \|V(c(t))\|^2$ by the definition of the metric g^T, it follows that $\|\dot{c}(t)\|$ remains bounded on finite t-intervals. Therefore $\{c(t_n)\}$ is Cauchy and Proposition **8.1.20** implies that $c(t)$ can be continuously extended to $c(b)$. ∎

Remark Note that completeness of Q, an estimate on the potential V, and conservation of the enrgy E, replaces "compactness" in Proposition **4.1.19** with "boundedness." ✍

Exercises

8.1A Let (M, ω) be a symplectic manifold with $\omega = \mathbf{d}\theta$ and $f : M \to M$ a local diffeomorphism. Prove that f is a symplectic iff for every compact oriented two-manifold B with boundary, $B \subset M$, we have

$$\int_{\partial B} \theta = \int_{f(\partial B)} \theta$$

8.1B (J. Moser) Use the method of proof of Darboux's theorem to prove that if M is a compact manifold, μ and ν are two volume forms with the same orientation, and

$$\int \mu = \int \nu ,$$

then there is a diffeomorphism $f : M \to M$ such that $f^*\nu = \mu$. (*Hint*: Use the Lie transform method. Since $\int \mu = \int \nu$, $\mu - \nu = \mathbf{d}\alpha$ (see Supplement **7.5B**); put $\nu_t = t\nu + (1-t)\mu$ and define X_t by letting the interior product of X_t with ν_t be α. Let φ_t be the flow of X_t and set $f = \varphi_1$.)

8.1C On $T^*\mathbb{R}^3$, consider the periodic three-dimensional ***Toda lattice*** Hamiltonian,

$$H(\mathbf{q}, \mathbf{p}) = \frac{1}{2}\|\mathbf{p}\|^2 + e^{q_1 - q_2} + e^{q_2 - q_3} + e^{q_3 - q_1} .$$

(i) Write down Hamilton's equations.

(ii) Show that

$$f_1(\mathbf{q}, \mathbf{p}) = p_1 + p_2 + p_3, \quad f_2 = H, \text{ and}$$

$$f_3(\mathbf{q}, \mathbf{p}) = \frac{1}{3}(p_1^3 + p_2^3 + p_3^3) + p_1(\exp(q^1 - q^2) + \exp(q^3 - q^1))$$
$$+ p_2(\exp(q^1 - q^2) + \exp(q^2 - q^3)) + p_3(\exp(q^1 - q^2) + \exp(q^2 - q^3))$$

are in involution and are independent everywhere.

(iii) Prove the same thing for

$$g_1 = f_1, \quad g_2(q, p) = \exp(q^1 - q^2) + \exp(q^2 - q^3) + \exp(q^3 - q^1), \text{ and}$$
$$g_3(q, p) = p_1 p_2 p_3 - p_1 \exp(q^2 - q^3) - p_2 \exp(q^3 - q^1) - p_3 \exp(q^1 - q^2).$$

(iv) Can you establish (iii) without explicitly computing the Poisson brackets? (*Hint* : Express g_1, g_2, g_3 as polynomials of f_1, f_2, f_3.)

8.1D A ***second-order equaton*** on a manifold M is a vector field X on TM such that $T\tau_M \circ X = \mathrm{Id}_{TM}$. Show that

(i) X is a second-order equation iff for all integral curves c of X in TM we have $(\tau_M \circ c)' = c$. One calls $\tau_M \circ c$ a ***base integral curve***.

(ii) X is a second-order equation iff in every chart the local representative of X has the form $(u, e) \mapsto (u, e, e, V(u, e))$.

(iii) If M is finite dimensional and X is a second-order equation, then the base integral curves satisfy

$$\frac{d^2x(t)}{dt^2} = V(x(t), \dot{x}(t)),$$

where $(x, \dot{x})$ denotes standard coordinates on TM.

8.1E Noether theorem Prove the following result for Lagrangian systems.
Let Z *be a Lagrangian vector field for* $L : TM \to \mathbb{R}$ *and suppose* Z *is a second-order equation. Let* Φ_t *be a one-parameter group of diffeomorphisms of* M *generated by the vector field* $Y : M \to TM$. *Suppose that for each real number* t, $L \circ T\Phi_t = L$. *Then the function* $P(Y) : TM \to \mathbb{R}$, *defined by* $P(Y)(v) = \mathbb{F}L(v) \cdot Y$ *is constant along integral curves of* Z.

8.1F Use Exercise **8.1E** to show conservation of linear (resp., angular) momentum for the motion of a particle in $\mathbb{R}^3$ moving in a potential that has a translation (resp., rotational) symmetry.

8.1G Consider $\mathbb{R}^{2n+2}$ with coordinates $(q^1, ..., q^n, E, p_1, ..., p_n, t)$ and define the symplectic form $\omega = dq^i \wedge dp_i + dE \wedge dt$. Consider a function $P(q, p, E, t) = H(q, p, t) - E$. Show that the vector field $X = \dot{q}^i \partial/\partial q^i + \dot{p}_i \partial/\partial p_i + \dot{E}\partial/\partial t + \dot{t}\partial/\partial E$ defined by $i_X\omega = dP$ reproduces familiar equations for $\dot{q}$, $\dot{p}$, $\dot{t}$ and $\dot{E}$.

8.1H Show that the wave equation (see supplement **8.1A**) may be derived as a Lagrangian system.

8.1I Refer to Example **8.1.13** on the Schrödinger equation. Let A and B be self adjoint operators on $\mathcal{H}$ and let $f_A: \mathcal{H} \to \mathbb{R}$ be given by $f_A(\psi) = \langle \psi, A\psi \rangle$ (the expectation value of A in the state ψ). Show that Poisson brackets and commutators are related by

$$f_{i[A, B]} = \{f_A, f_B\}.$$

8.1J Show that the geodesic flow of a compact Riemannian manifold is complete. [Warning: Compact pseudo-Riemannian manifolds need not be complete; see Wolf [1982] and Marsden [1973]].

8.1K Show that any isometry of a weak pseudo-Riemannian manifold maps geodesics to geodesics. (A map $\varphi : Q \to Q$ is called an *isometry* if $\varphi^* g = g$, where g is the weak pseudo-Riemannian metric on Q.)

8.1L Let (Q, g) be a weak Riemannian manifold.

(i) If F_t is the flow of the spray of g show that $\tau(F_t(sv)) = \tau(F_{st}(v))$, where $\tau : TQ \to Q$ is the projection.

(ii) Let U be any bounded set in T_qQ. Show that there is an $\varepsilon > 0$ such that for any $v \in U$, the integral curve of the spray with initial condition v exists for a time $\geq \varepsilon$. (*Hint:* Let $V \subset U$ be an open neighborhood of **v** such that all integral curves starting in V exist for time $\geq \delta$. Find $R > 0$ such that $R^{-1}U \subset V$ and use (i).)

8.1M A weak pseudo-Riemannian manifold (Q, g) is called *homogeneous* if for any $x, y \in Q$ there is an isometry φ such that $\varphi(x) = y$. Show that homogeneous weak Riemannian manifolds are complete by using **8.1K** and **8.1L**. (*Hint:* Put the initial condition v in a ball B and choose ε as in **8.1L** (ii). Let v(t) be the integral curve of S through v and let q(t) be the corresponding geodesic. The geodesic starting at $q(\varepsilon)$ in the direction $v(\varepsilon)$, is φ applied to the geodesic through $q = \tau(v)$ in the direction $T\varphi^{-1}(v)(\varepsilon))$; φ is the isometry sending q to $q(\varepsilon)$. The latter geodesic lies in the ball B, so it exists for time $\geq \varepsilon$.)

§8.2 *Fluid Mechanics*

We present a few of the basic ideas concerning the motion of an ideal fluid from the point of view of manifolds and differential forms. This is usually done in the context of Euclidean space using vector calculus. For the latter approach and additional details, the reader should consult one of the standard texts on the subject such as Batchelor [1967], Chorin and Marsden [1979], or Gurtin [1981]. The use of manifolds and differential forms can give additional geometric insight.

The present section is for expository reasons somewhat superficial and is intended only to indicate how to use differential forms and Lie derivatives in fluid mechanics. Once the basics are understood, more sophisticated questions can be asked, such as: in what sense is fluid mechanics an infinite - dimensional Hamiltonian system? For the answer, see Arnold [1978], Abraham and Marsden [1978], Marsden and Weinstein [1983], and Marsden, Ratiu, and Weinstein [1984]. For analogous topics in elasticity, see Marsden and Hughes [1983], and for plasmas, see **§8.4** and Marsden and Weinstein [1982].

Let M be a compact, oriented finite-dimensional Riemannian n-manifold, possibly with boundary. Let the Riemannian volume form be denoted $\mu \in \Omega^n(M)$, and the corresponding volume element $d\mu$. Usually M is a bounded region with smooth boundary in two - or three - dimensional Euclidean space, oriented by the standard basis, and with the standard Euclidean volume form and inner product.

Imagine M to be filled with fluid and the fluid to be in motion. Our object is to describe this motion. Let $x \in M$ be a point in M and consider the particle of fluid moving through x at time t = 0. For example, we can imagine a particle of dust suspended in the fluid; this particle traverses a trajectory which we denote $\varphi_t(x) = \varphi(x, t)$. Let $u(x, t)$ denote the velocity of the particle of fluid moving through x at time t. Thus, for each fixed time, u is a vector field on M. See Figure **8.2.1**. We call u the ***velocity field of the fluid***. Thus the relationship between u and φ_t is

$$\frac{d\varphi_t(x)}{dt} = u(\varphi_t(x), t);$$

that is, u is a time-dependent vector field with evolution operator φ_t in the same sense as was used in **§4.1**.

For each time t, we shall assume that the fluid has a well-defined mass density and we write $\rho_t(x) = \rho(x, t)$. Thus if W is any subregion of M, we assume that the mass of fluid in W at time t is given by

$$m(W, t) = \int_W \rho_t d\mu .$$

Our derivation of the equations is based on three basic principles, which we shall treat in turn:

1 Mass is neither created nor destroyed.

2 (Newton's second law) The rate of change of momentum of a portion of the fluid equals the force applied to it .

3 Energy is neither created nor destroyed.

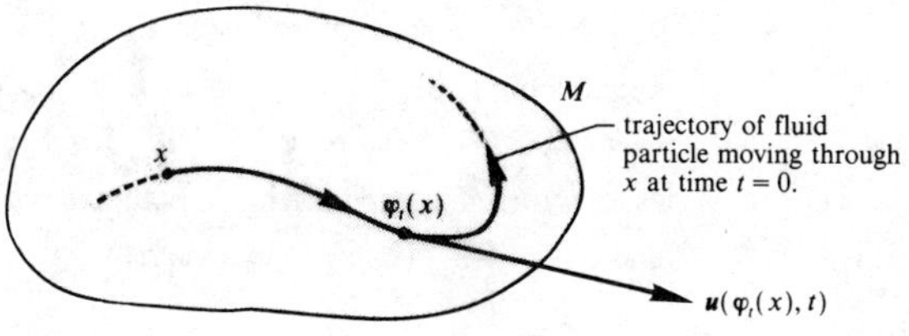

Figure 8.2.1

1. Conservation of mass. This principle says that the total mass of the fluid, which at time t = 0 occupied a nice region W, remains unchanged after time t; i.e.,

$$\int_{\varphi_t(W)} \rho_t d\mu = \int_W \rho_0 d\mu .$$

(We call a region W "nice" when it is an open subset of M with smooth enough boundary to allow us to use Stokes' theorem.) Let us recall how to use the transport theorem **7.1.12** to derive the continuity equation. Using the change-of-variables formula, conservation of mass may be rewritten as

$$\int_W \varphi_t^*(\rho_t\mu) = \int_W \rho_0\mu$$

for any nice region W in M, which is equivalent to

$$\varphi_t^*(\rho_t\,\mu) = \rho_0\,\mu\,, \quad \text{or} \quad (\varphi_t^*\rho_t)J(\varphi_t) = \rho_0\,,$$

where $J(\varphi_t)$ is the Jacobian of φ_t. This in turn is equivalent to

$$0 = \frac{d}{dt}\varphi_t^*(\rho_t\mu) = \varphi_t^*\left(\mathcal{L}_u(\rho_t\mu) + \frac{\partial\rho}{\partial t}\mu\right) = \varphi_t^*\left\{\left(u[\rho_t] + \rho_t \text{div } u + \frac{\partial\rho}{\partial t}\right)\mu\right\} = \varphi_t^*\left\{\left(\text{div}(\rho_t u) + \frac{\partial\rho}{\partial t}\right)\mu\right\}$$

by the Lie derivative formula and **6.5.17**. Thus

$$\frac{\partial\rho}{\partial t} + \text{div } (\rho_t u) = 0$$

is the differential form of the law of conservation of mass, also known as the ***continuity equation***.

Because of shock waves that could be present, ρ and u may not be smooth enough to justify the steps leading to the differential form of this law; the integral form will then be the one to use. Also note that the Riemannian metric has as yet played no role; only the volume element of M was needed.

2. **Balance of momentum**. Newton's second law asserts that the rate of change of momentum of a portion of the fluid equals the total force applied to it. To see how to apply this principle on a general manifold, let us discuss the situation $M \subset \mathbb{R}^3$ first. Here we follow the standard vector calculus conventions and write the velocity fields in boldface type. The momentum of a portion of the fluid at time t that at time $t = 0$ occupied the region W is

$$\int_{\varphi_t(W)} \rho \mathbf{u} d\mu .$$

Here and in what follows the integral is $\mathbb{R}^3$-valued, so we apply all theorems on integration componentwise.

For any continuum, forces acting on a piece of material are of two types. First there are ***forces of stress***, whereby the piece of material is acted on by forces across its surface by the rest of the continuum. Second, there are external, or ***body forces***, such as gravity or a magnetic field, which exert a force per unit volume on the continuum. The clear formulation of surface stress forces in a continuum is usually attributed to Cauchy. We shall assume that the body forces are given by a given force density $\mathbf{b}$, i.e., the total body forces acting on W are $\int_W \rho \mathbf{b}\, d\mu$. In continuum mechanics the forces of stress are assumed to be of the form $\int_{\partial W} \sigma(x, t) \cdot \mathbf{n}\, da$, where da is the induced volume element on the boundary, $\mathbf{n}$ is the outward unit normal, and $\sigma(x, t)$ is a time-dependent contravariant symmetric two-tensor, called the ***Cauchy stress tensor***. The contraction $\sigma(t, x) \cdot \mathbf{n}$ is understood in the following way: if σ has components σ^{ij} and $\mathbf{n}$ has components n^k, then $\sigma \cdot \mathbf{n}$ is a vector with components $(\sigma \cdot \mathbf{n})^i = g_{jk}\, \sigma^{ij} n^k$, where g is the metric (in our case $g_{jk} = \delta_{jk}$). The vector $\sigma \cdot \mathbf{n}$, called the ***Cauchy traction vector***, measures the force of contact (per unit area orthogonal to $\mathbf{n}$) between two parts of the continuum. (A theorem of Cauchy states that if one postulates the existence of a continuous Cauchy traction vector field $\mathbf{T}(\mathbf{x}, t, \mathbf{n})$ satisfying balance of momentum, then it must be of the form $\sigma \cdot \mathbf{n}$, for a two-tensor, σ; moreover if balance of moment of momentum holds, σ must be symmetric. See Chorin and Marsden [1979], Gurtin [1981], or Marsden and Hughes [1983] for details.) ***Balance of momentum*** is said to hold when

$$\frac{d}{dt}\int_{\varphi_t(W)} \rho \mathbf{u} d\mu = \int_{\varphi_t(W)} \rho \mathbf{b} d\mu + \int_{\partial\varphi_t(W)} \sigma \cdot \mathbf{n}\, da$$

for any nice region W in $M = \mathbb{R}^3$. If $\operatorname{div} \sigma$ denotes the vector with components $(\operatorname{div}(\sigma^{1i}), \operatorname{div}(\sigma^{2i}), \operatorname{div}(\sigma^{3i}))$, then by Gauss' theorem

$$\int_{\partial\varphi_t(W)} \sigma \cdot \mathbf{n}\, da = \int_{\varphi_t(W)} (\mathrm{div}\, \sigma)\, d\mu.$$

By the change-of-variables formula and Lie derivative formula, we get

$$\frac{d}{dt}\int_{\varphi_t(W)} \rho u^i\, d\mu = \int_W \frac{d}{dt}\, \varphi_t^*(\rho u^i\, d\mu) = \int_{\varphi_t(W)} \left(\frac{\partial(\rho u^i)}{\partial t} + (\mathbf{L}_u \rho) u^i + \rho \mathbf{L}_u u^i + \rho u^i \mathrm{div}\, u \right) d\mu,$$

so that the balance of momentum is equivalent to

$$\frac{\partial \rho}{\partial t} u^i + \rho \frac{\partial u^i}{\partial t} + (d\rho \cdot \mathbf{u}) u^i + \rho \mathbf{L}_u u^i + \rho u^i \mathrm{div}\, \mathbf{u} = \rho b^i + (\mathrm{div}\, \sigma)^i.$$

But $d\rho \cdot \mathbf{u} + \rho\, \mathrm{div}\, \mathbf{u} = \mathrm{div}(\rho \mathbf{u})$ and by conservation of mass, $\partial\rho/\partial t + \mathrm{div}(\rho\mathbf{u}) = 0$. Also, $\mathbf{L}_\mathbf{u} u^i = (\partial u^i/\partial x^j) u^j = (\mathbf{u} \cdot \nabla) u^i$, so we get

$$\frac{\partial \mathbf{u}}{\partial t} + (\mathbf{u} \cdot \nabla)\mathbf{u} = \mathbf{b} + \frac{1}{\rho} \mathrm{div}\, \sigma,$$

which represents the basic ***equations of motion***. Here the quantity $\partial\mathbf{u}/\partial t + (\mathbf{u} \cdot \nabla)\, \mathbf{u}$ is usually called the ***material derivative*** and is denoted by $D\mathbf{u}/dt$. These equations are for any continuum, be it elastic or fluid.

An ***ideal fluid*** is by definition a fluid whose Cauchy stress tensor σ is given in terms of a function $p(x, t)$ called the ***pressure***, by $\sigma^{ij} = -pg^{ij}$. In this case, balance of momentum in differential form becomes the ***Euler equations for an ideal fluid***:

$$\frac{\partial \mathbf{u}}{\partial t} + (\mathbf{u} \cdot \nabla)\mathbf{u} = \mathbf{b} - \frac{1}{\rho} \mathrm{grad}\ p.$$

The assumption on the stress σ in an ideal fluid means that if S is any fluid surface in M with outward unit normal $\mathbf{n}$, then the force of stress per unit area exerted across a surface element S at $\mathbf{x}$ with normal $\mathbf{n}$ at time t is $-p(\mathbf{x}, t)\mathbf{n}$ (see Figure 8.2.2).

Let us return to the context of a Riemannian manifold M. First, it is not clear what the vector-valued integrals should mean. But even if we could make sense out of this, using, say parallel transport, there is a more serious problem with the integral form of balance of momentum as stated. Namely, if one changes coordinates, then balance of momentum *does not look the same*. One says that the integral form of balance of momentum is *not covariant*. Therefore we shall concentrate on the differential form and from now on we shall deal only with ideal fluids. (For a detailed discussion of how to formulate the basic integral balance laws of continuum mechanics covariantly, see Marsden and Hughes [1983]. A genuine difficulty with shock wave theory is that the notion of weak solution is not a coordinate independent concept.)

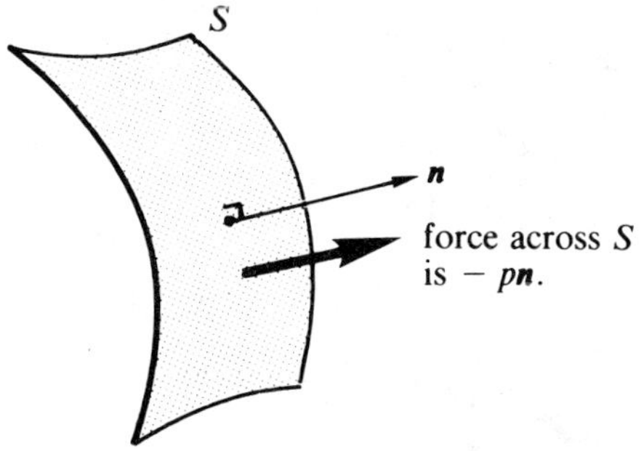

Figure 8.2.2

Rewrite Euler's equations in $\mathbb{R}^3$ with indices down; i.e., take the flat of these equations. Then the i-th equation, $i = 1, 2, 3$ is

$$\frac{\partial u_i}{\partial t} + u_1 \frac{\partial u_i}{\partial x^1} + u_2 \frac{\partial u_i}{\partial x^2} + u_3 \frac{\partial u_i}{\partial x^3} = b_i - \frac{1}{\rho}\frac{\partial p}{\partial x^i}.$$

We seek an invariant meaning for the sum of the last three terms on the left-hand side. For fixed i this expression is

$$u_j \frac{\partial u_i}{\partial x^j} = u_j \frac{\partial u_i}{\partial x^j} + u_j \frac{\partial u_j}{\partial x^i} - u \frac{\partial u_j}{\partial x^i} = (\mathbf{L}_{\mathbf{u}}\mathbf{u}^\flat)_i - \left(\frac{1}{2}\mathbf{d}\|\,\mathbf{u}\,\|^2\right)_i.$$

That is, Euler's equations can be written in the invariant form

$$\frac{\partial \mathbf{u}^\flat}{\partial t} + \mathbf{L}_{\mathbf{u}}\mathbf{u}^\flat - \frac{1}{2}\,\mathbf{d}(\mathbf{u}^\flat(\mathbf{u})) = -\frac{1}{\rho}\mathbf{d}p + \mathbf{b}^\flat.$$

We postulate this equation as the balance of momentum in M for an ideal fluid. The reader familiar with Riemannian connections (see Supplement **8.1B**) can prove that this form is equivalent to the form

$$\frac{\partial \mathbf{u}}{\partial t} + \nabla_{\mathbf{u}}\mathbf{u} = -\frac{1}{\rho}\operatorname{grad} p + \mathbf{b}$$

by showing that

$$\mathbf{L}_{\mathbf{u}}\mathbf{u}^\flat = (\nabla_{\mathbf{u}}\mathbf{u})^\flat + \frac{1}{2}\mathbf{d}(\mathbf{u}^\flat(\mathbf{u})),$$

where $\nabla_{\mathbf{u}}\mathbf{u}$ is the covariant derivative of $\mathbf{u}$ along itself, with ∇ the Riemannian connection given by g.

The boundary conditions that should be imposed come from the physical significance of ideal fluid: namely, no friction should exist between the fluid and ∂M; i.e., $\mathbf{u}$ is tangent to ∂M at points of ∂M. Summarizing, the *equations of motion* of an ideal fluid on a compact Riemannian manifold M with smooth boundary ∂M and outward unit normal n are

$$\frac{\partial u^\flat}{\partial t} + \mathbf{L}_u u^\flat - \frac{1}{2}d(u^\flat(u)) = -\frac{1}{\rho}dp + b^\flat \quad \text{and} \quad \frac{\partial \rho}{\partial t} + \operatorname{div}(\rho u) = 0.$$

We also have the ***boundary conditions***

$$u \,\|\, \partial M, \text{ i.e., } u \cdot n = 0 \text{ on } \partial M;$$

and ***initial conditions***

$$u(x, 0) = u_0(x) \text{ given on } M.$$

We shall assume $b = 0$ from now on for simplicity.

3. Conservation of energy A basic problem of ideal fluid dynamics is to solve the initial-boundary-value problem. The unknowns are u, ρ, and p, i.e., $n + 2$ scalar unknowns. We have, however, only $n + 1$ equations. Thus one might suspect that to specify the fluid motion, one more equation is needed. This is in fact true and the law of conservation of energy will supply the necessary extra equation in fluid mechanics. (The situation is similar for general continua; see Marsden and Hughes [1983].)

For a fluid moving in M with velocity field u, the ***kinetic energy*** of the fluid is

$$E_{\text{kinetic}} = \frac{1}{2}\int_M \rho \,\| u \|^2 \, d\mu$$

where $\| u \|^2 = \langle u, u \rangle$ is the square length of the vector function u. We assume that the total energy of the fluid can be written

$$E_{\text{total}} = E_{\text{kinetic}} + E_{\text{internal}},$$

where E_{internal} is the energy that relates to energy we cannot "see" on a macroscopic scale and derives from sources such as intermolecular potentials and molecular vibrations. If energy is pumped into the fluid or if we allow the fluid to do work, E_{total} will change. We describe two particular examples of energy equations that are useful.

A Assume that E_{internal} = constant. Then we ought to have E_{kinetic} as a constant of the motion; i.e.,

$$\frac{d}{dt}\left(\frac{1}{2}\int_M \rho \,\| u \|^2 d\mu\right) = 0.$$

To deal with this equation it is convenient to use the following.

8.2.1 Transport Theorem with Mass Density *Let* f *be a time-dependent smooth function on* M. *Then if* W *is any (nice) open set in* M,

$$\frac{d}{dt}\int_{\varphi_t(W)} \rho f \, d\mu = \int_{\varphi_t(W)} \rho \frac{Df}{dt} d\mu,$$

where $Df/dt = \partial f/\partial t + \mathbf{L}_u f$.

Proof By the change-of-variables formula, the Lie derivative formula, $\operatorname{div}(\rho u) = u[\rho] + \rho \operatorname{div}(u)$, and conservation of mass, we have

$$\begin{aligned}
\frac{d}{dt}\int_{\varphi_t(W)} \rho f \, d\mu &= \frac{d}{dt}\int_W \varphi_t^*(\rho f \mu) = \int_W \varphi_t^*\left(\frac{\partial(\rho f)}{\partial t}\mu + \mathbf{L}_u(\rho f \mu)\right) \\
&= \int_{\varphi_t(W)} \left(\frac{\partial \rho}{\partial t} f\mu + \rho \frac{\partial f}{\partial t}\mu + u[\rho] f\mu + \rho(\mathbf{L}_u f)\mu + \rho \mathbf{L}_u \mu\right) \\
&= \int_{\varphi_t(W)} \left[\left(\frac{\partial \rho}{\partial t} + u[\rho] + \rho \operatorname{div} u\right) f\mu + \rho\left(\frac{\partial f}{\partial t} + \mathbf{L}_u f\right)\mu\right] \\
&= \int_{\varphi_t(W)} \left[f\left(\frac{\partial \rho}{\partial t} + \operatorname{div}(\rho u)\right) + \rho\left(\frac{\partial f}{\partial t} + \mathbf{L}_u f\right)\right]\mu = \int_{\varphi_t(W)} \rho\left(\frac{\partial f}{\partial t} + \mathbf{L}_u f\right)\mu . \quad \blacksquare
\end{aligned}$$

Making use of $\mathbf{L}_u(\| u \|^2) = \mathbf{L}_u(u^\flat(u)) = (\mathbf{L}_u u^\flat)(u) = \mathbf{d}(u^\flat(u))(u)$, the transport lemma, and Euler's equations, we get

$$\begin{aligned}
0 &= \frac{d}{dt}\left(\frac{1}{2}\int_M \rho \| u \|^2 d\mu\right) = \frac{1}{2}\int_M \rho\left(\frac{\partial \| u \|^2}{\partial t} + \mathbf{L}_u \| u \|^2\right) d\mu = \int_M \rho \frac{\partial u^\flat}{\partial t} \cdot u \, d\mu + \frac{1}{2}\int_M (\mathbf{L}_u u^\flat) \cdot u \, d\mu \\
&= \int_M \rho \frac{\partial u^\flat}{\partial t} \cdot u \, d\mu + \int_M \rho(\mathbf{L}_u u^\flat) \cdot u \, d\mu - \frac{1}{2}\int_M \rho \mathbf{d}(u^\flat(u)) \cdot u \, d\mu \\
&= -\int_M \mathbf{d}p \cdot u \, d\mu = \int_M \{(\operatorname{div} u) p\mu - \mathbf{L}_u(p\mu)\} \qquad \text{(by the Leibniz rule for } \mathbf{L}_u\text{)} \\
&= \int_M \{(\operatorname{div} u) p\mu - \mathbf{d}(\mathbf{i}_u p\mu)\} = \int_M (\operatorname{div} u) p\mu.
\end{aligned}$$

The last equality is obtained by Stokes' theorem and the boundary conditions $0 = (u \cdot n) da = \mathbf{i}_u \mu$. If we imagine this to hold for the same fluid in all conceivable motions, we are forced to postulate one of the additional equations

$$\operatorname{div} u = 0 \quad \text{or} \quad p = 0.$$

The case $\operatorname{div} u = 0$ is that of an ***incompressible fluid***. Thus in this case the Euler equations are

$$\frac{\partial u^\flat}{\partial t} + \mathbf{L}_u u^\flat - \frac{1}{2}\mathbf{d}\,\|u\|^2 = -\frac{1}{\rho}\mathbf{d}p$$

$$\frac{\partial \rho}{\partial t} + \operatorname{div}(\rho u) = 0$$

$$\operatorname{div} u = 0$$

with the boundary condition $\mathbf{i}_u\mu = 0$ on ∂M and initial condition $u(x, 0) = u_0(x)$. The case $p = ($ is also possible but is less interesting.

For a homogeneous incompressible fluid, with *constant* density ρ, Euler's equations can be reformulated in terms of the Hodge decomposition theorem (see Section **7.5**). Nonhomogeneous incompressible flow requires a weighted Hodge decomposition (see Marsden [1976]). Recall that any one-form α can be written in a unique way as $\alpha = \mathbf{d}\beta + \gamma$, where $\delta\gamma = 0$. Define the linear operator $\mathbb{P} : \Omega^1(M) \to \{\gamma \in \Omega^1(M) \mid \delta\gamma = 0\}$ by $\mathbb{P}(\alpha) = \gamma$.

We are now in a position to reformulate Euler's equations. Let $\Omega^1{}_{\delta=0}$ be the set of C^∞ one-forms γ with $\delta\gamma = 0$ and γ tangent to ∂M; i.e., $*\gamma\,|_{\partial M} = 0$. Let $T : \Omega^1{}_{\delta=0} \to \Omega^1{}_{\delta=0}$ be defined by

$$T(u^\flat) = \mathbb{P}(\mathbf{L}_u u^\flat).$$

Thus Euler's equations can be written as $\partial u^\flat/\partial t + T(u^\flat) = 0$, which is in the "standard form" for an evolution equation. Note that T is nonlinear. Another important feature of T is that it is *nonlocal*; this is because $\mathbb{P}(\alpha)(x)$ depends on the values of α on all of M and not merely those in the neighborhood of $x \in M$.

B We postulate an internal energy over the region W to be of the form

$$E_{\text{internal}} = \int_W \rho w \, d\mu ,$$

where the function w is the internal energy density per unit mass.

We assume that ***energy is balanced*** in the sense that the rate of change of energy in a region equals the work done on it:

$$\frac{d}{dt}\left(\int_{\varphi_t(W)} \frac{1}{2}\|u\|^2 d\mu + \int_{\varphi_t(W)} \rho w \, d\mu\right) = -\int_{\partial\varphi_t(W)} p u \cdot n \, da.$$

By the transport theorem and arguing as in our previous results, this reduces to

$$0 = \int_{\varphi_t(W)} \left(p \operatorname{div} u + \rho \frac{Dw}{dt}\right) d\mu.$$

Since W is arbitrary,

$$p \operatorname{div} u + \rho \frac{Dw}{dt} = 0.$$

Now assume that w depends on the fluid motion through the density; i.e., the internal energy depends only on how much the fluid is compressed. Such a fluid is called ***ideal isentropic*** or ***barotropic***. The preceding identity then becomes

$$0 = p \operatorname{div} u + \rho\left(\frac{\partial w}{\partial t} + \mathbf{d}w \cdot u\right) = p \operatorname{div} u + \rho \frac{\partial w}{\partial \rho}\frac{\partial \rho}{\partial t} + \rho \frac{\partial w}{\partial \rho}\mathbf{d}\rho \cdot u = p \operatorname{div} u + \rho \frac{\partial w}{\partial \rho}(-\rho \operatorname{div} u)$$

using the equation of continuity. Since this is an identity and we are not restricting div u, we get

$$p = \rho^2 w'(\rho).$$

If p is a given function of ρ note that $w = -\int p\, d(1/\rho)$. Note that $\mathbf{d}p/\rho = \mathbf{d}(w + p/\rho)$. This follows from $p = \rho^2 w'$ by a straightforward calculation in which p and w are regarded as functions of ρ. The quantity $w + p/\rho = w + \rho w'$ is called the ***enthalpy*** and is often denoted h.

Thus Euler's equations for compressible ideal isentropic flow are

$$\frac{\partial u^\flat}{\partial t} + \mathbf{L}_u u^\flat - \frac{1}{2}\mathbf{d}(u^\flat(u)) + \frac{\mathbf{d}p}{\rho} = 0$$

$$\frac{\partial \rho}{\partial t} + \operatorname{div}(\rho u) = 0$$

$$u(x, 0) = u_0(x) \text{ on } M \text{ and } u \cdot n = 0 \text{ on } \partial M.$$

where $p = \rho^2 w'(\rho)$ is a function of ρ, called an ***equation of state***, which depends on the particular fluid. It is known that these equations lead to a well-posed initial value problem (i.e., there is a local existence and uniqueness theorem) only if $p'(\rho) > 0$. This agrees with the common experience that increasing the surrounding pressure on a vlume of fluid causes a decrease in occupied volume and hence an increase in density. Many gases can often be viewed as satisfying our hypotheses, with $p = A\rho^\gamma$ where A and γ are constants and $\gamma \geq 1$.

Cases **A** and **B** above are rather opposite. For instance, if $\rho = \rho_0$ is a constant for an incompressible fluid, then clearly p cannot be an invertible function of ρ. However, the case $\rho =$ constant may be regarded as a limiting case $p'(\rho) \to \infty$. In case **B**, p is an explicit function of ρ. In case **A**, p is implicitly determined by the condition div u = 0. Finally, notice that in neither case **A** or **B** is the possibility of a loss of total energy due to friction taken into account. This leads to the subject of ***viscous fluids***, not dealt with here.

Given a fluid flow with velocity field **u**(x, t), a ***streamline*** at a fixed time t is an integral curve of u; i.e., if x(s) is a streamline parametrized by s at the instant t, then x(s) satisfies

$$\frac{dx}{ds} = u(x(s), t), \quad t \text{ fixed.}$$

On the other hand, a ***trajectory*** is the curve traced out by a particle as time progresses, as explained at the beginning of this section; i.e., is a solution of the differential equation

$$\frac{dx}{dt} = u(x(t), t)$$

with given initial conditions. If u is independent of t (i.e., $\partial u/\partial t = \mathbf{0}$), then, streamlines and trajectories coincide. In this case, the flow is called ***stationary*** or ***steady***. This condition means that the "shape" of the fluid flow is not changing. Even if each particle is moving under the flow, the global configuration of the fluid does not change. The following criteria for steady solutions for homogeneous incompressible flow is a direct consequence of Euler's equations, written in the form $\partial u^\flat/\partial t + \mathbb{P}(\mathbf{L}_u u^\flat) = 0$, where $\mathbb{P}$ is the Hodge projection to the co-closed 1-forms.

8.2.2 Proposition *Let* u_t *bea solution to the Euler equations for homogeneous incompressible flow on a compact manifold* M *and* φ_t *its flow. The following are equivalent:*

(i) u_t *is a steady flow* (i.e., $(\partial u/\partial t) = \mathbf{0}$).

(ii) φ_t *is a one-parameter group*: $\varphi_{t+s} = \varphi_t \circ \varphi_s$.

(iii) $\mathbf{L}_{u_0} u_0^\flat$ *is an exact* 1–form.

(iv) $\mathbf{i}_{u_0} \mathbf{d}u_0^\flat$ *is an exact 1–form.*

It follows from (iv) that if u_0 is a *harmonic* vector field; i.e., u_0 satisfies $\delta u_0^\flat = 0$ and $\mathbf{d}u_0^\flat = \mathbf{0}$, then it yields a stationary flow. Also, it is known that there are other steady flows. For example, on a closed two-disk, with polar coordinates (r, θ), $u = f(r)(\partial/\partial\theta)$ is the velocity field of a steady flow because

$$(u \cdot \nabla)u = -\nabla p, \quad \text{where} \quad p(r, \theta) = \int_0^r f^2(s) s \, ds \; .$$

Clearly such a u need not be harmonic.

We saw that for compressible ideal isentropic flow, the total energy $\int_M (\|u\|^2/2 + \rho w) \, d\mu$ is conserved. We can refine this a little for stationary flows as follows.

8.2.3 Bernoulli's Theorem *For stationary compressible ideal isentropic flow, with* p *a function of* ρ,

$$\frac{1}{2}\|u\|^2 + \int \frac{dp}{\rho} = \frac{1}{2}\|u\|^2 + w + \frac{p}{\rho}$$

is constant along streamlines where the enthalpy $\int dp/\rho = w + p/\rho$ *denotes a potential for the one form* dp/ρ. *The same holds for stationary homogeneous* (ρ *= constant in space* $= \rho_0$) *incompressible flow with* $\int dp/\rho$ *replaced by* p/ρ_0. *If body forces deriving from a potential* U *are present, i.e.,* $b^\flat = -\mathbf{d}U$, *then the conserved quantity is*

$$\frac{1}{2}\|u\|^2 + \int \frac{dp}{\rho} = \frac{1}{2}\|u\|^2 + w + \frac{p}{\rho} + U.$$

Proof Since $\mathbf{L}_u(u^\flat)\cdot u = \mathbf{d}(u^\flat(u))\cdot u$, for stationary ideal compressible or incompressible homogeneous flows we have

$$0 = \frac{\partial u^\flat}{\partial t}\cdot u = -\left(\mathbf{L}_u u^\flat)\cdot u + \frac{1}{2}\mathbf{d}(u^\flat)\right)\cdot u - \frac{\mathbf{d}p}{\rho}\cdot u = -\frac{1}{2}(\mathbf{d}\,\|u\|^2)\cdot u - \frac{1}{\rho}\mathbf{d}p\cdot u\,,$$

so that

$$\left(\frac{1}{2}\|u\|^2 + \int \mathbf{d}p/\rho\right)\Bigg|_{x(s_1)}^{x(s_2)} = \int_{s_1}^{s_2} \mathbf{d}\left(\frac{1}{2}\|u\|^2 + \int\frac{\mathbf{d}p}{\rho}\cdot u\right)\cdot x'(s)ds = \int_{s_1}^{s_2}\frac{\partial u^\flat}{\partial s}\cdot u(x(s))ds = 0$$

since $x'(s) = u(x(s))$. ■

The two-form $\omega = \mathbf{d}u^\flat$ is called ***vorticity***, which, in $\mathbb{R}^3$ can be identified with curl $\mathbf{u}$. Our assumptions so far have precluded any tangential forces and thus any mechanism for starting or stopping rotation. Hence, intuitively, we might expect rotation to be conserved. Since rotation is intimately related to the vorticity, we can expect the vorticity to be involved. We shall now prove that this is so.

Let C be a simple closed contour in the fluid at $t = 0$ and let C_t be the contour carried along the flow. In other words, $C_t = \varphi_t(C)$ where φ_t is the fluid flow map. (See Fig. 8.2.3.) The ***circulation*** around C_t is defined to be the integral

$$\Gamma_{C_t} = \int_{C_t} u^\flat,$$

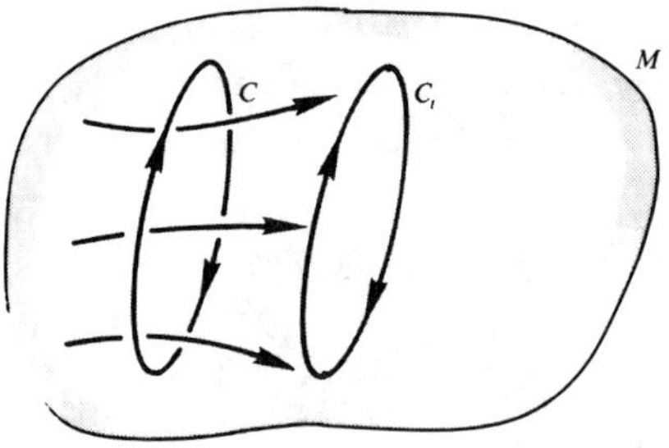

Figure 8.2.3

8.2.4 Kelvin Circulation Theorem *Let* M *be a manifold and* $\ell \subset M$ *a smooth closed loop, i.e., a compact one-manifold. Let* u_t *solve the Euler equations on* M *for ideal isentropic compressible or homogeneous incompressible flow and* $\ell(t)$ *be the image of* ℓ *at time* t *when each particle moves under the flow* φ_t *of* u_t*; i.e.,* $\ell(t) = \varphi_t(\ell)$*. Then the circulation is constant* *in time; i.e.,*

$$\frac{d}{dt}\int_{\ell(t)} u_t^\flat = 0$$

Proof Let φ_t be the flow of u_t. Then $\ell(t) = \varphi_t(\ell)$, and so changing variables,

$$\frac{d}{dt}\int_{\varphi_t(\ell)} u_t^\flat = \int_\ell \left[\varphi_t^*(\mathbf{L}_u u^\flat) + \varphi_t^*\left(\frac{\partial u^\flat}{\partial t}\right)\right].$$

However, $\mathbf{L}_u u^\flat + \partial u^\flat/\partial t$ is exact from the equations of motion and the integral of an exact form over a closed loop is zero. ■

We now use Stokes' theorem, which will bring in the vorticity. If Σ is a surface (a two-dimensional submanifold of M) whose boundary is a closed contour C, then Stokes' theorem yields

$$\Gamma_C = \int_C u^\flat = \int_\Sigma du^\flat = \int_\Sigma \omega.$$

See Fig. 8.2.4.

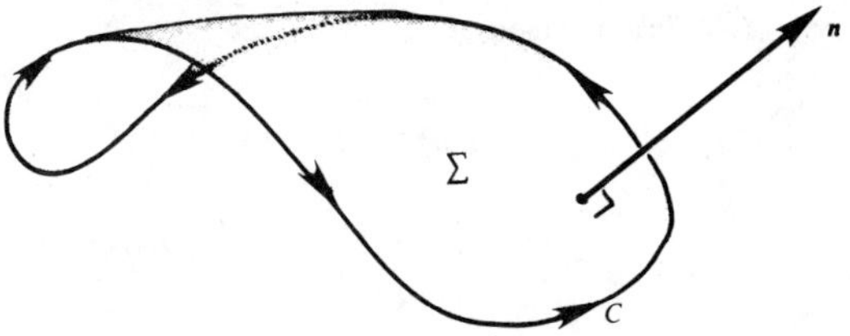

Figure 8.2.4

Thus, as a corollary of the circulation theorem, we can conclude:

8.2.5 Helmholtz' Theorem *Under the hypotheses of* **8.2.4**, *the flux of vorticity across a surface moving with the fluid is constant in time.*

We shall now show that ω and $\eta = \omega/\rho$ are *Lie propagated by the flow.*

8.2.6 Proposition *For isentropic or homogeneous incompressible flow, we have*

(i) $$\frac{\partial \omega}{\partial t} + \mathbf{L}_u\omega = 0 \quad \text{and} \quad \frac{\partial \eta}{\partial t} + \mathbf{L}_u\eta - \eta\,\text{div}\,u = 0$$

called the ***vorticity-stream equation*** *and*

(ii) $$\varphi_t^*\omega = \omega_0 \text{ and } \varphi_t^*\eta_t = J(\varphi_t)\eta_0$$

where $\eta_t(x) = \eta(x, t)$ *and* $J(\varphi_t)$ *is the Jacobian of* φ_t.

Proof Applying $\mathbf{d}$ to Euler's equations for the two types of fluids we get the ***vorticity equation***:

$$\frac{\partial\omega}{\partial t} + \mathbf{L}_u\omega = 0.$$

Thus

$$\begin{aligned}\frac{\partial\eta}{\partial t} + \mathbf{L}_u\eta &= \frac{1}{\rho}\left(\frac{\partial\omega}{\partial t} + \mathbf{L}_u\omega\right) - \frac{\omega}{\rho^2}\left(\frac{\partial\rho}{\partial t} + \mathbf{d}\rho\cdot u\right) \\ &= -\frac{\eta}{\rho}\left(\frac{\partial\rho}{\partial t} + \mathbf{d}\rho\cdot u + \rho \operatorname{div} u\right) + \eta \operatorname{div} u = \eta \operatorname{div} u\end{aligned}$$

by conservation of mass.

From $\partial\omega/\partial t + \mathbf{L}_u\omega = 0$ it follows that $(\partial/\partial t)(\varphi_t^*\omega_t) = 0$ and so $\varphi_t^*\omega_t = \omega_0$. Since $\varphi_t^*\rho_t = \rho_0 / J(\varphi_t)$ we also get $\varphi_t^*\eta_t = J(\varphi_t)\eta_0$. ■

In three dimensions we can associate to η the vector field $\zeta = *\eta$ (or equivalently $\mathbf{i}_\zeta\mu = \eta$). Thus $\zeta = \operatorname{curl} \mathbf{u}/\rho$, if M is embedded in $\mathbb{R}^3$.

8.2.7 Corollary *If* dim M = 3, *then* ζ *is transported as a* ***vector*** *by* φ_t; *i.e.*,

$$\zeta_t = \varphi_{t*}\,\zeta_0 \quad \text{or} \quad \zeta_t(\varphi_t(x)) = T_x\,\varphi_t(\zeta_t(x)).$$

Proof $\varphi_t^*\eta_t = J(\varphi_t)\eta_0$ by **8.2.6**, so $\varphi_t^*\mathbf{i}_{\zeta_t}\mu = J(\varphi_t)\mathbf{i}_{\zeta_0}\mu$. But $\varphi_t^*\mathbf{i}_{\zeta_t}\mu = \mathbf{i}_{\varphi_t^*\zeta_t}\varphi_t^*\mu = \mathbf{i}_{\varphi_t^*\zeta_t} J(\varphi_t)\mu$. Thus $\mathbf{i}_{\varphi_t^*\zeta_t}\mu = \mathbf{i}_{\zeta_0}\mu$, which gives $\varphi_t^*\zeta_t = \zeta_0$. ■

Notice that the vorticity as a *two-form* is Lie transported by the flow but as a *vector field* it is vorticity/ρ, which is Lie transported. Here is another instance where distinguishing between forms and vector fields makes an important difference.

The flow φ_t of a fluid plays the role of a configuration variable and the velocity field u plays the role of the corresponding velocity variable. In fact, to understand fluid mechanics as a Hamiltonian system in the sense of §**8.1**, a first step is to set up its phase space using the set of all diffeomorphisms $\varphi : M \to M$ (volume preserving for incompressible flow) as the configuration space. The references noted at the beginning of this section carry out this program (see also Exercise **8.2I** and §**8.4**).

Exercises

8.2A In classical texts on fluid mechanics, the identity

$$(\mathbf{u} \cdot \nabla)\mathbf{u} = (1/2)\, \nabla(\mathbf{u} \cdot \mathbf{u}) + (\nabla \times \mathbf{u}) \times \mathbf{u}$$

is often used. To what identity does this correspond in this section?

8.2B A flow is called ***potential flow*** if $u^\flat = d\varphi$ for a function φ. For (not necessarily stationary) homogeneous incompressible or isentropic flow prove ***Bernoulli's law*** in the form

$$\partial\varphi/\partial t + (1/2)\, \|\, u \,\|^2 + \int dp/\rho = \textit{constant} \text{ on a streamline.}$$

8.2C Complex variables texts "show" that the gradient of $\varphi(r, \theta) = (r + 1/r)\cos\theta$ describes stationary ideal incompressible flow around a cylinder in the plane. Verify this in the context of this section.

8.2D Translate **8.2.2** into vector analysis notation in $\mathbb{R}^3$ and give a direct proof.

8.2E Let $\dim M = 3$, and assume the vorticity ω has a one-dimensional kernel. (i) Using Frobenius' theorem, show that this distribution is integrable. (ii) Identify the one-dimensional leaves with integral curves of ζ (see **8.2.7**)--these are called ***vortex lines***. (iii) Show that vortex lines are propagated by the flow.

8.2F Assume $\dim M = 3$. A ***vortex tube*** T is a closed oriented two-manifold in M that is a union of vortex lines. The ***strength*** of the vortex tube is the flux of vorticity across a surface Σ inside T whose boundary lies on T and is transverse to the vortex lines. Show that vortex tubes are propagated by the flow and have a strength that is constant in time.

8.2G Let $f: \mathbb{R}^3 \to \mathbb{R}$ be a linear function and $g : S^2 \to \mathbb{R}$ be its restriction to the unit sphere. Show that $\mathbf{d}g$ gives a stationary solution of Euler's equations for flow on the two-sphere.

8.2H Stream Functions

(i) For incompressible flow in $\mathbb{R}^2$, show that there is a function ψ such that $u^1 = \partial\psi/\partial y$ and $u^2 = -\partial\psi/\partial x$. One calls φ the ***stream function*** (as in Batchelor [1967]).

(ii) Show that if we let $*\psi = \psi\, dx \wedge dy$ be the associated two form, then $\mathbf{u}^\flat = \delta*\psi$.

(iii) Show that $\mathbf{u}$ is a Hamiltonian vector field (see Section **8.1**) with energy ψ directly in $\mathbb{R}^2$ and then for arbitrary two-dimensional Riemannian manifolds M.

(iv) Do stream functions exist for arbitrary fluid flow on $\mathbb{T}^2$? On S^2?

(v) Show that the vorticity is $\omega = \Delta^*\psi$.

8.2I Clebsch Variables; Clebsch [1859]. Let $\mathcal{F}$ be the space of functions on a compact manifold M with the dual space $\mathcal{F}^*$, taken to be densities on M; the pairing between $f \in \mathcal{F}$ and $\rho \in \mathcal{F}^*$ is $\langle f, \rho \rangle = \int_M f\rho$.

(i) On the symplectic manifold $\mathcal{F} \times \mathcal{F}^* \times \mathcal{F} \times \mathcal{F}^*$ with variables $(\alpha, \lambda, \mu, \rho)$, show that Hamilton's equations for a given Hamiltonian H are

$$\dot{\alpha} = \frac{\delta H}{\delta \lambda}, \quad \dot{\mu} = \frac{\delta H}{\delta \rho}, \quad \dot{\lambda} = -\frac{\delta H}{\delta \alpha}, \quad \dot{\rho} = -\frac{\delta H}{\delta \mu},$$

where $\delta H/\delta\lambda$ is the ***functional derivative*** of H defined by

$$\langle \frac{\delta H}{\delta \lambda}, \dot{\lambda} \rangle = \mathbf{D}H(\lambda)\cdot\dot{\lambda}\,.$$

(ii) In the ideal isentropic compressible fluid equations, set $\mathbf{M} = \rho u^\flat$, the ***momentum density***, where dx denotes the Riemannian volume form on M. Identify the density $\sigma(x)dx \in \mathcal{F}^*$ with the function $\sigma(x) \in \mathcal{F}$ and write $\mathbf{M} = -(\rho \mathbf{d}\mu + \lambda \mathbf{d}\alpha)dx$. For momentum densities of this form show that Hamilton's equations written in the variables $(\alpha, \lambda, \mu, \rho)$ imply Euler's equation and the equation of continuity.

§8.3 *Electromagnetism*

Classical electromagnetism is governed by Maxwell's field equations. The form of these equations depends on the physical units chosen, and changing these units introduces factors like 4π, c = the speed of light, ε_0 = the dielectric constant and μ_0 = the magnetic permeability. This discussion assumes that ε_0, μ_0 are constant; the choice of units is such that the equations take the simplest form; thus $c = \varepsilon_0 = \mu_0 = 1$ and factors 4π disappear. We also do not consider Maxwell's equations in a material, where one has to distinguish **E** from **D**, and **B** from **H**.

Let **E**, **B**, and **J** be time dependent C^1-vector fields on $\mathbb{R}^3$ and $\rho : \mathbb{R}^3 \times \mathbb{R} \to \mathbb{R}$ a scalar. These are said to satisfy ***Maxwell's equations*** with ***charge density*** ρ ***and current density*** **J** when the following hold:

$$\operatorname{div} \mathbf{E} = \rho \quad \textit{(Gauss's law)} \tag{1}$$

$$\operatorname{div} \mathbf{B} = 0 = 0 \quad \textit{(no magnetic sources)} \tag{2}$$

$$\operatorname{curl} \mathbf{E} + \frac{\partial \mathbf{B}}{\partial t} = \mathbf{0} \quad \textit{(Faraday's law of induction)} \tag{3}$$

$$\operatorname{curl} \mathbf{B} - \frac{\partial \mathbf{E}}{\partial t} = \mathbf{J} \quad \textit{(Ampère's law)}. \tag{4}$$

E is called the ***electric field*** and **B** the ***magnetic field.***

The quantity $\int_\Omega \rho \, dV = Q$ is called the *charge* of the set $\Omega \subset \mathbb{R}^3$. By Gauss' theorem, (1) is equivalent to

$$\int_{\partial\Omega} \mathbf{E} \cdot \mathbf{n} \, dS = \int_\Omega \rho \, dV = Q \tag{5}$$

for any (nice) open set $\Omega \subset \mathbb{R}^3$; i.e., the *electric flux out of a closed surface equals the total charge inside the surface.* This generalizes Gauss' law for a point charge discussed in Section 7.3. By the same reasoning, (2) is equivalent to

$$\int_{\partial\Omega} \mathbf{B} \cdot \mathbf{n} \, dS = 0. \tag{6}$$

That is, the *magnetic flux out of any closed surface is zero.* In other words there are no magnetic sources inside any closed surface.

By Stokes' theorem, (3) is equivalent to

$$\int_{\partial S} \mathbf{E} \cdot d\mathbf{s} = \int_S (\operatorname{curl} \mathbf{E}) \cdot \mathbf{n} \, dS = -\frac{\partial}{\partial t} \int_S \mathbf{B} \cdot \mathbf{n} \, dS \tag{7}$$

for any closed loop ∂S bounding a surface S. The quanity $\int_{\partial S} \mathbf{E} \cdot d\mathbf{s}$ is called the ***voltage*** around

∂S. Thus, Faraday's law of induction (3) says that *the voltage around a loop equals the negative of the rate of change of the magnetic flux through the loop.*

Finally, again by the classical Stokes' theorem, (4) is equivalent to

$$\int_{\partial S} \mathbf{B} \cdot d\mathbf{s} = \int_S (\text{curl } \mathbf{B}) \cdot \mathbf{n}\, dS = \frac{\partial}{\partial t}\int_S \mathbf{E} \cdot \mathbf{n}\, dS + \int_S \mathbf{J} \cdot \mathbf{n}\, dS. \tag{8}$$

Since $\int_S \mathbf{J} \cdot \mathbf{n}\, dS$ has the physical interpretation of *current,* this form states that *if* $\mathbf{E}$ *is constant in time, then the magnetic potential difference* $\int_{\partial S} \mathbf{B} \cdot d\mathbf{s}$ *around a loop equals the current through the loop.* In general, if $\mathbf{E}$ varies in time, Ampère's law states that the *magnetic potential difference around a loop equals the total current in the loop plus the rate of change of electric flux through the loop.*

We now show how to express Maxwell's equations in terms of differential forms. Let $M = \mathbb{R}^4 = \{(x, y, z, t)\}$ with the Lorentz metric g on $\mathbb{R}^4$ of diagonal form $(1, 1, 1, -1)$ in standard coordinates (x, y, z, t).

8.3.1 Proposition *There is a unique two-form* F *on* $\mathbb{R}^4$, *called the* ***Faraday two-form*** *such that*

$$\mathbf{E}^\flat = -\mathbf{i}_{\partial/\partial t} F; \tag{9}$$

$$\mathbf{B}^\flat = -\mathbf{i}_{\partial/\partial t} * F. \tag{10}$$

(Here the $^\flat$ *is Euclidean in* $\mathbb{R}^3$ *and the* $*$ *is Lorentzian in* $\mathbb{R}^4$.)

Proof If

$$F = F_{xy}\, dx \wedge dy + F_{zx}\, dz \wedge dx + F_{yz}\, dy \wedge dz + F_{xt}\, dx \wedge dt + F_{yt}\, dy \wedge dt + F_{zt}\, dz \wedge dt,$$

then (see Example **6.2.14E**),

$$* F = F_{xy}\, dz \wedge dt + F_{zx}\, dy \wedge dt + F_{yz}\, dx \wedge dt - F_{xt}\, dy \wedge dz - F_{yt}\, dz \wedge dx - F_{zt}\, dx \wedge dy$$

and so

$$-\mathbf{i}_{\partial/\partial t} F = F_{xt}\, dx + F_{yt}\, dy + F_{zt}\, dz$$

and

$$-\mathbf{i}_{\partial/\partial t} * F = F_{xy}\, dz + F_{zx}\, dy + F_{yz}\, dx.$$

Thus, $\mathbf{F}$ is uniquely determined by (9) and (10), namely

$$F = E^1\, dx \wedge dt + E^2\, dy \wedge dt + E^3\, dz \wedge dt + B^3\, dx \wedge dy + B^2\, dz \wedge dx + B^1\, dy \wedge dz. \quad \blacksquare$$

We started with $\mathbf{E}$ and $\mathbf{B}$ and used them to construct F, but one can also take F as the primitive object and construct $\mathbf{E}$ and $\mathbf{B}$ from it using (9) and (10). Both points of view are useful.

Similarly, out of ρ and $\mathbf{J}$ we can form the ***source one-form*** $j = -\rho\, dt + J_1\, dx + J_2\, dy +$

$J_3\,dz$; i.e., j is uniquely determined by the equations $-\mathbf{i}_{\partial/\partial t}\,j = \rho$ and $\mathbf{i}_{\partial/\partial t}{*}j = *\mathbf{J}^\flat$; in the last relation, **J** is regarded as being defined on $\mathbb{R}^4$.

8.3.2 Proposition *Maxwell's equations* (1)-(4) *are equivalent to the equations*

$$\mathbf{d}F = 0 \quad \textit{and} \quad \delta F = j$$

on the manifold $\mathbb{R}^4$ *endowed with the Lorentz metric.*

Proof A straightforward computation shows that

$$\begin{aligned}\mathbf{d}F = {} & \left(\operatorname{curl}\mathbf{E} + \frac{\partial \mathbf{B}}{\partial t}\right)_x dy\wedge dz\wedge dt + \left(\operatorname{curl}\mathbf{E} + \frac{\partial \mathbf{B}}{\partial t}\right)_y dz\wedge dx\wedge dt \\ & + \left(\operatorname{curl}\mathbf{E} + \frac{\partial \mathbf{B}}{\partial t}\right)_z dx\wedge dy\wedge dt + (\operatorname{div}\,\mathbf{B})dx\wedge dy\wedge dz.\end{aligned}$$

Thus $\mathbf{d}F = 0$ is equivalent to (2) and (3).

Since the index of the Lorentz metric is 1, we have $\delta = *\,\mathbf{d}\,*$. Thus

$$\begin{aligned}\delta F &= *\,\mathbf{d}*F = *\,\mathbf{d}(-E^1 dy\wedge dz - E^2 dz\wedge dx - E^3 dx\wedge dy \\ &\qquad + B^1 dx\wedge dt + B^2 dy\wedge dt + B^3 dz\wedge dt) \\ &= *\Bigg[-(\operatorname{div}\mathbf{E})dx\wedge dy\wedge dz + \left(\operatorname{curl}\mathbf{B} - \frac{\partial \mathbf{E}}{\partial t}\right)_x dy\wedge dz\wedge dt + \\ &\qquad + \left(\operatorname{curl}\mathbf{B} - \frac{\partial \mathbf{E}}{\partial t}\right)_y dz\wedge dx\wedge dt + \left(\operatorname{curl}\mathbf{B} - \frac{\partial \mathbf{E}}{\partial t}\right)_z dx\wedge dy\wedge dt\Bigg] \\ &= \left(\operatorname{curl}\mathbf{B} - \frac{\partial \mathbf{E}}{\partial t}\right)_x dx + \left(\operatorname{curl}\mathbf{B} - \frac{\partial \mathbf{E}}{\partial t}\right)_y dy + \left(\operatorname{curl}\mathbf{B} - \frac{\partial \mathbf{E}}{\partial t}\right)_z dz - (\operatorname{div}\mathbf{E})dt\end{aligned}$$

Thus $\delta F = j$ iff (1) and (4) hold. ■

As a skew matrix, we can represent F as follows

$$F = \begin{array}{c} \begin{array}{cccc} x & y & z & t \end{array} \\ \begin{bmatrix} 0 & B^3 & -B^2 & E^1 \\ -B^3 & 0 & B^1 & E^2 \\ B^2 & -B^1 & 0 & E^3 \\ -E^1 & -E^2 & -E^3 & 0 \end{bmatrix} \begin{array}{c} x \\ y \\ z \\ t \end{array} \end{array}$$

Recall from Section **6.5** and Exercise **7.5G**, the formula

$$(\delta F)^i = |\det[g_{\ell j}]|^{-1/2}\,(F^{ik}\,|\det[g_{\ell j}]|^{1/2})_{,k}$$

Since $|\det[g_{k\ell}]| = 1$, Maxwell's equations can be written in terms of the Faraday two-form F in components as

$$F_{ij,k} + F_{jk,i} + F_{ki,j} = 0 \tag{11}$$

and

$$F^{ik}{}_{,k} = -j^i, \tag{12}$$

where $F_{ij,k} = \partial F_{ij}/\partial x^k$, etc. Since $\delta^2 = 0$, we obtain

$$0 = \delta^2 F = \delta j = * \,\mathbf{d} * j = * \,\mathbf{d}(-\rho\, dx \wedge dy \wedge dz + (*\, \mathbf{J}^\flat) \wedge dt)$$

$$= * \left[\left(\frac{\partial \rho}{\partial t} + \operatorname{div} \mathbf{J}\right) dx \wedge dy \wedge dz \wedge dt\right] = \frac{\partial \rho}{\partial t} + \operatorname{div} \mathbf{J};$$

i.e., $\partial\rho/\partial t + \operatorname{div} \mathbf{J} = 0$, which is the ***continuity equation*** (see Section **8.2**). Its integral form is, by Gauss' theorem

$$\frac{dQ}{dt} = \frac{d}{dt}\int_\Omega \rho\, dV = \int_{\partial\Omega} \mathbf{J} \cdot \mathbf{n}\, dS$$

for any bounded open set Ω. Thus the continuity equation says that the *flux of the current density out of a closed surface equals the rate of change of the total charge inside the surface.*

Next we show that Maxwell's equations are Lorentz invariant, i.e., are special-relativistic. The ***Lorentz group*** $\mathcal{L}$ is by definition the orthogonal group with respect to the Lorentz metric g, i.e.,

$$\mathcal{L} = \{A \in \mathbb{GL}(\mathbb{R}^4) \mid g(Ax, Ay) = g(x, y) \text{ for all } x, y \in \mathbb{R}^4\}.$$

Lorentz invariance means that F satisfies Maxwell's equations with j iff A^*F satisfies them with A^*j, for any $A \in \mathcal{L}$. But due to Proposition **8.3.1** this is clear since pull-back commutes with $\mathbf{d}$ and orthogonal transformations commute with the Hodge operator (see Exercise **6.2D**) and thus they commute with δ.

As a 4×4 matrix, the Lorentz transformation A acts on F by $F \mapsto A^*F = AFA^T$. Let us see that the action of $A \in \mathcal{L}$ mixes up **E**'s and **B**'s. (This is the source of statements like: "A moving observer sees an electric field partly converted to a magnetic field.")

Proposition **8.3.1** defines **E** and **B** intrinsically in terms of F. Thus, if one performs a Lorentz transformation A on F, the new resulting electric and magnetic fields $\mathbf{E}'$ and $\mathbf{B}'$ with respect to the Lorentz unit normal $A^*(\partial/\partial t)$ to the image $A(\mathbb{R}^3 \times 0)$ in $\mathbb{R}^4$ are given by

$$(\mathbf{E}')^\flat = -\mathbf{i}_{A^*\partial/\partial t} A^*F, \quad (\mathbf{B}')^\flat = -\mathbf{i}_{A^*\partial/\partial t} A^*F.$$

For a Lorentz transformation of the form

$$x' = \frac{x - vt}{\sqrt{1 - v^2}}, \quad y' = y, \quad z' = z, \quad t' = \frac{t - vx}{\sqrt{1 - v^2}}$$

(the special-relativistic analogue of an observer moving uniformly along the x-axis with velocity v) we get

$$\mathbf{E}' = \left(E^1, \frac{E^2 - vB^3}{\sqrt{1-v^2}}, \frac{E^3 + vB^2}{\sqrt{1-v^2}} \right)$$

and

$$\mathbf{B}' = \left(B^1, \frac{B^2 - vE^3}{\sqrt{1-v^2}}, \frac{B^3 + vE^2}{\sqrt{1-v^2}} \right).$$

We leave the verification to the reader.

By the way we have set things up, note that Maxwell's equations make sense on any ***Lorentz manifold***; i.e., a four-dimensional manifold with a pseudo-Riemannian metric of signature (+, +, +, −).

Maxwell's vacuum equations (i.e., $j = 0$) will now be shown to be ***conformally invariant*** on any Lorentz manifold (M, g). A diffeomorphism $\varphi : (M, g) \to (M, g)$ is said to be ***conformal*** if $\varphi^* g = f^2 g$ for a nowhere vanishing function f. (See Fulton, Rohrlich, and Witten [1962] for a review of conformal invariance in physics and the original literature references.)

8.3.3 Propositon *Let* $F \in \Omega^2(M)$ *where* (M, g) *is a Lorentz manifold, satisfy* $\mathbf{d}F = 0$ *and* $\delta F = j$. *Let* φ *be a conformal diffeomorphism. Then* $\varphi^* F$ *satisfies* $\mathbf{d}\varphi^* F = 0$ *and* $\delta \varphi^* F = f^2 \varphi^* j$. *Hence Maxwell's vacuum equations (with* $j = 0$*) are conformally invariant; i.e., if* F *satisfies them, so does* $\varphi^* F$.

Proof Since φ^* commutes with $\mathbf{d}$, $\mathbf{d}F = 0$ implies $\mathbf{d}\varphi^* F = 0$. The second equation implies $\varphi^* \delta F = \varphi^* j$. By Exercise **7.5H**, we have $\delta_{\varphi^* g} \varphi^* \beta = \varphi^* \delta \beta$. Hence $\delta F = j$ implies $\delta_{\varphi^* g} \varphi^* F = = \varphi^* j = \delta_{f^2 g} \varphi^* F$ since φ is conformal. The local formula for δF, namely

$$(\delta F)_i = |\det[g_{ks}]|^{-1/2} g_{ir} \frac{\partial}{\partial x^t} (g^{ra} g^{tb} F_{ab} |\det[g_{ks}]|^{1/2})$$

shows that when one replaces g by $f^2 g$, we get

$$\delta_{f^2 g} \varphi^* F = f^{-2} \varphi^* F,$$

and so

$$\delta \varphi^* F = f^2 \varphi^* j. \qquad \blacksquare$$

Let us now discuss the energy equation for the electromagnetic field. Introduce the ***energy density*** of the field $\mathcal{E} = (\mathbf{E} \cdot \mathbf{E} + \mathbf{B} \cdot \mathbf{B})/2$ and the *Poynting energy-flux vector* $\mathbf{S} = \mathbf{E} \times \mathbf{B}$. *Poynting's theorem* states that

$$-\frac{\partial \mathcal{E}}{\partial t} = \operatorname{div} \mathbf{S} + \mathbf{E} \cdot \mathbf{J}.$$

This is a straightforward computation using (3) and (4). We shall extend this result to $\mathbb{R}^4$ and, at the same time, shall rephrase it in the framework of forms.

Introduce the ***stress-energy-momentum tensor*** (or the ***Maxwell stress tensor***) T by

$$T^{ij} = F^{ik}F^{j}_{\ k} - \frac{1}{4}g^{ij}F_{pq}F^{pq} \tag{13}$$

(or intrinsically, $T = F \cdot F - (1/4)\langle F, F\rangle g$, where $F \cdot F$ denotes a single contraction of F with itself). A straightforward computation shows that the divergence of T equals

$$T^{ij}_{\ ,j} = F^{ik}_{\ \ ,j}\, F^{j}_{\ k} + F^{ik}F^{j}_{\ k,j} - \frac{1}{2}F_{pq}{}^{,i}F^{pq} \tag{19}$$

where $F_{pq}{}^{,i} = (\partial F_{pq}/\partial x^k)g^{ik}$. Taking into account $\delta F = j$ written in the form (12), it follows that

$$T^{i\ell}_{\ ,\ell} = F^{ik}j_k\,. \tag{14}$$

For $i = 4$, the relation (14) becomes Poynting's theorem. [Poynting's theorem can also be understood in terms of a Hamiltonian formulation; see **8.4.2** below. The Poynting energy-flux vector is the Noether conserved quantity for the action of the diffeomorphism group of $\mathbb{R}^3$ on $T^*\mathcal{A}$, where $\mathcal{A}$ is the space of vector potentials $\mathbf{A}$ defined in the following paragraph, and Poynting's theorem is just conservation of momentum (Noether's theorem). We shall not dwell upon these considerably more advanced aspects and refer the interested reader to Abraham and Marsden [1978].] It is clear that T is a symmetric 2-tensor. As a symmetric matrix,

$$T = \begin{bmatrix} \sigma & \mathbf{E}\times\mathbf{B} \\ (\mathbf{E}\times\mathbf{B})^T & \mathcal{E} \end{bmatrix}$$

where σ is the ***stress tensor*** and $\mathcal{E}$ is the energy density. The symmetric 3×3 matrix σ has the following components

$$\begin{aligned}
\sigma^{11} &= \frac{1}{2}[-(E^1)^2 - (B^1)^2 + (E^2)^2 + (B^2)^3 + (E^3)^2 + (B^3)^2] \\
\sigma^{22} &= \frac{1}{2}[(E^1)^2 + (B^1)^2 - (E^2)^2 - (B^2)^2 + (E^3)^3 + (B^3)^2] \\
\sigma^{33} &= \frac{1}{2}[(E^1)^2 + (B^1)^2 + (E^2)^2 + (B^2)^2 - (E^3)^2 - (B^3)^2] \\
\sigma^{12} &= E^1E^2 + B^1B^2 \\
\sigma^{13} &= E^1E^3 + B^1B^3 \\
\sigma^{23} &= E^2E^3 + B^2B^3.
\end{aligned}$$

We close this section with a discussion of Maxwell's equations in terms of vector potentials. We first do this directly in terms of $\mathbf{E}$ and $\mathbf{B}$. Since $\operatorname{div}\mathbf{B} = 0$, if $\mathbf{B}$ is smooth on all of $\mathbb{R}^3$, there exists a vector field $\mathbf{A}$, called the ***vector potential***, such that $\mathbf{B} = \operatorname{curl}\mathbf{A}$, by the Poincaré lemma. This vector field $\mathbf{A}$ is not unique and one could use equally well $\mathbf{A}' = \mathbf{A} + \operatorname{grad} f$ for some (possibly time-dependent) function $f: \mathbb{R}^3 \to \mathbb{R}$. This freedom in the choice of $\mathbf{A}$ is called

gauge freedom. For any such choice of $\mathbf{A}$ we have by (3)

$$0 = \text{curl } \mathbf{E} + \frac{\partial \mathbf{B}}{\partial t} = \text{curl } \mathbf{E} + \frac{\partial}{\partial t}\text{curl } \mathbf{A} = \text{curl}\left(\mathbf{E} + \frac{\partial \mathbf{A}}{\partial t}\right),$$

so that again by the Poincaré lemma there exists a (time-dependent) function $\varphi : \mathbb{R}^3 \to \mathbb{R}$ such that

$$\mathbf{E} + \frac{\partial \mathbf{A}}{\partial t} = -\text{ grad } \varphi. \tag{8}$$

Recall that the Laplace-Beltrami operator on functions is defined by $\nabla^2 f = \text{div}(\text{grad } f)$. On vector fields in $\mathbb{R}^3$ this operator may be defined componentwise. Then it is easy to check that

$$\text{curl}(\text{curl } \mathbf{A}) = \text{grad}(\text{div } \mathbf{A}) - \nabla^2 \mathbf{A}.$$

Using this identity, (8), and $\mathbf{B} = \text{curl } \mathbf{A}$ in (4), we get

$$\begin{aligned}\mathbf{J} = \text{curl } \mathbf{B} - \frac{\partial \mathbf{E}}{\partial t} &= \text{curl}(\text{curl } \mathbf{A}) - \frac{\partial}{\partial t}\left(-\frac{\partial \mathbf{A}}{\partial t} - \text{grad } \varphi\right)\\ &= \text{grad}(\text{div } \mathbf{A}) - \nabla^2 \mathbf{A} + \frac{\partial^2 \mathbf{A}}{\partial t^2} + \frac{\partial}{\partial t}(\text{grad } \varphi),\end{aligned}$$

and thus

$$\nabla^2 \mathbf{A} - \frac{\partial^2 \mathbf{A}}{\partial t^2} = -\mathbf{J} + \text{grad}\left(\text{div } \mathbf{A} + \frac{\partial \varphi}{\partial t}\right). \tag{16}$$

From (1) we obtain as before

$$\rho = \text{div } \mathbf{E} = \text{div}\left(-\frac{\partial \mathbf{A}}{\partial t} - \text{grad } \varphi\right) = -\nabla^2 \varphi - \frac{\partial}{\partial t}(\text{div } \mathbf{A}),$$

that is,

$$\nabla^2 \varphi = -\rho - \frac{\partial}{\partial t}(\text{div } \mathbf{A}),$$

or, subtracting $\partial^2 \varphi / \partial t^2$ from both sides,

$$\nabla^2 \varphi - \frac{\partial^2 \varphi}{\partial t^2} = -\rho - \frac{\partial}{\partial t}\left(\text{div } \mathbf{A} + \frac{\partial \varphi}{\partial t}\right). \tag{17}$$

It is apparent that (16) and (17) can be considerably simplified if one could choose, using the gauge freedom, the vector potential $\mathbf{A}$ and the function φ such that

$$\text{div } \mathbf{A} + \frac{\partial \varphi}{\partial t} = 0.$$

So, assume one has $\mathbf{A}_0, \varphi_0$ and one seeks a function f such that $\mathbf{A} = \mathbf{A}_0 + \text{grad } f$ and $\varphi = \varphi_0 - \partial f/\partial t$ satisfy $\text{div } \mathbf{A} + \partial\varphi/\partial t = 0$. This becomes, in terms of f,

$$0 = \text{div}(\mathbf{A}_0 + \text{grad } f) + \frac{\partial}{\partial t}\left(\varphi_0 - \frac{\partial f}{\partial t}\right) = \text{div } \mathbf{A}_0 + \frac{\partial \varphi_0}{\partial t} + \nabla^2 f - \frac{\partial^2 f}{\partial t^2};$$

i.e.,

$$\nabla^2 f - \frac{\partial^2 f}{\partial t^2} = -\left(\text{div } \mathbf{A}_0 + \frac{\partial \varphi_0}{\partial t}\right). \tag{18}$$

This equation is the calssical ***inhomogeneous wave equation***. The homogeneous wave equation (right-hand side equals zero) has solutions $f(t, x, y, z) = \psi(x - t)$ for any function ψ. This solution propagates the graph of ψ like a wave--hence the name wave equation.

Now we we can draw some conclusions regarding Maxwell's equations. In terms of the vector potential $\mathbf{A}$ and the function φ, (1) and (4) become

$$\nabla^2 \varphi - \frac{\partial^2 \varphi}{\partial t^2} = -\rho, \quad \nabla^2 \mathbf{A} - \frac{\partial^2 \mathbf{A}}{\partial t^2} = -\mathbf{J}, \tag{19}$$

which again are inhomogeneous wave equations. Conversely, if $\mathbf{A}$ and φ satisfy the foregoing equations and $\text{div } \mathbf{A} + \partial\varphi/\partial t = 0$, then $\mathbf{E} = -\text{grad } \varphi - \partial \mathbf{A}/\partial t$ and $\mathbf{B} = \text{curl } \mathbf{A}$ satisfy Maxwell's equations.

Thus in $\mathbb{R}^4$, this procedure reduces the study of Maxwell's equations to the wave equation, and hence solutions of Maxwell's equations can be expected to be wavelike.

We now repeat the foregoing constructions on $\mathbb{R}^4$ using differential forms. Since $\mathbf{d}F = 0$, on $\mathbb{R}^4$ we can write $F = \mathbf{d}G$ for a one-form G. Note that F is unchanged if we replace G by $G + \mathbf{d}f$. This again is the *gauge freedom* Substituting $F = \mathbf{d}G$ into $\delta F = j$ gives $\delta\mathbf{d}G = j$. Since $\Delta = \mathbf{d}\delta + \delta\mathbf{d}$ is the Laplace-deRham operator in $\mathbb{R}^4$, we get

$$\Delta G = j - \mathbf{d}\delta G. \tag{20}$$

Suppose we try to choose G so that $\delta G = 0$ (a gauge condition). To do this, given an initial G_0, we can let $G = G_0 + \mathbf{d}f$ and demand that

$$0 = \delta G = \delta G_0 + \delta\mathbf{d}f = \delta G_0 + \Delta f$$

so f must satisfy $\Delta f = -\delta G_0$. Thus, if the gauge condition

$$\Delta f = -\delta G_0 \tag{21}$$

holds, then Maxwell's equations become

$$\Delta G = j. \tag{22}$$

Equation (21) is equivalent to (18) and (22) to (19) by choosing $G = A^\flat + \varphi\, dt$ (where $^\flat$ is Euclidean in $\mathbb{R}^3$).

Exercises

8.3A Assume that the Faraday two-form F depends only on $t - x$.

(i) Show that $\mathbf{d}F = 0$ is then equivalent to $B^3 = E^2$, $B^2 = -E^3$, $B^1 = 0$.

(ii) Show that $\delta F = 0$ is then equivalent to $B^3 = E^2$, $B^2 = -E^3$, $E^1 = 0$.

These solutions of Maxwell's equations are called *plane electromagnetic waves*; they are determined only by E^2, E^3 or B^2, B^3, respectively.

8.3B Let $u = \partial/\partial t$. Show that the Faraday two-form $F \in \Omega^2(\mathbb{R}^4)$ is given in terms of E and $B \in \mathfrak{X}(\mathbb{R}^4)$ by $F = u^\flat \wedge E^\flat - *(u^\flat \wedge B^\flat)$.

9.3C Show that the Poynting vector satisfies

$$S^\flat = *(B^\flat \wedge E^\flat \wedge u^\flat)$$

where $u = \partial/\partial t$ and $E, B \in \mathfrak{X}(\mathbb{R}^4)$.

8.3D Let (M, g) be a Lorentzian four-manifold and $u \in \mathfrak{X}(M)$ a timelike unit vector field on M; i.e., $g(u, u) = -1$.

(i) Show that any $\alpha \in \Omega^2(M)$ can be written in the form

$$\alpha = (i_u\alpha) \wedge u^\flat - *((i_u * \alpha) \wedge u^\flat).$$

(ii) Show that if $i_u\alpha = 0$, where $\alpha \in \Omega^2(M)$ ("α is orthogonal to u"), then $*\alpha$ is decomposable, i.e., $*\alpha$ is the wedge product of two one-forms. Prove that α is also locally decomposable. (*Hint:* Use the Darboux theorem.)

8.3E The field of a stationary point charge is given by

$$\mathbf{E} = \frac{e\mathbf{r}}{4\pi r^3}, \quad \mathbf{B} = 0,$$

where $\mathbf{r}$ is the vector $x\mathbf{i} + y\mathbf{j} + z\mathbf{k}$ in $\mathbb{R}^3$ and r is its length. Use this and a Lorentz transformation to show that the electromagnetic field produced by a charge e moving along the x-axis with velocity $\mathbf{v}$ is

$$\mathbf{E} = \frac{e}{4\pi}\,\frac{(1-v^2)\mathbf{r}}{[x^2+(1-v^2)(y^2+z^2)]^{3/2}}$$

and, using spherical coordinates with the x-axis as the polar axis,

$$B_r = 0,\ \ B_\theta = 0,\ \ B_\varphi = \frac{e(1-v^2)v\sin\theta}{4\pi r^2(1-v^2\sin^2\theta)^{3/2}}$$

(the magnetic field lines are thus circles centered on the polar axis and lying in planes perpendicular to it).

8.3F (Misner, Thorne, and Wheeler [1973].) The following is the Faraday two-form for the field of an electric dipole of magnitude p_1 oscillating up and down parallel to the z-axis.

$$\mathbf{F} = \mathrm{Re}\left\{ p_1 e^{i\omega r - i\omega t}\left[2\cos\theta\left(\frac{1}{r^2} - \frac{i\omega}{r^2}\right)dr\wedge dt + \sin\theta\left(\frac{1}{r^3} - \frac{i\omega}{r^2} - \frac{\omega^2}{r}\right) r\,d\theta\wedge dt \right.\right.$$
$$\left.\left. + \sin\theta\left(-\frac{i\omega}{r^2} - \frac{\omega^2}{r}\right)dr\wedge r\,d\theta \right]\right\}.$$

Verify that $\mathbf{dF} = 0$ and $\delta\mathbf{F} = 0$, except at the origin.

8.3G Let the Lagrangian for electromagnetic theory be

$$\mathcal{L} = |\mathbf{F}|^2 = -\frac{1}{2}F_{ij}F_{k\ell}\,g^{ik}g^{j\ell}\sqrt{-\det g}\ .$$

Check that $\partial\mathcal{L}/\partial g_{ij}$ is the stress-energy-momentum tensor T^{ij} (see Hawking and Ellis [1973, sec. 3.3]).

§8.4 *The Lie-Poisson Bracket in Continuum Mechanics and Plasma Physics*

This section studies the equations of motion for some Hamiltonian systems in Poisson bracket formation. As opposed to §**8.1**, the emphasis is placed here on the Poisson bracket rather than on the underlying symplectic structure. This naturally leads to a generalization of Hamiltonian mechanics to systems whose phase space is a "Poisson manifold". We do not intend to develop here the theory of Poisson manifolds but only to illustrate it with the most important example, the Lie-Poisson bracket. See Marsden and Ratiu [1988] for further details.

If (P, ω) is a (weak) symplectic manifold, $H : P \to \mathbb{R}$ a smooth Hamiltonian with Hamiltonian vector field $\mathbf{X}_H \in \mathfrak{X}(P)$ whose flow is denoted by φ_t, recall from **8.1.11** that

$$\frac{d\varphi_t}{dt}(p) = X_H(\varphi_t(p)) \tag{1}$$

is equivalent to

$$\frac{d}{dt}(F \circ \varphi_t) = \{F \circ \varphi_t, H \circ \varphi_t\} \tag{2}$$

for any smooth locally defined function $f : U \to \mathbb{R}$, U open in P. In (2), $\{\ ,\ \}$ denotes the Poisson bracket defined by ω, i.e.,

$$\{F, G\} = \omega(X_F, X_G) = X_G[F] = -X_F[G]. \tag{3}$$

Finally, recall that the Poisson bracket is an antisymmetric bilinear operation on $\mathcal{F}(P)$ which satisfies the ***Jacobi identity***

$$\{\{F, G\}, H\} + \{\{G, H\}, F\} + \{\{H, F\}, G\} = 0, \tag{4}$$

i.e., $(\mathcal{F}(P), \{\ ,\ \})$ is a Lie algebra. In addition, the multiplicative ring structure and the Lie algebra structure of $\mathcal{F}(P)$ are connected by the ***Leibniz rule***

$$\{FG, H\} = F\{G, H\} + G\{F, H\}, \tag{5}$$

i.e., $\{\ ,\ \}$ is a derivation in each argument. These observations naturally lead to the following generalization of the concept of symplectic manifolds.

8.4.1 ***Definition*** *A smooth manifold* P *is called a* ***Poisson manifold*** *if* $\mathcal{F}(P)$, *the ring of functions on* P, *admits a Lie algebra structure which is a derivation in each argument. The bracket* *operation on* $\mathcal{F}(P)$ *is called a* ***Poisson bracket*** *and is usually denoted by* $\{\ ,\ \}$.

From the remarks above, we see that any (weak) symplectic manifold is a Poisson manifold. One of the purposes of this section is to show that there are physically important Poisson manifolds which are not symplectic. But even in the symplectic context it is sometimes easier to compute the Poisson bracket than the symplectic form, as the following example shows.

8.4.2 Maxwell's Vacuum Equations as an Infinite Dimensional Hamiltonian System We shall indicate how the dynamical pair of Maxwell's vacuum equations (3) and (4) of the previous section with current $\mathbf{J} = 0$ are a Hamiltonian system.

As the configuration space for Maxwell's equations, we take the space $\mathcal{A}$ of vector potentials. (In more general situations, one should replace $\mathcal{A}$ by the set of connections on a principal bundle over configuration space.) The corresponding phase space is then the cotangent bundle $T^*\mathcal{A}$ with the canonical symplectic structure. Elements of $T^*\mathcal{A}$ may be identified with pairs $(\mathbf{A}, \mathbf{Y})$ where $\mathbf{Y}$ is a vector field density on $\mathbb{R}^3$. (We do not distinguish $\mathbf{Y}$ and $\mathbf{Y}\, d^3x$.) The pairing between $\mathbf{A}$'s and $\mathbf{Y}$'s is given by integration, so the canonical symplectic structure ω on $T^*\mathcal{A}$ is

$$\omega((\mathbf{A}_1, \mathbf{Y}_1), (\mathbf{A}_2, \mathbf{Y}_2)) = \int_{\mathbb{R}^3} (\mathbf{Y}_2 \cdot \mathbf{A}_1 - \mathbf{Y}_1 \cdot \mathbf{A}_2) d^3x, \tag{6}$$

with associated Poisson bracket

$$\{F, G\}(\mathbf{A}, \mathbf{Y}) = \int_{\mathbb{R}^3} \left(\frac{\delta F}{\delta \mathbf{A}} \cdot \frac{\delta G}{\delta \mathbf{Y}} - \frac{\delta F}{\delta \mathbf{Y}} \cdot \frac{\delta G}{\delta \mathbf{A}} \right) d^3x, \tag{7}$$

where $\delta F/\delta \mathbf{A}$ is the vector field defined by

$$\mathbf{D_A} F(\mathbf{A}, \mathbf{Y}) \cdot \mathbf{A}' = \int \frac{\delta F}{\delta \mathbf{A}} \cdot \mathbf{A}'\, d^3x .$$

with the vector field $\delta F/\delta \mathbf{Y}$ defined similarly. With the Hamiltonian

$$H(\mathbf{A}, \mathbf{Y}) = \frac{1}{2} \int \| \mathbf{Y} \|^2 d^3x + \frac{1}{2} \int \| \operatorname{curl} \mathbf{A} \|^2 d^3x, \tag{8}$$

Hamilton's equations are easily computed to be

$$\frac{\partial \mathbf{Y}}{\partial t} = -\operatorname{curl} \operatorname{curl} \mathbf{A} \quad \text{and} \quad \frac{\partial \mathbf{A}}{\partial t} = \mathbf{Y}. \tag{9}$$

If we write $\mathbf{B}$ for $\operatorname{curl} \mathbf{A}$ and $\mathbf{E}$ for $-\mathbf{Y}$, the Hamiltonian becomes the field energy

$$\frac{1}{2}\int \| \mathbf{E} \|^2 d^3x + \frac{1}{2}\int \| \mathbf{B} \|^2 d^3x. \tag{10}$$

Equation (9) implies Maxwell's equations

$$\frac{\partial \mathbf{E}}{\partial t} = \operatorname{curl} \mathbf{B} \quad \text{and} \quad \frac{\partial \mathbf{B}}{\partial t} = -\operatorname{curl} \mathbf{E}, \tag{11}$$

and the Poisson bracket of two functions $F(\mathbf{A}, \mathbf{E})$, $G(\mathbf{A}, \mathbf{E})$ is

$$\{F, G\}(\mathbf{A}, \mathbf{E}) = -\int_{\mathbb{R}^3} \left(\frac{\delta F}{\delta \mathbf{A}} \cdot \frac{\delta G}{\delta \mathbf{E}} - \frac{\delta G}{\delta \mathbf{A}} \cdot \frac{\delta F}{\delta \mathbf{E}} \right) d^3x. \tag{12}$$

We can express this Poisson bracket in terms of $\mathbf{E}$ and $\mathbf{B} = \operatorname{curl} \mathbf{A}$. To do this, we consider functions $\mathcal{F} : \mathcal{V} \times \mathfrak{X}(\mathbb{R}^3) \to \mathbb{R}$, where $\mathcal{V} = \{\operatorname{curl} \mathbf{Z} \mid \mathbf{Z} \in \mathfrak{X}(\mathbb{R}^3)\}$. We pair $\mathcal{V}$ with itself relative to the L^2-inner product. This is a weakly non-degenerate pairing since by the Hodge - Helmholtz decomposition

$$\int_{\mathbb{R}^3} \operatorname{curl} \mathbf{Z}_1 \cdot \operatorname{curl} \mathbf{Z}_2 \, d^3x = 0$$

for all $\mathbf{Z}_2 \in \mathfrak{X}(\mathbb{R}^3)$ implies that $\operatorname{curl} \mathbf{Z}_1 = \nabla f$, whence $\Delta f = \operatorname{div} \nabla f = \operatorname{div} \operatorname{curl} \mathbf{Z}_1 = 0$, i.e., $f =$ constant by Liouville's theorem. Therefore $\operatorname{curl} \mathbf{Z}_1 = \nabla f = 0$, as was to be shown.

We compute $\delta F/\delta \mathbf{A}$ in terms of the functional derivative of an arbitrary extension F of $\mathcal{F}$ to $\mathfrak{X}(\mathbb{R}^3)$, where $F(\mathbf{A}, \mathbf{E}) = \mathcal{F}(\mathbf{B}, \mathbf{E})$, for $\mathbf{B} = \operatorname{curl} \mathbf{A}$. Let L be the linear map $L(\mathbf{A}) = \operatorname{curl} \mathbf{A}$ so that $F = \mathcal{F} \circ (L \times \text{Identity}) = \mathsf{F} \circ (L \times \text{Identity})$. By the chain rule, we have for any $\delta \mathbf{A} \in \mathfrak{X}(\mathbb{R}^3)$,

$$\begin{aligned} \int_{\mathbb{R}^3} \frac{\delta F}{\delta \mathbf{A}} \cdot \delta \mathbf{A} \, d^3x &= \mathbf{D}F(\mathbf{A}) \cdot \delta \mathbf{A} = (\mathbf{D}\mathsf{F}(\mathbf{B}) \circ \mathbf{D}L(\mathbf{A})) \cdot \delta \mathbf{A} \\ &= \int_{\mathbb{R}^3} \frac{\delta \mathsf{F}}{\delta \mathbf{B}} \cdot \operatorname{curl} \delta \mathbf{A} \, d^3x = \int_{\mathbb{R}^3} \operatorname{curl} \frac{\delta \mathsf{F}}{\delta \mathbf{B}} \cdot \delta \mathbf{A} \, d^3x, \end{aligned}$$

since $\mathbf{D}L(\mathbf{A}) = L$ and $\int_{\mathbb{R}^3} \mathbf{X} \cdot \operatorname{curl} \mathbf{Y} \, d^3x = \int_{\mathbb{R}^3} \mathbf{Y} \cdot \operatorname{curl} \mathbf{X} \, d^3x$. Therefore

$$\frac{\delta F}{\delta \mathbf{A}} = \operatorname{curl} \frac{\delta \mathsf{F}}{\delta \mathbf{B}}. \tag{13}$$

This formula seems to depend on the extension F of $\mathcal{F}$. However, this is not the case. More precisely, let $K : \mathfrak{X}(\mathbb{R}^3) \times \mathfrak{X}(\mathbb{R}^3) \to \mathbb{R}$ be such that $K \mid \mathcal{V} \times \mathfrak{X}(\mathbb{R}^3) \equiv 0$. We claim that if $\mathbf{B} \in \mathcal{V}$, then $\delta K/\delta \mathbf{B}$ is a gradient. Granting this statement, this shows that (13) is independent of the extension, since any two extensions of $\mathcal{F}$ coincide on $\mathcal{V} \times \mathfrak{X}(\mathbb{R}^3)$ and since $\operatorname{curl} \circ \operatorname{grad} = 0$. To prove the claim, note that for any $\mathbf{Z} \in \mathfrak{X}(\mathbb{R}^3)$,

$$0 = \mathbf{DK}(\mathbf{B}) \cdot \operatorname{curl} \mathbf{Z} = \int_{\mathbb{R}^3} \frac{\delta \mathrm{K}}{\partial \mathbf{B}} \cdot \operatorname{curl} \mathbf{Z}\, d^3x = \int_{\mathbb{R}^3} \operatorname{curl} \frac{\delta \mathrm{K}}{\delta \mathbf{B}} \cdot \mathbf{Z}\, d^3x$$

whence curl $\delta\mathrm{K}/\delta\mathbf{B} = 0$, i.e., $\delta\mathrm{K}/\delta\mathbf{B}$ is a gradient. Thus (13) implies

$$\frac{\delta \mathrm{F}}{\delta \mathbf{A}} = \operatorname{curl} \frac{\delta F}{\delta \mathbf{B}}, \tag{14}$$

where on the right-hand side $\delta F/\delta\mathbf{B}$ is understood as the functional derivative relative to $\mathbf{B}$ of an arbitrary extension of F to $\mathfrak{X}(\mathbb{R}^3)$. Since $\delta F/\delta\mathbf{E} = \delta\mathrm{F}/\delta\mathbf{E}$, the Poisson bracket (12) becomes

$$\{F, G\}(\mathbf{B}, \mathbf{E}) = \int_{\mathbb{R}^3} \left(\frac{\delta F}{\delta \mathbf{E}} \cdot \operatorname{curl} \frac{\delta G}{\delta \mathbf{B}} - \frac{\delta G}{\delta \mathbf{E}} \cdot \operatorname{curl} \frac{\delta F}{\delta \mathbf{B}} \right) d^3x \tag{15}$$

This bracket was found by Born and Infeld [1935] by a different method.

Using the Hamiltonian

$$H(\mathbf{B}, \mathbf{E}) = \frac{1}{2} \int_{\mathbb{R}^3} (\| \mathbf{B} \|^2 + \| \mathbf{E} \|^2)\, d^3x, \tag{16}$$

equations (11) are equivalent to the Poisson bracket equations

$$\dot{F} = \{F, H\}. \tag{17}$$

Indeed, since $\delta H/\delta\mathbf{E} = \mathbf{E}$, $\operatorname{curl}(\delta H/\delta\mathbf{B}) = \operatorname{curl} \mathbf{B}$, we have

$$\begin{aligned} \{F, H\} &= \int_{\mathbb{R}^3} \left(\frac{\delta F}{\delta \mathbf{E}} \cdot \operatorname{curl} \mathbf{B} - \mathbf{E} \cdot \operatorname{curl} \frac{\delta F}{\delta \mathbf{B}} \right) d^3x \\ &= \int_{\mathbb{R}^3} \left(\frac{\delta F}{\delta \mathbf{E}} \cdot \operatorname{curl} \mathbf{B} - \mathbf{E} \operatorname{curl} \cdot \frac{\delta F}{\delta \mathbf{B}} \right) d^3x \end{aligned}$$

Moreover,

$$\begin{aligned} F &= \mathbf{D}F(\mathbf{B}) \cdot \dot{\mathbf{B}} + \mathbf{D}F(\mathbf{E}) \cdot \dot{\mathbf{E}} \\ &= \int_{\mathbb{R}^3} \left(\frac{\delta' F}{\delta \mathbf{B}} \cdot \dot{\mathbf{B}} + \frac{\delta F}{\delta \mathbf{E}} \cdot \dot{\mathbf{E}} \right) d^3x \\ &= \int_{\mathbb{R}^3} \left(\frac{\delta F}{\delta \mathbf{B}} \cdot \dot{\mathbf{B}} + \frac{\delta F}{\delta \mathbf{E}} \cdot \dot{\mathbf{E}} \right) d^3x, \end{aligned}$$

where $\delta' F/\delta\mathbf{B}$ denotes the functional derivative of F relative to $\mathbf{B}$ in $\mathcal{V}$, i.e.,

$$\mathbf{D}F(\mathbf{B}) \cdot \delta\mathbf{B} = \int_{\mathbb{R}^3} \frac{\delta' F}{\delta \mathbf{B}} \cdot \delta\mathbf{B}\, d^3x. \tag{18}$$

The last equality in the formula for $\dot{F}$ is proved in the following way. Recall that $\delta F/\delta\mathbf{B}$ is the functional derivative of an arbitrary extension of F computed at $\mathbf{B}$, i.e.,

$$\mathbf{D}F(\mathbf{B})\cdot\mathbf{Z} = \int \frac{\delta F}{\delta \mathbf{B}} \cdot \mathbf{Z}\, d^3x \quad \text{for any} \quad \mathbf{Z} \in \mathfrak{X}(\mathbb{R}^3);$$

therefore since $\delta\mathbf{B}$ is a curl, this implies $\delta' F/\delta\mathbf{B}$ and $\delta F/\delta\mathbf{B}$ differ by a gradient which is L^2-orthogonal to $\dot{\mathbf{B}}$, since $\dot{\mathbf{B}}$ is divergence free (again by the Helmholtz-Hodge decomposition). Therefore (11) holds if and only if (17) does. ♦

We next turn to the most important example of a Poisson manifold which is not symplectic. Let $\mathfrak{g}$ denote a Lie algebra i.e., as on page 609, a vector space with a pairing $[\xi, \eta]$ of elements of $\mathfrak{g}$ that is bilinear, antisymmetric and satisfies Jacobi's identity. Let $\mathfrak{g}^*$ denote its "dual", i.e., a vector space weakly paired with $\mathfrak{g}$ via $\langle\ ,\ \rangle : \mathfrak{g}^* \times \mathfrak{g} \to \mathbb{R}$. If $\mathfrak{g}$ is finite dimensional, we take this pairing to be the usual action of forms on vectors.

9.4.3 Definition *For* $F, G : \mathfrak{g}^* \to \mathbb{R}$, *define the* $(\pm)$ ***Lie-Poisson brackets*** *by*

$$\{F, G\}_{\pm}(\mu) = \pm \left\langle \mu, \left[\frac{\delta F}{\delta \mu}, \frac{\delta G}{\delta \mu}\right]\right\rangle \tag{19}$$

where $\mu \in \mathfrak{g}^*$ *and* $\delta F/\delta\mu$, $\delta G/\delta\mu \in \mathfrak{g}$ *are the functional derivatives of* F *and* G, *that is,* $\mathbf{D}F(\mu)\cdot\delta\mu = \langle \delta\mu, \delta F/\delta\mu\rangle$.

If $\mathfrak{g}$ is finite dimensional with a basis ξ_i, and the structure constants are defined by

$$[\xi_i, \xi_j] = c_{ij}^{\ k}\, \xi_k\,,$$

the Lie-Poisson bracket is

$$\{F, G\} = \pm\, \mu_i^{\ j}\, c_{jk}^{\ i}\, \frac{\delta F}{\delta \mu_j}\, \frac{\delta G}{\delta \mu_k}.$$

8.4.4 Lie-Poisson Theorem $\mathfrak{g}^*$ *with the* $(\pm)$*Lie-Poisson bracket is a Poisson manifold.*

Proof Clearly $\{\ ,\ \}_{\pm}$ is bilinear and skew symmetric. To show $\{\ ,\ \}_{\pm}$ is a derivation in each argument, we show that

$$\frac{\delta(FG)}{\delta\mu} = F(\mu)\,\frac{\delta G}{\delta\mu} + G(\mu)\,\frac{\delta F}{\delta\mu}\,. \tag{20}$$

To prove (20), let $\delta\mu \in \mathfrak{g}^*$ be arbitrary. Then

$$\begin{aligned}
\left\langle \delta\mu, \frac{\delta(FG)}{\delta\mu}\right\rangle &= \mathbf{D}(FG)(\mu)\cdot\delta\mu \\
&= F(\mu)\mathbf{D}G(\mu)\cdot\delta\mu + G(\mu)\,\mathbf{D}F(\mu)\cdot\delta\mu \\
&= \left\langle \delta\mu, F(\mu)\,\frac{\delta G}{\delta\mu} + G(\mu)\,\frac{\delta F}{\delta\mu}\right\rangle.
\end{aligned}$$

Finally, we prove the Jacobi identity. We start by computing the derivative of the map $\mu \in \mathfrak{g}^* \mapsto \delta F/\delta\mu \in \mathfrak{g}$. We have for every $\lambda, \nu \in \mathfrak{g}^*$

$$\mathbf{D}\left(\langle \nu, \frac{\delta F}{\delta\mu}\rangle\right)(\mu)\cdot \lambda = \mathbf{D}(\mathbf{D}F(\cdot)\cdot \nu)(\mu)\cdot \lambda = \mathbf{D}^2F(\mu)(\nu, \lambda),$$

i.e.,

$$\mathbf{D}\left(\frac{\delta F}{\delta\mu}\right)(\mu)\cdot \lambda = \mathbf{D}^2F(\mu)(\lambda, \cdot\,). \tag{21}$$

Therefore, the derivative of $\mu \mapsto \left[\frac{\delta F}{\delta\mu}, \frac{\delta G}{\delta\mu}\right]$ is

$$\mathbf{D}\left[\frac{\delta F}{\delta\mu}, \frac{\delta G}{\delta\mu}\right](\mu)\cdot\nu = \left[\mathbf{D}^2F(\mu)(\nu, \cdot), \frac{\delta G}{\delta\mu}\right] + \left[\frac{\delta F}{\delta\mu}, \mathbf{D}^2G(\mu)(\nu, \cdot)\right] \tag{22}$$

where $\mathbf{D}^2F(\mu)(\nu, \cdot\,) \in L(\mathfrak{g}^*, \mathbb{R})$ is assumed to be represented via $\langle\ ,\ \rangle$ by an element of $\mathfrak{g}$. Therefore by (19) and (22)

$$\begin{aligned}\langle \nu, \frac{\delta}{\delta\mu}\{F, G\}\rangle &= \mathbf{D}\{F, G\}(\mu)\cdot \nu\\ &= \langle \nu, \left[\frac{\delta F}{\delta\mu}, \frac{\delta G}{\delta\mu}\right]\rangle + \langle \mu, \left[\mathbf{D}^2F(\mu)(\nu,\cdot), \frac{\delta G}{\delta\mu}\right]\rangle + \langle \mu, \left[\frac{\delta F}{\delta\mu}, \mathbf{D}^2G(\mu)(\nu,\cdot)\right]\rangle\\ &= \langle \nu, \left[\frac{\delta F}{\delta\mu}, \frac{\delta G}{\delta\mu}\right]\rangle + \langle \mathrm{ad}\left(\frac{\delta G}{\delta\mu}\right)^* \mu, \mathbf{D}^2F(\mu)(\nu,\cdot)\rangle - \langle \mathrm{ad}\left(\frac{\delta F}{\delta\mu}\right)^* \mu, \mathbf{D}^2G(\mu)(\nu,\cdot)\rangle,\end{aligned}$$

where $\mathrm{ad}(\xi)\colon \mathfrak{g} \to \mathfrak{g}$ is the linear map $\mathrm{ad}(\xi)\cdot \eta = [\xi, \eta]$ and $\mathrm{ad}(\xi)^* : \mathfrak{g}^* \to \mathfrak{g}^*$ is its dual defined by

$$\langle \mathrm{ad}(\xi)^*\mu, \eta\rangle = \langle \mu, [\xi, \nu]\rangle, \quad \eta \in \mathfrak{g},\ \mu \in \mathfrak{g}^*.$$

Therefore,

$$\langle \nu, \frac{\delta}{\delta\nu}\{F, G\}\rangle = \langle \nu, \left[\frac{\delta F}{\delta\mu}, \frac{\delta G}{\delta\mu}\right]\rangle - \langle \nu, \mathbf{D}^2F(\mu)\left(\mathrm{ad}\left(\frac{\delta G}{\delta\mu}\right)^* \mu, \cdot\right)\rangle + \langle \nu, \mathbf{D}^2G(\mu)\left(\mathrm{ad}\left(\frac{\delta F}{\delta\mu}\right)^* \mu, \cdot\right)\rangle$$

$$\frac{\delta}{\delta\nu}\{F, G\} = \left[\frac{\delta F}{\delta\mu}, \frac{\delta G}{\delta\mu}\right] - \mathbf{D}^2F(\mu)\left(\mathrm{ad}\left(\frac{\delta G}{\delta\mu}\right)^* \mu, \cdot\right) + \mathbf{D}^2G(\mu)\left(\mathrm{ad}\left(\frac{\delta F}{\delta\mu}\right)^* \mu, \cdot\right), \tag{23}$$

which in turn implies

$$\begin{aligned}\{F, G\}(\mu) = \langle \mu, \left[\frac{\delta}{\delta\mu}\{F, G\}, \frac{\delta H}{\delta\mu}\right]\rangle &= \langle \mu, \left[\left[\frac{\delta F}{\delta\mu}, \frac{\delta G}{\delta\mu}\right], \frac{\delta H}{\delta\mu}\right]\rangle\\ &+ \mathbf{D}^2F(\mu)\left(\mathrm{ad}\left(\frac{\delta G}{\delta\mu}\right)^* \mu, \mathrm{ad}\left(\frac{\delta H}{\delta\mu}\right)^* \mu\right) - \mathbf{D}^2G(\mu)\left(\mathrm{ad}\left(\frac{\delta F}{\delta\mu}\right)^* \mu, \mathrm{ad}\left(\frac{\delta H}{\delta\mu}\right)^* \mu\right).\end{aligned}$$

The two cyclic permutations in F, G, H added to the above formula sum up to zero: all six terms involving second derivatives cancel and the three first terms add up to zero by the Jacobi identity for the bracket of $\mathfrak{g}$. ■

8.4.5 The Free Rigid Body The equations of motion of the free rigid body described by an observer fixed on the moving body are given by ***Euler's equation***

$$\dot{\Pi} = \Pi \times \omega, \tag{24}$$

where $\Pi, \omega \in \mathbb{R}^3$, $\Pi_i = I_i \omega_i$, $i = 1, 2, 3$, $I = (I_1, I_2, I_3)$ are the principal moments of inertia, the coordinate system in the body is chosen so that the axes are the principal axes, ω is the angular velocity in the body, and Π is the angular momentum in the body. It is straightforward to check that the kinetic energy

$$H(\Pi) = \frac{1}{2}\Pi \cdot \omega \tag{25}$$

is a conserved quantity for (24).

We shall prove below that (24) *are Hamilton's equations with Hamiltonian* (25) *relative to a* (–) *Lie-Poisson structure on* $\mathbb{R}^3$.

The vector space $\mathbb{R}^3$ is in fact a Lie algebra with respect to the bracket operation given by the cross product, i.e. $[\mathbf{x}, \mathbf{y}] = \mathbf{x} \times \mathbf{y}$. (This is the structure that it inherits from the rotation group.) We pair $\mathbb{R}^3$ with itself using the usual dot-product, i.e., $\langle \mathbf{x}, \mathbf{y} \rangle = \mathbf{x} \cdot \mathbf{y}$. Therefore, if $F : \mathbb{R}^3 \to \mathbb{R}$, $\delta F/\delta\Pi = \nabla F(\Pi)$. Thus, the (–) Lie-Poisson bracket is given via (19) by the triple product

$$\{F, G\}(\Pi) = -\Pi \cdot (\nabla F(\Pi) \times \nabla G(\Pi)). \tag{26}$$

Since $\delta H/\delta\Pi = \omega$, we see that for any $F : \mathbb{R}^3 \to \mathbb{R}$,

$$\frac{d}{dt}(F(\Pi)) = DF(\Pi) \cdot \dot{\Pi} = \dot{\Pi} \cdot \nabla F(\Pi) = -\Pi \cdot (\nabla F(\Pi) \times \omega = \nabla F(\Pi) \cdot (\Pi \times \omega)$$

so that $\dot{F} = \{F, H\}$ for any $F: \mathbb{R}^3 \to \mathbb{R}$ is equivalent to Euler's equations of motion (24).

8.4.6 Proposition (Pauli [1953], Martin [1959], Arnold [1966], Sudarshan and Mukunda [1974]) *Euler's equations* (24) *for a free rigid body are a Hamiltonian system in* $\mathbb{R}^3$ *relative to the* (–) *Lie Poisson bracket* (24) *and Hamiltonian function* (25).

8.4.7 Ideal Incompressible Homogeneous Fluid Flow In §**8.2** we have shown that the equations of motion for an ideal incompressible homogeneous fluid in a region $\Omega \subset \mathbb{R}^3$ with smooth boundary $\partial\Omega$ are given by ***Euler's equations of motion***

$$\left.\begin{aligned} &\frac{\partial \mathbf{v}}{\partial t} + (\mathbf{v} \cdot \nabla)\mathbf{v} = -\nabla p \\ &\operatorname{div} \mathbf{v} = 0 \\ &\mathbf{v}(t, x) \in T_x(\partial\Omega) \;\text{ for }\; x \in \partial\Omega \end{aligned}\right\} \tag{27}$$

with initial condition $\mathbf{v}(0, x) = \mathbf{v}_0(x)$, a given vector field on Ω. Here $\mathbf{v}(t, x)$ is the Eulerian or spatial velocity, a time dependent vector field on Ω. The pressure p is a function of $\mathbf{v}$ and is uniquely determined by $\mathbf{v}$ (up to a constant) by the Neumann problem (take div and the dot product with $\mathbf{n}$ of the first equation in (27))

$$\left.\begin{aligned} &\Delta p = -\operatorname{div}((\mathbf{v} \cdot \nabla)\mathbf{v}) \\ &\frac{\partial p}{\partial \mathbf{n}} = \nabla p \cdot \mathbf{n} = -((\mathbf{v} \cdot \nabla)\mathbf{v}) \cdot \mathbf{n} \;\text{ on }\; \partial\Omega, \end{aligned}\right\} \tag{28}$$

where $\mathbf{n}$ is the outward unit normal to $\partial\Omega$. The kinetic energy

$$H(\mathbf{v}) = \frac{1}{2} \int_\Omega \| \mathbf{v} \|^2 \, d^3x \tag{29}$$

has been shown in §**8.2** to be a conserved quantity for (27). We shall prove below that the first equation in (27) is Hamiltonian relative to a (+) Lie-Poisson bracket with Hamiltonian function given by (29).

Consider the Lie algebra $\mathfrak{X}_{\mathrm{div}}(\Omega)$ of divergence free vector fields on Ω tangent to $\partial\Omega$ with bracket given by *minus* the bracket of vector fields, i.e., for $\mathbf{u}, \mathbf{v} \in \mathfrak{X}_{\mathrm{div}}(\Omega)$ define

$$[\mathbf{u}, \mathbf{v}] = (\mathbf{v} \cdot \nabla)\mathbf{u} - (\mathbf{u} \cdot \nabla)\mathbf{v}. \tag{30}$$

(The reason for this strange choice comes from the fact that the usual Lie bracket for vector fields is the right Lie algebra bracket of the diffeomorphism group of Ω.) Now pair $\mathfrak{X}_{\mathrm{div}}(\Omega)$ with itself via the L^2-pairing. As in **8.4.2**, using the Hodge-Helmholtz decomposition, it follows that this pairing is weakly non-degenerate. In particular

$$\frac{\delta H}{\delta \mathbf{v}} = \mathbf{v}\,. \tag{31}$$

The (+) Lie-Poisson bracket on $\mathfrak{X}_{\mathrm{div}}(\Omega)$ is

$$\{F, G\}(\mathbf{v}) = \int_\Omega \mathbf{v} \cdot \left[\left(\frac{\delta G}{\delta \mathbf{v}} \cdot \nabla\right)\frac{\delta F}{\delta \mathbf{v}} - \left(\frac{\delta F}{\delta \mathbf{v}} \cdot \nabla\right)\frac{\delta G}{\delta \mathbf{v}}\right] d^3x. \tag{32}$$

Therefore, for any $\mathfrak{X}_{\mathrm{div}}(\Omega) \to \mathbb{R}$, we have

$$\frac{d}{dt}(F(v)) = DF(v) \cdot \dot{v} = \int_\Omega \frac{\delta F}{\delta v} \cdot \dot{v}\, d^3x = \int_\Omega v \cdot \left[(v \cdot \nabla) \frac{\delta F}{\delta v} - \left(\frac{\delta F}{\delta v} \cdot \nabla \right) v \right] d^3x$$
$$= \int_\Omega v \cdot \left((v \cdot \nabla) \frac{\delta F}{\delta v} \right) d^3x - \int_\Omega \frac{\partial F}{\partial v} \cdot \nabla\left(\frac{1}{2} \| v \|^2 \right) d^3x.$$

To handle the first integral, observe that if

$$f, g : \Omega \to \mathbb{R}, \text{ then } \operatorname{div}(fgv) = f \operatorname{div}(gv) + gv \cdot \nabla f,$$

so that by Stokes' theorem and $v \cdot \mathbf{n} = 0$, $\operatorname{div} v = 0$, we get

$$\int_\Omega gv \cdot \nabla f\, d^3x = \int_{\partial\Omega} fg\, v \cdot \mathbf{n}\, dS - \int_\Omega f \operatorname{div}(gv)\, d^3x$$
$$= - \int_\Omega fv \cdot \nabla g\, d^3x .$$

Applying the above relation to $g = v^i$, $f = \dfrac{\delta F}{\delta v^i}$, and summing over $i = 1, 2, 3$ we get

$$\int_\Omega v \cdot \left((v \cdot \nabla) \frac{\delta F}{\delta v} \right) d^3x = - \int_\Omega \frac{\delta F}{\delta v} \cdot ((v \cdot \nabla) v) d^3x$$

so that $\dot{F} = \{F, H\}$ reads

$$\int_\Omega \frac{\delta F}{\delta v} \cdot \dot{v}\, d^3x = - \int_\Omega \frac{\delta F}{\delta v} \cdot \left[(v \cdot \nabla) v + \frac{1}{2} \nabla \| v \|^2 \right] d^3x \tag{33}$$

for any $F : \mathfrak{X}_{\text{div}}(\Omega) \to \mathbb{R}$. One would like to conclude from here that the coefficients of $\delta F/\delta v$ on both sides of (33) are equal. This conclusion, however, is incorrect, since $(v \cdot \nabla) v + (1/2) \nabla \| v \|^2$ is not divergence free. Thus, applying the Hodge-Helmholtz decomposition, write

$$(v \cdot \nabla) v + \frac{1}{2} \nabla \| v \|^2 = \mathbf{X} - \nabla f \tag{34}$$

where $X \in \mathfrak{X}_{\text{div}}(\Omega)$ and f is determined by

$$\left.\begin{aligned} \Delta\left(f + \frac{1}{2} \| v \|^2 \right) &= - \operatorname{div}((v \cdot \nabla) v) \\ \frac{\partial}{\partial \mathbf{n}}\left(f + \frac{1}{2} \| v \|^2 \right) &= - ((v \cdot \nabla) \cdot v) \cdot \mathbf{n} \end{aligned}\right\}$$

which coincides with (28), i.e.,

$$f + \frac{1}{2} \| \mathbf{v} \|^2 = \mathbf{p} + \text{constant} \tag{35}$$

Moreover, since

$$\int_\Omega \frac{\delta F}{\delta \mathbf{v}} \cdot \nabla f \, d^3x = \int_{\partial\Omega} f \frac{\partial F}{\partial \mathbf{v}} \cdot \mathbf{n} \, dS - \int_\Omega f \operatorname{div} \frac{\delta F}{\delta \mathbf{v}} \, d^3x = 0$$

we have from (33), (34), (35)

$$\frac{\partial \mathbf{v}}{\partial t} = -\mathbf{X} = -(\mathbf{v} \cdot \nabla)\mathbf{v} - \nabla p$$

which is the first equation in (27). We have thus proved

8.4.8 Proposition (Arnold [1966], Marsden and Weinstein [1983].) *Euler's equations* (27) *are a Hamiltonian system on* $\mathfrak{X}_{\text{div}}(\Omega)$ *relative to the* (+) *Lie-Poisson bracket* (32) *and Hamiltonian function given by* (29).

8.4.9 The Poisson-Vlasov Equation We consider a collisionless plasma consisting (for notational simplicity) of only one species of particles with charge q and mass m moving in Euclidean space $\mathbb{R}^3$ with positions **x** and velocities **v**. Let $f(\mathbf{x}, \mathbf{v}, t)$ be the plasma density in the plasma space at time t. In the Coulomb or electrostatic case in which there is no magnetic field, the motion of the plasma is described by the Poisson-Vlasov equations which are the (collisionless) Boltzmann equations for the density function f and the Poisson equation for the scalar potential φ_f :

$$\frac{\partial f}{\partial t} + \mathbf{v} \cdot \frac{\partial f}{\partial \mathbf{x}} - \frac{q}{m} \frac{\partial \varphi_f}{\partial \mathbf{x}} \cdot \frac{\partial f}{\partial \mathbf{v}} = 0 \tag{36}$$

$$\Delta \varphi_f = -q \int f(\mathbf{x}, \mathbf{v}) \, d^3v = \rho_f , \tag{37}$$

where $\partial/\partial\mathbf{x}$ and $\partial/\partial\mathbf{v}$ denote the gradients in $\mathbb{R}^3$ relative to the **x** and **v** variables, ρ_f is the charge density in physical space, and Δ is the Laplacian. Equation (36) can be written in "Hamiltonian" form

$$\frac{\partial f}{\partial t} + \{f, \mathcal{H}\} = 0, \tag{38}$$

where { , } is the canonical Poisson bracket on phase space,

$$\{f, g\} = \frac{\partial f}{\partial \mathbf{x}} \cdot \frac{\partial g}{\partial \mathbf{p}} - \frac{\partial f}{\partial \mathbf{p}} \cdot \frac{\partial f}{\partial \mathbf{x}} = \frac{1}{m}\left[\frac{\partial f}{\partial \mathbf{x}} \cdot \frac{\partial g}{\partial \mathbf{v}} - \frac{\partial f}{\partial \mathbf{v}} \cdot \frac{\partial g}{\partial \mathbf{x}}\right] \tag{39}$$

$\mathbf{p} = m\mathbf{v}$ and $\mathcal{H}_f = m \| \mathbf{v} \|^2 + q\varphi_f$ is the single particle energy, called the *self - consistent Hamiltonian.* Indeed,

$$\{ \mathcal{H}_f, f \} = \frac{1}{m}\left(\frac{\partial \mathcal{H}_f}{\partial \mathbf{x}} \cdot \frac{\partial f}{\partial \mathbf{v}} - \frac{\partial \mathcal{H}_f}{\partial \mathbf{v}} \cdot \frac{\partial f}{\partial \mathbf{x}}\right) = \frac{1}{m}\left(q\,\frac{\partial \varphi_f}{\partial \mathbf{x}} \cdot \frac{\partial f}{\partial \mathbf{v}} - m\mathbf{v} \cdot \frac{\partial f}{\partial \mathbf{x}}\right)$$
$$= \frac{q}{m}\frac{\partial \varphi_f}{\partial \mathbf{x}} \cdot \frac{\partial f}{\partial \mathbf{v}} - \mathbf{v} \cdot \frac{\partial f}{\partial \mathbf{x}} = \frac{\partial f}{\partial t}$$

according to (36). There is another very useful way to think of the evolution of f. If F(f) is any functional of the density function f and f evolves according to the Poisson-Vlasov equations (36) (or (38) equivalently) then F evolves in time by

$$\dot{F} = \{F, H\}_+$$

where $\{\ ,\ \}_+$ is a (+) Lie-Poisson bracket (to be defined) of functionals and H is the total energy. Let us state this more precisely. Let $V = \{f \in C^k(\mathbb{R}^6) \mid f \to 0 \text{ as } \|\mathbf{x}\| \to \infty, \ \|\mathbf{v}\| \to \infty\}$ with the L^2-pairing $\langle\ ,\ \rangle : V \times V \to \mathbb{R}$; $\langle f, g\rangle = \int f(\mathbf{x}, \mathbf{v})\, g(\mathbf{x}, \mathbf{v})\, d^3x\, d^3v$. If $F: V \to \mathbb{R}$ is differentiable at $f \in V$, the functional derivative $\frac{\delta F}{\delta f}$ is, by definition, the unique element $\frac{\delta F}{\delta f} \in V$ such that

$$DF(f) \cdot g = \left\langle \frac{\delta F}{\delta f}, g \right\rangle = \int \frac{\delta F}{\delta f}(\mathbf{x}, \mathbf{v})\, g(\mathbf{x}, \mathbf{v})\, d^3x\, d^3v.$$

The vector space V is a Lie algebra relative to the canonical Poisson bracket (39) on $\mathbb{R}^6$. For two functionals $F, G : V \to \mathbb{R}$ their (+) Lie-Poisson bracket $\{F, G\}_+ : V \to \mathbb{R}$ is then given by

$$\{ F, G\}_+ (\mathbf{x}, \mathbf{v}) = \int f(\mathbf{x}, \mathbf{v}) \left\{ \frac{\delta F}{\delta f}, \frac{\delta G}{\delta f} \right\} (\mathbf{x}, \mathbf{v})\, d^3x\, d^3v,$$

where $\{\ ,\ \}$ is the canonical bracket (39). For any $f, g, h \in V$ we have the formula

$$\int f\{g, h\}\, d^3x\, d^3v = \int g\{h, f\}\, d^3x\, d^3v\,. \tag{40}$$

Indeed by integration by parts, we get

$$\begin{aligned}\int f\{g, h\}\, d^3x\, d^3v &= \frac{1}{m}\int f\,\frac{\partial g}{\partial \mathbf{x}} \cdot \frac{\partial h}{\partial \mathbf{v}}\, d^3x\, d^3v - \frac{1}{m}\int f\,\frac{\partial h}{\partial \mathbf{x}} \cdot \frac{\partial g}{\partial \mathbf{v}}\, d^3x\, d^3v \\ &= -\frac{1}{m}\int \frac{\partial f}{\partial \mathbf{x}} \cdot g\,\frac{\partial h}{\partial \mathbf{v}}\, d^3x\, d^3v + \frac{1}{m}\int g\,\frac{\partial f}{\partial \mathbf{v}} \cdot \frac{\partial h}{\partial \mathbf{x}}\, g\, d^3x\, d^3v \\ &= \frac{1}{m}\int g\left(\frac{\partial h}{\partial \mathbf{x}} \cdot \frac{\partial f}{\partial \mathbf{v}} - \frac{\partial f}{\partial \mathbf{x}} \cdot \frac{\partial h}{\partial \mathbf{v}}\right) d^3x\, d^3v = \int g\{h, f\}\, d^3x\, d^3v\,.\end{aligned}$$

8.4.10 Proposition (Iwinski and Turski [1976], Kaufman and Morrison [1980] and Morrison [1980]) *Densities* $f \in V$ *evolve according to the Poisson-Vlasov equation* (38) *if and only if any* *differentiable function* $F: V \to \mathbb{R}$ *having functional derivative* $\delta F/\delta f$ *evolves by the* (+) *Lie-Poisson equation*

$$\dot{F}(f) = \{F, H\}_+(f) \tag{41}$$

with the Hamiltonian $H: V \to \mathbb{R}$ *equal to the total energy*

$$H(f) = \frac{1}{2}\int m \, \| \mathbf{v} \|^2 f(\mathbf{x}, \mathbf{v})\, d^3x\, d^3v + \int \frac{1}{2}\varphi_f(\mathbf{x})\, d^3x.$$

Proof First we compute $\frac{\delta H}{\delta f}$ using the definition $DH(f) \cdot \delta f = \int \frac{\delta H}{\delta f}\, \delta f$. Note that the first term of $H(f)$ is linear in f and the second term is $\frac{1}{2}\int \varphi_f \rho_f\, d^3x = \frac{1}{2}\int \| \nabla \varphi_f \|^2\, d^3x$ since $\Delta \varphi_f = -\rho_f$. Then

$$\begin{aligned} DH(f) \cdot \delta f &= \frac{1}{2}\int m \, \| \mathbf{v} \|^2\, \delta f\, d^3x\, d^3v + \int (\nabla \varphi_f)(\mathbf{D}\,(\nabla \varphi_f))\delta f\, d^3x \quad \text{(chain rule)} \\ &= \frac{1}{2}\int m \| \mathbf{v} \|^2 \delta f\, d^3x\, d^3v - \int \varphi_f \mathbf{D}\,(\Delta \varphi_f))\delta f\, d^3x \quad \text{(integration by parts)} \\ &= \frac{1}{2}\int m \, \| \mathbf{v} \|^2 \delta f\, d^3x\, d^3v + \int \varphi_f \left(\mathbf{D}\left(q \int f\, d^3v\right)\right)(f)\, \delta f\, d^3x \\ &= \frac{1}{2}\int m \, \| \mathbf{v} \|^2 \delta f\, d^3x\, d^3v + \int \varphi_f q \delta f\, d^3v. \quad \text{(linearity in f)} \end{aligned}$$

Therefore $\frac{\delta H}{\delta f} = \frac{1}{2} m \, \| \mathbf{v} \|^2 + q\varphi_f = \mathcal{H}_f$. We have

$$\dot{F}(f) = DF(f) \cdot \dot{f} = \int \frac{\delta F}{\delta f}\, \dot{f}\, d^3x\, d^3v$$

and

$$\int \frac{\delta F}{\delta f}\{\mathcal{H}_f, f\} d^3x\, d^3v = \int \frac{\delta F}{\delta f}\left\{\frac{\delta H}{\delta f}, f\right\} d^3x\, d^3v = \int f\left\{\frac{\delta F}{\delta f}, \frac{\delta H}{\delta f}\right\} d^3x\, d^3v = \{F, H\}_+(f).$$

by (40). Thus (41), for any F having functional derivatives, is equivalent to (38). ■

8.4.11 The Maxwell-Vlasov Equations We consider a plasma consisting of particles with charge q_1 and mass m moving in Euclidean space $\mathbb{R}^3$ with positions $\mathbf{x}$ and velocities $\mathbf{v}$. For simplicity we consider only one species of particle; the general case is similar. Let $f(\mathbf{x}, \mathbf{v}, t)$ be the plasma density at time t, $\mathbf{E}(\mathbf{x}, t)$ and $\mathbf{B}(\mathbf{x}, t)$ be the electric and magnetic fields. The ***Maxwell-Vlasov equations*** are:

$$\frac{\partial f}{\partial t} + \mathbf{v} \bullet \frac{\partial f}{\partial \mathbf{x}} + \frac{q}{m}\left(\mathbf{E} + \frac{\mathbf{v} \times \mathbf{B}}{c}\right) \bullet \frac{\partial f}{\partial \mathbf{v}} = 0, \tag{42}$$

$$\frac{1}{c}\frac{\partial \mathbf{B}}{\partial t} = -\operatorname{curl} \mathbf{E}, \tag{43a}$$

$$\frac{1}{c}\frac{\partial \mathbf{E}}{\partial t} = \operatorname{curl} \mathbf{B} - \frac{q}{c}\int \mathbf{v} f(\mathbf{x}, \mathbf{v}, t) d^3v\,, \tag{43b}$$

together with the non evolutionary equations

$$\operatorname{div} \mathbf{E} = \rho_f, \quad \text{where} \quad \rho_f = q \int f(\mathbf{x}, \mathbf{v}, t) d^3v, \tag{44a}$$

$$\operatorname{div} \mathbf{B} = 0. \tag{44b}$$

Letting $c \to \infty$ leads to the *Poisson-Vlasov equation* (36)

$$\frac{\partial f}{\partial t} + \mathbf{v} \cdot \frac{\partial f}{\partial \mathbf{x}} - \frac{q}{m} \frac{\partial \varphi_f}{\partial \mathbf{x}} \cdot \frac{\partial f}{\partial \mathbf{v}} = 0,$$

where $\Delta\varphi_f = -\rho_f$. In what follows we shall set $q = m = c = 1$.

The Hamiltonian for the Maxwell-Vlasov system is

$$H(f, \mathbf{E}, \mathbf{B}) = \int \frac{1}{2} \| \mathbf{v} \|^2 f(\mathbf{x}, \mathbf{v}, t) dx\, dv + \int \frac{1}{2} \left[\|\mathbf{E}(\mathbf{x}, t)\|^2 + \| \mathbf{B}(\mathbf{x}, t)\|^2 \right] d^3x. \tag{45}$$

Let $\mathcal{V} = \{\operatorname{curl} \mathbf{Z} \mid \mathbf{Z} \in \mathfrak{X}(\mathbb{R}^3)\}$.

8.4.12 Theorem (Iwinski and Turski [1976], Morrison [1980], Marsden and Weinstein [1982].)
(i) *The manifold* $\mathcal{F}(\mathbb{R}^6) \times \mathfrak{X}(\mathbb{R}^3) \times \mathcal{V}$ *is a Poisson manifold relative to the bracket*

$$\{F, G\}(f, E, B) = \int f\left\{\frac{\delta F}{\delta f}, \frac{\delta G}{\delta f}\right\} d^3x\, d^3v + \int \left(\frac{\delta G}{\delta \mathbf{B}} \cdot \operatorname{curl} \frac{\delta G}{\delta \mathbf{B}} - \frac{\delta G}{\delta \mathbf{E}} \cdot \operatorname{curl} \frac{\delta F}{\delta \mathbf{B}} \right) d^3x$$

$$+ \int \left(\frac{\delta F}{\delta \mathbf{E}} \cdot \frac{\partial f}{\partial \mathbf{v}} \frac{\delta G}{\delta f} - \frac{\delta G}{\delta \mathbf{E}} \cdot \frac{\partial f}{\partial \mathbf{v}} \frac{\delta F}{\delta f} \right) d^3x\, d^3v + \int f\mathbf{B} \cdot \left(\frac{\partial}{\partial \mathbf{v}} \frac{\delta F}{\delta f} \times \frac{\partial}{\partial \mathbf{v}} \frac{\delta G}{\delta f} \right) d^3x\, d^3v. \tag{46}$$

(ii) *The equations of motion* (42) *and* (43) *are equivalent to*

$$\dot{F} = \{F, H\}$$

where F *is any locally defined function with functional derivatives and* $\{\ ,\ \}$ *is given by* (46).

Part (a) follows from general considerations on reduction (see Marsden and Weinstein [1982]). The direct verification is laborious but straightforward, if one recognizes that the first two terms are the Poisson bracket for the Poisson-Vlasov equation and the Born-Infeld bracket respectively.

Proof of (b) Since $\delta H/\delta f = \| \mathbf{v} \|^2/2$, $\delta H/\delta \mathbf{E} = \mathbf{E}$, $\operatorname{curl} \delta H/\delta \mathbf{B} = \operatorname{curl} \mathbf{B}$, we have, by (40) and integration by parts in the fourth integral,

$$\{F, H\} = \int f\left\{\frac{\delta F}{\delta f}, \frac{1}{2}\|\mathbf{v}\|^2\right\} d^3x\, d^3v + \int\left(\frac{\delta F}{\delta \mathbf{B}} \cdot \operatorname{curl} \mathbf{B} - \mathbf{E} \cdot \operatorname{curl} \frac{\delta F}{\delta \mathbf{B}}\right) d^3x$$
$$+ \int\left(\frac{\delta F}{\delta \mathbf{E}} \cdot \frac{\partial f}{\partial \mathbf{v}} \frac{1}{2}\|\mathbf{v}\|^2 - \mathbf{E} \cdot \frac{\partial f}{\partial \mathbf{v}} \frac{\delta F}{\delta \mathbf{E}}\right) d^3x\, d^3v$$
$$+ \int f\mathbf{B}\left(\frac{\partial}{\partial \mathbf{v}} \frac{\delta F}{\delta f} \times \frac{\partial}{\partial \mathbf{v}}\left(\frac{1}{2}\|\mathbf{v}\|^2\right)\right) d^3x\, d^3v$$
$$= \int \frac{\delta F}{\delta f}\left[\left\{\frac{1}{2}\|\mathbf{v}\|^2, f\right\} - \mathbf{F} \cdot \frac{\delta f}{\delta \mathbf{v}} - \operatorname{div}_{\mathbf{v}}(\mathbf{v} \times f\mathbf{B})\right] d^3x\, d^3v$$
$$+ \int\left(\frac{\delta F}{\delta \mathbf{E}} \cdot \left[\operatorname{curl} \mathbf{B} + \int \frac{\partial f}{\partial \mathbf{v}} \frac{1}{2}\|\mathbf{v}\|^2 d^3v\right] d^3x - \int \mathbf{E} \cdot \frac{\delta F}{\delta \mathbf{B}} \operatorname{curl} d^3x$$

where $\operatorname{div}_{\mathbf{v}}$ denotes the divergence only with respect to the $\mathbf{v}$-variable. Since

$$\left\{\frac{1}{2}\|\mathbf{v}\|^2, f\right\} = -\mathbf{v} \cdot \frac{\partial f}{\partial \mathbf{x}}, \quad \operatorname{div}_{\mathbf{v}}(\mathbf{v} \times f\mathbf{B}) = \frac{\partial f}{\partial \mathbf{v}} \cdot (\mathbf{v} \times \mathbf{B}),$$
$$\int \frac{\partial f}{\partial \mathbf{v}} \frac{1}{2}\|\mathbf{v}\|^2 d^3x = -\int \mathbf{v}\, f\, d^3x,$$

and

$$\int \mathbf{E} \cdot \operatorname{curl} \frac{\delta F}{\delta \mathbf{B}} d^3x = \int \frac{\delta F}{\delta \mathbf{B}} \cdot \operatorname{curl} \mathbf{E}\, d^3x,$$

we get

$$\{F, H\} = \int \frac{\delta F}{\delta f}\left[-\mathbf{v} \cdot \frac{\partial f}{\partial \mathbf{x}} - (\mathbf{E} + \mathbf{v} \times \mathbf{B}) \cdot \frac{\partial f}{\partial \mathbf{v}}\right] d^3x\, d^3v$$
$$+ \int \frac{\delta F}{\delta \mathbf{E}} \cdot \left(\operatorname{curl} \mathbf{B} - \int \mathbf{v} f(\mathbf{x}, \mathbf{v}, t)\, d^3v\right) d^3v - \int\left(\frac{\delta F}{\delta \mathbf{B}} \cdot \operatorname{curl} \mathbf{E}\right) d^3x, \tag{48}$$

and since $\dot{F} = \int \frac{\delta F}{\delta f} \cdot \dot{f}\, d^3x\, d^3v + \int \frac{\delta F}{\delta \mathbf{E}} \cdot \dot{\mathbf{E}}\, d^3x + \int \frac{\delta' F}{\delta \mathbf{B}} \cdot \dot{\mathbf{B}}\, d^3x$ taking into account that $\frac{\delta F}{\delta \mathbf{B}}, \frac{\delta' F}{\delta \mathbf{B}}$ differ by a gradient (by (15)) which is L^2-orthogonal to $\mathcal{V}$ (of which both $\operatorname{curl} \mathbf{E}$ and $\mathbf{B}$ are a member), it follows from (48) that the equations (42)-(44) (with $q = c = m = 1$) are equivalent to (47). ■

Exercises

8.4A Find the symplectic form equivalent to the Born-Infeld bracket (16) on $\mathcal{V} \times \mathfrak{X}(\mathbb{R}^3)$.

8.4B Show that the Hamiltonian vector field $X_H \in \mathfrak{X}(\mathfrak{g}^*)$ relative to the (±) Lie-Poisson bracket is given by $X_H(\mu) = \mp \operatorname{ad}\left(\frac{\delta H}{\delta \mu}\right)^* \mu$.

8.4C (Gardner [1970]). Let $V = \left\{ f: \mathbb{R} \to \mathbb{R} \;\middle|\; f \text{ is } C^\infty,\ \lim_{|x| \to \infty} f(x) = 0 \right\}$.

(i) Show that the prescription

$$\{F, G\}(f) = \int_{-\infty}^{+\infty} \frac{\delta F}{\delta f} \frac{d}{dx} \frac{\delta G}{\delta f} \, dx$$

defines a Poisson bracket on V for appropriate functions F and G (be careful about what hypotheses you put on F and G).

(ii) Show that the Hamiltonian vector field of $H : V \to \mathbb{R}$ is given by

$$X_H(f) = \frac{d}{dx} \frac{\delta H}{\delta f} .$$

(iii) Let $H(f) = \int_{-\infty}^{+\infty} \left(f^3 + \frac{1}{2} f_x^2 \right) dx$. Show that the differential equation for X_H is the Korteweg-deVries equation

$$f_t - 6ff_x + f_{xxx} = 0.$$

8.4D (Ratiu [1980]). Let $\mathfrak{g}$ be a Lie algebra and $\varepsilon \in \mathfrak{g}^*$ be fixed. Show that the prescription

$$\{F, G\}_\varepsilon (\mu) = \left\langle \varepsilon, \left[\frac{\delta F}{\delta \mu}, \frac{\delta G}{\delta \mu} \right] \right\rangle$$

defines a Poisson bracket on $\mathfrak{g}^*$. (*Hint:* Look at the formulas in the proof of Theorem 8.4.4.)

8.4E (i) (Pauli [1953], R. Jost [1964]). Let P be a finite dimensional Poisson manifold satisfying the following condition: If $\{F, G\} = 0$ for any *locally* defined F implies G = constant. Show that there exists an open dense set U in P such that the Poisson bracket restricted to U comes from a symplectic form on U. (*Hint*: Define $B : T^*P \times T^*P \to \mathbb{R}$ by $B(\mathbf{d}F, \mathbf{d}G) = \{F, G\}$. Show first that $U = \{p \in P \mid B_p(\alpha, \beta) = 0$ for all $\alpha \in T_p{}^*P$ implies $\beta_p = 0\}$ is open and dense in P. Then show that B can be inverted at points in U.)

(ii) Show that, in general, $U \neq P$ by the following example. On $\mathbb{R}^2$ define

$$\{F, G](x, y) = y\left(\frac{\partial F}{\partial x} \frac{\partial G}{\partial y} - \frac{\partial F}{\partial y} \frac{\partial G}{\partial x} \right)$$

Show that U in (i) is $\mathbb{R}^2 \setminus \{Ox\text{-axis}\}$. Show that on U, the symplectic form generating the above Poisson bracket is $dx \wedge dy/y$.

§8.5 *Constraints and Control*

The applications in this final section all involve the Frobenius theorem. Each example is necessarily treated briefly, but hopefully in enough detail so the interested reader can pursue the subject further by utilizing the given references.

We start with the subject of *holonomic constraints in Hamiltonian systems*. A Hamiltonian system as discussed in §**8.1** can have a condition imposed that limits the available points in phase space. Such a condition is a *constraint*. For example, a ball tethered to a string of unit length in $\mathbb{R}^3$ may be considered to be constrained only to move on the unit sphere S^2 (or possibly interior to the sphere if the string is collapsible). If the phase space is T^*Q and the constraints are all derivable from constraints imposed only on the configuration space (the q's), the constraints are called *holonomic*. For example, if there is one constraint $f(q) = 0$ for $f : Q \to \mathbb{R}$, the constraints on T^*Q can be simply obtained by differentiation: $\mathbf{d}f = 0$ on T^*Q. If the phase space is TQ, then the constraints are holonomic iff the constraints on the velocities are saying that the velocities are tangent to some constraint manifold of the positions. A constraint then can be thought of in terms of velocities as a subset $E \subset TM$. If it is a subbundle, this *constraint is thus holonomic iff it is integrable in the sense of Frobenius' theorem.*

Constraints that are not holonomic, called *nonholonomic constraints*, are usually difficult to handle. Holonomic constraints can be dealt with in the sense that one understands how to modify the equations of motion when the constraints are imposed, by adding *forces of constraint*, such as centrifugal force. See, for example Goldstein [1980, Ch. 1], and Abraham and marsden [1978, §3.7]. We shall limit ourselves to the discussion of two examples of nonholonomic constraints.

A classical example of a nonholonomic system is a disk rolling without slipping on a plane. The disk of radius a is constrained to move without slipping on the (x, y)-plane. Let us fix a point P on the disk and call θ the angle between the radius at P and the contact point Q of the disk with the plane, as in Fig. 8.5.1. Let (x, y, a) denote the coordinates of the center of the disk. Finally, if φ denotes the angle between the tangent line to the disk at Q and the x-axis, the position of the disk in space is completely determined by (x, y, θ, φ). These variables form elements of our configuration space $M = \mathbb{R}^2 \times S^1 \times S^1$. The condition that there is no slipping at Q means that the velocity at Q is zero; i.e.,

$$\frac{dx}{dt} + a\frac{d\theta}{dt}\cos\varphi = 0, \quad \frac{dy}{dt} + a\frac{d\theta}{dt}\sin\varphi = 0$$

(total velocity = velocity of center plus the velocity due to rotation by angular velocity $d\theta/dt$).

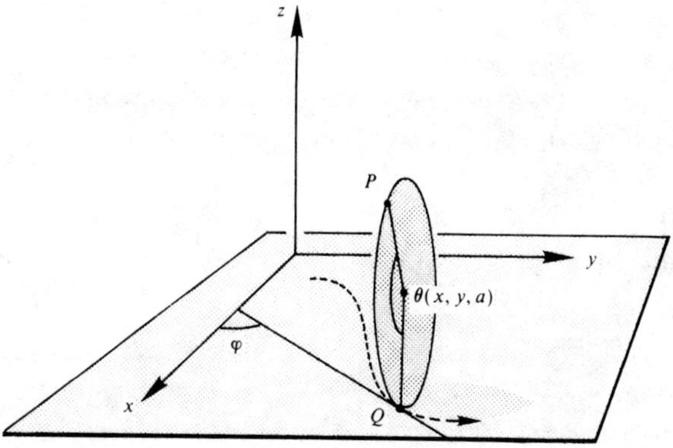

Figure 8.5.1

The expression of these constraints in terms of differential forms is $\omega_1 = 0$, $\omega_2 = 0$ where $\omega_1 = dx + a\cos\varphi\, d\theta$ and $\omega_2 = dy + a \sin\varphi\, d\theta$. We compute that

$$\omega = \omega_1 \wedge \omega_2 = dx \wedge dy + a\cos\varphi\, d\theta \wedge dy + a \sin\varphi\, dx \wedge d\theta.$$

$$d\omega_1 = -a\sin\varphi\, d\varphi \wedge d\theta, \qquad d\omega_2 = a\cos\varphi\, d\varphi \wedge d\theta.$$

$$d\omega_1 \wedge \omega = -a\sin\varphi\, d\varphi \wedge d\theta \wedge dx \wedge dy, \quad d\omega_2 \wedge \omega = a\cos\varphi\, d\varphi \wedge d\theta \wedge dx \wedge dy.$$

These do not vanish identically. Thus, according to Theorem **6.4.20**, this system is not integrable and hence these constraints are nonholonomic.

A second example of constraints is due to E. Nelson [1967]. Consider the motion of a car and denote by (x, y) the coordinates of the center of the front axle, φ the angle formed by the moving direction of the car with the horizontal, and θ the angle formed by the front wheels with the car (Fig. 8.5.2).

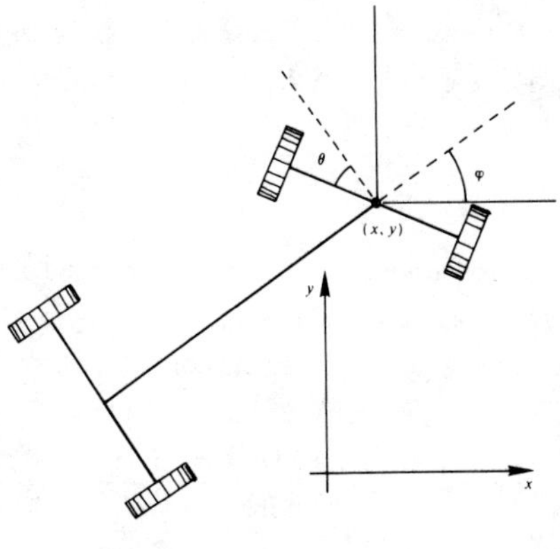

Figure 8.5.2

The configuration space of the car is $\mathbb{R}^2 \times \mathbb{T}^2$, parametrized by (x, y, φ, θ). We shall prove that the constraints imposed on this motion are nonholonomic. Call the vector field $X = \partial/\partial\theta$ ***steer.*** We want to compute a vector field Y corresponding to ***drive.*** Let the car be at the configuration point (x, y, φ, θ) and assume that it moves a small distance h in the direction of the front wheels. Notice that the car moves foward and simultaneously turns. Then the next configuration is

$$\big(x + h\cos(\varphi+\theta) + o(h),\ y + h\sin(\varphi+\theta) + o(h),\ \ \varphi + h\sin\theta + o(h),\ \theta\big).$$

Thus the "drive" vector field is

$$Y = \cos(\varphi+\theta)\frac{\partial}{\partial x} + \sin(\varphi+\theta)\frac{\partial}{\partial y} + \sin\theta\frac{\partial}{\partial\varphi}.$$

A direct computation shows that the vector field ***wriggle,***

$$W = [X, Y] = -\sin(\varphi+\theta)\frac{\partial}{\partial x} + \cos(\varphi+\theta)\frac{\partial}{\partial y} + \cos\theta\frac{\partial}{\partial\varphi},$$

and ***slide,***

$$S = [W, Y] = -\sin\varphi\frac{\partial}{\partial x} + \cos\varphi\frac{\partial}{\partial y},$$

satisfy

$$[X, W] = -Y,\ \ [S, X] = 0,\quad [S, Y] = \sin\theta\cos\varphi\frac{\partial}{\partial x} + \sin\theta\sin\varphi\frac{\partial}{\partial y},$$

and

$$[S, W] = \cos\theta\cos\varphi\frac{\partial}{\partial x} + \cos\theta\sin\varphi\frac{\partial}{\partial y}.$$

Define the vector fields Z_1, Z_2 by

$$Z_1 = [S, Y] = -W + (\cos\theta)S + \cos\theta\frac{\partial}{\partial\varphi},\quad Z_2 = [S, W] = Y - (\sin\theta)S - \sin\theta\frac{\partial}{\partial\varphi}.$$

A straightforward calculation shows that

$$[X, Z_1] = Z_2,\ [X, Z_2] = -Z_1,\ [S, Z_1] = 0, [S, Z_2] = 0, [Z_1, Z_2] = 0,$$

i.e., $\{X, Z_1, Z_2, S\}$ span a four dimensional Lie algebra $\mathfrak{g}$ with one dimensional center spanned by S. In addition, its derived Lie algebra $[\mathfrak{g}, \mathfrak{g}] \subset \mathfrak{g}$, equals span $\{Z_1, Z_2\}$ and is therefore abelian and two dimensional. Thus $\mathfrak{g}$ has no nontrivial non-abelian Lie subalgebras.

In particular the subbundle of $T(\mathbb{R}^2 \times \mathbb{T}^2)$ spanned by X and Y is not involutive and thus not integrable. By the Frobenius theorem, the field of two-dimensional planes spanned by X and Y is not tangent to a family of two-dimensional integral surfaces. Thus the motion of the car, subjected *only* to the constraints of "steer" and "drive" is nonholonomic. On the other hand, the motion of the car subjected to the constraints of "steer", "drive" and "wriggle" is holonomic.

Moreover, since the Lie algebra generated by these three vector fields is abelian, the motion of the car with these constraints can be described by applying these three vector fields in any order.

Next we turn our attention to *some elementary aspects of control theory*. We shall restrict our attention to a simple version of a local controllability theorem. For extensions and many additional results, we recommend consulting a few of the important papers and notes such as Brocket [1970, 1983], Sussmann [1977], Hermann and Krener [1977], Russell [1979], Hermann [1980], and Ball, Marsden, and Slemrod [1982] and references therein.

Consider a system of differential equations of the form

$$\dot{w}(t) = X(w(t)) + p(t)Y(w(t)) \tag{1}$$

on a time interval $[0, T]$ with initial conditions $w(0) = w_0$ where w takes values in a Banach manifold M, X and Y are smooth vector fields on M and $p\colon [0, T] \to \mathbb{R}$ is a prescribed function called a *control.*

The existence theory for differential equations guarantees that (1) has a flow that depends smoothly on w_0 and on p lying in a suitable Banach space Z of maps of $[0, T]$ to $\mathbb{R}$, such as the space of C^1 maps. Let the flow of (1) be denoted

$$F_t(w_0, p) = w(t, p, w_0). \tag{2}$$

We consider the curve $w(t, 0, w_0) = w_0(t)$; i.e., an integral curve of the vector field X. We say that (1) is ***locally controllable*** (at time T) if there is a neighborhood U of $w_0(T)$ such that for any point $h \in U$, there is a $p \in Z$ such that $w(T; p, w_0) = h$. In other words, we can alter the endpoint of $w_0(t)$ in a locally arbitrary way by altering p. (Fig. 8.4.3).

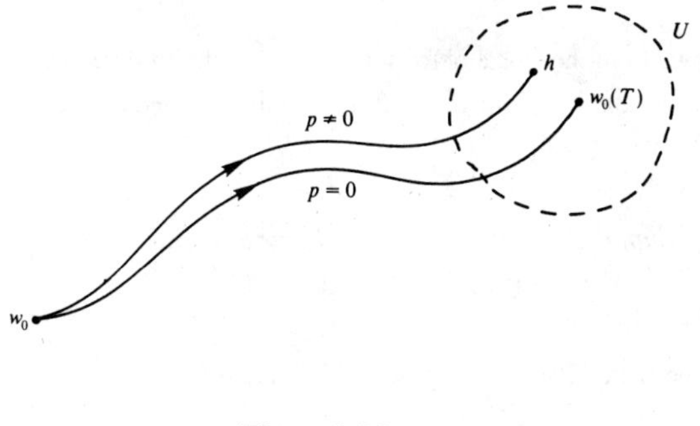

Figure 8.5.3

To obtain a condition under which local controllability can be guaranteed, we fix T and w_0

and consider the map

$$P : Z \to M; \quad p \mapsto w(T, p, w_0). \tag{3}$$

The strategy is to apply the inverse function theorem to P. The derivative of $F_t(w_0, p)$ with respect to p in the direction $\rho \in Z$ is denoted

$$D_pF_t(w_0, 0) \cdot \rho = L_t\rho \in T_{F_t(w_0, 0)}M.$$

Differentiating

$$\frac{d}{dt} w(t, p) = X(w(t, p)) + p(t)Y(w(t, p))$$

with respect to p at $p = 0$, we find that in T^2M

$$\frac{d}{dt} L_t\rho = X(w_0(t)) \cdot L_t\rho + (\rho Y(w_0(t)))_{\text{vertical lift}} . \tag{4}$$

To simplify matters, let us assume $M = E$ is a Banach space and that X is a *linear operator*, so (4) becomes

$$\frac{d}{dt}L_t\rho = X \cdot L_t\rho + \rho Y(w_0(t)). \tag{5}$$

Equation (5) has the following solution given by the variation of constants formula

$$L_T\rho = \int_0^T e^{(T-s)X}\rho(s)Y(e^{sX}w_0)ds \tag{6}$$

since $w_0(t) = e^{tX}w_0$ for linear equations.

8.5.1 Proposition *If the linear map* $L_T : Z \to E$ *given by* (6) *is surjective, then the* (1) *is locally controllable* (*at time* T).

Proof This follows from the "local onto" form of the implicit function theorem (see **2.5.9**) applied to the map P. Solutions exist for time T for small p since they do for $p = 0$; see corollary **4.1.25.** ■

8.5.2 Corollary *Suppose* $E = \mathbb{R}^n$ *and* Y *is linear as well. If* dim span $\{Y(w_0), [X, Y](w_0), [X,[X,Y]](w_0), ...\} = n$, *then* (1) *is locally controllable.*

Proof We have the Baker-Campbell-Hausdorff formula

$$e^{-sX}Ye^{sX} = Y + s[X, Y] + \frac{s^2}{2}[X, [X, Y]] + ...,$$

obtained by expanding $e^{sX} = I + sX + (s^2/2)X^2 + \ldots$ and gathering terms. Substitution into (6) shows that L_T is surjective. ∎

For the case of nonlinear vector fields and the system (1) on finite-dimensional manifolds, controllability hinges on the dimension of the space obtained by replacing the foregoing commutator brackets by Lie brackets of vector fields, n being the dimension of M. This is related to what are usually called *Chow's theorem* and *Hermes' theorem* in control theorey (see Chow [1947]).

To see that some condition involving brackets is necessary, suppose that the span of X and Y forms an involutive distribution of TM. Then by the Frobenius theorem, w_0 lies in a unique maximal two-dimensional leaf $\mathcal{L}(w_0)$ of the corresponding foliation. But then the solution of (1) can never leave $\mathcal{L}(w_0)$, no matter how p is chosen. Hence in such a situation, (1) would not be locally controllable; rather, one would only be able to move in a two-dimensional subspace. If repeated bracketing with X increases the dimension of vectors obtained then the attainable states increase in dimension accordingly.

Exercises

8.5A Check that the system in Fig. 8.5.1 is nonholonomic by verifying that there are two vector fields X, Y on M spanning the subset E of TM defined by the constraints

$$\dot{x} = a\,\dot{\theta}\cos\varphi = 0 \qquad \text{and} \qquad \dot{y} + a\,\dot{\theta}\sin\varphi = 0$$

such that [X, Y] is not in E; i.e., use Frobenius' theorem directly rather than using Pfaffian systems.

8.5B Justify the names *wriggle* and *slide* for the vector fields W and S in the example of Fig. 8.5.2 using the product formula in Exercise **4.2D**. Use these formulas to explain the following statement of Nelson [1967, p. 35]: "the Lie product of "steer" and "drive" is equal to "slide" and "rotate" ($= \partial/\partial\varphi$) on $\theta = 0$ and generates a flow which is the simultaneous action of sliding and rotating. This motion is just what is needed to get out of a tight parking spot."

8.5C The word *holonomy* arises not only in mechanical constraints as explained in this section but also in the theory of connections (Kobayashi-Nomizu [1963, II, §7,8]). What is the relation between the two uses, if any?

8.4D In linear control theory (1) is replaced by

$$\dot{w}(t) = X \cdot w(t) + \sum_{i=1}^{N} p_i(t) Y_i ,$$

where X is a linear vector field on $\mathbb{R}^n$ and Y_i are *constant* vectors. By using the methods used to prove **8.5.1**, rediscover for yourself the *Kalman criterion* for local controllability, namely, $\{X^k Y_i \mid k = 0, 1, ..., n-1, i = 1, ..., N\}$ spans $\mathbb{R}^n$.

References

R. Abraham [1963] *Lectures of Smale on Differential Topology,* Notes, Columbia University.

R. Abraham and J. Marsden [1978] *Foundations of Mechanics,* Second Edition, Addison-Wesley, Reading Mass.

R. Abraham and J. Robbin [1967] *Transversal Mappings and Flows,* Addison-Wesley, Reading Mass.

R.A. Adams [1975] *Sobolev Spaces,* Academic Press, New York.

V. I. Arnol'd [1966] Sur la geometrie differentielle des groupes de Lie de dimenson infinie et ses applications a L'hydrodynamique des fluids parfaits, *Ann. Inst. Fourier. Grenoble,* **16,** 319 - 361.

V.I. Arnol'd [1978] *Mathematical Methods of Classical Mechanics,* Springer Graduate Texts in Mathematics, **60**, Springer-Veriag, New York.

V.I. Arnol'd and A. Avez [1967] *Theorie ergodique des systemes dynamiques,* Gauthier-Villars, Paris (English ed., Addison-Wesley, Reading, Mass., 1968)

J. M. Ball and J.E. Marsden [1984] Quasiconvexity at the boundary, positivity of the second variation and elastic stability, *Arch.Rat. Mech. An.* **86**, 251-277.

J.M. Ball, J.E. Marsden and M. Slemrod [1982] Controllability for distributed bilinear systems, *SIAM J. Control and Optim.*, **20**, 575-597.

G.K. Batchelor [1967] *An Introduction to Fluid Dynamics,* Cambridge Univ. Press, Cambridge, England.

M. Berger and D. Ebin [1969] Some decompositions of the space of symmetric tensors on a Riemannian manifold, *J. Diff. Geom.*, **3**, 379-392.

G.D. Birkhoff [1931] Proof of the ergodic theorem, *Proc. Nat Acad. Sci.,* **17**, 656-660.

D. Bleecker [1981] *Gauge Theory and Variational Principles,* Global Analysis: Pure and Applied, **1**, Addison- Wesley, Reading, Mass.

R. Bonic and F. Reis [1966] A characterization of Hilbert space, *Acad. Bras. de Cien,* **38**, 239-241.

R. Bonic and J. Frampton [1966] Smooth functions on Banach manifolds, *J. Math. Mech.* **16**, 877-898.

J.M. Bony [1969] Principe du maximum, inegalité de Harnack et unicité du problémè de Cauchy pour les operateurs elliptiques dégenerés, *Ann. Inst. Fourier Grenoble* **19,** 277-304.

M. Born and L. Infeld [1935] On the quantization of the new field theory, *Proc. Roy. Soc. A.* **150,** 141-162.

R. Bott [1970] On a topological obstruction to integrability, *Proc. Symp. Pure Math.* **16,** 127-131.

N. Bourbaki [1971] Variétés differentielles et analytiqes, *Fascicule de résultats* **33,** Hermann.

J.P. Bourguigon [1975] Une stratification de l'espace des structures riemanniennes, *Comp. Math.* **30,** 1-41.

J. P. Bourguignon and H. Brezis [1974] Remarks on the Euler equation, *J. Funct. An.* **15,** 341-363.

R. Bowen [1975] *Equilibrium States and the Ergrodic Theory of Anosov Diffeo - morphisms,* Springer Lecture Notes in Mathematics, **470.**

G. E. Bredon [1972] *Introduction to Compact Transformation Groups,* Academic Press, New York.

H. Brezis [1970] On Characterization of Flow Invariant Sets, *Commun. Pure Appl. Math* **23,** 261-263.

R.W. Brockett [1970] *Finite Dimensional Linear Systems,* Wiley, New York.

R. W. Brockett [1983] *A Geometrical Framework for Nonlinear Control and Estimation,* CBMS Conference Series, SIAM.

M. Buchner and J. Marsden and S. Schecter [1983] Applications of the blowing-up construction and algebraic geometry to bifurcation problems, *J. Diff. Eqs.* **48,** 404-433.

M. Buchner and J. Marsden and S. Schecter [1983] Examples for the infinite dimensional Morse Lemma, *SIAM J. Math An.* **14,** 1045-1055.

D. Burghelea, A. Albu, and T. Ratiu [1975] *Compact Lie Group Actions* (in Romanian), Monografii Matematice **5,** Universitatea Timisoara.

W.L. Burke [1980] *Spacetime, Geometry, Cosmology,* University Science Books, Mill Valley, Ca.

C. Camacho, A.L. Neto [1985] *Geometric Theory of Foliations,* Birkhäuser, Boston, Basel, Stuttgart.

M. Cantor [1981] Elliptic operators and the decomposition of tensor fields, *Bull. Am. Math. Soc.* **5,** 235-262.

C. Caratheodory [1909] Untersuchungen über die Grundlagen der Thermodynamik, *Math. Ann.* **67,** 355-386.

C. Caratheodory [1965] *Calculus of Variation and Partial Differential Equations,* Holden-Day, San Francisco.

E. Cartan [1945] *Les systemes differentiels exterieurs et leur applications geometriques,* Hermann, Paris.

F.F. Chen [1974] *Introduction to Plasma Physics,* Plenum

P. Chernoff [1974] *Product Formulas, Nonlinear Semigroups and Addition of Unbounded Operators,* Memoirs of Am. Math. Soc. **140.**

P. Chernoff and J. Marsden [1974] *Properties of Infinite Dimensional Hamiltonian Systems,* Springer Lecture Notes in Mathematics, **425.**

C. Chevalley [1946] *Theory of Lie groups,* Princeton University Press, Princeton, N.J.

G. Choquet [1969] *Lectures on Analysis,* 3 vols. Addison-Wesley, Reading, Mass.

Y. Choquet-Bruhat, C. DeWitt-Morette and M. Dillard-Blieck [1982] *Analysis, Manifolds, and Physics*, Rev. ed., North-Holland, Amsterdam.

A. Chorin and J. Marsden [1979] *A Mathematical Introduction to Fluid Mechanics*, Springer Universitext.

A. Chorin, T.J.R. Hughes, M.F. McCracken and J. Marsden [1978]. Product formulas and numerical algorithms, *Commun. Pure Appl. Math.* **31,** 205-256.

S.N. Chow and J.K. Hale [1982] *Methods of Bifucation Theory*, Springer, New York.

S.N. Chow, J. Mallet-Paret and J. Yorke [1978] Finding zeros of maps: hompotopy methods that are constuctive with probability one. *Math. Comp.* **32,** 887-899.

W. L. Chow [1947] Über Systeme von linearen partiellen Differentialgleichungen. *Math. Ann.* **117,** 89-105.

A. Clebsch [1859] Über die Integration der hydrodynamischen Gleichungen, *J. Reine. Angew. Math.* **56,** 1-10.

Clemmow and Daugherty [1969] *Electrodynamics of Particles and Plasmas*, Addison Wesley

J.M. Cook [1966] Complex Hilbertian structures on stable linear dynamical systems, *J. Math. Mech.* **16,** 339-349.

M. Craioveanu and T. Ratiu [1976] *Elements of Local Analysis*, Vol. 1, 2 (in Romanian), Monografii Mathematice **6,7.** Universitatea Timisoara.

R.C. Davidson [1972] *Methods in Nonlinear Plasma Theory*, Academic Press

G. deRham [1984] *Differentiable Manifolds: Forms, Currents, Harmonic Forms,* Grundlehren der Math. Wissenschaften, **266**, Springer-Verlag, New York.

P.A.M. Dirac [1964] Lectures on Quantum Mechanics, *Belfer Graduate School of Sci., Monograph Series*, **2**, Yeshiva University

G. Duff and D. Spencer [1952] Harmonic tensors on Riemannian manifolds with boundary, *Ann. Math.* **56,** 128-156.

G. Duffing [1918] *Erzwungene Schwingungen bei veränderlichen Eigenfrequenz,* Vieweg u. Sohn, Braunschweig.

I.E. Dzyaloshinskii and G.E. Volovick [1980] Poisson Brackets in Condensed Matter Physics, *Ann. of Phys.*, **125**, 67-97

D. Ebin [1970] On completeness of Hamiltonian vector fields, *Proc. Am. Math. Soc.* **26,** 632-634.

D. Ebin and J. Marsden [1970] Groups of diffeomorphisms and the motion of an incompressible fluid, *Ann Math.* **92,** 102-163.

J. Eells [1958] On the geometry of function spaces, in *Symposium de Topologia Algebrica,* Mexico UNAM, Mexico City, pp. 303-307.

H. Elliasson [1967] Geometry of manifolds of maps, *J. Diff. Geom.* **1,** 169-194.

D. Elworthy and A. Tromba [1970a] Differential structures and Fredholm maps on Banach manifolds, *Proc. Symp. Pure Math.* **15,** 45-94.

D. Elworthy and A. Tromba [1970b] Degree theory on Banach manifolds, *Proc. Symp. Pure Math.* **18,** 86-94.

A. Fischer [1970] A theory of superspace, in *Relativity*, M. Carmelli et al. (Eds). Plenum, New York.

A. Fischer and J. Marsden [1975] Deformations of the scalar curvature, *Duke Math. J.***42,** 519-547.

A. Fischer and J. Marsden [1979] Topics in the dynamics of general relativity, in *Isolated Gravitating Systems in General Relativity,* J. Ehlers (ed.) Italian Physical Society, North-Holland, Amsterdam, pp. 322-395.

H. Flanders [1963] *Differential Forms*, Academic Press, New York.

L.E. Fraenkel [1978] Formulae for high derivatives of composite functions, *Math. Proc. Camb. Phil. Soc.* **83,** 159-165.

J. Frampton and A. Tromba [1972] On the classification of spaces of Hölder continuous functions, *J. Funct. An.* **10,** 336-345.

T. Fulton, F. Rohrlich and L. Witten [1962] Conformal invariance in physics, *Rev. Mod. Phys.* **34,** 442-457.

M.P. Gaffney [1954] A special Stokes's theorem for complete Riemannian manifolds, *Ann. Math.* **60,** 140-145.

C.S. Gardener, J.M. Greene, M.D. Kruskal and R.M. Muira [1974] Korteweg-deVries Equation and Generalizations. VI. Methods for Exact Solution, *Comm. PUre Appl. Math.*, **27**, 97-133

G. Glaeser [1958] Étude de quelques algèbres Tayloriennes, *J. Anal. Math.* **11,** 1-118.

H. Goldstein [1980] *Classical Mechanics*, 2nd ed. Addison-Wesley, Reading, Mass.

M. Golubitsky and J. Marsden [1983] The Morse lemma in infinite dimensions via singularity theory, *SIAM. J. Math. An.* **14,** 1037-1044.

M. Golubitsky and V. Guillemin [1974] *Stable Mappings and their Singularities,* Graduate Texts in Mathematics, Vol **14,** Springer - Verlag, New York.

M. Golubitsky and D. Schaeffer [1985] *Singularities and Groups in Bifurcation Theory*, I, Springer, Verlag, New York.

L. Graves [1950] Some mapping theorems, *Duke Math. J.* **17,** 111-114.

H. Grauert [1958] On Levi's problem and the imbedding of real-analytic manifolds, *Am. J. of. Math.* **68,** (2), 460-472.

J. Guckenheimer and P. Holmes [1983] *Nonlinear Oscillations, Dynamical Systems and Bifurcations of Vector Fields*, Springer Applied Math. Sciences, Vol. **43.**

V. Guillemin and A Pollack [1974] *Differential Topology,* Prentice-Hall, Englewood Cliffs, N.J.

V. Guillemin and S. Sternberg [1977] *Geometric Asymptotics,* American Mathematical Society Surveys, Vol **14.**

V. Guillemin and S. Sternberg [1984] *Symplectic Techniques in Physics,* Cambridge University Press.

M. Gurtin [1981] *An Introduction to Continuum Mechanics,* Academic Press, New York.

J.K. Hale [1969] *Ordinary Differential Equations,* Wiley-Interscience, New York.

P. R. Halmos [1956] *Lectures on Ergodic Theory,* Chelsea, New York.

R. Hamilton [1982] The inverse function theorem of Nash and Moser, *Bull. Am. Math. Soc.* **7,** 65-222.

P. Hartman [1972] On invariant sets and on a theorem of Wazewski, *Proc. Am. Math. Soc.* **32,** 511-520.

P. Hartman [1973] *Ordinary Differential Equations,* 2nd ed. Reprinted by Birkhäuser, Boston.

S. Hawking and G.F.R. Ellis [1973] *The Large Scale Structure of Space-Time,* Cambridge Univ.Press, Cambridge, England.

C. Hayashi [1964] *Nonlinear Oscillations in Physical Systems,* McGraw-Hill, New York.

R. Hermann [1973] *Geometry, Physics and Systems,* Marcel Dekker, New York.

R. Hermann [1977] *Differential Geometry and the Calculus of Variations,* 2nd ed. Math. Sci. Press, Brookline, Mass.

R. Hermann [1980] *Cartanian Geometry, Nonlinear Waves and Control Theory,* Part B. Math. Sci. Press, Brookline, Mass.

R. Hermann and A.J. Krener [1977] Nonlinear conrollability and observability, *IEEE Trans. on Auto. Control.* **22,** 728-740.

T.H. Hildebrandt and L.M.Graves [1927] Implicit functions and their differentials in general analysis, *Trans. Am. Math. Soc.* **29,** 127-53.

M.W. Hirsch [1976] *Differential Topology.* Graduate Texts in Mathematics, Vol **33,** Springer - Verlag. New York.

M. W. Hirsch and S. Smale [1974] *Differential Equations, Dynamical Systems and Linear Algebra,* Academic Press, New York.

V.W.D. Hodge[1952] *Theory and Applications of Harmonic Integrals,* 2nd ed. Cambridge University Press. Cambridge. England.

D.D. Holm, J.E. Marsden, T. Ratiu and A. Weinstein [1985] Nonlinear Stability of Fluid and Plasma Equilibria, *Physics Reports,* **123**, 1-116

H. Holmann and H. Rummler [1972] *Alternierende Differentialformen.* BI- Wissen - schaftsverlag, Zürich.

P. Holmes [1979a] A nonlinear oscillator with a strange attractor, *Phil. Trans. Roy. Soc. London.*

P. Holmes [1979b] Averaging and chaotic motions in forced oscillations, *SIAM J. on Appl. Math.* **38,** 68-80, and **40,** 167-168.

V.H. Hopf [1931] Über die Abbildungen der Dreidimensionalen Sphäre auf die Kugelfäche, *Math. Annalen.* **104,** 637-665.

W. Huebsch [1955] On the covering homotopy theorem, *Annals. Math.* **61,** 555-563.

T.J.R. Hughes and J. Marsden [1977] Some Applications of Geometry in Continuum Mechanics, *Reports on Math. Phys.* **12,** 35-44.

D. Husemoller [1975] *Fibre Bundles,* 2nd ed. Graduate Texts in Mathematics Vol. **20,** Springer-Verlag, New York.

M.C. Irwin [1980] *Smooth Dynamical Systems,* Academic Press.

Z.R. Iwi´nski and K.A. Turski [1976] Canonical Theories of systems Interacting Electromagnetically, *Letters in Applied and Engineering Sciences,* **4**, 179-191

F. John [1975] *Partial Differential Equations,* 2nd ed. Applied Mathematical Sciences, Vol **1.** Springer-Verlag, New York.

L. Karp [1981] On Stokes' theorem for non-compact manifolds, *Proc. Am. Math. Soc.* **82.** 487-490.

T. Kato [1966] *Perturbation Theory for Linear Operators,* Springer. (Second edition, 1977).

A.N. Kaufman and P.J. Morrison [1982] Algebraic Structure of the Plasma Quasilinear Equations, *Phys. Lett.*, **88**, 405-406

R. Kaufmann [1979] A singular map of a cube onto a square, *J. Diff. Equations* **14,** 593-594.

J. Kelley [1975] *General Topology.* Graduate Texts in Mathematics **27**, Springer-Verlag, New York.

W.Klingenberg [1978] *Lectures on Closed Geodesics.* Grundlehren der Math. Wissenschaften **230,** Springer-Verlag, New York.

G. Knowles [1981] *An Introduction to Applied Optimal Control,* Academic Press, New York.

S. Kobayashi and K Nomizu [1963] *Foundations of Differential Geometry,* Wiley, New York.

K. Kodaira [1949] Harmonic fields in Riemannian manifolds, *Ann. of Math.* **50,** 587-665.

B.O. Koopman [1931] Hamiltonian systems and transformations in Hilbert space, *Proc. Nat. Acad. Sci.* **17,** 315-318.

S. Lang [1972] *Differential Manifolds*, Addison-Wesley, Reading, Mass.

H.B. Lawson [1977] *The Qualitative Theory of Foliations,* American Mathematical Society CBMS Series, Vol. **27.**

P.D. Lax [1973] *Hyperbolic Systems of Conservative Laws and the Mathematical Theory of Shock Waves,* SIAM, CBMS Series, Vol. **11.**

E. Leonard and K.Sunderesan [1973] A note on smooth Banach spaces, *J. Math. Anal. Appl.* **43,** 450-454.

D. Lewis, J.E. Marsden and T. Ratiu [1986b] The Hamiltonian Structure for Dynamic Free Boundary Problems, *Physica*, **18D**, 391-404

J. Lindenstrauss and L. Tzafriri [1971] On the complemented subspace problem, *Israel J. Math.* **9**, 263-269.

L. Loomis and S. Sternberg [1968] *Advanced Calculus*, Addison-Wesley, Reading Mass.

D. G. Luenberger [1969] *Optimization by Vector Space Methods*, John Wiley, New York.

G.W. Mackey [1963] *Mathematical Foundations of Quantum Mechanics*, Addison-Wesley, Reading, Mass.

G.W. Mackey [1962] Point realizations of transformation groups, *Illinois J. Math.* **6**, 327-335.

J. Marcinkiewicz and A. Zygmund [1936] On the differentiability of functions and summability of trigonometric series, *Fund. Math.* **26**, 1-43.

J.E. Marsden [1968a] Genralized Hamiltonian mechanics, *Arch. Rat. Mech. An.* **28**, 326-362.

J.E. Marsden [1968b] Hamiltonian one parameter groups, *Arch. Rat. Mech. An.* **28**, 362-396.

J.E. Marsden [1972] Darboux's theorem fails for weak symplectic forms, *Proc. Am. Math. Soc.*, **32**, 590-592.

J.E. Marsden [1973] A proof of the Calderón extension theorem, *Can. Math. Bull.* **16**, 133-136.

J.E. Marsden [1974a] *Elementary Classical Analysis*, W.H. Freeman, San Francisco.

J.E. Marsden [1974b] *Applications of Global Analysis in Mathematical Physics*, Publish or Perish, Waltham, Mass.

J. E. Marsden [1976] Well-posedness of equations of non-homogeneous perfect fluid, *Comm.* PDE **1**, 215-230.

J.E. Marsden [1981] *Lectures on Geometric Methods in Mathematical Physics*, CBMS Vol. **37**, SIAM, Philadelphia.

J.E. Marsden and T. Hughes [1976] *A Short Course in Fluid Mechanics*, Publish or Perish

J.E. Marsden and T.J.R. Hughes [1983] *Mathmematical Foundations of Elasticity*, Prentice-Hall, Redwood City, Calif.

J.E. Marsden, T. Ratiu and A. Weinstein [1982] Semi-direct products and reduction in mechanics, *Trans. Am. Math. Soc.* **281**, 147-177.

J.E. Marsden and A. Tromba [1981] *Vector Calculus*, W.H. Freeman, San Francisco.

J.E. Marsden and A. Weinstein [1974] Reduction of symplectic manifolds with symmetry, *Rep. Math. Phys.*, **5**, 121-130

J.E. Marsden and A. Weinstein [1982] The Hamiltonian structure of the Maxwell-Vlasov equations, *Physica D.* **4**, 394-406.

J.E. Marsden and A. Weinstein [1983] Coadjoint orbits, vortices and Clebsch variables for incompressible fluids, *Physica D.* **7**, 305-323

J.E. Marsden and J.Scheurle [1987] The construction and smoothness of invariant manifolds by the deformation method, *SIAM J. Math. An.* (to appear).

J.L. Martin [1959] Generalized Classical Dynamics and teh 'Classical Analogue' of a Fermi Oscillation, *Proc. Roy. Soc.*, **A251**, 536.

R.H. Martin [1973] Differential equations on closed subsets of a Banch space. *Trans. Am. Math. Soc.* **179,** 339-414.

W.S. Massey [1977] *Algebraic Topology. An Introduction*, Graduate Texts in Math. **56**, Springer-Verlag, New York.

A. Mayer [1872] Über unbeschränkt integrable Systeme von linearen Differential - gleichungen, *Math. Ann.* **5,** 448-470.

J. Milnor [1965] *Topology from the Differential Viewpoint,* University of Virginia Press, Charlottesville, Va.

C. Misner, K. Thorne and J.A. Wheeler [1973] *Gravitation*, W.H. Freeman, San Francisco.

C.B. Morrey [1966] *Multiple Integrals in the Calculus of Variations,* Springer-Verlag, New York.

P.J. Morrison [1980] The Maxwell-Vlasov Equations as a Continuous Hamiltonian System, *Phys. Lett.*, **80A**, 383-386

P.J. Morrison and J.M. Greene [1980] Noncanonical Hamiltonian Density formulation of Hydrodynamics and Ideal Magnetohydrodyanics, *Phys. Rev. Lett.*, **45**, 790-794

J. Moser [1965] On the volume elements on a manifold, *Trans. Am. Math. Soc.* **120,** 286-294.

M. Nagumo [1942] Über die Lage der Integralkurven gewöhnlicher Differential - gleichungen, *Proc. Phys. - Math. Soc. Jap.* **24,** 551-559.

E. Nelson [1959] Analytic vectors, *Ann. Math.* **70,** 572-615.

E. Nelson [1967] *Tensor Analysis,* Princeton University Press, Princeton, N.J.

E. Nelson [1969] *Topics in Dynamics I: Flows,* Priceton University Press, Princeton, N.J.

L. Nirenberg [1974] *Topics in Nonlinear Analysis,* Courant Institute Lecture Notes.

S.P. Novikov [1965] Topology of Foliations, *Trans. Moscow Math. Soc.* 268-304.

G.F. Oster and A.S. Perelson [1973] Systems, circuits and thermodynamics, *Israel J. Chem.* **11,** 445-478, and *Arch. Rat. Mech. An.* **55,** 230-274, and **57,** 31-98.

R. Palais [1954] Definition of the exterior drivative in terms of the Lie derivative, *Proc. Am. Math. Soc.* **5,** 902-908.

R. Palais [1963] Morse theory on Hilbert manifolds, Topology **2,** 299-340.

R. Palais [1965a] *Seminar on the the Atiyah - Singer Index Theorem,* Princeton University Press, Princeton, N.J.

R. Palais [1965b] *Lectures on the Differential Topology of Infinite Dimensional Manifolds,* Notes by S. Greenfield, Brandeis University.

R. Palais [1968] *Foundations of Global Nonlinear Analysis,* Addison-Wesley, Reading, Mass.

R. Palais [1969] The Morse lemma on Banach spaces, *Bull .Am. Math. Soc.* **75,** 968-971.

W. Pauli [1953] On the Hamiltonian Structure of Non-local Field Theories, *Il Nuovo Cimento*, **X**, 648-667

N.H. Pavel [1984] *Differential Equations, Flow Invariance and Applications,* Research Notes in Mathematics, Pitman, Boston-London.

J-P. Penot [1970] Sur le théorème de Frobenius, *Bull. Math. Soc. France,* **98,** 47-80.

A. Pfluger [1957] *Theorie der Riemannschen Flächen,* Grundlehren der Math. Wissenshaften **89,** Springer-Verlag, New York.

A. Povzner [1966] A global existence theorem for a nonlinear system and the defect index of a linear operator, *Transl. Am. Math. Soc.* **51,** 189-199.

M.M. Rao [1972] Notes on characterizing Hilbert space by smoothness and smoothness of Orlicz spaces, *J. Math. Anal. Appl.* **37,** 228-234.

B. Rayleigh [1887] *The Theory of Sound,* 2 vols., (1945 ed.).Dover, New York.

G. Reeb [1952] Sur certains proprietés topologiques des variétés feuilletées, *Actual. Sci. Ind. No.* **1183.** Hermann, Paris.

R.M. Redheffer [1972] The theorems of Bony and Brezis on flow-invariant sets, *Am. Math. Monthly.* **79,** 740-747.

M. Reed, and B. Simon [1974] *Methods on Modern Mathematical Physics,* Vol. 1: *Functional Analysis,* Vol. 2: *Self-adjointness and Fourier Analysis,* Academic Press, New York.

G. Restrepo [1964] Differentiable norms in Banach spaces, *Bull. Am. Math. Soc.* **70,** 413-414.

F. Riesz [1944] Sur la théorie ergodique, *Comm. Math. Helv.* **17,** 221-239.

J. Robbin [1968] On the existence theorem for differential equations, *Proc. Am. Math. Soc.* **19,** 1005-1006.

E. H. Roth [1986] *Various Aspects of Degree Theory in Banach Spaces,* AMS Surveys and Monographs, **23.**

H. Royden [1968] *Real Analysis,* 2nd ed., Macmillan, New York.

W.Rudin [1966] *Real and Complex Analysis,* McGraw - Hill, New York.

W.Rudin [1973] *Functional Analysis,* McGraw - Hill, New York.

W.Rudin [1976] *Principles of Mathematical Analysis,* 3rd ed. McGraw-Hill, New York.

D. Russell [1979] *Mathematics of Finite Dimensional Control Systems, Theory and Design,* Marcel Dekker, New York.

A. Sard [1942] The measure of the critical values of differentiable maps, *Bull. Am. Math. Soc.* **48,** 883-890.

B. Schutz [1980] *Geometrical Methods of Mathematical Physics,* Cambridge University Press, Cambridge, England.

J.T. Schwartz [1967] *Nonlinear Functional Analysis,* Gordon and Breach, New York.

J. P. Serre [1965] *Lie Algebras and Lie Groups,* W. A. Benjamin, Inc. Reading, Mass.

I. Singer and J Thorpe [1976] *Lecture Notes on Elementary Topology and Geometry,* Undergraduate Texts in Mathematics, Springer-Verlag, New York.

S. Smale [1964] Morse theory and a nonlinear generalization of the Dirichlet problem, *Ann. Math.* **80,** 382-396.

S. Smale [1967] Differentiable dynamical systems, *Bull. Am. Math. Soc.* **73,** 747-817.

S. Smale [1965] An infinite-dimensional version of Sard's theorem, *Amer. J. Math.* **87,** 861-866.

J. Smoller [1983] *Mathematical Theory of Shock Waves and Reaction Diffusion Equations,* Grundlehren der Math. Wissenschaffen **258**, Springer-Verlag, New York.

S.S. Sobolev [1939] On the theory of hyperbolic partial differential equations, *Mat. Sb.* **5,** 71-99.

A. Sommerfeld [1964] *Thermodynamics and Statistical Mechanics,* Lectures on Theoretical Physics, Vol. 5. Academic Press, New York.

J.M. Souriau [1970] *Structure des Systemes Dynamiques,* Dunod, Paris

M. Spivak [1979] *Differential Geometry,* Vols 1-5. Publish or Perish, Waltham, Mass.

E. Stein [1970] *Singular Integrals and Differentiability Properties of Functions,* Princeton Univ. Press, Princeton, N.J.

S. Sternberg [1983] *Lectures on Differential Geometry,* 2nd ed., Chelsea, New York.

S. Sternberg [1969] *Celestial Mechanics,* Vols. 1,2. Addison-Wesley, Reading, Mass.

J.J. Stoker [1950] *Nonlinear Vibrations,* Wiley, New York.

M. Stone [1932a] Linear transformations in Hilbert space, *Am. Math. Soc. Colloq. Publ.* **15.**

M. Stone [1932b] On one-parameter unitary groups in Hilbert space. *Ann. of. Math.* **33,** 643-648.

K. Sundaresan [1967] Smooth Banach spaces, *Math. Ann.* **173,** 191-199.

H. J. Sussmann [1975] A generalization of the closed subgroup theorem to quotients of arbitrary manifolds, *J. Diff. Geom.* **10,** 151-166.

H.J. Sussmann [1977] Existence and uniqueness of minimal realizations of nonlinear systems, *Math. Systems Theory* **10,** 263-284.

F. Takens [1974] Singularities of vector fields, *Publ. Math. IHES.* **43,** 47-100.

H.F. Trotter [1958] Approximation of semi-groups of operators, *Pacific. J. Math.* **8,** 887-919.

V.T. Tuan and D.D. Ang [1979] A representation theorem for differentiable functions, *Proc. Am. Math. Soc.* **75,** 343-350.

Y. Ueda [1980] Explosion of strange attractors exhibited by Duffing's equation, *Ann. N.Y. Acad. Sci.* **357,** 422-434.

V. S. Varadarajan [1974] *Lie Groups, Lie Algebras and Their Representations,* Graduate Texts in Math. **102**, Springer-Verlag, New York.

J. von Neumann [1932] Zur Operatorenmethode in der klassischen Mechanik, *Ann. Math.* **33**, 587-648, 789.

C. von Westenholz [1981] *Differential Forms in Mathematical Physics*, North-Holland, Amsterdam.

F. Warner [1983] *Foundations of Differentiable Manifolds and Lie Groups,* Graduate Texts in Math. **94**, Springer-Verlag, New York.

A. Weinstein [1969] Symplectic structures on Banach manifolds, *Bull. Am. Math. Soc.* **75**, 804-807.

A. Weinstein [1977] *Lectures on Symplectic Manifolds,* CBMS Conference Series Vol. **29**, American Mathematical Society.

J.C. Wells [1971] C^1-partitions of unity on non-separable Hilbert space, *Bull. Am. Math. Soc.* **77**, 804-807.

J.C. Wells [1973] Differentiable functions on Banach spaces with Lipschitz derivatives, *J. Diff. Geometry* **8**, 135-152.

R. Wells [1980] *Differential Analysis on Complex Manifolds,* 2nd ed., Graduate Texts in Math. **65**, Springer-Verlag, New York.

H. Whitney [1935] A function not constant on a connected set of critical points, *Duke Math. J.* **1**, 514-517.

H. Whitney [1943a] Differentiability of the remainder term in Taylor's formula, *Duke Math. J.* **10**, 153-158.

H. Whitney [1943b] Differentiable even functions, *Duke Math. J.* **10**, 159-160.

H. Whitney [1944] The self intersections of a smooth n-manifold in 2n-space, *Ann. of. Math.* **45**, 220-246.

F. Wu and C.A. Desoer [1972] Global inverse function theorem, *IEEE Trans. CT.* **19**, 199-201.

J.L. Wyatt, L.O. Chua, and G.F. Oster [1978] Nonlinear n-port decomposition via the Laplace operator, *IEEE Trans. Circuits Systems* **25**, 741-754.

S. Yamamuro [1974] *Differential Calculus in Topological Linear Spaces*, Springer Lecture Notes, **374**.

S.T. Yau [1976] Some function theoretic properties of complete Riemannian manifolds and thier applications to geometry, *Indiana Math J.* **25**, 659-670.

J.A. Yorke [1967] Invariance for ordinary differential equations, *Math. Syst. Theory.* **1**, 353-372.

Index

Applied Mathematical Sciences

cont. from page ii

55. *Yosida:* Operational Calculus: A Theory of Hyperfunctions.
56. *Chang/Howes:* Nonlinear Singular Perturbation Phenomena: Theory and Applications.
57. *Reinhardt:* Analysis of Approximation Methods for Differential and Integral Equations.
58. *Dwoyer/Hussaini/Voigt (eds.):* Theoretical Approaches to Turbulence.
59. *Sanders/Verhulst:* Averaging Methods in Nonlinear Dynamical Systems.
60. *Ghil/Childress:* Topics in Geophysical Dynamics: Atmospheric Dynamics, Dynamo Theory and Climate Dynamics.
61. *Sattinger/Weaver:* Lie Groups and Algebras with Applications to Physics, Geometry, and Mechanics.
62. *LaSalle:* The Stability and Control of Discrete Processes.
63. *Grasman:* Asymptotic Methods of Relaxation Oscillations and Applications.
64. *Hsu:* Cell-to-Cell Mapping: A Method of Global Analysis for Nonlinear Systems.
65. *Rand/Armbruster:* Perturbation Methods, Bifurcation Theory and Computer Algebra.
66. *Hlaváček/Haslinger/Nečas/Lovíšek:* Solution of Variational Inequalities in Mechanics.
67. *Cercignani:* The Boltzmann Equation and Its Applications.
68. *Temam:* Infinite Dimensional Dynamical System in Mechanics and Physics.
69. *Golubitsky/Stewart/Schaeffer:* Singularities and Groups in Bifurcation Theory, Vol. II.
70. *Constantin/Foias/Nicolaenko/Temam:* Integral Manifolds and Inertial Manifolds for Dissipative Partial Differential Equations.
71. *Catlin:* Estimation, Control, and the Discrete Kalman Filter.
72. *Lochak/Meunier:* Multiphase Averaging for Classical Systems.
73. *Wiggins:* Global Bifurcations and Chaos.
74. *Mawhin/Willem:* Critical Point Theory and Hamiltonian Systems.
75. *Abraham/Marsden/Ratiu:* Manifolds, Tensor Analysis, and Applications, 2nd ed.
76. *Lagerstrom:* Matched Asymptotic Expansions: Ideas and Techniques.